STRESS BIOLOGY Of CYANOBACTERIA

Molecular Mechanisms to Cellular Responses

STRESS BIOLOGY of CYANOBACTERIA

Molecular Mechanisms to Cellular Responses

Edited by **Ashish Kumar Srivastava**
Amar Nath Rai • Brett A. Neilan

CRC Press
Taylor & Francis Group
Boca Raton London New York

CRC Press is an imprint of the
Taylor & Francis Group, an **informa** business

CRC Press
Taylor & Francis Group
6000 Broken Sound Parkway NW, Suite 300
Boca Raton, FL 33487-2742

First issued in paperback 2017

ISBN-13: 978-1-4665-0478-3 (hbk)
ISBN-13: 978-1-138-19874-6 (pbk)

Library of Congress Cataloging-in-Publication Data

Stress biology of cyanobacteria : molecular mechanisms to cellular responses / [edited by] Ashish Kumar Srivastava, Amar Nath Rai, Brett A. Neilan.
pages cm
Includes bibliographical references and index.
ISBN 978-1-4665-0478-3 (hardback)
1. Cyanobacteria. 2. Cyanobacteria--Effect of stress on--Molecular aspects. I. Srivastava, Ashish Kumar editor of compilation. II. Rai, A. N. (Amar Nath), 1944- editor of compilation. III. Neilan, Brett A., editor of compilation.

QR99.63.S77 2013
579.3'9--dc23 2012045687

Visit the Taylor & Francis Web site at
http://www.taylorandfrancis.com

and the CRC Press Web site at
http://www.crcpress.com

Contents

Preface

This book is a compilation of the holistic responses of cyanobacteria, ranging from ecological and physiological to the modern aspects of their molecular biology, genomics, and biochemistry. Cyanobacteria are one of the best prokaryotic phyla to study bioenergetics and stress biology due to their cosmopolitan occurrence, archaic origin, and relatively simple bacterial architecture with a physiology ancestral to higher plants. They are also one of the most fascinating groups of microorganisms possessing both nitrogen fixation and oxygenic photosynthesis, even though both are temporally or spatially separated, which are major contributors to the present geochemical status of planet Earth.

With the current global climate change scenario, it is of prime importance to assess cyanobacterial ecophysiologies to have a broader knowledge of climate impacts and possible remediation. This is relevant and timely since cyanobacteria, along with algae, produce almost half of the global oxygen, especially from the oceans. They occupy almost every niche on the Earth, including fresh and marine waters, rice fields, hot springs, arid deserts, and the polar regions. The rice agroecosystem assumes special significance due to its wide occurrence across the world, mainly in tropical countries like India and China, and is a daily staple in the diet of billions of people. In water-logged conditions of rice fields, cyanobacteria have a significant role in nitrogen fixation. In addition, harmful bloom formation is another significant ecophysiological phenomenon associated with cyanobacteria. Toxins produced by certain cyanobacteria have been widely explored at the chemical, toxicological, and ecological levels. Cyanotoxins such as microcystin, anatoxin, saxitoxin, and cylindrospermopsin have been intensively investigated, and their effects on human and animal health have been assessed. These are but a few of the potentials of cyanobacteria, which make them one of the most significant components and primary producers of many different ecosystems globally.

Recent advances in cyanobacterial genome sequencing and proteomics have also opened many new and exciting avenues of research. Molecular biological research on gene expression under different environmental conditions, proteomics and metabolomics of cyanobacteria and transgenic cyanobacteria are emerging fields that are providing previously unattainable insights into the physiology and evolution of this bacterial phylum. Our current in-depth understanding of photosynthesis has been defined in cyanobacteria using classical physiology, and now also at the level of structural and functional genomics.

This book is divided into two parts: Part I, Bioenergetics and Molecular Mechanisms of Stress Tolerance, and Part II, Cellular Responses and Ecophysiology. Both parts collectively cover almost every aspect of cyanobacterial stress biology. Chapters 1 through 3 mainly focus on the basics of molecular bioenergetics of photosynthesis and respiration in cyanobacteria. These chapters are compiled in such a way as to provide a detailed understanding of bioenergetics and a readily comprehensible perspective of different stress tolerance mechanisms in cyanobacteria with regards to this unifying metabolic process. From Chapter 4 onward, we have emphasized the effect of particular stresses on a number of unique and vital cyanobacterial processes. These stresses include heavy metal, high and low temperature, salt, osmotic, and UV-B. The chapters include stress-induced changes in a wide array of physiological, biochemical, and molecular processes of cyanobacteria. Part II deals with cellular responses and ecophysiology. It describes mechanisms of symbiosis, stress-induced bioproducts, the role of environmental factors on nitrogen fixation, mutation and cyanobacterial adaptation, and the most widely studied cyanotoxin microcystin.

The initial seed from which this book grew was sown during our discussions at Mizoram University and through research exchanges by the editors at the University of New South Wales and Banaras Hindu University. The real work started during a fruitful conference in Spain.

The idea behind writing this book has been a decade-long gap between the publication of any book dedicated to cyanobacterial stress biology. Furthermore, we wanted to compile the work of these researches from across the globe into a book that deals with stress responses of cyanobacteria under both controlled laboratory conditions and in the natural environment.

Ashish Kumar Srivastava
Tehri Garhwal, Uttarakhand, India

Amar Nath Rai
Shillong, Meghalaya, India

Brett A. Neilan
Sydney, New South Wales, Australia

Acknowledgments

We are grateful to Govindjee, University of Illinois, United States, for his valuable suggestions and help in organizing these chapters. We would like to acknowledge all the contributors who spared their valuable time to write chapters within the given time frame. Each chapter has been delivered in a very professional manner. We would also like to express our gratitude to the many reviewers of the chapters who diligently ensured the scientific standard of the chapters (and probably learnt a lot in the process). We are thankful to Poonam Bhargava for her valuable suggestions during the editing of chapters. Support from Banaras Hindu University, North Eastern Hill University, Mizoram University, and the University of New South Wales is gratefully acknowledged. Ashish is very thankful to Lal Bahadur Shastri National Academy of Administration, Mussoorie, and District Administration of Tehri Garhwal for allowing him to use facilities essential for editing and compilation work. We also appreciate the help and support from John Sulzycki and Kathryn Everett, Taylor & Francis Group, which made this project a reality. Last but not the least, we are grateful to our families and friends who supported us in every possible way over the last year.

Editors

Dr. Ashish Kumar Srivastava completed his MSc in botany and PhD in molecular biology of cyanobacteria from Banaras Hindu University, Varanasi, India. His area of research is molecular assessment of diversity and stress responses of cyanobacteria inhabiting different ecosystems. He for the first time characterized whole photorespiratory pathway in *Anabaena* using the proteomics and transcript analyses. He established a rapid as well as phenol- and lysozyme-free DNA isolation method, which is equally applicable to both cultures and soil samples. He also provided first-hand data on the role of salinity in distribution of nitrogen-fixing cyanobacteria in rice fields using the 16S rRNA gene and denaturing gradient gel electrophoresis. He received junior and senior research fellowship from the Council of Scientific and Industrial Research, New Delhi, India. He has served in Mizoram University and Banaras Hindu University as assistant professor. He is the recipient of the Endeavour Research Award 2009 from the Australian Government and was a visiting scientist in the University of New South Wales, Sydney, Australia. He received the Young Scientist Award from the Department of Science and Technology, Government of India, and the BPS Project Award of British Psychological Society. He is also a life member of the National Academy of Sciences, India. He has published many research papers in journals of international repute and did many research projects funded by the Department of Science and Technology, Ministry of Environment and Forests and Council of Scientific and Industrial Research, Government of India. He joined the prestigious Indian Administrative Services in 2011 and is currently posted in Tehri Garhwal, Uttarakhand, India.

Professor Amar Nath Rai, after having served Mizoram University as the vice-chancellor, joined NEHU as vice-chancellor in 2010. Born in 1955 at Parsa (Ghazipur, Uttar Pradesh, India), he completed his MSc from B.H.U. (gold medalist) and PhD from the University of Dundee (United Kingdom). He joined NEHU as a reader in 1984 after having served the University of Hyderabad as a lecturer. Having excellent academic record and publications in highly rated journals, he was appointed as professor in 1987, being the youngest professor in the university at that time. He is an internationally recognized scientist in the field of nitrogen fixation and plant–microbe interactions. He has written books published by renowned publishers like CRC Press (United States) and Kluwer Academic Publisher (the Netherlands) and published his research work in journals with high-impact factor.

Rai had been visiting professor at Uppsala and Stockholm Universities for over a decade (1985–2002). He was elected fellow of the Linnean Society of London and member of the Institute of Biology, London, in 1982. He is also the recipient of Amity Academic Excellence Award and Young Scientist Award.

His continuous contribution to NEHU since 1984 is well recognized for developing the infrastructure and academic standards. His contributions to Mizoram University during his tenure as vice-chancellor have been well acknowledged by the faculty, staff, students, and general public of Mizoram for developing necessary infrastructure and creating facilities within a short period of time.

He was a member of the Education Reform Commission for Mizoram that submitted its report within one year recommending comprehensive reform measures for primary to college education.

He is also one of the members of the following: Central Advisory Board on Education (CABE), GOI; Scientific Advisory Committee to the Cabinet, GOI; Governing Council of the National Assessment & Accreditation Council (NAAC), Bangalore; and Visitor's Nominee (The President of India) to the Executive Council, Academic Council & Court of various central universities.

Brett A. Neilan is a molecular biologist and an expert in the study of toxic cyanobacteria that affect water quality globally. Today, his research group at the University of New South Wales (UNSW) comprises 30 researchers, including 18 graduate students. The main topic of their work is microbial and molecular diversity, specifically the genetics of toxic cyanobacteria (blue-green algae). The research has led to an understanding of the biochemical pathways that are responsible for the production of toxins in our water supplies. He obtained his PhD in microbial and molecular biology from UNSW in 1995. Prior to his PhD training, Brett obtained a bachelor of applied science degree in biomedical science (1985) from the University of Technology, Sydney, and then worked as a medical researcher, hospital scientist, and forensic biologist. His postdoctoral position as an Alexander von Humboldt Fellowship in Berlin was on nonribosomal peptide biosynthesis. The continuation of this and his earlier postdoctoral work at Stanford University on a NASA fellowship has become the basis for current studies regarding the search for microbial natural products in novel environments, including that of symbionts of medicinal plants.

The research has been communicated in more than 200 peer-reviewed publications and was awarded three Australian Museum Eureka Prizes and the Australian Academy of Science Fenner Medal. The research has been funded by more than $25 million of government and industry grants. He has been a visiting scientist in Germany, Japan, Brazil, the United States, China, and Korea. He is currently professor of microbial chemistry, Australian Research Council Federation Fellow, deputy director of the Australian Centre for Astrobiology, co-director of the Environmental Microbiology Initiative at UNSW, member of the Australian Society for Microbiology and Sydney Institute for Marine Science, and adjunct professor at the Chinese Academy of Science.

Brett's future plans are to characterize the genetics of marine and freshwater toxins and gain a better understanding of the factors that influence the production of toxins. Brett will also be looking at the mechanisms responsible for the complex biosynthesis of a range of pharmacologically active compounds to assist in the design and synthesis of novel bioactive compounds.

Contributors

Marina Aboal
Faculty of Biology
Laboratory of Algology
Department of Plant Biology
University of Murcia
Murcia, Spain

M. Leigh Ackland
Centre for Cellular and Molecular Biology
Deakin University
Melbourne, Victoria, Australia

Chhavi Agrawal
Molecular Biology Section
Laboratory of Algal Biology
Centre for Advanced Studies in Botany
Banaras Hindu University
Varanasi, India

Poonam Bhargava
Department of Biotechnology
Birla Institute of Scientific Research
Jaipur, India

Eduardo Costas
Veterinary Faculty
Department of Animal Production (Genetics)
Complutense University of Madrid
Madrid, Spain

Lutz A. Eichacker
Department of Mathematics and Natural Science
Centre for Organelle Research
University of Stavanger
Stavanger, Norway

Camino García-Balboa
Veterinary Faculty
Department of Animal Production (Genetics)
Complutense University of Madrid
Madrid, Spain

Attila Glatz
Institute of Biochemistry
Biological Research Centre
Hungarian Academy of Sciences
Szeged, Hungary

Raquel Gonzalez
Veterinary Faculty
Department of Animal Production (Genetics)
Complutense University of Madrid
Madrid, Spain

Govindjee
Department of Plant Biology
and
Department of Biochemistry
and
Center of Biophysics and Computational Biology
University of Illinois at Urbana-Champaign
Urbana, Illinois

and

School of Life Sciences
Jawaharlal Nehru University
New Delhi, India

Donat P. Häder
Department of Biology
Friedrich-Alexander University
Erlangen, Germany

Martin Hagemann
Department of Plant Physiology
Institute of Biosciences
University of Rostock
Rostock, Germany

Ibolya Horváth
Institute of Biochemistry
Biological Research Centre
Hungarian Academy of Sciences
Szeged, Hungary

Lee Hudek
Centre for Cellular and Molecular Biology
Deakin University
Melbourne, Victoria, Australia

Aran Incharoensakdi
Faculty of Science
Laboratory of Cyanobacterial Biotechnology
Department of Biochemistry
Chulalongkorn University
Bangkok, Thailand

Pilla Sankara Krishna
Department of Biotechnology
School of Life Sciences
University of Hyderabad
Hyderabad, India

Ashok Kumar
Microbial Biotechnology Unit
School of Biotechnology
Banaras Hindu University
Varanasi, India

Victoria Lopez-Rodas
Veterinary Faculty
Department of Animal Production (Genetics)
Complutense University of Madrid
Madrid, Spain

Surya Kant Mehta
Laboratory of Algal Physiology
and Biochemistry
Department of Botany
Mizoram University
Aizawl, India

Agnes Michalczyk
Centre for Cellular and Molecular Biology
Deakin University
Melbourne, Victoria, Australia

Pushpendra Kumar Mishra
Laboratory of Algal Physiology
and Biochemistry
Department of Botany
Mizoram University
Aizawl, India

Atefeh Nemati Moghaddam
Department of Chemistry
Institute for Advanced Studies in Basic
Sciences
Zanjan, Iran

Mohammad Mahdi Najafpour
Department of Chemistry
and
Center of Climate Change and Global Warming
Institute for Advanced Studies in Basic Sciences
Zanjan, Iran

Hitoshi Nakamoto
Department of Biochemistry and Molecular
Biology
and
Institute for Environmental Science
and Technology
Saitama University
Saitama, Japan

Brett A. Neilan
School of Biotechnology and Biomolecular
Science
University of New South Wales
Sydney, New South Wales, Australia

Christian Obinger
Division of Biochemistry
Department of Chemistry
BOKU-University of Natural Resources
and Applied Life Sciences
Vienna, Austria

Nadin Pade
Department of Plant Physiology
Institute of Biosciences
University of Rostock
Rostock, Germany

Sarita Pandey
Molecular Biology Section
Laboratory of Algal Biology
Centre for Advanced Studies in Botany
Banaras Hindu University
Varanasi, India

Leanne Andrea Pearson
School of Biotechnology and Biomolecular
Science
University of New South Wales
Sydney, New South Wales, Australia

Günter A. Peschek
Institute of Physical Chemistry
University of Vienna
Vienna, Austria

Roman Y. Pishchalnikov
Wave Research Centre
Prokhorov General Physics Institute
Russian Academy of Sciences
Moscow, Russia

Jogadhenu S.S. Prakash
Department of Biotechnology
School of Life Sciences
University of Hyderabad
Hyderabad, India

Amar Nath Rai
Department of Biochemistry
North-Eastern Hill University
Shillong, India

Ashutosh Kumar Rai
Microbial Biotechnology Unit
School of Biotechnology
Banaras Hindu University
Varanasi, India

Lal Chand Rai
Molecular Biology Section
Laboratory of Algal Biology
Centre for Advanced Studies in Botany
Banaras Hindu University
Varanasi, India

Snigdha Rai
Molecular Biology Section
Laboratory of Algal Biology
Centre for Advanced Studies in Botany
Banaras Hindu University
Varanasi, India

Richa
Centre of Advanced Study in Botany
Banaras Hindu University
Varanasi, India

John L. Sailo
Laboratory of Algal Physiology
and Biochemistry
Department of Botany
Mizoram University
Aizawl, India

Jian-Ren Shen
Division of Bioscience
Graduate School of Natural Science
and Technology
and
Faculty of Science
Department of Biology
Okayama University
Okayama, Japan

Dmitriy Shevela
Department of Chemistry
Chemistry Biology Centre
Umeå University
Umeå, Sweden

and

Department of Mathematics and Natural
Science
Centre for Organelle Research
University of Stavanger
Stavanger, Norway

Sisinthy Shivaji
Molecular Biodiversity Group
Center for Cellular and Molecular Biology
Hyderabad, India

Alok Kumar Shrivastava
Molecular Biology Section
Laboratory of Algal Biology
Centre for Advanced Studies in Botany
Banaras Hindu University
Varanasi, India

Prashant Kumar Singh
Molecular Biology Section
Laboratory of Algal Biology
Centre for Advanced Studies in Botany
Banaras Hindu University
Varanasi, India

Rajeshwar P. Sinha
Centre of Advanced Study in Botany
Banaras Hindu University
Varanasi, India

Ashish Kumar Srivastava
Institute of Environment and Sustainable Development
Banaras Hindu University
Varanasi, India

Lucas J. Stal
Department of Marine Microbiology
Netherlands Institute of Sea Research
Yerseke, the Netherlands

and

Department of Aquatic Microbiology
Institute for Biodiversity and Ecosystem Dynamics
University of Amsterdam
Amsterdam, the Netherlands

Mayashree B. Syiem
Department of Biochemistry
North-Eastern Hill University
Shillong, India

Zsolt Török
Institute of Biochemistry
Biological Research Centre
Hungarian Academy of Sciences
Szeged, Hungary

László Vígh
Institute of Biochemistry
Biological Research Centre
Hungarian Academy of Sciences
Szeged, Hungary

Rungaroon Waditee-Sirisattha
Faculty of Science
Department of Microbiology
Chulalongkorn University
Bangkok, Thailand

Nishikant V. Wase
Department of Biochemistry
University of Nebraska-Lincoln
Lincoln, Nebraska

Phillip C. Wright
Department of Chemical and Biological Engineering
Chemical Engineering at the Life Science Interface Institute
The University of Sheffield
Sheffield, United Kingdom

Saw Ow Yen
Department of Chemical and Biological Engineering
Chemical Engineering at the Life Science Interface Institute
The University of Sheffield
Sheffield, United Kingdom

Part I

Bioenergetics and Molecular Mechanisms of Stress Tolerance

1 Oxygenic Photosynthesis in Cyanobacteria

Dmitriy Shevela, Roman Y. Pishchalnikov, Lutz A. Eichacker, and Govindjee**

CONTENTS

1.1 INTRODUCTION

Oxygenic photosynthesis started about 3 billion years ago, when ancient cyanobacteria-like organisms evolved an apparatus capable of capturing and utilizing visible solar radiation (300–700 nm). By using electrons extracted from H_2O, the reduction of CO_2 to energy-rich carbohydrates with concomitant release of O_2 had become possible (for recent reviews on evolution of photosynthesis, see Refs. [1–7]). The unique advent of O_2 released by the first cyanobacteria and its subsequent accumulation in the Earth's atmosphere was, undoubtedly, the *biological Big Bang* [8] for the evolution of the whole biosphere. It created an aerobic condition and the requisite background for the development and sustenance of aerobic metabolism and more-advanced forms of life [9–13]. Another great input of cyanobacteria is the evolutionary event of *endosymbiosis* [14]. Cyanobacteria are the photosynthetic ancestors of plastids in algae and plants (for reviews, see [5,15,16]).

* Corresponding authors: dmitriy.shevela@alumni.tu-berlin.de; gov@illinois.edu

Cyanobacteria (formerly classified as "blue-green algae") are one of the largest and versatile groups of prokaryotes of enormous biological importance. About 20%–30% of global primary photosynthetic productivity originates from cyanobacteria [17]. This corresponds to the yearly fixation of about 20–30 Gt of CO_2 into biomass and release of about 50–80 Gt of O_2 in the atmosphere by these oxygenic prokaryotes [18,19]. In addition, many cyanobacteria can fix atmospheric N_2 into a biologically accessible form and thereby play a key role in the nitrogen cycle of biosphere [20,21]. Cyanobacteria are highly adaptable; they exhibit wide ecological tolerance and gliding mobility: they can be found almost in any environment, including extreme ones (e.g., benthos, plankton, cold and hot deserts, antarctic dry valleys, tropical rain forests) [22,23]. Although all existing cyanobacteria (except recently discovered oceanic unicellular N_2-fixing cyanobacteria from UCYN-A group [24], discussed in Section 1.3.1) have the ability to perform oxygenic photosynthesis (they use H_2O as electron donor), some are able to grow as anaerobic photoautotrophs using H_2S as an alternative electron donor [25]. This represents unique additional capability of anoxygenic photosynthesis in these organisms, while anoxygenic photobacteria are not able to utilize H_2O as a substrate and produce O_2.

The mechanism of oxygenic photosynthesis in cyanobacteria remarkably resembles that of oxygenic eukaryotes (algae and higher plants). This allows us to use cyanobacteria as a suitable model to study different aspects of oxygenic photosynthesis and its regulation that is often difficult to study in higher plants or algae. Similar to all photosynthetic eukaryotic organisms, cyanobacteria share the use of unique reaction centers (RCs), photosystem I (PSI) and photosystem II (PSII), to drive light-induced electron transfer from H_2O to $NADP^+$ (the oxidized form of nicotinamide adenine dinucleotide phosphate); its reduced form, NADPH, is used to power the synthesis of carbohydrates. Like algae and plants, cyanobacteria have two light reactions that work in series, as known from experiments referred to as the Emerson's enhancement effect (see, e.g., [26,27]), and from antagonistic effects of light I (absorbed by PSI) and light II (absorbed by PSII) on specific components of electron transfer (see, e.g., [28,29]).

Like all other oxygenic photosynthesizers, cyanobacteria contain the photosynthetic pigment *chlorophyll* (Chl) *a* [30] (for an overview on why Chl *a* was chosen by nature, see [31]). In addition to Chl *a*, most cyanobacteria contain *carotenoids* (Cars) [32,33] and *phycobilins* (phycocyanin, allophycocyanin, and, in some species, phycoerythrin) [34]. Phycobilins are not present in plants, but in cyanobacteria they are organized in large light-harvesting multiprotein complexes called *phycobilisomes* (PBSs) [35–41]. The ultrastructure of PBSs may vary among cyanobacteria and is dependent on the growing conditions. There are some cyanobacteria, e.g., prochlorophytes, that contain Chl *b* in addition to Chl *a* [42], while Chl *d* is known to be dominant in an apparently widespread *Acaryochloris*-like organisms [43–46]. Moreover, some cyanobacteria can be transformed to contain Chl *b*, thereby representing their great flexibility [47].

Cyanobacterial cells are surrounded by two membranes: an outer one, which forms *the cell wall* (made of murein), and an inner one, *the cytoplasmic membrane*, which separates *the cytoplasm* from *the periplasm* (see Figure 1.1 and its legend). The light reaction of oxygenic photosynthesis takes place in the so-called *thylakoid membranes* that occur in pairs; the space between the pair is called *the lumen*, and the space between two pairs is contiguous with *the cytoplasm*. One of the major differences between cyanobacteria and photosynthetic eukaryotes is that in cyanobacteria respiratory and photosynthetic redox-active protein complexes share a *common* thylakoid membrane (see [48] and Figure 1.1). The thylakoid membrane in cyanobacteria does not form grana as it does in plants and algae. Moreover, there is even a cyanobacterium *Gloeobacter violaceus* that lacks thylakoids, and the photosynthetic pigments are associated with the cytoplasmic membrane [49]. Nevertheless, despite some minor differences in the composition of the redox-active complexes, the photosynthetic electron transport chain of cyanobacteria is very similar to that of plants and algae. In all oxygenic organisms (both prokaryotic and eukaryotic), the result of the light-driven electron transport is the oxidation of H_2O coupled with evolution of O_2, reduction of $NADP^+$ to NADPH, and phosphorylation of ADP to ATP (see Figure 1.2).

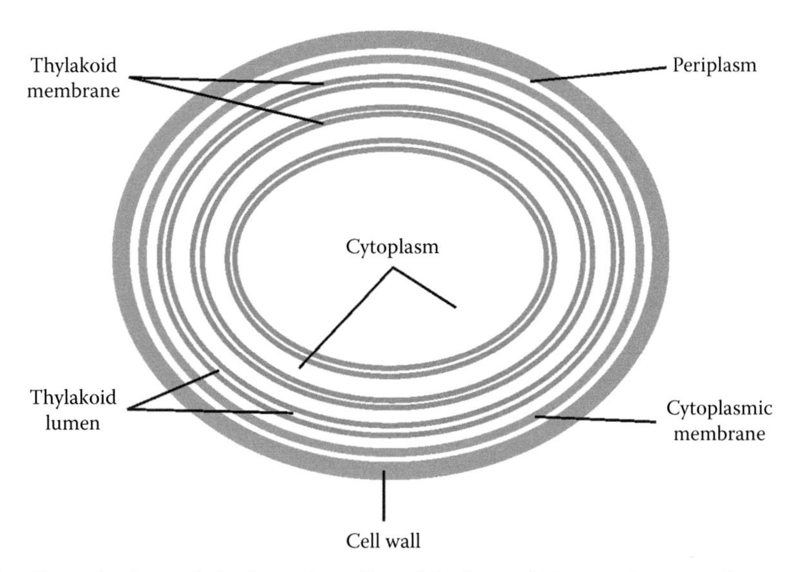

FIGURE 1.1 **(See color insert.)** A schematic outline of the intracellular membranes and compartments in a cyanobacterial cell. The thylakoid membranes (green) contain chlorophyll *a* and perform both photosynthetic and respiratory electron transport, while the cytoplasmic membrane system (yellowish), which contains carotenoids, is involved only in respiration. As a consequence of photosynthetic and respiratory electron transport in thylakoid membranes, protons are brought into the thylakoid lumen, the space between a pair of thylakoid membranes. The resulting proton gradient across the thylakoid membrane is utilized for the synthesis of ATP. (Modified and adapted from Vermaas, W.F.J., *Encyclopedia of Life Sciences (ELS)*, John Wiley & Sons, Ltd, London, U.K., 2001. Copyright Wiley-VCH Verlag GmbH & Co. KGaA.)

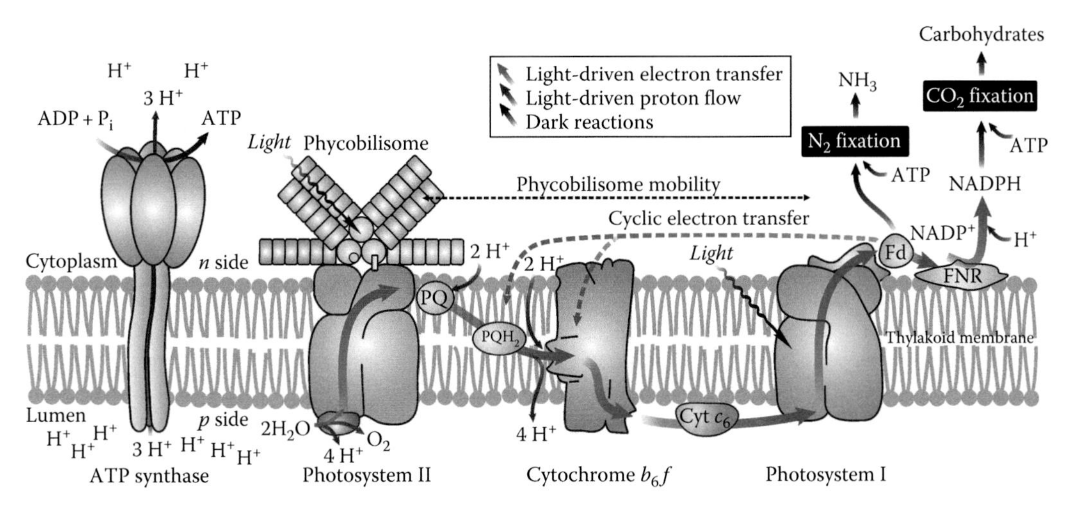

FIGURE 1.2 **(See color insert.)** A schematic representation of the protein complexes involved in light-induced electron and proton transfer reactions of oxygenic photosynthesis in cyanobacteria. The arrows indicating the light-driven electron transfer and proton flow as well as some dark reactions are colored individually, as shown in the figure. Looking at the abbreviated components from the left of the diagram: PQ, plastoquinone; PQH_2, plastoquinol; Cyt c_6, cytochrome c_6 (also known as Cyt c_{553}); Fd, ferredoxin; FNR, ferredoxin-NADP reductase. Note that although the diagram does not show PBSs that are attached to PSI, PBSs can be redistributed to PSI due to their mobility (indicated by black dashed arrow). Also note that cyanobacteria use Cyt c_6 (as shown) or plastocyanin (not shown) to transfer electrons from Cyt b_6f to PSI. For the sake of simplicity, respiratory complexes (type 1 *NADPH dehydrogenase* that oxidizes NADPH to $NADP^+$; *Succinate dehydrogenase* that oxidizes succinate to fumarate and reduces PQ to PQH_2; and a *(terminal) oxidase* that reduces O_2 to water) are not shown. For further details, see text.

The structure of the O_2-evolving apparatus in the cyanobacterial membranes is highly conserved throughout evolution. Many fundamental questions of oxygenic photosynthesis that are often difficult to approach in plants or other eukaryotic photosynthetic organisms are therefore investigated in cyanobacterial model systems. Cyanobacteria show rapid growth in nature and under laboratory culture conditions; they owe a large part of their evolutionary success to their tremendous metabolic flexibility. The simplicity of a single cell system and ease to manipulate them genetically as compared to the multicellular higher plant models has also contributed to cyanobacteria becoming important model systems to study responses to abiotic stress.

In this chapter, we provide a basic introduction to the light-induced reactions of photosynthesis in cyanobacteria, on the two photosystems, their structure and function leading to NADP production, and ATP synthesis. A glimpse of our views on evolution of cyanobacteria is also presented. More detailed information on the photosynthetic and respiratory processes in cyanobacteria and on the historical discoveries in this field can be obtained from the extensive available literature (e.g., see [48,50–59]).

1.2 OVERVIEW OF PHOTOSYNTHETIC ENERGY CONVERSION IN CYANOBACTERIA

1.2.1 The Light-Induced Reactions and Photosynthetic Electron Transport Chain

The thylakoid membrane in most cyanobacteria is the site of the photosynthetic light reactions (see Figures 1.1 and 1.2). However, whereas in algae and plants the thylakoid membrane is located in a special organelle (chloroplast), in cyanobacteria the membrane is within the cytoplasm [48].

The initial event in photosynthetic light reactions of cyanobacteria begins with the absorption of light (photons) by large antenna systems, PBSs, attached to the cytoplasmic surface of photosynthetic membrane. PBSs are made of the *phycobiliproteins* that covalently bind phycobilins (open-chain tetrapyrroles) in a special geometric arrangement (Figure 1.2; for further details, see Section 1.3.4 and [38,60–62]). Phycobilins deliver the energy of absorbed light (excitation energy) (discussed in Section 1.4.1) to the large pigment-protein RC complexes, PSII and PSI (for details, see Section 1.3.3; for recent reviews, see [63–67] and references therein), integrated into the thylakoid membrane. Cyanobacteria contain relatively low amount of PSII complexes as compared to PSI (the PSI/PSII ratio may vary from ~3 up to 5.8) [68–71], and PBSs are primarily associated with PSII (as the external antenna). However, under certain conditions, PBSs are redistributed to PSI, thereby regulating the efficiency of excitation energy transfer between the two photosystems [70,72,73]. In a similar way, Chl *a*-/Chl *b*-containing membrane-integral light-harvesting complexes (LHCs) in plants show an energy equilibration between the photosystem complexes [74]; however, LHCs are absent in thylakoids of cyanobacteria. Normally, the PBS system covers absorption in the wavelength range of 300–700 nm. However, *Acaryochloris*-like cyanobacteria with Chl *d*-dominated photosynthetic antenna system do extend the absorption wavelength range up to 775 nm [43,75].

Conversion of solar energy into chemical forms of energy is the result of the two photochemical RCs (PSII and PSI) acting in tandem. Delivery of excitation energy to the RC-Chl *a* molecules, referred to as P680 (special Chl *a*-complex in PSII; for alternative definition of P680, see [76]) and P700 ("heterodimeric" Chl *a*/*a*′-complex in PSI), initiates the energy conversion process. Due to the redox-active cofactors embedded into a protein matrix of both photosystems (see Sections 1.3.3.1 and 1.3.3.2, and references therein), the photochemical acts within RCs involve fast, sequential electron transfers that result in stabilized charge separation, stepwise "extraction" of electrons from water and their transfer to $NADP^+$. Figure 1.3 illustrates the linear (noncyclic) electron transfer pathway in a way that all redox-active cofactors that make this transfer possible are arranged according to their redox potentials. This arrangement, known as *the Z-scheme*, was proposed by Hill and Bendall in 1960 ([77]; for historical perspective on the evolution of the Z-scheme over the last 50 years, see [78]).

The primary photochemical reaction in PSII involves formation of the singlet excited state of P680, P680* (Figure 1.3), delocalized among the ensemble of four RC Chl *a* molecules, either

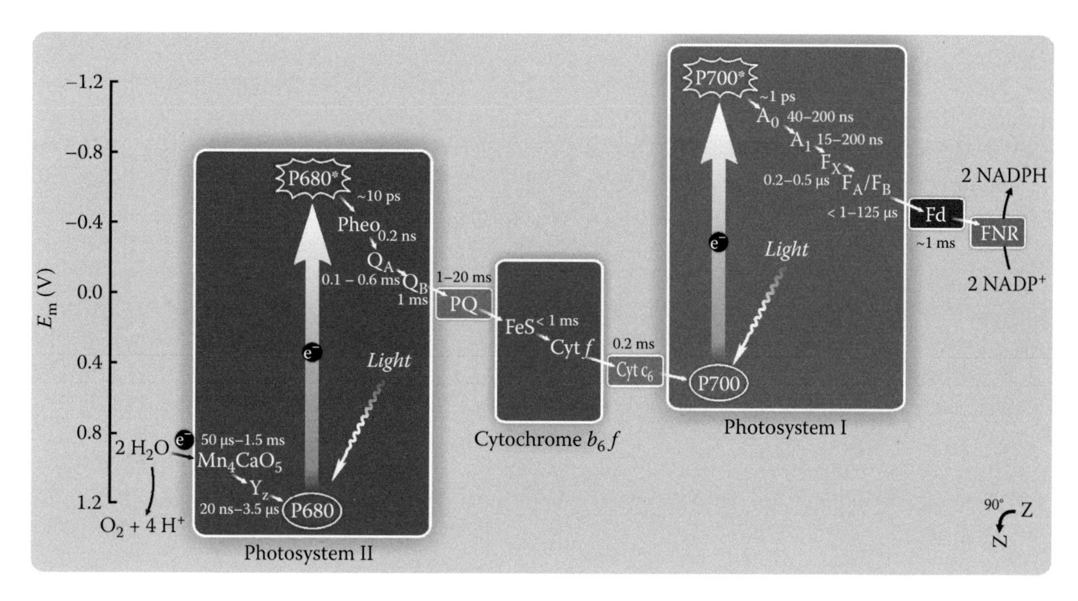

FIGURE 1.3 The Z-scheme of oxygenic photosynthesis in cyanobacteria representing the energetics of linear electron transfer from water to $NADP^+$ plotted on redox midpoint potential (E_m, at pH 7) scale. The two white upward arrows symbolize the transition of the RC Chl *a* molecules (P680 in PSII and P700 in PSI) from the ground to the (singlet) excited state (P*) attained mostly after excitation energy transfer from their respective light-harvesting antenna systems and to a lesser extent by direct absorption of photon ($h\nu$) (see Sections 1.3.4 and 1.4.1). The numbers 680 and 700 are the wavelengths (in nm) of the red absorption maxima for the Chl *a* molecules of the RCs of PSII and PSI. The diagram also shows half-times of several electron transfer steps. In all likelihood, the "bottleneck" reaction is at the PQ/PQH_2 level (1–20 ms). Abbreviations of the components involved in the electron transfer (from the left of the diagram): Mn_4CaO_5, tetra-nuclear manganese-calcium cluster with five oxygen atoms; Y_Z, redox-active tyrosine residue; Pheo, pheophytin; Q_A, primary quinone electron acceptor of PSII; Q_B, secondary quinone electron acceptor of PSII; PQ, a pool of mobile PQ molecules; FeS, an iron-sulfur protein (known as Rieske FeS protein); Cyt *f*, cytochrome *f*; A_0, a special Chl *a* molecule; A_1, a phylloquinone molecule; F_X, F_A, F_B, iron-sulfur clusters. Other abbreviations are as in Figure 1.2. For further details, see text.

upon excitation energy transfer from antenna molecules or upon absorption of a photon with a wavelength of <680 nm. The photochemistry-driving energy per each photon absorbed by PSII is equal to the energy of a 680 nm photon (E_{680}), i.e., 1.83 eV. This results in stable and directed charge separation (see Section 1.4.1 and references therein) between the Chl *a* molecules and the nearby pheophytin (Pheo) molecule, followed by the formation of the radical pair $P680^{\bullet+}Pheo^{\bullet-}$ with less than 10% energy loss from E_{680} [79,80]. The created cation radical $P680^{\bullet+}$, which has one of the highest known oxidizing potential in nature (~1.25 V) [81,82], is strong enough to energetically drive the sequential oxidation of water, *via* a redox-equivalent-accumulating catalyst, a Mn_4CaO_5 cluster, which links the one-electron photochemistry with the four-electron oxidation chemistry of two water molecules. The result of water oxidation is the evolution of molecular oxygen and the release of protons into the thylakoid lumen (Figure 1.2). The free energy stored by PSII due to the transfer of one electron from water to the mobile electron acceptor plastoquinone (PQ) is about 50% of E_{680} [79,80]. The electron transfer in PSII and its composition is further discussed in Section 1.3.3.1. Photosynthetic water splitting is briefly outlined in Section 1.4.2 and in more details in Chapter 2.

PSI, which has some similarities to PSII, captures (upto) <700 nm light energy (E_{700} = 1.77 eV) to drive the redox reactions of the electron transfer in that system (Figure 1.3). The redox potential for the primary electron donor P700 in cyanobacteria was estimated to be 400–450 mV, while in higher plants it has a slightly higher value (~470 mV) [83,84]. Although, this oxidizing potential is not high enough to extract electrons directly from water, it is enough to drive transmembrane electron transfer between the external electron donor (cytochrome c_6 (Cyt c_6) and/or plastocyanin (PC)) in the thylakoid lumen and acceptor (ferredoxin (Fd)) in the cytoplasm (Figures 1.2 and 1.3). We will come back to the structural organization of PSI and the electron transfer pathways in Section 1.3.2.2.

The integral membrane cytochrome b_6f (Cyt b_6f) complex plays a key role in mediating the photo-induced electron transfer between the two photosystems and in increasing the number of protons pumped across the membrane into the lumen (in addition to that associated with the water-splitting PSII (Figure 1.2)). It is a large multi-subunit protein with several prosthetic groups, which has high structural and functional similarity between those from cyanobacteria and plants. Structure and function of this complex has thoroughly been discussed in reviews [85–90] and is not included in our chapter. The mobile lipophilic electron carrier PQ, in the membrane, connects PSII with Cyt b_6f, and the mobile soluble electron carrier Cyt c_6 (also known as Cyt c_{553}) in the thylakoid lumen connects Cyt b_6f with PSI [91–93]. Under copper-replete conditions, many cyanobacterial species substitute the iron-containing Cyt c_6 by the copper-containing redox carrier, PC [94]. Interestingly, in the absence of both Cyt c_6 and PC, *Synechocystis* sp. PCC 6803 still shows photoautotrophic growth indicating the high flexibility of this prokaryotic photoautotroph to cope with stress [95]. We note that cyanobacteria share the use of PQ pool, Cyt b_6f complex, and PC/Cyt c_6 for both photosynthetic and respiratory electron transport pathways (the latter is not discussed in this chapter) (for further details, see [48,96]). Finally, production of NADPH from NADP is catalyzed by the membrane-associated flavoprotein ferredoxin-NADP$^+$ reductase (FNR) (see Figure 1.2 and Section 1.4.3; for recent reviews, see [97,98]).

Photosynthetic electron transfer, energized by the two photosystems, also involves light-induced proton flow (Figure 1.2)), which establishes a proton electrochemical potential difference ($\Delta\psi$) across the thylakoid membrane (the cytoplasmic side being called the *n* (negative) side; and the luminal side being called the *p* (positive) side). This proton gradient, more precisely *the proton motive force* (PMF), which includes membrane potential gradient, is utilized by the ATP synthase for the phosphorylation of ADP to ATP. This occurs when the protons return to the cytoplasm through the protein complexes of the ATP synthase (for further details, see Section 1.4.4 and [99–102]). Under certain conditions, cyclic electron flow occurs from the reducing side of PSI, through PQ pool and/or Cyt b_6f complex and back to PSI (depicted in Figure 1.2) [103–105]. This cyclic electron flow increases proton pumping into the lumen, thereby increasing the synthesis of ATP.

Thus, the light-driven electron transport from H_2O to NADP$^+$ catalyzed by PSII and PSI results in the formation of the energy-rich compounds NADPH and ATP and leads to the evolution of O_2. The energy stored in NADPH and ATP is subsequently used to drive the reactions of CO_2 fixation. This production of energy-rich carbohydrates by the metabolic C-3 or the Calvin–Benson cycle does not directly require light (reviewed in [106–109]; it is not discussed further in this chapter). For a background in CO_2 fixation and concentration mechanism in cyanobacteria, see [110]. We also note here that in cyanobacteria, ATP can also be utilized for N_2 fixation and other cell processes.

1.2.2 Time Sequence of Light Reactions: From Picoseconds to Milliseconds

Here we will discuss the time sequence of events of photosynthetic light reactions that lead to evolution of O_2 and production of NADPH in cyanobacteria and other oxygenic photosynthesizers.

Photosynthetic electron transfer from water to NADP$^+$ is like a *bucket brigade* from one intermediate to the other. One can also imagine this process as a *relay race*. It begins simultaneously by charge separation at P680 (in PSII) and at P700 (in PSI), and the process is over almost simultaneously within a few ps (see a Z-scheme in Figure 1.3; most of the times given are essentially half-times for single-electron transfers). In PSII, P680 is oxidized and Pheo is reduced; and in PSI, P700 is oxidized and a specific Chl *a* molecule A_0 (first stable electron acceptor) is reduced [65,80,111–115]. Light energy is, thus, converted to chemical energy! Within ~200 ps, the electron on reduced Pheo (in PSII) moves to Q_A (a tightly bound primary quinone electron acceptor of PSII), and the electron on reduced A_0 (in PSI) moves to A_1 (phylloquinone of PSI) and to F_X within 200 ns (the first iron-sulfur cluster of PSI) (see [67,116] and references therein). In PSII this is followed by reduction of $P680^{•+}$ by Y_Z (redox-active tyrosine residue) within the time range of 20 ns–35 μs [67,81,117]. Then, in the 200 μs range, several steps occur: an electron on the reduced Q_A moves to Q_B (secondary quinone acceptor loosely bound to PSII), and the oxidized Y_Z is reduced by the Mn_4CaO_5 cluster in the so-called oxygen-evolving complex (OEC) [118]; if Cyt c_6/PC was in the reduced state, then during this time (~200 μs), $P700^{•+}$ would also be reduced (see, e.g., [119]).

The Q_B has a very long life, and a proton from His-252 on the D1 protein stabilizes its negative charge [120–122]. There is a "two-electron gate" at the Q_B-site [123]. After a second light reaction, electrons arriving from the excited P680, *via* Pheo, and Q_A reduce Q_B^- to Q_B^{2-}; this occurs in 400–600 μs; however, the reaction is "slow" because of electrostatic repulsion between negative charges on Q_A^- and Q_B^- [124]. The negative charge on Q_B^{2-} is further stabilized by a proton, arriving, *via* several amino acids, from a bicarbonate ion (hydrogen carbonate), bound on the non-heme-iron (NHI) between Q_A and Q_B (see [125] and references cited in [122]). As discussed in Section 1.4.2, oxygen evolution is a period four clock. After four light reactions, oxygen is evolved (from two H_2O) and two molecules of mobile PQ are reduced to plastoquinol (PQH_2). These last reactions (1) the last step of oxygen evolution and (2) reduction of PQ to PQH_2 in addition the electron flow from reduced F_X to Fd in PSI all take place in the ms time range (see [67,116] and references therein).

Overall the slowest reactions of electron transfer from water to $NADP^+$ involve (1) formation of PQH_2 from PQ, (2) uptake of protons from the stromal side (the *n* side), (3) diffusion of PQH_2 to the Cyt b_6f complex, and (4) the release of protons to the luminal side (the *p* side). These aforementioned processes take ~20 ms and are the *bottleneck* reactions of the entire electron transfer (see, e.g., [126,127]). Still, it is difficult to decipher which of the steps is the slowest because this depends upon other conditions of the system. Further, the time taken for the reduction of $NADP^+$ to NADPH depends upon various conditions and the status of the enzyme FNR. Apparently, in continuous strong light, the entire process of electron transfer from water to $NADP^+$ requires about ~500 ms, when all the PSII and PSI electron acceptors are reduced (see, e.g., [128]). This time is also seen by the appearance of the peak "P" in Chl *a* fluorescence transient as in *Acaryochloris marina* [129]. However, superimposition of other phenomena, such as *state changes*, and the redox status of the PQ pool (due to overlap of respiratory and photosynthesis reactions) do not allow us to observe this time through Chl *a* fluorescence measurements in most cyanobacteria. For reviews on the relation of Chl *a* fluorescence transient and photosynthesis, we refer readers to reviews [130–132].

1.3 PHOTOSYNTHETIC UNITS OF CYANOBACTERIA

1.3.1 Photosynthetic Reaction Centers: Evolutionary Perspective

In every photosynthetic organism, the primary events of energy conversion involve the so-called *photosynthetic unit* (reviewed in [133]). This concept includes a certain RC protein complex performing transmembrane charge separation and the light-harvesting pigment-containing system delivering the energy of the captured photons to the RC. All photosynthetic RCs (both prokaryotic and eukaryotic) share structural similarities; they all contain an integral membrane-protein complex of homodimeric or heterodimeric nature to which pigments (Cars and Chls) and redox-active cofactors (such as Chls, quinones, or iron-sulfur clusters) are bound [64,66,134,135]. However, the differences in the nature of the terminal electron acceptors allow classification of all existing RCs into two types (for reviews, see [1,7,136,137]): *FeS-type RCs* and *Q-type RCs*. Those that have iron-sulfur clusters as the terminal acceptor (e.g., the RC of green sulfur bacteria and PSI of cyanobacteria and oxygenic eukaryotes) belong to the FeS-type RCs (also called *type I RCs*), while those with quinones as final acceptor belong to the Q-type RCs (or *type II RCs*) (e.g., the RCs of purple bacteria and PSII of cyanobacteria and other oxygenic organisms) (Figure 1.4). Although structural

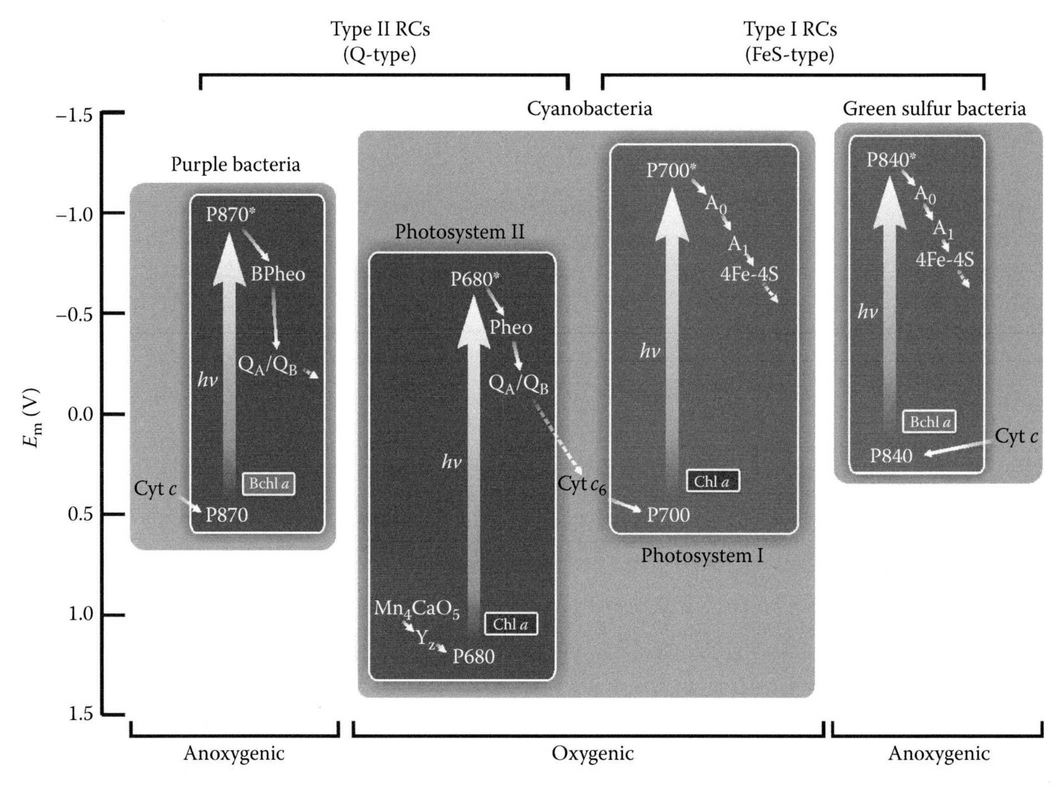

FIGURE 1.4 Photosynthetic RCs in anoxygenic (purple bacteria and green sulfur bacteria) and oxygenic (cyanobacteria) organisms. For each RC the electron donors, as well as the electron acceptors, are shown and plotted on the redox midpoint potential scale (E_m, at pH 7). Depending on the final electron acceptors (quinones (Q_A/Q_B) or iron-sulfur clusters (4Fe-4S)), the RCs are classified as Q- or FeS-type (also called as type II and type I RC, respectively). The numbers after P represent the long-wavelength absorption maxima of the RC pigments and are inversely related to the absorbed energy (length of straight white arrows). For the sake of simplicity, some redox components of the electron transfer chain are omitted (indicated by dashed arrows). Cyt *c* in purple bacteria symbolizes all the four different forms of cytochrome *c* used by these bacteria. A_0 denotes an electron acceptor: bacteriochlorophyll (in green sulfur bacteria) or chlorophyll (in PSI of cyanobacteria). Other specific abbreviations are BPheo, bacteriopheophytin; 4Fe-4S, RC-associated iron-sulfur centers: F_X, F_A, and F_B; Q_A/Q_B, the primary (Q_A) and secondary (Q_B) quinone electron acceptors. See text and [7,56,134] for further details. (Based on the data in Hohmann-Marriott, M.F. and Blankenship, R.E., *Annu. Rev. Plant Biol.*, 62, 515, 2011; Govindjee and Shevela, D., *Front. Plant Sci.*, 2, 28, 2011; Wraight, C., Reaction centers, electron flow, and energy transduction, in *Photosynthesis*, Govindjee, Ed., Academic Press, New York, p. 17, 1982.)

protein and cofactor similarities between the two types of RCs strongly support the idea that they all have a common evolutionary origin, there is presently no clear agreement on whether the earliest photosynthetic RC was the type I (similar to homodimeric RC of current green sulfur bacteria *Chlorobiaceae*), the type II (similar to one of purple bacteria), or an intermediate between these two types (called *type 1.5*) (see e.g., [1,66,138–144]).

The evolutionary origin of direct ancestors of the O_2-evolving cyanobacteria is even more mysterious and not yet resolved. We know that oxygenic cyanobacteria (as well as all other O_2-evolving organisms) have both types of RCs (PSII and PSI), whereas all anoxygenic photosynthetic bacteria have just one of these types (either type I or type II RC) (Figure 1.4). This indicates that the origin of oxygenic photosynthesis begins with the origin of the functionally linked RCs (PSII and PSI) in a direct ancestor of O_2-evolving cyanobacteria-like organisms. Two different hypotheses have been proposed to explain how both type I and II RCs ended up in the first cyanobacteria (for details, see [1,3,7,11,144]). Briefly, *the selective loss hypothesis* assumes that the two types of RCs evolved separately in one ancestor of cyanobacteria and that the anoxygenic bacteria with just one RC are derived from this organism by a loss of either type I or type II RCs. Later, two RCs in a primitive cyanobacterium became linked and achieved the ability to oxidize water and evolve O_2 and to reduce $NADP^+$. According to *the fusion hypothesis*, type I and type II RCs developed separately in distinct anoxygenic bacteria and later were brought together in one cyanobacteria-like organism by a large-scale lateral gene transfer.

What is unique in cyanobacterial flexibility is that it may reflect their position in evolutionary history. Examples are shown by Vermaas et al. [145], who observed that electrons from PSII could be used by an oxidase in the absence of PSI, and by Wang et al. [146], who reported that several PSI-minus mutants of *Synechocystis* sp. PCC 6803 were able to evolve O_2 in the light in the presence of glucose for up to 30 min. On the other hand, some "anoxygenic" cyanobacteria among the so-called UCYN-A group of oceanic unicellular N_2-fixing cyanobacteria have been recently discovered [24,147]. These cyanobacteria are not able to perform oxygenic photosynthesis because they completely lack a functional PSII apparatus, although PSI was found to be intact (it is not known yet, however, how PSI functions in the absence of PSII in these organisms). These two "opposite" examples clearly reveal global importance of water-splitting PSII for oxygenic photosynthesis.

How and when the ability to oxidize water was added to type II RC remains a mystery. Undoubtedly, the development of a strongly oxidizing RC (PSII), together with the invention of the catalytic site for water oxidation (Mn_4CaO_5-protein complex) that is capable of collecting and storing oxidizing equivalents, formed the central stage in the transition from anoxygenic to oxygenic photosynthesis [3,148–150].

1.3.2 Primary Reaction Center Pigments in Cyanobacteria

Which invention was required for the type II RC (primitive PSII) to become so highly oxidizing? It is assumed that the evolutionary development of a specific protein environment of the RC that contained a photoactive pigment with a redox potential (E_m) higher (more positive) than +0.82 V for oxidation of H_2O (at pH 7.0) was that "invention" in a direct progenitor of cyanobacteria [1,31,150]. Unfortunately, we do not know the E_{m} value of the RC photoactive pigment in these ancient organisms. What we know is that in modern water-oxidizing photoautotrophs, the E_{m} value of the RC photoactive Chl *a* (cation radical $P680^{\bullet+}$) is about +1.25 V [81,82] and that this value exceeds the oxidizing power of Chl *a* (Chl $a^{\bullet+}$) *in vitro* (in acetonitrile) by about 0.4 V [151]. This clearly indicates that such energetics (E_{m} value) for P680 *in vivo* (in PSII) is achieved not only due to the chemical nature of Chl *a* but to a great extent due to the specific structural arrangement of the surrounding protein matrix of the PSII RC [80,152,153]. Although the basic principles of the photosynthetic RCs are very similar and highly conserved throughout evolution, there are no other natural RCs that contain photoactive pigments (cation radicals) with such strong oxidizing potential as $P680^{\bullet+}$ of PSII. Thus, the E_{m} of $P680^{\bullet+}$ is at least 0.5 V above the cation radicals in the RCs of all anoxygenic bacteria that contain bacteriochlorophyll (BChl) *a* (Figure 1.4) and other forms of BChl (*b* and *g*) [154]; for an overview of possible scenarios for the evolution of Chls and BChls, see [11,56,150,155–158]). The evolutionary replacement of BChl *a* by Chl *a* in oxygenic photosynthetic organisms must have been "inspired" by the unique physicochemical property of Chl *a* to form a sufficiently stable radical pair $P680^{\bullet+}Q_A^{\bullet-}$ [153]. Such functional compatibility of Chl *a* with neighboring cofactors of electron transfer and protein surroundings makes this pigment to be unique and essential in the RCs of all water-splitting photosynthesizers [31,153].

Are photoactive RC pigments in oxygenic organisms always a Chl *a*-complex? For many years this was believed to be true. However, the unique role of Chl *a* for P680 became a matter of debate after 1996, when Miyashita et al. [43] discovered the Chl *d*-dominant O_2-evolving cyanobacterium *A. marina*. This marine cyanobacterium was found to contain more than 95% Chl *d* and only trace amount of Chl *a* (~3% of the total Chl) [43]. Such an unusual pigment composition results in a unique constitution of the photosystems in cyanobacteria from the genus *Acaryochloris*. Thus far, these cyanobacteria are the only known organisms that perform oxygenic photosynthesis by using Chl *d* as the RC photopigment for both photosystems (PSI and PSII) [159,160]. Thus, the photoactive RC pigment of PSI, P700, which is Chl *a*/Chl *a′* heterodimer in all oxygenic photosynthesizers, was shown to be replaced by a Chl *d*/Chl *d′* heterodimer in *A. marina* [161]. Since this pigment shows a flash-induced absorbance difference maximum at ~740 nm, the primary donor of PSI is termed P740 [159], corresponding to P700 in other O_2-evolving organisms. The E_m of P740/P740$^{\bullet+}$ was found to be +335 mV [159], which is 70–120 mV lower than the E_m of P700/P700$^{\bullet+}$ of the usual cyanobacterial Chl *a*-containing PSI [83]. It is still an open question whether the special Chl pair (P_{D1}/P_{D2}) of the photoactive pigment in *Acaryochloris* PSII RC is a Chl *d*/Chl *d* homodimer [160,162,163] or a Chl *a*/Chl *d* heterodimer [164–166]. At the same time, there is an agreement that the primary electron acceptor of PSII in *A. marina* is Pheo *a*, not Pheo *d* [162,167,168]. However, not all RC Chls *a* are substituted by Chl *d* in photosystems of this unusual cyanobacterium (reviewed in [163]) indicating that Chl *a* is still the unique RC photoactive pigment for driving oxygenic photosynthesis (for a viewpoint on why Chl *a* is unique among the other Chl species for serving RCs of oxygenic organisms, see [31]). Despite this, the Chl *d*-containing *A. marina* provides an interesting system to further explore the energetics of the primary reactions of photosynthesis and water-splitting reaction by PSII (see e.g., [166,168–172]).

1.3.3 Cyanobacterial Photosystems: Structure and Function

1.3.3.1 Photosystem II

1.3.3.1.1 General Structural Organization

PSII is a large multimeric pigment-protein complex that exists as a dimer with a total weight of ~700 kDa. Figure 1.5 shows general arrangement of cyanobacterial dimeric PSII based on the recent PSII crystal structure from *Thermosynechococcus vulcanus* at 1.9 Å resolution [125]. Each monomer of PSII from cyanobacteria comprises 17 integral membrane-protein subunits, 3 peripheral protein subunits on the luminal side of the complex, and close to 90 cofactors (Figures 1.5 and 1.6) [63,64,67,125,173–176].

All redox cofactors required for photochemical charge separation and for water splitting are located on the D1 (PsbA) and D2 (PsbD) polypeptides that form the D1/D2 heterodimer. Each polypeptide has five transmembrane α-helices (TMHs) (Figure 1.5A). The D1 and D2 proteins show sequence homology with the L and M subunits of the photosynthetic RC of anoxygenic purple bacteria [177]. The D1/D2 heterodimer of cyanobacteria coordinates 6 Chl *a* molecules (among which are the pair of P_{D1}/P_{D2} and 2 accessory Chls (Chl_{D1} and Chl_{D2}) now included in the symbol P680), 2 Pheo *a* molecules ($Pheo_{D1}$ and $Pheo_{D2}$), 2 or 3 quinones (Q_A on the D2 side and Q_B on the D1 side; the third quinone Q_c identified in PSII crystal structure at 2.9 Å resolution [178] was not seen in the recent PSII structure at 1.9 Å resolution [125]), at least 2 β-Cars and 1 NHI (between Q_A and Q_B) with an associated (bi)carbonate ion (HCO_3^-/CO_3^{2-}) [125,174]. Although the symmetrically related cofactors located on the D2-branch do not participate in electron transfer through PSII ("inactive" branch) (Figure 1.6), some of them are known to play protective roles against photo-induced damage of PSII. In fact, the overall structure and arrangement of the cofactors on the reducing side of the D1/D2 heterodimer closely resembles that of their counterparts from purple bacteria [179]. The only clear exception here is that in PSII, the NHI is bound to (bi)carbonate and not to Glu as in anoxygenic bacterial RC (recently reviewed in [73,122]) and that the Q_B site is slightly larger and in a closer contact with the cytoplasmic surface than in anoxygenic bacteria.

Two redox-active Tyr residues, Y_Z (D1-Tyr161) and Y_D (D2-Tyr161), are located on the electron-donor side of the D1/D2 heterodimer (see Figure 1.6). The Pheo *a* molecules (discovery of Pheo is reviewed in [180]) and the Tyr residues on D1 and D2 are homologous. However, they are not

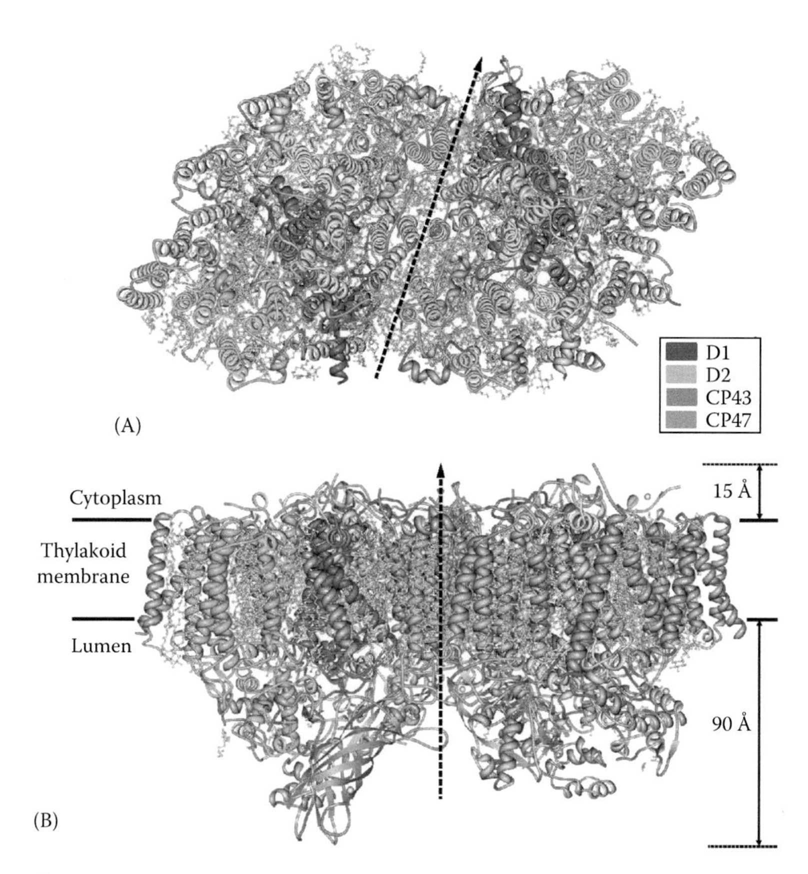

FIGURE 1.5 (See color insert.) Overall structure of the cyanobacterial PSII dimer from *T. vulcanus*. (A) View of the dimer from the cytoplasmic side. (B) Side-on view (perpendicular to the membrane normal). The PSII core proteins D1, D2, CP43, and CP47 are colored individually, as shown in the figure, whereas others are colored in light gray. PSII extends to ~15 Å on the cytoplasmic side but to a much larger distance, ~90 Å, on the lumen side. Approximate boundary between the monomeric subunits of the homodimer is indicated by dashed arrows. See text for further details. The PSII model was generated with *RCSB Protein Workshop Viewer* using x-ray crystallographic coordinates deposited at Protein Data Bank (PDB) with ID 3ARC. (By permission from Macmillan Publishers Ltd. *Nature*, Umena, Y. et al., Crystal structure of oxygen-evolving photosystem II at a resolution of 1.9 Å, 473, 55, 2011. Copyright 2011.)

equivalent, since the electron transfer of PSII proceeds only through the $Pheo_{D1}$ and Y_Z on D1 protein (Figure 1.6). Nevertheless, although Y_D (located on D2) is not involved in electron flow of PSII, it plays an important role in the redox processes in PSII (see e.g., [181–183]), and it is probably also involved in the photoactivation of the OEC [184].

The D1 protein not only holds the redox-active cofactors of the "active" electron transfer branch (Figure 1.6) but also provides most of the ligands to the catalytic site of oxidative water splitting [185], a cluster of four Mn atoms, one Ca atom, and five bridging oxygen atoms (the Mn_4CaO_5 cluster) [125,186]. Together with its protein ligands [185] and at least one Cl^- ion as cofactor (two Cl^- ions according to [125]), the Mn_4CaO_5 cluster forms a functional unit, the OEC (also called the water-oxidizing complex). The Mn_4CaO_5 cluster is surrounded by three extrinsic proteins on the lumenal side of PSII: PsbO (33 kDa), PsbV (also known as cytochrome c_{550} (Cyt c_{550}); 17 kDa) and PsbU (12 kDa) (Figure 1.6). The latter two smaller proteins are present only in the cyanobacterial OEC; in case of higher plants, they are substituted by PsbP (23 kDa) and PsbQ (17 kDa). PsbO is often called the *Mn-stabilizing protein* for its key role in stabilization of the Mn_4CaO_5 cluster and hence the O_2-evolving activity. These three extrinsic proteins are highly important for shielding the

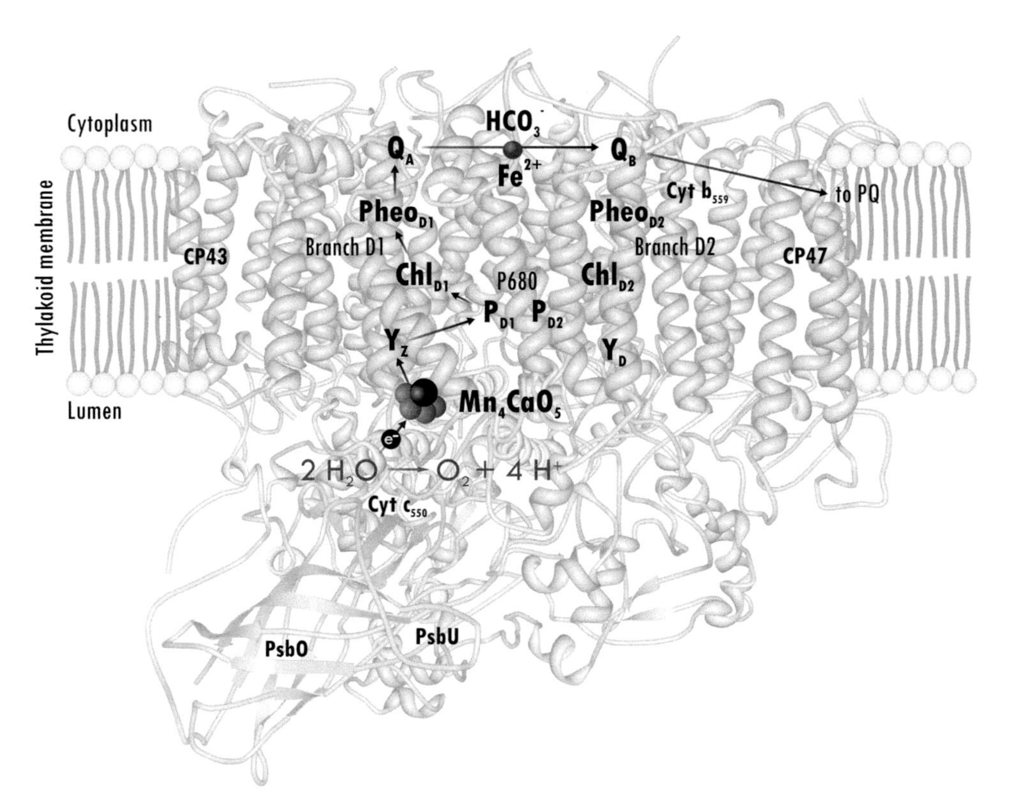

FIGURE 1.6 (See color insert.) Schematic arrangement of electron transfer cofactors in cyanobacterial PSII monomer. Figure shows side-on view from the direction parallel to the membrane plane. All cofactors (dark blue labels) of the monomer are arranged in two branches on the D1 and D2 protein subunits (see Figure 1.5). The light-induced single electron transfer occurs mainly on the D1 protein of the PSII RC (so-called active branch). The direction of electron transfer is indicated by dark blue arrows. The location of some protein subunits is shown (black labels). Note that primary electron donor P680 (traditional definition) refers to a pair of Chl *a* molecules (P_{D1} and P_{D2}), and two accessory Chls (Chl_{D1} and Chl_{D2}). For further details and abbreviations, see text. The protein background of PSII monomer was generated with the *RCSB Protein Workshop viewer* using coordinates deposited at PDB under ID 3KZI.

Mn_4CaO_5 cluster and for the optimization of water oxidation at physiological concentrations of Ca^{2+} and Cl^- ions (for recent reviews, see [187–189]).

The D1-D2 heterodimer is closely associated with two internal antenna subunits, the Chl *a*-containing proteins CP47 (PsbB) and CP43 (PsbC) (Figures 1.5 and 1.6), which bind 16 and 13 Chls *a* molecules, respectively [174]. These Chls funnel excitation energy from light-harvesting antenna system (PBSs) to PSII RC Chls. However, pigments in CP43 and CP47 are not only Chls but also Cars (reviewed in [190]). The CP47 and CP43 proteins are structurally homologous, each having large membrane-extrinsic loops interacting with the extrinsic subunits on the lumenal side. Both proteins are indispensable for establishing the water-splitting site. CP43 contributes a ligand (Glu354) to the first coordination sphere of the Mn_4CaO_5 cluster [125,191], while the large loop of CP47 is known to be essential for stabilization of the entire OEC [192].

In addition to the aforementioned proteins of the PSII RC, there are also 13 small membrane-intrinsic protein subunits, referred to as *low molecular weight* (LMW) *proteins* (PsbE, PsbF, PsbH, PsbI, PsbJ, PsbK, PsbL, PsbM, PsbT, PsbX, PsbY, PsbZ, and Ycf12) (for reviews, see [175,193]. Among them, only PsbZ has two TMHs, whereas all others have just one TMH. Two of the LMW proteins, namely, PsbE (or α-subunit) and PsbF (or β-subunit), provide His ligands for the high-potential heme of cytochrome b_{559} (Cyt b_{559}) (Figure 1.6). The Cyt b_{559} and the nearby β-Car (not shown in Figure 1.6) are located on the D2 side. Their function is to protect the system against photo-induced damage [193,194].

Cyanobacterial PSII complex is known to have the highest content of lipids as essential cofactors among the others membrane-bound photosynthetic complexes. Thus, in total, 25 lipids per PSII monomer were revealed by x-ray structural models at 3.0 Å [173] and 2.9 Å [178] resolution. Lipids are mainly bound at the interface of the D1/D2 heterodimer and the antenna proteins CP43 and CP47, playing a significant role in the structure and function of PSII (for a review, see Ref. [195]).

All of the aforementioned protein subunits and cofactors of PSII have been physically seen in the latest x-ray structures of dimeric PSII complexes from thermophilic cyanobacteria at 1.9 Å [125] (also see Figure 1.5) and 2.9 Å [178] resolution. The crystal structure of cyanobacterial monomeric PSII complexes is also available, though at a lower resolution (3.6 Å) [196].

1.3.3.1.2 Electron Transfer Pathway and Function of PSII

The time sequence of the light-induced reactions within PSII is discussed in Section 1.2.2, and the electron transfer pathway is shown in Figure 1.6. An important point here is to realize that among the 4 RC Chl *a* molecules that form P680 and are mentioned earlier, Chl_{D1} is the Chl molecule where the very first light-induced charge separation begins, followed by the transfer of the positive charge to P_{D1} soon thereafter and formation of the $P_{D1}^{\bullet+}Pheo_{D1}^{\bullet-}$ pair (see, e.g., [197] and Section 1.4.1 for primary photochemistry). Each electron passes rapidly (see Section 1.2.2 for time sequence) from $Pheo^{\bullet-}$ to a permanently bound Q_A, leading to the stabilized radical ion pair $P_{D1}^{\bullet+}Pheo_{D1}Q_A^{\bullet-}$, before finally arriving at the loosely bound Q_B in the Q_B-pocket. The resulting $P_{D1}^{\bullet+}$ ($P680^{\bullet+}$) serves as the driving force for oxidative splitting of two H_2O molecules to O_2 and four protons. The electrons from water are transferred to $P_{D1}^{\bullet+}$ *via* the Mn_4CaO_5 cluster and the redox-active Y_Z (see Figure 1.6). Transfer of the second electron reduces Q_B^- to Q_B^{2-}, and the reduced Q_B^{2-} picks up two protons, ultimately, from the cytoplasmic side of the medium, yielding PQH_2 (Figure 1.3) [121]. A bicarbonate ion, bound to the NHI between Q_A and Q_B, plays a key role in this protonation event (reviewed in [122,198]). Upon protonation, PQH_2 leaves the PSII complex and diffuses in the membrane to the Cyt b_6f complex. Thereafter, another PQ molecule from the PQ pool immediately fills the empty Q_B-pocket of PSII.

Thus, PSII functions as a water:plastoquinone oxidoreductase (see recent reviews, see [65,67,199] and relevant chapters in [200]; it utilizes light energy to remove electrons from H_2O (water oxidation) and adds these electrons, as well as protons, to PQ (reduction of PQ pool). The role of the OEC in this process is indispensable: it couples the successive one-electron oxidations of $P_{D1}^{\bullet+}$ ($P680^{\bullet+}$) to the four-electron oxidation of water to molecular oxygen (for further details, see Section 1.4.2 and Chapter 2 and references therein).

1.3.3.2 Photosystem I

1.3.3.2.1 General Structural Organization

Cyanobacterial PSI is the largest photosynthetic membrane-bound protein complex; its oligomeric form predominantly exists as a trimer and has a molecular weight of ~1100 kDa (for recent overviews, see [63,201,202] and relevant chapters in [203]). The structure from the thermophilic cyanobacterium *T. elongatus* was determined at 2.5 Å resolution (see Figure 1.7 and Refs. [204,205]). Under certain environmental conditions, especially under high light intensities, PSI in cyanobacteria can also become a monomer (as it always exists in higher plants) [206,207]. Each monomeric PSI complex consists of 12 protein subunits that harbor 127 noncovalently bound cofactors (96 Chls *a*, 22 β-Cars, 2 phylloquinones, 3 [4Fe-4S] clusters, and 4 lipids) [208,209].

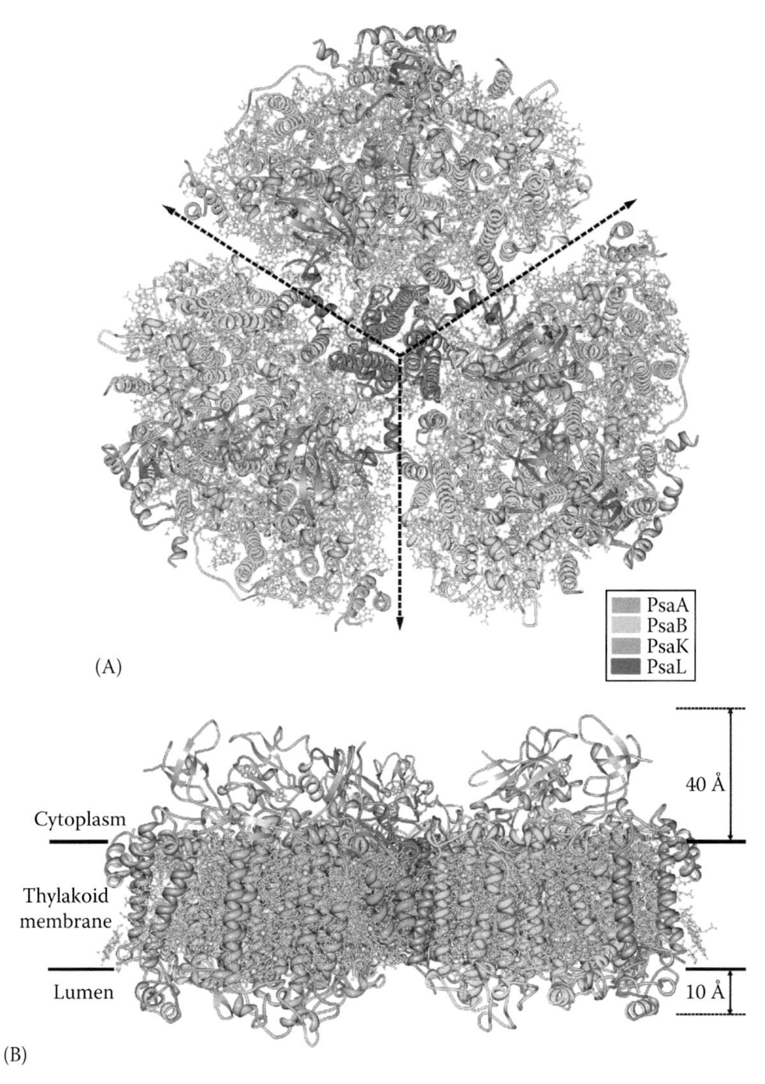

FIGURE 1.7 **(See color insert.)** Overall structure of the cyanobacterial PSI trimer from *T. elongatus*. (A) View of the trimer from the cytoplasmic side. (B) Side-on view (from the direction perpendicular to the membrane normal). PSI core proteins PsaA, PsaB, PsaK, and PsaL are colored individually, as shown in the figure, whereas the others are colored in light gray. PSI extends to ~40 Å on the cytoplasmic side but to a lesser distance, ~10 Å, on the lumen side. Approximate boundary between monomeric RCs is indicated by dashed arrows. For details, see text. The PSI model was generated with the *RCSB Protein Workshop viewer* using coordinates from PDB with ID 3PCQ. (By permission from Macmillan Publishers Ltd. *Nature*, Jordan, P. et al., Three-dimensional structure of cyanobacterial photosystem I at 2.5 A resolution, 411, 909, 2001. Copyright 2001; *Nature*, Chapman, H.N. et al., Femtosecond X-ray protein nanocrystallography, 470, 73, 2011. Copyright 2011.)

Two main protein subunits, PsaA and PsaB (the PsaA/PsaB heterodimer), form the core of PSI. Although these proteins are larger (each has 11 TMHs) than the core components of PSII (D1 and D2) and the anoxygenic RCs, they are homologous. The similarities of the TMHs arrangements are thought to indicate an evolutionary evolvement of PSI and PSII from a common ancestor [142]. Not all, but the majority of cofactors of the electron transfer chain of PSI are coordinated by the PsaA and PsaB. These cofactors are the primary electron donor (P700) that is the "special pair" of Chl *a* and Chl *a′* heterodimer, P_A and P_B, according to a nomenclature suggested by Redding and van der Est [210]; the

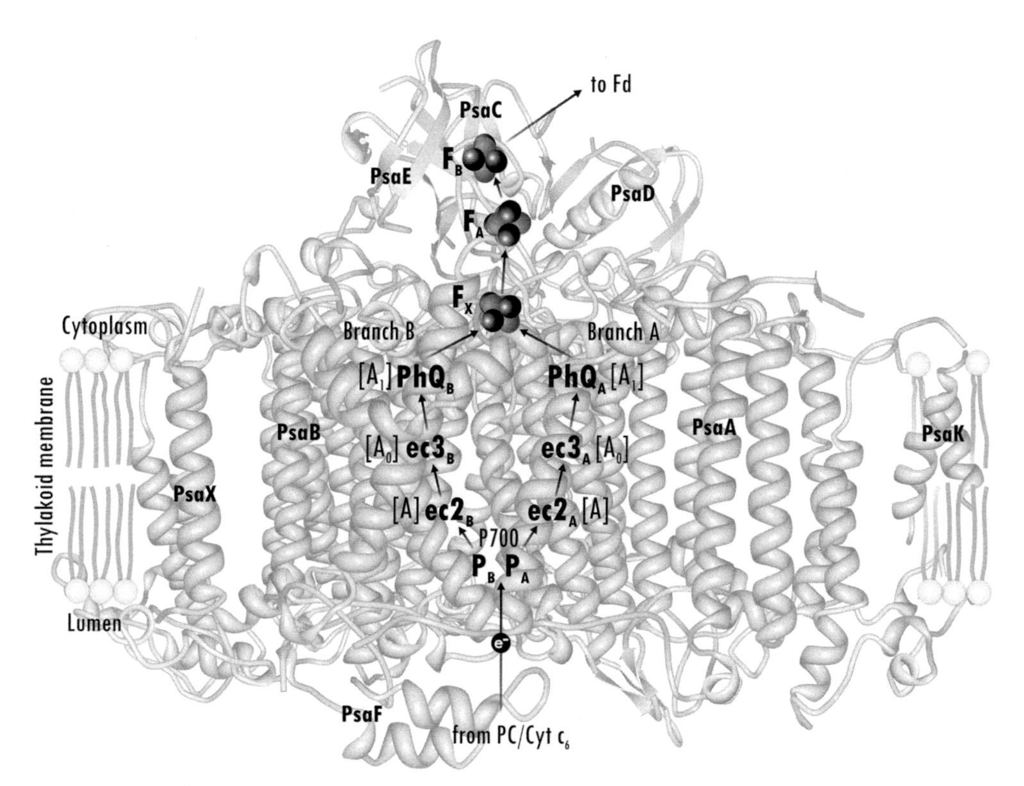

FIGURE 1.8 (See color insert.) Schematic arrangement of electron transfer cofactors in cyanobacterial PSI monomer. This figure is a side view of the monomer from the direction of the membrane exposed periphery of PSI along the membrane plane into the center of the trimer. As in PSII, the electron transfer cofactors (dark blue labels) in PSI are organized in two parallel branches denoted as A and B (named after the protein subunits PsaA and PsaB, respectively; see Figure 1.7). The arrows indicate the direction of the electron transfer and the two possible electron transfer pathways with almost equal probability (for overviews of bidirectional electron transfer in PSI and its possible advantages for photosynthetic organism, see [210,214]). The electron transfer cofactors of two branches are indicated according to the nomenclature suggested by Redding and van der Est [210], while in brackets we show the commonly used (traditional) names that refer to the spectroscopic signatures of these cofactors. Note that the primary electron donor P700 (P_A and P_B) is used for a "special pair" of Chl *a* and Chl *a′* heterodimer. The location of some protein subunits is indicated by black labels. For further details and cofactor abbreviations, see text. The protein background of PSI monomer was modeled by the *RCSB Protein Workshop viewer* using coordinates deposited at PDB with ID 1JB0. (By permission from Macmillan Publishers Ltd. *Nature*, Jordan, P. et al., Three-dimensional structure of cyanobacterial photosystem I at 2.5 A resolution, 411, 909, 2001. Copyright 2001.)

initial electron acceptor, A, that consists of two Chls *a* ($ec2_A$ and $ec2_B$); the first stable electron acceptor A_0 that is also formed by two Chls *a* ($ec3_A$ and $ec3_B$); two phylloquinone molecules A_1, PhQ_A and PhQ_B; and the first [4Fe-4S] cluster termed as F_X (see Figure 1.8). Moreover, most of the antenna Chls of PSI (79 of the 90) as well as most of the Cars are bound to the PsaA/PsaB heterodimer. This represents a unique feature of PSI to form strong structural and functional cohesiveness of the RC and the integral antenna system. Another important function of PsaA and PsaB is that they are involved in the docking of the mobile electron donors to P700 (Cyt c_6/PC) on the lumenal side of PSI [211].

The PsaA/PsaB heterodimer is surrounded by seven small membrane-integral protein subunits (PsaF, PsaI, PsaJ, PsaK, PsaL, PsaM, and PsaX), each containing from 1 up to 3 TMHs. Four of the subunits (PsaF, PsaJ, PsaK, and PsaX) are located at the membrane exposed surface of the PSI trimer (Figures 1.7 and 1.8) and are involved in the stabilization of the core-antenna system of PSI. Other three proteins (PsaI, PsaL, and PsaM) are located in the interface of the neighboring monomers in the trimeric PSI complex, thereby forming "trimerization domain" [209]. With the exception of one subunit, PsaX, which is not found in higher plants [212], all other cyanobacterial protein subunits are present in higher plants.

Three proteins (PsaC, PsaD, and PsaE) on the cytoplasmic side do not contain TMHs (Figures 1.7 and 1.8). They form the docking site of PSI for the mobile electron acceptor of PSI, Fd. PsaC is the most important among these proteins, because it binds the [4Fe-4S] clusters F_A and F_B, the two terminal cofactors of the electron transfer chain of PSI. This protein has high sequence homology and identical structure in cyanobacteria and higher plants [204,212].

1.3.3.2.2 Electron Transfer Pathway and Function of PSI

The electron transfer chain in PSI is arranged in two quasi-symmetrical branches (A and B) that contain 6 Chls and 2 phylloquinones A_1 (Figure 1.8). Both branches are found to be active [213,214]. However, the electron transfer in cyanobacteria takes place mostly on the branch A [215,216], while in photosynthetic eukaryotes, branch B is thought to be more active than A [217–219].

Recent data suggest that the primary charge separation in PSI may begin not from P700, as it was initially thought, but from the A/A_0 pair (Figure 1.8), followed by the generation of a primary $A^{\bullet+}A_0^{\bullet-}$ radical pair, rapid reduction by P700 and the formation of the subsequent radical pair $P700^{\bullet+}A_0^{\bullet-}$ [220–222]. The oxidized $P700^{\bullet+}$ is then reduced by the electron provided by Cyt c_6/PC. The electron available at the $A_0^{\bullet-}$ moves though the remaining intermediates across the membrane to reduce water-soluble Fd. For its important function to power the electron transfer from Cyt c_6/PC to Fd by using energy of light, PSI may be called as light-driven plastocyanin:ferredoxin oxidoreductase.

1.3.4 Light-Harvesting Complexes: Phycobilisomes

In Sections 1.3.3.1 and 1.3.3.2, we have already mentioned membrane-bound antenna systems associated with the two photosystems in cyanobacteria. For further information on these integral (core) antennas and their structures, see [38,61,133]. Here, we briefly focus on a distinctive feature of cyanobacteria—the giant multidimensional extraneous antenna system on the cytoplasmic side of their thylakoid membranes, the PBSs (Figure 1.2), which give the specific blue-green color to cyanobacteria (for overviews on PBSs, see [38,61,62,223]). However, cyanobacteria are not the sole photosynthetic organisms that have PBSs; red algae also employ the PBSs for harvesting sunlight (see, e.g., [40,224,225]).

PBSs are organized into 3 core cylinders (two in the base and one on the top) and 6 (sometimes 8 or 10) peripheral stacked disk rods (Figure 1.2). The core cylinders, and the peripheral rods of the PBSs, are formed by water-soluble proteins, the *phycobiliproteins*, which contain covalently bound (*via* thioester bonds) open-chain tetrapyrroles, the *phycobilins*. In most cyanobacteria, phycobilins include *phycocyanins* (with absorption maxima in the 615–640 nm range) located in the peripheral rods, and *allophycocyanins* (with absorption at 650–655 nm) located in the core rods (see Figure 1.9A). There are some cyanobacteria that, in addition to phycocyanins and allophycocyanins, contain *phycoerythrins* (with absorption at 495 and/or at 565–575 nm) in their PBSs (see, e.g., [60,226–228]).

The absorption of light by phycobilins in the wavelength range 495–655 nm gives cyanobacteria an advantage at ocean depths where mainly green light is available. Excitation energy transfer is efficient and the cascade of energy transfer starts at the outer rods downhill toward the phycobilins localized at the core cylinders that absorb at longer wavelength [62]. The terminal long-wavelength allophycocyanins of the core cylinders interact with Chls *a* of internal antenna subunits of PSII, CP43 and CP47 (see Figures 1.5 and 1.6 and Section 1.3.3.1). One of these allophycocyanins, namely, allophycocyanin B (with absorption maxima at 670 nm) [229], is located on the PBS core-membrane linker protein ApcE, also called an "anchor" polypeptide for its mediating function with thylakoid membrane and/or the RCs [223,230]. Cyanobacterial PBSs are much larger in size than the photosystems. Thus, their most abundant hemidiscoidal morphological form normally has a diameter in the range of 300–800 Å [133,231]. The size of PBSs may vary depending on the light conditions: under

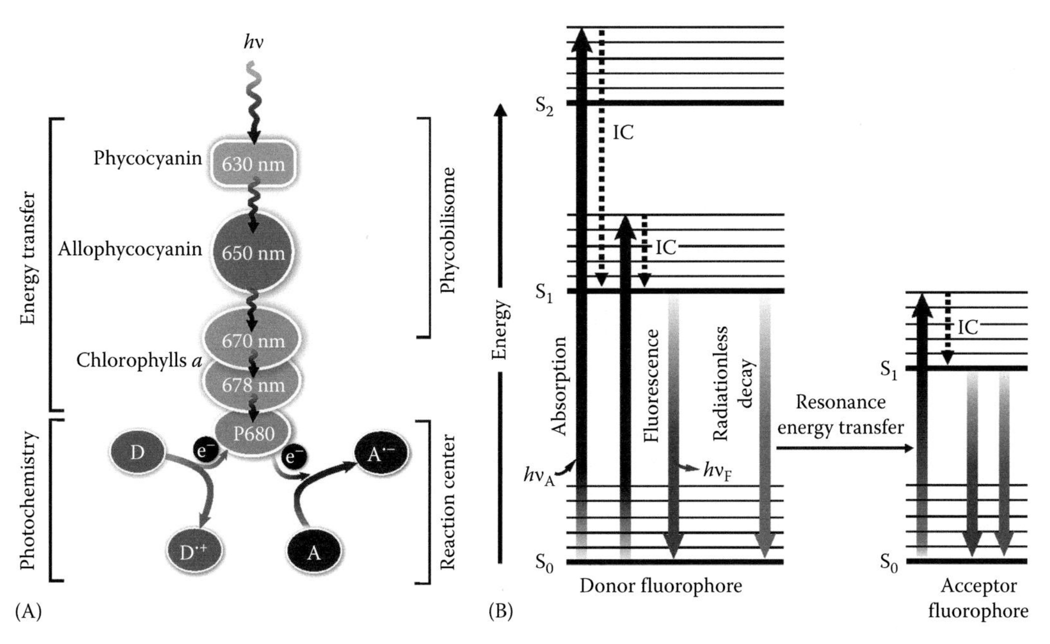

FIGURE 1.9 Excitation energy transfer in cyanobacteria. (A) Energy transfer steps in PBSs including primary photochemistry (charge separation) at the PSII RC of cyanobacteria. The energy of absorbed photon ($h\nu$) passes through a number of phycobilin pigments (phycocyanin and allophycocyanin) in PBSs and Chl *a* molecules until it reaches the RC Chls *a* (P_{D1}). The excited P_{D1} donates its electron to an acceptor (A) leading to a primary charge separation and formation of $P_{D1}^{\bullet+}$ and $A^{\bullet-}$ ($Pheo^{\bullet-}$). The electron vacancy in $P_{D1}^{\bullet+}$ is filled by the electron from an electron donor (D). The wavelength numbers (nm) represent pigments corresponding to the long wavelength absorption maxima of these molecules. (B) Perrin–Jablonski diagram illustrating the transfer of excitation energy between antenna molecules by Förster Resonance Energy Transfer (FRET). Absorption of energy in the form of electromagnetic radiation (photon) by an antenna molecule induces a very fast (10^{-13} s) transition from the lowest vibration level (only a few vibrational energy levels are shown) of a ground electronic singlet state (S_0) to an excited electronic states (S_1 or S_2). The magnitude of the absorbed energy ($h\nu_A$) determines which vibrational level of S_1 (or S_2) becomes populated. In the next 10^{-13} s, due to the process called internal conversion (IC), the antenna molecule relaxes to the lowest vibrational level of the first excited singlet state (S_1). One of the routes by which molecule can return to S_0 is fluorescence—photon emission (with the magnitude $h\nu_F$) that occurs between states S_1 and S_0. In most cases, emitted light will have a longer wavelength and therefore lower energy than the absorbed radiation. However, because of the proximity of other antenna molecules with a near (or the same) energy levels, the excited singlet state energy has a very high probability to be transferred to a neighboring antenna molecule *via* FRET. Since the S_1 state of Chl *a* is energetically lower than that of phycobilins, excitation energy is rapidly localized on the Chl *a* molecules. See text for further details.

low light intensities, PBSs increase in size, whereas they decreased in size under high light intensities (in some extreme cases, the outer rods may even disappear and only core allophycocyanins remain) [61].

For a long time, PBSs were thought to be the light-harvesting antennas of only PSII [62]. However, numerous investigations clearly revealed that effective excitation energy transfer may occur directly from PBSs to PSI RC (see e.g., [70,232,233]), indicating that PBSs are mobile complexes and that depending on the light conditions, they can be attached either to PSII or to PSI [72]. In a few cyanobacterial species, phycobiliproteins act as extrinsic antenna system as phycobiliproteins aggregates and are not organized into PBSs. The representatives of such PBS-lacking cyanobacteria are the Chl *d*-dominating cyanobacterium *A. marina* [171,234,235] and Chl *b*-containing cyanobacterium *Prochlorococcus marinus* [236–238].

The light-harvesting PBSs are highly important for photoautotrophic growth and directly mirror the physical parameters of the environment, especially light. Light can vary with respect to quality and amount, and the organism senses and responds to the light environment by tuning its physiology. With respect to light harvesting, size, composition, number, and location of PBSs are optimized to maximize yield of physiologically usable light energy while minimizing damage to the photosynthetic apparatus and the cell [239]. During the so-called *state transitions* (also termed as State 1–State 2 (inversely, State 2–State 1) transition), cyanobacteria balance the light excitation between the two RCs on a timescale of seconds to minutes. Upon over-excitation of PSII (State 2), excitation energy channeled to PSI is increased, while upon over-excitation of PSI (State 1), excitation energy channeled to PSII is increased. In higher plants and green algae, state transitions are regulated *via* phosphorylation/dephosphorylation and redistribution of an LHC species within thylakoid membranes and between PSII and PSI [240]. In cyanobacteria, PBSs are required and rpaC has been found essential to establish State 2 (see, e.g., [72,241,242]). In addition, cyanobacteria respond with alterations in the composition of PBSs *via* a long-term physiological process termed *chromatic adaptation* (*acclimation*) upon changes in the quality (light color) of the light environment [243–246]. Cyanobacteria showing maximal rates of photosynthesis in one particular light intensity and quality can also acclimate and operate with less effective light conditions through a process termed *photoacclimation* [247,248]. Finally, physiology of cyanobacteria copes with excess light energy utilizing a number of photoprotective mechanisms [249].

1.4 LIGHT-INDUCED REACTIONS IN CYANOBACTERIA

1.4.1 Light Absorption, Excitation Energy Transfer, and Primary Photochemistry

Three types of pigment molecules, Chls, phycobilins, and Cars, are responsible for the light absorption properties of the cyanobacterial antenna complexes. Membrane-integral antenna complexes of PSII and PSI, together with PBSs, serve the RCs with excitation energy needed for the primary photochemistry processes. Here, we briefly outline the reaction sequence of excitation energy transfer that leads to primary photochemistry in the RCs of cyanobacteria and other oxygenic organisms (for reviews, see [80,183,250]).

The reaction sequence initiated by absorption of photon ($h\nu$) and excitation energy transfer within the PBSs and the RC integral antenna complexes can be represented as follows (also see Figure 1.9A for a simplified schematic drawing of these processes):

$$\text{Pbl} + h\nu \rightarrow \text{Pbl}^* \quad (1.1)$$

$$\text{Pbl}^*\,\text{Pbl} \rightarrow \text{Pbl}\,\text{Pbl}^* \quad (1.2)$$

$$\text{Pbl}^*\,\text{Chl}\,a \rightarrow \text{Pbl}\,\text{Chl}\,a^* \quad (1.3)$$

$$\text{Chl}\,a^*\,\text{Chl}\,a \rightarrow \text{Chl}\,a\,\text{Chl}\,a^* \quad (1.4)$$

where

Pbl represents phycobilins

Chl *a* represents Chls of the RC-integrated antenna complexes

$h\nu$ is a photon (quantum)

the asterisk indicates an electronically excited state

After absorption of photons, antenna pigment molecules go into specific excited states, depending upon the wavelength of light. The light energy is converted into excitation energy of the molecules that absorbed the light. Finally, light energy absorbed by different phycobilins (reaction 1.1) is transferred (reaction 1.2), as shown in Figure 1.9A, to the antenna Chls (reaction 1.3). This is followed by excitation energy transfer among Chl *a* molecules (reaction 1.4). The nature of the excitation energy transfer in photosynthetic units has been debated for a long time. The *Förster hopping model* and the *delocalized (coherent) exciton model* are two mechanisms for the description of excitation energy transfer. Both models can be derived from quantum mechanics principles and, in fact, the applicability of these theories strongly depend on the interaction energies between different chromophores (pigments) and the protein environment: The Förster theory, the Redfield theory, the Modified Redfield theory, and the Generalized Förster theory are the result of direct application of the known physical concepts, which include the *exciton concept*, to the photosynthetic system [251–253]. Figure 1.9B shows the Perrin–Jablonski diagram illustrating the transfer of excitation energy transfer between antenna molecules by Förster Resonance Energy Transfer (FRET) (also see legend of this figure for description). In general, the Förster theory appears applicable to energy transfer from say phycocyanin to allophycocyanin and then to Chls *a*. The important feature of exciton dynamics is a cooperative mechanism of the excitation energy transfer. Such mechanism has a very clear physical interpretation: due to nonzero coupling energy between Chls in the antenna complexes, molecules lose their individual properties and the system behaves like a "super molecule" with a set of exciton energies which we observe in the optical experiments.

Taking PSII RC complex as an example, the excitation energy ("exciton" in the language of many) from the antenna Chls is trapped by the photoactive accessory pigment (Chl_{D1}) (see Figure 1.6); this is followed by primary charge separation (for simplicity, some steps have been left out):

$$\text{Chl } a^{*}\ \text{Chl}_{D1} \rightarrow \text{Chl } a\ \text{Chl}_{D1}^{*} \tag{1.5}$$

$$\text{Chl}_{D1}^{*}\ \text{P}_{D1}\ \text{Pheo}_{D1} \rightarrow \text{Chl}_{D1}^{\bullet+}\ \text{P}_{D1}\ \text{Pheo}_{D1}^{\bullet-} \rightarrow \text{Chl}_{D1}\ \text{P}_{D1}^{\bullet+}\ \text{Pheo}_{D1}^{\bullet-} \tag{1.6}$$

where the dot symbolizes a radical (unpaired electron).

There is still a debate as to the definition of P680 and the detailed steps involved in the primary photochemistry; we do not discuss it here (for further details, see Refs. [65,111,254,255]). Thus, the end result of the excitation energy transfer to Chl_{D1} (reaction 1.5), subsequent charge separation and the transfer of the positive charge to P_{D1} is the generation of the charge separated state (Chl_{D1}) $P_{D1}^{\bullet+}$ $Pheo_{D1}^{\bullet-}$ (reaction 1.6). This state is further stabilized by the following reactions on the acceptor side of PSII:

$$\text{Pheo}_{D1}^{\bullet-}\ \text{Q}_A \rightarrow \text{Pheo}_{D1}\ \text{Q}_A^{\bullet-} \tag{1.7}$$

$$\text{Q}_A^{\bullet-}\ \text{Q}_B \rightarrow \text{Q}_A\ \text{Q}_B^{\bullet-} \tag{1.8}$$

The absorption of the second photon ($h\nu$) initiates a second turnover of the reaction sequences (1.1) through (1.7), leading to the following reaction sequence on the Q_B-site of PSII:

$$\text{Q}_A^{\bullet-}\ \text{Q}_B^{\bullet-} \rightarrow \text{Q}_A\ \text{Q}_B^{2-} + 2\text{H}^+_{\text{cytoplasm}} \rightarrow \text{Q}_A\ \text{Q}_B\,\text{H}_2 \tag{1.9}$$

$$\text{Q}_A\ \text{Q}_B\,\text{H}_2 + \text{PQ} \rightarrow \text{Q}_A\ \text{Q}_B + \text{PQH}_2 \tag{1.10}$$

$$\text{PQH}_2 \rightarrow \text{PQ} + 2\text{H}^+_{\text{lumen}} \tag{1.11}$$

Thus, after the reduction of Q_B to Q_B^{2-}, followed by its protonation with two protons from cytoplasm (reaction 1.9), plastoquinone Q_B forms plastoquinol PQH_2 (reaction 1.10), which then diffuses towards the Cyt b_6f complex finally resulting in delivery of two protons into the lumen (reaction 1.11) and two electrons to the redox-cofactors of the Cyt b_6f complex. As it has been already discussed in Section 1.2.2, reaction (1.11) is the slowest reaction event among the all light-induced reactions of the electron transfer chain of thylakoid membrane.

1.4.2 Light-Induced Water Oxidation

Mechanism of water oxidation in oxygenic organisms is discussed in Chapter 2. Therefore, in this section, we briefly mention the basic principles of this process.

The basis for the understanding of the mechanism of photosynthetic water oxidation, known to be identical in both prokaryotic and eukaryotic O_2-evolving organisms, was set more than 40 years ago. Illuminating dark-adapted algae and chloroplasts by short ("single turnover") saturating flashes, Joliot and coworkers [256] discovered that O_2 evolved with a characteristic periodicity of four. The periodicity of four was readily explained by the four-electron chemistry of water splitting. On the other hand, the fact that the first maximum of O_2 evolution occurred after the third rather than the fourth flash, and that O_2 oscillation was damped after a few cycles, indicated an unexpected level of complexity in the mechanism of water oxidation by PSII. Based on these findings, Kok et al. [257] developed an elegant model of photosynthetic water oxidation (also called as Kok cycle or the "oxygen clock") (see Figure 1.10 and also [258] for various kinetic models of water oxidation). It assumes that each OEC cycles through five different redox states during oxidation of two water molecules, named S_i states ($i = 0, \ldots, 4$), where i is the number of oxidizing equivalents stored within the Mn_4CaO_5 cluster of the OEC. The formation of the four oxidizing equivalents occurs during repeated oxidation (one electron at a time) of the OEC by P680$^{\bullet+}$ *via* a redox-active tyrosine, $Y_Z^{\bullet}$

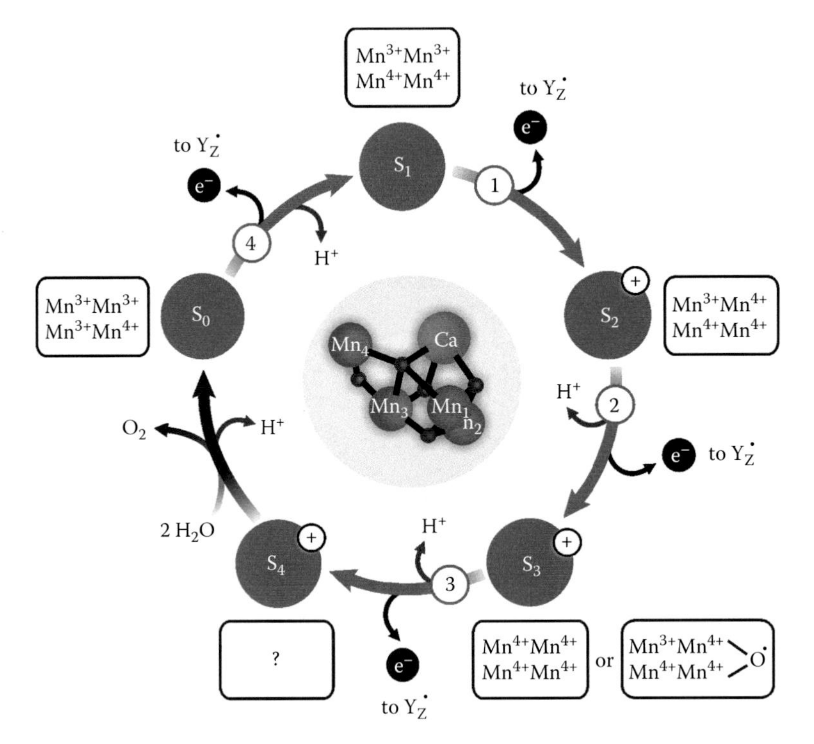

FIGURE 1.10 (See color insert.) The Kok cycle (also known as the "oxygen clock") that illustrates the stepwise process of photosynthetic water oxidation by the Mn_4CaO_5 cluster of oxygenic organisms. Blue arrows indicate light-induced S-state transitions and the numbers in circles on the arrows indicate the number of light flashes required for that transition, assuming that in the dark the Mn_4CaO_5 cluster is mostly in the S_1 state. The $S_4 \rightarrow S_0$ transition does not require light and is shown in black. Note that depending on the S-state transition, either the electron or the proton is thought to be removed first. Currently discussed oxidation states of the four Mn ions of the Mn_4CaO_5 cluster in various S states are shown. For further details on photosynthetic water splitting, see text (this chapter), Chapter 2 (this book), and Refs. [266,267,320]. For the original version of the Kok cycle and other models, see [256–258]. The structural model of the Mn_4CaO_5 cluster in the center is as derived from the recent x-ray crystallographic PSII structure.

(see Figures 1.6 and 1.10). In order to explain the maximum O_2 evolution after the third flash, Kok et al. assumed that in long dark-adapted samples, practically all (almost 100%) PSII centers are in the S_1 state, and not in the S_0. Therefore, the first S cycle begins from the S_1 state and is completed after three given flashes, while each next cycle is completed after four flashes. Indeed, later studies showed that the S_0 state is slowly (tens of minutes) oxidized to the S_1 state by the oxidized form of tyrosine D, $Y_D^{\bullet}$, of polypeptide D2 (Figure 1.6) [182,259,260]. The S_2 and S_3 states are reduced within seconds to minutes into the S_1 state by the reduced tyrosine Y_D *via* a fast decay [259,261,262] or a slower decay due to electron donation from the reduced acceptor side quinones [263–265]. The $S_4 \rightarrow S_0$ transition does not require a light flash. Therefore, the S_4 state is thought to spontaneously decay into the S_0 state releasing O_2; this is accompanied by the binding of at least one new substrate water molecule (reviewed in Refs. [67,266,267]). Lifetime measurements of the S_2 and S_3 states clearly indicate that the redox potentials and kinetics within the OEC do not differ between various species of cyanobacteria and higher plants [170,268]. In spite of numerous attempts, the S_4 state has not yet been detected because of its very short lifetime [269–271].

The oxidation state changes in the Mn_4CaO_5 cluster are mostly Mn-based. Several spectroscopic studies have been made to record the oxidation states of Mn ions in each S state (see reviews [67,272–275] and references therein). The results of these studies are summarized in Figure 1.10. Important insight about the structure of the Mn_4CaO_5 complex was provided by x-ray diffraction crystallography studies and extended x-ray absorption fine structure spectroscopy (see e.g., [173,174,186]). Recently, x-ray crystallography investigations on the PSII at 1.9 Å resolution have provided important new structural information about the Mn_4CaO_5 cluster in cyanobacteria (see [125] and Figure 1.10).

1.4.3 Some Comments on Production of the Reducing Power (NADPH) in Cyanobacteria

As mentioned earlier, the cyanobacterial system is not only unique but more flexible than those of higher plants and algae (see Section 1.1). Cyanobacteria are unique in the sense that both the reduction of $NADP^+$ to NADPH (for photosynthesis) and its reverse, the oxidation of NADPH (for respiration) take place on the same thylakoid membrane, although respiration also occurs on cytoplasmic membranes (see, e.g., [48] and Figure 1.1). In addition to PSI, PSII and Cyt b_6f complex (Figure 1.2), needed for the production of NADPH, thylakoid membrane also harbors (1) type 1 NADPH dehydrogenase (NDH-1) that oxidizes NADPH to $NADP^+$; (2) a terminal oxidase that reduces O_2 to H_2O; and (3) succinate dehydrogenase (SDH) that oxidizes succinate to fumarate providing reducing power to the PQ pool [48]. This complicates the kinetics of the production of NADPH. Another uniqueness lies in the fact that, as already mentioned in Section 1.2.1, the number of PSI to PSII is not 1:1 as is the case in higher plants and algae, but ranges from 3 to 5:1 (see, e.g., [68–70])—clearly favoring, in principle, a higher "cyclic" reaction around PSI (see e.g., [103]), over noncyclic electron flow from water to $NADP^+$.

1.4.4 Production of ATP

The basic steps preceding the energy-requiring ATP synthesis from ADP and inorganic phosphate (P_i) are (1) charge separation in both PSII and PSI leading to a negative charge on the thylakoid membrane side facing the cytoplasmic side (the *n* side), and a positive charge on the thylakoid membrane side facing the lumen side (the *p* side); (2) deposition of four protons into the lumen when two H_2O molecules are oxidized to O_2 (see Section 1.4.2); (3) utilization of four electrons available from water oxidation for the conversion of two PQ molecules to two molecules of PQH_2, with the four protons, needed for this reaction, being taken from the cytoplasmic side (see Figure 1.2); this step being followed by proton release into the lumen, while the electrons on PQH_2 are being transferred to the Cyt c_6 (or PC) *via* the Cyt b_6f complex; and (4) in addition to the noncyclic pathway of electron transfer from H_2O to $NADP^+$, and also a cyclic electron flow around PSI, there is a "Q-cycle" (see, e.g., [198,276]). These cycles also lead to proton translocation from the cytoplasmic side to the lumen side. The sum of the two components (ΔpH and membrane potential, $\Delta\psi$) forms the PMF (see Section 1.2.1). Peter Mitchell ([277]; Nobel Prize in Chemistry in 1978) suggested that dissipation of the PMF through ATP synthase provides energy for ATP synthesis. About 42 kJ of converted light energy in this reaction is stored in each mole of the high-energy phosphate, ATP.

The mechanism of ATP formation is very different from the electron transfer processes described in Section 1.2. Jagendorf and Uribe [278] had observed that if chloroplasts (experiments were not done in cyanobacteria then) were first suspended in an acidic medium and then transferred to an alkaline medium in the presence of ADP and P_i, ATP formation occurred in darkness. Junge and coworkers (reviewed in [100,102]) found that valinomycin that dissipates the electric field inhibited phosphorylation. These experiments clearly support the Mitchell's chemiosmotic hypothesis.

The ATP synthase is shown schematically in Figures 1.2 and 1.11 (for more detailed models and structures, see the figures in [100–102]); this enzyme is ~15 nm long and ~12 nm wide and it has a molecular weight of ~600 kDa. It has two basic parts, one hydrophobic part (F_o) that, is embedded in the thylakoid membrane and another hydrophilic part (F_1) protruding into the stroma (see Figure 1.11A and B). The F_o of cyanobacteria contains several subunits: a (one copy), b (two copies; sometimes b and b′; [279]), and c (10–15 copies), whereas F_1 contains α (three copies), β (three copies), γ (one copy), δ (one copy), and ε (one copy), subunits. The F_o-enzyme part acts as a rotary motor, and most of the membrane-associated subunits rotate (Figure 1.11C), while H^+'s are translocated across the membrane (reviewed in [101]). Mechanical energy in this rotary motion is coupled to the formation of "high-energy phosphate bonds" in ATP at the stator (α, β, and δ) F_1 part of the enzyme (Figure 1.11C). Paul Boyer and John Walker received, in 1997, a Nobel Prize in Chemistry for their explanation of how the ΔpH energy is converted into mechanical energy on F_o followed by this rotation energy being converted into chemical energy, during ATP formation from ADP and P_i on the α and β subunits of F_1 (for historical perspective see [102]).

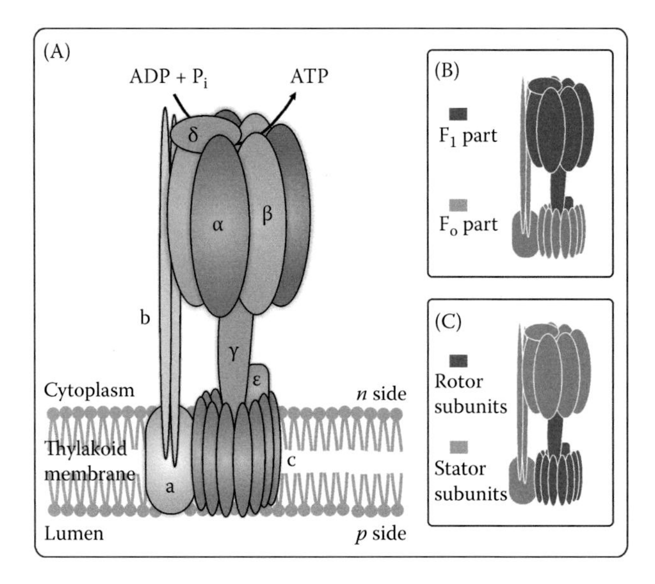

FIGURE 1.11 Schematic representation of ATP synthase and its subunits. (A) Overall subunit scheme of ATP synthase. (B) Schematic view of two basic parts of ATP synthase: a membrane-embedded F_o part and a peripheral (cytoplasmic) F_1 part (shaded individually as shown in the figure). (C) Schematic view of rotor and stator subunits of ATP synthase (shaded differently as shown in the figure).

Protons on the *p* side bind to the "c subunits." Pogoryelov et al. [280] had examined c subunits in several cyanobacteria species; although they had >80% sequence identity, one species had 13 copies; four species had 14 copies and two species had 15 copies. Again, this may be interpreted to show a high degree of flexibility of the cyanobacterial biology. There are many reports that deal with the number of protons needed per ATP produced. Van Walvaren et al. [281] had found this number to be 4 in two cyanobacterial strains (also see [282]).

In summary, the PMF (mainly proton gradient, just mentioned) is used by F_o, particularly the c subunit, to be converted into a mechanical rotation energy (a torque), which is then used by F_1, particularly its α/β unit, that converts this rotation energy into chemical energy, resulting in the formation of ATP from ADP and P_i. Further, both ε and γ subunits are involved in the regulation of this ATP synthase activity [283,284]. For details of mechanism, see Ref. [198].

1.5 COMMENTS ON EVOLUTION

One of the great enigmas in the evolution of life is the question when and how the first cyanobacteria initiated oxygenic photosynthesis and hence became capable of carrying out the thermodynamically and chemically demanding reaction of water oxidation into O_2? This evolutionary mystery remains unsolved. We know that before the event of oxygenic photosynthesis, photosynthetic organisms were anoxygenic (not O_2-producing). For reduction of CO_2 (carbon fixation) an oxygenic phototrophs used electrons from available reductants, such as H_2, H_2S, CH_4, and/or $Fe(OH)^+$ [7,285,286]. Although H_2O was another potential and the most abundant electron donor present in the environment, its utilization as source of electrons for carbon fixation required significantly higher investment in energy than for other electron donors. However, a constantly decreasing supply of H_2 and of other substrates on the surface of the planet, and the presence of an enormous (almost unlimited) H_2O pool, and the necessity of the photobacteria to gain energy for sustaining essential cellular processes may have resulted in constant evolutionary pressure towards the development of photosynthetic apparatus capable of extracting electrons from water. During the period between 3.2 and 2.4 billion years ago (the exact date is not known but is intensively discussed in the literature [157,287–296]), the problem of water splitting was solved: using the energy of sunlight, the emerged photosynthetic machinery in the first cyanobacteria successfully bridged the large potential barrier between water and $NADP^+$ [3,144,148,297,298]. Unlike anoxygenic photosynthetic apparatus, the new water-splitting photosynthetic apparatus contained the following inventions: (i) two types of *functionally linked RCs* (PSI and PSII) each comprising a new photosynthetic pigment, Chl *a*, and the unique RC, where the primary photochemistry takes place (see [3,31,35]; also see Sections 1.3.1 and 1.3.2); (ii) the *catalytic site of water oxidation* (OEC) functioning as a charge-accumulating system coupled to a strongly oxidizing RC (PSII) (see [149] and Section 1.4.2); and (iii) characteristic *light-harvesting antenna complexes* (PBSs) for the capture of light energy and its delivery, in the form of excitation energy, to the RCs (see [299] and Section 1.3.4). Since the metabolic waste product of the water-splitting reaction was O_2, the first oxygenic organisms had the great advantage of being able "to poison" their competitors with "toxic" dioxygen. As result, only those organisms survived that either developed protective mechanisms against O_2 or found O_2-free ecological niches. The origin of cyanobacteria and the levels of atmospheric O_2 is well depicted in the relationship between important events in the evolution of life and photosynthesis (see Figure 1.12A and its legend for further details and references). Most of the successful survived organisms, including cyanobacteria, managed to develop highly efficient respiratory processes which utilized O_2 as the terminal electron acceptor for "biochemical burning" which allowed the release of at least 10–15 times more free energy from organic substances than through anoxygenic processes [13,300]. Geochemical data indicate that already about 2.3 billion years ago, the amount of O_2 in the atmosphere was

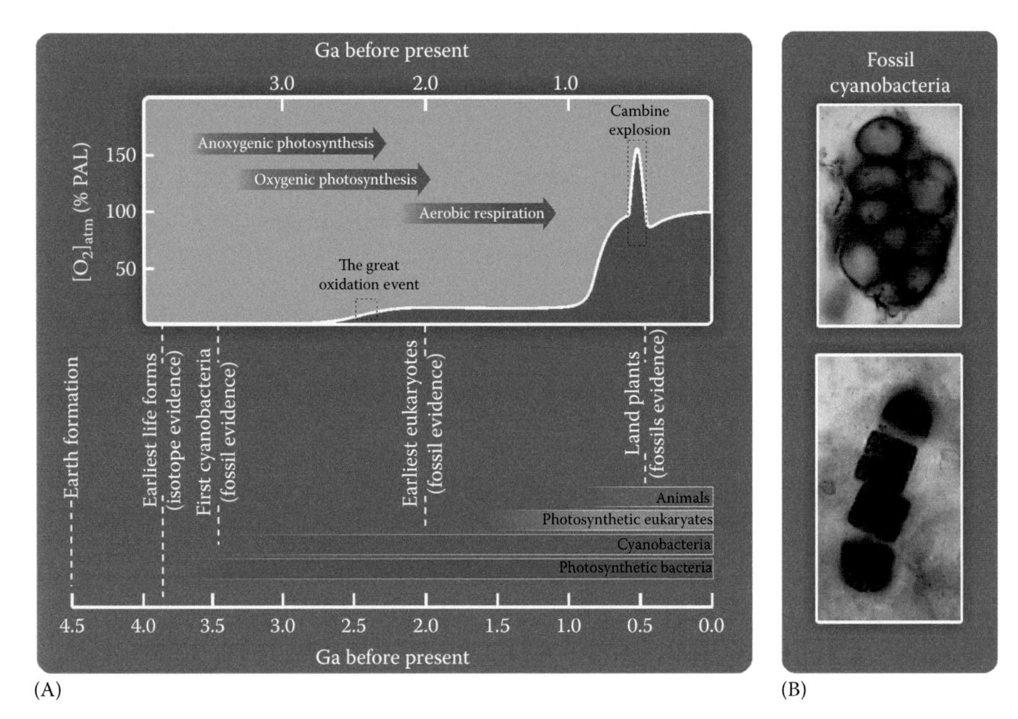

FIGURE 1.12 **(See color insert.)** The role of cyanobacteria in the evolution of life and evolution of metabolic pathways during Earth's history. (A) Schematic view on the relationships between important events in the evolution of life and photosynthesis, origin of cyanobacteria (highlighted in blue), and atmospheric O_2 concentration (in % of present atmospheric level (PAL) of O_2) through geological times in billions of years (Ga). These relationships and the dates in the figure are approximations based on numerous literature data [1,7,9,157,289,290,292, 296,298,302,303]. Only selected events of evolutionary diversification and the emergence of some organisms are shown. There is no firm conclusion as to when oxygenic photosynthesis was invented by primitive cyanobacteria. The earliest known fossil record of cyanobacteria occurring at 3.45 Ga [291,293] has been questioned [286,287]. Although there are other ample indications for the presence of primitive morphological forms of cyanobacteria in stomatolites about 3.5 Ga ago [294,295], they do not rule out the possibility that some of the earliest cyanobacteria behaved like anoxygenic photosynthetic bacteria. Concentration curve of atmospheric O_2 over time is displayed between a lower and an upper limits of PAL values provided in [290]. (B) Two representatives of ancient fossil cyanobacteria from the ~0.85-Ga-old Bitter Springs Chert of central Australia. On the top is a nonmobile colonial chroococcacean cyanobacterium (Coccoidal cyanobacteria), and on the bottom is the filamentous cyanobacterium *Palaeolyngbya* (Oscillatoriaceae). The Oscillatoriaceaen cyanobacteria have changed little or not at all over the last thousands of millions years [292]. The pictures of fossil cyanobacteria were kindly provided to the authors by J. William Schopf. (Based on Blankenship, R.E., *Plant Physiol.*, 154, 434, 2010; Hohmann-Marriott, M.F. and Blankenship, R.E., *Annu. Rev. Plant Biol.*, 62, 515, 2011; Falkowski, P.G., *Science*, 311, 1724, 2006; Raymond, J. and Blankenship, R.E., *Coord. Chem. Rev.*, 252, 377, 2008; Jagendorf, A.T. and Uribe, E., *Proc. Natl. Acad. Sci. USA*, 55, 170, 1966; Dunn, S.D. et al., *Biochemistry*, 40, 187, 2000; Van Walraven, H.S. et al., *FEBS Lett.*, 379, 309, 1996; Olson, J.M., *Photosynth. Res.*, 88, 109, 2006; Brasier, M.D. et al., *Nature*, 416, 76, 2002; Schopf, J.W., *Science*, 260, 640, 1993; Schopf, J.W., *Photosynth. Res.*, 107, 87, 2011.)

high enough to form an ozone layer, which began to absorb a large part of the highly damaging UV radiation from the sun [11,290,301] (Figure 1.12). Consequently, this allowed organisms to make better use of the terrestrial environment and led to a significant increase of genomic and metabolic complexity of life, as we observe it today [9,302,303].

Two representatives of ancient fossil cyanobacteria are shown in Figure 1.12B (also see legend and [292] for further details). Given the high degree of conservation of the photosynthetic apparatus, it is well possible, that the first cyanobacteria which learned how to split water and produce O_2, and thus changed the world, looked very similar to one of these two fossil representatives.

1.6 ABIOTIC STRESS ADAPTATION IN CYANOBACTERIA

Cyanobacteria live in many different environmental conditions that often include extremes of temperature, light, and nutrient status. Clearly, abiotic stress affects light-induced reactions of oxygenic photosynthesis in cyanobacteria. However, cyanobacteria have evolved many strategies for survival and adaptation (for recent review, see [304]). We provide here a glimpse of two of the many abiotic stresses and how cyanobacteria deal with them: nutrient stress and light stress. For a general description of abiotic stress adaptation in plants, see various chapters in [305].

An example of nutrient stress adaptation is the lack or excess of copper. Lack of copper leads to Cyt c_6 (Cyt c_{553}) replacing PC as electron donor to $P700^{\bullet+}$; on the other hand, the presence of copper in the medium leads to induction of the PC gene [306]. There is indeed an exchange of Cyt c_{553} with PC [93]; and copper mediates regulation of the two interchangeable intermediates [94,95]. Another nutrient stress that cyanobacteria experience is iron stress. Falk et al. [307] discussed the production of a CP43′ polypeptide of PSII under iron deficiency conditions. Recently, Ivanov et al. [308] demonstrated that iron stress induces the production of CP43′ and monomerization of PSI trimers and reduces the capacity for state transition. Leonhardt and Strauss [309] described the iron stress operon involved in dealing with electron transport and, thus, how cyanobacteria cope with this stress. Excess light can lead to irreversible photoinhibition and destroy the organism. However, cyanobacteria do tolerate high light and protect themselves by what is termed "non-photochemical quenching" of the first singlet excited state of Chl molecules; this entails de-excitation of the excited state through release of heat. In addition, excess of light in PSII causes the formation of low-fluorescent state II, whereas excess of light in PSI causes the formation of high-fluorescent state I; this indicates that there is reorganization of pigments and PBSs between PSI and II (state transitions). For a description and understanding of these regulatory mechanisms, we refer readers to some of the available literature (see, e.g., [132,310–318]). Recent data suggest that state transitions in cyanobacteria are important physiological adaptation mechanism to maximize the efficiency of light harvesting at very low light intensities, and that they play no role in protection from photoinhibition [72]. For further details on abiotic stress responses in cyanobacteria, see the chapters of this book.

1.7 CONCLUDING REMARKS

There is tremendous need and interest in solving problems of dwindling resources, global climate changes, and sustainability for our future. We need to develop artificial photosynthesis to capture the all abundant solar energy as well as improve photosynthesis efficiency. To achieve this goal, we must exploit the knowledge of natural photosynthesis [319–327] (also see Chapter 2). We believe that cyanobacteria can help us in this respect. We hope to exploit the flexibility of cyanobacteria in learning how to use them to produce hydrogen [328–331] and fix nitrogen [20,21] and for the synthesis of biopharmaceuticals [332,333]. Information provided in this chapter on the structure and function of the antenna system, and the photosystems, and on evolution is crucial, in our opinion, in designing future directions for using cyanobacteria to serve society.

ACKNOWLEDGMENTS

We are grateful to Gernot Renger for reading the manuscript and for his valuable suggestions and to J. William Schopf for providing us pictures of ancient cyanobacteria that has been used in Figure 1.12 and for checking the accuracy of this figure. D.S. and L.A.E. acknowledge the financial support by *The Research Council of Norway* (grant 192436 to D.S. and 197119 to L.A.E.). G. acknowledges the financial support of the Department of Plant Biology at the University of Illinois. Lastly, we thank Mythili Gopi, project manager at SPi Global for patience with our corrections.

REFERENCES

1. Blankenship, R.E., Early evolution of photosynthesis, *Plant Physiol.* 154, 434, 2010.
2. Buick, R., When did oxygenic photosynthesis evolve? *Philos. Trans. R. Soc. Lond. B* 363, 2731, 2008.
3. Blankenship, R.E., Sadekar, S., and Raymond, J., The evolutionary transition from anoxygenic to oxygenic photosynthesis, in *Evolution of Aquatic Photoautotrophs*, Falkowski, P. and Knoll, A.N., Eds., Academic Press, New York, 2007, p. 21.
4. Dismukes, G.C. and Blankenship, R.E., The origin and evolution of photosynthetic oxygen production, in *Photosystem II. The Light-Driven Water:Plastoquinone Oxidoreductase*, Wydrzynski, T. and Satoh, K., Eds., Springer, Dordrecht, the Netherlands, 2005, p. 683.
5. Björn, L.O. and Govindjee, The evolution of photosynthesis and chloroplasts, *Curr. Sci.* 96, 1466, 2009.
6. Drews, G., The evolution of cyanobacteria and photosysthesis, in *Bioenergetic Processes of Cyanobacteria: From Evolutionary Singularity to Ecological Diversity*, Peschek, G.A., Obinger, C., and Renger, G., Eds., Springer, Dordrecht, the Netherlands, 2011, p. 265.
7. Hohmann-Marriott, M.F. and Blankenship, R.E., Evolution of photosynthesis, *Annu. Rev. Plant Biol.* 62, 515, 2011.
8. Barber, J., Photosynthetic generation of oxygen, *Philos. Trans. R. Soc. Lond. B* 363, 2665, 2008.
9. Falkowski, P.G., Tracing oxygen's imprint on Earth's metabolic evolution, *Science* 311, 1724, 2006.
10. Des Marais, D.J., Evolution: When did photosynthesis emerge on Earth? *Science* 289, 1703, 2000.
11. Olson, J.M. and Blankenship, R.E., Thinking about the evolution of photosynthesis, *Photosynth. Res.* 80, 373, 2004.
12. Lane, N., *Oxygen: The Molecule That Made the World*, Oxford University Press, Oxford, U.K., 2004.
13. Renger, G., Biological energy conservation, in *Biophysics*, Hoppe, W. et al., Eds., Springer, Berlin, Germany, 1983, p. 347.
14. Margulis, L., *Origin of Eukaryotic Cells*, Yale University Press, New Haven, CT, 1970.
15. Bhattacharya, D., Yoon, H.S., and Hackett, J.D., Photosynthetic eukaryotes unite: Endosymbiosis connects the dots, *BioEssays* 26, 50.60, 2004.
16. Archibald, J.M., The puzzle of plastid evolution, *Curr. Biol.* 19, R81, 2009.
17. Waterbury, J.B. et al., Widespread occurrence of a unicellular, marine, planktonic, cyanobacterium, *Nature* 277, 293, 1979.
18. Field, C.B. et al., Primary production of the biosphere: Integrating terrestrial and oceanic components, *Science* 281, 237, 1998.
19. Barber, J., Water, water everywhere, and its remarkable chemistry, *Biochim. Biophys. Acta* 1655, 123, 2004.
20. Zehr, J.P., Nitrogen fixation by marine cyanobacteria, *Trends Microbiol.* 19, 162, 2011.
21. Bothe, H. et al., Nitrogen fixation and hydrogen metabolism in cyanobacteria, *Microbiol. Mol. Biol. Rev.* 74, 529, 2010.
22. Cohen, Y. and Gurevitz, M., The cyanobacteria—Ecology, physiology and molecular genetics, in *The Prokaryotes*, Dworkin, M. et al., Eds., Springer, New York, 2006, p. 1074.
23. Stal, L.J., Cyanobacteria: Diversity and versality, clues to life in extreme environments, in *Algae and Cyanobacteria in Extreme Environments*, Seckbach, J., Ed., Springer, Dordrecht, the Netherlands, 2007, p. 659.
24. Zehr, J.P. et al., Globally distributed uncultivated oceanic N_2-fixing cyanobacteria lack oxygenic photosystem II, *Science* 322, 1110, 2008.
25. Cohen, Y. et al., Adaptation to hydrogen sulfide of oxygenic and anoxygenic photosynthesis among cyanobacteria, *Appl. Environ. Microbiol.* 51, 398, 1986.
26. Govindjee and Rabinowitch, E., Action spectrum of the second Emerson effect, *Biophys. J.* 1, 73, 1960.
27. Owens, O.H. and Hoch, G., Enhancement and de-enhancement effect in *Anacystis nidulans*, *Biochim. Biophys. Acta* 75, 183, 1963.
28. Kok, B., Light induced absorption changes in photosynthetic organisms. II. A split-beam difference spectrophotometer, *Plant Physiol.* 34, 184, 1959.
29. Amesz, J. and Duysens, L.N., Action spectrum, kinetics and quantum requirement of phosphopyridine nucleotide reduction and cytochrome oxidation in the blue-green alga *Anacystis nidulans*, *Biochim. Biophys. Acta* 64, 261, 1962.
30. Scheer, H., An overview of chlorophylls and bacteriochlorophylls: Biochemistry, biophysics, functions and applications, in *Chlorophylls and Bacteriochlorophylls*, Grimm, B. et al., Eds., Springer, Dordrecht, the Netherlands, 2006, p. 1.
31. Björn, L.O. et al., A viewpoint: Why chlorophyll *a*? *Photosynth. Res.* 99, 85, 2009.

32. Hirschberg, J. and Chamovitz, D., Carotenoids in cyanobacteria, in *The Molecular Biology of Cyanobacteria*, Bryant, D.A., Ed., Springer, Dordrecht, the Netherlands, 1994, p. 559.
33. Govindjee, Carotenoids in photosynthesis: A historical perspective, in *The Photochemistry of Carotenoids: Applications in Biology*, Frank, H.A. et al., Eds., Kluwer Academic Publishers, Dordrecht, the Netherlands, 1999, p. 1.
34. Beale, S.I., Biosynthesis of cyanobacterial tetrapyrrole pigments hemes, chlorophylls, and phycobilins, in *The Molecular Biology of Cyanobacteria*, Bryant, D.A., Ed., Springer, Dordrecht, the Netherlands, 2004, p. 519.
35. Blankenship, R.E., *Molecular Mechanisms of Photosynthesis*, Blackwell Publishing, Oxford, U.K., 2002.
36. Govindjee and Mohanty, P., Photochemical aspects of photosynthesis in blue-green algae, in *Biology and Taxonomy of Blue-Green Algae*, Desikachary, T., Ed., University of Madras, Madras, India, 1972, p. 171.
37. Glazer, A.N., Comparative biochemistry of photosynthetic light-harvesting systems, *Annu. Rev. Biochem.* 52, 125, 1983.
38. Collins, A.M., Wen, J., and Blankenship, R.E., Photosynthetic light-harvesting complexes, in *Molecular Solar Fuels*, Wydrzynski, T. and Hillier, W., Eds., RSC Publishing, Cambridge, U.K., 2012, p. 85.
39. Gray, B.H., Lipschultz, C.A., and Gantt, E., Phycobilisomes from a blue-green alga *Nostoc* species, *J. Bacteriol.* 116, 471, 1973.
40. Tandeau de Marsac, N., Phycobiliproteins and phycobilisomes: The early observations, in *Discoveries in Photosynthesis*, Govindjee et al., Eds., Springer, Dordrecht, the Netherlands, 2005, p. 443.
41. Wildman, R.B. and Bowen, C.C., Phycobilisomes in blue-green algae, *J. Bacteriol.* 117, 866, 1974.
42. Matthijs, H.C.P., van der Staay, G.W.M., and Mur, L.R., Prochlorophytes: The 'other' cyanobacteria, in *The Molecular Biology of Cyanobacteria*, Bryant, D.A., Ed., Kluwer Academic Publishers, Dordrecht, the Netherlands, 1994, p. 49.
43. Miyashita, H. et al., Chlorophyll *d* as a major pigment, *Nature* 383, 402, 1996.
44. Larkum, A.W.D. and Kuhl, M., Chlorophyll *d*: The puzzle resolved, *Trends Plant. Sci.* 10, 355, 2005.
45. Miller, S.R. et al., Discovery of a free-living chlorophyll *d*-producing cyanobacterium with a hybrid proteobacterial/cyanobacterial small-subunit rRNA gene, *Proc. Natl. Acad. Sci. USA* 102, 850, 2005.
46. Murakami, A. et al., Chlorophyll *d* in an epiphytic cyanobacterium of red algae, *Science* 303, 1633, 2004.
47. Vavilin, D. et al., Energy and electron transfer in photosystem II of a chlorophyll *b*-containing *Synechocystis sp.* PCC 6803 mutant, *Biochemistry* 42, 1731, 2003.
48. Vermaas, W.F.J., Photosynthesis and respiration in cyanobacteria, in *Encyclopedia of Life Sciences (ELS)*, John Wiley & Sons, Ltd, London, U.K., 2001.
49. Rippka, R., Waterbury, J., and Cohen-Bazire, G., A cyanobacterium which lacks thylakoids, *Arch. Microbiol.* 100, 419, 1974.
50. Carr, N.G. and Whitton, B.A., Eds., *The Biology of Cyanobacteria*, vol. 19, University of California Press, Berkeley and Los Angeles, CA, 1982.
51. Packer, L. and Glazer, A.N., Eds., *Cyanobacteria*, vol. 167, Academic Press, Inc., San Diego, CA, 1988.
52. Bryant, D.A., Ed., *The Molecular Biology in Cyanobacteria*, vol. 1, Kluwer Academic Publishers, Dordrecht, the Netherlands, 1994.
53. Herrero, A., Flores, E., and Flores, F.G., *The Cyanobacteria: Molecular Biology, Genomics, and Evolution*, Horizon Scientific Press/Caister Academic Press, Norwich, U.K., 2008.
54. Gault, P.M. and Marler, H.J., *Handbook on Cyanobacteria: Biochemistry, Biotechnology and Applications*, Nova Science Publishers, New York, 2009.
55. Peschek, G.A., Obinger, C., and Renger, G., *Bioenergetic Processes of Cyanobacteria: From Evolutionary Singularity to Ecological Diversity*, Springer, Dordrecht, the Netherlands, 2011.
56. Govindjee and Shevela, D., Adventures with cyanobacteria: A personal perspective, *Front. Plant Sci.* 2, 28, doi:10.3389/fpls.2011.00028, 2011.
57. Govindjee, Amesz, J., and Fork, D.C., *Light Emission by Plants and Bacteria*, Academic Press, Orlando, FL, 1986.
58. Govindjee and Krogmann, D., Discoveries in oxygenic photosynthesis (1727–2003): A perspective, *Photosynth. Res.* 80, 15, 2004.
59. Schmetterer, G. and Pils, D., Cyanobacterial respiration, in *Respiration in Archaea and Bacteria: Diversity of Prokaryotic Respiratory Systems*, Zannoni, D., Ed., Springer, Dordrecht, the Netherlands, 2004, p. 261.
60. Mimuro, M. and Kikuchi, H., Antenna systems and energy transfer in Cyanophyta and Rhodophyta, in *Light-Harvesting Antennas in Photosynthesis*, Green, B.R. and Parson, W.W., Eds., Springer, Dordrecht, the Netherlands, 2003, p. 281.

61. Mimuro, M. et al., Oxygen-evolving cyanobacteria, in *Primary Processes of Photosynthesis, Part 1 Principles and Apparatus*, Renger, G., Ed., RSC Publishing, Cambridge, U.K., 2008, p. 261.
62. Sidler, W.A., Phycobilisome and phycobiliprotein structures, in *The Molecular Biology of Cyanobacteria*, Bryant, D.A., Ed., Springer, Dordrecht, the Netherlands, 1994, p. 139.
63. Fromme, P. and Grotjohann, I., Structure of cyanobacterial photosystems I and II, in *Bioenergetic Processes of Cyanobacteria: From Evolutionary Singularity to Ecological Diversity*, Peschek, G.A., Obinger, C., and Renger, G., Eds., Springer, Dordrecht, the Netherlands, 2011, p. 285.
64. Kargul, J. and Barber, J., Structure and function of photosynthetic reaction centres, in *Molecular Solar Fuels*, Wydrzynski, T. and Hillier, W., Eds., RSC Publishing, Cambridge, U.K., 2012, p. 107.
65. Renger, T., Photophysics of photosynthetic reaction centres, in *Molecular Solar Fuels*, Wydrzynski, T. and Hillier, W., Eds., RSC Publishing, Cambridge, U.K., 2012, p. 143.
66. Nelson, N., Photosystems and global effects of oxygenic photosynthesis, *Biochim. Biophys. Acta* 1807, 856, 2011.
67. Govindjee et al., Photosystem II, in *Encyclopedia of Life Sciences (ELS)*, John Wiley & Sons, Ltd., Chichester, U.K., 2010.
68. Melis, A., Spectroscopic methods in photosynthesis: Photosystem stoichiometry and chlorophyll antenna size, *Philos. Trans. R. Soc. Lond. B* 323, 397, 1989.
69. Murakami, A. and Fujita, Y., Steady state of photosynthesis in cyanobacterial photosynthetic systems before and after regulation of electron transport composition: Overall rate of photosynthesis and PSI/PS II composition, *Plant Cell Physiol.* 29, 305, 1988.
70. Rakhimberdieva, M.G. et al., Interaction of phycobilisomes with photosystem II dimers and photosystem I monomers and trimers in the cyanobacterium *Spirulina platensis*, *Biochemistry* 40, 15780, 2001.
71. Shen, G., Boussiba, S., and Vermaas, W., Synechocystis sp PCC 6803 strains lacking photosystem I and phycobilisome function, *Plant Cell* 5, 1853, 1993.
72. Mullineaux, C.W. and Emlyn-Jones, D., State transitions: An example of acclimation to low-light stress, *J. Exp. Bot.* 56, 389, 2005.
73. McConnell, M.D. et al., Regulation of the distribution of chlorophyll and phycobilin-absorbed excitation energy in cyanobacteria. A structure-based model for the light state transition, *Plant Physiol.* 130, 1201, 2002.
74. Lemeille, S. and Rochaix, J.-D., State transitions at the crossroad of thylakoid signalling pathways, *Photosynth. Res.* 106, 33, 2010.
75. Schiller, H. et al., Light-harvesting in *Acaryochloris marina*—Spectroscopic characterization of a chlorophyll *d*-dominated photosynthetic antenna system, *FEBS Lett.* 410, 433, 1997.
76. Renger, G. and Renger, T., Photosystem II: The machinery of photosynthetic water splitting, *Photosynth. Res.* 98, 53, 2008.
77. Hill, R. and Bendall, F., Function of the 2 cytochrome components in chloroplasts—Working hypothesis, *Nature* 186, 136, 1960.
78. Govindjee and Björn, L.O., Dissecting oxygenic photosynthesis: The evolution of the "Z"-scheme for thylakoid reactions, in *Photosynthesis: Overviews on Recent Progress and Future Perspectives*, Itoh, S., Mohanty, P., and Guruprasad, K.N., Eds., IK Publishers, New Delhi, India, 2012, p. 1.
79. Dau, H. and Zaharieva, I., Principles, efficiency, and blueprint character of solar-energy conversion in photosynthetic water oxidation, *Acc. Chem. Res.* 42, 1861, 2009.
80. Renger, G. and Holzwarth, A.R., Primary electron transfer, in *Photosystem II. The Light-Driven Water:Plastoquinone Oxidoreductase*, Wydrzynski, T.J. and Satoh, K., Eds., Springer, Dordrecht, the Netherlands, 2005, p. 139.
81. Rappaport, F. and Diner, B.A., Primary photochemistry and energetics leading to the oxidation of the (Mn)4Ca cluster and to the evolution of molecular oxygen in Photosystem II, *Coord. Chem. Rev.* 252, 259, 2008.
82. Ishikita, H. et al., Redox potentials of chlorophylls in the photosystem II reaction center, *Biochemistry* 44, 4118, 2005.
83. Nakamura, A. et al., Species dependence of the redox potential of the primary electron donor P700 in photosystem I of oxygenic photosynthetic organisms revealed by spectroelectrochemistry, *Plant Cell Physiol.* 52, 815, 2011.
84. Nakamura, A. et al., Significant species-dependence of P700 redox potential as verified by spectroelectrochemistry: Comparison of spinach and *Theromosynechococcus elongatus*, *FEBS Lett.* 579, 2273, 2005.
85. Berry, E.A. et al., Structure and function of cytochrome bc complexes, *Annu. Rev. Biochem.* 69, 1005, 2000.

86. Bernat, G. and Rögner, M., Center of the cyanobacteria electron transport network: The cytochrome b_6f complex, in *Bioenergetic Processes of Cyanobacteria: From Evolutionary Singularity to Ecological Diversity*, Peschek, G.A., Obinger, C., and Renger, G., Eds., Springer, Dordrecht, the Netherlands, 2011, p. 573.
87. Cramer, W.A. et al., Transmembrane traffic in the cytochrome b_6f complex, *Annu. Rev. Biochem.* 75, 769, 2006.
88. Allen, J.F., Cytochrome b_6f: Structure for signalling and vectorial metabolism, *Trends Plant. Sci.* 9, 130, 2004.
89. Cramer, W.A. et al., Structure-function of the cytochrome b_6f complex: A design that has worked for three billion years, in *Primary Processes of Photosynthesis, Part 2: Principles and Apparatus*, Renger, G., Ed., RSC Publishing, Cambridge, U.K., 2008, p. 417.
90. Baniulis, D. et al., Structure–function of the cytochrome b_6f complex, *Photochem. Photobiol.* 84, 1349, 2008.
91. De la Rosa, M.A., Navarro, J.A., and Hervás, M., The convergent evolution of cytochrome c_6 and plastocyanin has been driven by geochemical changes, in *Bioenergetic Processes of Cyanobacteria: From Evolutionary Singularity to Ecological Diversity*, Peschek, G.A., Obinger, C., and Renger, G., Eds., Springer, Dordrecht, the Netherlands, 2011, p. 607.
92. Bendall, D.S., Schlarb-Ridley, B.G., and Howe, C.J., Transient interactions between soluble electron transfer proteins. The case of plastocyanin and cytochrome c_6, in *Bioenergetic Processes of Cyanobacteria: From Evolutionary Singularity to Ecological Diversity*, Peschek, G.A., Obinger, C., and Renger, G., Eds., Springer, Dordrecht, the Netherlands, 2011, p. 541.
93. Sandmann, G., Formation of plastocyanin and cytochrome c-553 in different species of blue-green algae, *Arch. Microbiol.* 145, 76, 1986.
94. Zhang, L. et al., Copper-mediated regulation of cytochrome *c*553 and plastocyanin in the cyanobacterium *Synechocystis* 6803, *J. Biol. Chem.* 267, 19054, 1992.
95. Zhang, L., Pakrasi, H.B., and Whitmarsh, J., Photoautotrophic growth of the cyanobacterium *Synechocystis* sp. PCC 6803 in the absence of cytochrome c_{553} and plastocyanin, *J. Biol. Chem.* 269, 5036, 1994.
96. Nicholls, P., History and function: The respiratory and photosynthetic electron transport chains, in *Bioenergetic Processes of Cyanobacteria: From Evolutionary Singularity to Ecological Diversity*, Peschek, G.A., Obinger, C., and Renger, G., Eds., Springer, Dordrecht, the Netherlands, 2011, p. 189.
97. Aliverti, A. et al., Structural and functional diversity of ferredoxin-$NADP^+$ reductases, *Arch. Biochem. Biophys.* 474, 283, 2008.
98. Medina, M., Structural and mechanistic aspects of flavoproteins: Photosynthetic electron transfer from photosystem I to $NADP^+$, *FEBS J.* 276, 3942, 2009.
99. McCarty, R.E., Evron, Y., and Johnson, E.A., The chloroplast ATP synthase: A rotary enzyme? *Annu. Rev. Plant Physiol. Plant Mol. Biol.* 51, 83, 2000.
100. Junge, W., Sielaff, H., and Engelbrecht, S., Torque generation and elastic power transmission in the rotary F_OF_1-ATPase, *Nature* 459, 364, 2009.
101. Bald, D., ATP synthase: Structure, function and regulation of a complex machine, in *Bioenergetic Processes of Cyanobacteria: From Evolutionary Singularity to Ecological Diversity*, Peschek, G.A., Obinger, C., and Renger, G., Eds., Springer, Dordrecht, the Netherlands, 2011, p. 239.
102. Junge, W., Protons, proteins and ATP, *Photosynth. Res.* 80, 197, 2004.
103. Bendall, D.S. and Manasse, R.S., Cyclic photophosphorylation and electron transport, *Biochim. Biophys. Acta* 1229, 23, 1995.
104. Van Thor, J.J. et al., Salt shock-inducible photosystem I cyclic electron transfer in *Synechocystis* PCC6803 relies on binding of ferredoxin:$NADP^+$ reductase to the thylakoid membranes *via* its CpcD phycobilisome-linker homologous N-terminal domain, *Biochim. Biophys. Acta* 1457, 129, 2000.
105. Joliot, P., Joliot, A., and Johnson, G., Cyclic electron transfer around photosystem I, in *Photosystem I: The Light-Driven Plastocyanine:Ferredoxin Oxidoreductase*, Golbeck, J.H., Ed., Springer, Dordrecht, the Netherlands, 2006, p. 639.
106. Martin, W., Scheibe, R., and Schnarrenberger, C., The Calvin cycle and its regulation, in *Photosynthesis: Physiology and Metabolism*, Leegood, R.C., Shakey, T.D., and von Caemmerer, S., Eds., Kluwer Academic Publishers, Dordrecht, the Netherlands, 2000, p. 9.
107. Tabita, F.R., The biochemistry and molecular regulation of carbon dioxide metabolism in cyanobacteria, in *The Molecular Biology of Cyanobacteria*, Bryant, D.A., Ed., Springer, Dordrecht, the Netherlands, 1994, p. 437.
108. Fukuzawa, H., Ogawa, T., and Kaplan, A., The uptake of CO_2 by cyanobacteria and microalgae, in *Photosynthesis: Plastid Biology, Energy Conversion and Respiration*, Eaton-Rye, J.J., Tripathy, B.C., and Sharkey, T.D., Eds., Springer, Dordrecht, the Netherlands, 2012, p. 625.

109. Sharkey, T.D. and Weise, S.E., Autotrophic carbon dioxide fixation, in *Photosynthesis: Plastid Biology, Energy Conversion and Carbon Assimilation*, Eaton-Rye, J.J., Tripathy, B.C., and Sharkey, T.D., Eds., Springer, Dordrecht, the Netherlands, 2012, p. 651.
110. Badger, M.R. and Spalding, M.H., CO_2 acquisition, concentration and fixation in cyanobacteria and algae, in *Photosynthesis: Physiology and Metabolism*, Leegood, R.C., Sharkey, T.D., and Cammerer, S., Eds., Kluwer Academic Publishers, Dordrecht, the Netherlands, 2000, p. 369.
111. Renger, G., Photosynthetic water splitting: Apparatus and mechanism, in *Photosynthesis: Plastid Biology, Energy Conversion and Carbon Assimilation*, Eaton-Rye, J.J., Tripathy, B.C., and Sharkey, T.D., Eds., Springer, Dordrecht, the Netherlands, 2012, p. 359.
112. Shuvalov, V.A. et al., Primary charge separation between P700* and the primary electron acceptor complex A-A_0: A comparison with bacterial reaction centers, in *Photosystem I. The Light-Driven Plastocyanin:Ferredoxin Oxidoreductase*, Golbeck, J.H., Ed., Springer, Dordrecht, the Netherlands, 2006, p. 291.
113. DiMagno, L. et al., Energy transfer and trapping in photosystem I reaction centers from cyanobacteria, *Proc. Natl. Acad. Sci. USA* 92, 2715, 1995.
114. Savikhin, S., Ultrafast optical spectroscopy of photosystem I, in *Photosystem I. The Light-Driven Plastocyanin:Ferredoxin Oxidoreductase*, Golbeck, J.H., Ed., Springer, Dordrecht, the Netherlands, 2006, p. 155.
115. Renger, G. et al., Fluorescence and spectroscopic studies of exciton trapping and electron transfer in photosystem II of higher plants, *Aust. J. Plant Physiol.* 22, 167, 1995.
116. Brettel, K. and Leibl, W., Electron transfer in photosystem I, *Biochim. Biophys. Acta* 1507, 100, 2001.
117. Kühn, P. et al., Analysis of the $P680^{+\bullet}$ reduction pattern and its temperature dependence in oxygen-evolving PSII core complexes from a thermophilic cyanobacteria and higher plants, *Phys. Chem. Chem. Phys.* 6, 4838, 2004.
118. Åhrling, K., Pace, R., and Evans, M.C.W., The catalytic manganese cluster: Implications from spectroscopy, in *Photosystem II. The Light-Driven Water:Plastoquinone Oxidoreductase*, Wydrzynski, T., Satoh, K., and Freeman, J., Eds., Springer, Dordrecht, the Netherlands, 2005, p. 285.
119. Medina, M. et al., A comparative laser-flash absorption spectroscopy study of *Anabaena* PCC 7119 plastocyanin and cytochrome c_6 photooxidation by photosystem I particles, *Eur. J. Biochem.* 213, 1133, 1993.
120. Padden, S. et al., Site specific mutagenesis reveals a critical role for histidine 252 of the D1 subunit in the two-electron gate of photosystem II, Presented at the *28th Annual Eastern Regional Photosynthesis Conference*, Marine Biological Laboratory, Woods Hole, MA, April 1–3, 2011.
121. Petrouleas, V. and Crofts, A.R., The iron-quinone acceptor complex, in *Photosystem II. The Light-Driven Water:Plastoquinone Oxidoreductase*, Wydrzynski, T. and Satoh, K., Eds., Springer, Dordrecht, the Netherlands, 2005, p. 177.
122. Shevela, D. et al., Photosystem II and the unique role of bicarbonate: A historical perspective, *Biochim. Biophys. Acta* 1817, 1134, 2012.
123. Velthuys, B.R. and Amesz, J., Charge accumulation at the reducing side of system 2 of photosynthesis, *Biochim. Biophys. Acta* 333, 85, 1974.
124. Bowes, J.M. and Crofts, A.R., Binary oscillations in the rate of reoxidation of the primary acceptor of Photosystem II, *Biochim. Biophys. Acta* 590, 373, 1980.
125. Umena, Y. et al., Crystal structure of oxygen-evolving photosystem II at a resolution of 1.9 Å, *Nature* 473, 55, 2011.
126. Haehnel, W., Electron transport between plastoquinone and chlorophyll A_1 in chloroplasts. II. Reaction kinetics and the function of plastocyanin in situ, *Biochim. Biophys. Acta* 459, 418, 1977.
127. Siggel, U. et al., Investigation of absorption changes of plastoquinone system in broken chloroplasts: Effect of bicarbonate depletion, *Biochim. Biophys. Acta* 462, 196, 1977.
128. Schreiber, U. and Klughammer, C., New NADPH/9-AA module for the DUAL-PAM-100: Description, operation and examples of application, *PAM Application Notes* 2, 1, 2009.
129. Papageorgiou, G.C., Tsimilli-Michael, M., and Stamatakis, K., The fast and slow kinetics of chlorophyll *a* fluorescence induction in plants, algae and cyanobacteria: A viewpoint, *Photosynth. Res.* 94, 275, 2007.
130. Stirbet, A. and Govindjee, On the relation between the Kautsky effect (chlorophyll *a* fluorescence induction) and Photosystem II: Basics and applications of the OJIP fluorescence transient, *J. Photochem. Photobiol. B Biol.* 104, 236, 2011.
131. Papageorgiou, G. and Govindjee, Photosystem II fluorescence: Slow changes—Scaling from the past, *J. Photochem. Photobiol. B Biol.* 104, 258, 2011.
132. Papageorgiou, G.C., The photosynthesis of cyanobacteria (blue bacteria) from the perspective of signal analysis of chlorophyll *a* fluorescence, *J. Sci. Ind. Res. India* 55, 596, 1996.

133. Boichenko, V.A., Photosynthetic units of phototrophic organisms, *Biochemistry (Moscow)* 69, 471, 2004.
134. Wraight, C., Reaction centers, electron flow, and energy transduction, in *Photosynthesis,* Govindjee, Ed., Academic Press, New York, 1982, p. 17.
135. Allen, J.P. and Williams, J.C., Photosynthetic reaction centers, *FEBS Lett.* 438, 5, 1998.
136. Blankenship, R.E., Origin and early evolution of photosynthesis, *Photosynth. Res.* 33, 91, 1992.
137. Olson, J.M., 'Evolution of Photosynthesis' (1970), re-examined thirty years later, *Photosynth. Res.* 68, 95, 2001.
138. Baymann, F. et al., Daddy, where did (PS)I come from? *Biochim. Biophys. Acta* 1507, 291, 2001.
139. Nelson, N. and Ben-Shem, A., The structure of photosystem I and evolution of photosynthesis, *BioEssays* 27, 914, 2005.
140. Rutherford, A.W. and Nitschke, W., Photosystem II and the quinone-iron-containing reaction centers: Comparison and evolutionary perspectives, in *Origin and Evolution of Biological Energy Conversion,* Batscheffky, H., Ed., VCH, New York, 1996, p. 143.
141. Nitschke, W., Mattioli, T., and Rutherford, A.W., The Fe-S-type photosystems and the evolution of photosynthetic reaction centers, in *Origin and Evolution of Biological Energy Conversion*, Baltscheffky, H., Ed., VCH, New York, 1996, p. 177.
142. Schubert, W.D. et al., A common ancestor for oxygenic and anoxygenic photosynthetic systems: A comparison based on the structural model of photosystem I, *J. Mol. Biol.* 280, 297, 1998.
143. Sadekar, S., Raymond, J., and Blankenship, R.E., Conservation of distantly related membrane proteins: Photosynthetic reaction centers share a common structural core, *Mol. Biol. Evol.* 23, 2001, 2006.
144. Xiong, J. and Bauer, C.E., Complex evolution of photosynthesis, *Annu. Rev. Plant Biol.* 53, 503, 2002.
145. Vermaas, W.F.J., Shen, G., and Styling, S., Electrons generated by photosystem II are utilized by an oxidase in the absence of photosystem I in the cyanobacterium *Synechocystis* sp. PCC 6803, *FEBS Lett.* 337, 103, 1994.
146. Wang, Q.J. et al., Net light-induced oxygen evolution in photosystem I deletion mutants of the cyanobacterium *Synechocystis sp.* PCC 6803, *Biochim. Biophys. Acta* 1817, 792, 2012.
147. Moisander, P.H. et al., Unicellular cyanobacterial distributions broaden the oceanic N_2 fixation domain, *Science* 327, 1512, 2010.
148. Allen, J.F. and Martin, W., Evolutionary biology—Out of thin air, *Nature* 445, 610, 2007.
149. Raymond, J. and Blankenship, R.E., The origin of the oxygen-evolving complex, *Coord. Chem. Rev.* 252, 377, 2008.
150. Blankenship, R.E. and Hartman, H., The origin and evolution of oxygenic photosynthesis, *Trends Biochem. Sci.* 23, 94, 1998.
151. Kobayashi, M. et al., Redox potential of chlorophyll *d in vitro*, *Biochim. Biophys. Acta* 1767, 596, 2007.
152. Ishikita, H. et al., How photosynthetic reaction centers control oxidation power in chlorophyll pairs P680, P700, and P870, *Proc. Natl. Acad. Sci. USA* 103, 9855, 2006.
153. Renger, G., The light reactions of photosynthesis, *Curr. Sci.* 98, 1305, 2010.
154. Blankenship, R.E., Madigan, M.T., and Bauer, C.E., Eds., *Anoxygenic Photosynthetic Bacteria*, vol. 2, Kluwer Academic Publishers, Dordrecht, the Netherlands, 1995.
155. Burke, D.H., Hearst, J.E., and Sidow, A., Early evolution of photosynthesis—Clues from nitrogenase and chlorophyll iron proteins, *Proc. Natl. Acad. Sci. USA* 90, 7134, 1993.
156. Raymond, J. et al., Evolution of photosynthetic prokaryotes: A maximum-likelihood mapping approach, *Philos. Trans. R. Soc. Lond. B* 358, 223, 2003.
157. Larkum, A.W.D., The evolution of photosynthesis, in *Primary Processes of Photosynthesis, Part 2: Principles and Apparatus*, Renger, G., Ed., RSC Publishing, Cambridge, U.K., 2008, p. 491.
158. Lockhart, P.J. et al., Evolution of chlorophyll and bacteriochlorophyll: The problem of invariant sites in sequence analysis, *Proc. Natl. Acad. Sci. USA* 93, 1930, 1996.
159. Hu, Q. et al., A photosystem I reaction center driven by chlorophyll *d* in oxygenic photosynthesis, *Proc. Natl. Acad. Sci. USA* 95, 13319, 1998.
160. Tomo, T. et al., Identification of the special pair of photosystem II in a chlorophyll *d*-dominated cyanobacterium, *Proc. Natl. Acad. Sci. USA* 104, 7283, 2007.
161. Akiyama, M. et al., Quest for minor but key chlorophyll molecules in photosynthetic reaction centers—Unusual pigment composition in the reaction centers of the chlorophyll *d*-dominated cyanobacterium *Acaryochloris marina*, *Photosynth. Res.* 74, 97, 2002.
162. Itoh, S. et al., Function of chlorophyll *d* in reaction centers of photosystems I and II of the oxygenic photosynthesis of *Acaryochloris marina*, *Biochemistry* 46, 12473, 2007.
163. Tomo, T., Allakhverdiev, S.I., and Mimuro, M., Constitution and energetics of photosystem I and photosystem II in the chlorophyll *d*-dominated cyanobacterium *Acaryochloris marina*, *J. Photochem. Photobiol. B Biol.* 104, 333, 2011.

164. Kobayashi, M. et al., Minor but key chlorophylls in photosystem II, *Photosynth. Res.* 84, 201, 2005.
165. Schlodder, E. et al., Both chlorophylls *a* and *d* are essential for the photochemistry in photosystem II of the cyanobacteria, *Acaryochloris marina, Biochim. Biophys. Acta* 1767, 589, 2007.
166. Renger, T. and Schlodder, E., The primary electron donor of photosystem II of the cyanobacterium *Acaryochloris marina* is a chlorophyll *d* and the water oxidation is driven by a chlorophyll a/chlorophyll d heterodimer, *J. Phys. Chem. B* 112, 7351, 2008.
167. Chen, M. et al., Structure of a large photosystem II supercomplex from *Acaryochloris marina, FEBS Lett.* 579, 1306, 2005.
168. Allakhverdiev, S.I. et al., Redox potentials of primary electron acceptor quinone molecule (Q_A^-) and conserved energetics of photosystem II in cyanobacteria with chlorophyll *a* and chlorophyll *d, Proc. Natl. Acad. Sci. USA* 108, 8054, 2011.
169. Allakhverdiev, S.I. et al., Redox potential of pheophytin *a* in photosystem II of two cyanobacteria having the different special pair chlorophylls, *Proc. Natl. Acad. Sci. USA* 107, 3924, 2010.
170. Shevela, D. et al., Characterization of the water oxidizing complex of photosystem II of the Chl *d*-containing cyanobacterium *Acaryochloris marina via* its reactivity towards endogenous electron donors and acceptors, *Phys. Chem. Chem. Phys.* 8, 3460, 2006.
171. Boichenko, V.A. et al., Functional characteristics of chlorophyll *d*-predominating photosynthetic apparatus in intact cells of *Acaryochloris marina, Photosynth. Res.* 65, 269, 2000.
172. Cser, K. et al., Energetics of Photosystem II charge recombination in *Acaryochloris marina* studied by thermoluminescence and flash-induced chlorophyll fluorescence measurements, *Photosynth. Res.* 98, 131, 2008.
173. Loll, B. et al., Towards complete cofactor arrangement in the 3.0 Å resolution structure of photosystem II, *Nature* 438, 1040, 2005.
174. Guskov, A. et al., Recent progress in the crystallographic studies of photosystem II, *ChemPhysChem* 11, 1160, 2010.
175. Shi, L.-X. et al., Photosystem II, a growing complex: Updates on newly discovered components and low molecular mass proteins, *Biochim. Biophys. Acta* 1817, 13, 2012.
176. Ferreira, K.N. et al., Architecture of the photosynthetic oxygen-evolving center, *Science* 303, 1831, 2004.
177. Deisenhofer, J. and Michel, H., The photosynthetic reaction center from the purple bacterium *Rhodopseudomonas viridis, Biosci. Rep.* 9, 383, 1989.
178. Guskov, A. et al., Cyanobacterial photosystem II at 2.9-Angstrom resolution and the role of quinones, lipids, channels and chloride, *Nat. Struct. Mol. Biol.* 16, 334, 2009.
179. Feher, G. et al., Structure and function of bacterial photosynthetic reaction centers, *Nature* 339, 111, 1989.
180. Klimov, V.V., Discovery of pheophytin function in the photosynthetic energy conversion as the primary electron acceptor of Photosystem II, *Photosynth. Res.* 76, 247, 2003.
181. Faller, P. et al., Tyrosyl radicals in photosystem II: The stable tyrosyl D and the catalytic tyrosyl Z, *J. Inorg. Biochem.* 86, 214, 2001.
182. Styring, S. and Rutherford, A.W., In the oxygen evolving complex of photosystem II the S_0 state is oxidized to the S_1 state by Y_D^+ (signal II_{slow}), *Biochemistry* 26, 2401, 1987.
183. Diner, B.A. and Rappaport, F., Structure, dynamics, and energetics of the primary photochemistry of photosystem II of oxygenic photosynthesis, *Annu. Rev. Plant Biol.* 53, 551, 2002.
184. Ananyev, G.A. et al., A functional role for tyrosine-D in assembly of the inorganic core of the water oxidase complex of photosystem II and the kinetics of water oxidation, *Biochemistry* 41, 974, 2002.
185. Debus, R., The catalytic manganese cluster: Protein ligation, in *Photosystem II. The Light Driven Water:Plastiquinone Oxidoreductase*, Wydrzynski, T. and Satoh, K., Eds., Springer, Dordrecht, the Netherlands, 2005, p. 261.
186. Yano, J. et al., Where water is oxidized to dioxygen: Structure of the photosynthetic Mn_4Ca cluster, *Science* 314, 821, 2006.
187. Roose, J.L., Wegener, K.M., and Pakrasi, H.B., The extrinsic proteins of photosystem II, *Photosynth. Res.* 92, 369, 2007.
188. Fagerlund, R.D. and Eaton-Rye, J.J., The lipoproteins of cyanobacterial photosystem II, *J. Photochem. Photobiol. B Biol.* 104, 191, 2011.
189. Bricker, T.M. et al., The extrinsic proteins of Photosystem II, *Biochim. Biophys. Acta* 1817, 121, 2012.
190. Bricker, T.M. and Frankel, L.K., The structure and function of CP47 and CP43 in Photosystem II, *Photosynth. Res.* 72, 131, 2002.
191. Debus, R.J., Protein ligation of the photosynthetic oxygen-evolving center, *Coord. Chem. Rev.* 252, 244, 2008.

192. Gleiter, H.M. et al., Involvement of the CP47 protein in stabilization and photoactivation of a functional water oxidizing complex in the cyanobacterium *Synechocystis sp* PCC 6803, *Biochemistry* 34, 6847, 1995.
193. Shi, L.X. and Schröder, W.P., The low molecular mass subunits of the photosynthetic supracomplex, photosystem II, *Biochim. Biophys. Acta* 1608, 75, 2004.
194. Stewart, D.H. and Brudvig, G.W., Cytochrome b_{559} of photosystem II, *Biochim. Biophys. Acta* 1367, 63, 1998.
195. Dörmann, P. and Hölzl, G., The role of glycolipids in photosynthesis, in *Lipids in Photosynthesis: Essential and Regulatory Functions*, Wada, H. and Murata, N., Eds., Springer, Dordrecht, the Netherlands, 2009, p. 265.
196. Broser, M. et al., Crystal structure of monomeric photosystem II from *Thermosynechococcus elongatus* at 3.6-Å resolution, *J. Biol. Chem.* 285, 26255, 2010.
197. Raszewski, G. et al., Spectroscopic properties of reaction center pigments in photosystem II core complexes: Revision of the multimer model, *Biophys. J.* 95, 105, 2008.
198. Spetzler, D. et al., Energy transduction by the two molecular motors of the F_1F_0 ATP synthase, in *Photosynthesis: Plastid Biology, Energy Conversion and Carbon Assimilation*, Eaton-Rye, J.J., Tripathy, B.C., and Sharkey, T.D., Eds., Springer, Dordrecht, the Netherlands, 2012, p. 561.
199. Renger, G., Functional pattern of photosystem II, in *Primary Processes of Photosynthesis, Part 2 Principles and Apparatus*, Renger, G., Ed., RSC Publishing, Cambridge, U.K., 2008, p. 237.
200. Wydrzynski, T. and Satoh, K., Eds., *Photosystem II. The Light-Driven Water:Plastoquinone Oxidoreductase*, vol. 22, Springer, Dordrecht, the Netherlands, 2005.
201. Nelson, N. and Yocum, C.F., Structure and function of photosystems I and II, *Annu. Rev. Plant Biol.* 57, 521, 2006.
202. Amunts, A. and Nelson, N., Plant photosystem I design in the light of evolution, *Structure* 17, 637, 2009.
203. Golbeck, J.H., Ed., *Photosystem I. The Light-Driven Plastocyanin:Ferredoxin Oxidoreductase*, vol. 24, Springer, Dordrecht, the Netherlands, 2006.
204. Jordan, P. et al., Three-dimensional structure of cyanobacterial photosystem I at 2.5 A resolution, *Nature* 411, 909, 2001.
205. Chapman, H.N. et al., Femtosecond X-ray protein nanocrystallography, *Nature* 470, 73, 2011.
206. Rögner, M. et al., Mono-, di- and trimeric PS I reaction center complexes isolated from the thermophilic cyanobacterium *Synechococcus* sp.: Size, shape and activity, *Biochim. Biophys. Acta* 1015, 415, 1990.
207. Kruip, J. et al., Evidence for the existence of trimeric and monomeric Photosystem I complexes in thylakoid membranes from cyanobacteria, *Photosynth. Res.* 40, 279, 1994.
208. Fromme, P., Jordan, P., and Krauß, N., Structure of photosystem I, *Biochim. Biophys. Acta* 1507, 5, 2001.
209. Fromme, P. and Grotjohann, I., Structural analysis of cyanobacterial Photosystem I, in *Photosystem I. The Light-Driven Plastocyanin:Ferredoxin Oxidoreductase*, Golbeck, J.H., Ed., Springer, Dordrecht, the Netherlands, 2006, p. 47.
210. Redding, K. and van der Est, A., The directionality of electron transport in photosystem I, in *Photosystem I: The Light-Driven Plastocyanin:Ferredoxin Oxidoreductase*, Golbeck, J.H., Ed., Springer, Dordrecht, the Netherlands, 2006, p. 413.
211. Grotjohann, I., Jolley, C., and Fromme, P., Evolution of photosynthesis and oxygen evolution: Implications from the structural comparison of Photosystems I and II, *Phys. Chem. Chem. Phys.* 6, 4743, 2004.
212. Ben-Shem, A., Frolow, F., and Nelson, N., Crystal structure of plant photosystem I, *Nature* 426, 630, 2003.
213. Guergova-Kuras, M. et al., Evidence for two active branches for electron transfer in photosystem I, *Proc. Natl. Acad. Sci. USA* 98, 4437, 2001.
214. Rutherford, A.W., Osyczka, A., and Rappaport, F., Back-reactions, short-circuits, leaks and other energy wasteful reactions in biological electron transfer: Redox tuning to survive life in O_2, *FEBS Lett.* 586, 603, 2012.
215. Cohen, R.O. et al., Evidence for asymmetric electron transfer in cyanobacterial photosystem I: Analysis of a methionine-to-leucine mutation of the ligand to the primary electron acceptor A_0, *Biochemistry* 43, 4741, 2004.
216. Dashdorj, N. et al., Asymmetric electron transfer in cyanobacterial Photosystem I: Charge separation and secondary electron transfer dynamics of mutations near the primary electron acceptor A_0, *Biophys. J.* 88, 1238, 2005.
217. Fairclough, W.V. et al., Bidirectional electron transfer in photosystem I: Electron transfer on the PsaA side is not essential for phototrophic growth in *Chlamydomonas*, *Biochim. Biophys. Acta* 1606, 43, 2003.

218. Pushkar, Y.N. et al., Recruitment of a foreign quinone into the A_1 site of photosystem I. Consecutive forward electron transfer from A_0 to A_1 to F_X with anthraquinone in the A_1 site as studied by transient EPR, *J. Biol. Chem.* 280, 12382, 2005.
219. Berthold, T. et al., Exploring the electron transfer pathways in photosystem I by high-time-resolution electron paramagnetic resonance: Observation of the B-Side radical pair $P_{700}{}^+A_{1B}{}^-$ in whole cells of the deuterated green alga *Chlamydomonas reinhardtii* at cryogenic temperatures, *J. Am. Chem. Soc.* 134, 5563, 2012.
220. Müller, M.G. et al., Independent initiation of primary electron transfer in the two branches of the photosystem I reaction center, *Proc. Natl. Acad. Sci. USA* 107, 4123, 2010.
221. Holzwarth, A.R. et al., Ultrafast transient absorption studies on photosystem I reaction centers from *Chlamydomonas reinhardtii*. 2: Mutations near the P700 reaction center chlorophylls provide new insight into the nature of the primary electron donor, *Biophys. J.* 90, 552, 2006.
222. Giera, W. et al., Effect of the P700 pre-oxidation and point mutations near A_0 on the reversibility of the primary charge separation in Photosystem I from *Chlamydomonas reinhardtii*, *Biochim. Biophys. Acta* 1797, 106, 2010.
223. MacColl, R., Cyanobacterial phycobilisomes, *J. Struct. Biol.* 124, 311, 1998.
224. Arteni, A. et al., Structure and organization of phycobilisomes on membranes of the red alga *Porphyridium cruentum*, *Photosynth. Res.* 95, 169, 2008.
225. Glazer, A.N., Light harvesting by phycobilisomes, *Annu. Rev. Biophys. Biophys. Chem.* 14, 47, 1985.
226. Wilbanks, S.M., de Lorimier, R., and Glazer, A.N., Phycoerythrins of marine unicellular cyanobacteria. III. Sequence of a class II phycoerythrin, *J. Biol. Chem.* 266, 9535, 1991.
227. Bryant, D.A., Phycoerythrocyanin and phycoerythrin: Properties and occurrence in cyanobacteria, *J. Gen. Microbiol.* 128, 835, 1982.
228. Rodriguez, H. et al., Nitrogen-fixing cyanobacterium with a high phycoerythrin content, *Appl. Environ. Microbiol.* 55, 758, 1989.
229. Lundell, D.J. and Glazer, A.N., Allophycocyanin B. A common b subunit in *Synechococcus* allophycocyanin B (l_{max} 670 nm) and allophycocyanin (l_{max} 650 nm), *J. Biol. Chem.* 256, 12600, 1981.
230. Mullineaux, C.W., Phycobilisome-reaction centre interaction in cyanobacteria, *Photosynth. Res.* 95, 175, 2008.
231. Grossman, A.R. et al., The phycobilisome, a light-harvesting complex responsive to environmental conditions, *Microbiol. Rev.* 57, 725, 1993.
232. Mullineaux, C.W., Excitation energy transfer from phycobilisomes to Photosystem I in a cyanobacterium, *Biochim. Biophys. Acta* 1100, 285, 1992.
233. Glazer, A.N. et al., Selective disruption of energy flow from phycobilisomes to Photosystem I, *Photosynth. Res.* 40, 167, 1994.
234. Marquardt, J. et al., Isolation and characterization of biliprotein aggregates from *Acaryochloris marina*, a Prochloron-like prokaryote containing mainly chlorophyll *d*, *FEBS Lett.* 410, 428, 1997.
235. Chen, M., Quinnell, R.G., and Larkum, A.W.D., The major light-harvesting pigment protein of *Acaryochloris marina*, *FEBS Lett.* 514, 149, 2002.
236. Lokstein, H., Steglich, C., and Hess, W.R., Light-harvesting antenna function of phycoerythrin in *Prochlorococcus marinus*, *Biochim. Biophys. Acta* 1410, 97, 1999.
237. Hess, W.R. et al., Coexistence of phycoerythrin and a chlorophyll a/b antenna in a marine prokaryote, *Proc. Natl. Acad. Sci. USA* 93, 11126, 1996.
238. Wiethaus, J. et al., Phycobiliproteins in *Prochlorococcus marinus*: Biosynthesis of pigments and their assembly into proteins, *Eur. J. Cell Biol.* 89, 1005, 2010.
239. Gutu, A. and Kehoe, D.M., Emerging perspectives on the mechanisms, regulation, and distribution of light color acclimation in cyanobacteria, *Mol. Plant* 5, 1, 2012.
240. Rochaix, J.-D., Role of thylakoid protein kinases in photosynthetic acclimation, *FEBS Lett.* 581, 2768, 2007.
241. Kondo, K., Mullineaux, C.W., and Ikeuchi, M., Distinct roles of CpcG1-phycobilisome and CpcG2-phycobilisome in state transitions in a cyanobacterium *Synechocystis* sp. PCC 6803, *Photosynth. Res.* 99, 217, 2009.
242. Li, H. and Sherman, L.A., A redox-responsive regulator of photosynthesis gene expression in the cyanobacterium *Synechocystis* sp. Strain PCC 6803, *J. Bacteriol.* 182, 4268, 2000.
243. Tandeau de Marsac, N., Occurrence and nature of chromatic adaptation in cyanobacteria, *J. Bacteriol.* 130, 82, 1977.
244. Postius, C. et al., N_2-fixation and complementary chromatic adaptation in non-heterocystous cyanobacteria from Lake Constance, *FEMS Microbiol. Ecol.* 37, 117, 2001.

245. Kehoe, D.M. and Gutu, A., Responding to color: The regulation of complementary chromatic adaptation, *Annu. Rev. Plant Biol.* 57, 127, 2006.
246. Duxbury, Z. et al., Chromatic photoacclimation extends utilisable photosynthetically active radiation in the chlorophyll *d*-containing cyanobacterium, *Acaryochloris marina*, *Photosynth. Res.* 101, 69, 2009.
247. MacIntyre, H.L. et al., Photoacclimation of photosynthesis irradiance response curves and photosynthetic pigments in microalgae and cyanobacteria, *J. Phycol.* 38, 17, 2002.
248. Walters, R.G., Towards an understanding of photosynthetic acclimation, *J. Exp. Bot.* 56, 435, 2005.
249. Bailey, S. and Grossman, A.R., Photoprotection in cyanobacteria: Regulation of light harvesting, *Photochem. Photobiol.* 84, 1410, 2008.
250. Renger, G., Ed., *Primary Processes of Photosynthesis: Principles and Apparatus*, vols. 8–9, RSC Publishing, Cambridge, U.K., 2008.
251. Voigt, J. et al., Excitonic effects in the light-harvesting Chl a/b–protein complex of higher plants, *Phys. Status Solidi B* 194, 333, 1996.
252. Renger, T. and May, V., Multiple exciton effects in molecular aggregates: Application to a photosynthetic antenna complex, *Phys. Rev. Lett.* 78, 3406, 1997.
253. Yang, M. and Fleming, G.R., Influence of phonons on exciton transfer dynamics: Comparison of the Redfield, Förster, and modified Redfield equations, *Chem. Phys.* 275, 355, 2002.
254. Saito, K. et al., Distribution of the cationic state over the chlorophyll pair of the photosystem II reaction center, *J. Am. Chem. Soc.* 133, 14379, 2011.
255. Prokhorenko, V.I. and Holzwarth, A.R., Primary processes and structure of the photosystem II reaction center: A photon echo study, *J. Phys. Chem. B* 104, 11563, 2000.
256. Joliot, P., Barbieri, G., and Chabaud, R., Un nouveau modele des centres photochimiques du systeme II, *Photochem. Photobiol.* 10, 309, 1969.
257. Kok, B., Forbush, B., and McGloin, M., Cooperation of charges in photosynthetic O_2 evolution, *Photochem. Photobiol.* 11, 457, 1970.
258. Mar, T. and Govindjee, Kinetic models of oxygen evolution in photosynthesis, *J. Theoret. Biol.* 36, 427, 1972.
259. Messinger, J. and Renger, G., Generation, oxidation by the oxidized form of the tyrosine of polypeptide D2, and possible electronic configuration of the redox States S_0, S_{-1} and S_{-2} of the water oxidase in isolated spinach thylakoids, *Biochemistry* 32, 9379, 1993.
260. Vass, I. et al., The accessory electron-donor tyrosine D of photosystem II is slowly reduced in the dark during low-temperature storage of isolated thylakoids, *Biochim. Biophys. Acta* 1018, 41, 1990.
261. Vermaas, W.E.J., Renger, G., and Dohnt, G., The reduction of the oxygen evolving system in chloroplasts by thylakoid components, *Biochim. Biophys. Acta* 764, 194, 1984.
262. Vermaas, W.F.J., Rutherford, A.W., and Hansson, O., Site directed mutagenesis in photosystem II of the cyanobacterium *Synechocystis sp.* PCC 6803: Donor D is a tyrosine residue in the D2 protein, *Proc. Natl. Acad. Sci. USA* 85, 8477, 1988.
263. Diner, B.A., Dependence of deactivation reactions of photosystem II on redox state of plastoquinone pool A varied under anaerobic conditions. Equilibria on the acceptor side of photosystem II, *Biochim. Biophys. Acta* 460, 247, 1977.
264. Nugent, J.H.A., Demetriou, C., and Lockett, C.J., Electron donation in photosystem II, *Biochim. Biophys. Acta* 894, 534, 1987.
265. Rutherford, A.W. and Inoue, Y., Oscillation of delayed luminescence from PS II: Recombination of $S_2Q_B^-$ and $S_3Q_B^-$, *FEBS Lett.* 165, 163, 1984.
266. Messinger, J. and Renger, G., Photosynthetic water splitting, in *Primary Processes of Photosynthesis, Part 2 Principles and Apparatus*, Renger, G., Ed., RSC Publishing, Cambridge, U.K., 2008, p. 291.
267. Messinger, J., Noguchi, T., and Yano, J., Photosynthetic O_2 evolution, in *Molecular Solar Fuels*, Wydrzynski, T. and Hillier, W., Eds., RSC Publishing, Cambridge, U.K., 2012, p. 163.
268. Isgandarova, S., Renger, G., and Messinger, J., Functional differences of photosystem II from *Synechococcus elongatus* and spinach characterized by flash-induced oxygen evolution patterns, *Biochemistry* 42, 8929, 2003.
269. Shevela, D. et al., Membrane-inlet mass spectrometry reveals a high driving force for oxygen production by photosystem II, *Proc. Natl. Acad. Sci. USA* 108, 3602, 2011.
270. Kolling, D.R.J. et al., Photosynthetic oxygen evolution is not reversed at high oxygen pressures: Mechanistic consequences for the water-oxidizing complex, *Biochemistry* 48, 1381, 2009.
271. Haumann, M. et al., Photosynthetic water oxidation at elevated dioxygen partial pressure monitored by time-resolved X-ray absorption measurements, *Proc. Natl. Acad. Sci. USA* 105, 17384, 2008.

272. Debus, R.J., The manganese and calcium ions of photosynthetic oxygen evolution, *Biochim. Biophys. Acta* 1102, 269, 1992.
273. Noguchi, T., Light-induced FTIR difference spectroscopy as a powerful tool toward understanding the molecular mechanism of photosynthetic oxygen evolution, *Photosynth. Res.* 91, 59, 2007.
274. Yachandra, V.K., The catalytic manganese cluster: Organisation of the metal ions, in *Photosystem II. The Light-Driven Water:Plastoquinone Oxidoreductase*, Wydrzynski, T. and Satoh, K., Eds., Springer, Dordrecht, the Netherlands, 2005, p. 235.
275. Meyer, T.J., Huynh, M.H.V., and Thorp, H.H., The possible role of proton-coupled electron transfer (PCET) in water oxidation by photosystem II, *Angew. Chem. Int. Ed.* 46, 5284, 2007.
276. Crofts, A.R., The Q-cycle—A personal perspective, *Photosynth. Res.* 80, 223, 2004.
277. Mitchell, P., Chemiosmotic coupling in oxidative and photosynthetic phosphorylation, *Biol. Rev.* 41, 445, 1966.
278. Jagendorf, A.T. and Uribe, E., ATP formation caused by acid-base transition of spinach chloroplasts, *Proc. Natl. Acad. Sci. USA* 55, 170, 1966.
279. Dunn, S.D., Kellner, E., and Lill, H., Specific heterodimer formation by the cytoplasmic domains of the b and b′ subunits of cyanobacterial ATP synthase, *Biochemistry* 40, 187, 2000.
280. Pogoryelov, D. et al., The oligomeric state of c rings from cyanobacterial F-ATP synthases varies from 13 to 15, *J. Bacteriol.* 189, 5895, 2007.
281. Van Walraven, H.S. et al., The H^+/ATP coupling ratio of the ATP synthase from thiol-modulated chloroplasts and two cyanobacterial strains is four, *FEBS Lett.* 379, 309, 1996.
282. Ferguson, S.J., ATP synthase: From sequence to ring size to the P/O ratio, *Proc. Natl. Acad. Sci. USA* 107, 16755, 2010.
283. Konno, H. et al., The regulator of the F1 motor: Inhibition of rotation of cyanobacterial F1-ATPase by the e subunit, *EMBO J.* 25, 4596, 2006.
284. Krenn, B.E. et al., ATP synthase from a cyanobacterial *Synechocystis* 6803 mutant containing the regulatory segment of the chloroplast gamma subunit shows thiol modulation, *Biochem. Soc. Trans.* 23, 757, 1995.
285. Olson, J.M., Photosynthesis in the Archean Era, *Photosynth. Res.* 88, 109, 2006.
286. Tice, M.M. and Lowe, D.R., Photosynthetic microbial mats in the 3,416-Myr-old ocean, *Nature* 431, 549, 2004.
287. Brasier, M.D. et al., Questioning the evidence for Earth's oldest fossils, *Nature* 416, 76, 2002.
288. Cavalier-Smith, T., Brasier, M., and Embley, T.M., Introduction: How and when did microbes change the world? *Philos. Trans. R. Soc. Lond. B* 361, 845, 2006.
289. Falkowski, P., The biological and geological contingencies for the rise of oxygen on Earth, *Photosynth. Res.* 107, 7, 2011.
290. Kump, L.P., The rise of atmospheric oxygen, *Nature* 451, 277, 2008.
291. Schopf, J.W., Microfossils of the early Archean apex chert: New evidence of the antiquity of life, *Science* 260, 640, 1993.
292. Schopf, J.W., The paleobiological record of photosynthesis, *Photosynth. Res.* 107, 87, 2011.
293. Schopf, J.W. et al., Laser-Raman imagery of Earth's earliest fossils, *Nature* 416, 73, 2002.
294. Allwood, A.C. et al., Stromatolite reef from the Early Archaean era of Australia, *Nature* 441, 714, 2006.
295. Bosak, T. et al., Morphological record of oxygenic photosynthesis in conical stromatolites, *Proc. Natl. Acad. Sci. USA* 106, 10939, 2009.
296. Farquhar, J., Zerkle, A., and Bekker, A., Geological constraints on the origin of oxygenic photosynthesis, *Photosynth. Res.* 107, 11, 2011.
297. Oparin, I.A., The origin of life and the origin of enzymes, in *Advances in Enzymology and Related Areas of Molecular Biology*, Nord, F.F., Ed., Interscience Publishers, New York, 1965, p. 347.
298. Tomitani, A. et al., The evolutionary diversification of cyanobacteria: Molecular-phylogenetic and paleontological perspectives, *Proc. Natl. Acad. Sci. USA* 103, 5442, 2006.
299. Green, B.R., The evolution of light-harvesting antennas, in *Light-Harvesting Antennas in Photosynthesis*, Green, B.R. and Parson, W.W., Eds., Springer, Dordrecht, the Netherlands, 2003, p. 129.
300. Peschek, G.A. et al., Life inplies work: A holistic account of our microbial biosphere focussing on the bioenergetic processes of cyanobacteria, the ecologically most successful organism on our Earth, in *Bioenergetic Processes of Cyanobacteria: From Evolutionary Singularity to Ecological Diversity*, Peschek, G.A., Obinger, C., and Renger, G., Eds., Springer, Dordrecht, the Netherlands, 2011, p. 3.
301. Bekker, A. et al., Dating the rise of atmospheric oxygen, *Nature* 427, 117, 2004.

302. Payne, J. et al., The evolutionary consequences of oxygenic photosynthesis: A body size perspective, *Photosynth. Res.* 107, 37, 2011.
303. Gantt, E., Oxygenic photosynthesis and the distribution of chloroplasts, *Photosynth. Res.* 107, 1, 2011.
304. Zorina, A. et al., Regulation systems for stress responses in cyanobacteria, *Russ. J. Plant Physiol.* 58, 749, 2011.
305. Pareek, A. et al., Eds., *Abiotic Stress Adaptation in Plants. Physiological, Molecular and Genomic Foundation*, 1st edn., Springer, Dordrecht, the Netherlands, 2010.
306. Briggs, L.M., Pecoraro, V.L., and Mcintosh, L., Copper-induced expression, cloning, and regulatory studies of the plastocyanin gene from the cyanobacterium *Synechocystis* sp. PCC 6803, *Plant Mol. Biol.* 15, 633, 1990.
307. Falk, S. et al., Functional analysis of the iron-stress induced CP 43′ polypeptide of PS II in the cyanobacterium *Synechococcus* sp. PCC 7942, *Photosynth. Res.* 45, 51, 1995.
308. Ivanov, A.G. et al., Iron deficiency in cyanobacteria causes monomerization of photosystem I trimers and reduces the capacity for state transitions and the effective absorption cross section of photosystem I *in vivo*, *Plant Physiol.* 141, 1436, 2006.
309. Leonhardt, K. and Straus, N.A., An iron stress operon involved in photosynthetic electron transport in the marine cyanobacterium *Synechococcus* sp. PCC 7002, *J. Gen. Microbiol.* 138, 1613, 1992.
310. Campbell, D. and Oquist, G., Predicting light acclimation in cyanobacteria from nonphotochemical quenching of photosystem II fluorescence, which reflects state transitions in these organisms, *Plant Physiol.* 111, 1293, 1996.
311. Clarke, A.K. et al., Dynamic responses of photosystem II and phycobilisomes to changing light in the cyanobacterium *Synechococcus* sp. PCC 7942, *Planta* 197, 553, 1995.
312. Fork, D.C. and Satoh, K., State I-State II transitions in the thermophilic blue-green alga (cyanobacterium) *Synechococcus lividus*, *Photochem. Photobiol.* 37, 421, 1983.
313. Kaňa, R. et al., The slow S to M fluorescence rise in cyanobacteria is due to a state 2 to state 1 transition, *Biochim. Biophys. Acta* 1817, 1237, 2012.
314. Lüttge, U. et al., Photosynthesis of terrestrial cyanobacteria under light and desiccation stress as expressed by chlorophyll fluorescence and gas exchange, *J. Exp. Bot.* 46, 309, 1995.
315. Mullineaux, C.W. and Allen, J.F., State 1-State 2 transitions in the cyanobacterium *Synechococcus* 6301 are controlled by the redox state of electron carriers between Photosystems I and II, *Photosynth. Res.* 23, 297, 1990.
316. Schubert, H., Forster, R.M., and Sagert, S., *In situ* measurement of state transition in cyanobacterial blooms: Kinetics and extent of the state change in relation to underwater light and vertical mixing, *Mar. Ecol. Progr.* 128, 99, 1995.
317. Vernotte, C., Astier, C., and Olive, J., State 1-state 2 adaptation in the cyanobacteria *Synechocystis* PCC 6714 wild type and *Synechocystis* PCC 6803 wild type and phycocyanin-less mutant, *Photosynth. Res.* 26, 203, 1990.
318. Karapetyan, N., Non-photochemical quenching of fluorescence in cyanobacteria, *Biochemistry (Moscow)* 72, 1127, 2007.
319. McConnell, I., Li, G.H., and Brudvig, G.W., Energy conversion in natural and artificial photosynthesis, *Chem. Biol.* 17, 434, 2010.
320. Najafpour, M.M. et al., Biological water oxidation: Lessons from Nature, *Biochim. Biophys. Acta* 1817, 1110, 2012.
321. Styring, S., Artificial photosynthesis for solar fuels, *Faraday Discuss.* 155, 357, 2012.
322. Kalyanasundaram, K. and Graetzel, M., Artificial photosynthesis: Biomimetic approaches to solar energy conversion and storage, *Curr. Opin. Biotechnol.* 21, 298.
323. Antal, T. et al., Use of near-infrared radiation for oxygenic photosynthesis via photon up-conversion, *Int. J. Hydrogen Energ.* 37, 8859, 2012.
324. Barber, J., Photosynthetic energy conversion: Natural and artificial, *Chem. Soc. Rev.* 38, 185, 2009.
325. Blankenship, R.E. et al., Comparing photosynthetic and photovoltaic efficiencies and recognizing the potential for improvement, *Science* 332, 805, 2011.
326. Gust, D. et al., Engineered and artificial photosynthesis: Human ingenuity enters the game, *MRS Bulletin* 33, 383, 2008.
327. Messinger, J. and Shevela, D., Principles of photosynthesis, in *Fundamentals of Materials and Energy and Environmental Sustainability*, Ginley, D. and Cachen, D., Eds., Cambridge University Press, Cambridge, U.K., 2012, p. 302.

328. Ghirardi, M.L. et al., Photobiological hydrogen-producing systems, *Chem. Soc. Rev*. 38, 52, 2009.
329. Lubitz, W., Reijerse, E.J., and Messinger, J., Solar water-splitting into H_2 and O_2: Design principles of photosystem II and hydrogenases, *Energy Environ. Sci*. 1, 15, 2008.
330. Allakhverdiev, S.I. et al., Hydrogen photoproduction by use of photosynthetic organisms and biomimetic systems, *Photochem. Photobiol. Sci*. 8, 148, 2009.
331. Lee, H.-S., Vermaas, W.F.J., and Rittmann, B.E., Biological hydrogen production: Prospects and challenges, *Trends Biotechnol*. 28, 262, 2010.
332. Abed, R.M.M., Dobretsov, S., and Sudesh, K., Applications of cyanobacteria in biotechnology, *J. Appl. Microbiol*. 106, 1, 2009.
333. Lem, N.W. and Glick, B.R., Biotechnological uses of cyanobacteria, *Biotechnol. Adv*. 3, 195, 1985.

2 Water Oxidation and Water-Oxidizing Complex in Cyanobacteria

Mohammad Mahdi Najafpour, Atefeh Nemati Moghaddam, Jian-Ren Shen, and Govindjee**

CONTENTS

2.1 INTRODUCTION

Cyanobacteria, or blue-green algae as they were called, are a group of bacteria that obtain their energy through oxygenic photosynthesis (for a perspective, see [1]; for evolution, see [2]).Cyanobacteria converted the early reducing atmosphere into an oxidizing one and changed the composition of life forms on Earth. The consensus is that chloroplasts in plants and eukaryotic algae have evolved from cyanobacterial ancestors via endosymbiosis [3]. In this chapter, we will discuss the structure and function of the water-oxidizing complex (WOC) in cyanobacteria [4,5].

* Corresponding authors: mmnajafpour@iasbs.ac.ir; gov@illinois.edu

Using the energy from sunlight, photosynthesis converts CO_2 into organic compounds [6,7]. In plants, algae, and cyanobacteria, photosynthesis uses CO_2 and water, and releases oxygen as a waste product. In these oxygenic photosynthetic organisms, a linear electron-transport system is used for the conversion of nicotinamide adenine dinucleotide phosphate ($NADP^+$) to its reduced form (NADPH); water is the ultimate source of electrons and is oxidized to oxygen in the process [8]. Oxygenic photosynthetic organisms catalyze photosynthetic water oxidation, and are therefore responsible for the presence of oxygen in the earth's atmosphere. This process requires two photosystems—photosystem I (PSI) and photosystem II (PSII)—and two light reactions (I and II), working in series, using what is commonly known as the Z-Scheme [9] (see Chapter 1 for a description of the overall steps in oxygenic photosynthesis.). PSII (water–plastoquinone oxido-reductase) uses light (photons) to energize specific reaction center chlorophyll molecules; this leads to electron transfer from water, through several intermediates (coenzymes and cofactors), to plastoquinone [10]; water is oxidized to hydronium ions and molecular oxygen [11,12]. The resulting protons generated by the oxidation of water are used to create a proton gradient that is used by ATP synthase to generate ATP [13]. The reduced plastoquinone (plastoquinol) transfers its electrons to PSI [14] via a cytochrome b_6f (Cyt b_6f) complex [15] where, again, protons are released into the lumen and a proton gradient is produced across the thylakoid membrane and used for ATP synthesis. The electrons (from PSII) transferred to plastoquinone are ultimately used, by PSI, to reduce $NADP^+$ to NADPH or are used in cyclic photophosphorylation around PSI [13]. In Cyt b_6f complex, electrons pass through several intermediates (cytochrome f, Rieske iron center) to plastocyanin (or cytochrome c_6 in some cyanobacteria), which is the electron donor to PSI, in its reduced form.

The WOC, also referred to as oxygen-evolving complex (OEC), in PSII is the protein complex that oxidizes water [3,16–19]. PSII may serve as a model to split water by sunlight, which is a prerequisite for a sustainable hydrogen economy [18]. In this chapter, we will review water oxidation and the WOC in natural photosynthesis (see Wydrzynski and Hillier [19] for reviews that deal with both natural and artificial photosynthesis and their relationship to "Solar Fuels").

2.2 A BIT OF HISTORY OF PHOTOSYNTHESIS

Joseph Priestley (1733–1804) described the ability of plants to generate "phlogiston" (the power to store the air which had been injured by the burning of candles); this was the discovery of oxygen evolution by plants [6,20]. During this period of "New Chemistry," Carle Wilhelm Scheele (1742–1786) and Antoine Laurent Lavoisier (1743–1794) identified this gas as oxygen. Jan Ingenhousz (1730–1799) discovered the role of light and the importance of the green color (later established as chlorophyll) of plants, and Jean Senebier (1742–1809) discovered the role of CO_2 in photosynthesis. Nicholas Theodore de Saussure (1767–1845) established the role of water, and finally Julius Robert Mayer (1814–1878) provided the concept that in photosynthesis light energy is converted into chemical energy. Robert Hill (1899–1991) discovered that when "chloroplasts" were exposed to light in the presence of an artificial electron acceptor, oxygen evolution was observed; this "Hill reaction" shows that carbon assimilation and oxygen evolution are not obligatorily linked and two distinct systems may exist [21]. (For a timeline of photosynthesis see [22].)

There are two parts to photosynthesis (Figure 2.1) [7]:

1. The reactions that depend directly on light take place in specific pigment–protein complexes in the thylakoid membranes; they are called the "light reactions." Here, light energy is converted into chemical energy. The end product of this set of reactions, which includes many dark reactions as well, is the production of oxygen, of the reducing power (NADPH) and of the ATP. Production of oxygen is the focus of this chapter.
2. The so-called *dark reactions* that do not depend directly on light take place in the stroma or cytoplasmic region; here CO_2 is converted to sugars. The dark reactions involve a cycle called the Calvin–Benson cycle, in which CO_2 and energy from NADPH and ATP are used to form sugars.

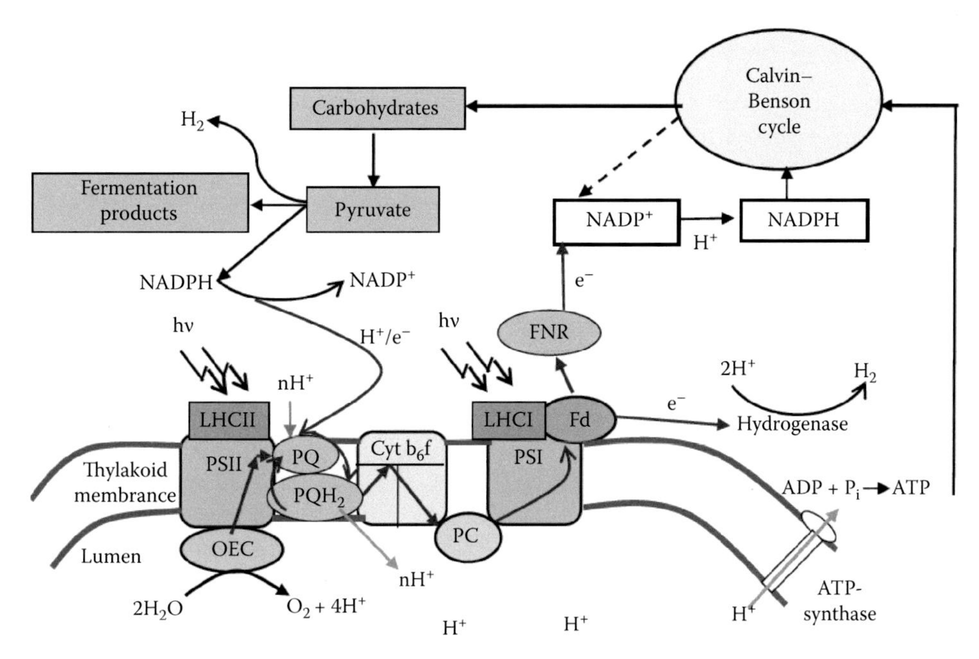

FIGURE 2.1 Carbon fixation and oxygen evolution take place in two distinct spaces in oxygenic photosynthesis. The diagram shows a schematic view of light-powered hydrogen production during oxygenic photosynthesis, as well as carbohydrate synthesis that can also be followed by hydrogen production. The photosynthetic processes are driven by light energy captured by the light-harvesting complexes (LHCII and LHCI) of PSII and PSI. Electrons are ultimately derived from H_2O by its oxidation at the water-oxidizing complex (WOC) of PSII; these electrons are passed along the photosynthetic electron-transport chain via plastoquinone (PQ), the cytochrome b_6/f complex (Cyt b_6/f), plastocyanin (PC) (or cytochrome c_6), PSI, and to ferredoxin (Fd). Then, ferredoxin–$NADP^+$ oxidoreductase (FNR) transfers the electrons to $NADP^+$ with the final production of NADPH. Protons (H^+ ions) are released into the thylakoid lumen by the WOC as water is oxidized, as well as when PQH_2 delivers electrons to Cyt b_6/f complex. The proton gradient across the thylakoid membrane is used by ATP synthase to produce ATP. The ATP and NADPH generated during the primary photosynthetic processes are consumed during CO_2 fixation in the Calvin–Benson cycle, which produces sugars and ultimately starch. Under anaerobic conditions, hydrogenase can accept electrons from the reduced Fd molecules and use them to reduce protons to molecular hydrogen. Anaerobic conditions also allow the use of starch as a source of protons and electrons for H_2 production (via NADPH, PQ, Cyt b_6/f, PC, and PSI) using a hydrogenase enzyme. (From *J. Photochem. Photobiol. B*, 104, Allakhverdiev, S.I., Recent progress in the studies of structure and function of photosystem II, 1–8, 2011, Copyright 2011, with permission from Elsevier.)

In the following sections, we describe the structure and function of the manganese–calcium cluster that performs one of the most important reactions in Nature, *water oxidation*.

2.3 WATER OXIDATION AND WATER-OXIDIZING COMPLEX IN NATURAL PHOTOSYNTHESIS

Water oxidation is one of the most important reactions on the Earth since it is the source of nearly all the atmosphere's oxygen. The WOC is a manganese–calcium cluster that oxidizes water with modest driving force and with a turnover of up to 50 molecules of O_2 released per second [23]. The structure is expected to be the same in plants, algae, and cyanobacteria. In this section, we present the available atomic level structure and the most accepted mechanism of water oxidation by the Mn–Ca cluster. (We refer the readers to a small, but wonderful, book that deals with oxygen itself, its history, and its role in the evolution of life [24].)

2.3.1 Structure of the Water-Oxidizing Complex

The first pioneering paper dealing with the structure of PSII was on the cyanobacterium *Synechococcus elongatus* from the research groups of Horst Witt and Horst Saenger [25], which was followed by structural analysis of PSII from *Thermosynechococcus vulcanus* by Kamiya and Shen [26]. The first clear evidence for the cubane model came from the laboratories of James Barber and So Iwata in 2004 [27]. These authors also provided, for the first time, information on Ca in the WOC: here was a Mn_3Ca-cubane with the fourth Mn attached a bit far away [27].

The atomic level structure of the Mn_4Ca cluster is important for the understanding of the mechanism of water oxidation. Both extended x-ray absorption fine structure (EXAFS) and x-ray diffraction (XRD) studies have been successfully used to determine the structure of the WOC in PSII, particularly from cyanobacteria [25–30].

The most accepted model based on XRD, EXAFS, hyperfine splitting, and other physical constraints is the *dangler model*, where three Mn ions are strongly coupled and one "dangling" Mn interacts with the trimer [25–30]. These XRD, EXAFS, and other methods provided the arrangement of all of the protein subunits and the location of the chlorophylls and other cofactors, and formed a basis for further investigations on PSII [25–30]. However, the early investigations did not provide enough details for the structure of the Mn_4Ca cluster, the location of the substrate water molecules, or the precise arrangement of the amino acid side chains and cofactors that may have significant mechanistic consequences in water oxidation.

In 2011, the research groups of Jian-Ren Shen and Nobuo Kamiya significantly improved the resolution of the PSII crystals from the thermophilic cyanobacterium *T. vulcanus* down to a high resolution of 1.9 Å; further, the authors analyzed their structure in details [29,30]. Their investigation has provided many more details of the structure of the WOC containing the number and location of the bridged oxygen, the location of substrate water molecules, and the precise arrangement of the amino acid side chains [29,30] (for a historical account, see [31]).

In this latest structure of the WOC, Umena et al. [29,30] found four manganese ions, one calcium ion, and five oxygen atoms that serve as oxo bridges linking the five metal ions (four manganese and one calcium ion) (Figure 2.2). In addition, four terminal water ligands were found, two of which were coordinated to Ca and two to the dangling Mn(Mn(4)).

The aforementioned structure suggests that the manganese–calcium cluster could be described as $Mn_4CaO_5(H_2O)_4$. Of these five metal ions and five oxygen atoms, the calcium and three manganese ions occupy four corners and four oxygen atoms form the other four corners of the cubane-like structure. Regarding the Ca–O and Mn–O bond lengths, the cubane-like structure is not an ideal and symmetric one. Another manganese ion is located outside the cubane and is linked to two manganese ions within the cubane by one oxygen of the cubane and the fifth oxygen by a di-μ-oxo bridge (an oxygen atom bridged between two or three metal ions) [29,30]. The location of possible substrate water molecules is very important for the understanding of the mechanism of water oxidation by the WOC.

A few amino acids with carboxylate and imidazole groups are coordinated to the $Mn_4CaO_5(H_2O)_4$ cluster (Table 2.1) [29,30]. Generally, the carboxylate ion may coordinate to a metal ion in different modes (Figure 2.3). In the WOC, only one monodentate mode of carboxylate is observed and other carboxylate groups serve as bidentate modes [29,30]. Each of the four manganese ions has six ligands, whereas the calcium has seven ligands (Table 2.1).

In the following sections, we describe the detailed structure of the WOC revealed at a resolution of 1.9 Å.

2.3.1.1 Manganese Ions

Manganese is a trace mineral that participates in many enzymes [33]. It is found widely in *Nature*, but occurs only in trace amounts in human tissues. Mn(II) or (III) ions function as cofactors for a number

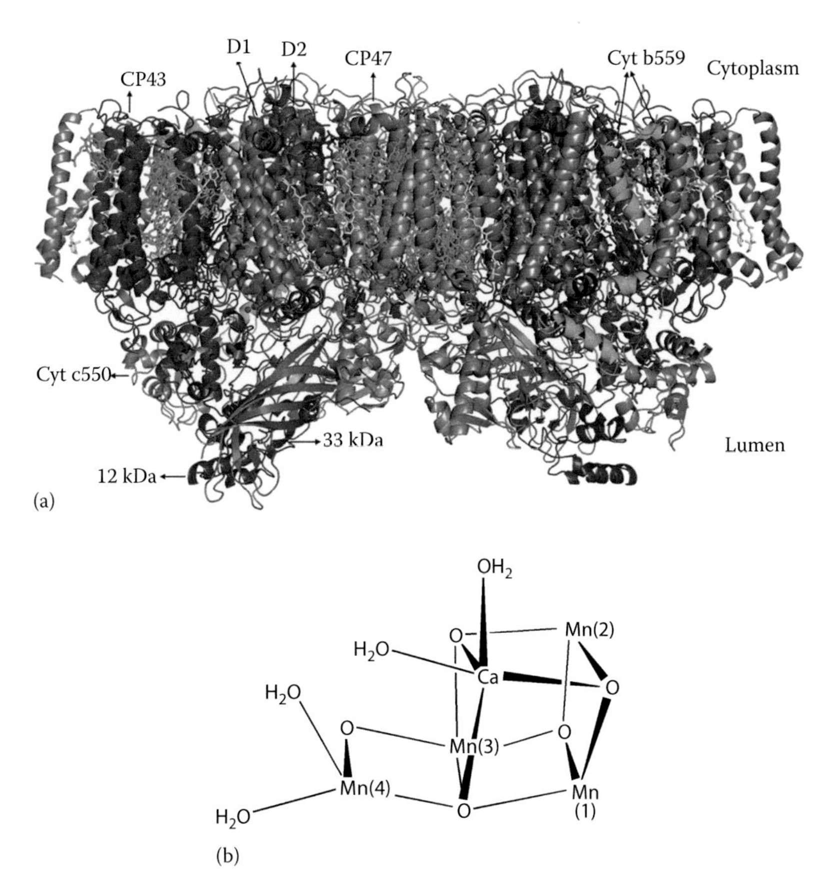

FIGURE 2.2 **(See color insert.)** (a) Structure of a cyanobacterial PSII dimer [29]. View from a direction perpendicular to the membrane normal. Molecules in green, yellow, and blue represent chlorophylls, β-carotenes, lipids, and detergent molecules, respectively. Red and yellow balls at the lumenal surface represent Mn and Ca ions, respectively. Protein subunits are labeled in the figure. For clarity, water molecules are omitted. (b) The entire structure of the Mn_4CaO_5 cluster resembles a distorted chair, with the asymmetric cubane. (From Umena, Y. et al., *Nature*, 473, 55–60, 2011; Kawakami, K. et al., *J. Photochem. Photobiol. B*, 104, 9–18, 2011.)

TABLE 2.1
Ligands for Manganese and Calcium Ions in the WOC

Ion	Ligands
Mn(1)	3(μ_3-O), 1(monodentate COO^-), 1(bridging COO^-), 1(imidazole)
Mn(2)	3(μ_3-O), 3 (bridging COO^-)
Mn(3)	3(μ_3-O), 1(μ_2-O), 2 (bridging COO^-)
Mn(4)	1(μ_4-O), 1(μ_2-O), 2 (bridging COO^-), 2(H_2O)
Ca	3(μ_3-O), 2 (bridging COO^-), 2(H_2O)

Sources: From Umena, Y. et al., *Nature*, 473, 55–60, 2011; Kawakami, K. et al., *J. Photochem. Photobiol. B*, 104, 9–18, 2011.

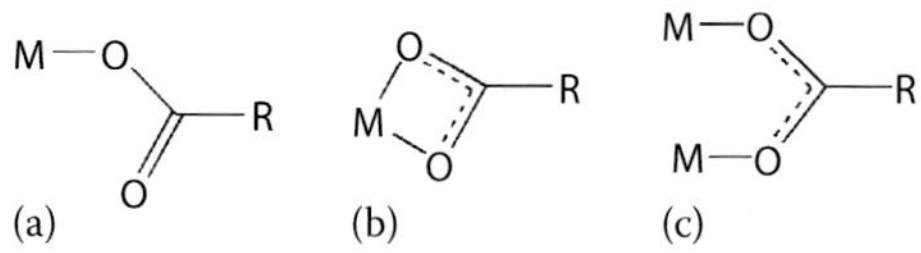

FIGURE 2.3 Unidentate (a), bidentate (b), and bridging carboxylate modes (c) [32]. In the structure of the WOC, one monodentate mode of carboxylate is observed and other carboxylate groups serve as bidentate modes.

of enzymes in higher organisms, where they are essential for detoxification of free radicals [33]. This element is required as a trace mineral for all known living organisms. The human body contains a total of 15–20 mg of manganese, most of which is located in the bones, with the remainder found in the kidneys, liver, pancreas, pituitary glands, and adrenal glands. In larger amounts, manganese can cause a poisoning syndrome in mammals, with neurological damage, which is sometimes irreversible. The most common oxidation states of manganese in biological systems are (II), (III), and (IV). Mn(II) often competes with Mg(II) in biological systems. Manganese compounds with oxidation states V, VI, and VII are strong oxidizing agents and are vulnerable to undergo disproportionation reactions. The most stable oxidation state for manganese in many mononuclear manganese enzymes and also manganese catalase is (II). In water, Mn(III) ion is unstable and prone to disproportionate to Mn(II) and Mn(IV), but this oxidation state could be stabilized with many "hard" ("hard" applies to chemical species that are small, have high charge states, and are weakly polarizable [34]) ligands in enzymes [34]. Mn(IV) is a usual oxidation state in minerals and could be stabilized by many hard ligands in biological systems. As shown in Figure 2.2b, there are four manganese ions in the WOC in PSII. In this section, we discuss the details of coordination chemistry of metal ions and a few important groups near the manganese–calcium cluster. However, it is worth mentioning that in the manganese–calcium cluster, there is charge distribution and charge on each ion is lower than suggested by its oxidation state. In other words, the $Mn_4CaO_5(H_2O)_4$ cluster is a delocalized system and each ion should not be studied completely separately, but rather in an integrated manner.

We describe next what is known about the four individual Mn atoms.

2.3.1.1.1 Manganese(1)

The ligands around Mn(1) are similar to one of the manganese ions in manganese catalase enzymes (Figure 2.4a). As usual, the coordination number for Mn(III) or (IV) is 6. Three μ_3-O as hard ligands and two carboxylate and one imidazole group as a borderline ligand could stabilize the oxidation state of III or IV for the manganese ion. As shown in Table 2.1 and in Figure 2.4a, in the WOC of PSII, Mn(1) has six ligands: two μ_3-O, one μ_4-O (μ_n-O means an oxo bridge linking n atoms together), one monodentate carboxylate, one bridging carboxylate, and one imidazole ligand [29,30].

2.3.1.1.2 Manganese(2)

The six ligands around this ion are three μ_3-O and three bridging COO^- [29,30] (Figure 2.4b). These ligands could stabilize oxidation state of III or IV for the ion. The coordination number of the ion is 6. The ion is connected to calcium and two manganese ions with a bridging carboxylate and three oxo groups.

2.3.1.1.3 Manganese(3)

The six ligands around this ion are two μ_3-O, one μ_4-O, one μ_2-O, and two bridging COO^-. Four hard μ-O could stabilize Mn(IV) than Mn(III) (Figure 2.4c).

2.3.1.1.4 Manganese(4)

The ligands around this ion are one μ_4-O, one μ_2-O, two bridging COO^-, and two H_2O (Figure 2.4d) [29,30]. These two water molecules are very important and one of them may serve as one of the substrates for water oxidation [29,30].

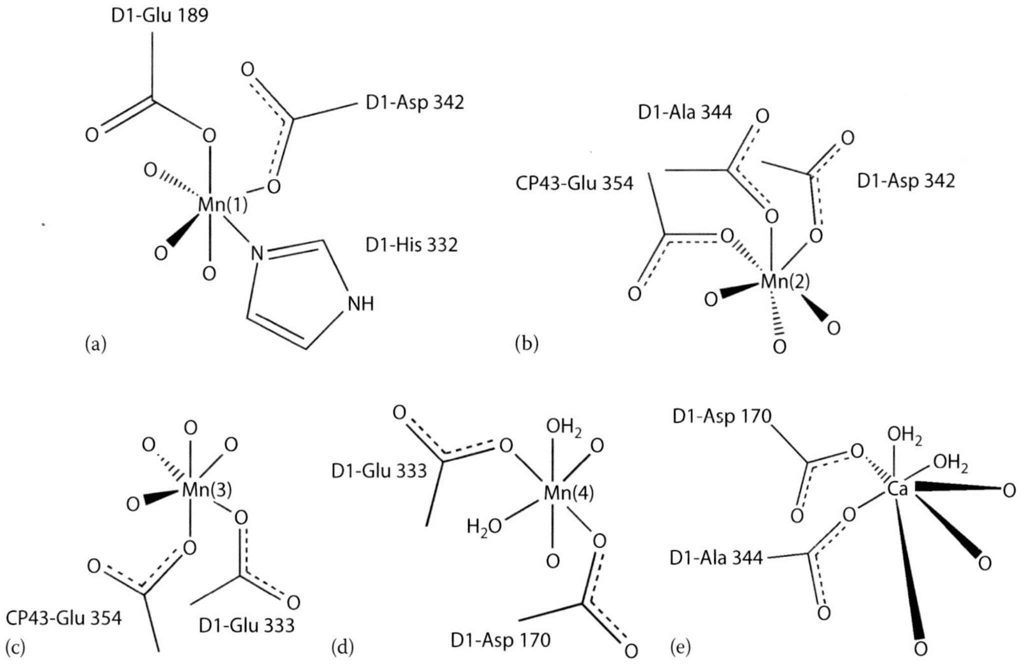

FIGURE 2.4 Mn(1) (a), Mn(2) (b), Mn (3) (c), Mn(4) (d), and Ca (e) and their surrounding ligands. (From Umena, Y. et al., *Nature*, 473, 55–60, 2011; Kawakami, K. et al., *J. Photochem. Photobiol. B*, 104, 9–18, 2011.)

Regarding these ligands, the oxidation state of Mn(III) could be stabilized for the ion, but deprotonation of water molecules could stabilize the oxidation state of Mn(IV) as well as higher oxidation states (e.g., Mn(V)).

2.3.1.2 Calcium Ion

Calcium is an essential ion for many organisms, particularly in cell physiology, where the movement of the calcium ion into and out of the cytoplasm functions as a signal for many cellular processes. Calcium is also a major structural element in bones and teeth. The usual role of this ion is structural and it is important for the stabilization of a number of proteins and enzymes. Calcium has been identified as an essential cofactor in water oxidation, and the calcium-binding sites in PSII have been previously studied by several methods. Strontium (II) is the only cation that can functionally substitute for calcium in the WOC [35]. In the 1.9 Å structure of the WOC, calcium has seven ligands, two μ_3-O, one μ_4-O, two bridging COO^-, and two H_2O molecules (Figure 2.4e) [29,30]. Similar to water molecules coordinated to Mn(4), these two water molecules are very important and one of them may serve as the substrate for water oxidation. The *coordination number of calcium* ions varies from 6 to 10 in different compounds. Thus, a ligand may coordinate to or decoordinate from this ion in different states of water oxidation.

2.3.1.3 Water

The location of the substrate water binding sites on the inorganic Mn_4Ca core has been an important question in the study of the mechanism of water oxidation. Hillier and Wydrzynski [36] used ^{18}O exchange kinetics of the substrate water molecules in PSII to examine the interactions of calcium and strontium with substrate water and to probe a number of point mutations surrounding the catalytic site. The most direct approach to follow water ligand exchange is by using mass spectrometry. This involves the addition of ^{18}O water followed by time-dependent sampling of the products. In this technique, two kinetic phases at m/e = 34, representing separate ^{18}O exchange rates for the two substrate water molecules, were detected [36]: the slow and fast phases that show the exchange of the two nonequivalent substrate sites. Since four water molecules are coordinated in the structure of the WOC, two of them may serve as the substrate for water oxidation [29,30]. Other suggested substrates for water oxidation are μ-O groups [29,30]. Another water molecule, also found around WOC, is hydrogen-bonded to one of the μ-O and one carboxylate group in this structure. This water molecule, although less likely, could also serve as a substrate for water oxidation (Figure 2.5) [29,30].

In PSII, there are three types of channels: for oxygen, water, and protons [29,30,37–39]; they lead from the WOC to the lumenal side of PSII. The functional assignment of these channels has been based on electrostatic, structural, and orientation grounds. This strategy of having separate specific channels is expected to avoid the interaction of unwanted chemicals with the WOC and to increase the catalytic activity of the enzyme (Figure 2.5).

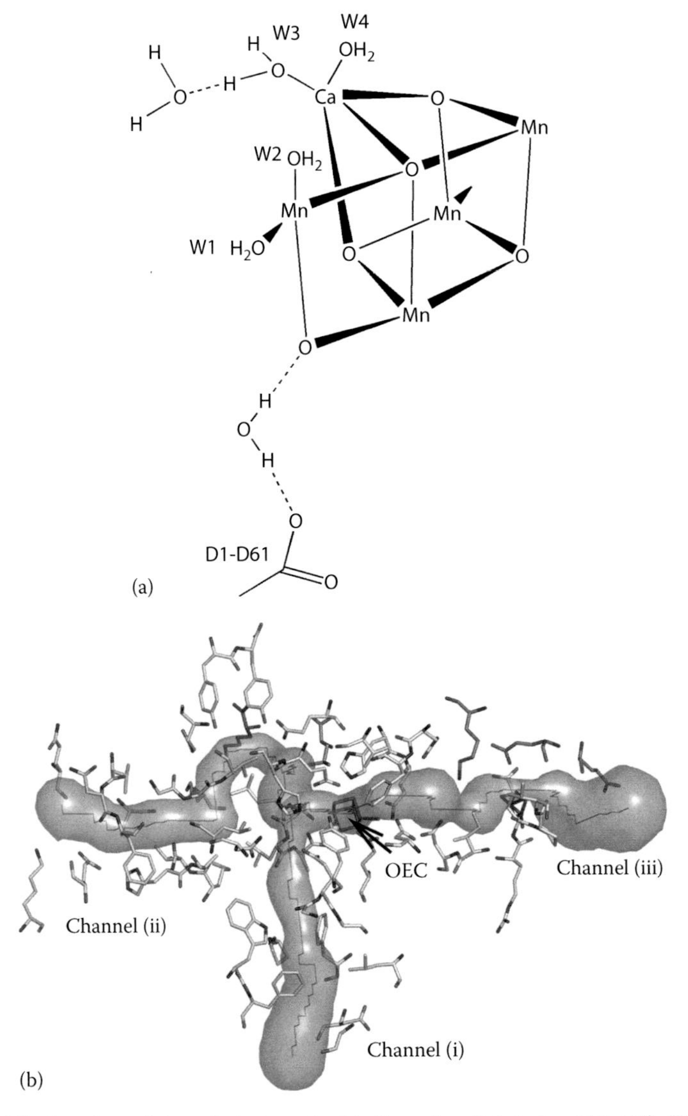

FIGURE 2.5 (a) The location of the substrate water binding sites (labeled as W1, W2, W3 and W4) on the WOC [29,30]. (b) The figure shows channels for hydrophobic oxygen (i), water (ii), and protons (iii), all leading to the catalytic Mn_4Ca cluster of PSII. OEC stands for oxygen-evolving complex. (Reproduced from Barber, J., *Inorg. Chem.*, 47(6), 1700, 2008.)

2.3.1.4 Amino Acids in the Second Coordination Sphere (D1-His 337 and CP47-Arg 357)

The imidazole nitrogen of D1-His 337 is hydrogen-bonded to one of the μ-O. The role of this hydrogen bond may be as a stabilizer for the WOC (Figure 2.6) [29,30].

There is an arginine in the second coordination sphere of WOC, CP43-Arg 357, and this residue may have an important role in maintaining the structure of the metal cluster, in stabilizing the cubane structure, and/or in providing partial positive charges to compensate for the negative charges induced by the oxo bridges and carboxylate ligands of the WOC [29,30]. One of the guanidinium nitrogens of CP43-Arg 357 is hydrogen-bonded to both μ-O manganese-calcium clusters, whereas the other is hydrogen-bonded to the carboxylate oxygen of D1-Asp 170 and to that of D1-Ala 344 (see Figure 2.6). The structure shows that the distances between the nitrogens of the arginine side chain and Ca^{2+} are 4.2 and 4.4 Å. Also, the distances between the nitrogen atoms of the arginine side chain and Mn(4) are 4.7 and 6.0 Å. The side chain of arginine may stabilize the structure of the WOC as it is hydrogen-bonded to two μ-O bridges and one carboxylate group bridging between Ca^{2+} and Mn(2).

2.3.1.5 Chloride

The function of chloride in biology, in general, could be to contribute negative charges in the formation of membrane potentials, responsible for the regulation of osmotic pressures in cells, to halogenate aromatic amino acids or to produce reactive species that are bactericidal and to act, as a bridging ligand, between heme *a* and Cu_B in the oxidized form of cytochrome oxidase. We know that chloride ion is a native anion and is required for electron donor reactions in the WOC [29,30,40–45]. Umena et al. [29,30] have identified two chloride ions in the structure of the WOC. Both Cl^- ions are surrounded by water molecules and amino acids. Umena et al. [29,30] have suggested that the two chloride anions may function to maintain the coordination environment of the $Mn_4CaO_5(H_2O)_4$, allowing the water oxidation reaction to proceed properly (Figure 2.7) (also see [45]).

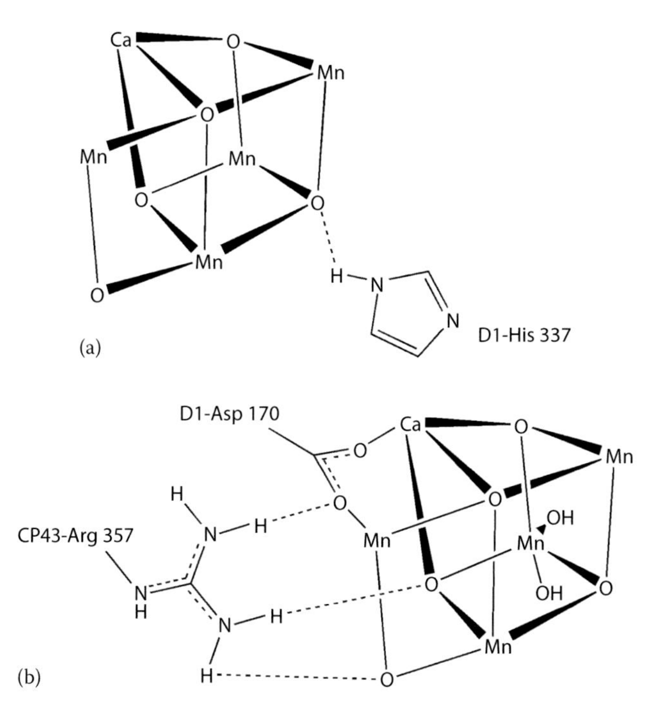

FIGURE 2.6 (a) The localization of D1-His 337 that is hydrogen-bonded to one of the μ-O in the WOC and (b) the localization of Arg 357. (From Umena, Y. et al., *Nature*, 473, 55–60, 2011; Kawakami, K. et al., *J. Photochem. Photobiol. B*, 104, 9–18, 2011.)

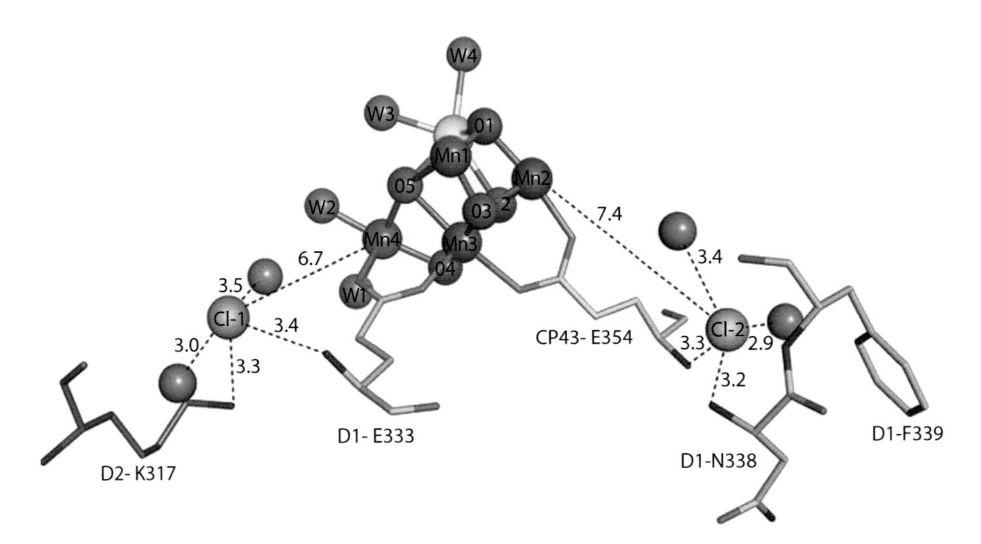

FIGURE 2.7 Structure of the two Cl^- binding sites near the $Mn_4CaO_5(H_2O)_4$ cluster. Hydrogen-bond distances are given in Å. (Reproduced from *J. Photochem. Photobiol. B*, 104, Kawakami, K. et al., Structure of the catalytic, inorganic core of oxygen-evolving photosystem II at 1.9 Å resolution, 9–18, 2011, Copyright 2011, with permission from Elsevier.)

2.3.1.6 Tyrosine 161

In PSII, photons are absorbed by light-harvesting pigment-protein complexes, and excitation energy is transferred to the reaction center chlorophyll P_{680}; here, primary charge separation occurs: oxidized P_{680} (P_{680}^+) and reduced pheophytin (Phe^-) are formed [10,46,47], for details, see Chapter 1. Then, P_{680}^+ is reduced by electron transfer from a tyrosine (Tyrosine 161) residue (Y_z), residing on the D1 protein, to form a tyrosine radical ($Y_z^{\bullet+}$) [4,5]. Electrons for the reduction of $Y_z^{\bullet+}$ are extracted from the WOC. As shown in Figure 2.8, this group forms a strong hydrogen bond with

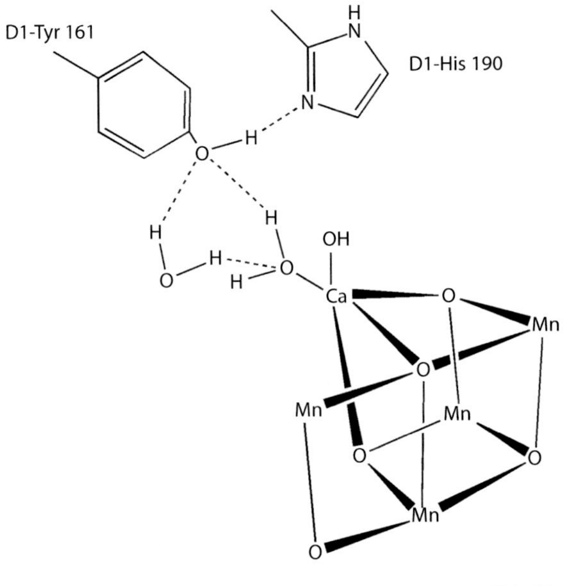

FIGURE 2.8 The location of Tyrosine 161. (From Umena, Y. et al., *Nature*, 473, 55–60, 2011; Kawakami, K. et al., *J. Photochem. Photobiol. B*, 104, 9–18, 2011.)

one water molecule coordinated to calcium [29,30]. Another hydrogen bond is observed between Y_z and the ε-nitrogen of a histidine (D1-His 190). This histidine is further hydrogen-bonded to other amino acids or water molecules to form a hydrogen-bond network suggested as an exit channel for protons [29,30].

2.3.2 Mechanism of Oxygen Evolution

2.3.2.1 Four-Electron Water Oxidation

In thermodynamic terms, a four-electron water oxidation is certainly easier than four-sequential one-electron oxidation or two-sequential two-electron oxidation because in those cases the first steps (H_2O to hydrogen peroxide and hydroxyl radical) are more endergonic than the four-electron water oxidation, and result in low over-voltage for practical operations. As described later, *Nature* uses a four-electron oxidation mechanism for water oxidation with lower activation energy than other known mechanisms [5,48,49].

2.3.2.2 Flash-Induced Oxygen Evolution Pattern: The Joliot Experiment and the Kok Cycle

An elegant method to study oxygen evolution in biological systems is to activate a photosynthetic system with short and intense light flashes, with appropriate dark periods, and measure the oxygen yield on each flash. Pierre Joliot's experiments in 1969 showed that flash illumination produced an oscillating pattern in the oxygen evolution and a maximum occurred on every *fourth* flash [50–56]. These patterns are very interesting because splitting of two water molecules to produce one oxygen molecule requires the removal of *four* electrons. In 1970, Kok et al. [56] proposed an explanation for the observed oscillation of the oxygen evolution pattern. The Kok et al. [56] hypothesis was that in a cycle of water oxidation a succession of oxidizing equivalents is stored on each separate and independent WOC, and when four oxidizing equivalents accumulate one by one, an oxygen is spontaneously evolved. Each oxidation state of the WOC is known as an "S-state," with S_0 being the most reduced state and S_4 the most oxidized state in the catalytic

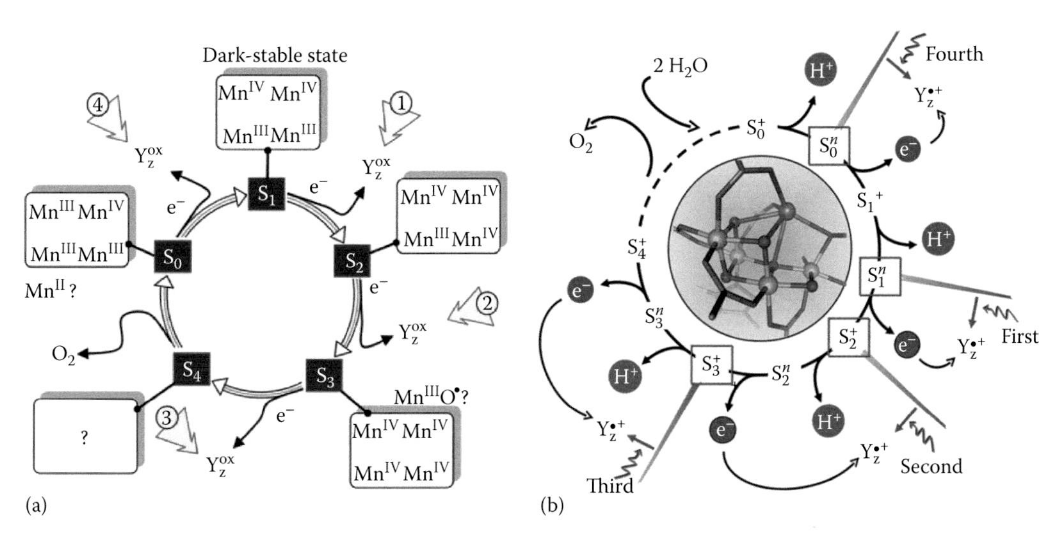

SCHEME 2.1 Classical S-state cycle of photosynthetic water oxidation [50–58]. Absorption of a photon causes charge separation at the reaction center P_{680} of PSII that leads to the formation of $Y_z^{\bullet+}$ (oxidized tyrosine-161 on the D1 protein) within less than 1 μs. Reduction of $Y_z^{\bullet+}$ by electron transfer (ET) from the manganese complex results in $S_i \rightarrow S_{i+1}$ transition. There are several similar S-state cycle schemes. Here, we show a plausible oxidation state of the four Mn ions in the different S-states (a). The extended S-state cycle including not only four oxidation but also four deprotonation steps is also shown in (b). (From Grundmeier, A. and Dau, H., *Biochim. Biophys. Acta*, 1817, 88, 2012.)

cycle (Scheme 2.1) [56]. It is essential to recognize that to explain the fact that the first maximum of oxygen evolution is after the 3rd flash, and then after 7th and 11th flashes, the S_1 state must be dark-stable. The $S_4 \rightarrow S_0$ transition is light independent and in this state oxygen is evolved. All other S-state transitions are initiated by the photochemical oxidation of P_{680} P_{680}^{+} [56].

2.3.2.3 Oxidation States of Manganese Ions in the Kok Cycle

It is well known that redox changes in the $S_0 \rightarrow S_1$ and $S_1 \rightarrow S_2$ transitions for the WOC are manganese-based [48]. The $S_2 \rightarrow S_3$ and the $S_3 \rightarrow S_4$ transitions are still controversial as to whether a metal-centered or a ligand-centered oxidation occurs [48]. In the $S_4 \rightarrow S_0$ transition (Scheme 2.1), rapid oxidation of two substrate water molecules occurs [48]. Based on the experimental evidence accumulated thus far, the four Mn ions in the S_1-state are believed to be Mn_2(III)Mn_2(IV), but a lower valence combination may also be possible.

2.3.2.4 Mechanism of Water Oxidation in Nature

Detailed physico-chemical mechanism of water oxidation by the WOC of PSII is still not resolved [58]. There are many proposals for the mechanism of water oxidation by the WOC in PSII [48]. The most important models are the following (Scheme 2.2):

1. *Nucleophile–electrophile reaction*: Pecoraro et al. [59] have proposed that a terminal Mn(V)=O undergoes a nucleophilic attack by a Ca^{2+} bound hydroxide ligand to form a Mn-bound hydroperoxide. Brudvig et al. [60] have proposed a mechanism in which a Ca^{2+} ion plays a role as a weak Lewis acid. In this mechanism, a water molecule bound to calcium reacts with a Mn(V)=O species to form the O=O bond through a nucleophilic attack. Lee and Brudvig [61] provided direct support for the proposal that Ca^{2+} plays a structural role in the early S-state transitions, which may also be fulfilled by other cations with similar ionic radius. Umena et al. [29,30] suggested that one of the water molecules that is coordinated with calcium, and is near Tyr 161, may serve as one of the substrate molecules, and the water molecule coordinated to Mn(4) may serve as the second substrate in the O–O formation.
2. *Coupling of an oxyl radical and a manganese-bound oxo-ligand*: Siegbahn [62], based on extensive DFT calculations by his research group, has suggested that a Mn(IV)$-O^{\bullet}$ may react with a manganese-bound oxo-ligand to form oxygen. In this mechanistic hypothesis, spin alignment of the reactive oxygen atoms is important [62].

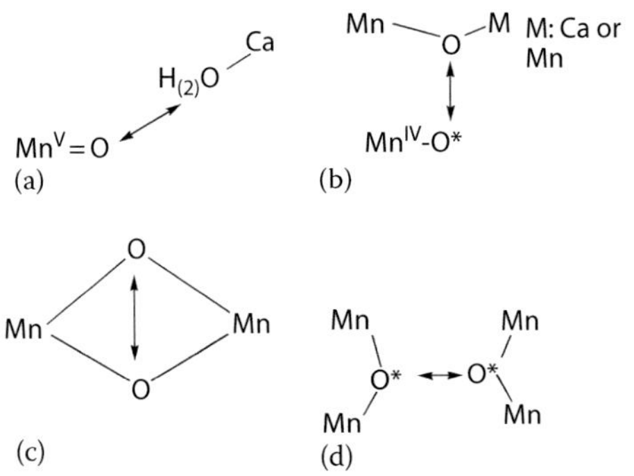

SCHEME 2.2 The most important proposed mechanisms of oxygen evolution by the WOC (a–d). (From Springer Science+Business Media: Burnap, R.L. and Vermaas, W.F.J., Eds., *Advances in Photosynthesis and Respiration Functional Genomics and Evolution of Photosynthetic Systems*, Probing functional diversity of thermophilic cyanobacteria in microbial mats, Vol. 33, Chap. 2, 2012, pp. 17–46, Bhaya, D.)

3. *Reductive elimination of two bridging oxo-ligands*: Rüttinger et al. [63] in the research group of Charles Dismukes proposed that a high-valent Mn-oxo-cluster collapses to form oxygen from two bridging oxo-ligands.
4. *Radical coupling mechanism*: In this mechanism, two oxyl radicals ($O^•$) are formed followed by a radical coupling to generate the oxygen–oxygen bond [64].

Umena et al. [29,30] suggested that one oxygen that bridges between Ca, Mn(4), Mn(3), and Mn(1) may exist as a hydroxide ion in the S_1 state and that it may provide one of the substrates for dioxygen evolution. One of the water molecules coordinated to calcium or Mn(4) may provide another substrate for oxygen evolution (Scheme 2.2).

Research on many WOC mutants has provided information on the role of specific amino acids in the mechanism of oxygen evolution (see, e.g., Ref. [48]). We need to be aware that although the basic mechanisms of oxygen evolution must be the same in all systems, it needs to be studied and checked even within different diverse groups of cyanobacteria. We already know that just even within thermophilic cyanobacteria there is functional diversity in natural populations [65].

2.4 POSSIBLE EVOLUTIONARY ORIGIN OF THE WATER-OXIDIZING COMPLEX IN PHOTOSYSTEM II

The high concentration of oxygen in Earth's atmosphere is one of the most geologically important signatures of life. Accumulation of oxygen began after the evolution of oxygenic photosynthesis in cyanobacteria around 3 billion years ago [66]. The increasing concentration of oxygen may have many biological patterns, among them the evolution of "body size" [66]. However, it is an enigma as to when and how oxygen-producing photosynthetic cyanobacteria evolved from their photosynthetic bacterial precursors [1]. It has been suggested that two of the bacterial reaction centers and PSII are evolutionarily related [67]. However, PSII must have provided a very strong oxidation potential to oxidize water because water is a stable molecule and to oxidize water a molecule must have a midpoint potential greater than ~ 0.82 V versus the standard hydrogen electrode at pH 7 [67]. P_{865}, reaction center of an anoxygenic bacteria, has only a moderate E_m (midpoint potential) value of 0.5 V and thus cannot oxidize water or even tyrosine [67] (Table 2.2).

The WOC in PSII found in cyanobacteria and in the thylakoid membranes of plant chloroplasts are believed to have evolved from a single common ancestor [68,69]. There are several hypotheses for the origin of the WOC. One hypothesis suggests that the WOC originated in binuclear manganese active sites, including ribonucleotide reductase, catalase, and arginase [70,71]. Perhaps, Mn catalase could have been a key intermediate en route to oxygenic photosynthesis [70,71]. Blankenship and Hartman [71] proposed that a primitive Mn catalase was the original template upon which the modern WOC was structured. Raymond and Blankenship [72] developed an approach for determining the optimal superposition of the atoms concentrated around the active sites of PSII and binuclear-manganese proteins. These observations support a common structural core in the WOC and in distinct manganese binuclear enzymes. It is also possible that

TABLE 2.2
Midpoint Potential at pH = 7 Relative to the Standard Hydrogen Electrode

Compound	Midpoint Potential
P_{680}/P_{680}^+	1.1–1.4
H_2O/O_2	0.82
Tyr/Tyr^+	1–1.1
P_{865}/P_{865}^+	0.5

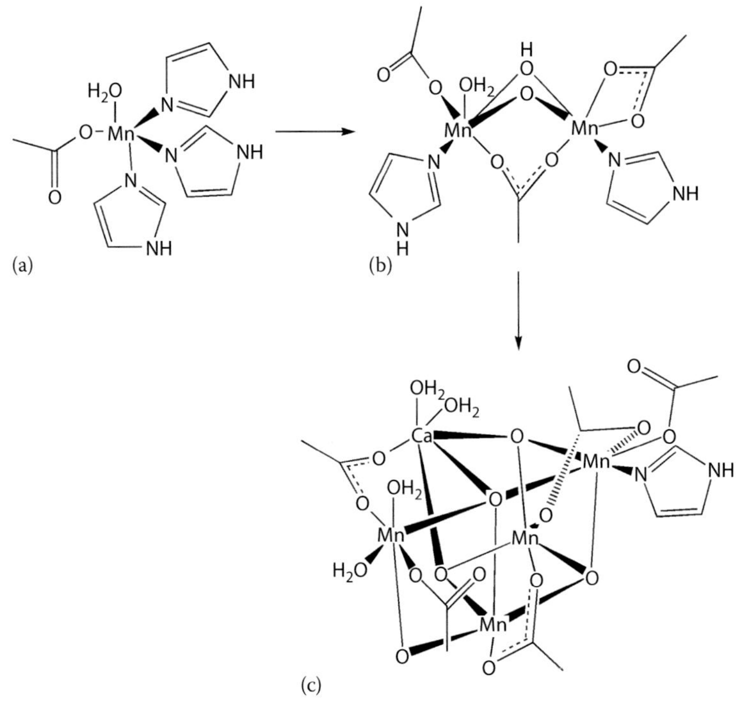

FIGURE 2.9 A mononuclear enzyme similar to manganese superoxide dismutase (a) might have served as an origin for catalase (b) and then the WOC (c). (With permission from Springer Science + Business Media: *Origins Life Evol. Biosphere*, A possible evolutionary origin for the Mn_4 cluster in photosystem II: From manganese superoxide dismutase to oxygen evolving complex, 39, 2009, 151–163, Najafpour, M.M.)

the development of oxygenic photosynthesis occurred in steps, the first of which involved only mononuclear manganese enzymes—the mononuclear manganese enzymes could have been a key intermediate en route to catalase and then to the WOC (Figure 2.9) [73].

Recently, Allen et al. [74] demonstrated that modifications of bacterial reaction centers can produce a highly oxidizing protein with a tight Mn-binding site that is redox active. Allen et al. [74] further showed that after light-induced electron transfer from the primary donor to the electron acceptors, the bound Mn is oxidized and can react with superoxide to produce molecular oxygen. This interesting system could serve as a useful model for understanding the involvement of intermediates in the evolutionary development of PSII [74].

Williamson et al. [68] have proposed that manganese ions may have replaced iron in iron-binding site of an enzyme and formed the precursor to the WOC. One of us (MMN) (see [75]) has also proposed a novel origin for the WOC from the manganese-oxidizing bacteria. The Archaean ocean may have sufficient Mn and Ca, ions with high affinity for interaction with the Mn oxide, and alkaline conditions may have enabled protocyanobacteria to assemble mineral oxides as functional compliments of early active site of PSII. Thus, those bacteria did not need to do large amount of manganese oxidation any more, since few manganese ions were enough to oxidize a large amount of water [75]. It is highly likely that those manganese-oxidizing bacteria may have changed to become the water-oxidizing bacteria. Water oxidation may have been an advantage for water-oxidizing bacteria because the amount of water on Earth was huge and water-oxidizing bacteria could reproduce and survive more easily than the other bacteria. These water-oxidizing bacteria may be the origin for cyanobacteria, and thus for algae and plants. In this regard, it is interesting that manganese oxides, in the form of dispersed powders, have been tested as catalysts for the four-electron oxidation of water to oxygen in the presence of different oxidants [76]. More interestingly, it has been shown that incorporation of calcium into mixed-valence manganese oxides produces a structure similar to the WOC and greatly improves the water oxidation activity of these manganese oxides toward water oxidation [77–80]. Further, these manganese oxides, with calcium, but without any additional groups, show a structure similar to the WOC in PSII [77–80].

2.5 STRESS AND OXYGEN EVOLUTION

Cyanobacteria, just as algae and plants, are prone to different stresses. They have, for example, cold stress, heat stress, salt stress, water stress, and light stress. It is instructive to learn from all the systems to understand the impact of stress on cyanobacteria. We recommend the readers to consult Refs. [81–86]. Among the photosynthetic apparatus, the WOC is known to be one of the most susceptible sites of inactivation induced by various stresses. This is largely due to the unstable nature of the protein components that constitute the WOC, as well as the vulnerable structure of the distorted Mn_4CaO_5 structure. For example, heating is known to release extrinsic proteins, indispensable components for an intact WOC, of cyanobacteria and higher plant PSII, thereby inhibiting oxygen evolution [87–89]. The binding of extrinsic proteins in the WOC is weak and can also be affected by cold stress [90] and salt stress [91], leading to their dissociation from PSII and the inactivation of oxygen evolution. One of the primary targets of photoinhibition has been proposed to be the Mn site [92,93], as Mn absorbs ultraviolet light, which may cause the destruction of the WOC. Thus, engineering a system with the aim to improve stability of the WOC complex might lead to cyanobacteria and plants that can better cope with various stress conditions. In this regard, photoinhibition by visible light and UV radiation on PSII is an important issue. Hakala et al. [94] showed that the release of Mn ions to the thylakoid lumen is the earliest detectable step in both UV- and visible-light-induced photoinhibition. After Mn release from the OEC, oxidative damage to the PSII reaction center occurs because the Mn-depleted OEC cannot reduce P_{680}^+ normally. As discussed by Vass et al. [95–98], PSII has mechanisms to protect itself from photodamage by light, by nonradiative charge recombination, and repair of damaged reaction center complexes.

2.6 WATER OXIDATION IN ARTIFICIAL PHOTOSYNTHESIS

The goal of artificial photosynthesis is to make different useful material or high-energy chemicals for energy storage using sunlight [99]. Hydrogen production by water splitting may be one of the most important goals of artificial photosynthesis [99]. To evolve hydrogen efficiently in a sustainable manner, it is necessary to first synthesize a "super catalyst" for water oxidation, which is the more challenging half reaction of water splitting [100]. There is an efficient system for water oxidation in cyanobacteria, algae, and plants (see e.g. Refs. [22–27]). Published data on the Mn–Ca cluster have provided details on the mechanism and structure of the WOC [29,30]. To design an efficient WOC for artificial photosynthesis, we must learn and use wisely the knowledge about water oxidation and the WOC in the natural system [101–103].

In the end, we refer the readers to the web site of Royal Society of Chemistry for a collection of articles : A Comment on "Artificial Photosynthesis, titled "Running on sun" is online at Chemistry World, Royal Society of Chemistry: http://rsc.li/PCKq86; it is a part of a special collection http://blogs.rsc.org/cs/2012/09/25/a-centenary-for-solar-fuels/ to mark the centenary of Ciamician's paper "The Photochemistry of the Future".

ACKNOWLEDGMENTS

M.M. Najafpour and A. Nemati Moghaddamare are grateful to the Institute for Advanced Studies in Basic Sciences, in Iran, for financial support. Govindjee is grateful to the School of Life Sciences, Jawaharlal Nehru University, New Delhi, India, for support since this review was completed while he was a visiting professor there in 2012.

REFERENCES

1. Govindjee and Shevela, D., Adventures with cyanobacteria: A personal perspective, *Front. Plant Sci.*, 2, 1–17, 2011.
2. Drews, G., The evolution of cyanobacteria and photosynthesis, in *Bioenergetic Processes of Cyanobacteria*, Peschek, G.A., Obinger, C., and Renger, G., Eds., Springer, Dordrecht, the Netherlands, 2011, Chap. 11, pp. 265–284.
3. Björn, L.O. and Govindjee, The evolution of photosynthesis and chloroplasts, *Curr. Sci.*, 96, 1466–1474, 2009.
4. Fromme, P. and Grotjohann, I., Structure of cyanobacterial photosystems I and II, in *Bioenergetic Processes of Cyanobacteria from Evolutionary Singularity to Ecological Diversity*, Peschek, G.A., Obinger, C., and Renger, G., Eds., Springer, Dordrecht, the Netherlands, 2011, Chap. 12, pp. 285–335.
5. Renger, G. and Ludwig, B., Mechanism of photosynthetic production and respiratory reduction of molecular dioxygen, A biophysical and biochemical comparison, in *Bioenergetic Processes of Cyanobacteria from Evolutionary Singularity to Ecological Diversity*, Peschek, G.A., Obinger, C., and Renger, G., Eds., Springer, Dordrecht, the Netherlands, 2011, Chap. 13, pp. 337–394.
6. Rabinowitch, E. and Govindjee, *Photosynthesis*, John Wiley, New York, 1969, available free at: http://www.life.illinois.edu/govindjee/photosynBook.html
7. Allakhverdiev, S.I., Recent progress in the studies of structure and function of photosystem II, *J. Photochem. Photobiol. B*, 104, 1–8, 2011.
8. Govindjee, Kern, J.F., Messinger, J., and Whitmarsh, J., Photosystem II, in *Encyclopedia of Life Sciences (ELS)*, John Wiley & Sons, Ltd, Chichester, U.K., 2010, pp. 1–15, doi: 10.1002/9780470015902.a0000669.pub2 http://www.els.net.
9. Govindjee and Björn, L.O., Dissecting oxygenic photosynthesis: The evolution of the "Z"-scheme for thylakoid reactions, in *Photosynthesis: Overviews on Recent Progress and Future Perspectives*, Itoh, S., Mohanty, P., and Guruprasad, K.N., Eds., I.K. Publishers, New Delhi, India, 2012, Chap. 1, pp. 1–28.
10. Wydrzynski, T.J. and Satoh, K., Eds., *Photosystem II: The Light-driven Water: Plastoquinone Oxidoreductase*, Advances in Photosynthesis and Respiration, Vol. 22, Springer, Dordrecht, the Netherlands, 2005.
11. Renger, G., Photosynthetic water splitting: Apparatus and mechanism, in *Photosynthesis: Plastid Biology, Energy Conversion and Carbon Assimilation*, Vol. 34, Eaton-Rye, J.J., Tripathy, B.C., and Sharkey, T.D., Eds., Advances in Photosynthesis and Respiration, Springer, Dordrecht, the Netherlands, 2012, Chap. 17, pp. 359–412.
12. Satoh, K., Wydrzynski, T.J., and Govindjee, Introduction to photosystem II, in *Advances in Photosynthesis and Respiration Photosystem II: The Light-Driven Water: Plastoquinone Oxidoreductase*, Vol. 22, Wydrzynski, T.J. and Satoh, K., Eds., Springer, Dordrecht, the Netherlands, 2005, Chap. 1, pp. 11–22.
13. Spetzler, D. et al., Energy transduction by the two molecular motors of the F1F0 ATP synthase, in *Advances in Photosynthesis and Respiration Photosynthesis: Plastid Biology, Energy Conversion and Carbon Assimilation*, Vol. 34, Eaton-Rye, J.J., Tripathy, B.C., and Sharkey, T.D., Eds., Springer, Dordrecht, the Netherlands, 2012, Chap. 22, pp. 561–590.
14. Golbeck, J.H., Ed., *Advances in Photosynthesis and Respiration Photosystem I: The Light-Driven Plastocyanin: Ferredoxin Oxidoreductase*, Vol. 24, Springer, Dordrecht, the Netherlands, 2006.
15. Kallas, T., Cytochrome b_6f complex at the heart of energy transduction and redox signalling, in *Advances in Photosynthesis and Respiration Photosynthesis: Plastid Biology, Energy Conversion and Carbon Assimilation*, Vol. 34, Eaton-Rye, J.J., Tripathy, B.C., and Sharkey, T.D., Eds., Springer, Dordrecht, the Netherlands, 2012, Chap. 21, pp. 501–560.
16. Govindjee, Kambara, T., and Coleman, W., The electron donor side of photosystem: The oxygen evolving complex, *Photochem. Photobiol.*, 42, 187–210, 1985.
17. Kambara, T. and Govindjee, Molecular mechanism of water oxidation in photosynthesis based on the functioning of manganese in two different environments, *Proc. Natl Acad. Sci. U.S.A.*, 82, 6119–6123, 1985.
18. Luber, S. et al., S_1-state model of the O_2-evolving complex of photosystem II, *Biochemistry*, 50, 6308–6311, 2011.
19. Wydrzynski, T.J. and Hillier, W., Eds., *Molecular Solar Fuels*, Royal Society of Cambridge, Cambridge, U.K., 2011.
20. Hill, J.F., Early pioneers of photosynthesis research, in *Advances in Photosynthesis and Respiration Photosynthesis: Plastid Biology, Energy Conversion and Carbon Assimilation*, Vol. 34, Eaton-Rye, J.J., Tripathy, B.C., and Sharkey, T.D., Eds., Springer, Dordrecht, the Netherlands, 2012, Chap. 30, pp. 771–800.

21. Walker, D.A., 'And whose bright presence' an appreciation of Robert Hill and his reaction, in *Advances in Photosynthesis and Respiration Discoveries in Photosynthesis*, Vol. 20, Govindjee, Beatty, J.T, Gest, H., and Allen, J.F., Eds., Springer, Dordrecht, the Netherlands, 2005, pp. 109–112.
22. Govindjee and Krogmann, D., Discoveries in oxygenic photosynthesis (1727–2003): A perspective, in *Discoveries in Photosynthesis*, Vol. 20, Govindjee, Beatty, J.T, Gest, H., and Allen, J.F., Eds., Springer, Dordrecht, the Netherlands, 2005, pp. 63–105.
23. Moore, G.F. and Brudvig, G.W., Energy conversion in photosynthesis: A paradigm for solar fuel production, *Annu. Rev. Condens. Matter Phys.*, 2, 303–327, 2011.
24. Decker, H. and Van Holde, K.E., *Oxygen and the Evolution of Life*, Springer, Heidelberg, Germany, 2011.
25. Zouni, A. et al., Crystal structure of photosystem II from *Synechococcus elongatus* at 3.8 Å resolution, *Nature*, 409, 739–743, 2001.
26. Kamiya, N. and Shen, J.R., Crystal structure of oxygen-evolving photosystem II from *Thermosynechococcus vulcanus* at 3.7 Å resolution, *Proc. Natl Acad. Sci. U.S.A.*, 100, 98–103, 2003.
27. Ferreira, K.N. et al., Architecture of the photosynthetic oxygen evolving centre, *Science*, 303, 1831–1838, 2004.
28. Yano, J. et al., Where water is oxidized to dioxygen: Structure of the photosynthetic Mn_4Ca cluster, *Science*, 314, 821–825, 2006.
29. Umena, Y. et al., Crystal structure of oxygen-evolving photosystem II at a resolution of 1.9 Å, *Nature*, 473, 55–60, 2011.
30. Kawakami, K. et al., Structure of the catalytic, inorganic core of oxygen-evolving photosystem II at 1.9 Å resolution, *J. Photochem. Photobiol. B*, 104, 9–18, 2011.
31. Witt, H.T., Steps on the way to building blocks, topologies, crystals and x-ray structural analysis of photosystems I and II of water oxidizing photosynthesis, in *Discoveries in Photosynthesis*, Vol. 20, Govindjee, Beatty, J.T, Gest, H., and Allen, J.F., Eds., Springer, Dordrecht, the Netherlands, 2005, pp. 237–259.
32. Nakamoto, K., *Infrared and Raman Spectra of Inorganic and Coordination Compounds*, 4th edn., Wiley-Interscience Publication, Hoboken, NJ, 1986, pp. 231–233.
33. Pecoraro, V.L., *Manganese Redox Enzymes*, VCH, New York, 1992.
34. Pearson, R.G., Hard and soft acids and bases, *J. Am. Chem. Soc.*, 85, 3533–3539, 1963.
35. Boussac, A. et al., Biosynthetic Ca^{2+}/Sr^{2+} exchange in the photosystem II oxygen evolving enzyme of *Thermosynechococcus elongatus*, *J. Biol. Chem.*, 279, 22809–22819, 2004.
36. Hillier, W. and Wydrzynski, T., ^{18}O-water exchange in photosystem II: Substrate binding and intermediates of the water splitting reaction, *Coord. Chem. Rev.*, 252, 306–317, 2008.
37. Barber, J., Crystal structure of the oxygen-evolving complex of photosystem II, *Inorg. Chem.*, 47(6), 1700–1710, 2008.
38. Barber, J. and Murray, J.W., Revealing the structure of the Mn-cluster of photosystem II by x-ray crystallography, *Coord. Chem. Rev.*, 252, 233–243, 2008.
39. Guskov, A. et al., Cyanobacterial photosystem II at 2.9-Å resolution and the role of quinones, lipids, channels and chloride, *Nat. Struct. Mol. Biol.*, 16, 334–342, 2009.
40. Baianu, I.C. et al., NMR study of chloride-ion interactions with thylakoid membranes, *Proc. Natl Acad. Sci. U.S.A.*, 81, 3713–3717, 1984.
41. Coleman, W.J. and Govindjee, A model for the mechanism of chloride activation of oxygen evolution in photosystem II, *Photosynth. Res.*, 13, 199–223, 1987.
42. Popelkova, H. et al., Mutagenesis of basic residues R151 and R161 in manganese-stabilizing protein of photosystem II causes inefficient binding of chloride to the oxygen-evolving complex, *Biochemistry*, 45, 3107–3115, 2006.
43. Yocum, C.F., The calcium and chloride requirements of the O_2 evolving complex, *Coord. Chem. Rev.*, 252, 296–305, 2008.
44. Kawakami, K. et al., Location of chloride and its possible functions in oxygen-evolving photosystem II revealed by x-ray crystallography, *Proc. Natl Acad. Sci. U.S.A.*, 106, 8567–8572, 2009.
45. Rivalta, I. et al., Structural functional role of chloride in photosystem II, *Biochemistry*, 50, 6312–6315, 2011.
46. Govindjee and Seibert, M., Picosecond spectroscopy of the isolated reaction centers from the photosystems of oxygenic photosynthesis—Ten years (1987–1997) of fun. A tribute to Michael R. Wasielewski on his 60th birthday, *Photosynth. Res.*, 103, 1–6, 2010.
47. Renger, G. Eds., *Primary Processes of Photosynthesis: Principles and Apparatus, Parts 1 and 2*, The Royal Society of Chemistry, Cambridge, U.K., 2008.
48. McEvoy, J. and Brudvig, G.W., Water-splitting chemistry of photosystem II, *Chem. Rev.*, 106, 4455–4482, 2006.

49. Ruttinger, W. and Dismukes, G.C., Synthetic water oxidation catalysts for artificial photosynthetic water oxidation, *Chem. Rev.*, 97, 1–24, 1997.
50. Joliot, P., Barbieri, G., and Chabaud, R., Un nouveau modele des centres photochimiques du systeme II, *Photochem. Photobiol.*, 10, 309–329, 1969.
51. Mar, T. and Govindjee, Kinetic models of oxygen evolution, *J. Theor. Biol.*, 36, 427–446, 1972.
52. Joliot, P. and Kok, B., Oxygen evolution in photosynthesis, in *Bioenergetics of Photosynthesis*, Govindjee, Ed., Academic Press, New York, 1975, pp. 387–412.
53. Grundmeier, A. and Dau, H., Structural models of the manganese complex of photosystem II and mechanistic implications, *Biochim. Biophys. Acta*, 1817, 88–105, 2012.
54. Joliot, P., Period-four oscillations of the flash-induced oxygen formation in photosynthesis, *Photosynth. Res.*, 20, 371–378, 2005.
55. Whitmarsh, J. and Govindjee, Photosystem II, in *Encyclopedia of Life Sciences*, MacMillan Reference Ltd, London, U.K., 2001.
56. Kok, B., Forbush, B., and McGloin, M., Cooperation of charges in photosynthetic O_2 evolution: I. A linear four-step mechanism, *Photochem. Photobiol.*, 11, 457–475, 1970.
57. Renger, G., Mechanism of light induced water splitting in photosystem II of oxygen evolving photosynthetic organisms, *Biochim. Biophys. Acta*, 1817(8), 1164–1176, 2012.
58. Bader, K.P., Thibault, P., and Schmid, G.H., A study on oxygen evolution and on the S-state distribution in thylakoid preparations of the filamentous blue-green alga *Oscillatoria chalybea*, *Z Naturforsch*, 38c, 778–792, 1983.
59. Pecoraro, V. L. et al., A proposal for water oxidation in photosystem II, *Pure Appl. Chem.*, 70, 925–929, 1998.
60. Limburg, J., Szalai, A., and Brudvig, G.W., A mechanistic and structural model for the formation and reactivity of a Mn(V) = O species in photosynthetic water oxidation, *J. Chem. Soc., Dalton Trans.*, 1353–1362, 1999.
61. Lee, C., Brudvig, G.W., and Lakshmi, K.V., Probing the functional role of Ca^{2+} in the oxygen-evolving complex of photosystem II by metal ion inhibition, *Biochemistry*, 46(11), 3211–3223, 2007.
62. Siegbahn, P.E.M., Studies of O-O bond formation in photosystem II, *Inorg. Chem.*, 47, 1779–1786, 2008.
63. Rüttinger, W. et al., Theoretical O_2 evolution from the manganese-oxo cubane core $Mn_4O_4^{6+}$: A molecule mimic of the photosynthetic water oxidation enzyme?, *J. Am. Chem. Soc.*, 122, 10353–10357, 2000.
64. Yachandra, V.K., Sauer, K., and Klein, M.P., Where plants oxidize water to dioxygen, *Chem. Rev.*, 96, 2927–2950, 1996.
65. Bhaya, D., Probing functional diversity of thermophilic cyanobacteria in microbial mats, in *Advances in Photosynthesis and Respiration Functional Genomics and Evolution of Photosynthetic Systems*, Vol. 33, Burnap, R.L. and Vermaas, W.F.J., Eds., Springer, Dordrecht, the Netherlands, 2012, Chap. 2, pp. 17–46.
66. Payne, J.L. et al., The evolutionary consequences of oxygenic photosynthesis: A body size perspective, *Photosynth. Res.*, 107, 7–10, 2011.
67. Allen, J.P. and Williams, J.C., The evolutionary pathway from anoxygenic to oxygenic photosynthesis examined by comparison of the properties of photosystem II and bacterial reaction centers, *Photosynth. Res.*, 107, 59–69, 2011.
68. Williamson, A. et al., The evolution of photosystem II: Insights into the past and future, *Photosynth. Res.*, 107, 71–86, 2011.
69. Murray, J.W., Sequence variation at the oxygen-evolving centre of photosystem II: A new class of 'rogue' cyanobacterial D1 proteins, *Photosynth. Res.*, 110, 177–184, 2011.
70. McKay, C. and Hartman, H., Hydrogen peroxide and the evolution of oxygenic photosynthesis, *Origins Life Evol. Biosphere*, 21, 157–163, 1991.
71. Blankenship, R.E. and Hartman, H., The origin and evolution of oxygenic photosynthesis, *Trends. Biochem. Sci.*, 23, 94–97, 1998.
72. Raymond, J. and Blankenship, R.E., The origin of the oxygen-evolving complex, *Coord. Chem. Rev.*, 252, 377–383, 2008.
73. Najafpour, M.M., A possible evolutionary origin for the Mn_4 cluster in photosystem II: From manganese superoxide dismutase to oxygen evolving complex, *Origins Life Evol. Biosphere*, 39, 151–163, 2009.
74. Allen, J.P. et al., Light-driven oxygen production from superoxide by Mn-binding bacterial reaction centers, *Proc. Natl Acad. Sci. U.S.A.*, 109, 2314–2318, 2012.
75. Najafpour, M.M., Amorphous manganese-calcium oxides as a possible evolutionary origin for the $CaMn_4$ cluster in photosystem II, *Origins Life Evol. Biosphere*, 41, 237–247, 2011.
76. Harriman, A. et al., Metal oxides as heterogeneous catalysts for oxygen evolution under photochemical condition, *J. Chem. Soc. Faraday*, 84, 2795–2806, 1988.

77. Najafpour, M.M. et al., Calcium-manganese (III) oxides ($CaMn_2O_4.xH_2O$) as biomimetic oxygen-evolving catalysts, *Angew. Chem. Int. Ed.*, 49, 2233–2237, 2010.
78. Najafpour, M.M., Mixed-valence manganese calcium oxides as efficient catalysts for water oxidation, *Dalton Trans.*, 40, 3793–3795, 2011.
79. Najafpour, M.M., Calcium manganese oxides as structural and functional models for the active site in the oxygen evolving complex in photosystem II: Lessons from simple models, *J. Photochem. Photobiol. B*, 104, 111–117, 2011.
80. Zaharieva, I. et al., Synthetic manganese-calcium oxides mimic the water-oxidizing complex of photosynthesis functionally and structurally, *Energy Environ. Sci.*, 4, 2400–2408, 2011.
81. Szalontai, B., Domonkos, I., and Gombos, Z., The role of membrane structure in acclimation to low-temperature stress, in *Photosynthesis: Plastid Biology, Energy Conversion and Carbon Assimilation*, Vol. 34, Eaton-Rye, J.J., Tripathy, B.C., and Sharkey, T.D., Eds., Advances in Photosynthesis and Respiration, Springer, Dordrecht, the Netherlands, 2012, Chap. 11, pp. 233–250.
82. Mohanty, P. et al., Heat stress: Susceptibility, recovery and regulation, in *Photosynthesis: Plastid Biology, Energy Conversion and Carbon Assimilation*, Vol. 34, Eaton-Rye, J.J., Tripathy, B.C., and Sharkey, T.D., Eds., Advances in Photosynthesis and Respiration, Springer, Dordrecht, the Netherlands, 2012, Chap. 12, pp. 251–274.
83. Allahverdiyev, S.I. and Aro, E.M., Photosynthetic responses of plants to excess light: Mechanisms and conditions for photoinhibition, excess energy dissipation and repair, in *Photosynthesis: Plastid Biology, Energy Conversion and Carbon Assimilation*, Vol. 34, Eaton-Rye, J.J., Tripathy, B.C., and Sharkey, T.D., Eds., Advances in Photosynthesis and Respiration, Springer, Dordrecht, the Netherlands, 2012, Chap. 13, pp. 275–298.
84. Heddad, M., Engelken, J., and Adamska, I., Light stress proteins in viruses, cyanobacteria and photosynthetic eukaryote, in *Photosynthesis: Plastid Biology, Energy Conversion and Carbon Assimilation*, Vol. 34, Eaton-Rye, J.J., Tripathy, B.C., and Sharkey, T.D., Eds., Advances in Photosynthesis and Respiration, Springer, Dordrecht, the Netherlands, 2012, Chap. 14, pp. 299–318.
85. Danon, A., Environmentally-induced oxidative stress and its signaling, in *Photosynthesis: Plastid Biology, Energy Conversion and Carbon Assimilation*, Vol. 34, Eaton-Rye, J.J., Tripathy, B.C., and Sharkey, T.D., Eds., Advances in Photosynthesis and Respiration, Springer, Dordrecht, the Netherlands, 2012, Chap. 15, pp. 319–330.
86. Kanesaki, Y. et al., Sensors and signal transducers of environmental stress in cyanobacteria, in *Abiotic Stress Adaptation in Plants: Physiological, Molecular and Genomic Foundation*, Pareek, A., Sopory, S.K., Bohnert, H.J., and Govindjee, Eds., Springer, Dordrecht, the Netherlands, 2010, pp. 15–31.
87. Enami, I. et al., Is the primary cause of thermal inactivation of oxygen evolution in spinach PSII membranes release of the extrinsic 33 kDa protein or of Mn?, *Biochim. Biophys. Acta*, 1186, 52–58, 1994.
88. Nishiyama, Y. et al., Photosynthetic oxygen evolution is stabilized by cytochrome c550 against heat inactivation in *Synechococcus* sp PCC 7002, *Plant Physiol.*, 105, 1313–1319, 1994.
89. Nishiyama, Y., Los, D.A., and Murata, N., PsbU, a protein associated with photosystem II, is required for the acquisition of cellular thermotolerance in *Synechococcus* species PCC 7002, *Plant Physiol.*, 120, 301–308, 1999.
90. Shen, J.R., Terashima, I., and Katoh, S., Cause for dark, chilling-induced inactivation of photosynthetic oxygen-evolving system in cucumber leaves, *Plant Physiol.*, 93(4), 1354–1357, 1990.
91. Allakhverdiev, S.I. and Murata, N., Salt stress inhibits photosystems II and I in cyanobacteria, *Photosynth. Res.*, 98, 529–539, 2008.
92. Takahashi, S., and Murata, N., How do environmental stresses accelerate photoinhibition?, *Trends Plant Sci.*, 13, 178–182, 2008.
93. Ohnishi, N. et al., Two-step mechanism of photodamage to photosystem II: Step 1 occurs at the oxygen-evolving complex and step 2 occurs at the photochemical reaction center, *Biochemistry*, 44(23), 8494–8499, 2005.
94. Hakala, M. et al., Evidence for the role of the oxygen-evolving manganese complex in photoinhibition of photosystem II, *Biochim. Biophys. Acta*, 1706, 68–80, 2005.
95. Renger, G. et al., On the mechanism of photosystem II deterioration by UV-B irradiation, *Photochem. Photobiol.*, 49, 97–105, 1989.
96. Vass, I., Molecular mechanisms of photodamage in the photosystem II complex, *Biochim. Biophys. Acta*, 1817, 209–217, 2012.
97. Vass, L. et al., UV-B induced inhibition of photosystem II electron transport studied by EPR and chlorophyll fluorescence. Impairment of donor and acceptor side components, *Biochemistry*, 35, 8964–8973, 1996.

98. Vass, I. et al., The mechanism of UV-A radiation-induced inhibition of photosystem II electron transport studied by EPR and chlorophyll fluorescence, *Biochemistry*, 41, 10200–10208, 2002.
99. Najafpour, M.M., Ed., *Artificial Photosynthesis*, In Tech Publications, Rijeka, Croatia, ISBN: 979-953-307-665-1, 2012.
100. Bockris, J.O.M., *Energy-the Solar Hydrogen Alternative*, Wiley & Sons, New York, 1977.
101. Najafpour, M.M. et al., Biological water oxidation: Lessons from nature, *Biochim. Biophys. Acta*, 1817, 1110–1121, 2012.
102. Najafpour, M.M. and Govindjee, Oxygen evolving complex in photosystem II: Better than excellent, *Dalton Trans.*, 40, 9076–9084, 2011.
103. Scholes, G.D. et al., Lessons from nature about solar light harvesting, *Nat. Chem.*, 3, 763–774, 2011.

3 Origin, Evolution, and Interaction of Bioenergetic Processes in Cyanobacteria under Normal and Stressful Environments

Günter A. Peschek, Christian Obinger, and Surya Kant Mehta*

CONTENTS

3.1 INTRODUCTION

Prokaryotic and eukaryotic cells, the two types of cell architecture (structure), are the only really different types on earth since even Carl Woese's "archaebacteria" are nothing but prokaryotes! There is not a single transitional form ("missing link") between prokaryotes and eukaryotes. It is worth noting that the *all-decisive biochemical or functional features* that make up a living cell are strikingly uniform in all existing types of living cells. There is not a single biochemical trait in any eukaryote that would not also be found in one or the other of the prokaryotes. Rather, in the opposite

* Corresponding author: cyano1@aon.at

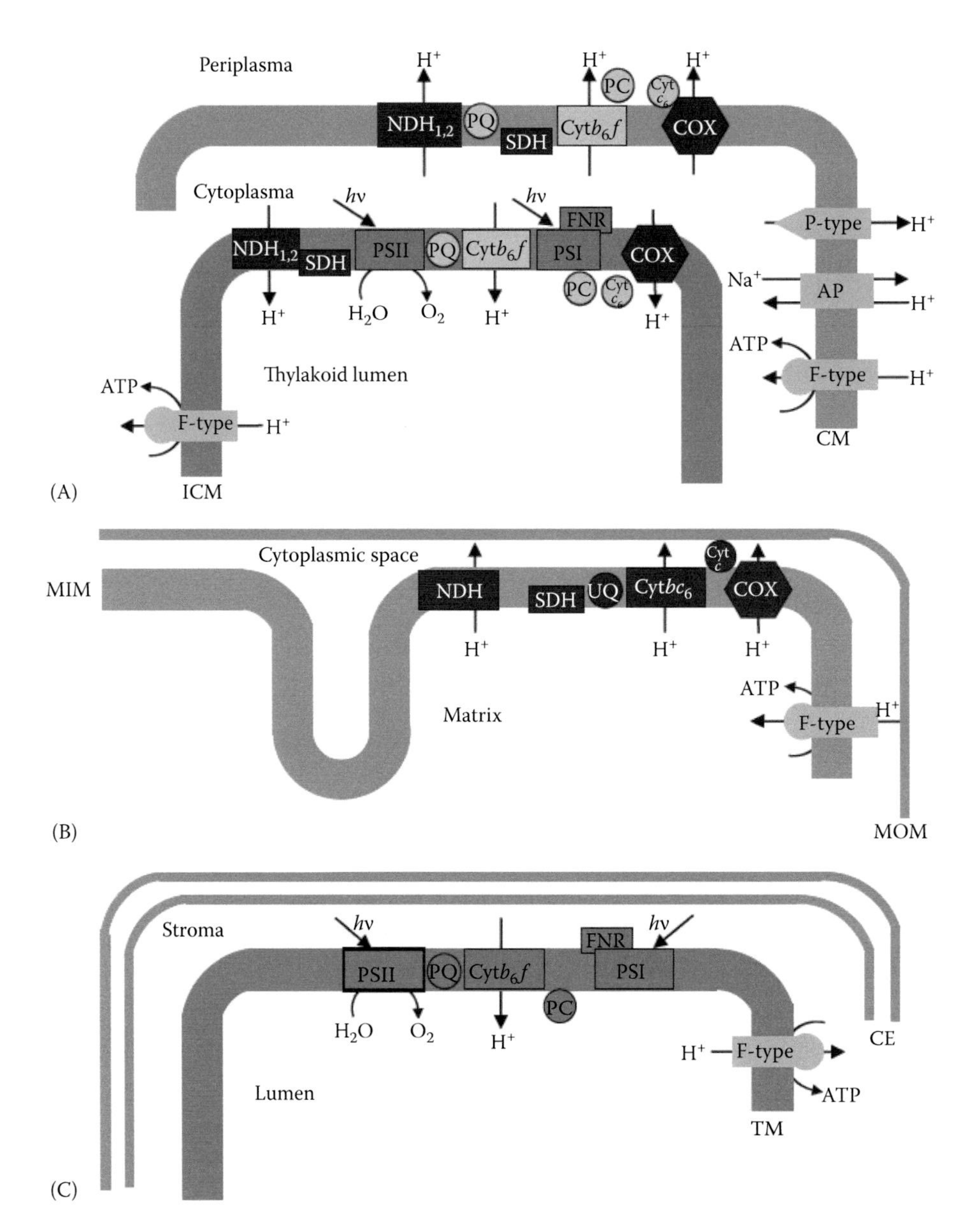

FIGURE 3.1 Comparison of bioenergetic membrane functions in a cyanobacterium (A), a mitochondrion (B), and a chloroplast (C). CM, plasma or cytoplasmic membrane; ICM or TM, intracytoplasmic or thylakoid membrane; CE, chloroplast envelope; MOM, mitochondrial outer membrane; MIM, mitochondrial inner membrane. Anastomoses between CM and ICM, which, though clearly demonstrated in anoxyphototrophs and also implicated for cyanobacteria and chloroplasts by both phylogenetic and functional considerations, could so far not be *physically* detected in cyanobacteria by electron microscopy or similar techniques [2,3]. Further abbreviations: AP, Na^+/H^+-antiporter; COX, aa_3-type cytochrome *c* oxidase; Cyt*c*, soluble *c*-type cytochrome; Cytc_6, cytochrome c_6; PC, plastocyanin; Cytb_6f, cytochrome b_6f; PQ, plastoquinol; UQ, ubiquinol; PSI, photosystem I; PSII, photosystem II; FNR, ferredoxin $NADP^+$ oxidoreductase; NDH, NAD[P]H dehydrogenase; SDH, succinate dehydrogenase; P-type, proton-pumping P-type ATPase; F-type ATPase, that is, ATP synthase. Note the mutually opposite polarity (sidedness) of thylakoid and inner mitochondrial membranes with an otherwise functionally and structurally quite similar electron transport assembly. For more details about cyanobacterial structure–function relationships, see [4].

direction, there are a few specifically prokaryotic functions that had apparently been lost in *all* eukaryotes, for example, nonmitotic proliferation or N_2 fixation.

The classical endosymbiont hypothesis that describes the endosymbiotic origin of a green plant or algal chloroplast from an engulfed, formerly free-living cyanobacterium [1], is almost unequivocally accepted [2]; however, modern molecular biology has added much to this striking relatedness, for example, with respect to the inhibition profiles of protein synthesis in free-living prokaryotes (viz., *cyanobacteria* but also, e.g., α-proteobacteria and others) and in *chloroplasts* (and mitochondria).

Another hypothesis, "orthodox endosymbiont hypothesis," assumes two separate and consecutive events: first, the "uptake" of a respiring purple bacterium [3,4] (e.g., *Paracoccus denitrificans* or another, *Rickettsia*-like, bacterium) and, secondly, a symbiosis between this proto-mitochondriate eu-oid-karyote and an ordinary (unicellular) cyanobacterium leading to a "full-fledged eukaryote" with mitochondrion and chloroplast. Now it was recently proposed that both chloroplast and mitochondrion have originated from a single endosymbiont event between an Archaeon and a cyanobacterium [5–7]. A comparison of genome signatures [8–10] and proteomes of *Rickettsiae*, mitochondria, chloroplasts, and cyanobacteria [11] has supported the aforementioned "generalized," endosymbiont hypothesis.

Electron transfer, proton movement, and ATP synthesis are the three bioenergetic machines. Figure 3.1 compares the membrane *functions*, viz., electron transfer, proton movement, and ATP synthesis of a cyanobacterium (top), a mitochondrion (middle), and a chloroplast (below). There is not only accidental *analogy* but also evolutionary *homology*, that is, a common monophyletic evolutionary origin of all three bioenergetic processes. Such monophyletic origin of respiration and photosynthesis is being amply discussed by the so-called *conversion hypothesis* [12–14], which now receives much support from the analysis of genome signatures [11]. Cyanobacteria, being photoautotrophs, are unique living cells equipped with oxygenic photosynthesis, on the one hand, and nitrogen fixation, on the other hand. Although cyanobacterial cell lacks membrane bound cell organelles, they have a unique special separation of photosynthesis and nitrogen fixation processes. This chapter deals with the three basic processes of energy metabolism: photosynthesis, respiration, and ATP synthesis.

3.2 ENERGY AND LIFE

There is no life without constant input of environmental energy (energy from outside), which, in turn, according to fundamental thermodynamical principles, is used to maintain the characteristic low-entropy ("high-order") state synonymous with a living cell. Three bioenergetic processes (fermentation, photosynthesis, and respiration) can uniquely be found in a single cyanobacterial cell. Life depends on the continuous input of *environmental energy*, which, in the living cell, is immediately and permanently transformed into biologically useful forms of energy, viz., ATP and other nucleoside triphosphates [15,16] (Figure 3.2). Energy is taken up from the environment by the living cell and given off as work (Figure 3.2). Energy conversion in both photosynthesis and respiration is impossible without *chemiosmosis*, that is, without the *membrane principle*. Except the retinal-based halobacterial photosynthesis, which is based on a purely conformational chemiosmotic proton pump, chemiosmosis implies membrane-bound ion-translocating electron transport [17,18]. Only in fermentative energy conversions through substrate level phosphorylation (SLP) [19,20], membranes are mechanistically not involved.

3.2.1 ATP, Adenylate Kinase, and Energy Charge

ATP (Figure 3.2A) is called the freely convertible *biological energy quantum* and is synthesized from ADP and inorganic phosphate, P_i (Equation 3.1), with K being the equilibrium constant, R the universal gas constant, T the absolute temperature (K), and $\Delta G^{o\prime}$ the difference between Gibbs energies [free enthalpies] *after* and *before* the reaction, under physiological standard conditions ($T = 298$ K, pressure = 1 atm = 10^5 Pa, concentration of each reactant, except H^+, = 1 M [1 mol/L], but $[H^+] = 10^{-7}$ M [pH = 7]).

$$\text{ADP} + P_i \uparrow \text{ATP} + \text{H}_2\text{O} K = 10^{-5}; \quad \Delta G^{o\prime} = -R.T.\ln K = +28.5 \text{ kJ/mol} \tag{3.1}$$

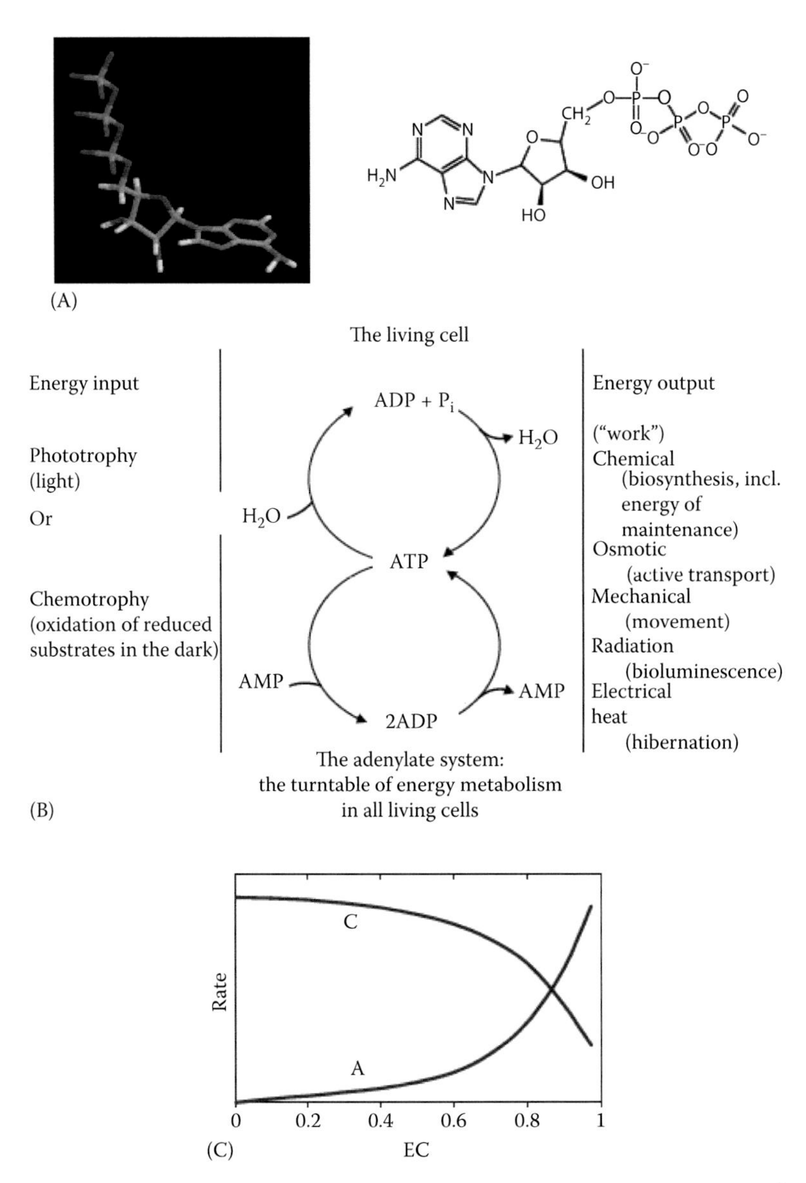

FIGURE 3.2 ATP, energy flow, and energy charge. (A) Structural formula of ATP (adenosine 5′-triphosphate). (B) The flow of energy from the environment through the adenylate system (see Figure 3.5) in a living cell and, in the form of work, back to the environment again. During this flow, the energy is "degraded" (it becomes less and less reversible, thermodynamically speaking), that is, its entropy content increases steadily. The entropy is taken up from the living cell, which is thereby left behind in the typical low-entropy (high-order) state characteristic of a living cell [12]. (C) Generalized representation of enzyme activity responses to energy charge (EC). C, enzymes involved in ATP regeneration (*catabolism*); A, biosynthetic enzymes that use ATP (*anabolism*).

TABLE 3.1
Phosphate Group Transfer Potentials ($-\Delta G^{\circ\prime}$) of Various Metabolically Important Phosphorylated High-Energy Intermediates in Physicochemical Standard Conditions (1 M concentration, 25°C, and at pH 7)

Phosphorylated Compound	$-\Delta G^{\circ\prime}$ (kJ/mol)
Phosphoenolpyruvate	62.2
1,3-Bisphosphoglycerate	49.6
Phosphocreatine	43.2
Acetylphosphate	42.4
Phosphoarginine	32.2
ATP (γ-phosphate)	28.5
GTP (γ-phosphate)	28.5
UTP (γ-phosphate)	28.5
Glucose 1-phosphate	21.0
Glucose 6-phosphate	13.9
Fructose 6-phosphate	16.0
Glycerol 1-phosphate	9.2
Carbamyl phosphate (at pH 9.5)	52

Water is taken as the standard phosphate acceptor in ordinary hydrolysis reactions. The compounds shown are of particular relevance to SLPs. $\Delta G^{\circ\prime}$ values mean $\Sigma G_i^{\circ\prime}$ of products minus $\Sigma G_i^{\circ\prime}$ of reactants, terminal state minus initial state (see Equations 3.1 and 3.4).

Enzyme adenylate kinase catalyzes the extremely fast (k_{cat} of several millions per min) and energetically neutral *equilibrium* between the three adenylates according to

$$\text{ADP} + \text{ADP} \uparrow \text{ATP} + \text{AMP} \tag{3.2}$$

Thus, the "*ATP cycle*" that describes the chemistry of "phosphorylation" is represented by Equations 3.1 and 3.2. Together with other nucleoside triphosphate interconversions (Table 3.1) ATP serves as *energy buffer* in living cells [15,16,21–23]. Equation 3.2 demonstrates that 50% of intracellular ADP is still energetically equivalent to ATP. The *energy charge* (EC) is a complete and sufficient measure of the entire cellular free energy content or working capacity of an integral living cell (Figure 3.2C):

$$\text{EC} = \frac{\{[\text{ATP}] + 0.5 \times [\text{ADP}]\}}{\{[\text{ATP}]+[\text{ADP}]+[\text{AMP}]\}} \tag{3.3}$$

Equation 3.3 immediately shows that the (intra)cellular EC only can assume values between 0.0 (totally de-energized cell: all adenylates present as AMP) and 1.0 (fully or over-energized cell, all adenylates present as ATP). Both states are unrealistic, of course. But most interestingly, due to the many and multifarious allosterically regulated enzymes in a living cell [24,25], the overall metabolic activity crucially and sensibly responds to the EC value of the cell.

ATP, through a multitude of various (soluble) transphosphorylases (phosphotransferases), is in rapid (isoenergetic) metabolic equilibrium with other "energy-rich" [26] nucleoside (tri-) phosphates such as GTP, CTP, UTP, etc. (Table 3.1), as well as with an energizable biomembrane [13,27].

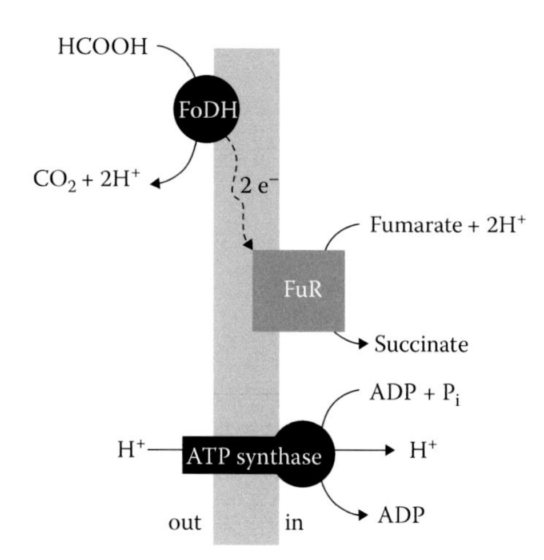

FIGURE 3.3 Energy-conserving pre-respiratory transmembrane electron transfer. Electron transfer from formate dehydrogenase (FoDH) to fumarate reductase (FuR), *chemically* coupled to the formation of a p.m.f. as "fuel" for the ATP synthase: model of the evolutionary origin of chemiosmotic energy conversion in the absence of a more advanced membrane-bound electron transport chain [12,13,14,27]. The scheme strongly resembles the extant fumarate respiration (Table 3.3), that is, the transmembrane formate-fumarate reductase complex.

It is biochemically significant that *all enzymatic* reactions of nucleoside phosphates (particularly adenylates) are absolutely dependent on Mg^{2+}, which is complexed by the nucleoside and nucleotide phosphates [25]. However, it is *always* and *irrevocably* ATP that forms the primary and exclusively *first* biochemically utilizable energy pool in all living cells. And the reason for this unique role of ATP (in contrast to GTP, CTP, or UTP, which do store practically the same amount of chemical energy in their anhydride γ- and β-phosphate bonds) must be sought not in a possibly peculiar constitution of ATP itself but in *evolving life* as such, viz., in the entirety of a living cell that has evolved according to the Darwinian principles of *struggle for life* and *survival of the fittest.* And the fittest cell that "had finally overcome" and evolved further on, that is, handed over its possibly even improved (genetic) identity to the next generation, for whatever—be it even a more or less statistical, population-dynamical—reason in detail, *by chance* was the one that used ATP as the metabolic energy source. This "decision- and pace-making power" of evolving life itself had already been pointed out by George Wald many decades ago and similar reasoning must apply, by the way, to other *unexplained biological dichotomies* ("one-sidednesses") such as the preponderance of L-amino acids (over D-amino acids) in natural proteins and of D-sugars (over L-sugars) in natural polymeric carbohydrates. This type of reasoning—which might almost be seen to ascribe some "*intelligent*" or "*teleological*" capacity to evolving life—does, of course, imply a strictly *monophyletic origin of life* in general. Irrespective of *lateral gene transfer* etc., hardly any biologist nowadays would like to question a monophyletic origin of life, which, by the way, is also elegantly supported by the *LUCA* theorem of Carl Woese and associates ("LUCA" = last universal common ancestor) (Figure 3.3) (see [28–35]). At any rate, the decision-making "switch-points" in the evolution of living cells, leading to the major unexplained *dichotomies* of Nature, most evidently must have been "governed" by the famous "chance" of Jacques Monod [36]. They bear convincing witness of the undeniable and all-embracing *contingency of our world and our lives* [37,38].

3.2.2 Biological Energy Conversion

The *Gibbs–Helmholtz equation* (Equation 3.4) dealt with the *driving force* of a (chemical or physical) reaction. The total or *free* environmental enthalpy difference, in physicochemical terms, the maximum reversible reaction work either performed by (–), or needed to effect (+), the reaction, is commonly symbolized by $\pm\Delta G$ and called the *Gibbs energy* or free enthalpy difference between the two states in question, viz., *after* and *before* the reaction. The ΔG tapped by living cells in order to sustain their life processes is determined by the Gibbs–Helmholtz equation, which comprises two terms: the difference of *heat contents* ΔH and the difference of the states of *disorder* = *entropy*, ΔS, *with* Δ being "*after* minus *before*," and it reads as follows:

$$\Delta G = \Delta H - T\Delta S \tag{3.4}$$

At thermodynamic equilibrium (no reaction) $\Delta G = 0$. In case of $\Delta G < 0$, the process (reaction) is *exergonic*, that is, energy-yielding, and in case of $\Delta G > 0$, it is *endergonic*, that is, energy-requiring.

ΔS represent the difference of degree of disorder, that is, the thermodynamical probability.

The flow of energy through a living cell, from an environmental energy *source* to the *work* given off, can best be seen from Figure 3.2B. Only two forms of environmental energy can be directly used by living cells, either radiation energy ($h\nu$) from sunlight (*photosynthesis*) or the energy released in the oxidation or "biological combustion" of food stuff. The primary energy from the environment is inevitably and universally first converted into, and temporarily stored in, ATP and the adenylate system (Figure 3.2). In each and all living cells, the biologically utilizable energy from the environment is always *first* captured in ATP through "phosphorylation," that is, the endergonic synthesis of ATP from ADP and P_i (Equation 3.1). Phosphorylation can be classified as (1) substrate-level phosphorylation (SLP) in the completely soluble and isotropic cytosolic system of a cell and (2) energetically efficient chemiosmotic electron transport phosphorylation (ETP) [17,18].

3.2.2.1 Substrate-Level Phosphorylation

The essence of SLP is *intramolecular redox reactions*. In SLP, the "missing link" between metabolic redox energy and ATP is various (soluble) phosphorylated "high-energy" intermediates [26] from which the P_i finally disembarks *exergonically* (Equation 3.1) on ADP to yield ATP (Table 3.1). Thus, membrane-bound processes are not needed for SLP. The well-established and almost colloquial term "high-energy or energy-rich compound" is always applied to organic high-energy *phosphates*. High-energy compounds show a strongly negative value for the Gibbs energy difference ($\Delta G^{o\prime}$) during their hydrolysis (Table 3.1). High-energy compounds are compiled in Table 3.1, with special reference to SLP [20]. The SLP is the evolutionarily most ancient mode of biological energy conversion as it functions, without participation of "advanced" membrane-bound ("vectorial") enzymes, in the soluble ("isotropic") system of the cytosol, in the dark and without obligate participation of electron transport processes [12,18,20]. No energy conversion is possible without *oxidation* and as no oxidation is possible without conjugated *reduction*, also SLP does imply *redox reactions* in the absence of light or an external oxidant (viz., e.g., O_2!). The SLP redox reactions are *intramolecular* as most clearly seen in glycolysis [19].

Cyanobacteria, combining all of the three bioenergetic processes—fermentation, respiration, and (oxygenic) photosynthesis—together in a single prokaryotic (bacterial) cell [39–41], are absolutely unique among living cells. But SLP, though potentially active in cyanobacteria in the form of, for example, lactic acid fermentation [39,42], plays a very marginal role. Not a single cyanobacterium is known that is able to grow in the absence of both light and oxygen. More than half of all cyanobacteria that have been successfully cultured so far [43] are obligate *photo*trophs, mostly even photo*auto*trophs. A large number of cyanobacterial species exclusively need light as an external energy source for growth and proliferation [44,45] while organic matter, primarily carbohydrates, may at least partly serve as carbon source. This *obligate phototrophy* [44–46] is to be compared with obligate lithotrophs and methylotrophs that need specific inorganic (or 1-carbon-) reductants to be taken up from the environment [45,47,48]. Chemo- and photo-litho*auto*trophs depend on CO_2 for carbon supply [45,47–49] metabolic reductants for CO_2, that is, ferredoxin [$E^{o\prime} = -420$ mV] and NAD[P]H [$E^{o\prime} = -320$ mV] (Tables 3.2 and 3.3).

TABLE 3.2
Energetics of (Mostly Aerobic) Chemolithoautotrophic Bacteria

Redox Couple	$E^{\circ\prime}$ (mV)	Chemical Reaction	$\Delta G^{\circ\prime}$ (kJ)	Bacteria
CO_2/CO	–513	$CO + 0.5O_2 \rightarrow CO_2$	–257.2	Carboxydo-
$2H^+/H_2$	–420	$H_2 + 0.5O_2 \rightarrow H_2O$	–239.3	Knallgas-
CO_2/CH_4	–244	$CH_4 + 2O_2 \rightarrow CO_2 + 2H_2O$	–818.0	Methylotrophs
SO_{42-}/H_2S	–223	$H_2S + 2O_2 \rightarrow H_2SO_4$	–805.3	Thiobacilli
SO_{42-}/S°	–205	$S + 1.5O_2 + H_2O \rightarrow H_2SO_4$	–593.6	Thiobacilli
NO_{2-}/NH_{4+}	+334	$NH_3 + 1.5\ O_2 \rightarrow HNO_2 + H_2O$	–281.5	Nitroso-
NO_{3-}/NO_2	+425	$NO_{2-} + 0.5O_2 \rightarrow NO_{3-}$	–76.1	Nitro-
Fe^{3+}/Fe^{2+}	+762	$Fe^{2+} + 0.25\ O_2 + H^+ \rightarrow Fe^{3+} + 0.5\ H_2O$	–4.35 (–44.3 at pH 0)	Thio-(ferro-)
NO_{3-}/N_2	+752	$N_2 + 2.5O_2 + H_2O \rightarrow 2NO_{3-} + 2H^+$	–65.6	Hypothetical

Molecular oxygen was taken as the standard acceptor throughout [$E^{\circ\prime}(O_2/H_2O)$ = +820 mV]. $E^{\circ\prime}$, standard reduction potential; $\Delta G^{\circ\prime}$, change in free enthalpy (Equation 3.4). For calculations: $\Delta G = -n.F.\Delta E$, ΔE, standard reduction potential of acceptor minus donor; n, number of transferred electrons; F = 96.5 kJ $mol^{-1}.V^{-1}$.

TABLE 3.3
Energetics of Several Types of (Aerobic and Anaerobic) Respiration with H_2 Taken as the Standard Donor Throughout [$E^{\circ\prime}(2H^+/H_2)$ = –420 mV]

Redox Couple	$E^{\circ\prime}$ (mV)	Chemical Reaction	$\Delta G^{\circ\prime}$ (kJ)	Bacteria
O_2/H_2O	+820	$O_2 + 2H_2,\ldots, 2H_2O$	–478.6	Knallgas
NO_{3-}/N_2	+740	$NO_{3-} + 2.5H_2 + H^+\ 0.5N_2 + 3H_2O$	–560.0	Nitrate resp.
NO_{3-}/NO_{2-}	+425	$NO_{3-} + H_2\ NO_2 + H_2O$	–163.2	Denitrification
Fumarate/succinate	+33	Fumarate + H_2 succinate	–87.4	Fumarate resp.
SO_{42-}/H_2S	–223	$SO_{42-} + 4H_2\ S^{2-} + 4H_2O$	–152.2	Sulfate resp.
S°/H_2S	–271	$S^\circ + H_2\ H_2S$	–28.8	Sulfur resp.
CO_2/CH_4	–244	$CO_2 + 4H_2\ CH_4 + 2H_2O$	–135.0	Carbonate resp.
CO_2/CH_3COO-	–285	$CO_2 + 2H_2\ 0.5\ AcOH + H_2O$	–52.3	Clostridium aceticum
N_2/NH_{4+}	–284	$N_2 + 3H_2 + 2H^+\ 2NH_{4+}$	–78.7	Hypothetical

For calculations: $\Delta G = -n.F.\Delta E$, ΔE, standard reduction potential of acceptor minus donor; n, number of transferred electrons; F = 96.5 kJ.$mol^{-1}.V^{-1}$.

3.2.2.2 Chemiosmotic ("Electron Transport") Phosphorylation

Compared to SLP, chemiosmotic ETP is energetically much more efficient [17,18]. Its essence is *membrane energization* (Figures 3.3 and 3.4) by an electrochemical ion potential gradient across the membrane, either $\Delta\mu_H^+$ or proton motive force (p.m.f.) [50–53]. ETP can be classified either as *photophosphorylation* (photosynthesis) or *oxidative phosphorylation* (respiration). Both types of phosphorylation are in essence ETPs [18]. However, in the former, the primary oxidative step, proper, for example, oxidation of H_2S or H_2O, is *endergonic*, viz., light-dependent (*photooxidations*), while in the latter the oxidation of the primary electron donor, for example, of NADH, is

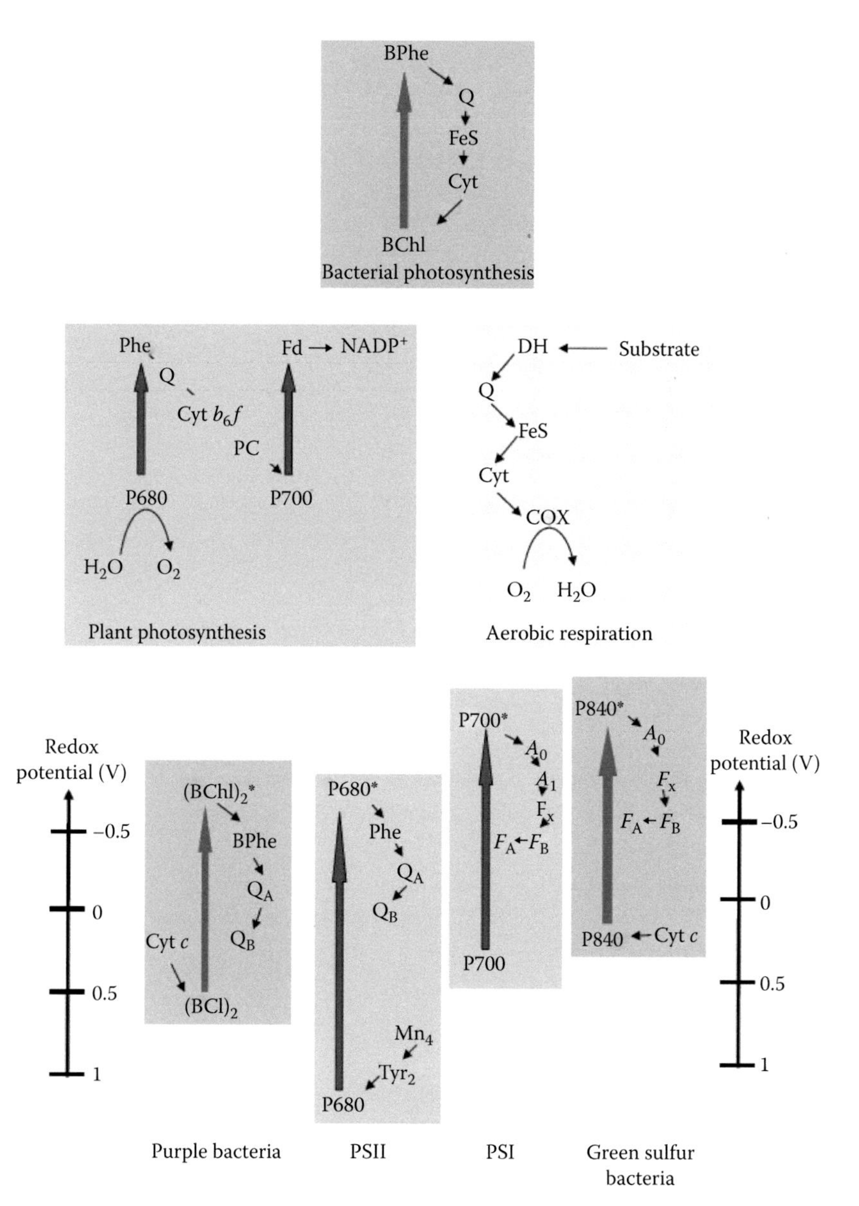

FIGURE 3.4 Evolutionary transition from anoxygenic photosynthesis to oxygenic photosynthesis and/or to aerobic respiration. Top: simplified linear scheme for the evolutionary transition from anoxygenic ("bacterial") photosynthesis as found in purple and green sulfur bacteria to oxygenic (cyanobacterial or plant-type) photosynthesis and/or to aerobic respiration as envisaged by the generalized endosymbiont hypothesis. Bottom: quantitative illustration of the redox potential "split-and-shift," which occurred upon transition from the primordially separated two bacterial photosystems to the two extant photosystems I (low potential, FeS-type RC) and II (high-potential, pheophytin-quinone type RC), hand in hand with an independent (convergent?) evolution of a WOC (OEC) as the "electron-switch" from water, from a di-Mn-containing catalase (see text and [35]). PSI, photosystem I; PSII, photosystem II; BChl, bacterial chlorophyll; Phe, pheophytine; BPhe, bacterial pheophytine; Q, quinol; Cyt, cytochrome; Cytb_6f, cytochrome b_6f complex; PC, plastocyanin; A_0 and A_1, electron acceptors of P700*; FeS, iron-sulfur cluster; F_x, F_A, F_B, types of 4Fe-4S clusters; Fd, ferredoxin; $NADP^+$, nicotinamide adenine dinucleotide phosphate; DH, dehydrogenase; COX, cytochrome c oxidase; Cyt$c$, cytochrome c.

exergonic and occurs also in darkness [18]. There is a single type of photophosphorylation that occurs without electron transport in the membrane. This is the retinal-based photophosphorylation in the purple membrane of halobacteria [54–57]. Halobacteria are photosynthetically capable of making some ATP in the light, however, are *non-phototrophic* as they are devoid of an autotrophic CO_2-reducing mechanism [55].

Both chlorophyll-based and halobacterial photosynthesis depend on a *chemiosmotic* type of ATP synthesis [18,51,52], needing an anisotropic, reversible, membrane-bound ATPase [27,58,59] (Figure 3.3). The simple vectorial transmembrane electron transfer, for example, from a membrane-bound formate dehydrogenase facing the outside to a fumarate reductase facing the inside of the cell, that is, "Mitchellian" reaction, is coupled to outside acidification and inside alkalinization, creating a p.m.f. across the membrane that serves as the driving force of a membrane-bound ATP synthase. Such mechanism represents a feasible primordial chemiosmotic system for energy conversion [12–14]. In general, the *free enthalpy* ΔG (kJ/mol) stored in a transmembrane electrochemical potential gradient is represented by Equation 3.5, with $\Delta\psi$ being the electrical membrane potential (defined as p-side minus n-side) and ΔpH being the pH difference (again p-side minus n-side). In the specific case of proton translocation, ΔG is usually expressed as the proton electrochemical gradient $\Delta\mu_H{}^+$ (with $\Delta\mu_H{}^+ = -F\Delta\psi + 2.3RT\Delta\text{pH}$):

$$\Delta G = -nF\Delta\psi + 2.3RT\Delta\text{pH} \qquad (3.5)$$

Mitchell has defined the term proton motive force (p.m.f. or Δp) in units of voltage (Equation 3.6). Using Δp and substituting values for R and T at 25°C gives Equation 3.7:

$$\Delta p[\text{mV}] = \frac{-[\Delta\mu H^+]}{F} \qquad (3.6)$$

$$\Delta p[\text{mV}] = \Delta\psi - 59\Delta\text{pH} \qquad (3.7)$$

Equation 3.7 shows the energetic equivalence of membrane potential [$\Delta\psi$] and ΔpH. A comprehensive survey of all existent chemiosmotic reactions is presented in Tables 3.2 and 3.3.

3.2.2.3 Oxidative Phosphorylation and Photophosphorylation

The largest amount of biologically synthesized ATP in our biosphere is the result of respiration and photosynthesis. Equation 3.8 shows the chemical reaction schemes of aerobic respiration and oxygenic photosynthesis revealing that both processes are represented by the same chemical equation, just to be read in opposite directions:

$$12H_2O + 6CO_2 \uparrow C_6H_{12}O_6 + 6O_2 + 6H_2O \quad \Delta G^{o\prime} = \pm 2821.5 \text{ kJ/mol} \qquad (3.8)$$

Equation 3.8, also known as water–water cycle [60], is kind of "life's token" on earth. This equation shows that oxidation of H_2O to O_2 shares important steps in common with the reduction of O_2 to H_2O [61,62].

Ultimately, all primary production on earth is energetically driven by (sun) light. Photosynthetically an estimated 10^{11} tons of carbon per year is converted into biomass by plant-type photosynthesis (to which the primordial cyanobacteria have given birth to, approximately 3.2 billion years ago [60,63,64]; and the equivalent amount of O_2 is thereby released from water according to Equation 3.8). Recent estimates assign between 20% and 30% of this worldwide primary production to cyanobacteria, particularly to small unicellular marine *Synechococcus* species [65] and likewise planktonic prochlorophytes [66], which are nevertheless extremely widespread in *all* oceans. On a more global scale the total biological and geological O_2-cycle on earth, together with respective pool sizes (H_2O, CO_2, and O_2) is described in more detail in [154], also see [12,35, and 92].

3.2.3 Photosynthetic Oxygen Evolution

It had been the cyanobacteria or their immediate ancestors that had "invented" the *oxygenic photosynthesis*. More or less simultaneously, the water-oxidizing or oxygen-evolving complex (WOC or OEC) together with a photosystem II (pheophytin-quinone-type reaction center [RC]) must have evolved. They provided for the so-called *biological Big-Bang* in evolution [67,68] and thus paved the path for all other "higher forms" of life, from simple eukaryotes (protozoa) up to "*Homo sapiens sapiens*."

3.2.4 Evolution of Photosynthetic Reaction Centers

In parallel to the evolution of the WOC, the evolution of a two-photosystem- (2-RC-) photosynthetic apparatus [12] had occurred (Figure 3.4). Under the impact of an increasing oxygen partial pressure in the atmosphere, the primordial single photosynthetic system of purple and of green photosynthetic bacteria underwent a further redox potential split or *disproportionation* into a *high-potential* and into a *low-potential* photosystem in a single cell.

In the course of further evolution, the two individual photosystems, "pre-PSI and pre-PSII," have merged into a full-fledged cyanobacterial (plant-type) Z-scheme [21,69–71]. At the same time, by the same path, a "respiratory chain" might have "split off" the common branch between ("pre") PSII and ("pre") PSI, now with an "alternative" cytochrome *c* oxidase (COX) as a terminal electron acceptor (schematically described in Figure 3.4). A similarly *convergent* evolutionary path may lead the evolution of cyanobacterial cytochrome c_6 and plastocyanin [72,73], the cyanobacterial C-phycocyanin family [74], and several other soluble cyanobacterial electron transport catalysts [73].

3.3 CYANOBACTERIAL PHOTOSYNTHETIC AND RESPIRATORY ELECTRON TRANSPORT

The much higher energetic efficiency and mechanistic versatility of ETP have by far "won the race" [18] from SLP. Versatile membrane-bound electron transport assemblies have freed the organisms from the slavish dependence on particular phosphorylatable substrates (Table 3.1). *Cyanobacteria* were among the first that have supplemented and streamlined the primordial chemiosmotic transmembrane electron transfer system as shown in Figures 3.4 and 3.5 with additional membrane-bound electron carriers such as, first of all, an electron/proton-transporting quinone [12–14,27,75].

3.3.1 Photosynthetic Electron Transport

With the exception of *Gloeobacter violaceus*, all cyanobacteria contain two physically separated intracellular membrane systems, the (yellow) chlorophyll-free cytoplasmatic or plasma membrane and the green chlorophyll-containing thylakoid or intracytoplasmic membrane (ICM) (Figure 3.5). Either of the two encloses an osmotically autonomous compartment, the cytosol (surrounded by the CM) and the intrathylakoid or luminal space (surrounded by the ICM).

In cyanobacteria, all components of PET are located at or associated with ICM (Figure 3.5). Photosynthetic electron transport starts with capturing of the light by antenna systems, which transfer the excitation energy to RCs that, in turn, catalyze light-induced electron transfer across the thylakoid membrane. The highly efficient job of capturing light energy by chlorophyll *a* molecules is not enough to supply all the needed light. Cyanobacteria have partially solved this problem by

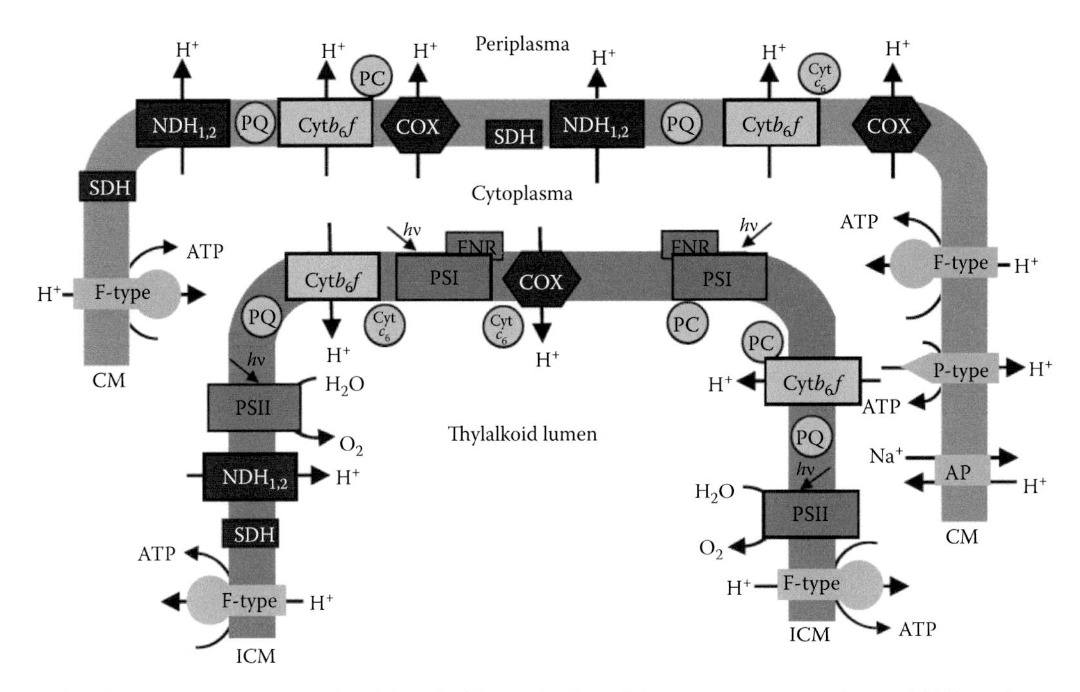

FIGURE 3.5 General structural and functional organization of electron transport carriers and ATP synthases in the membrane systems of a model cyanobacterium. Electron donor and acceptor as well as protein complexes involved exclusively in PET are depicted by PSI, PSII, and FNR, those found only in RET by SDH, NDH, and COX. Electron components used in both PET and RET. CM, cytoplasmic (plasma) membrane(s); ICM, intracytoplasmic (thylakoid) membrane(s); PSII and PSI, photosystems II and I; PQ, plastoquinone; Cytb_6f, cytochrome b_6f; PC, plastocyanin; Cytc_6, cytochrome c_6; FNR, ferredoxin-NADP$^+$ reductase; NDH_1, (mitochondrial-like) NDH-1 enzyme and NDH_2, (nonproton-pumping) NDH-2 enzyme; SDH, succinate dehydrogenase; COX, cytochrome *c* oxidase; proton-pumping P-type ATPase (ATP hydrolase) and F-type ATPase (ATP synthase) and AP, the Na^+/H^+ antiporter; are also shown. Note that neither in cyanobacteria nor in mature chloroplasts a plain structural continuity between CM and ICM could so far be discovered.

utilizing the phycobilisome (PBS) located on the stromal side of ICM. The PBS is water-soluble antenna complex of fluorescent, pigmented proteins associated with PSII. PBS is responsible for as much as 50% of protein mass within a cyanobacterial cell.

The PBS consists of a base consisting of three cylinders attached to ICM, and several rods that protrude from it in a fan-like arrangement. The rods are trimers or hexamers of phycoerythrin, phycocyanin, and allophycocyanin. Although each protein directly absorbs light (within 450 and 660 nm), it will either pass on or absorb energy to or from one of the other proteins with an established order of energy transfer: phycoerythrin ∀ phycocyanin ∀ allophycocyanin ∀ photosystem chlorophyll. Phycoerythrocyanin may replace phycoerythrin in certain organisms [76]. In addition to the membrane extrinsic PBSs, cyanobacteria also have membrane intrinsic antenna complexes belonging to the Pcb family [77].

Photosystem II (PSII) catalyzes light-driven electron transfer from water to PQ and has the same dimeric oligomeric structure in plants and cyanobacteria. It consists of 17 protein subunits, which bind 35–40 chlorophylls and 8–12 carotenoids [78–82]. The subunits are named according to their genes PsbA to PsbO, PsbU, PsbV, and PsbX. Light for the photoreactions in PSII is captured by the PBS and the core antenna of PSII [79]. Excitation energy is transferred from the antenna into the center of the complex. Each PSII monomer harbors the electron transfer chain consisting of PsbA (D1) and PsbD (D2). The structural similarity between the central core of PSI and PSII is remarkable, and clearly suggests that all photo-RCs have evolved from a common ancestor as previously proposed [42,83]. The cofactors of the ETC are four chlorophyll *a* molecules, two pheophytins, two PQs, two tyrosines (TyrZ and TyrD), and the manganese cluster.

P680 has the highest redox potential reported for any cofactor in biological systems and this is the thermodynamic prerequisite for water oxidation. The four chlorophylls are arranged in two symmetrically related pairs and all may be able to perform the initial charge separation across ICM. In each charge separation event, $P680^+$ is re-reduced by extracting one electron from the OEC (Mn cluster, WOC or OEC) consisting of four manganese ions and a calcium ion. The structure of the Mn cluster is still unsolved. A redox active tyrosine functions as the intermediate electron carrier between $P680^+$ and the Mn cluster. One molecule of O_2 is evolved after four positive charges have been accumulated. During this process, four protons are released into the lumen of the thylakoids. From P680, the electron is transferred to pheophytin and via tightly bound PQ_A to mobile PQ_B at the Q_B-side that serves as the terminal electron acceptor in PSII. After two charge separation events, PQ is doubly reduced, takes up two protons from the stroma, leaves PSII as PQH_2, and diffuses through the membrane to the b_6f complex. A PQ from the PQ-pool re-fills the Q_B site (Figure 3.5).

Electrons from mobile Q_B are transferred to the cytochrome b_6f complex that is also located in ICM within proximity to PSII and PSI. Electrons move internally through the cofactors of the cytochrome b_6f complex to one of the luminal mobile electron carriers, such as plastocyanin (PC) or cytochrome c_6 (Cytc_6) (Figures 3.4 and 3.5). Cytochrome b_6f complex (Cytb_6f) is a dimeric multiprotein complex with each monomer consisting of eight subunits and several cofactors ligated by the protein. The cofactors are cytochrome *f*, a 2Fe-2S-cluster (Rieske iron-sulfur protein) and cytochrome b_6. Electrons are transferred from mobile PQ having two protons added in its doubly reduced form. Unlike in the photosystem proteins, the two electrons do not take the same path. The first electron moves through the high-potential transfer chain including the Rieske protein and cytochrome *f* both located at the lumenal side of the ICM. The second electron is transferred by two *b*-type hemes (referred to by their absorbance maxima b_{562} and b_{565} or b_n and b_p, with the subscript referring to whether the particular heme is closer to the *n*-side [stromal] or *p*-side [luminal] of ICM) to the stromal side of the membrane. Alternatively, it may also be moved to the anionic plastosemiquinol if cyclic electron transfer is taking place. In either case, an extra proton is drawn from the stromal side and either released to the lumen or taken up by plastoquinol. This “Q-cycle” may be compared to the electron transfer cycle in cytochrome bc_1; however, in spite of the functional similarity, Cytb_6f and cytochrome bc_1 are only distantly related in structural terms. In any case, in cyanobacterial membranes, the b_6f-complex shares immunological as well as inhibition characteristics with the mitochondrial bc_1 complex [84–87].

In connection with Cytb_6f, the surprising discovery of a novel semi-*c*-type cytochrome, termed cytochrome *x* or c_n, deserves special attention. Its heme group is covalently attached to the protein via only one thioether bond and the fifth and sixth coordination sites of its iron ion are H_2O-only and free, respectively. Cytochrome c_x resides between heme b_n and PQ on the stromal side of ICM in close vicinity to a conspicuous positively charged amino acid region of the b_6f complex. Cytochrome c_x is a highly plausible candidate for the long-sought-for ferredoxin:PQ oxidoreductase necessary for cyclic endogenous PET around PSI [88]. Cytochrome *f* is the largest subunit of Cytb_6f and has the largest membrane-extrinsic domain projecting into the lumen. This domain contains the *c*-type heme ligated by a histidine and a tyrosine and covalently bound with the protein. Cytochrome *f* donates electrons to either plastocyanin (PC) or cytochrome c_6 (Cytc_6), both being small soluble electron carriers present on the luminal side of ICM (Figure 3.5).

While higher plants and many cyanobacteria utilize solely PC, in some green algae and cyanobacteria PC may be functionally substituted by cytochrome c_6 [72]. Despite their very different structures, both proteins share a number of physicochemical properties: their molecular masses are about 10 kDa and their midpoint redox potential values are around 350 mV at pH 7, work as electron shuttle between cytochrome b_6f and PSI [89].

Both PC and Cytc_6 approach PSI through either electrostatic or hydrophobic interaction and docks in a groove on the luminal side of PSI and donates its electron to the photooxidized chlorophyll dimer, P700$^+$ [88]. PSI is a light-driven oxidoreductase that actually transfers the electron from the luminal electron carrier to ferredoxin on the stromal side. That particular process supplies electrons for subsequent reduction of NADP$^+$ to NADPH by the ferredoxin-NADP$^+$ reductase (FNR) (Figure 3.5). It also serves generation of a proton electrochemical gradient across the thylakoid membrane. PSI has 12 subunits (PsaA-PsaF, PsaI-PsaM, PsaX) and 127 cofactors (chlorophylls *a*, phylloquinones, iron-sulfur clusters, carotinoids). PsaA and PsaB are the largest subunits making up the core of the (heterodimeric) RC and provide the binding side for PC or Cytc_6 as well as two conserved tryptophans important for electron transfer to P700H$^+$. The entire functional PSI is usually found in the ICM as a trimer of these individual heterodimeric RCs. In total, each of the PSI trimer's monomers contains 95 Chl*a* molecules with six actually belonging to the ETC.

The final cofactor in PSI that actually transfers electrons to the electron acceptor proteins, ferredoxin or flavodoxin, is F_B. Ferredoxin is a small (10.6 kDa) soluble, globular, and acidic protein that approaches PSI from the stromal side of the ICM. In cyanobacteria, ferredoxin (encoded by light-induced *fedI* gene) is characterized by a 2Fe-2S cluster of −420 mV, making reduced ferredoxin one of the strongest soluble natural reducing agents, also still not as strong as PSI. Ferredoxin or flavodoxin finally transfers the electron to the ferredoxin:NADP$^+$ oxidoreductase (FNR), a low-potential flavoprotein, which reduces NADP$^+$ to NADPH [90].

3.3.2 Respiratory Electron Transport

3.3.2.1 Origin of Aerobic Respiration

Striking molecular sequence similarity between aerobic COX and certain electron transport components of anaerobic nitrate respiration has given foundation of "respiration-early" hypothesis (see Table 3.3), including the presence of Cu_A in *Pseudomonas stutzeri* N_2O reductase. However, molecular similarities of aerobic and anaerobic respiratory electron transport (RET) components do not necessarily mean that the former have evolved from the latter. The same similarities would be found if the latter had evolved from the former [35,86]. "Reverse evolution" of aerobic respiration also seems to be valid for many representatives of Archaea [32,33,54–57]. Geological evidence lets it appear highly unlikely that before the advent of bulk O_2 in the biosphere substantial quantities of potential electron acceptors such as N−O compounds could have been available at all [91]. In contrast to oxygenated N-compounds, fumarate, CO_2, and SO_4^{2-}, rather than NO_3^-, would have been readily available for tinkering up some anaerobic RET system.

Comparisons between signature-based whole genome sequences of mitochondria, chloroplasts, and cyanobacteria strongly support the so-called conversion hypothesis. The similarity of the photosynthetic apparatus of a free living cyanobacterium and a chloroplast is more than striking in both functional and structural respects. Additionally, there is close mechanistic and structural similarity between respiration and photosynthesis in general (cf. Equation 3.8), as also analyzed in detail by the conversion hypothesis [12–14]. This has led to the conclusion that both respiration and photosynthesis might share a common evolutionary (viz., a *monophyletic*) origin. Also, the discovery of thylakosomes in several protists such as *Psalteriomonas lanterna* and representatives of the parasitic genus *Apicomplexa* [92] is in line with a "generalized endosymbiont hypothesis" according to which both chloroplasts and mitochondria might have started as endosymbiotically engulfed cyanobacteria.

3.3.2.2 Cyanobacterial Respiratory Electron Transport

Table 3.4 presents a survey of COX and reductase activities of cytoplasmic and intracytoplasmic membranes. It seems appropriate here to compare these rates with whole cell *photosynthetic* rates (O_2 release = CO_2 fixation). The physiological significance of electron transport rates measured in isolated membranes compared with intact cells may be assessed from the following typical rates

TABLE 3.4
Survey of Cytochrome c Oxidase (COX) and Reductase Activities of Cytoplasmic (Plasma, CM) and Intracytoplasmic Membrane (ICM) Separated and Purified from 19 Cyanobacterial Species Followed Spectrophotometrically

	COX		Cytochrome c Reductase				Apparent Molar Mass from SDS-PAGE (kDa)		Molar Mass Calculated from Genes (kDa)		Number of Amino Acids/cIP	
			CM		ICM							
Cyanobacteria	CM	ICM	+NAD	+NADPH	+NADH	+NADPH	SUI	SUII	SUI	SUII	SUI	SUII
Synechococcus sp. PCC7003[a]	30	16	110	38	5	0.1	40	25	—	—	—	—
Synechococcus sp. PCC6715[b]	32	35	—	5	1	10	34	14–19	—	—	—	—
Synechococcus elongatus PCC6301 (*Anacystis nidulans*)	60	3	20	3	5	3	46–55	29–32	60	34	541/8.6	315/5.6
S. elongatus PCC7942 (*Anacystis* R2)	52	8	18	10	3		45–52	30–35				
Synechocystis sp. PCC6714	2	81			20	23	34–36	21–23				
Synechocystis sp. PCC6803	3	90	553	10	20	16	34–36	21–23	61	6	551/5.8	332/4.6
Microcystis sp. PCC7005 (*Synechocystis* sp. PCC7005)	5	85	43	28	22	18	34–38	22–25				
Gloeobacter violaceus PCC7421	20	—	0.3	0.1	—	—	52	25	62	33	560/6.6	308/6.4
Spirulina sp. PCC6313	7	78	93	34	17	13	46–50	44	—	—	—	—

(*continued*)

TABLE 3.4 (continued)
Survey of Cytochrome c Oxidase (COX) and Reductase Activities of Cytoplasmic (Plasma, CM) and Intracytoplasmic Membrane (ICM) Separated and Purified from 19 Cyanobacterial Species Followed Spectrophotometrically

	COX		Cytochrome c Reductase				Apparent Molar Mass from SDS-PAGE (kDa)		Molar Mass Calculated from Genes (kDa)		Number of Amino Acids/cIP	
			CM		ICM							
Cyanobacteria	CM	ICM	+NAD	+NADPH	+NADH	+NADPH	SUI	SUII	SUI	SUII	SUI	SUII
Oscillatoria sancta PCC7515	66	9	23	9	32	11	49	35				
Prochlorothrix	145	20	23	0.1	0.1	0.1	50–51	26				
Chlorogloeopsis fritschii PCC6718	48	12	33	13	9	19	40	33				
Leptolyngbya PCC73110 (*Plectonema boryanum* IAM M-129	72	5	19	2	2	4	48–50	42–43				
Calothrix desertica PCC7102	58	13	33	13	44	14	48–51	19–25				
Fischerella muscicola PCC7414	83	34	22	6	10	6	43–47	14–18				
Anabaena variabilis PCC7937/ ATCC29413 Veg	35	46	430	17	5	0.1	48–49	36	64	39	575/6.1	355/4.8
Het	350	2400	170	26	20	0.1			62	35	559/6.8	324/4.9

Nostoc	Veg	35	42	380	68	4	0.1	48–51	36	64	39	580/6.0	361/7.8
punctiforme	Het	530	4200	28	16	13	5			62	35	562/7.3	329/5.0
PCC73102													
Nostoc sp.	Veg	38	50	450	20	8	0.1	50	36	64	39	575/6.0	355/4.8
PCC7120	Het	420	3500	195	48	17	—			62	35	559/6.8	327/5.0
Nostoc sp.	Veg	5	33	370	42	—	3	50	36				
PCC8009	Het	62	320	12	53	39							

Oxidase and reductase activities are expressed as nmol horse heart ferro(ferri)cytochrome *c* oxidized (reduced) per min per mg membrane protein. Rates of O_2-reduction were measured also polarographically being exactly four times the corresponding rates of spectrophotometrically determined rates. All cytochrome c reductase activity was abolished after extraction (by *n*-pentane or *n*-octane) or destruction (by appropriate UV irradiation) of the electron transport quinone, viz., of the PQ-9. In addition, apparent molar masses (kDa) of subunits I and II of COX in isolated and purified CM and ICM are shown. Apparent molar masses were obtained from SDS-PAGE/Western Blot analysis and identification of the proteins by means of polyclonal antisera raised against subunits I and II (and the holoenzyme) of COX from *Paracoccus denitrificans* and beef heart mitochondria. Moreover, the Table depicts molar masses calculated from *cox* genes (Table 3.6), number of amino acids, and calculated isoelectric points, cIPs.

In *Nostocales* (i.e., heterocysts forming diazotrophic cyanobacteria), cells were also grown with N_2 as the sole N-source, heterocysts were isolated and purified and CM and ICM thus were individually obtained from both vegetative cells and heterocysts and used separately for COX and reductase assays. Data are from all pertinent experiments performed in our laboratory since 1971 (at least 10 independent membrane preparations and measurements per strain, standard deviations always ranging within 20% of the corresponding mean). Reproducible detection by both spectral and polarographic techniques of rates <0.1 nmol/min/mg protein is rather unreliable. "–", not determined; PCC, Pasteur Culture Collection, Paris, France; SAG, Sammlung von Algenkulturen, Göttingen, Germany; ATCC, American Type Culture Collection.

[a] *Coccochloris elabens*, sometimes also named *Agmenellum quadruplicatum*; halophilic, grown in 0.5 M NaCl.

[b] *Synechococcus lividus*; thermophilic, grown at 53°C A functional COX in terms of cytochrome c oxidizing activity and immunological cross reaction has been determined in addition in *Synechococcus* WH7803, *Anabaena (cylindrica)* PCC7122, *Plectonema* ATCC73110, *Oscillatoria brevis*, *Microcystis* PCC7813, *Gloeothece* PCC6501, *Gloeothece* PCC6909, *Cyanothece* PCC7822, *Dermocarpa* PCC7438 (*Chroococcidiopsis*), *Nostoc* ATCC8009 (*Mac*), *Nostoc* (*muscorum*) PCC6314, *Nostoc circinalis*, *Phormidium* PCC6409, *Phormidium laminosum*, *Prochlorothrix hollandica* (endocyanelle of *Cyanophora paradoxa*), *Anabaenopsis siamensis* (SAG), *Chlorogloeopsis fritschii* (SAG), *Dermocarpella* sp. L. (SAG), *Cylindrospermum* sp. K. (SAG) (PCC7326), *Spirulina* (*Arthrospira*) *platensis* (SAG), *Aphanizomenon flos-aquae* (SAG).

of *Anacystis nidulans* (*Synechococcus* sp. PCC6301: 15–35 nmol NAD[P]H or succinate oxidized per min/mg membrane protein (i.e., 2× nmol horse heart) ferricytochrome *c* reduced per min/mg membrane protein). Concomitant O_2 uptake by these membranes measured polarographically in the dark was determined to be 10–20 nmol O_2/mg protein, which compares with an O_2 uptake by intact cells of 130–150 nmol O_2/h.mg dry wt. under identical conditions (approximate correlations: 1 mg dry wt. of cells = 0.5 mg protein = 0.25 mg (crude) membrane protein = 0.02 mg chlorophyll). These data clearly show that rates of RET in isolated cyanobacterial membranes (measured spectrophotometrically or polarographically) are fully compatible with rates of *in vivo* dark oxygen uptake by intact cells (30°C–35°C).

Cyanobacterial RET comprises the following five invariant components (from low (–) to high (+) potential): one or more dehydrogenases, an electron transport quinone (lipid-soluble mobile carrier), a Cyt (*bc*-type) b_6f complex, PC and/or Cytc_6 (water-soluble mobile carrier), and a terminal respiratory oxidase (TRO) as the final electron acceptor ("electron sink").

For chemiosmotic energy conversion [50,93], the membranes (i.e., CM and ICM) form a closed, osmotically autonomous compartment and the same membrane also possesses a reversible F_0F_1-ATPase or ATP synthase (F-type ATPase in Figure 3.5) of appropriate orientation to catalyze the endergonic synthesis of ATP from ADP and P_i. This enzyme is the most strictly conserved biochemical device in the whole biosphere and certainly was present even before any sort of electron transport was "invented" [13,14,27,94]. Figure 3.5 depicts the topology of ATP synthase in ICM and CM underlying its important role of energy transduction in both PET and RET. Structurally ATP synthase is very similar in all organisms though eukaryotic mitochondrial ATP synthases are much more complex than the prokaryotic and chloroplast enzymes [95–97] comprising two structurally and functionally distinct domains: a membrane-intrinsic proton translocation system (the F_0 part) and a membrane-extrinsic catalytic domain (the F_1 part). F_1 consists of a hexamer of three α- (with three regulatory nucleotide-binding sites) and three β-subunits (with three catalytic binding sites of different nucleotide occupancy), as well as a γ-subunit located asymmetrically in the middle of the hexamer. The F_0 part consists of three different proteins: one subunit *a*, two subunits *b*, and 10–15 subunits *c*. F_1 and F_0 are structurally and functionally coupled by at least two stalks. The enzyme works as a molecular motor. Ion flux through the F_0 part drives rotation of the rings of subunit *c*, coupled by the central stalk to the rotation of the γ-subunit, which drives ATP synthesis at the catalytic sites.

As to the low-potential end of RET, the physiological functions of both a photosynthetic and a respiratory hydrogenase had been described in non-nitrogen-fixing cyanobacterium *A. nidulans* (*Synechococcus* sp. PCC6301) many years ago already [98–103], this was later confirmed by genome analysis. Hydrogenases are extremely widespread and quite randomly distributed among all bacteria including cyanobacteria (for reviews, see [104–106]). In dinitrogen-fixing cyanobacteria, hydrogenases are useful for the recycling of the electrons inevitably going to H^+ instead of to N_2 [107]. It should be noted in this context that hydrogenases *sensu stricto* metabolize H_2 without hydrolysis of ATP while H_2 production by nitrogenase needs on an average two ATP/e^- transferred from the Fe- to the MoFe-protein. It is suggested that a "primordial" energy-requiring hydrogenase might, in the course of evolution, have evolved into a nitrogenase [108–110]. Characteristically, both hydrogenase and nitrogenase are low-potential iron-sulfur enzymes operating at around $E^{o\prime} = -420$ mV and both are severely inactivated in the presence of free O_2.

The occurrence of a "mitochondrial" energy-transducing, multi-subunit NADH dehydrogenase in both CM and ICM of cyanobacteria as well as in chloroplasts (i.e., in *oxyphototrophs*) was first described by Steinmüller and associates [111,112]. However, in all oxyphototrophs, three of the 14 NDH-1 subunits that form the minimal functioning complex I [112,113], viz., NuoE, F, and G [112], are not known. Cyanobacteria synthesize two quite different types of NADH dehydrogenases: the classical multi-subunit NDH-1 enzyme pumps protons, uses FMN and several FeS clusters as co-enzymes, oxidizes both deamino-NADH and NAD[P]H, and is strongly inhibited by rotenone or piericidin A (this enzyme is marked NDH_1 in Figure 3.5). The alternative NDH-2 enzyme (see NDH_2 in Figure 3.5) usually consists of one subunit only, does not pump protons, utilizes FAD instead of FMN, and is devoid of FeS clusters. The NDH-2 enzyme does not oxidize NADPH [114] nor deamino-NADH while the cyanobacterial NDH-1 enzyme may oxidize *both* NADH and NADPH [115,116].

A further remarkable peculiarity of cyanobacterial RET is the practically exclusive role of PQ as the lipid-soluble mobile carrier (Figure 3.5). No ubiquinone or menaquinone has ever been detected. The next speciality of cyanobacterial RET is the cytochrome b_6f complex. As has been mentioned already, b_6f complexes from the cyanobacterium *Mastigocladus laminosus* [88] and from *Chlamydomonas reinhardtii* chloroplasts [117] have been structurally resolved in all details quite recently. However, surprisingly, the cyanobacterial b_6f complex occurs in both ICM and (chlorophyll-free!) CM [118,119] (Figure 3.5). It is strongly inhibited by the classical complex III inhibitor antimycin *A*, which normally does not at all affect canonical b_6f complexes [120,121]. Extensive reviews on structure–function relationships of Cyt*bc* and b_6f complexes can be found in Refs [122–125].

In RET downstream of Cytb_6f, extrinsic, water-soluble mobile carriers, either plastocyanin or cytochrome c_6, transport electrons to TRO[s]. Quinol oxidases are also found in cyanobacteria; however, their physiological role is still under discussion. Contrary to previous claims [126], at least either Cytc_6 or PC is absolutely indispensable for integral PET and RET [90,127], and in cyanobacteria both proteins are capable of reducing P700 of photosystem I and the COX [59,86,128–134]. Cyanobacterial genomes sequencing have given astonishing evidence for a variety of several different terminal respiratory oxidases (TROs), sometimes even up to three different TROs in one and the same species (Table 3.5) [135]. However, as most of those species are *obligate photoautotrophs,* they are incapable of making major ecological use of respiration for growth and proliferation.

Heme-copper oxidases play a key role in aerobic RET. They are redox-driven proton pumps that couple the four-electron reduction of molecular oxygen to water to the directional translocation of protons across the membrane. They contain a heme-copper binuclear RC comprising a Cu_B electronically coupled with a high-spin heme. Additionally, all heme-copper oxidases contain a low-spin heme in subunit I [136]. Depending on the nature of electron donors, the superfamily of heme-copper oxidases is divided into two branches, namely, cytochrome *c* oxidases (COX, aa_3-type cytochrome oxidase) and quinol oxidases (QOX, *bo*-type quinol oxidase) [137–139]. COX has been found in all cyanobacteria both at genomic and protein level. Thus, the occurrence of orthodox RET and of a proton-translocating aa_3-type COX in both CM and ICM of cyanobacteria is well established. COXs have been discovered spectrophotometrically in crude membrane preparations of *Anabaena* sp. [140], *Nostoc* MAC [141], and *A. nidulans* [14]. A gene cluster encoding subunits II, I, and III of COX from *Synechocystis* PCC6803 has been cloned [142–144]. Operons encoding subunits I–III in the same order have also been cloned from other species (for review, see Ref. [135]). Sequence alignment COX protein clearly demonstrates [135] that subunits I and II contain all redox cofactors and residues involved in O_2 reduction and proton pumping.

COX subunit II includes a typical hairpin-like structure comprising two transmembrane helices and a peripheral domain on the *p*-side of the membrane, which later can be either CM or ICM. Subunit II of COX participates in both electron transfer and proton pumping. Electrons from mobile donor (i.e., PC or Cytc_6) are transferred to the Cu_A center, which is composed of two electronically coupled, mixed-valence copper ions [145]. In all COXs (including cyanobacterial), this center is binuclear and the strictly conserved ligands are located near the C-terminus of subunit II. The role of both PC and cytochrome c_6 as mobile electron donors for COX is not fully known. A recent investigation of two deletion mutants of *Synechocystis* PCC6803 (each mutant lacking either *petE* or *petJ*) and their photoautotrophic and heterotrophic growth rate in copper-free and copper-supplemented medium clearly demonstrated that respiration requires the presence of either Cytc_6 or PC [89]. It is suggested that cyanobacterial RET can use both PC and Cytc_6 as electron donor. Deletion of the operon that encodes COX in *Synechocystis* PCC6803 resulted in a mutant strain that was still viable under both photoautotrophic and photomixotrophic conditions. The strain respired at near wild-type rates, and this respiration was cyanide-sensitive [146]. Isolated membranes from the mutant were unable to oxidize reduced horse heart cytochrome *c*. This was the first evidence for a second cyanide-sensitive terminal respiratory oxidase in cyanobacteria. Although COX appears to be the

TABLE 3.5

ORFs and Genes of Cyanobacterial Terminal Respiratory Heme-Copper Oxidases (i.e., Cytochrome *c* Oxidases [COX] and *bo*-Quinol Oxidases [QOX]) and Non-Heme-Copper Quinol Oxidases, *bd*-QOX

Heme-Copper Oxidases					Genome					
COX	*bo*-QOX	*bd*-QOX	Cyanobacteria	Classification	Size (Mb)	N_2 Fixation	Heterocysts	PC	Cytc6	CytcM
1	0	0	*Prochlorococcus marinus* str. MIT9301	Pleurocapsales	1.6	No	No	1	1	1
1	1		*P. marinus* str. MIT9515	Pleurocapsales	1.7	No	No	1	0	0
1	0.90%		10 × *P. marinus*[a]	Pleurocapsales	1.7–2.7	No	No	1	0%–40%	1
	1.10%							1	1%–40%	
								1	2%–20%	
1[#]	0		*G. violaceus* PCC7421	Chroococcales	4.6	No	No	2	2	1
1	0		*Microcystis aeruginosa* NIES-843	Chroococcales	5.8	No	No	1	1	1
1	0		*S. elongatus* PCC6301	Chroococcales	2.7	No	No	1	3	1
1	0		*S. elongatus* PCC7942	Chroococcales	2.7	No	No	1	3	1
1	0.30%		14 × *Synechococcus* sp.[b]	Chroococcales	2.2–3.0	No	No	1	1%–10%	1
	1.70%							0	2%–50%	
								1	3%–10%	
								1	4%–30%	
1	1		*Synechocystis* sp. PCC6803	Chroococcales	3.6	No	No	1	1	1
1	0		*Thermosynechococcus elongatus* BP-1	Chroococcales	2.6	No	No	0	1	1

1	0	*Acaryochloris marina* MBIC11017		6.5	No	No	1	2	1
1	1	*Crocosphaera watsonii* WH8501*	Chroococcales	6.2	Yes	No	1	3	1
1	1	*Cyanothece* sp. ATCC51142	Chroococcales	4.9	Yes	No	1	2	1
1	1	*Cyanothece* sp. CCY0110*	Chroococcales	5.9	Yes	No	1	2	1
1	1	*Cyanothece* sp. PCC7424*	Chroococcales	6.4	Yes	No	2	2	1
1	0	*Cyanothece* sp. PCC8801*	Chroococcales	4.6	Yes	No	1	2	1
1	1	*Lyngbya* sp. PCC8106*	Oscillatoriales	7.0	Yes	No	1	2	1
1	1	*Trichodesmium erythraeum* IMS101	Oscillatoriales	7.7	Yes	No	1	2	1
2	2#	*A. variabilis* ATCC29413	Nostocales	6.3	Yes	Yes	1	3	1
2	1	*Nodularia spumigena* CCY9414*	Nostocales	5.3	Yes	Yes	1	4	1
2	1#	*N. punctiforme* PCC73102*	Nostocales	9.0	Yes	Yes	1	3	1
2	1	*Nostoc* (*Anabaena*) sp. PCC7120	Nostocales	6.4	Yes	Yes	1	2	1

In addition, the (putative) electron donors for COX are given, that is, plastocyanin (PC), cytochrome *c*6 (Cyt*c*6), and cytochrome *c*M (Ccyt*c*M) 44 completely or partially (*) sequenced strains were analyzed. The genome size is included.

[a] The *P. marinus* genus includes following strains: Pro9601, Pro9211, Pro9215, Pro9303, Pro9312, Pro9313, ProNATL1A, ProNATL2A, Pro1375, Pro1986.

[b] Following strains are included in the *Synechococcus* group: Syn107*, Syn9311, Syn9605, Syn9902, SynJA23, SynJA33, Syn7002, Syn307, Syn9916*, Syn9917*, Syn5701*, Syn7803, Syn7805*, Syn8102. # Genome analysis shows the presence of additional but incomplete operons for COX or QOX (absence of subunit III).

main terminal oxidase, there is also evidence of the presence and function of the other heme-copper oxidase superfamily (QOX) or *bo*-type quinol oxidase in many cyanobacteria. QOX is encoded by a set of genes very similar in sequence to the operon that encodes COX. QOX subunit I contains all redox cofactors and is homologous to the corresponding subunit in COX. In contrast to COXs, QOX subunit II has neither a binuclear Cu_A center nor a cytochrome *c* binding site [135,147]. Instead, heme *b* receives electrons directly from a membrane solubilized quinol molecule (plastoquinol in case of cyanobacteria?). The protons produced upon quinol oxidation are released on *p*-side of the membrane. Table 3.5 shows that many unicellular cyanobacteria lack the genes encoding QOX, whereas all nitrogen-fixing species have usually one operon. Transcription of operons encoding QOX has been detected in several species, and studies using deletion mutants and inhibitors proposed QOX to be active and to contribute to energy metabolism (for a review, see Ref. [135]). Additionally, genome analysis has indicated the presence of genes or ORFs for another terminal oxidase, namely, the *bd*-quinol oxidase [135], which is not a member of the heme-copper protein superfamily. It does not contain a binuclear reaction site and is incapable of proton pumping. However, since up to the present day not a single terminal respiratory oxidase of any cyanobacterium, except the conventional aa_3-type cytochrome-*c* oxidase, could as yet be characterized or isolated as a functional enzyme but only traced in the form of elusive genes, the potential role of all these enzymes ("alternative oxidases") in cyanobacterial electron transport must remain open (also see Table 3.6).

3.3.2.3 Interaction between PET and RET in Cyanobacteria

Figure 3.6 summarizes the present knowledge of branched RET in cyanobacteria and its interrelationship with PET. The dual-function PET-RET pathways in ICM downstream of plastquinol (PQ) share common membrane protein complexes and mobile electron carriers involved in both pathways (PQ, cytochrome b_6f, plastocyanin, and cytochrome c_6). Nowadays, the respective identities and locations of the various electron transport components are fairly well documented and proven [119,140,148–151]. An immediate interaction between PET and RET can only be expected for the ICM-contained electron transport system. Figure 3.6 shows this intramembrane "electron transport bottleneck," which must hold for all uncompartmentalized dual-function membranes of all prokaryotes capable of both respiration and photosynthesis, irrespective of the peculiar electron transport components involved. Since, in cyanobacteria, the "production of electrons" in the light from water-oxidation is much more rapid than from substrate dehydrogenation, results in a well-known "light-inhibition of respiration" in illuminated prokaryotes. Since the competition between PET and RET in a dual-function biomembrane is mutual, an inhibition of RET by PET can also be measured in special conditions. However, as the inhibition of photosynthesis by oxygen (known as "Warburg effect" [53] may result, in an intact cell from widely different molecular effects, for example, photorespiration, ribulose-bisphosphate carboxylase/-oxygenase competition, glycolate oxidase action, other more or less unspecific oxygenases, etc.). The competition between PET and RET, proper, can only be recognized as such in experiments on isolated and separated cyanobacterial membranes as had already been demonstrated in our laboratory in the early eighties (see Ref. 18).

Interaction between RET and PET is obvious under environmental stress conditions as well as nitrogen fixation. Respiratory membrane energization and ATP supply are crucial to stressed cyanobacteria, which usually shut down PET first. In *Synechococcus* PCC6301 exposed to a variety of seemingly unrelated stress conditions (e.g., high salt or light limitation), RET in CM was shown to be significantly enhanced and known as salt respiration. ICM-bound respiratory chain is much less affected under stresses. The amount of COX in CM (but not in ICM) increases in stressed cyanobacterial cells [153]. The increased respiration generates a proton gradient to drive H^+/Na^+ exchange. H^+/Na^+ antiporters and proton-pumping P-type ATPases have been identified in *Synechocystis* PCC6803 (Figure 3.5) [154]. Recently, the interrelation between RET and PET under conditions of photoinhibition and high light tolerance has been demonstrated in *Synechococcus* PCC7002 [155]. Both COX and QOX were reported to be dispensable for growth under normal to moderately high light conditions, whereas at more stressful conditions COX seems to be the primary oxidase in RET.

TABLE 3.6
List of All Mentioned Cyanobacterial Strains, Abbreviations, and Access Codes of All Genomes

Abbreviations	Sequenced Cyanobacteria	Accession Numbers
Aca.mar	*A. marina* MBIC11017	NC_009925
Ana.var	*A. variabilis* ATCC29413	NC_007413
Cro.wat	*C. watsonii* WH8501	NZ_AADV00000000
Cya51142	*Cyanothece* sp. ATCC51142	NC_010546
Cya0110	*Cyanothece* sp. CCY0110	NZ_AAXW00000000
Cya7424	*Cyanothece* sp. PCC7424	NC_011729
Cya8801	*Cyanothece* sp. PCC8801	NC_011726
Glo.vio	*G. violaceus* PCC7421	NC_005125
Lyn8106	*Lyngbya* sp. PCC8106	NZ_AAVU00000000
Mic.aer	*M. aeruginosa* NIES-843	NC_010296
Nod.spu	*N. spumigena* CCY9414	NZ_AAVW00000000
Nos.pun	*N. punctiforme* PCC73102	NC_010628
Nos7120	*Nostoc* (*Anabaena*) sp. PCC7120	NC_003272
Pro9601	*P. marinus* str. AS9601	NC_008816
Pro9211	*P. marinus* str. MIT9211	NC_009976
Pro9215	*P. marinus* str. MIT9215	NC_009840
Pro9301	*P. marinus* str. MIT9301	NC_009091
Pro9303	*P. marinus* str. MIT9303	NC_008820
Pro9312	*P. marinus* str. MIT9312	NC_007577
Pro9313	*P. marinus* str. MIT9313	NC_005071
Pro9515	*P. marinus* str. MIT9515	NC_008817
ProNATL1A	*P. marinus* str. NATL1A	NC_008819
ProNATL2A	*P. marinus* str. NATL2A	NC_007335
Pro1375	*P. marinus* subsp. marinus str. CCMP1375	NC_005042
Pro1986	*P. marinus* subsp. marinus str. CCMP1986	NC_005072
Syn6301	*S. elongatus* PCC6301	NC_006576
Syn7942	*S. elongatus* PCC7942	NC_007604
Syn107	*Synechococcus* sp. BL107	NZ_AATZ00000000
Syn9311	*Synechococcus* sp. CC9311	NC_008319
Syn9605	*Synechococcus* sp. CC9605	NC_007516
Syn9902	*Synechococcus* sp. CC9902	NC_007513
SynJA23	*Synechococcus* sp. JA-2-3B′a(2–13)	NC_007776
SynJA33	*Synechococcus* sp. JA-3-3Ab	NC_007775
Syn7002	*Synechococcus* sp. PCC7002	NC_010475
Syn307	*Synechococcus* sp. RCC307	NC_009482
Syn9916	*Synechococcus* sp. RS9916	NZ_AAUA00000000
Syn9917	*Synechococcus* sp. RS9917	NZ_AANP00000000
Syn5701	*Synechococcus* sp. WH5701	NZ_AANO00000000
Syn7803	*Synechococcus* sp. WH7803	NC_009481
Syn7805	*Synechococcus* sp. WH7805	NZ_AAOK00000000
Syn8102	*Synechococcus* sp. WH8102	NC_005070
Syc6803	*Synechocystis* sp. PCC6803	NC_000911
The.elo	*T. elongatus* BP-1	NC_004113
Tri.ery	*T. erythraeum* IMS101	NC_008

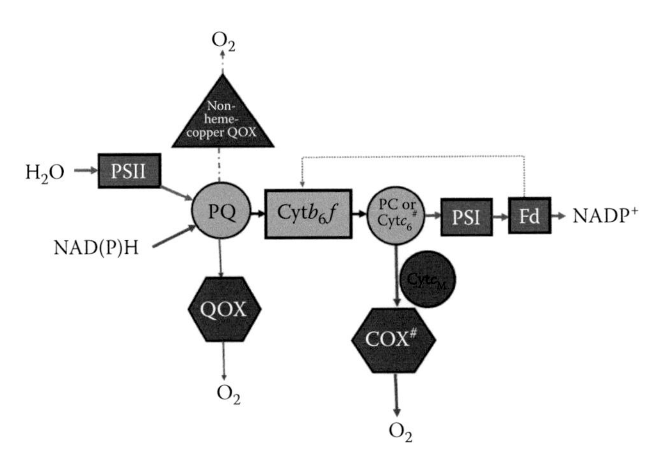

FIGURE 3.6 Scheme of the branched dual-function photosynthetic (PET)-respiratory (RET) electron transport chain(s) in cyanobacteria. Note that in all cyanobacterial species except *Gleobacter Violaceus* (that does not contain thylakoids at all) PET is only localized in intracytoplasmic membranes or thylakoids (ICM), whereas RET is active in both ICM and cytoplasmic membrane (CM). Electron donor and acceptor as well as protein complexes involved exclusively in PET are depicted by PSI, PSII, and Fd, those found only in RET by COX, QOX and CM. Electron components used in both PET and RET are shown by plastoquinone, the chimeric cyanobacterial cytochrome b.c complex (b_6f/bcl), and cytochrome c_6 or plastocyanin (PET and RET). In contrast to cytochrome *c* (aa_3-type) oxidase (COX, hexagon), which has been detected on a protein level in both CM and ICM, there is no unequivocal information about the localization of heme-copper (QOX, hexagon) and non-heme-copper quinol oxidases (triangle). Depending on copper-availability plastocyanin (PC) and/or cytochrome c_6 (Cytc_6) functions as electron donor for both, photosystem I (PSI) or COX. Another mobile electron carrier exclusively found in cyanobacteria (cytochrome c_M, Cytc_M) most probably is involved in electron shuttling between cytochrome b_6f (Cytb_6f) and COX during environmental stress. COX is the main terminal oxidase in RET, but—depending on growth conditions and nitrogen supply—can partially be substituted by QOX or non-heme-copper oxidases that both use plastoquinol-9 (PQ) as electron donor. In contrast to PC and Cytc_M, up to four paralogs for Cytc_6 are found in cyanobacterial genomes. In *Nostocales*, two paralogs for COX are found. Cyclic electron flow around PSI is shown in dashed line.

COX (and RET) plays an important role in protection against excessive reductive stress via diverting electrons from the reduced PQ pool to TROs (Figure 3.6).

A small *c*-type membrane-associated cytochrome c_M (Cytc_M in Figure 3.6) can reduce Cu_A in COX [156] and exclusively participate in RET. Under stress conditions, cyanobacteria not only shut down PET but also suppress the expression of both Cytc_6 and PC, whereas expression of *cytM* (that is scarcely expressed under normal growth conditions) is enhanced [157]. The interrelation between PET and RET is also evident during nitrogen fixation. Cyanobacteria are the only diazotrophs that produce oxygen as a by-product of PET. The biological reduction of N_2 is absolutely restricted to prokaryotes and is catalyzed by a multimeric metalloenzyme (nitrogenase), which is irreversibly inhibited by molecular oxygen and reactive oxygen species [111]. The rates of RET increase when cyanobacteria grow on atmospheric N_2 as the sole source. Moreover, RET efficiently lowers the concentration of dissolved O_2 and maintains nitrogenase in a virtually oxygen-free cellular environment with simultaneous satisfaction of the severe energy (ATP) requirement [148,158,159].

During the course of planetary evolution, cyanobacteria have co-evolved with the changing oxidation state of the ocean and atmosphere to accommodate the machinery of PET and RET and oxygen-sensitive N_2 fixation within the same cell and/or colony of cells. These strategies have been generally described as temporal and/or spatial separation of photosynthesis and N_2 fixation (for review, see [135]). The simplest adaptation is seen for the genus *Trichodesmium* (order Oscillatoriales) that fixes nitrogen during the day [160]. Here, oxygen protection is a complex interaction between spatial and temporal segregation of PET, RET, and N_2 fixation within the photoperiod.

In *Trichodesmium*, nitrogenase is compartmentalized in a fraction of the cells (10%–20%) that are often arranged consecutively along the trichome. However, active photosynthetic components are found in all cells, even those harboring nitrogenase [161]. Light initiates PET, providing energy and reductants for carbohydrate synthesis and storage, stimulating cyclic and pseudocyclic (Mehler) electron flow through PSI, and poising the PQ pool at reduced levels. Interestingly, RET has been shown to be active also early in the photoperiod at high rates supplying carbon skeletons for amino acid synthesis (the primary sink for fixed nitrogen) [161]. RET reduces the PQ pool further, sending negative feedback to linear PET. As a consequence, PSII becomes down-regulated and oxygen consumption exceeds oxygen production. This opens a window of opportunity for N_2 fixation during the photoperiod. As the carbohydrate pool is consumed, PET through the PQ pool diminishes. The PQ pool becomes increasingly oxidized, net O_2 production exceeds consumption, and nitrogenase activity is lost during the following day.

A full temporal separation, in which nitrogen is fixed only at night, is evident in other filamentous genera of the order Oscillatoriales as well as in unicellular members of the order of Chroococcales. These organisms fix nitrogen at night and nitrogenase is typically found in all cells. These organisms show extensive metabolic periodicities of PET, RET, and nitrogen fixation when grown under N_2-fixing conditions. Finally, a highly refined specialization is found in heterocystous cyanobacteria (Nostocales). In these organisms, nitrogenase is confined to a micro-anaerobic cell, the heterocysts [162], which differentiates completely and irreversibly 12–20 h after combined nitrogen sources are removed from the medium. Heterocysts are enveloped by two superimposed layers, one consisting of polysaccharides and the other of glycolipids, the latter slowing the diffusion of O_2. Heterocysts have lost the capacity of cell division, exhibit high PSI activity (but lack PSII) and increased RET. They rely on vegetative cells as source of carbon and reductant and, in return, they supply the surrounding vegetative cells with fixed nitrogen. Cyclic electron flow around PSI and RET supplies ATP. Interestingly, heterocysts prefer ICM as the site of RET, while unicellular N_2-fixing species prefer CM [159].

3.3.2.4 Conclusion

In conclusion, cyanobacteria are paradigmatic organisms of oxygenic photosynthesis and aerobic respiration in bioenergetic, evolutionary, and ecological respects. They have uniquely accommodated both a photosynthetic and an RET chain within a single prokaryotic cell. Since many cyanobacteria in addition are also capable of N_2 fixation, they are often called the bioenergetic "nonplus-ultra" among living beings. Cyanobacteria not only were the first oxygenic (plant-type) photosynthesizers but also the first aerobic respirers, with COX being the first functional aerobic terminal oxidase [35].

ACKNOWLEDGMENTS

This work was supported by the Austrian Science Funds (FWF-project P17928 at present). Devoted and invaluable technical assistance has always been provided by Mr. Otto Kuntner. We are thankful to Ms. Poonam Bhargava for assistance in manuscript preparation.

REFERENCES

1. Mereschkowsky, C., Über Natur und Ursprung der Chromatophoren im Pflanzenreiche. *Biol. Zentralbl.*, 25, 593, 1905.
2. Nierzwicki-Bauer, S. A., Balkwill, D. L., and Stevens, S. E., Three-dimensional ultrastructure of a unicellular cyanobacterium, *J. Cell Biol.*, 97, 713, 1983; Schmetterer, G., Peschek, G. A., and Sleytr, U. B., Thylakoid degradation during photooxidative bleaching of the cyanobacterium *Anacystis nidulans, Protoplasma*, 115, 202, 1983.
3. Liberton, M., Howard Berg, R., Heuser, J., Roth, R., and Pakrasi, H., Ultrastructure of the membrane systems in the unicellular cyanobacterium *Synechocystis* sp. strain PCC 6803, *Protoplasma*, 227, 129, 2006; Peschek, G. A. and Sleytr, U. B., Thylakoid morphology of the cyanobacteria *Anabaena variabilis* and *Nostoc* MAC grown under light and dark conditions, *J. Ultrastruct. Res.*, 82, 233, 1983.

4. Peschek, G. A., Structure–function relationships in the dual-function photosynthetic-respiratory electron-transport assembly of cyanobacteria (blue-green algae), *Biochem. Soc. Trans.*, 24, 729, 1996.
5. Margulis, L., Archaeal-eubacterial mergers in the origin of Eukarya: Phylogenetic classification of life. *Proc. Natl. Acad. Sci. U S A*, 93, 1071, 1996.
6. Peschek, G. A., Cyanobacteria viewed as free-living chloromitochondria: The endosymbiont hypothesis revisited. *Plant. Physiol. Biochem.*, 38, 266, 2000.
7. Peschek, G. A., Cyanobacteria viewed as free-living proto-chloromitochondria, in *Photosynthesis: Fundamental Aspects to Global Perspectives*. van der Est, A. and Bruce, D. (Eds.), The international Society of Photosynthesis, Toronto, Canada, 2005, p. 746.
8. Gupta, R. S., Protein phylogenies and signature sequences: A reappraisal of evolutionary relationships among Archaebacteria, Eubacteria and Eukaryotes. *Microbiol. Mol. Biol. Rev.*, 62, 1435, 1998.
9. Karlin, S. and Burge, C., Dinucleotide relative abundance extremes: A genomic signature. *Trends Genet.*, 11, 283, 1995.
10. Wang, Y. et al., The spectrum of genomic signatures: From dinucleotides to chaos game representation. *Gene*, 346, 173, 2005.
11. Sasikumar, R., Jijoy, J., and Peschek. G. A., Genome signature comparison points to a common cyanobacterial ancestry for mitochondria and chloroplasts, in *International Colloquium on Endocytobiology and Symbiosis*, Gmunden, Austria. Galambos, C. (Ed.), International Society of Endocytobiology [ISE], 2007.
12. Broda, E., *The Evolution of the Bioenergetic Processes*. Pergamon Press, Oxford, U.K., 1975, p. 211.
13. Broda, E. and Peschek, G. A., Did respiration or photosynthesis come first? *J. Theor. Biol.*, 81, 201, 1979.
14. Peschek, G. A., Phylogeny of photosynthesis and the evolution of electron transport: The bioenergetic backbone. *Photosynthetica*, 15, 543, 1981.
15. Nitschmann, W. H., Chemiosmotic coupling in cyanobacteria: ATP synthesis in a two-membrane system. *J. Theor. Biol.* 122, 409, 1986.
16. Nitschmann, W. H. and Peschek, G. A., Oxidative phosphorylation and energy buffering in cyanobacteria. *J. Bacteriol.* 168, 1285, 1986.
17. Baltscheffsky, H. and Baltscheffsky, M., Electron transport phosphorylation. *Annu. Rev. Biochem.*, 43, 871, 1974.
18. Peschek, G. A., Electron transport chains in oxygenic cyanobacteria, in *Primary Processes of Photosynthesis: Principles and Applications*, Vol. 2. Renger, G. (Ed.), *European Society of Photobiology*. RSC Publishing, Cambridge, U.K., 2008, p. 283.
19. Racker, E., *Mechanisms in Bioenergetics*. Academic Press, New York, 1965.
20. Thauer, R. K., Jungermann, K., and Decker, K., Energy conservation in chemotrophic anaerobic bacteria. *Bacteriol. Rev.*, 41, 100, 1977.
21. Nitschke, W. and Rutherford, A. W., Photosynthetic reaction centres: Variations on a common theme? *Trends Biochem. Sci.*, 16, 241, 1991.
22. Nitschmann, W. H. and Peschek, G. A., Modes of proton translocation across the cell membrane of respiring cyanobacteria. *Arch. Microbiol.*, 141, 330, 1985.
23. Peschek, G. A., Nitschmann, W. H., and Czerny, T., Respiratory proton extrusion and plasma membrane energization. *Methods Enzymol.*, 167, 361, 1988.
24. Gross, T. et al., *Introductory Microbiology*. Chapman & Hall, London, U.K., 1995.
25. Stryer, H., *Biochemistry*. W. H. Freeman & Company, San Francisco, CA, 1981.
26. Lipmann, F. A., Metabolic generation and utilization of phosphate bond energy, *Adv. Enzymol.*, 1, 99, 1941.
27. Raven, J. A. and Smith, F. A., The evolution of chemiosmotic energy coupling, *J. Theor. Biol.*, 57, 301, 1976.
28. Doolittle, W. F. and Brown, J. R., Tempo, mode, the progenote, and the universal root, *Proc. Natl. Acad. Sci. USA*, 91, 6721, 1994.
29. Doolittle, W. F., Phylogenetic classification and the universal tree, *Science*, 284, 2124, 1999.
30. Doolittle, W. F., The nature of the universal ancestor and the evolution of the proteome, *Curr. Opin. Struct. Biol.*, 10, 355, 2000.
31. Forterre, P., Where is the root of the universal tree of life? *Bioessays*, 21, 871, 1999.
32. Woese, C. R. and Fox, G. E., The archaebacteria—A third kingdom of life? *Proc. Natl. Acad. Sci. U S A*, 74, 5088, 1977.
33. Woese, C. R., Magrum, L. J., and Fox, G. E., Archaebacteria, *J. Mol. Evol.*, 11, 245, 1978.
34. Woese, C., Kandler, O., and Wheelis, M. L., The tripartite system of living cells. *Proc. Natl. Acad. Sci. USA*, 87, 4576, 1990.
35. Lane, N., *Oxygen, The Molecule that Made the World*. Oxford University Press, Oxford, U.K., 2002, p. 171.
36. Monod, J., *Le hasard et la necessite*. Editions du Seuil, Paris, France, 1970.

37. Dawkins, R., *The Selfish Gene*. Oxford University Press, Oxford, U.K., 1989.
38. Tarnas, R., *The Passion of Western Mind*. Random House, Inc., New York, 1991, p. 369.
39. Peschek, G. A. and Broda, E., Utilization of fructose by a unicellular blue-green alga, *Anacystis nidulans*. *Naturwissenschaften*, 60, 479, 1973.
40. Peschek, G. A., Light inhibition of respiration and fermentation and its reversal by inhibitors and uncouplers of photophosphorylation in whole cells of a prokaryotic alga, *Anacystis nidulans*, in *Proceedings of the Third International Congress on Photosynthesis*. Avron, M. (Ed.), Elsevier, Amsterdam, the Netherlands, 1974, p. 921.
41. Peschek, G. A., The bioenergetic processes of the blue-green alga, *Anacystis nidulans*. Doctoral thesis, University of Vienna, Vienna, Austria [in German], 1975.
42. Stal, L. J. and Moezelaar, R., Fermentation in cyanobacteria, *FEMS Microbiol. Rev.*, 21, 179, 1997.
43. Rippka, R. et al., Generic assignments, strain histories and properties of pure cultures of cyanobacteria, *J. Gen. Microbiol.*, 111, 1, 1979.
44. Benedict, C. R., Nature of obligate photoautotrophy, *Annu. Rev. Plant Physiol.*, 29, 67, 1978.
45. Smith, A. J. and Hoare, D. S., Specialist phototrophs, lithotrophs, and methylotrophs: A unity among a diversity? *Bacteriol. Rev.*, 41, 419, 1977.
46. Whittenbury, R. and Kelly, S.P., Autotrophy: A conceptual phoenix, in *Microbial Energetics, Symposium 27*. Hoddock, B. A. and Hamilton, W. A. (Eds.), Soc. Gen. Microbiol., Cambridge University Press, Cambridge, U.K., 1977, p. 121.
47. Suzuki, I., Mechanisms of inorganic oxidation and energy coupling, *Annu. Rev. Microbiol.*, 28, 85, 1974.
48. Kelly, D. P., Autotrophy: Concepts of lithotrophic bacteria and their organic metabolism. *Ann. Rev. Microbiol.*, 25, 177, 1971.
49. Gest, H., Energy conversion and generation of reducing power in bacterial photosynthesis. *Adv. Microbiol. Physiol.*, 7, 243, 1972.
50. Mitchell, P., Coupling of phosphorylation to electron and hydrogen transfer by a chemiosmotic type of mechanism. *Nature*, 191, 144–148, 1961.
51. Mitchell, P., Chemiosmotic coupling in oxidative and photosynthetic phosphorylation. *Biol. Rev.*, 41, 455, 1966.
52. Nicholls, P. et al., Cytochrome oxidase in *Anacystis nidulans*: Stoichiometries and possible functions in cytoplasmic and thylakoid membranes. *Biochim. Biophys. Acta*, 1098, 184, 1992.
53. Skulachev, V. P., Chemiosmotic systems and the basic principles of cell energetics, in *Molecular Mechanisms in Bioenergetics*. Ernster, L. (Ed.), Elsevier, Amsterdam, the Netherlands, 1992, p. 37.
54. Oesterhelt, D. and Stoeckenius, W., Rhodopsin-like protein from the purple membrane of *Halobacterium halobium*, *Nat. New Biol.*, 233, 149, 1971.
55. Ovchinikov, Y. A. et al., The structural basis of the functioning of bacteriorhodopsin: An overview, *FEBS Lett.*, 100, 219, 1979.
56. Racker, E. and Stoeckenius, W., Reconstitution of purple membrane vesicles catalyzing light driven proton uptake and adenosine triphosphate formation, *J. Biol. Chem.*, 249, 662, 1974.
57. Renthal, R., Bacteriorhodopsin, in *Molecular Mechanisms in Bioenergetics*. Ernster, L. (Ed.), Elsevier, Amsterdam, the Netherlands, 1992, p. 75.
58. Abrahams, J. P. et al., Structure at 2.8 Å resolution of F_1-ATPase from bovine heart mitochondria, *Nature*, 370, 621, 1994.
59. Pedersen, P. L. and Amzel, M., ATP synthases—Structure, reaction center, mechanism, and regulation of one of nature's most unique machines, *J. Biol. Chem.*, 268, 9937, 1993.
60. Barghoorn, E. S. and Schopf, J. W., Microorganisms from the late precambrian of Central Australia, *Science*, 150, 337, 1965.
61. Babcock, G. T. and Wikström, M., Oxygen activation and the conservation of energy in cell respiration. *Nature*, 356, 301, 1992.
62. Hoganson, C. W. et al., From water to oxygen and back again: Mechanistic similarities in the enzymatic redox conversions between water and dioxygen, *Biochim. Biophys. Acta*, 1365, 170, 1998.
63. Barghoorn, E. S. and Schopf, J.W., Microorganisms three billion years old from the precambrian of South Africa. *Science*, 152, 758, 1966.
64. Schopf, J. W., Precambrian microorganisms and evolutionary events prior to the origin of vascular plants. *Biol. Rev.*, 45, 319, 1970.
65. Waterbury, J. B. et al., Widespread occurrence of a unicellular, marine, planktonic cyanobacterium. *Nature*, 277, 293, 1979.
66. Chisholm, S. W. et al., A novel free living Prochlorophyte abundant in the oceanic euphotic zone. *Nature*, 340, 340, 1988.

67. Barber, J., Water, water everywhere, and its remarkable chemistry. *Biochim. Biophys. Acta*, 1655, 123, 2004.
68. Des Marais, D. J., When did photosynthesis emerge on earth? *Science*, 289, 1703, 2000.
69. Hill, R., Oxygen evolved by isolated chloroplasts. *Nature*, 139, 281, 1937.
70. Hill, R. and Bendall, F., Function of the two cytochrome components in chloroplasts: A working hypothesis. *Nature*, 186, 136, 1960.
71. Xiong, J. et al., Molecular evidence for the early evolution of photosynthesis. *Science*, 289, 1724, 2000.
72. De la Rosa, M. A. et al., Convergent evolution of cytochrome c_6 and plastocyanin, in *The Light-Driven Plastocyanin: Ferredoxin Oxidoreductase*. Golbeck, J. H. (Ed.), Springer Verlag, Dordrecht, the Netherlands, 2006, p. 683.
73. Morand, L. Z. et al., Soluble electron transfer catalysts of cyanobacteria, in *The Molecular Biology of Cyanobacteria*. Byrant, D. A. (Ed.), Kluwer Academic, Dordrecht, the Netherlands, 1994, p. 381.
74. Sidler, W. A., Phycobilisome and phycobiliprotein structures, in *The Molecular Biology of Cyanobacteria*. Bryant, D. A. (Ed.), Kluwer Academic Publishers, Dordrecht, the Netherlands, 1994, p. 139.
75. Nicholls, P., The mitochondrial and bacterial respiratory chains. From MacMunn and Keilin to current concepts, in *Frontiers of Cellular Bioenergetics*. Papa, S. (Ed.), Kluwer Academic/Plenum Publishers, New York, 1999, p. 1.
76. Grossman, A. R., A molecular understanding of complementary chromatic adaptation, *Photosynth. Res.*, 76, 207, 2003.
77. Bibby, T. S. et al., Low-light-adapted *Prochlorococcus* species possess specific antennae for each photosystem, *Nature*, 424, 1051, 2003.
78. Biesiadka, J. et al., Crystal structure of cyanobacterial photosystem II at 3.2 Å resolution: A closer look at the Mn cluster, *Phys. Chem. Chem. Phys.*, 6, 4733, 2004.
79. DeRuyter, Y. S. and Fromme, P., Molecular structure of the photosynthetic apparatus, in *The Cyanobacteria: Molecular Biology and Evolution*. Herrero A. and Flores, E. (Eds.), Caister Academic Press, Norfolk, U.K., 2008.
80. Ferreira, K. N. et al., Architecture of the photosynthetic oxygen-evolving center. *Science*, 303, 1831, 2004.
81. Loll, B. et al., Towards complete cofactor arrangement in the 3.0 Å resolution structure of photosystem II. *Nature*, 438, 1040, 2005.
82. Zouni, A. et al., Crystal structure of photosystem II from *Synechococcus elongatus* at 3.8 Å resolution. *Nature*, 409, 739, 2001.
83. Blankenship, R. E., Origin and early evolution of photosynthesis, *Photosynth. Res.*, 33, 91, 1992.
84. Dworsky, A. et al., Functional and immunological characterization of both "mitochondria-like" and "chloroplast-like" electron/proton transport proteins in isolated and purified cyanobacterial membranes. *Bioelectrochem. Bioenerg.*, 38, 35, 1995.
85. Baniulis, D. et al., Purification and crystallization of the cyanobacterial cytochrome b6f complex. *Methods Mol. Biol.*, 684, 65, 2011.
86. Paumann, M. et al., The bioenergetic role of dioxygen and the terminal oxidase(s) in cyanobacteria. *Biochim. Biophys. Acta*, 1707, 231, 2005.
87. Peschek, G. A., Obinger, C., and Paumann, M., The respiratory chain of blue-green algae (cyanobacteria). *Physiol. Plant*, 120, 358, 2004.
88. Kurisu, G. et al., Structure of the cytochrome b_6f complex of oxygenic photosynthesis: Tuning the cavity. *Science*, 302, 1009, 2003.
89. Duran, V. et al., The efficient functioning of photosynthesis and respiration in *Synechocystis* sp. PCC 6803 strictly requires the presence of either cytochrome c_6 or plastocyanin. *J. Biol. Chem.*, 279, 7229, 2004.
90. Karplus, P. A., Daniels, M. J., and Herriott, J. R., Atomic structure of ferredoxin-$NADP^+$ reductase: Prototype for a structurally novel flavoprotein family. *Science*, 251, 60, 1991.
91. Gilbert, D. L. (Ed.), *Oxygen and Living Processes. An Interdisciplinary Approach*. Springer Verlag, New York, 1981.
92. Hackstein, J. H. P. et al., Plastid-like organelles in anaerobic mastigotes and parasitic Apicomplexans, in *Eukaryotism and Symbiosis. Intertaxonic combination versus symbiotic adaptation. Endocytobiology VI*. Schenk, H. E. A., Herrmann, R. G., Jeon, K. W., Müller, N. E., and Schwemmler, W. (Eds.), Springer Verlag, Berlin, Germany, 1997, p. 49.
93. Nicholls, D. G. and Ferguson, S. J., *Bioenergetics*, 2nd edn., Academic Press, London, U.K., 1992.
94. Peschek, G. A. and Broda, E., Fermentation-prerespiration-photosynthesis-respiration: A bioenergetic succession, in *An Interdisciplinary Study of the Origin and Evolution of Earth's Earliest Biosphere*. Schopf, J. W. (Ed.), UCLA, Los Angeles, CA, 1980, p. 23.

95. Hausrath, A. C., Capaldi, R. A., and Matthews, B. W., The conformation of the epsilon- and gamma-subunits with the *Escherichia coli* F1 ATPase. *J. Biol. Chem.*, 276, 47227, 2001.
96. Rubinstein, J. L., Walker, J. E., and Henderson, R., Structure of the mitochondrial ATP synthase by electron cryomicroscopy, *EMBO J.*, 22, 6182, 2003.
97. Walker, J. E. and Dickson, V. K., The peripheral stalk of the mitochondrial ATP synthase. *Biochim. Biophys. Acta*, 1757, 286, 2006.
98. Peschek, G. A., Reduced sulfur and nitrogen compounds and molecular hydrogen as electron donors for anaerobic CO_2 photoreduction in *Anacystis nidulans. Arch. Microbiol.*, 119, 313, 1978.
99. Peschek, G. A., Aerobic hydrogenase activity in *Anacystis nidulans*. The oxyhydrogen reaction. *Biochim. Biophys. Acta*, 548, 203, 1979.
100. Peschek, G. A., Anaerobic hydrogenase activity in *Anacystis nidulans*. H_2-dependent photoreduction and related reactions. *Biochim. Biophys. Acta*, 548, 187, 1979.
101. Peschek, G. A., Evidence for two functionally distinct hydrogenases in *Anacystis nidulans. Arch. Microbiol.*, 123, 81, 1979.
102. Peschek, G. A., The role of the Calvin cycle for anoxygenic CO_2 photoassimilation *Anacystis nidulans. FEBS Lett.*, 106, 34, 1979.
103. Peschek, G. A., Electron transport reactions in respiratory particles of hydrogenase-induced *Anacystis nidulans. Arch. Microbiol.*, 125, 123, 1980.
104. Bothe, H., Boison, G., and Schmitz, O., Hydrogenases in cyanobacteria, in *The Phototrophic Prokaryotes*. Peschek, G. A., Löffelhardt, W., and Schmetter G. (Eds.), Kluwer Academic/Plenum Publishers, New York, 1999, p. 589.
105. Houchins, P., The physiology and biochemistry of hydrogen metabolism in cyanobacteria. *Biochim. Biophys. Acta*, 768, 227, 1984.
106. Tamagnini, P. et al., Hydrogenases and hydrogen metabolism in cyanobacteria. *Microbiol. Mol. Biol. Rev.*, 66, 1, 2002.
107. Jensen, B. J., Energy requirement for diazotrophic growth of the cyanobacterium *Anabaena variabilis* determined from groth yields in the dark. *J. Gen. Microbiol.*, 129, 2633, 1983.
108. Broda, E. and Peschek, G. A., Evolutionary considerations on the thermodynamics of nitrogen fixation. *Biosystems*, 13, 47, 1980.
109. Broda, E. and Peschek, G. A., Nitrogen fixation as evidence for the reducing nature of the early biosphere. *Biosystems*, 16, 1, 1983.
110. Broda, E. and Peschek, G. A., The evolution of dinitrogen fixation. *Origins Life*, 14, 653, 1984.
111. Berger, S., Ellersiek, U., and Steinmüller, K., Cyanobacteria contain a mitochondrial complex I-homologous NADH-dehydrogenase, *FEBS Lett.*, 286, 129, 1991.
112. Friedrich, T., Steinmüller, K., and Weiss, H., The proton-pumping respiratory complex I of bacteria and mitochondria and its homologue in chloroplasts, *FEBS Lett.*, 367, 107, 1995.
113. Leif, H. et al., Isolation and characterization of the proton-translocating NADH:ubiquinone oxidoreductase from *Escherichia coli. Eur. J. Biochem.*, 230, 538, 1995.
114. Alpes, I., Scherer, S., and Böger, P., The respiratory NADH dehydrogenase of the cyanobacterium *Anabaena variabilis*: Purification and characterization. *Biochim. Biophys. Acta*, 973, 41, 1989.
115. Festetics, T., Kinetic investigations on the oxidation of succinate and NAD(P)H by plasma and thylakoid membranes isolated and purified from cyanobacteria. Diploma thesis, University of Vienna, Vienna, Austria, 2004.
116. Flasch, H., Spectrophotometric investigations on complexes I, II and III of the respiratory electron transport system in isolated and purified plasma and thylakoid membranes of cyanobacteria. Diploma thesis, University of Vienna, Vienna, Austria, 1997.
117. Stroebel, D. et al., An atypical heam in the cytochrome b_6f complex. *Nature*, 426, 413, 2003.
118. Kraushaar, H. et al., Immunologically cross-reactive and redox-competent cytochrome b_6/f-complexes in the chlorophyll-free plasma membrane of cyanobacteria. *FEBS Lett.*, 273, 227, 1990.
119. Sherman, D. M., Troyan, T. A., and Sherman, L. A., Localization of membrane proteins in the cyanobacterium *Synechococcus* sp. PCC 7942. *Plant Physiol.*, 106, 251, 1994.
120. Dzelzkalns, V. A. et al., Deletion of the structural gene for the NADH-dehydrogenase subunit 4 of *Synechocystis* 6803 alters respiratory properties. *Plant Physiol.*, 105, 1435, 1994.
121. Fay, P., Oxygen relations of nitrogen fixation in cyanobacteria. *Microbiol. Rev.*, 56, 340, 1992.
122. Berry, E. A. et al., Structure and function of cytochrome bc complexes. *Annu. Rev. Biochem.*, 69, 1005, 2000.
123. Malkin, R., Cytochrome bc_1 and b_6f-complexes of photosynthetic membranes. *Photosynth. Res.*, 33, 121, 1992.
124. Trumpower, B. L., Cytochrome bc_1 complexes of microorganisms. *Microbiol. Rev.*, 54, 101, 1990.

125. Trumpower, B. L., The protonmotive Q cycle. *J. Biol. Chem.*, 265, 11409, 1990.
126. Zhang, L., Pakrasi, H. B., and Whitmarsh, J., Photoautotrophic growth of the cyanobacterium *Synechocystis* sp. PCC 6803 in the absence of cytochrome c_{553} and plastocyanin. *J. Biol. Chem.*, 269, 5036, 1994.
127. Moser, D. et al., Acidic cytochrome c_6 of unicellular cyanobacteria is an indispensable and kinetically competent electron donor to cytochrome oxidase in plasma and thylakoid membranes. *Biochem. Int.*, 24, 757, 1991.
128. Lockau, W., Evidence for a dual role of cytochrome c-553 and plastocyanin in photosynthesis and respiration of the cyanobacterium, *Anabaena variabilis. Arch. Microbiol.*, 128, 336, 1981.
129. Paumann, M. et al., Soluble Cu_A domain of cyanobacterial cytochrome c oxidase. *J. Biol. Chem.*, 27, 10293, 2004.
130. Paumann, M. et al., Kinetics of electron transfer between plastocyanin and the soluble Cu_A domain of cyanobacterial cytochrome c oxidase. *FEMS Microbiol. Lett.*, 239, 301, 2004.
131. Paumann, M. et al., Kinetics of interprotein electron transfer between cytochrome c_6 and the soluble Cu_A domain of cyanobacterial cytochrome c oxidase. *FEBS Lett.*, 576, 101, 2004.
132. Peikert, R., Plastocyanin from *Synechocystis* 6803—Isolation, purification, properties and the role of ionic strength in electron transfer to cyanobacterial membranes. Diploma thesis, University of Vienna, Vienna, Austria, 1995.
133. Sandmann, G. and Böger, P., Copper-induced exchange of plastocyanin and cytochrome c-553 in cultures of *Anabaena variabilis. Plant Sci. Lett.*, 17, 417, 1980.
134. Siegelman, H. et al., Plastocyanin: Possible significance of quaternary structure. *Eur. J. Biochem.*, 64, 131, 1976.
135. Bernroitner, M. et al., Heme-copper oxidases and their electron donors in cyanobacterial respiratory electron transport. *Chem. Biodivers.*, 5, 1927, 2008.
136. Pereira, M. M. and Teixeira, M., Proton pathways, ligand binding and dynamics of the catalytic site in heam-copper oxygen reductases: A comparison between the three families. *Biochim. Biophys. Acta, Bioenerg.*, 1665, 340, 2004.
137. Hart, S. E. et al., Terminal oxidases of cyanobacteria. *Biochem. Soc. Trans.*, 33, 832, 2005.
138. Howitt, C. A. and Vermaas, W. F. J., Quinol and cytochrome oxidases in the cyanobacterium *Synechocystis* sp. PCC 6803. *Biochemistry*, 37,17944, 1998.
139. Musser, S. M., Stowell, M. H. B., and Chan S. I., Comparison of ubiquinol and cytochrome c terminal oxidases. An alternative view. *FEBS Lett.*, 327, 131, 1993.
140. Peschek, G. A. et al., Photosynthesis and respiration in blue-green algae: Electron-transporting functions of thylakoids and plasma membrane, and occurrence of cytochrome aa_3, in *Anacystis nidulans, in Photosynthesis V. Chloroplast Development.* Akoyunoglu, G. (Ed.), Balaban International Science Services, Philadelphia, PA, 1981, p. 707.
141. Peschek, G. A., Spectral properties of a cyanobacterial cytochrome oxidase: Evidence for cytochrome aa_3. *Biochem. Biophys. Res. Commun.*, 98, 72, 1981.
142. Alge, D. and Peschek, G. A., Characterization of a cta/CDE operon-like genomic region encoding subunits I-III of the cytochrome c oxidase of the cyanobacterium *Synechocystis* PCC 6803. *Biochem. Mol. Biol. Int.*, 29, 511, 1993.
143. Alge, D. and Peschek, G. A., Identification and characterization of the ctaC (coxB) gene as part of an operon encoding subunits I, II, and III of the cytochrome c oxidase (cytochrome aa_3) in the cyanobacterium *Synechocystis* PCC 6803. *Biochem. Biophys. Res. Commun.*, 191, 9, 1993.
144. Alge, D., Schmetterer, G., and Peschek, G. A., The gene encoding cytochrome-c oxidase subunit I from *Synechocystis* PCC6803. *Gene*, 138, 127, 1994.
145. Iwata, S. et al., Structure at 2.8 Å resolution of cytochrome c oxidase from *Paracoccus denitrificans. Nature*, 376, 660, 1995.
146. Schmetterer, G., Alge, D., and Gregor, W., Deletion of cytochrome c oxidase genes from the cyanobacterium *Synechocystis* sp. PCC6803: Evidence for alternative respiratory pathways. *Photosynth. Res.*, 42, 43, 1994.
147. Abramson, J. et al., The structure of the ubiquinol oxidase from *Escherichia coli* and its ubiquinone binding site, *Nat. Struct. Biol.*, 7, 910, 2000.
148. Bergman, B. et al., Cytochrome oxidase: Subcellular distribution and relationship to nitrogenase expression in the nonheterocystous marine cyanobacterium *Trichodesmium thiebautii. Appl. Environ. Microbiol.*, 59, 3239, 1993.
149. Peschek, G. A. et al., Correlation between immuno-gold labels and activites of the cytochrome c oxidase (aa_3 type) in membranes of salt-stressed cyanobacteria. *FEMS Microbiol. Lett.*, 124, 431, 1994.

150. Peschek, G. A. et al., Immunochemical localization of the cytochrome c oxidase in a cyanobacterium, *Synechococcus* PCC 7942 (*Anacystis nidulans*). *Biochim. Biophys. Acta,* 1187, 369, 1994.
151. Serrano, A. et al., Cellular localization of cytochrome c_{553} in the N_2-fixing cyanobacterium *Anabaena variabilis. Arch. Microbiol.*, 154, 614, 1990.
152. Turner, J. S. and Brittain, E. G., Oxygen and photosynthesis, *Biol. Rev.*, 37, 170, 1962.
153. Peschek, G. A. and Zoder, R., Temperature stress and basic bioenergetic strategies for stress defence, in *Algal Adaptation to Environmental Stresses. Physiological, Biochemical and Molecular Mechanisms*, Rai, L. C. and Gaur, J. P. (Eds.), Springer Verlag, Berlin, Germany, 2001, p. 203.
154. Inaba, M., Sakamoto, A., and Murata, N., Functional expression in *Escherichia coli* of low-affinity and high-affinity Na(+)(Li(+))/H(+) antiporters of *Synechocystis. J. Bacteriol.*, 183, 1376, 2001.
155. Nomura, C. T. et al., Characterization of two cytochrome oxidase operons in the marine cyanobacterium *Synechococcus* sp. PCC7002: Inactivation of ctaDI affects the PS I:PS II ratio. *Photosynth. Res.*, 87, 215, 2006.
156. Bernroitner, M. et al., Cyanobacterial cytochrome cM: Probing its role as electron donor for CuA of cytochrome c oxidase. *Biochim. Biophys. Acta (Bioenergetics)*, 1787, 135, 2009.
157. Malakhov, M. P., Malakhova, O. A., and Murata, N., Balanced regulation of expression of the gene for cytochrome cM and of genes for plastocyanin and cytochrome c_6 in *Synechocystis. FEBS Lett.*, 444, 281, 1999.
158. Peschek, G. A., Villgrater, K., and Wastyn, M., "Respiratory protection" of the nitrogenase in dinotrogen fixing cyanobacteria. *Plant Soil* 127, 17, 1991.
159. Wastyn, M. et al., Respiratory activities and aa_3-type cytochrome oxidase in plasma and thylakoid membranes from vegetative cells and heterocysts of the cyanobacterium *Anabaena* ATCC 29413. *Biochim. Biophys. Acta*, 935, 217, 1988.
160. Zehr, J. P., Mellon, M. T., and Zani, S., New nitrogen-fixing microorganisms detected in oligotrophic oceans by amplification of Nitrogenase (nifH) genes. *Appl. Environ. Microbiol.*, 64, 3444, 1998.
161. Rai, A. N., Borthakur, M., and Bergman, B., Nitrogenase derepression, its regulation and metabolic changes associated with diazotrophy in the non-heterocystous cyanobacterium *Plectonema boryanum. J. Gen. Microbiol.*, 138, 481, 1992.
162. Jones, K. M., Buikema W. J., and Haselkorn R., Heterocyst-specific expression of patB, a gene required fro nitrogen fixation in *Anabaena* sp. strain PCC 7120. *J. Bacteriol.*, 185, 2306, 2003.

4 Understanding the Mechanisms of Abiotic Stress Management in Cyanobacteria with Special Reference to Proteomics

*Snigdha Rai, Sarita Pandey, Alok Kumar Shrivastava, Prashant Kumar Singh, Chhavi Agrawal, and Lal Chand Rai**

CONTENTS

4.1 INTRODUCTION

Cyanobacteria are thought to have evolved around 3.5 billion years ago and have played a key role in earth's transition from an anaerobic state to the aerobic one. They are the largest group of Gram-negative photosynthetic prokaryotes that has a close resemblance to higher plants in terms of lipid composition and protein assembly of the thylakoid membrane [1]. Cyanobacteria are ubiquitous in aquatic and terrestrial ecosystems including extreme habitats like hot springs, deserts, and polar regions. They occupy a central position in the nutrient cycling due to their unique capacity to fix atmospheric nitrogen into the assimilatory form (NH_4^+), thus contributing to the global nitrogen budget [2,3]. They are reported to play a key role in the maintenance of soil fertility [4] and have proved to be a potential biofertilizer in the form of plant–cyanobacterial symbiosis. Besides, many cyanobacterial species are reported to synthesize bioactive metabolites of potential therapeutic use [5]. It can therefore be concluded that

* Corresponding author: lcrbhu15@gmail.com

cyanobacteria are a powerful model system for studying the molecular mechanisms of the responses and the acclimation strategies to abiotic stresses [6–8]. These mechanisms may provide models that may be applicable to plants as well.

Abiotic stresses result due to a shift in environmental parameters that affect the cellular homeostasis of organisms including cyanobacteria. Acquired survival and tolerance strategies employed include differential expression of specific protein genes, the emergence of adaptation proteins, and a shift in metabolic pathways [9,10]. In general, all abiotic stresses such as temperature extremes, drought, salinity, heavy metals, metalloid, pesticides, and ultraviolet (UV) radiation initiate detrimental effects on cell growth and metabolism [11–15].

A list of molecular approaches at the levels of DNA, RNA, and protein are used to study cellular physiology and metabolism under stress. The question arises as to how closely do transcription and translation correlate with specific cellular networks [16]. A recent report showed incongruent experimental patterns between mRNA and proteins [17] and it has been accepted that the analysis of differential gene expression is actually an indirect approach to understanding the molecular mechanisms involved in stress response [18,19]. On the other hand, proteomics reflects the functionality of the cell, bringing the virtual life of genes to the real life of proteins [20]. It reflects the entire protein content of an organism/system at a particular time point under a given set of conditions. Acting as a bridge between transcriptome and metabolome, it offers a continually evolving set of novel techniques to study all facets of protein structure and function, reflecting the actual state of cell physiology. Moreover, long-term adaptation to a stress depends on the synthesis of "adaptation proteins," which can be looked for and evaluated in order to understand cell adaptation under such adverse conditions.

While earlier cyanobacterial protein research endeavors were hypothesis driven and studied specific enzymes and their isolated functions, the increasing advances in proteomics tools have facilitated the study of whole cell protein extracts on 2-DE gels followed by its analysis using matrix-assisted laser desorption ionization time-of-flight mass spectrometry (MALDI-TOF/MS). In addition, high-throughput proteomics tools like liquid chromatography-mass spectrometry (LC-MS), differential gel electrophoresis (DIGE), isobaric tag for relative and absolute quantitation (iTRAQ), isotope coded affinity tag (iCAT), and stable isotope labeling with amino acid in cell culture (SILAC) are some advanced and refined approaches that enable the study of all the synthesized proteins under a single roof [21,22]. This provides a new dimension for a contemporary study. Current success in proteomic studies includes elucidating subcellular compartments such as plasma and outer membrane [23,24], periplasm [17] as well as highly purified PSII complex [25], and thylakoid assembly proteins [26]. Comparative proteomic studies can identify strain-specific variations and can spot core proteins of the target organism [27]. Besides identification and characterization of multiprotein complexes by blue native PAGE (BN-PAGE), it provides an integrative view of protein–protein interaction networks [28,29]. Proteomics offers a precise understanding of the effects of environmental perturbations in a system.

4.2 CYANOBACTERIAL RESPONSES

4.2.1 Temperature Stress

Temperature fluctuation in the environment has been a major concern over the decades due to inconsistent and evolving phenomena like global warming. A temperature rise of 0.2°C/decade is projected for the next couple of years [30]. Temperature affects a broad spectrum of cellular components and metabolism, and temperature extremes impose stresses of variable severity that depend on the rate of temperature change, intensity, and duration [31]. It is well known that oxygenic photosynthetic organisms acclimate to changes in the environmental temperature, although there are high- and low-temperature limits for these phenomena [32].

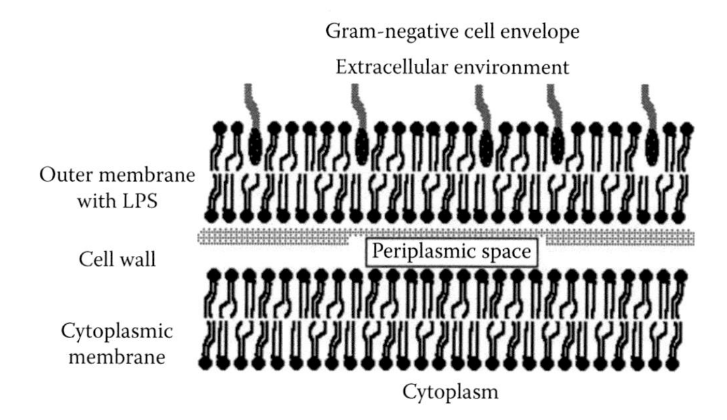

FIGURE 4.1 Schematic representation of cytoplasmic membrane. (Adapted from http://mansfield.ohio-state.edu/-sabedon/biol11080.htm)

The most significant change in response to temperature corresponds to alteration in membrane compositions [33]. A schematic presentation of cytoplasmic membrane depicts the involvement of lipid bilayer and protein (Figure 4.1). Temperature-induced change in membrane fluidity is one of the immediate consequences in all organisms including cyanobacteria. Temperature fluctuations affect the degree of lipid saturation and protein integrity of membranes. High temperature results in excessive fluidity of membrane lipids and alterations in bonds (hydrogen and electrostatic) between polar groups of proteins, thereby modifying their structure and thus causing ion leakage [34,35]. In contrast, at low temperature, membranes become less fluid and their protein components can no longer function normally. This results in inhibited H^+/ATPase activity and solute transport into and out of the cells [36,37]. Membrane lipid of cyanobacteria contains polyunsaturated fatty acids [37] that help them combat low temperature [38,39]. The importance of the desaturases and the expression of their genes in acclimation to cold in cyanobacteria have been well documented [6,34,38,40–42]. It is believed that the fluidity of thylakoid membranes in cyanobacteria is controlled so as to maintain the membranes in the optimal condition for the damaged D1 protein to be digested and then replaced by newly synthesized protein [43].

High temperature disturbs metabolic processes due to the narrow activity range of most cellular enzymes and modifications in protein folding [43]. Some thermophilic cyanobacterial species are reported to grow in hot springs and deserts, whereas mesophilic ones can tolerate temperatures up to 50°C [10,42]. The effect of elevated temperature at proteome level has been investigated in *Synechocystis* PCC 6803 [44,45], *Spirulina platensis* [46], and *Anabaena doliolum* [47] using 2-DE followed by MALDI-TOF analysis. High-temperature stress is known to cause protein aggregation and denaturation [43], thereby creating an imbalance in the overall protein trafficking and function. Heat shock proteins (HSPs) subsequently function as chaperonins and associated proteases that assist protein refolding [45]. According to Schmid et al. [48], elevated temperature may induce a general stress response and could lead to cross-protection against related stresses. This finds support from the fact that HSPs have been reported to express under diverse abiotic stresses. According to the literature survey, heat-exposed cyanobacteria exhibit differentially increased expression of two heat shock proteins—GroEL1 and GroEL2 (chaperones) [49–51], whose regulation involves the CIRCE/HrcA system [52]. Inactivation of several other genes, such as *clpB* (protease), *hspA*, *htpG*, *dnaK2* (chaperones), and *sigB* and *sigC* (sigma factors) have been reported to induce thermal sensitivity [53,54]. High temperature is further reported to induce expression of DNA-binding proteins of 72 and 32 kDa observed using gel mobility shift assays [55]. Similarly, *ClpB* (hsp 100) molecular chaperon was reported to be induced in *Synechococcus* sp. PCC 7942 for the acquisition of thermotolerance [56].

Adaptive mechanisms to low temperature include (1) evolution of cold shock and antifreeze proteins, (2) modulation in kinetics of key proteins, and (3) accumulation of polyunsaturated fatty acids in membranes [35]. A 15-fold induction in ClpC1 (caseinolytic proteases)—a new family of bacterial molecular chaperons, was reported within 24 h of cold treatment [57]. Similarly, Chamot et al. [58] identified two RNA helicase protein genes *crhB* and *crhC* in *Anabaena* PCC 7120, where *crhC* was suggested to overcome cold-induced blockage of the initiation of translation occurring at low temperature. Cyanobacterial protein s21, a component of small subunit of ribosome, was observed to accumulate in *Synechocystis* in response to cold [37], suggesting that this protein might be involved in regulating translational apparatus.

4.2.2 Salt Stress

Salinity affects 7% of the world's total land area and 20% of agricultural land and has been a major cause of annual crop productivity loss [59,60]. The physiological and molecular mechanisms of tolerance to the osmotic and ionic components of salinity have been addressed by many researchers. High salt concentration affects water potential inside cells causing loss of water, which in turn leads to osmotic stress. Further, elevated ion concentration (Na^+ and Cl^-) within cells destabilizes ion homeostasis, leading to ionic stress condition [61,62].

In general, salinity and osmotic stresses produce very similar effects on living cells [63]. Interestingly, however, Kanesaki et al. [64] suggested that *Synechocystis* is capable of discriminating salt and hyperosmotic stresses as different stimuli. A comparative analysis of the two showed that expression of genes like *hspA*, *dnaK*, *dnaJ*, *groEL*, *ggpS* were commonly induced by both, while genes like *rpl2*, *rpl4*, *rpl23* and *fabG*, *rplA*, *repA* were specific to salt and hyperosmotic stress, respectively (Figure 4.2a and b).

Cyanobacteria, in general, are considerably more tolerant to salt stress as compared to higher plants. Other than some marine species, freshwater inhabitants like *A. doliolum* can survive a salt concentration of 150 mM [65] in contrast to higher plants that generally can tolerate up to 50 mM [66]. Cyanobacterial species like *Nostoc*, *Scytonema*, *Microcoleus*, *Aulosira*, and *Anabaena* have been used in reclaiming alkaline and sodic lands [67,68]. Modern stromatolites in Hamelin pool, Western Australia can be another such remarkable example of cyanobacterial tolerance to high salt concentrations. Stromatolites, considered as living fossils, are hard rock-like structures created by the accumulation of photosynthesizing cyanobacteria along with solid sediments. The water of Hamelin pool is twice as saline as typical seawater, rendering it inhospitable to animals like snails from grazing on them and thus supporting their undisturbed growth.

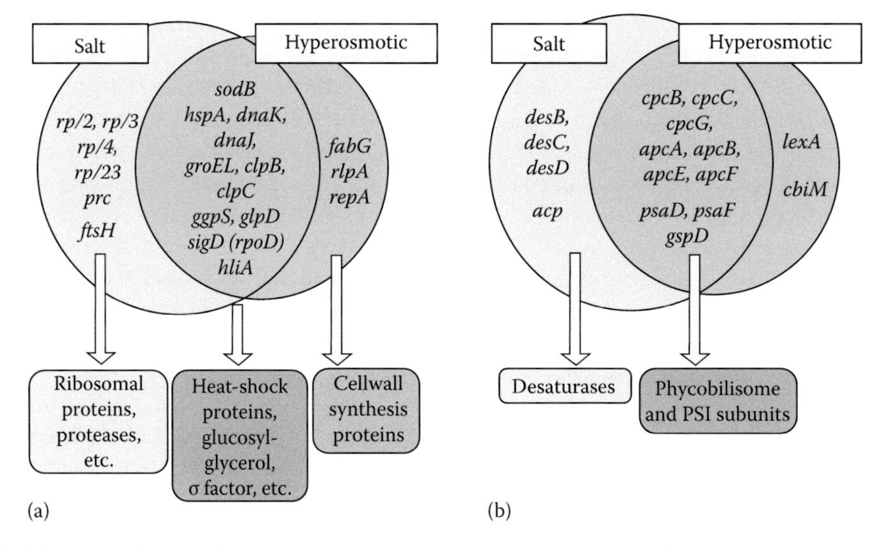

FIGURE 4.2 Identification of salt and hyperosmotic-stress-regulated genes in *Synechocystis*. (a) Up-regulated candidates, (b) down-regulated candidate. (Modified from Kanesaki, Y. et al., *Biochem. Biophys. Res. Commun.*, 290, 339, 2002.)

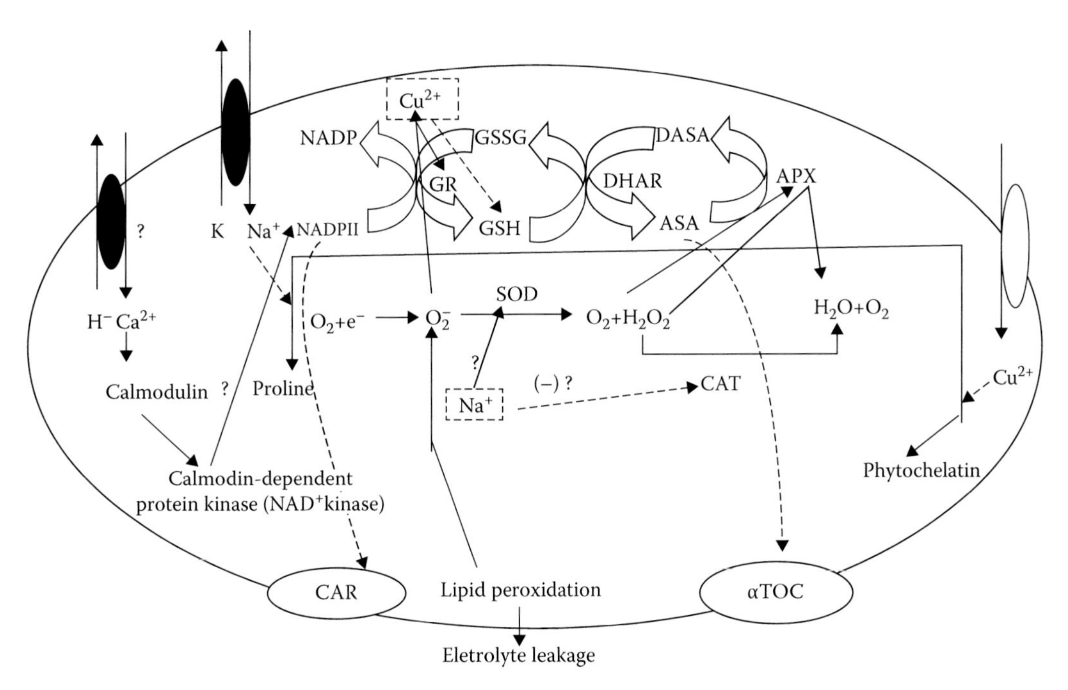

FIGURE 4.3 Hypothetical model for the induction of antioxidative defence system of A. doliolum under salinity and copper stress. Dotted line shows involvement in biosynthesis; box with dotted line as an inducer. The '?' depicts the effects with unknown reasons. α-TOC: α-tocopherol; APX: ascorbate peroxidase; ASA: ascorbate; CAR: carotenoid; CAT: catalase; DASA: dehydroascorbate; DHAR: dehydroascorbate reductase; GSH: glutathione reduced; GR: glutathione reductase; SOD: superoxide dismutase. (Modified from Srivastava, A.K. et al., *World J. Microbiol. Biotechnol.*, 21, 1291, 2005.)

Enhanced accumulation of Na^+ ions in the cell cytoplasm of cyanobacteria is prevented by two mechanisms, namely, (1) limited uptake of sodium and (2) its active efflux via Na^+/H^+ antiport proteins coupled with H^+-ATPase and respiratory cytochrome oxidase [69,70], whereas high sodium ion toxicity is compensated by (1) accumulating organic compounds to maintain the osmoticum [71], (2) inducing antioxidative defense system to detoxify the reactive oxygen species as shown in Figure 4.3 [65], and (3) expressing a set of salt-inducible proteins [17,72].

A review of the literature regarding sense and perception of salt shock indicates that it involves a cascade of molecular events and expression of specific stress genes and metabolites [73]. Salt overlay sensitive (SOS) pathway explains sensing mechanism under high salt concentrations (Figure 4.4). The extracellular signal is perceived by the membrane receptors, activating a signaling cascade involving mainly three genes: SOS 1, SOS 2, and SOS 3. In this context, most cyanobacterial genomes analyzed (27 out of 33 sequenced genomes) demonstrated the presence of *lexA* gene, which appears to function as the transcriptional regulator of the key SOS response genes [74].

Plasma membrane that regulates intracellular trafficking also plays a key role in salt adaptation. A change in membrane permeability was observed in *Synechocystis* owing to the increase in the proportion of long-chain saturated fatty acids and in the content of the membrane protein [16]. Much work has been done to understand the changes in membrane dynamics at an elevated concentration of salt by blue native SDS-PAGE followed by MALDI-TOF/MS. Studies in *Synechocystis* revealed the proteins that are involved in cell wall biosynthesis [16]. These alterations probably produced a strong diffusion barrier reducing the influx of inorganic ions into the periplasm.

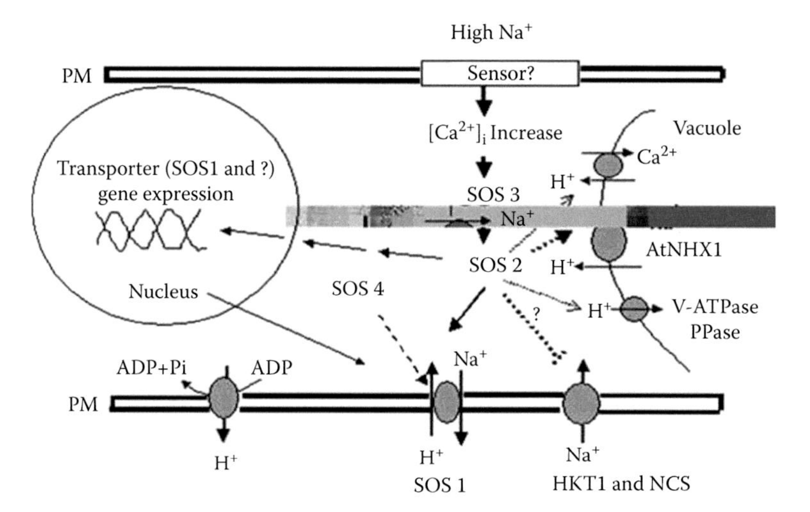

FIGURE 4.4 Overview of SOS signaling pathways.

The damage to cellular biomolecules primarily proteins during salt stress is due to effect on dehydration shell or surface change [73]. To keep up with this, accumulation and synthesis of nontoxic organic compounds (compatible solutes) occurs [75]. The role of GGP (glucosyl glycerol phosphate) synthase has been reported in *Synechocystis* sp. PCC 6803 [76]. The enzyme belongs to the biosynthetic pathway of osmoprotective compound, glucosylglycerol (GG), which proceeds from ADP glucose and glycerol 3-phosphate via glucosylglycerol phosphate in a two-step reaction.

Photosynthesis and respiration are among the key processes affected by salt stress [77]. Increase in Na^+ concentration causes two stabilizing proteins: cytochrome C550 and PsbU to dissociate from PSII complex and cytochrome C553 protein from PSI, thereby inhibiting photosynthetic electron transport. Recently, Allakhaverdiev et al. [78] demonstrated that salt stress inhibits the repair of the photodamaged PSII by inhibiting the synthesis of D1 protein in *Synechocystis*. The interrupted electron transport eventually results in the production of reactive oxygen species, which could be another reason for damage due to oxidation.

Cyanobacterial nitrogen fixation is very sensitive to ionic stress *per se*. Diversion of cellular energy from nitrogen fixation to Na^+ efflux during exposure to NaCl severely inhibits nitrogenase activity [79,80]. Besides, it also produces change in amino acid and carbohydrate synthesis [81,82]. Chen et al. [83] reported an increase in sucrose phosphate synthase activity in *Microcoleus vaginatus*. Differential induction of the enzymatic and nonenzymatic components of the antioxidative defense system has been reported in species like *Anabaena* and *Synechocystis* with respect to hypersaline conditions. A decline in the expression pattern of the enzyme superoxide dismutase was observed in *A. doliolum* after 24 h of 150 mM salt treatment [65].

Among the major identified metabolic changes, salinity has also been linked to glycolate metabolism-mediated photorespiration and glucosyl glycerol biosynthesis by many researchers. Hagemann et al. reported an enhanced activity of the enzyme glucosyl glycerol phosphate synthase [84] and the role of 2-phosphoglycolate metabolism in *Synechocystis* PCC 6803 [85] under salt stress condition. Gene-based and mutational approaches were used for extrapolating 2-phosphoglycolate pathway (Figure 4.5). Srivastava et al. [86] reported photorespiration in *A. doliolum* based on 2-DE and MALDI-TOF analyses. A time-dependent significant increase of 1.45- and 2.82-fold in glycolate oxidase activity observed at 150 mM NaCl treatment after 1 and 24 h, respectively, in *A. doliolum* clearly indicated a correlation between glycolate metabolism and salinity. Protein phosphorylation, however, is a general response to salt stress as identified in *Synechocystis* sp. PCC 7120 [87].

Classical proteomic approaches as well as high-throughput studies with respect to salt stress have led to the identification of some interesting membrane proteins. Huang et al. [16] have investigated

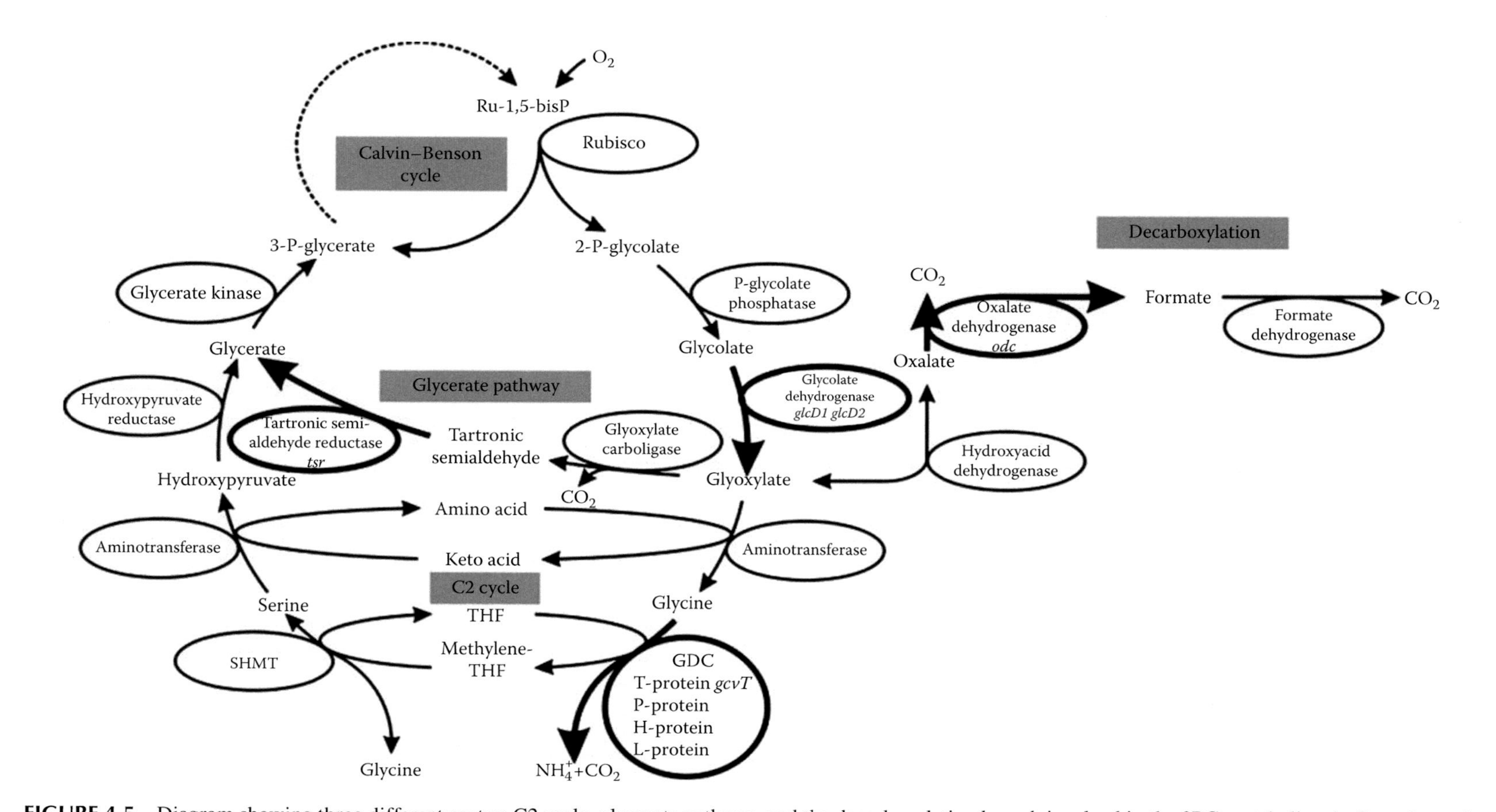

FIGURE 4.5 Diagram showing three different routes: C2 cycle, glycerate pathway, and the decarboxylating branch involved in the 2PG metabolism in *Synechocystis* sp. strain PCC 6803. Enzymatic steps mutated in strains used for this study have been highlighted. (Modified from Eisenhut, M. et al., *Plant Physiol.*, 142(1), 333, 2006; Chemot, D. et al., *J. Bacteriol.*, 181, 1728, 1999.)

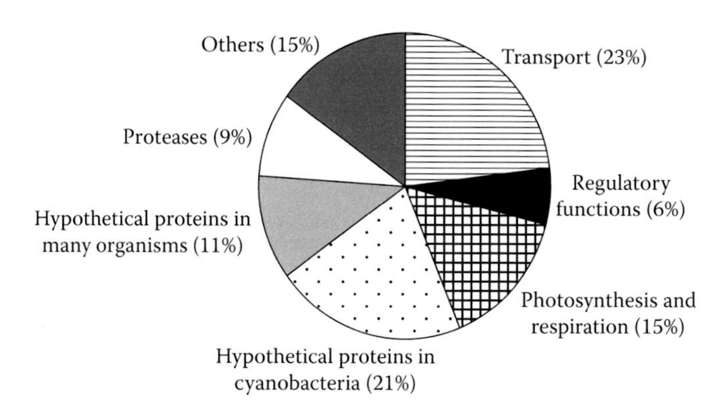

FIGURE 4.6 Piechart showing pattern of proteins from the plasma membrane preparations of *Synechocystis* into functional groups. (Modified from Huang, F. et al., *Proteomics*, 6, 910, 2006.)

how plasma membrane proteins of *Synechocystis* are affected by salt stress treatment, since this membrane represents the barrier of the cell toward the changing external salinity. Twenty-five proteins were observed to show change in expression pattern under salt stress based on differential and quantitative analysis of 2D gel profiles using PDQuest software (Figure 4.6). A substantial number of proteins, however, are still annotated as "hypothetical protein" due to the lack of sequence similarity to other proteins with known function.

Thus, it can be said that high salt concentration not only destabilizes structural and functional components of the cell but irreversibly damages and slows down the key processes within cells. We still need to explore the critical proteins involved in signaling, repair, and replenishment during salt stress.

4.2.3 UV Stress

Outside the visible spectrum, the only stressful radiation likely to be encountered under natural conditions is UV. With a reduction in ozone layer, organisms are exposed to an enhanced level of UV-B radiation (280–320 nm) [88,89]. This is particularly important to photosynthetic cyanobacteria, which require exposure to sunlight for survival and proliferation. In order to evade UV-B damage, cyanobacteria migrate to the regions of low UV-B flux. Cyanobacterial cells optimize their position in terrestrial mats or water column finding a compromise between optimal photosynthesis and pigment bleaching [90]. Besides, they also avoid UV-B-induced photooxidative damage by synthesis of screening metabolites like scytonemin and mycosporin like amino acids. Thus, they are found to possess mechanisms to avoid UV-B rays, defend against them, and repair any damage that UV-B might cause to their cellular components [91].

Direct damage of key proteins, enzymes, and DNA by UV-B and indirect oxidative damage of biomolecules are known to decrease the photosynthetic quantum yield, growth, and survival [5,7,8]. Uncommonly enhanced level of UV produces high amount of ROS. As a result, a number of physiological and biochemical processes of cyanobacteria such as growth, survival, pigmentation, photosynthetic oxygen production, phycobiliproteins composition, and $^{14}CO_2$ uptake have been reported to be susceptible to UV-B [92,93]. The photosynthetic pigments especially phycobiliproteins are readily bleached and destroyed [94]. Until now, the molecular mechanisms of diversified physiological response in cyanobacteria under UV-B stress have not been fully identified and knowledge about the effects of UV radiation at proteome level is very limited. Proteome-based reports on UV-B are available for *Synechocystis* PCC 6803 [20] and *Nostoc commune* [95]. Study on *Synechocystis* showed altered expression of several proteins upon UV-B exposure for 8, 72, 84, and 96 h [20]. The altered proteins were categorized in seven functional groups, including amino acid biosynthesis, photosynthesis, respiration, energy metabolism, translation, cellular processes, hypothetical, biosynthesis of cofactors, and prosthetic group (Figure 4.7). The down-regulated proteins under

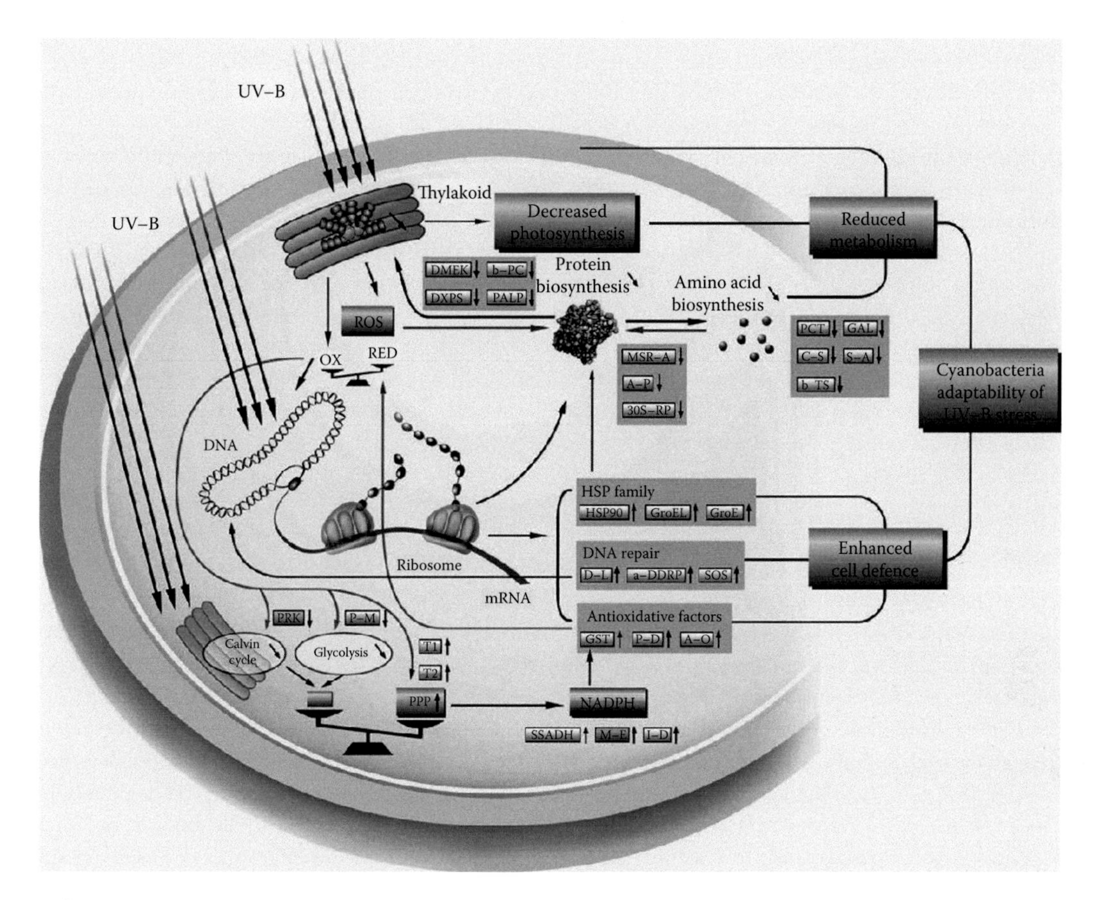

FIGURE 4.7 Hypothetical model of the long-term UV-B stress-responses in *Synechocystis sp.* PCC 6803. OX, oxidation; RED, reduction; DMEK, 4-diphosphocytidyl-2-C-methyl-D-erythritol kinase; b-PC, phycocyanin b subunit; DXPS, 1-deoxy-D-xylulose-5-phosphate synthase; PALP, phycocyanin associated linker protein; A-P, aminopeptidase P; MSR-A, methionine sulfoxide reductase A; b-TS, tryptophan synthase subunit beta; C-S, cysteine synthase; 30S-RP, 30S ribosomal protein S2; GAL, glutamate–ammonia ligase; D-L, DNA ligase; PCT, 3-phosphoshikimatel-carboxyvinyltransferase; S-A, sulfate adenylyltransferase; AO, amine oxidase; a-DDRP, DNA-directed RNA polymerase alpha subunit; P-D, phytoene desaturase; T1, transketolase; T2, transaldolase; GST, glutathione S-transferase; P-M, phosphoglycerate mutase; PRK, phosphoribulokinase; M-E, malic enzyme; I-D, isocitrate dehydrogenase; SSADH, succinate semialdehyde dehydrogenase. (Modified from Gao, Y. et al., *J. Exp. Bot.*, 60, 1141, 2009.)

long-term exposure were mainly associated with three functions: amino acid biosynthesis, photosynthesis, and protein biosynthesis. In contrast, up-regulated proteins were mainly divided into four functional categories: (1) DNA repair, (2) HSP family, (3) NADPH generation, and (4) cellular antioxidative reactions. Increased amount of the enzymes involved in DNA repair–DNA ligase, SOS function regulatory protein, and DNA-directed RNA polymerase alpha subunit suggested that DNA is the main target of UV-B stress. In some of the reports, up-regulation of enzymes comprising succinate–semialdehyde dehydrogenase, transketolase, transaldolase, malic enzyme, and citrate dehydrogenase and the down-regulation of phosphoglycerate mutase, indicated generation of high level of NADPH [96–98].

Similarly, in a study on *N. commune*, changes in the proteome during UV-B shock and UV acclimation, as revealed by subtractive high-resolution 2D gel electrophoresis (differential 2D display), have been studied [95]. Based on exposure time, the influence of UV-B on protein synthesis was grouped in three categories: (1) transient induction or repression of proteins after short-term exposure, (2) durable induction or repression of proteins after short-term exposure, and (3) continuous increase or decrease of proteins after prolonged exposure. Furthermore, partial sequences of a set of UV-B-stimulated proteins showed that these can be grouped into three functional classes: (1) proteins involved in lipid metabolism, (2) proteins involved in carbohydrate metabolism, and (3) proteins with regulatory function. Transient UV-B shock response had no overlap with the later UV-B acclimation response, thus indicating that these two are remarkably complex, but completely different strategies of the organism are required to protect itself against UV-B radiation.

Nitrogen-fixing enzyme nitrogenase is inhibited by UV-B even after a few minutes of exposure and a complete loss in activity was observed after 35–55 min depending upon the species. This may be due to the inhibition of ATP synthesis by UV-B [99]. Other adaptive strategies include stimulation of nitrate reductase [100] and inhibition of ammonia-assimilating enzyme glutamine synthetase [20]. It can be concluded that much alterations at the level of proteins and metabolites provide a basis to the cyanobacterial survival and tolerance under UV-B. There is still a need to explore and understand the integrated network of responses operating in the cell under UV-B stress.

4.2.4 Metals and Metalloid Stress

Cyanobacteria perform oxygenic photosynthesis and thus require far more metals than nonphotosynthetic organisms [10]. Metals act as cofactors for photosynthetic protein supercomplexes: iron for iron–sulfur clusters, cytochromes, and nonheme irons; manganese for the water-splitting complex; copper for plastocyanin, and magnesium for chlorophylls [101]. Living organisms sense metal-ion concentration and consequently adapt their cell metabolism. There is a narrow range between deficiency and toxicity of essential transition metals and tightly controlled homeostatic network operates for their regulation. Redox active metal ions trigger the formation of hydroxyl radicals via Haber–Weiss and Fenton reaction [102]. These radicals possess a high-affinity binding to S, N, O containing functional groups in biomolecules causing their inactivation and damage [103]. Along with its field counterparts, cyanobacteria too are exposed to elevated concentration of transition metals as soil and water contaminants. Cadmium, an example of toxic metal, has been studied as an industrial pollutant causing adverse effects on cyanobacteria [104]. Biochemical studies in *A. doliolum* by Bhargava et al. [105] showed that excess cadmium stimulates both enzymatic and nonenzymatic antioxidants in the cell. Zinc is another potentially toxic ion. When *Synechocystis* grows at maximum permissive concentration of zinc, transcription of the *ziaA* operator–promoter becomes greatly increased. Manganese is the essential cofactor for several enzymes, such as catalase and superoxide dismutase, and is an integral component of the oxygen-evolving complex of PSII. Using a DNA microarray approach, Yamaguchi et al. [106] identified *ManS* (a His kinase for the sensing of extracellular manganese) and *ManR* (a response regulator for the regulation and expression of the mntCAB operon) to be the possible participants in manganese homeostasis. Similarly, the cells require copper as an essential micronutrient for the synthesis of plastocyanin and cytochrome C oxidase, but copper becomes toxic at high doses. Cyanobacteria respond to copper deficiency as well as to excess copper [102]. Actually, copper is used as an effective algicide, and copper contamination is an important concern for algae and cyanobacteria population [104]. Likewise, most of the work related to metal stress has been done by using genomic tools like targeted mutagenesis, knockout libraries of specific sets of genes, and DNA microarrays—rather than employing proteomic studies [107], and very limited proteome-based studies are on record. In a 2-DE based study on the effect of copper [102], operation of anoxygenic photosynthesis was reported in copper acclimated *A. doliolum* growing in 250-fold excess Cu (Figure 4.8). Proteomic analysis of Cu acclimated strain under excess Cu, exhibited significant accumulation of both isoforms of SOD, plastocyanin, phosphoribulokinase, and transketolase, which counteract oxidative damage, serve as an alternate electron carrier from cytochrome b6/f complex to photosystem I, and fulfill the NADPH and ATP requirements, respectively, under copper stress (Figure 4.9).

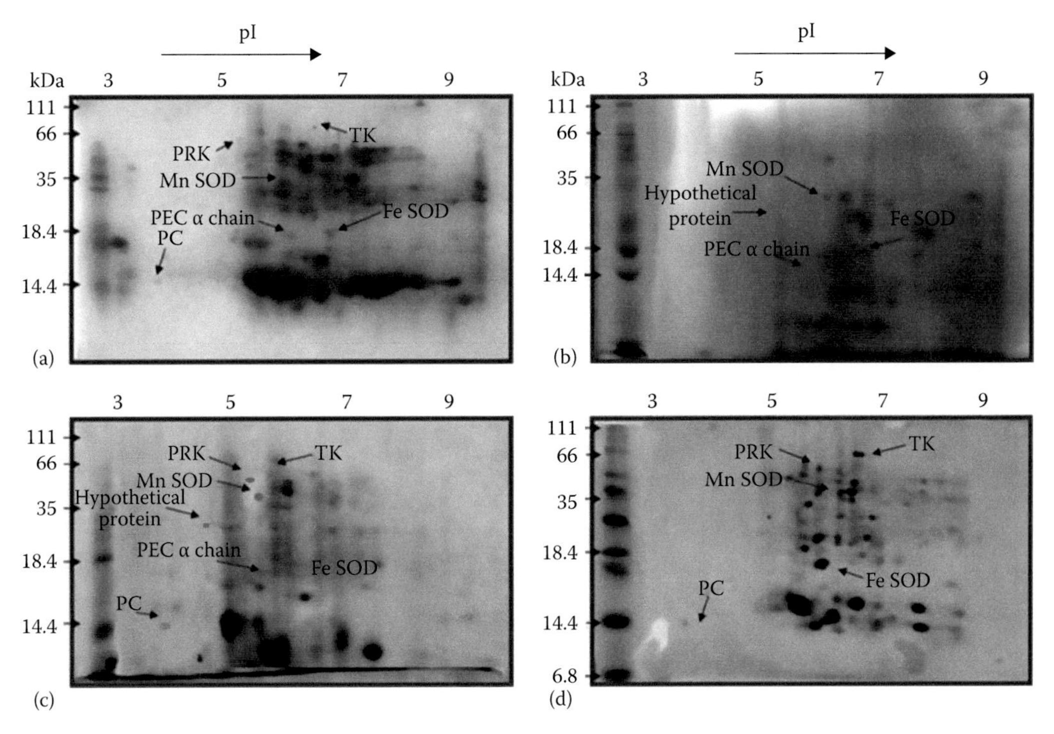

FIGURE 4.8 The 2-DE protein profiles of *A. doliolum*: control (a) exposed to 20 µM Cu for 10 (b) and 15 (c) days and Cu-acclimated (d). Altered expression of seven proteins was observed. (Modified from Bhargava, P. et al., *Photosynth. Res.*, 96, 61, 2008.)

Copper deprivation has also been studied at proteome level in *Synechocystis* [108]. To analyze the effect of Cu, comparative proteome of *Synechocystis* wild-type strain and the two knockout mutants, namely, Δ*petE* (lacking the Pc-coding gene) and Δ*petJ* (lacking the Cyt-coding gene), were done by 2-DE for their cytoplasmic soluble fractions, cultured with or without copper. Proteomic analysis of the wild-type strain revealed overexpression of 19 proteins belonging to different functional categories like amino acid synthesis, carbohydrate metabolism, nitrogen and carbon fixation, as well as GroEL1 chaperone and of ClpP protease. However, under Cu deficiency, ABC transporter's expression, which is a major contributor to the ferric iron transport across the plasma membrane, was enhanced [109]. In the presence of Cu, proteome of wild type and Δ*petE* showed differential expression in 43 proteins involved in amino acid and carbohydrate metabolisms, nitrogen cycle, carbon fixation, and cofactors synthesis, particularly changes in proteins of photosynthesis were observed in mutants. Likewise, comparison of wild type and Δ*petJ* cells grown in the absence of copper revealed changes in 29 proteins involved in some major metabolic pathways. Carbohydrate metabolism proteins, antenna proteins, and enzymes participating in the synthesis of different pigments were significantly changed in the mutant cells.

A proteome-based study of *Anabaena* sp. PCC 7120 subjected to LC_{50} dose of arsenic ($Na_2AsO_4 \cdot 7H_2O$) was performed in control and treated samples at 1 and 15 days by 2-DE and identified using MALDI-TOF/MS and LC-MS. Of the 45 proteins, belonging to different categories like antioxidative defense system, photosynthesis, respiration, amino acid metabolism, cofactor biosynthesis, protein synthesis and folding, and regulatory and hypothetical proteins, 31 were significantly up-regulated, while others including the DnaK and antioxidative enzymes like SOD, Fe-SOD, and AhpC/TSA were down-regulated [110]. While induction of catalase, peroxiredoxin, thioredoxin, and oxidoreductase provides testimony to their roles in protecting the cells from

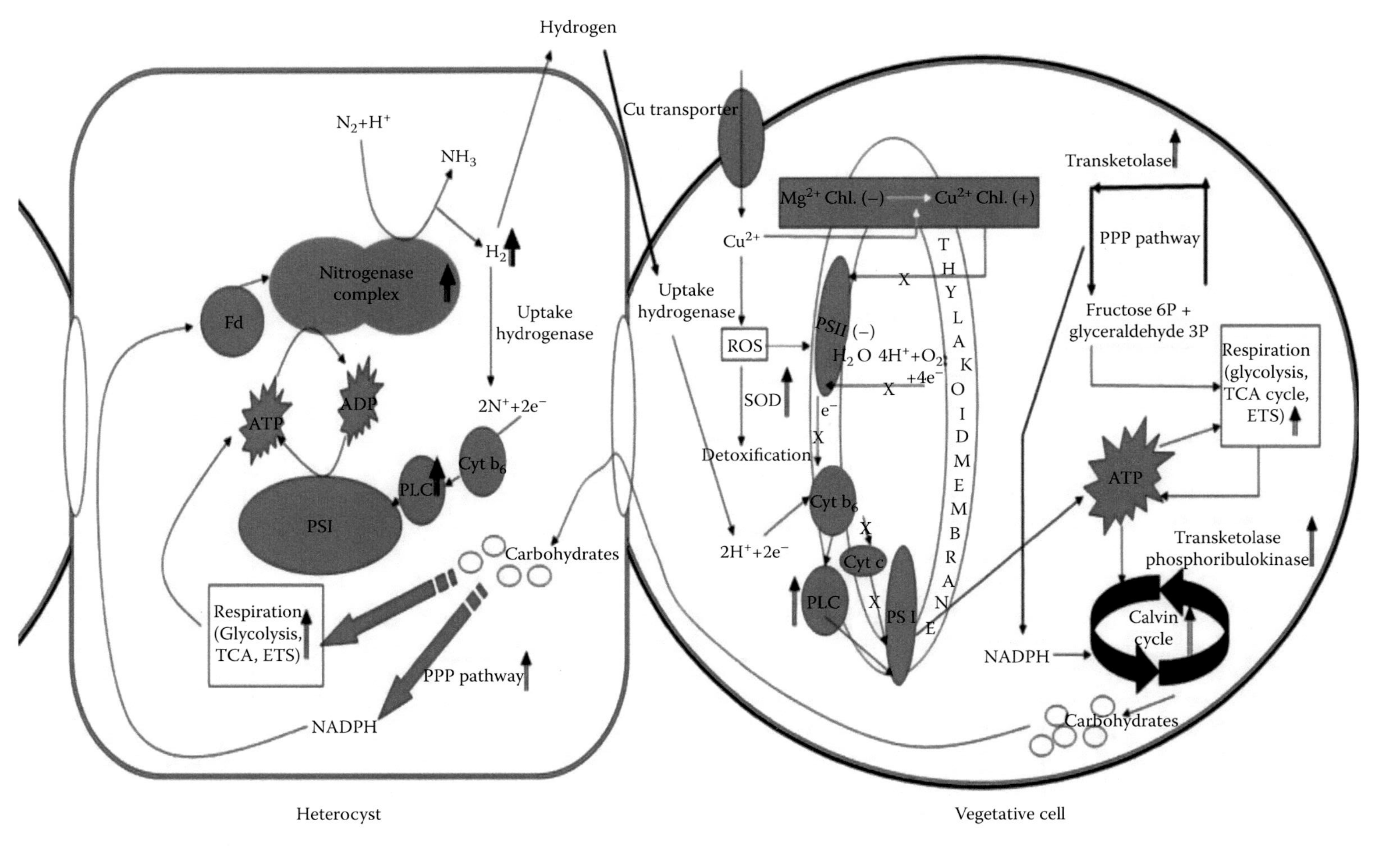

FIGURE 4.9 Diagrammatic sketch for the survival strategy of *A. doliolum* under excess Cu; "X" represents inhibition, and thick arrows indicate enhancement. (Modified from Bhargava, P. et al., *Photosynth. Res.*, 96, 61, 2008.)

oxidative stress, the down-regulation of the enzymes of energy metabolism like phosphoglycerate kinase, fructose bisphosphate aldolase II, fructose 1,6 bisphosphatase, transketolase, alpha aminolevulinic acid, uroporphyrinogen decarboxylase, and ATP synthase on day 1 and their enhanced expression on 15th day went hand-in-hand with reduction and/or enhancement in carbon fixation, photosynthesis, ATP content, and nitrogenase activity on day 1 and 15, respectively. Furthermore, increase in phytochelatin content and activity of glutathione S-transferase suggested their role in arsenic sequestration and protection of the cyanobacterium from arsenic-induced oxidative damage. Thus, it was demonstrated that arsenate reductase, arsenite reductase efflux genes, phytochelatin, Glutathion-S transferase, peroxiredoxin, and thioredoxin are the major players in cell survival under arsenic stress [110].

The best-studied case of metal stress response corresponds to iron deficiency, as it is an essential metal for all photosynthetic organisms. Iron is abundant in nature, but it is usually a limiting factor for photoautotrophic growth because of its poor solubility in aerobic environments. In *Synechocystis*, iron limitation enhances the expression of (1) *isiA*, a gene that codifies for an additional antenna system around photosystem I (PSI); (2) *isiB*, a gene coding for flavodoxin, a flavin-containing protein that replaces ferredoxin [111]; and (3) *idiA*, a gene coding for an integral subunit of PSII [112]. In *Synechococcus*, the *dpsA* gene encodes a stress-inducible DNA-binding protein whose transcription slightly increases during the stationary phase but as much as 12-fold under low iron conditions. Also the functioning of *dpsA* is essential for cell growth when iron is a limiting factor [113]. Similarly, *all3940* (a *dps* family gene) of *Anabaena* PCC 7120 was found to play a key role in combating nutrients limitation and multiple abiotic stresses [114]. An increased microcystin production in iron-starved condition has been reported in *Microcystis aeruginosa* strains suggesting that a highly efficient system for coping with short-term iron limitation is in place in these organisms [115].

4.2.5 pH Stress

Cyanobacteria prefer neutral to slightly alkaline pH for optimum growth. Alterations in pH can evoke a wide spectrum of changes in their cell behavior. There are few proteome-based studies related to pH response in cyanobacteria. *Synechocystis* PCC 6803 has been shown to exhibit physiological changes to acid stress [23]. Recently, it was reported that in *Synechocystis* hundreds of genes were up-regulated above 1.5-fold upon transfer of cells from pH 7.5 to 10.2 [116]. Likewise, Ohta et al. [117] have reported differential expression of a number of genes in *Synechocystis* when exposed to acid stress for varying time intervals. However, in proteomic analysis very few spots were detected in comparison to high-pH-regulated genes [118]. A study on cytoplasmic and periplasmic proteins showed altered expression at different external pH, that is, 9.0, 7.5, 6.0, and 5.5. In comparison to cytoplasmic fraction, periplasmic fractions showed remarkable changes in protein expression and is not surprising because the periplasm is more exposed to bear the brunt of environmental fluctuations. This suggests the importance of this soluble compartment in cyanobacterial adaptation to abiotic stresses. Oxalate decarboxylase and carbonic anhydrase, already known for their role in pH homeostasis, were induced upon pH stress. Several proteins of unknown functions were also induced upon pH stress [118]. Plasma membrane related studies by Zhang et al. [119] identified 39 high-pH-responsive proteins. The largest group of proteins enhanced by high pH consists of transport- and substrate-binding proteins of ABC transporters as well as proteins with unknown function. Up-regulation of proteins belonging to substrate- and ATP-binding category suggests that phosphate uptake system is highly susceptible to the elevated media pH. MinD involved in cell division, Cya2 in signaling, PsaF and CoxB proteins involved in photosynthesis and respiration, and several regulatory proteins were also identified. Among these proteins eight were found to be hypothetical ones. Although such complementary approaches have provided a new insight into pH-based cellular responses, a complete understanding of the effect of pH on basic cellular processes like carbon fixation, nitrogen status, photosynthesis, and respiration is still lacking.

4.2.6 Pesticide Stress

Pesticides are of equal importance for the farmer in controlling weeds, undesirable plants in monocultures of crop plants [120], and for the plant physiologists in trying to understand their mode of action, on the one hand, and, in addition, to using pesticides of known specificity and their mode of action as tools in the elucidation of physiological mechanisms, on the other hand [121]. Besides controlling pests, these pesticides damage a wide variety of nontarget species including cyanobacterial mats in paddy fields. An undesirable side effect from the use of pesticides is that they enter into freshwater ecosystems by spray, drift, leaching, runoff, or accidental spills [122] and persist in there [123] causing toxicity. Several reviews and research reports have discussed the effect of pesticides on algae, cyanobacteria, and other organisms [124–126]. Differential tolerance of blue-green algae to pesticides has also been reported [127]. Pesticides include mainly three classes: organic phosphate compounds, chlorinated hydrocarbons, and carbamates [128–130].

Pesticide application is known to affect negatively the soil microflora [131]. Microbial degradation is considered to be the primary mechanism for their dissipation from soil [132]. The pesticide-degrading bacteria produce a variety of enzymes with different specificities for their degradation. Such bacterial enzyme extracts have been characterized with respect to the rate and specificity of pesticides [133,134]. Since the degrading processes are directly executed by proteins, the state of degrading bacteria is essentially reflected in their proteome rather than genome, which is much more stable [135]. Cyanobacterial studies are particularly important in this because they have been reported to accumulate and detoxify these pesticides. Investigations pertaining to different pesticides have indicated a direct effect on the PSII activity, although there are very few proteome-based approaches in this area so far.

Among several pesticide studies, protein-based data are available for a chloroacetamide herbicide named butachlor, which can inhibit the very-long-chain fatty acid synthase [136]. It is a preemergence herbicide for the control of annual grasses and certain broad-leaved weeds in both seeded and transplanted rice. Two-dimensional gel electrophoresis based studies in butachlor-treated *Aulosira fertilissima* were performed at temporal isolation of 3, 6, and 15 days, respectively, by Kumari et al. [131]. This study showed 2.53-fold up-regulation of NusB protein, while proteins like fructose 1,6-bisphosphate aldolase class II (FBP aldolase), ATP synthase β-chain, C-phycocyanin α-subunit, GroES (10 kDa chaperonin), allophycocyanin β-chain, peroxiredoxin (Prx), and phycocyanin α-chain showed a severe down-regulation. Of the eight proteins characterized, six were found to be functionally associated with photosynthesis, respiration, and antioxidative defense system. In addition to this, down-regulation of GroES protein in the proteome map also indicated a negative effect on PSI and PSII and rubisco assembly. The up-regulation of Nus B and the down-regulation of GroES proteins together were thought to be responsible for the death of *A. fertilissima*.

4.3 CONCLUSION

Abiotic-stress-elicited responses in cyanobacteria have been studied widely and deeply. Proteins are the part mostly affected and are the real effectors in abiotic-stress-induced metabolic hustle and bustle. They encounter the stresses either directly as metal chelators (phytochelatin and metallothionine) and as antioxidative enzymes (superoxide dismutase, alkyl hydroxyl peroxide reductase, catalase, etc.) or indirectly via osmolyte synthesis and detoxification. Osmoprotectants like glucosyl glycerol, glycine betain, and proline favor the native form of the majority of proteins [137]. Molecular chaperonins such as GroEL are now widely regarded as essential components for the stabilization of integral membrane or secretory proteins before membrane insertion or translocation as well as for the assembly of macromolecular complexes like ribulose bis phosphate carboxylase oxygenase [138]. They also play key role in protein folding and as a catalyst of biomolecular events. Besides, conserved proteins like PII are reported to transmit signals of the cellular N and C status through phosphorylation of a serine residue [23]. In a nutshell, they work for system processing and as restorer under altered cellular homeostasis.

An overall assessment of protein data studied to date reflects two imperative edges. Despite the difference in the primary target in different stresses, due to interlinked metabolic web and series of interdependent reactions, a common effect on physiological attributes like carbon fixation, respiration, and nitrogen metabolism is known. Photosynthetic apparatus, especially PSII-associated D1 and D2 proteins, are most prone to damage. The role of HSPs in nearly all stresses demonstrates an overlap in the complex web of proteins. Stresses like cold, salt, drought, metals, and metalloids show a somewhat similar pattern of damage. Nearly all stresses are reported to show enhanced expression of proteins associated in oxidative damage. In contrast, some highlighting differences in the form of major areas affected and the overall mode of action can be observed. Photorespiration is mainly triggered in response to salt and UV radiation in some cases. UV-B directly causes damage at the level of DNA. While metals affect redox status of the system, temperature and pesticides mainly target membrane dynamics.

Second, a major chunk of proteins studied so far in cyanobacterial stress biology show similarity to that in higher plants, which further reiterates the close phylogenetic relationship of the two. The area of proteomics in cyanobacterial is only just burgeoning. Much of what we know is confined mainly to cytosolic fractions and in some most-opted models for study like *Anabaena, Synechocystis, Microcystis*, etc.

Although a wide range of studies have been conducted so far, we still find gaps in an overall understanding of stress responses in cyanobacterial proteomics. There is a need to further explore the complicated web of proteins and metabolites under combined stress conditions in order to cover the entire range and types of proteins and their role in survival and sustenance of cyanobacteria.

ACKNOWLEDGMENTS

L.C. Rai is thankful to DST for project and J.C. Bose National Fellowship. Snigdha Rai, Sarita Pandey, and Alok Kumar Shrivastava are thankful to CSIR-UGC for JRF and SRF. Chhavi Agrawal is thankful for JRF from DBT and Prashant Kumar Singh to DAE-BRNS for financial support.

REFERENCES

1. Los, D.A. and Murata, N., Responses to cold shock in cyanobacteria, *J. Mol. Microbiol. Biotechnol.*, 1, 221, 1999.
2. Capone, D.G. et al., *Trichodesmium*, a globally significant marine cyanobacterium, *Science*, 276, 1221, 1997.
3. Karl, D.M., Nutrient dynamics in the deep blue sea, *Trends Microbiol.*, 10, 410, 2002.
4. Singh, R.N., *Role of Blue Green Algae in Nitrogen Economy of Indian Agriculture*. New Delhi, India: Indian Council for Agricultural Research, 1961, p. 175.
5. Yadav, S. et al., Cyanobacterial secondary metabolites, *Int. J. Pharm. Bio. Sci.*, 2, B144, 2011.
6. Murata, N. and Wada, H., Acyl-lipid desaturases and their importance in the tolerance and acclimatization to cold of cyanobacteria, *Biochem. J.*, 308, 1, 1995.
7. Glatz, A. et al., The *Synechocystis* model of stress: From molecular chaperones to membranes, *Plant Physiol. Biochem.*, 37, 1, 1999.
8. Los, D.A. and Murata, N., Membrane fluidity and its roles in the perception of environmental signals, *Biochem. Biophys. Acta*, 1666, 142, 2004.
9. Los, D.A. et al., Stress responses in *Synechocystis*: Regulated genes and regulatory systems, in *The Cyanobacteria: Molecular Biology, Genomics and Evolution*, Herrero, A. and Flores, E., Eds. Norfolk, U.K.: Caister Academic Press, 2008, p. 117.
10. Casteilli, O. et al., Proteomic analyses of the response of cyanobacteria to different stress condition, *FEBS Lett.*, 583, 1753, 2009.

11. Xiong, L., Schumaker, K.S., and Zhu, J.K., Cell signaling during cold, drought and salt stress, *Plant Cell*, 14, S165, 2002.
12. Amme, S. et al., Proteome analysis of cold stress response in *Arabidopsis thaliana* using DIGE-technology, *J. Exp. Bot.*, 57, 1537, 2006.
13. Donker, V.A. and Hader, D.P., Effects of ultraviolet irradiation on photosynthetic pigments in some filamentous cyanobacteria, *Aquat. Microb. Ecol.*, 11, 143, 1996.
14. Xia, J., Response of growth, photosynthesis, and photoinhibition of the edible cyanobacterium *Nostoc sphaeroides* colonies to thiobencarb herbicide, *Chemosphere*, 59, 561, 2005.
15. Dhankher, O.P., Arsenic metabolism in plants: An inside story, *New Phytol.*, 168, 503, 2005.
16. Huang, F. et al., Proteomic screening of salt-stress-induced changes in plasma membranes of *Synechocystis* sp. strain PCC 6803, *Proteomics*, 6, 910, 2006.
17. Fulda, S. et al., Proteomics of *Synechocystis* sp. strain PCC 6803: Identification of periplasmic proteins in cells grown at low and high salt concentrations, *Eur. J. Biochem.*, 267, 5900, 2000.
18. Fulda, S. et al., Proteome analysis of salt stress response in the cyanobacterium *Synechocystis* sp. strain PCC 6803, *Proteomics*, 6, 2733, 2006.
19. De Souza, G.A. and Wiker, H.G., The impact of proteomic advances on bacterial gene annotation, *Curr. Proteomics*, 6, 84, 2009.
20. Gao, Y. et al., Identification of the proteomic changes in *Synechocystis* sp. PCC 6803 following prolonged UV-B irradiation, *J. Exp. Bot.*, 60, 1141, 2009.
21. Graves, P.R. and Haystead, T.A.J., Molecular biologist's guide to proteomics, *Microbiol. Mol. Biol. Rev.*, 66, 39, 2002.
22. Monteoliva, L. and Albar, J.P., Differential proteomics: An overview of gel and non gel based approaches, *Brief Funct. Genomic Proteomic*, 3, 220, 2004.
23. Huang, F. et al., Proteomics of *Synechocystis* sp. Strain PCC 6803: Identification of plasma membrane proteins, *Mol. Cell Proteomics*, 1, 956, 2002.
24. Huang, F. et al., Isolation of outer membrane of *Synechocystis* sp. PCC 6803 and its proteomic characterization, *Mol. Cell Proteomics*, 3, 586, 2004.
25. Kashino, Y. et al., Proteomic analysis of a highly active photosystem II preparation from the cyanobacterium *Synechocystis* sp. PCC 6803 reveals the presence of novel polypeptides, *Biochemistry*, 41, 8004, 2002.
26. Wang, Y.C., Sun, J., and Chitnis, P.R., Proteomic study of the peripheral proteins from thylakoid membranes of the cyanobacterium *Synechocystis* sp. PCC 6803, *Electrophoresis*, 21, 1746, 2000.
27. Alexova, R. et al., Comparative protein expression in different strains of the bloom-forming cyanobacterium *Microcystis aeruginosa*, *Mol. Cell Proteomics*, 10(9), M110, 2011.
28. Eubel, H., Braun, H.P., and Millar, A.H., Blue-native PAGE in plants: A tool in analysis of protein-protein interactions, *Plants Methods*, 16, 11, 2005.
29. Peng, Y. et al., A blue native PAGE analysis of membrane protein complexes in *Clostridium thermocellum*, *BMC Microbiol.*, 11, 22, 2011.
30. Pachauri, R.K. and Reisinger, A., Eds. Intergovernmental panel on climatic change, Forth assessment report: Climate change 2007 (AR4). Geneva, Switzerland, 2007, p. 104.
31. Sung, D.Y. et al., Acquired tolerance to temperature extremes, *Trends Plant Sci.*, 8, 179, 2003.
32. Berry, J. and Bjorkman, O., Photosynthetic response and adaptation to temperature in higher plants, *Annu. Rev. Plant Physiol.*, 31, 491, 1980.
33. Brock, D.T. et al., *Biology of Microorganisms*, 7th edn. Englewood Cliffs, NJ: Prentice Hall, 1994, Chapters 18 and 19.
34. Los, D.A. and Murata, N., Structure and expression of fatty acid desaturases, *Biochem. Biophys. Acta*, 1394, 3, 1998.
35. Morgan-Kiss, R.M. et al., Adaptation and acclimation of photosynthetic microorganisms to permanently cold environments, *Microbiol. Mol. Biol. Rev.*, 70, 222, 2006.
36. Zhang, J.H. et al., Changes in membrane associated H^+/ATPase activities and amounts in young grape plants during the cross adaptation to temperature stresses, *Plant Sci.*, 170, 768, 2006.
37. Singh, S.C., Sinha, R.P., and Hader, D.P., Role of lipid and fatty acids in stress tolerance in cyanobacteria, *Acta Protozool.*, 41, 297, 2002.
38. Gombos, Z., Wada, H., and Murata, N., Unsaturation of fatty acids in membrane lipids enhances tolerance of the cyanobacterium *Synechocystis* PCC 6803 to low-temperature photoinhibition, *Proc. Natl Acad. Sci. U.S.A.*, 89, 1992, 9959.
39. Gombos, Z., Wada, H., and Murata, N., The recovery of photosynthesis from low temperature photoinhibition is accelerated by unsaturation of membrane lipids: A mechanism of chilling tolerance, *Proc. Natl Acad. Sci. U.S.A.*, 91, 1994, 8787.

40. Wada, H. and Murata, N., Temperature-induced changes in the fatty acid composition of the cyanobacterium, *Synechocystis* PCC 6803, *Plant Physiol.*, 92, 1062, 1990.
41. Wada, H., Gombos, Z., and Murata, N., Contribution of membrane lipids to the ability of the photosynthetic machinery to tolerate temperature stress, *Proc. Natl Acad. Sci. U.S.A.*, 91, 1994, 4273.
42. Tasaka, Y. et al., Targeted mutagenesis of acyl-lipid desaturases in *Synechocystis*: Evidence for the important roles of polyunsaturated membrane lipids in growth, respiration and photosynthesis, *EMBO J.*, 15, 6416, 1996.
43. Inoue, N. et al., Acclimation to the growth temperature and the high-temperature effects on photosystem II and plasma membranes in a mesophilic cyanobacterium, *Synechocystis* sp. PCC 6803, *Plant Cell Physiol.*, 42, 1140, 2001.
44. Suzuki, I., Simon, W.J., and Slabas, A.R., The heat shock response of *Synechocystis* sp. PCC 6803 analysed by transcriptomics and proteomics, *J. Exp. Bot.*, 57, 1573, 2006.
45. Slabas, A.R. et al., Proteomic analysis of the heat shock response in *Synechocystis* PCC 6803 and a thermally tolerant knockout strain lacking the histidine kinase 34 gene, *Proteomics*, 6, 845, 2006.
46. Hongsthong, A. et al., Subcellular proteomic characterization of the high-temperature stress response of the cyanobacterium *Spirulina platensis*, *Proteome Sci.*, 7, 1, 2009.
47. Mishra, Y. et al., Proteomic evaluation of the non-survival of *Anabaena doliolum* (Cyanophyta) at elevated temperatures, *Eur. J. Phycol.*, 44, 551, 2009.
48. Schmid, A.K. et al., Global whole-cell FTIR mass spectrometric proteomics analysis of the heat shock response in the radioresistant bacterium *Deinococcus radiodurans*, *J. Proteome Res.*, 4, 709, 2005.
49. Tanaka, N., Hiyama, T., and Nakamoto, H., Cloning, characterization and functional analysis of groESL operon from thermophilic cyanobacterium *Synechococcus vulcanus*, *Biochim. Biophys. Acta*, 1343, 335, 1997.
50. Kovács, E. et al., The chaperonins of *Synechocystis* PCC 6803 differ in heat inducibility and chaperone activity, *Biochem. Biophys. Res. Commun.*, 289, 908, 2001.
51. Sato, S., Ikeuchi, M., and Nakamoto, H., Expression and function of groEL paralog in the thermophilic cyanobacterium *Thermosynechococcus elongatus* under heat and cold stress, *FEBS Lett.*, 582, 3389, 2008.
52. Singh, A.K. et al., The heat shock response in the cyanobacterium *Synechocystis* sp. strain PCC 6803 and regulation of gene expression by HrcA and SigB, *Arch. Microbiol.*, 186, 273, 2006.
53. Eriksson, M.J. et al., Novel form of ClpB/Hsp100 protein in the cyanobacterium *Synechococcus*, *J. Bacteriol.*, 183, 7392, 2001.
54. Tanaka, N. and Nakamoto, H., HtpG is essential for the thermal stress management in cyanobacteria, *FEBS Lett.*, 458, 117, 1999.
55. Sato, N. and Nakamura, A., Involvement of 5′ untranslated region in cold regulated expression of the rbpA1 gene in the cyanobacterium *Anabaena variabilis* M3, *Nucleic Acids Res.*, 26, 2192, 1998.
56. Eriksson, M.J. and Clarke, A.K., The heat shock proteins ClpB mediates the development of thermotolerance in the cyanobacterium *Synechococcus* sp. strain PCC 7942, *J. Bacteriol.*, 178, 4839, 1996.
57. Porankiewicz, J., Sachelin, J., and Clark, A.K., The ATP dependent Clp proteases is essential for acclimation to UV-B and low temperature in cyanobacterium *Synechococcus*, *Mol. Microbiol.*, 29, 275, 1998.
58. Chemot, D. et al., A cold shock-induced cyanobacterial RNA helicase, *J. Bacteriol.*, 181, 1728, 1999.
59. Szabolcs, I., Soil and salinization, in *Hand Book of Plant and Crop Stress*, Pessarakali, M., Ed. New York: Marcel Dekker, 1994, p. 3.
60. FAO, FAO Land and Plant Nutrition Management Service, 2007. http://www.fao.org/ag/agl/agll/spush
61. Roberts, M.F., Organic compatible solutes of halotolerant and halophilic microorganisms, *Saline Systems*, 1, 5, 2005.
62. Bartels, D. and Souer, E., Molecular responses of higher plants to dehydration, in *Plant Responses to Abiotic Stress*, Vol. 4, Hirt, H. and Shinozaki, K., Eds. Berlin, Germany: Springer-Verlag, 2004.
63. Csonka, L.N., Physiological and genetic responses of bacteria to osmotic stress, *Microbiol. Rev.*, 53, 121, 1989.
64. Kanesaki, Y. et al., Salt stress and hyperosmotic stress regulate the expression of different sets of genes in *Synechocystis* sp. PCC 6803, *Biochem. Biophys. Res. Commun.*, 290, 339, 2002.
65. Srivastava, A.K., Bhargava, P., and Rai, L.C., Salinity and copper-induced oxidative damage and changes in the antioxidative defence systems of *Anabaena doliolum*, *World J. Microbiol. Biotechnol.*, 21, 1291, 2005.
66. Serrano, R. and Gaxiola, R., Microbial models and salt stress tolerance in plants, *Crit. Rev. Plant Sci.*, 13, 121, 1994.
67. Singh, R.N., Reclamation of user lands in India through blue green algae, *Nature*, 165, 325, 1950.

68. Thomas, J. and Apte, S.K., Sodium requirement and metabolism in nitrogen fixing cyanobacteria, *J. Biosci.*, 6, 771, 1984.
69. Blumwald, E., Wolosin, J.M., and Packer, L., Na^+/H^+ exchange in cyanobacterium *Synechococcus* 6311, *Biochem. Biophys. Res. Commun.*, 122, 452, 1984.
70. Molitor, V., Erber, W., and Peschek, G.A., Increased levels of cytochrome-oxidase and sodium-proton antiporter in the plasma membrane of *Anacystis nidulans* after growth in sodium-enriched media, *FEBS Lett.*, 204, 251, 1986.
71. Fulda, S. et al., Analysis of stress responses in the cyanobacterial strains *Synechococcus* sp. PCC 7942, *Synechocystis* sp. PCC 6803, and *Synechococcus* sp. PCC 7418: Osmolyte accumulation and stress protein synthesis, *J. Plant Physiol.*, 154, 240, 1999.
72. Bhagwat, A.A. and Apte, S.K., Comparative analysis of proteins induced by heat-shock, salinity and osmotic stress in the nitrogen-fixing cyanobacterium *Anabaena* sp. strain L-31, *J. Bacteriol.*, 171, 5187, 1989.
73. Hagemann, M., Molecular biology of cyanobacterial salt acclimation, *FEMS Microbiol. Rev.*, 35, 87, 2011.
74. Li, S., Xu, M., and Su, Z., Computational analysis of *LexA* regulon in cyanobacteria, *BMC Genom.*, 11, 527, 2010.
75. Joset, F., Jeanjean, R., and Hagemann, M., Dynamics of the response of cyanobacteria to salt stress: Deciphering the molecular events, *Physiol. Plant*, 96, 738, 1996.
76. Hagemann, M. and Erdmann, N., Activation and pathway of GG synthesis in the cyanobacterium *Synechocystis* sp. PCC 6803, *Microbiology*, 140, 1427, 1994.
77. Allakhverdiev, S.I. et al., Genetic engineering of the unsaturation of fatty acids in membrane lipids alters the tolerance of *Synechocystis* to salt stress, *Proc. Natl. Acad. Sci. U.S.A.*, 96, 1999, 5862.
78. Allakhverdiev, S.I. et al., Salt stress inhibits the repair of photodamaged photosystem II by suppressing the transcription and translation of psbA genes in *Synechocystis*, *Plant Physiol.*, 130, 1443, 2002.
79. Apte, S.K., Reddy, B.R., and Thomas, J., Relationship between Na^+ influx and salt tolerance of nitrogen fixing cyanobacteria, *Appl. Environ. Microbiol.*, 53, 1934, 1987.
80. Apte, S.K. et al., Expression and possible role of stress responsive proteins in *Anabaena*, *J. Biosci.*, 23, 399, 1998.
81. Kawasaki, S. et al., Gene expression profiles during the initial phase of salt stress in rice, *Plant Cell*, 13, 889, 2001.
82. Moisinder, P.H., McClinton, E., and Paer, H.W., Salinity effects on growth, photosynthetic parameters, and nitrogenase activity in estuarine planktonic cyanobacteria, *Microbiol. Ecol.*, 43, 432, 2002.
83. Chen, L.Z. et al., Effects of salt stress on carbohydrate metabolism in desert soil alga *Microcoleus vaginatus* Gom., *J. Integr. Plant Biol.*, 48(8), 914, 2006.
84. Klahn, S. et al., The gene *ssl3076* encodes a protein mediating the salt-induced expression of *ggps* for the biosynthesis of the compatible solute glucosylglycerol in *Synechocystis* sp. strain PCC 6803, *J. Bacteriol.*, 192, 4403, 2010.
85. Eisenhut, M. et al., The plant-like C_2 glycolate pathway and the bacterial-like glycerate cycle cooperate in phosphoglycolate metabolism in cyanobacteria, *Plant Physiol.*, 142, 333, 2006.
86. Srivastava, A.K. et al., Salinity-induced physiological and proteomic changes in *Anabaena doliolum*, *Environ. Exp. Bot.*, 64, 49, 2008.
87. Hagemann, M. et al., Salt-dependent protein phosphorylation in the cyanobacterium *Synechocystis* sp. PCC 6803, *FEMS Microbiol. Lett.*, 113, 205, 1993.
88. Tyagi, R. et al., Protective role of certain chemicals against UV-B induced damage in the nitrogen fixing cyanobacterium *Nostoc muscorum*, *J. Basic Microbiol.*, 43, 137, 2003.
89. Singh, S.P. et al., Photoprotective and biotechnological potentials of cyanobacterial potentials of cyanobacterial sheaths pigment, scytonemin, *Afr. J. Biotechnol.*, 9, 580, 2010.
90. Donker, V.A. and Hader, D.P., Effects of solar ultraviolet radiation on motility, photomovement and pigmentation in filamentous, gliding cyanobacterium, *FEMS Microbiol. Ecol.*, 86, 159, 1991.
91. Schulz, E.M. and Scherer, S., UV protection in cyanobacteria, *Eur. J. Phycol.*, 34, 329, 1999.
92. Hader, D.P. et al., Effects on aquatic ecosystems. UNEP Environmental Effects Panel Report, 86, 1998.
93. Donker, V.A. and Hader, D.P., Ultraviolet radiation effects on pigmentation in the cyanobacterium, *Acta Protozool.*, 36, 49, 1997.
94. Aratoz, R. and Hader, D.P., Ultravoilet radiation induces both degradation and synthesis of phycobilisomes in *Nostoc* sp.: A spectroscopic and biochemical approach, *FEMS Microbiol. Ecol.*, 23, 301, 1997.
95. Schulz, M.E. et al., The UV-B stimulon of the terrestrial cyanobacterium *Nostoc commune* comprises early shock proteins and late acclimation proteins, *Mol. Microbiol.*, 46, 827, 2002.

96. Kletzien, R.F., Harris, P.K., and Foellmi, L.A., Glucose-6-phosphate dehydrogenase: A 'housekeeping' enzyme subject to tissue-specific regulation by hormones, nutrients, and oxidant stress, *FASEB J.*, 8, 174, 1994.
97. Busch, K.B. and Fromm, H., Plant succinic semialdehyde dehydrogenase. Cloning, purification, localization in mitochondria, and regulation by adenine nucleotides, *Plant Physiol.*, 121, 589, 1999.
98. Pocsi, I., Prade, R.A., and Penninckx, M.J., Glutathione, altruistic metabolite in fungi, *Adv. Microbiol. Physiol.*, 49, 1, 2004.
99. Kumar, A., Sinha, R.P., and Häder, D.P., Effect of UV-B on enzymes of nitrogen metabolism in the cyanobacterium *Nostoc calcicola*, *J. Plant Physiol.*, 148, 86, 1996.
100. Sinha, R.P. et al., Effects of UV-B irradiation on growth, survival, pigmentation and nitrogen metabolism enzymes in cyanobacteria, *Acta Protozool.*, 34, 187, 1995.
101. Cavet, J.S., Borrelly, G.P., and Robinson, N.J., Zn, Cu and Co in cyanobacteria: Selective control of metal availability, *FEMS Microbiol. Rev.*, 27, 165, 2003.
102. Bhargava, P. et al., Excess copper induces anoxygenic photosynthesis in *Anabaena doliolum*: A homology based proteomic assessment of its survival strategy, *Photosynth. Res.*, 96, 61, 2008.
103. Clemens, S., Toxic metal accumulation, responses to exposure and mechanism of tolerance in plants, *Biochemie*, 88, 1707, 2006.
104. Surosz, W. and Palinska, K.A., Effects of heavy-metal stress on cyanobacterium *Anabaena flos-aquae*, *Arch. Environ. Contam. Toxicol.*, 48, 40, 2004.
105. Bhargava, P. et al., Cadmium mitigates ultraviolet-B stress in *Anabaena doliolum*: Enzymatic and non-enzymatic antioxidants, *Biol. Plantarum*, 51, 546, 2007.
106. Yamaguchi, K. et al., A two-component $Mn2^{+}$-sensing system negatively regulates expression of the mntCAB operon in *Synechocystis*, *Plant Cell*, 14, 2901, 2002.
107. Murata, N. and Suzuki, I., Exploitation of genomic sequences in a systematic analysis to access how cyanobacteria sense environmental stress, *J. Exp. Bot.*, 57, 235, 2006.
108. Katoh, H.I. et al., Genes essential to iron transport in the cyanobacterium *Synechocystis* sp. strain PCC 6803, *J. Bacteriol.*, 183, 2779, 2001.
109. De la Cerda, B. et al., A proteomic approach to iron and copper homeostasis in cyanobacteria, *Brief Funct. Genomic Proteomic*, 6, 322, 2008.
110. Pandey, S., Rai, R., and Rai, L.C., Proteomics combines morphological, physiological and biochemical attributes to unravel the survival strategy of *Anabaena* sp. PCC7120 under arsenic stress, *J. Proteomics*, 74, 921, 2012.
111. Kunert, A. et al., Repression by Fur is not the main mechanism controlling the iron-inducible isaAB operon in the cyanobacterium *Synechocystis* sp. PCC 6803, *FEMS Microbiol. Lett.*, 227, 255, 2003.
112. Lax, J.E.M. et al., Structural response of photosystem 2 to iron deficiency: Characterization of a new photosystem 2-IdiA complex from the cyanobacterium *Thermosynechococcus elongatus* BP-1, *Biochim. Biophys. Acta*, 1767, 528, 2007.
113. Sen, A. et al., Growth phase, metal dependent regulation of the *dpsA* gene in *Synechococcus* sp. strain PCC 7942, *Arch. Microbiol.*, 173, 352, 2000.
114. Narayan, O.P., Kumari, N., and Rai, L.C., Heterologous expression of *Anabaena* PCC 7120 all3940 (a Dps family gene) protects *Escherichia coli* from nutrient limitation and abiotic stresses, *Biochem. Biophys. Res. Commun.*, 394, 163, 2010.
115. Alexova, R. et al., Iron uptake and toxin synthesis in the bloom-forming *Microcystis aeruginosa* under iron limitation, *Environ. Microbiol.*, 13, 1064, 2011.
116. Summerfield, T.C. and Sherman, L.A., Global transcriptional response of the alkali-tolerant cyanobacterium *Synechocystis* sp strain PCC 6803 to a pH 10 environment, *Appl. Environ. Microbiol.*, 74, 5276, 2008.
117. Ohta, H. et al., Identification of genes expressed in response to acid stress in *Scynechocystis* sp. PCC6803 using DNA microarrays, *Photosynth. Res.*, 84, 225, 2005.
118. Kurian, R., Phadwal, K., and Maenpaa, P., Proteomic characterization of acid stress response in *Synechocystis* sp. PCC 6803, *Proteomics*, 6, 3614, 2006.
119. Zhang, L.F. et al., Proteomic analysis of plasma membranes of cyanobacterium *Synechocystis* sp. Strain PCC 6803 in response to high pH stress, *J. Proteome Res.*, 8, 2892, 2009.
120. Fedtke, C., *Biochemistry and Physiology of Herbicide Action*, Vol. 1. Berlin, Germany: Springer-Verlag, 1982, p. 38.
121. Trebst, A., The three dimensional structure of the herbicide binding niche on the reaction centre polypeptides of photosystem II, *Z. Natureforsch*, 42c, 742, 1987.

122. Van der Werf, H.M.G., Assessing the impact of pesticides on the environment, *Agric. Ecosyst. Environ.*, 60, 81, 1996.
123. Padhy, R.N., Cyanobacteria and pesticides, *Residue Rev.*, 95, 1, 1985.
124. Mallison, S.M. and Cannon, R.E., Effects of pesticides on cyanobacterium *Plectonema boryanum* and cyanophage LPP-1, *Appl. Environ. Microbiol.*, 47, 910, 1984.
125. Kapoor, K. and Arora, L., Influence of some pesticides on cyanobacteria in vitro condition, *Indian J. Environ. Ecoplann.*, 3, 219, 2000.
126. Ravindran, C.R., Suguna, S., and Shanmugasundaram S., Tolerance of *Oscillatoria* isolates to agrochemicals and pyrethroid components, *Indian J. Exp. Biol.*, 38, 402, 2000.
127. Sardeshpande, J.S. and Goyal, S.K., Effect of insecticides on growth and nitrogen fixation by blue-green algae, in *Proceedings of National Symposium on Biological Nitrogen Fixation*, Kaushik, B.D., Ed. New Delhi, India: IARI, 1982, p. 582.
128. Edwards, C.A., *Environmental Pollution by Pesticides*, 1st edn. London, U.K.: Plenum Press, 1973, p. 43.
129. Brown, A.W.A., Insecticides and fish, in *Ecology of Pesticides*. New York: John Wiley & Sons, Inc., 1978, p. 525.
130. Rand, G.M. and Sam, R.P., *Fundamentals of Aquatic Methods and Application*. Washington, DC: Hemisphere Publishing Corp., 1985.
131. Kumari, N., Narayan, O.P., and Rai, L.C., Understanding butachlor toxicity in *Aulosira fertilissima* using physiological, biochemical and proteomic approaches, *Chemosphere*, 77, 1501, 2009.
132. Sethunathan, N. and Yoshida T., Parathion degradation in submerged rice soils in the Philippines, *J. Agric. Food Chem.*, 21, 504, 1973.
133. Hoskin, F.C.G., Kirkish, M.A., and Steinmann, K.E., Two enzymes of the detoxification of organophosphorous compounds: Sources, similarities and significance, *Fundam. Appl. Toxicol.*, 4, S165, 1984.
134. Brown, H.M., Van, J.A.T., and Carski, T.H., Degradation of thifensulfuron methyl in soil: Role of microbial carboxyesterase activity, *J. Agric. Food Chem.*, 45, 955, 1997.
135. Mulbry, W.W., Karns, J.S., and Kearney, P.C., Identification of plasmid borne parathion hydrolysate gene from *Pseudomonas dimunita*, *Appl. Environ. Microbiol.*, 51, 926, 1986.
136. Gotz, T. and Boger, P., The very-long-chain fatty acid synthase is inhibited by chloroacetamides, *Z. Naturforsch.*, 59c, 549, 2004.
137. Da Costa, M.S., Santos, H., and Galinski, E.A., An overview of the role and diversity of compatible solutes in bacteria and archaea. *Adv. Biochem. Eng. Biotechnol.*, 61, 117, 1998.
138. Webb, R., Reddy, K.J., and Sherman, L.A., Regulation and sequence of the *Synechocystis* sp. strain PCC 7942 groESL operon, encoding a cyanobacterial Chaperonin, *J. Bacteriol.*, 172, 5079, 1990.

5 Molecular Chaperones and Stress Tolerance in Cyanobacteria

*Hitoshi Nakamoto**

CONTENTS

* Corresponding author: nakamoto@mail.saitama-u.ac.jp

5.1 INTRODUCTION

Molecular chaperones play essential roles in protein quality control in the cell. Among molecular chaperones, five protein families are recognized as highly conserved, and ubiquitously distributed [1–3]. They are small Hsp, chaperonin or Hsp60 (GroEL), Hsp70 (DnaK), Hsp90 (HtpG), and Hsp104 (ClpB) (Table 5.1). GroEL, DnaK, HtpG, and ClpB are terms normally used for prokaryotic members. The chaperone families include members whose expression is constitutive and others that are responsive to heat and/or other stresses.

The highly crowded environment of the cell promotes the aggregation of nonnative protein molecules [4]. Unfolded, partially folded, or misfolded polypeptides tend to aggregate because they typically expose hydrophobic amino acid residues and regions of unstructured polypeptide backbone

TABLE 5.1
Major Molecular Chaperones

Chaperone Families	Biochemical Functions	Homologs in Cyanobacteria
Small Hsp	Prevent protein aggregation	HspA
Hsp60	Prevent protein aggregation Assist protein folding	GroEL1, GroEL2
Hsp70	Prevent protein aggregation Assist protein folding Solubilize protein aggregates Pull translocating proteins across membranes	DnaK1, DnaK2, DnaK3
Hsp90	Prevent protein aggregation Binding and stabilization/regulation of proteins such as signaling molecules	HtpG
Hsp100	Solubilize protein aggregates	ClpB1, ClpB2

that are normally buried within the interior of the protein. The aggregates are thermodynamically stable but potentially toxic to the cell. Polypeptides take nonnative forms during de novo folding, subunit assembly, dissociation of macromolecular complexes, transport of proteins into organelles, and intracellular protein breakdown under normal conditions [1–3]. They are forced to take nonnative forms under stress conditions such as high temperature. Thus, chaperones are needed under both non-stress and stress conditions.

A molecular chaperone recognizes and binds to the exposed hydrophobic surface of a nonnative protein to prevent its aggregation [1–3]. It also brids to polypetide backbone. This molecular chaperone/nonnative protein complex is not a dead-end product but a productive intermediate since the nonnative protein is bound by the chaperone in a refoldable state. Thus, a chaperone can hold it in a folding/assembly competent state. This chaperone activity is sometimes called as "holdase activity." Function of chaperones such as GroEL and DnaK is to assist a nonnative protein acquire its native conformation, with concomitant ATP hydrolysis. GroEL, DnaK, HtpG, and ClpB have weak ATPase activity. Binding/hydrolysis of ATP is usually essential for the binding/release of a protein substrate from these molecular chaperones. Furthermore, chaperones such as DnaK and ClpB disaggregate/solubilize protein aggregates to rescue proteins from an aggregated state. The molecular chaperone acts like an enzyme because its interaction with a nonnative protein(s) is transient and it is not present in native proteins or protein complexes.

In this chapter, I will describe the expression and function of the aforementioned five representative molecular chaperones in relation to abiotic stresses, in particular heat stress. I will also describe DnaJ (Hsp40), a co-chaperone for DnaK as it is not possible to understand the function of DnaK without its partner DnaJ. A histidine kinase that is involved in the transduction of stress signals regulates the expression of molecular chaperones [5]. Transcription of genes encoding some molecular chaperones depends on alternative sigma factors [6–8]. Furthermore, their transcription is regulated by changes in the physical order of membrane lipids [9]. In this chapter, I will not cover these global gene regulation mechanisms since they will be treated in other chapters, but rather focus on specific transcription factors/elements that are involved in the expression of particular chaperone genes. In addition to the representative molecular chaperones, several "heat shock proteins" such as Orf7.5, IsiA, and IsiB play an essential role under heat and other stresses [10,11]. I will also describe Orf7.5 briefly in this chapter.

5.2 SMALL HEAT SHOCK PROTEIN

5.2.1 Introduction

The small heat shock proteins (small Hsps or sHSPs) whose monomeric molecular masses range from 12 to 42 kDa have the most diverse amino acid sequences among the major Hsps [12,13]. They are distinguished by their secondary and tertiary structure rather than their size and primary structure. They have a conserved, core α-crystallin domain of ~100 amino acids, which is flanked by a divergent N-terminal arm and a C-terminal extension (Figure 5.1). The α-crystallin domain, which is named after the α-crystallin of the vertebrate eye lens, is built from a β-sandwich, comprising two anti-parallel β-sheets [14,15]. This domain mediates dimer formation of individual protomers, which is the most stable suboligomeric form or building unit as the small Hsps form large oligomers. For example, small Hsp from *Methanococcus jannaschii*, a hyperthermophilic archaeon, consists of 24 identical subunits [14], and wheat HSP16.9 is a homo 12-mer [15].

What distinguishes small Hsp from other representative molecular chaperones is the fact that it neither has ATPase activity nor the ability to assist/facilitate the refolding of a nonnative protein on its own [12,13]. However, small Hsp can bind a nonnative protein to form a soluble complex, preventing its aggregation. Small Hsps maintain nonnative polypeptides in a refolding-competent state [16,17].

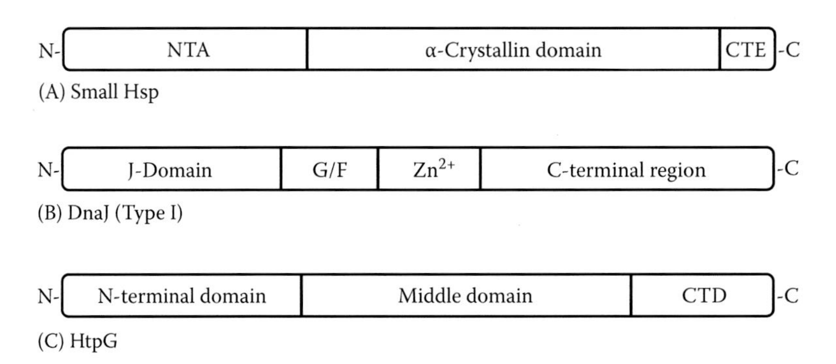

FIGURE 5.1 Linear representation of (A) small Hsp, (B) DnaJ, and (C) HtpG. The N-terminal arm and C-terminal extension of small Hsp are abbreviated as NTA and CTE, respectively. The glycine/phenylalanine-rich and cysteine-rich regions of DnaJ are abbreviated as G/F and Zn^{2+}, respectively. The C-terminal domain of HtpG is abbreviated as CTD.

The polypeptides can then be renatured with the assistance of ATP-dependent chaperones such as DnaK/Hsp70 and ClpB/Hsp104 [16–20].

Unlike DnaK and GroEL, which have defined substrate binding sites, in the case of small Hsp there is no specific surface involved in substrate binding, although the N-terminal arm is the major structural feature involved in substrate contacts [21]. Small Hsp relies on multiple contact sites distributed throughout the protein to protect substrates from irreversible aggregation. This may be the reason why the molar ratio of small Hsp to a substrate protein is variable.

Small Hsps have the ability to sense changes in temperature and can shift from an inactive to a chaperone-active state upon temperature upshift. There are two molecular mechanisms underlying this temperature regulation. In one mechanism, the small Hsp rearranges its quaternary organization at elevated temperatures. The small Hsp oligomer dissociates into sub-oligomeric species, which subsequently recombine with oligomeric species to form various higher-order oligomers [22]. The higher-order oligomers can bind nonnative proteins. In the other mechanism, a conformational rearrangement within the oligomer is responsible for the activation [23]. In this mechanism, subunit dissociation and/or subunit exchange may not be involved in the activation.

5.2.2 Cyanobacterial Small Hsp

5.2.2.1 Genes

The first cyanobacterial small Hsp gene, *hspA*, was cloned from the genomic library of the thermophilic cyanobacterium *Synechococcus vulcanus* [24]. Southern blot analysis indicated the absence of other *hspA* homologous genes in the genome. Complete genome sequencing also revealed the presence of a single small Hsp gene, sll1514, which is referred to as *hsp16.6*, *hsp17*, or *hspA* in *Synechocystis* sp. PCC 6803 [25]. In order to avoid confusion, I will use *hspA* to describe the *Synechocystis* gene encoding a small Hsp throughout this chapter. Some cyanobacterial genomes contain multiple genes encoding small Hsps. For example, the filamentous cyanobacterium *Anabaena* sp. PCC 7120 has two small Hsp, Hsp17.8 and Hsp17.1 [26].

5.2.2.2 Gene Expression under Various Stresses

DNA microarray analysis has shown that expression of *Synechocystis hspA* is enhanced by high light intensity, high salt concentration, hyperosmotic, high temperature, UV-B, peroxide, acid (pH 3.0), or alkali (pH 10) stress [27–33]. Northern blot analysis has shown that *Synechocystis hspA* is induced by high temperature, high salt concentration, or methylviologen stress [11,34–36].

According to proteome analysis the level of *Synechocystis*, HspA protein is enhanced by heat shock, or salt shock treatments [37,38]. Small Hsp as well as GroEL are the most abundantly synthesized cyanobacterial Hsps [39,40].

5.2.2.3 Regulatory Mechanism for Gene Expression

The heat-induction of the *hspA* gene expression is rapid, intense, and transient under light [24,35]. In the dark, the induction and the shut off of the heat shock response become slower [35].

Studies indicate that the expression of the cyanobacterial small Hsp gene is regulated at both the transcriptional and the posttranscriptional level as discussed below.

5.2.2.3.1 Gene Regulation at the Transcriptional Level

Transcription of *S. vulcanus hspA* is transiently heat-inducible at a promoter that is homologous to the *Escherichia coli* σ^{70}-dependent promoter. The promoter region lacks any known regulatory DNA elements [24], indicating that a unique regulatory mechanism suppresses the expression of *hspA* under non heat-shock conditions. Consistent with this, expression of the *hspA* gene was constitutive when the gene with its upstream regulatory region was introduced into *E. coli* [41]. In order to test the aforementioned hypothesis, we performed electrophoretic mobility shift assays with cell extracts to detect a (regulatory) protein that binds to a 5′-untranslated region of the *hspA* gene [42]. The results indicated the presence of such a protein(s) in the extract of unstressed cells of *Thermosynechococcus elongatus*, a thermophilic cyanobacterium closely related to *S. vulcanus*. The target sequence to which the putative DNA-binding protein binds was an AT-rich imperfect inverted-repeat sequence (ACAAgcAAA-TTTagTTGT). The DNA-binding activity in the cell, as well as in the cell extract, was lost much more quickly at a heat-shock temperature than at normal growth temperature. In cells, the activity was restored within 45 min after heat-shock by the heat-induced synthesis and stabilization of a DNA-binding protein. Thus, a plausible regulatory mechanism for the *hspA* expression is that the inverted repeat serves as a *cis*-active binding site, whereas the DNA-binding protein serves as a negative regulator (repressor). Upon heat shock, the *hspA* transcription takes place because the repressor becomes unstable and/or degraded transiently.

5.2.2.3.2 Gene Regulation at the Posttranscriptional Level

The first indication for the presence of a posttranscriptional regulatory mechanism for small Hsp gene expression came from a study in *E. coli* [41]. When *S. vulcanus hspA* was expressed in *E. coli*, the level of the HspA protein increased upon a temperature upshift without change in its mRNA level. Neither replacement of the native *hspA* promoter with the *lacZ* promoter nor the addition of a transcriptional inhibitor abolished the heat induction. Analyses of the expression of a series of the transcriptional and translational *hspA-lacZ* fusions demonstrated that the heat induction occurs only in the translational fusions. The expression analysis indicated that an element in the 5′-untranslated region and another one in the *hspA* coding region of the *hspA* mRNA are involved in regulation of the translation.

A 5′-untranslated region of the *Synechocystis hspA* mRNA has been studied in detail since a putative RNA thermometer was found [43]. RNA thermometers are translational control elements built into the 5′-untranslated region of bacterial heat shock or virulence genes [44]. Typically, they form a secondary structure that traps the Shine–Dalgarno (SD) sequence at low temperatures, resulting in inhibition of translation. An increase in temperature changes the structure so that the SD sequence can form the translation initiation complex. When the 5′-untranslated region of *hspA* was cloned between the pBAD promoter and the *bgaB* gene coding for a heat-stable β-galactosidase, the enzyme activity was heat-induced without change in the *bgaB* mRNA level. Evidence for the function of the RNA thermometer *in vivo* was provided by introduction of point mutations in the thermometer of the *Synechocystis* genome [43]. Their results obtained with *E. coli* are similar to our results [41], indicating that an RNA thermometer is also involved in the posttranscriptional regulation of *S. vulcanus hspA* (Figure 5.2). In addition, expression of the *hspA* gene is regulated by changes in the stability of the transcripts since *hspA* mRNA is greatly stabilized at heat shock temperature [24].

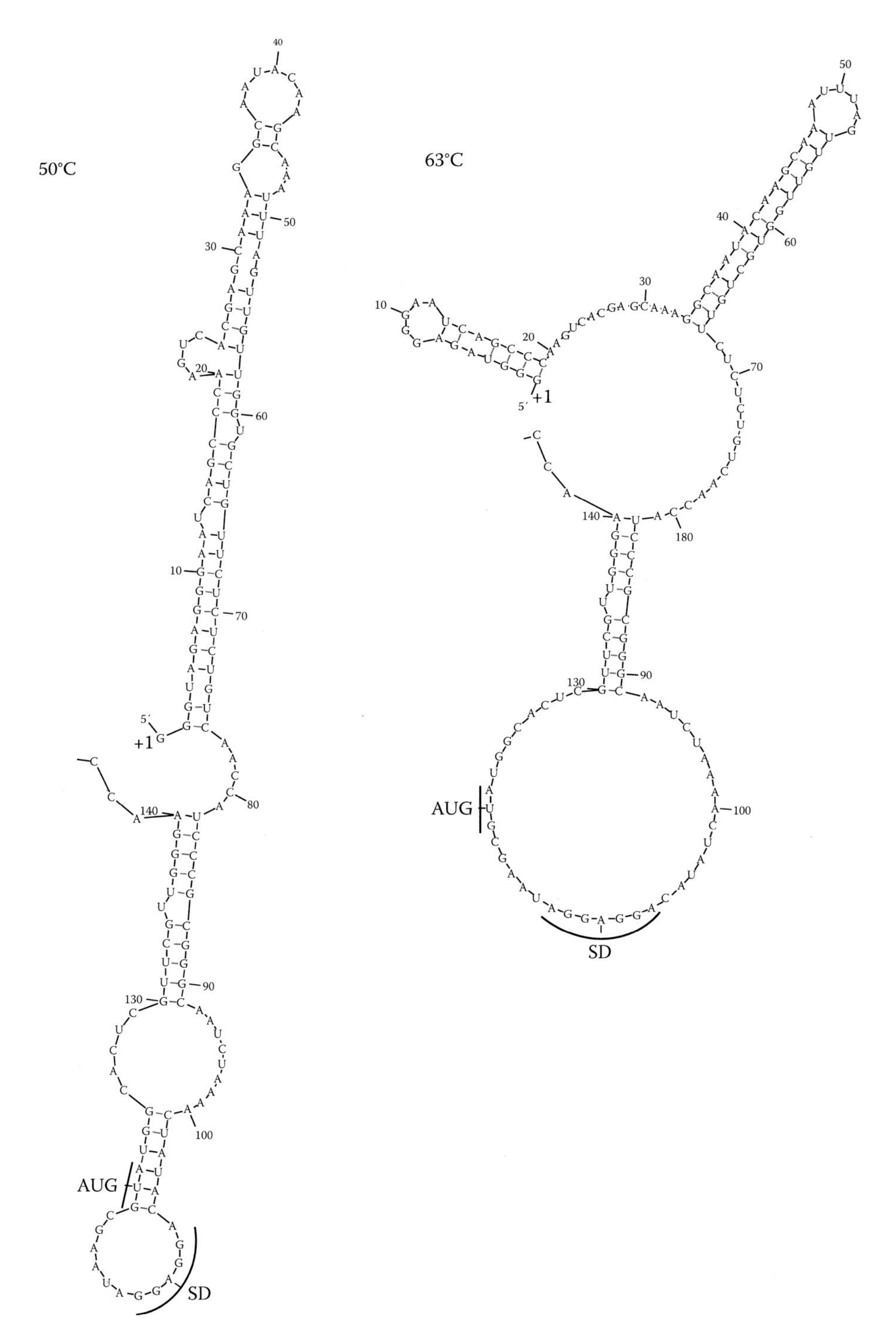

FIGURE 5.2 The secondary structure as predicted by the mfold program (http://www.bioinfo.rpi.edu/applications/mfold) of the entire *S. vulcanus hspA*-untranslated region is shown. The transcription start site (+1), the start codon (AUG), and the SD sequence are indicated. A putative RNA thermometer folds into a structure that traps SD and AUG at 50°C. An increase in temperature to 63°C destabilizes the structure, and liberates the SD and AUG sequences, which may result in the initiation of translation.

5.2.2.4 Biochemical Properties

The amino acid sequence alignment indicates that cyanobacterial small Hsps are homologous to plant cytoplasmic as well as to chloroplastic small Hsps [24,40]. They do not contain the methionine-rich region that is conserved in chloroplast small Hsps [45]. Size-exclusion chromatography (SEC) and nondenaturing gel electrophoresis (PAGE in the absence of SDS) demonstrated that native and recombinant HspA from *S. vulcanus* forms a homo-oligomer consisting of 24 subunits [40]. A recombinant HspA from the mesophilic cyanobacterium *Synechocystis* sp. PCC 6803 also elutes as a single peak at ≥400 kDa in SEC, which is consistent with an oligomer composed of 24 monomers [46]. Upon heating, this oligomer dissociates into suboligomeric species, indicating that the cyanobacterial small Hsp senses increases in temperature by the changes in its oligomer state. Size exclusion chromatography demonstrated that recombinant proteins of both *Anabaena* Hsp17.8 and Hsp17.1 form large oligomers approximately 420 and 410 kDa, respectively, which indicates that these small Hsps also form 24-mers [26].

The "anti-aggregation" activity of a cyanobacterial small Hsp was initially shown with HspA from *S. vulcanus* [40]. It prevented the heat-induced aggregation of model substrates such as malate dehydrogenase and citrate synthase. Like *Synechococcus* HspA, *Synechocystis* HspA, *Anabaena* Hsp17.1, and *Anabaena* Hsp17.8 protect model substrates from aggregation *in vitro* [26,46]. Recombinant *Synechocystis* HspA forms complexes with denatured malate dehydrogenase, transferring it to the *E. coli* DnaK/DnaJ/GrpE chaperone system for subsequent refolding [47]. Not only a model substrate, but also native substrates of *Synechocystis* cells bound to HspA are refolded by the *E. coli* DnaK chaperone system in an ATP-dependent fashion [48].

Oligomeric stability of HspA is required for cellular function [46]. Mutations in the α-crystallin domain that destabilize the HspA oligomer lead to loss of chaperone activity *in vitro.* These mutations reduce thermotolerance to the level of an *hspA* deletion mutant. A mutation in the C-terminal extension also destabilizes the HspA oligomer, and causes a defect in thermotolerance [49]. However, the mutated protein still shows some chaperone activity. Thus, there is no strict correlation between the thermo-protective function of the HspA mutants *in vivo* and their chaperone activity measured *in vitro.* Furthermore, mutations, located in the N-terminal arm of HspA, have little effect on *in vitro* chaperone activity or oligomeric stability, although they significantly decrease thermotolerance in *Synechocystis* [50]. Thus, there exist mutations that significantly impair the function of HspA *in vivo* without greatly disrupting oligomerization and/or chaperone function.

5.2.2.5 Cellular Localization

In both non-stressed and heat-stressed cells, small Hsp is localized in thylakoid membranes as well as the cytosol. When *S. vulcanus* HspA was overexpressed in *Synechococcus elongatus* PCC 7942, ~40% of the total cellular HspA was localized in the insoluble protein fraction of the cell extract, while DnaK and GroEL were present almost exclusively in the soluble fraction [51]. *Synechocystis* HspA was mainly detected in the soluble protein fraction in the extract of cells incubated at a normal or a moderately high temperature, however about one-half of the HspA was recovered in the pellet fraction when cells were treated at a lethal temperature at 50°C [48].

The membrane localization is partly due to association of small Hsp with surface localized proteins such as phycobilisome proteins [52]. In addition, small Hsp associates with synthetic and cyanobacterial lipids directly, and even penetrates into membranes, resulting in their stabilization [47]. Mutational studies have shown that small Hsp stabilizes thylakoid membranes [51,53,54].

Interestingly, distribution of HspA between cytosol and thylakoid membranes changes dynamically during heat stress [54]. In non-stressed *S. elongatus* cells, HspA is equally distributed in both compartments. Upon heat shock, the ratio of HspA in the cytosol increases transiently.

5.2.2.6 Cellular Function

Inactivation or deletion of the single *Synechocystis* small Hsp gene, *hspA*, in the genome does not have any effect on growth under non-stress conditions, but causes great loss of basal and acquired thermotolerance [46,53], indicating that small Hsp is dispensable under normal conditions, but plays a role under heat stress. Deficiency in small Hsp also results in a greatly reduced activity of photosynthetic oxygen evolution in heat-stressed *Synechocystis* cells [55]. These results indicate that HspA is essential for the survival of the organism under heat stress. The high-temperature sensitivity of the mutant may be even greater since a *Synechocystis hspA* mutant shows elevated levels of the *groESL1*, *groEL2*, and *htpG* mRNAs than the wild type [36]. The lack of small Hsp may induce higher levels of other major chaperones in order to compensate for the loss of chaperone function of small Hsp.

Constitutive overexpression of an exogenous *S. vulcanus* HspA in *S. elongatus* PCC 7942 results in greatly enhanced cell survival, and an increase in thermotolerance of whole-chain photosynthetic electron transport (from water to carbon dioxide), photosystem II(PSII), and phycocyanins at a high temperature [51]. The expression did not affect cellular levels of other major Hsps such as GroEL and DnaK. Overexpression of HspA stabilizes *Synechococcus* thylakoid membranes under heat or high-intensity light stress [54], while thylakoid membranes of heat-shocked *Synechocystis hspA* mutant cells disintegrate [53]. Thus, mutational studies with an *hspA* disruptant and an *hspA* overexpression strain indicate that small Hsp plays an essential role in the protection of cells and, in particular, PSII, phycocyanins, and thylakoid membranes against thermal stress.

It is generally thought that cyanobacteria and higher plants are sensitive to heat, with PSII being the most sensitive component. However, there is no strict correlation between the loss of PSII activity and the survival rate at a lethal high-temperature at 50°C in *S. elongatus* PCC 7942 [51]. For example, PSII activity of the *hspA* overexpressing strain was heat-inactivated much more in the dark than under illumination, but the survival rate was higher in the dark than under illumination. On the other hand, the extent of phycocyanin damage appears to correlate with the decrease in the survival rate. For example, both the survival rate and the phycocyanin bleaching (loss of light-harvesting function of phycocyanin) were similarly affected by light/dark conditions during the heat treatment. As shown in Figure 5.3A, phycobilisome proteins were aggregated or degraded in the light/heat-stressed *Synechococcus* cells, while *hspA* overexpression resulted in the protection of these proteins under the same conditions. Thus, the light/heat-dependent phycocyanin bleaching may be caused by the aggregation and/or degradation of phycocyanins in a cell. In terms of cell survival and cellular protein quality control in cyanobacteria, phycobilisomes may have to be well protected since phycobiliproteins such as phycocyanins are the most abundant proteins found in cyanobacteria. Phycobilisomes and their constituents are produced in massive amounts and may constitute 50% or more of the soluble protein in cyanobacteria [56].

We hypothesized that HspA recognizes phycocyanins as major cellular targets in order to protect them from irreversible denaturation and/or degradation, and therefore contributing toward the survival of *S. elongatus* PCC 7942 at a lethal high-temperature. To test this hypothesis, we analyzed the interaction between HspA and phycocyanins *in vitro* [52]. First, we worked out conditions under which phycocyanin bleaching takes place *in vitro*. Phycocyanin bleaching occurred when purified *Synechococcus* phycobilisomes were heat-treated in the presence of hydrogen peroxide, whereas isolated recombinant HspA suppressed their bleaching under the same conditions. Under the denaturing conditions, phycobilisomes were de-assembled to low molecular weight complexes and then aggregated. HspA associated physically with phycocyanins in the dissociated complexes, with concomitant suppression of aggregation. HspA interacted with the isolated phycocyanin $\alpha\beta$ heterodimer directly to protect it from denaturation.

The survival rate decreases greatly when *Synechococcus* cells are heat-treated under illumination [51]. Low irradiance is enough to cause the enhancement of cell death. At high temperatures, light

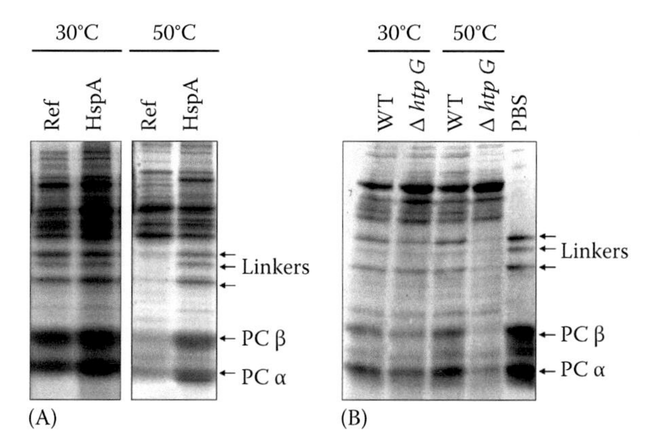

FIGURE 5.3 Stabilization of phycobilisome proteins and protection of proteome by small Hsp and HtpG. (A) An HspA expressing strain (HspA) and a reference strain (Ref) of *Synechococcus* [51] were grown at 30°C in the light (40 μE/m²/s), and directly shifted to 50°C for 15 min in the light. Soluble proteins that were isolated from the same number of cells either before or after the shift were analyzed by SDS-PAGE. (B) The wild-type *S. elongatus* (WT) and its *htpG* mutant (Δ*htpG*) [142] were grown at 30°C in the light (30 μE/m²/s), and shifted to 45°C for 180 min in the light and then to 50°C for 20 min in the light. Soluble proteins that were isolated from the same number of cells either before or after the shift were analyzed by SDS-PAGE. Linkers, PCβ, and PCα are the 33 kDa rod linker polypeptide, the 30 kDa rod linker polypeptide, the 27 kDa rod-core linker polypeptide, phycocyanin β subunit, and phycocyanin α subunit.

energy absorbed by photosynthetic pigments is not efficiently utilized in photosynthesis due to the heat-induced denaturation of enzymes/proteins, thus excess energy may be used to generate active oxygen in a cell. DNA microarray analysis has shown that active oxygen detoxifying enzymes such as superoxide dismutase are induced during heat shock in the cyanobacterium *Synechocystis* sp. PCC 6803 [30], indicating an increased level of cellular oxidative stress at high temperatures. Overexpression of HspA increased the survival rate under the light/heat conditions [51]. This may indicate that HspA protects cells under oxidative as well as heat stress. DNA microarray analysis has shown that *Synechocystis hspA* is the most highly induced Hsp gene under hydrogen peroxide stress [31]. Thus, we evaluated whether small Hsp plays a protective role under oxidative stress by analyzing the phenotype of a small Hsp deficient mutant and a small Hsp overexpressing strain in the presence of methylviologen or hydrogen peroxide. We found that a *Synechocystis hspA* deletion mutant is sensitive to methylviologen, while a *Synechococcus* strain overexpressing *hspA* shows an increased tolerance to hydrogen peroxide [57]. These results showed that small Hsp enables better growth, increases viability, and stabilizes membrane proteins such as the photosystems and phycobilisomes under oxidative stress. Thus, HspA plays an important role in the acclimation to oxidative stress in cyanobacteria.

HspA also plays a role in the acclimation to salt stress [36]. A *Synechocystis hspA* deletion mutant lost the ability to develop tolerance to a lethal salt concentration (2.0 M) by a moderate salt pretreatment (0.5 M NaCl), while the wild-type acquired the tolerance. Under various salt stresses, the mutant failed to undergo the ultrastructural changes characteristic of wild-type cells.

In addition to thylakoid membrane proteins (e.g., phycocyanins) and thylakoid membranes, cyanobacterial small Hsp protects a wide variety of proteins. Both immunoprecipitation and affinity chromatography have been employed to recover proteins that specifically interact with *Synechocystis* HspA *in vivo* during heat treatment [48]. Identification of the proteins by mass spectrometry revealed that they represent a wide range of cellular functions including transcription, translation, cell signaling, and secondary metabolism. In this analysis, phycocyanins were not identified as substrates for HspA. This may be due to the removal of phycobilisome proteins during the preparation of samples since they tend to precipitate by centrifugation.

5.3 DnaJ OR J-PROTEIN

5.3.1 Introduction

The DnaJ, Hsp40, or J-protein family is defined by the presence of J-domain of 70–80 amino acids. Members of the family can be divided into three subfamilies with respect to their additional conserved domains or regions [58,59]. Type I (or type A) members, such as *E. coli* DnaJ, possess a glycine and phenylalanine (G/F)-rich region adjacent to their N-terminal J-domain, a central cysteine-rich region containing two zinc binding sites, and a variable C-terminal domain (Figure 5.1). Type II (or type B) members contain all the domains listed earlier with the exception of the cysteine-rich region, and type III (or type C) DnaJ proteins contain only the J-domain.

The J-domain, which is absolutely essential for its cochaperone function (e.g., stimulation of ATP hydrolysis by DnaK), is composed of four helices [58,60]. The highly conserved J-domain HPD motif, which is essential for functional interaction with DnaK is located in an exposed loop between helices II and III. The G/F-rich region which contains the conserved DIF motif has been suggested to be critical for the release of a functional DnaK-substrate complex [58,61]. The cysteine-rich region is characterized by four CXXCXGXG repeats, which coordinate two zinc ions [62,63]. This region may be involved in substrate binding and/or locking-in substrate proteins to DnaK. The DnaJ C-terminal region is involved in the dimerization of DnaJ and substrate binding [64].

As described later, DnaJ (Hsp40) is an essential component of the DnaK (Hsp70) chaperone machinery that mediates protein folding [58,59]. Besides its cochaperone functions, DnaJ may bind unfolded substrates and prevent their aggregation in the absence of DnaK [65].

In the genomes from bacteria to human, many different DnaJ-encoding genes have been identified [59]. Some members may be primarily involved in cotranslational folding of proteins and/or transport of proteins across membranes, whereas other members may perform stress-related functions.

5.3.2 Cyanobacterial DnaJ

5.3.2.1 Genes

The first *dnaJ* gene cloned and characterized was *dnaJ3* from *S. elongatus* PCC 7942 [66]. Subsequently, it became evident that in the *Synechococcus* genome as well as other completely sequenced cyanobacterial genomes, *dnaJ* multigene families exist (CyanoBase). For example, in the genome of *Synechocystis* sp. PCC 6803, seven DnaJ proteins are encoded [67]. One protein belongs to type I, two belong to type II, and the others to type III. Type I DnaJ protein or DnaJ1 is strictly conserved in all cyanobacteria, while DnaJ2 (type II) is not encoded in the genome of *Gloeobacter violaceus*, and some marine cyanobacterial genomes [68]. Interestingly, Ftn2, a protein much larger than the DnaJs, and is involved in cell division, has an N-terminal DnaJ domain [69].

5.3.2.2 Gene Expression under Various Stresses

DNA microarray analysis has shown that expression of *Synechocystis* slr0093, which encodes a type II DnaJ protein, is enhanced by high salt concentration, hyperosmotic, high temperature, UV-B, peroxide, or acid (pH 3.0) stress [28–32], while that of *Synechocystis* sll0897, which encodes a type I DnaJ protein, is enhanced by high temperature, UV-B, or peroxide stress [8,29,31]. *Synechocystis* sll1666, which encodes a type III DnaJ protein, is induced by peroxide stress [31].

Northern blot analysis showed that among *S. elongatus dnaJ1*, *dnaJ2*, and *dnaJ3*, transcripts of *dnaJ2*, which encodes a type II DnaJ protein, accumulates significantly under high temperature or high light intensity stress [68].

5.3.2.3 Biochemical Properties and Cellular Localization

The anti-aggregation activity of *S. elongatus* DnaJ1, DnaJ2, and DnaJ3, which are classified into the three subtypes (type I, type II, and type III, respectively), has been characterized. DnaJ1 partially prevented aggregation of the model substrate luciferase when it was heat-treated, although more than 50% of the total substrate proteins were still detected in the insoluble fraction [68]. DnaJ2 and DnaJ3 did not prevent the aggregation significantly.

A major fraction of DnaJ3 and a fraction of DnaJ1 have been located in membrane fractions, while DnaJ2 is mainly present in the soluble/cytosol fraction [66,68].

5.3.2.4 Cellular Function

In *S. elongatus*, disruption of either one of *dnaJ1*, *dnaJ2*, and *dnaJ3* in all copies of the chromosome was impossible, suggesting that these genes are essential for normal growth. Only the *dnaJ4* gene that encodes a type III DnaJ could be disrupted [66,68]. In contrast, most of the *Synechocystis dnaJ* genes could be deleted individually [67]. Disruption of only one of the seven genes was impossible, which suggests an essential, but as yet unknown, function for this type II DnaJ homolog (Sll1933). Disruption of the type I *dnaJ* gene (sll0897) resulted in a growth defect at high (42°C) and low (20°C) temperatures, while the other mutants grew like the wild-type *Synechocystis* cells. Only the type I (Sll0897) and one of the type II DnaJ (Slr0093) proteins could complement for the temperature-sensitive growth of an *E. coli dnaJ* mutant, while the others were unable to complement [67]. Thus, the type I DnaJ is not essential under normal growth conditions, but plays an important role under temperature stress. The functions of other *Synechocystis* DnaJs are either not critical or are taken over by the remaining DnaJs. It is an unexpected finding that disruption of the gene encoding the type II DnaJ Slr0093 did not cause high- or low-temperature sensitivity because its gene expression is enhanced most among the *dnaJ* genes under cold and heat stresses [5,30]. DnaJ3 overexpression was toxic to *E. coli* and *Synechococcus* cells [66]. When this gene was expressed in *E. coli,* cells became filamentous in contrast to those expressing the *E. coli dnaJ* gene.

Yeast two-hybrid analysis has demonstrated that DnaJ2 interacts with both DnaK2 and RNase E [70]. RNase E is an endoribonuclease that is involved in the processing and degradation of RNA [71]. The C-terminal fragment of DnaJ2 interacted with the N-terminal fragment of RNase E. The presence of DnaJ2 and DnaK2 resulted in inhibition of the RNase E activity in an ATP-dependent manner [70].

Why do cyanobacteria contain a large number of DnaJ proteins? It is reasonable to surmise that the individual DnaJ proteins have specific functions and distinct substrate specificity, but little is known about this important issue. Furthermore, the DnaK partners of the different DnaJ proteins are currently unknown.

5.4 GroEL (Cpn60)

5.4.1 Introduction

GroEL, a prokaryotic member of the Hsp60 or chaperonin (cpn) family, forms a megadalton porous cylinder of 14 subunits made of two nearly sevenfold rotationally symmetrical rings stacked back-to-back. Inside the cylinder there is a cavity [72,73]. Each monomer consists of three domains: an equatorial domain, an intermediate domain, and an apical domain. The equatorial domain contains the N- and C-terminus and the ATP-binding pocket, and forms the foundation of the 14-mer assembly at its waist and holds the two heptameric rings together. The apical domain forms the opening of the central cavity and contains hydrophobic binding sites for a nonnative polypeptide and GroES. The intermediate domain connects the other two domains.

GroES, the co-chaperonin for GroEL, forms a heptameric dome [74] and closes the cavity of the GroEL cylinder, generating a hydrophilic, folding chamber in which an unfolded/folding polypeptide undergoes productive folding in isolation [75]. GroELs are ATPases, and the folding process is regulated by the ATP binding and its hydrolysis.

5.4.2 Cyanobacterial GroELs

5.4.2.1 Genes

The *groEL* gene was first cloned from the *S. elongatus* PCC 7942 genome [39], followed by cloning of the *cpn60* gene (sll0416) from the *Synechocystis* sp. PCC 6803 genome [76]. Unlike the *Synechococcus* gene and the *E. coli groESL* operon, the *Synechocystis cpn60* did not form an operon with the *groES* gene. This difference is not due to the species difference since the *groESL* operon was subsequently found in the *Synechocystis* genome [77], revealing the presence of two homologous *groEL* genes. The thermophilic cyanobacterium *S. vulcanus* also contained two copies of the *groEL* gene [78,79]. We designated one of the two genes as *groEL1* (the one that is arranged as an operon with *groES*), and the other as *groEL2*. We know now that cyanobacterial genomes contain two or (rarely) three *groEL* homologues (Cyanobase). In those cases with two homologues, only one has an upstream *groES* homologue like *Synechocystis* and *Synechococcus,* except *G. violaceus* PCC 7421, which precedes the diversification of cyanobacteria in the phylogenetic tree. Both *groEL* genes from this cyanobacterium have *groES* immediate upstream of each *groEL* gene. The phylogenetic tree shows much higher conservation in the GroEL1 amino acid sequences than in GroEL2 (see Figure S2 shown in Ref. [80]). The majority of bacterial species including *E. coli* have only one *groEL* gene (one *groESL* operon) that is essential for bacterial growth [81]. However, approximately one-third of fully sequenced bacterial genomes including cyanobacterial genomes contain more than one copy of the *groESL* operon or additional *groEL* genes. The functions of multiple GroELs are not yet fully understood.

5.4.2.2 Gene Expression under Various Stresses

DNA microarray analysis has shown that expression of *Synechocystis groEL1* (slr2076) and *groEL2/cpn60* (sll0416) is enhanced by high light intensity, high temperature, peroxide, or UV-B stress [27,29–31]. Only *groEL2* is induced by high salt concentration or hyperosmotic stress [28,36].

Northern blot analysis has shown that *Synechocystis groEL1* and *groEL2* are induced by high temperature [11,34–36,76,77,80,82], or low temperature stress [83]. Expression of both *Synechocystis groEL1* and *groEL2* is modulated by light under normal and heat shock conditions [35,80,82]. Furthermore, *Synechocystis groEL2* is induced by ethanol, nalidixic acid, hydrogen peroxide, or UV treatment [76].

The level of *Synechocystis* GroEL1 and GroEL2 proteins is elevated following heat shock [35,37,84] or high salt concentration [36,38]. *Synechocystis* GroEL1 protein also accumulates following prolonged UV-B irradiation [85].

For *groEL* genes from cyanobacteria other than *Synechocystis*, Northern blot analysis has shown that expression of both *groEL1* and *groEL2* genes from *S. vulcanus*, *T. elongatus*, and *S. elongatus* PCC 7942 increases greatly during high temperature treatment [39,78–80,86]. Transcription of both *groEL1* and *groEL2* genes from *T. elongatus* and *S. elongatus* is regulated by light [80]. Surprisingly, in a nitrogen-fixing *Anabaena* L-31, GroEL2 (Cpn60) disappears upon heat shock in the presence of a nitrogen source, while it is heat-induced under nitrogen-fixing conditions [87]. On the other hand, *groEL1* is induced under both nitrogen fixing and non-fixing conditions.

The level of *S. vulcanus* GroEL protein is elevated following heat shock [40]. The level of *S. elongatus* GroEL is enhanced following a low temperature treatment [88,89] or oxidative stress [90]. *T. elongatus* GroEL2 is also induced transcriptionally and translationally following a low-temperature treatment [86]. A UV-A-resistant marine cyanobacterium *Oscillatoria* sp. NKBG 091600 induces *groESL* operon transcriptionally and translationally after UV-A irradiation while a UV-A sensitive *S. elongatus* is unable to induce [91].

5.4.2.3 Regulatory Mechanism for Gene Expression

5.4.2.3.1 Regulation of Heat Shock Response by Light

Expression of both *Synechocystis groEL1* and *groEL2* genes is rapidly enhanced by heat shock (Figure 5.4). The expression is modulated by light. The heat shock induction of *groEL2* is abolished almost altogether in the dark, whereas that of *groEL1* is suppressed to some extent [35,82]. Heat shock induction of *groEL2* in the light is greatly reduced by the addition of 3-(3′,4′-dichlorophenyl)-1,1-dimethylurea (DCMU), whereas *groEL1* is less affected [35,82]. 2,5-dibromo-3-methyl-6-isopropyl-*p*-benzoquinone (DBMIB) does not inhibit the light-induced accumulation of the *groEL1* and *groEL2* mRNAs [35]. Basically the same results were obtained with the *groEL1* and *groEL2* genes of *S. elongatus* PCC 7942 and *T. elongatus* [80]. DCMU inhibits electron flow from PSII into plastoquinone, an electron carrier of the photosynthetic electron transport, while DBMIB inhibits oxidation of plastoquinone by PSI. Thus, we postulate that the redox state of plastoquinone is involved in the light-dependent heat induction of cyanobacterial *groEL* genes. This modulation takes place at the transcriptional level since the promoter activity of *Synechocystis groEL1* is greatly increased following light treatment with or without heat treatment, and this enhancement is abolished in the presence of DCMU [80]. Light also exerts a significant effect on the heat induction of *htpG* as shown in Figure 5.4 [35]. Thus, it is reasonable to expect that cells acquire more thermotolerance when they are pre-heat-treated under illumination than in the dark. Experimental evidence supports this expectation [35].

5.4.2.3.2 Negative Regulation of Cyanobacterial groEL Gene Expression by the CIRCE/HrcA System and Positive Regulation by the Novel DNA Element K-Box

The *groEL1* gene has its transcription initiation site within the CIRCE (*c*ontrolling *i*nverted *r*epeat of *c*haperone *e*xpression) sequences in *Synechocystis* sp. PCC 6803 [34,82], *S. elongatus* [39], and *T. elongatus* [86]. The *Synechocystis groEL2* gene has upstream CIRCE, and the –35/–10 regions of its promoter, but not the transcription initiation site, are located within the CIRCE sequences [34]. CIRCE is a perfect inverted repeat of 9 bp separated by a 9-bp spacer with the sequence TTAGCACTC-N9-GAGTGCTAA. It serves as a *cis*-active binding site for a negative regulator called HrcA (*h*eat shock *r*egulation at *C*IRCE) [92]. Cyanobacterial HrcA can physically bind to CIRCE [86,93] as proved by electrophoretic mobility shift assay. HrcA from *Anabaena* sp. strain L-31 forms a dimer [93]. Knocking out the *hrcA* gene in *Synechocystis* sp. PCC 6803 and *Anabaena* sp. strain PCC7120 derepresses transcription of both *groEL1* and *groEL2* (*cpn60*) genes,

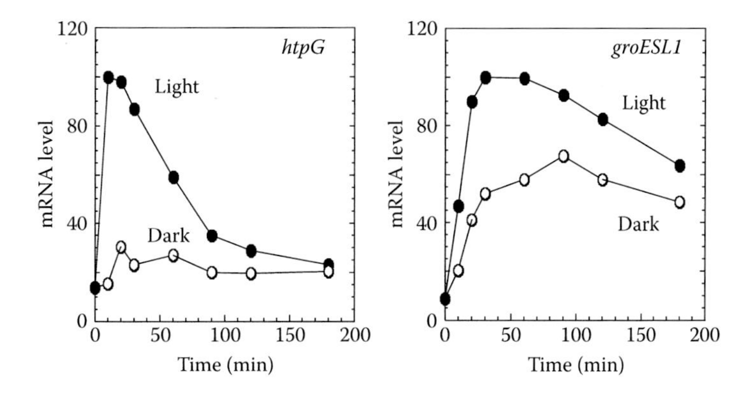

FIGURE 5.4 Modulation of cyanobacterial heat shock induction by light: time course for the accumulation of the *htpG* and *groESL1* mRNAs of *Synechocystis* sp. PCC 6803 after heat shock at 42°C in the presence (closed circles) or absence of light (open circles). Cells growing at 30°C under a light intensity of 35 μE/m²/s were subjected to the heat shock in the presence or absence of light for 3 h. Bands corresponding to the *htpG*, and *groESL1* mRNAs detected by Northern blot analysis [35] are quantified. The amount of RNA loaded was verified by rRNA staining (data not shown).

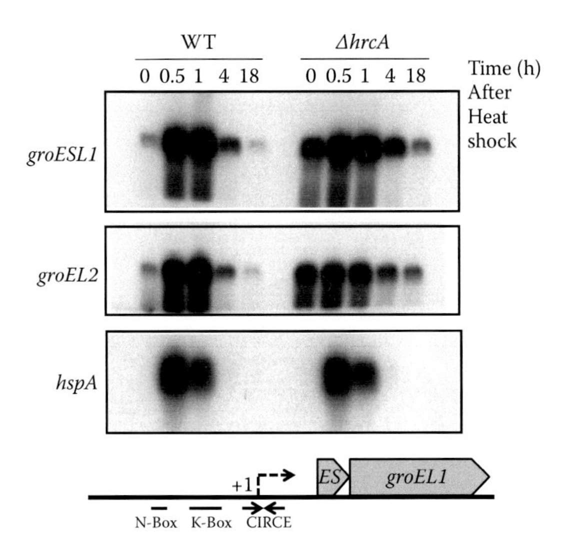

FIGURE 5.5 Induction and shut off of heat shock response in the wild-type *Synechocystis* sp. PCC 6803 and its *hrcA* mutant (Δ*hrcA*) [184]. Cells growing at 30°C under a light intensity of 30 μE/m^2/s were subjected to heat shock at 42°C in the presence of light for 18 h. Bands corresponding to the *groESL1*, *groEL2*, and *hspA* mRNAs were detected by Northern blot analysis. The operator sequence CIRCE to which the repressor HrcA binds is present upstream of the two *groEL* genes. Thus, de-repression of gene expression under normal conditions was observed in these genes. The amount of RNA loaded was verified by rRNA staining (data not shown). At the bottom, the *groESL1* operon and its upstream region are shown. This operon is regulated negatively by the CIRCE/HrcA system, and positively by the K- and N-boxes.

indicating that HrcA is involved with the regulation of the *groEL* gene expression. However, further induction can still occur on heat shock, suggesting that other mechanisms of regulation must also be present [34,93]. Figure 5.5 shows that heat-induction of the *Synechocystis groEL1* and *groEL2* gene expression takes place like the wild type although the basal levels of their transcripts are higher than in the wild type.

In order to elucidate the mechanism(s) for regulation of the *groEL* expression other than the CIRCE/HrcA system, we analyzed expression of the *Synechocystis groEL* genes in an *hrcA* mutant. The *groEL* expression in the *hrcA* mutant was modulated by light. Transcripts of both *groEL1* and *groEL2* decreased to an undetectable level in the dark, but increased to the wild-type level under high light without heat treatment, or under low light with heat treatment [80]. The rate and extent of the light induction appeared to be dependent on the light intensity. DCMU greatly suppressed the light-responsive induction of the *groEL1* and *groEL2* genes. Thus, we postulate that the redox state of plastoquinone is involved in a novel mechanism for the light and/or heat induction of the *groEL* genes. This novel mechanism may not play a role under normal growth conditions since HrcA effectively represses *groEL* gene expression. However, it may play a major role in inducing the *groEL* genes when HrcA is inactivated under stress. We performed *in silico* analysis to search for conserved sequences upstream of cyanobacterial *groEL1* and *groEL2* genes and found a highly conserved sequence located immediately upstream of the *groESL1* promoter sequences, which we designated as K-box. It is located from −63 to −48 and −62 to −47 (from the transcription start site) of the *Synechocystis groESL1* and *groEL2*, respectively, upstream sequences (Figure 5.5). The K-box has the consensus sequence 5′-GTTCGG-NNAN-CCNNAC-3′. The K-box and the CIRCE operator sequences are highly conserved in most *groESL1* operons, but not in cyanobacterial *groEL2* genes although *Synechocystis groEL2* contains both sequences [80]. Sequences similar to the K-box were also found upstream of the *dnaK2* genes [80,94]. Furthermore, specific binding of a protein(s) to the N-box, GATCTA, was detected by a gel mobility shift assay using cell extracts [80]. The N-box is located further upstream (−93 to −86) than the K-box (Figure 5.5). Analysis with *Synechocystis* reporter strains, in which the expression of reporter genes was driven by upstream DNA fragments of *groEL1,* clearly showed that K-box is essential not only for heat and/or high light stress induction, but also for the basal transcription of *groEL1*. The analysis also showed the N-box plays a role for enhancing the induction.

The *groEL2* genes from *S. elongatus* and *T. elongatus*, which lack CIRCE, K-box, and N-box naturally, are also induced by heat and/or light, and their induction is suppressed by DCMU [80]. Thus, the control mechanisms of the light/heat-responsive *groE* expression are highly diversified in cyanobacteria. What further complicates the regulatory mechanism for *T. elongatus groEL2* expression is that there are two transcription initiation sites for the gene [86].

5.4.2.3.3 Other Regulatory Factors

A heat shock gene, *orf 7.5*, which encodes a polypeptide of 63 amino acids, plays an indispensable role in growth and survival of *S. elongatus* under heat stress [10]. Disruption of the gene led to a marked inhibition of growth at a high temperature and a decrease in the basal and acquired thermotolerances in the mutants. The Orf7.5 protein may play a regulatory role for the *groESL1* expression since its heat induction was suppressed in the *orf7.5* mutant.

An RNA helicase, CrhR, regulates the low-temperature-inducible expression of *groEL1* and *groEL2* in *Synechocystis* sp. PCC 6803 [83]. Mutation of the *crhR* gene repressed the low-temperature-induction of the *groEL* genes at the transcript and protein levels. The mutation also stabilized these transcripts at an early phase during acclimatization to low temperatures. However, it did not affect the heat-induction of the *groEL2* gene.

5.4.2.4 Biochemical Properties

As described earlier, the GroEL-GroES cavity is indispensible for its chaperone function. Thus, it is quite reasonable to expect that the GroEL oligomer or the chaperonin double ring has a stable structure. This appears to be the case in *E. coli* GroEL. However, the GroEL tetradecamers of *Synechocystis* are unstable in the absence of Mg-ATP and glycerol during isolation and purification [84]. When the purified GroELs are subjected to native-PAGE in the absence of glycerol, they dissociate to monomers. Whether the instability was specific to one or the other of the two GroELs had not been tested. Thus, we compared the biochemical properties of GroEL1 from *S. elongatus* PCC 7942 with those of GroEL2 [95]. Various analyses showed that GroEL1 can form a tetradecamer like *E. coli* GroEL, while GroEL2 forms a heptamer or dimer, but both GroEL1 and GroEL2 oligomers are very unstable and liable to dissociate. In order to evaluate the oligomer states *in vivo*, the sizes of the two GroELs were analyzed by native-PAGE with cell extracts from the wild type and a *groEL2* deletion mutant. The results confirmed that *Synechococcus* GroEL1 can form a tetradecamer, while GroEL2 is unable to form. GroEL2 and to a lesser extent GroEL1, display enhanced hydrophobic exposure to the probe 1,1′-bis(4-anilino)naphthalene-5,5′-disulfonic acid (bis-ANS) as compared with *E. coli* GroEL. The increase may be related to the lower oligomer state as *E. coli* GroEL also exhibits significantly enhanced exposure when it is monomerized [96].

We examined the chaperone function of cyanobacterial GroELs *in vitro* for the first time with GroEL1 and GroEL2 of *S. elongatus* PCC 7942 [95]. In all these experiments, *E. coli* GroEL was used as a control. The k_{cat} (min^{-1}) for ATP hydrolysis by GroEL1, GroEL2, and *E. coli* GroEL were 1.15, 0.28, and 6.91, respectively. The low ATPase activities of cyanobacterial GroELs may be due to the instability of the oligomer assembly, since the *E. coli* GroEL monomer shows about one-seventh of the ATPase activity of the 14-mers [96]. Consistent with a previous report [97], the ATP hydrolysis by *E. coli* GroEL was inhibited by GroES. In contrast, the activities of both GroEL1 and GroEL2 were enhanced by GroES. This unusual property of cyanobacterial GroEL may be surmised to result from the instability of their oligomer assembly since the stimulating effect of GroES as well as low ATPase activity in the absence of GroES is observed in a single-ring mutant of *E. coli* GroEL [98]. Like cyanobacterial genomes, the *Mycobacterium tuberculosis* genome also carries a duplicate set of *groEL* genes, only one of which occurs on the *groESL* operon. The mycobacterial

GroELs are also reported to exist *in vivo* (in cell extracts) and *in vitro* as lower oligomers (dimer) and not tetradecamers [99]. The ATPase activities of the two GroELs are much lower than that of the *E. coli* GroEL. These authors argue that failure of the *M. tuberculosis* chaperonins to oligomerize can be attributed to amino acid mutations at the oligomeric interface.

Synechococcus GroEL1 and GroEL2 prevented aggregation of heat-denatured proteins in the same way as *E. coli* GroEL [95], indicating that they can bind nonnative proteins. However, their capability to assist protein folding was much lower than that of *E. coli* GroEL, regardless of the co-presence of GroES, which enhances the ATPase activity. Higher concentration of GroEL1, which stabilizes the tetradecamer in a folding reaction mixture, or utilization of different protein substrates, did not increase the folding activity [95]. Interestingly, the ATPase activity of GroEL1 was greatly enhanced in the presence of GroEL2, but the folding activities of GroEL1 and GroEL2 were still much lower than that of *E. coli* GroEL.

5.4.2.5 Cellular Localization

Synechococcus GroELs are present "almost" exclusively in soluble fractions of cell extracts [51]. However, this does not mean GroELs are not associated with membranes. It has been shown for *Synechocystis* sp. PCC 6803 that some of the GroEL protein associates with thylakoid membranes [100]. Both GroEL1 and GroEL2 of *Synechocystis* are detected in thylakoid membranes [101]. Furthermore, GroEL2 is identified in plasma membranes of *Synechocystis* [102]. GroEL in *Spirulina platensis* exposed to a low temperature is also detected in both thylakoid and plasma membranes [103]. *E. coli* GroEL can associate with model lipid membranes directly [104], suggesting that GroEL may associate with membrane *in vivo* as well to maintain membrane integrity under normal and stressed conditions.

5.4.2.6 Cellular Function

5.4.2.6.1 Differential Roles of GroEL1 and GroEL2

As described earlier, cyanobacterial genomes contain multiple *groEL* genes. What are the functions of multiple chaperonins? Do all of them play an essential house-keeping role like *E. coli* GroEL [105], or do they each have a specialized role?

The first indication for differential roles of GroEL1 and GroEL2 in their cellular functions came from complementation experiments. Introduction and overexpression of the *S. vulcanus groEL1* gene in a *groEL* defective mutant (*groEL44*) of *E. coli* resulted in the complementation of heat sensitivity [79], which contrasted with the result for *groEL2*. Although overexpressed in *E. coli* upon heat shock treatment, introduction of the *groEL2* gene of *S. vulcanus* into the *E. coli* mutant did not achieve complementation [78]. Our results indicate that GroEL1 has a function similar to *E. coli* GroEL, while GroEL2 may not. Similarly, overexpression of the *Synechocystis* GroEL1 in the *E. coli groEL* mutant strain allowed growth at elevated temperature [84]. Furthermore, the GroEL1 could substitute for *E. coli* GroEL in bacteriophage T4 morphogenesis as well as in the folding of the Rubisco enzyme. In contrast to GroEL1, *Synechocystis* GroEL2 only partially complemented the temperature-sensitive phenotype, and the Rubisco assembly defect, but did not promote the growth of the bacteriophage T4.

GroEL1, with GroES, seems to play a house-keeping, essential role in the cyanobacterial cell since a strain of *S. elongatus* PCC 7942 in which this gene is disrupted could not be constructed [94].

5.4.2.6.2 Low-Temperature Response of GroEL2

If only one GroEL is sufficient to fulfill the house-keeping role, an additional homolog may diverge to take on specialized roles. Our studies seem to support this idea. In contrast to the *Synechococcus groEL1* gene, the *groEL2* gene in *T. elongatus* can be deleted without effect under normal growth conditions at 50°C, although it is required for growth at both heat shock (62°C) and cold shock (40°C) temperatures [86]. In this thermophilic cyanobacterium, both GroEL1 and GroEL2 accumulate on heat shock, but only the accumulation of GroEL2 is enhanced by cold shock. GroEL1 does not change its cellular level significantly during cold shock [86]. It is noteworthy that the *groEL2* mutant shows temperature-sensitive phenotype even though wild type and mutant cells have the same level of GroEL1. Thus, we concluded that GroEL1 cannot replace GroEL2 function in response to high or low temperature stress. High- and low-temperature sensitive phenotype was also observed in a *Synechococcus groEL2* deletion mutant (S. Huq, 14. H. Nakamoyo, unpublished data, 2010). The level of *Synechococcus* GroEL is enhanced following a low temperature treatment [88,89]. CrhR may be involved in the regulation of the low-temperature-induction of the *Thermosynechococcus* and *Synechococcus groEL* genes as well as *Synechocystis groELs* [83]. In conclusion, considerable evidence supports the hypothesis that GroEL1 plays a house-keeping role and a role under stress, while GroEL2 plays a specialized and essential role under temperature-stress conditions. I propose that GroEL2 proteins may have evolved specialized functions in order to play a key role in cyanobacterial adaptation to various environmental stresses.

5.4.2.6.3 Multiple GroELs from Nitrogen-Fixing Cyanobacteria

In the nitrogen-fixing cyanobacterium *Anabaena* sp. strain L-31, both GroEL1 (GroEL) and GroEL2 (Cpn60) are induced upon heat shock under nitrogen-fixing (nitrogen deficient) conditions [87]. However, only GroEL1 is induced in the presence of a nitrogen source (NO_3^- and/or NH_4^+). The transcription of *groEL2* is repressed and GroEL2 is degraded under heat shock in the presence of nitrogen. Photosynthetic activity resists much more and recovers more rapidly under/from heat shock in the absence of nitrogen than in its presence. Overexpression of GroEL2 in *Anabaena* sp. strain PCC7120 confirmed the role of GroEL2 in the protection of photosynthesis and nitrate reductase under heat shock conditions. Overexpression of GroEL1 along with GroES enhances the heat and salinity stress tolerance of *Anabaena* sp. strain PCC7120 [106]. Thus, both GroEL1 and GroEL2 play a role under heat stress. It is not known whether GroEL2 also plays a role under salt stress in *Anabaena*.

5.5 DnaK

5.5.1 Introduction

Members of the DnaK family are characterized by an N-terminal ATPase domain of ~45 kDa [107], a less-conserved C-terminal substrate-binding domain of ~25 kDa, and a C-terminal extension [58,108]. The C-terminal domain is further divided into a β-sandwich subdomain (a substrate binding cavity) and an α-helical subdomain that is called a lid [109].

DnaK can form a binary complex with a nonnative protein and thus prevent its aggregation with other nonnative proteins. However, in the presence of DnaJ, GrpE, and ATP, DnaK can bind and release a substrate protein [58,108], keeping it within folding pathways that lead to its native structure. The binding and release of a substrate is controlled allosterically by the nucleotide occupancy and status in the ATPase domain. While the ATP-bound DnaK exhibits low affinity and fast exchange rate for its substrates, the ADP-bound form is characterized by high affinity and low exchange rate. In the latter form, the α-helical lid closes over the substrate binding cavity to stabilize the substrate-bound complex. Thus, ATP binding to the N-terminal ATPase domain triggers the transition to the low affinity state of the C-terminal substrate-binding domain, while ATP hydrolysis leads to the high affinity ADP-bound state.

The ATP-dependent DnaK chaperone machine is regulated by the cochaperone DnaJ and a nucleotide exchange factor GrpE [58,108]. DnaJ is thought to recruit a nonnative protein and transfer it to the ATP-bound state of DnaK. It also stimulates ATP hydrolysis by DnaK, leading to an ADP-bound DnaK in complex with its substrate. GrpE that interacts with the DnaK ATPase domain [110], promotes the exchange of ADP for ATP, and also augments substrate release from DnaK.

The DnaK chaperone machine binds to short extended hydrophobic polypeptide sequences present in its substrate proteins [111]. Because of this ability, it plays a role during the folding of nascent/nonnative polypeptides, translocation of polypeptides through membranes, oligomeric disassembly/assembly, degradation of proteins, and thermal or chemical stresses.

5.5.2 Cyanobacterial DnaKs

5.5.2.1 Genes

The DnaK genes form a multigene family in *S. elongatus* PCC 7942 [112,113]. The *Synechococcus* genome has three *dnaK* homologues (*dnaK1*, *dnaK2*, and *dnaK3*). Complete genome sequencing also revealed the presence of three *dnaK* homologs in *Synechocystis* sp. PCC 6803 [25]. We know now that cyanobacterial genomes generally contain multiple genes encoding DnaK. DnaK2 appears to be strictly conserved in all cyanobacterial lineages, and phylogenetic analysis of DnaK indicates that it represents the only DnaK retained in algal and plant plastids [114].

In *Synechocystis* genomes, *dnaK1* (sll0058) has *grpE* immediately upstream where it is also found in some low G + C Gram-positive bacteria and Chlamydia [115], whereas *dnaK3* (sll1932) has a *dnaJ* homolog (*dnaJ3*) immediately downstream, which is similar to the *E. coli dnaK/dnaJ* operon [115,116]. A halotolerant cyanobacterium *Aphanothece halophytica* has a *grpE-dnaK1-dnaJ* gene organization [117] like *Bacillus subtilis* [118].

5.5.2.2 Gene Expression under Various Stresses

DNA microarray analysis has shown that expression of *Synechocystis dnaK2* (sll0170) is enhanced by high light intensity, salt, hyperosmotic, heat, UV-B, peroxide, or acid (pH 3.0) stress [27–32]. Expression of *Synechocystis dnaK3* (sll1932) is enhanced by UV-B [29], while that of *dnaK1* (sll0058) is enhanced by peroxide, or alkali stress [31,33].

Northern blot analysis has shown that *Synechocystis dnaK2* is induced by ethanol, nalidixic acid, hydrogen peroxide, UV, heat, or salt stress [34,36,76,119]. The promoter activity of *Synechocystis dnaK2* increases following heat, cold, high light intensity, oxidative (H_2O_2), osmotic, or acid (pH 5.5) stress treatment [119]. High light intensity, osmotic shock, or acidic pH treatment also increases the *dnaK3* promoter activity. Among the *Synechocystis dnaK* genes, expression of *dnaK2* is most significant [119]. Quantitative RT-PCR analysis showed that the *dnaK1* as well as *dnaK2* transcripts increase upon a shift to high light in *Synechocystis* and a marine cyanobacterium *Prochlorococcus* MED4 [120].

Among the *Synechocystis* DnaKs, DnaK2 is the most abundant protein [119]. The level of *Synechocystis* DnaK2 protein is further enhanced following heat shock treatment [37], while salt stress increases the *Synechocystis* DnaK1 level [38]. Interestingly, *Aphanothece dnaK1* is up-regulated by both heat and salt stresses [117].

In *S. elongatus* PCC 7942, only the *dnaK2* gene responds to heat, high light intensity, high salt concentration, and osmotic stresses, but not to cold stress, which has been shown by Northern blot analysis [94]. Only the *dnaK2* promoter is significantly up-regulated under high light intensity or salt stress conditions. The *Synechococcus* DnaK2 protein shows a transient but significant increase in protein synthesis rate after heat shock, and the level of the DnaK2 protein is elevated following heat treatment [121].

5.5.2.3 Regulatory Mechanism for Gene Expression

Expression of the *Synechococcus dnaK2* gene is regulated by a 20-bp sequence element called *m*ultistress-Associated *r*egulatory Sequence (MARS) that is highly conserved among cyanobacterial *dnaK2s* [94]. The consensus sequence of MARS is 5′-NGGTTCGGN(A/T)(A/T)(A/T/C)C_4(C/A)C(T/A/C)(T/C/G)-3′. The sequence is similar to the K-box that is involved in the *groEL1* transcription [80] as described earlier. Like the K-box [80], MARS is essential not only for heat or high light stress induction, but also for the basal expression of *dnaK2*.

5.5.2.4 Biochemical Properties

The three DnaK homologs in *Synechococcus* or *Synechocystis* share the common DnaK domain structure, that is, the N-terminal ATPase and proximal C-terminal substrate binding domains. However, their C-terminal regions/extensions are variable. Both *Synechococcus* and *Synechocystis* DnaK2 proteins are mainly detected as a monomer, but some appears as a dimer when analyzed by native-PAGE (K. Sueoka, S. Narumi and H. Nakamoto, unpublished data, 2011).

Aphanothece DnaK1 exhibits an ATPase activity that is more resistant to inactivation under high salt concentration than *Synechococcus* DnaK1 [122]. When mature or precursor plastocyanin that had been urea-denatured was diluted in the presence of DnaK1 and copper, *Aphanothece* DnaK1 yielded copper-bound plastocyanin. This apparent holding activity of *Aphanothece* DnaK1 was detected in the presence of 150 mM KCl or NaCl, while the activity of *Synechococcus* DnaK1 was detected in the presence of KCl, but it was not detected in the presence of 150 mM NaCl. Thus, the *Aphanothece* DnaK1 may have evolved functions specifically directed to the protection of proteins under salt or NaCl stress.

DnaKs have highly diverse C termini as described earlier, which may explain its functional diversity. C-terminal truncation of approximately 60 amino acids of *Aphanothece* DnaK1 has no effect on its ATPase activity, but causes a significant decrease in the holding activity [122]. Among the three *Synechococcus* and *Synechocystis* DnaK homologs, DnaK3 is especially unique in that its C terminus is extended. Mutational studies showed that the C terminus of the *Synechococcus* DnaK3 protein can be deleted [123], while that (the last 78 amino acids) from *Synechocystis* cannot [119,124]. There may be differences between the *in vivo* functions of DnaK3 in the two cyanobacterial strains.

All three *Synechocystis* DnaKs can bind to the GrpE nucleotide exchange factor in the absence of ATP [119]. However, the cyanobacterial DnaK/DnaJ/GrpE chaperone system has not been established yet, so no folding activity of DnaK has been measured.

5.5.2.5 Cellular Localization

Synechococcus and *Synechocystis* DnaK1 and DnaK2 are mainly detected in the soluble protein fraction, although they are also partially, but strongly, bound to membranes peripherally [123–125]. *Synechocystis* DnaK2 and *S. platensis* DnaK is detected in thylakoid membranes [101,126]. *Synechococcus* and *Synechocystis* DnaK3 is almost exclusively attached to membranes, although it lacks membrane-spanning sequences. Specifically, *Synechococcus* DnaK3 is detected in membrane-bound polysomes [125]. The substrate-binding domain of DnaK3 has been suggested to be required for its association with thylakoid membranes [123].

5.5.2.6 Cellular Function

The *Synechococcus* and *Synechocystis dnaK1* genes are not essential for normal growth, or growth at high temperatures, which is consistent with the fact that some cyanobacterial genomes lack *dnaK1* homologues. On the other hand, *dnaK2* and *dnaK3* are essential under normal conditions [121,124]. Thus, DnaK2 and DnaK3 have some specific functions which cannot be compensated for by the remaining two DnaK proteins.

In *E. coli*, *dnaK* is not essential for growth, but deletion of the gene confers a high temperature-sensitive phenotype [127]. *Synechococcus* DnaK2 can suppress the temperature-sensitive growth of an *E. coli dnaK* mutant strain at 44.5°C, suggesting that DnaK2 is involved in cellular thermotolerance. On the other hand, DnaK1 and DnaK3 cannot; rather, overproduction of these proteins resulted in growth inhibition even at the permissive temperature 42°C [121]. In contrast to *Synechococcus*, none of the *Synechocystis dnaK* genes can complement the defect of an *E. coli dnaK* mutant [124].

In *E. coli*, overproduction of *E. coli* DnaK results in cell filamentation [128]. Similarly, cells overexpressing *Synechococcus* DnaK1 or DnaK2 become filamentous [121]. In contrast to *Synechococcus* DnaK1, overproduction of *Aphanothece* DnaK1 failed to produce the filamentation [114].

Analysis of suppressors of a *Synechococcus* mutant strain carrying a missense mutation in the substrate-binding domain of DnaK3 revealed that it interacts with the 50S ribosomal protein L24 [125]. In the missense mutant, the mutated DnaK3 changes its localization (from membrane and membrane-bound polysomes) to the cytosolic polysome fraction upon temperature upshift. This change in localization was suppressed in the suppressor mutant along with high-temperature sensitivity of the missense mutant. Furthermore, in the missense mutant, synthesis of the PSII reaction center protein D1 was impaired at the high temperature, which suggests that DnaK3 functions in the translational process on the thylakoid membrane [125].

In addition to suppressor analysis, yeast two-hybrid screening is another method to study protein-protein interaction. By this method, it has been shown that DnaJ2 interacts with both DnaK2 and RNase E, an essential endoribonuclease [70]. This interaction inhibits RNase E activity in an ATP-dependent manner, suggesting that DnaK2 and DnaJ2 are involved in RNA degradation through interaction with RNase E.

5.6 HtpG

5.6.1 Introduction

HtpG is a term for a prokaryotic homolog to Hsp90. Both eukaryotic Hsp90 and HtpG form a dimer at physiological temperatures [129,130]. Each monomer consists of three domains: N-terminal domain, middle domain, and C-terminal domain (Figure 5.1). The N-terminal domain possesses an ATP binding site [131]. Hsp90 is a very weak ATPase and *in vivo* function of Hsp90 is dependent on ATP binding and hydrolysis [132]. On the other hand, the C-terminal domain is essential for Hsp90 dimerization [133,134]. The middle domain is the major substrate binding site for HtpG [135]. Hsp90 can adopt a large number of structurally distinct conformational states. These states include the nucleotide-free (apo), ATP, ADP, and Grp94 like states [136–138]. The ATPase reaction involves large conformational changes of the Hsp90 dimer that are thought to drive structural changes of a substrate protein and its release [128,130,134]. Little is known about how these conformational states relate to chaperone function, but our work suggests that the substrate-binding affinity becomes weaker in the ADP-bound state of HtpG than the ATP-bound state [135]. Thus, a substrate may be loaded to the nucleotide-free or ATP-bound state of Hsp90/HtpG, and released when it is in the ADP-bound state.

Hsp90 affects the folding and activation of a wide variety of substrate proteins including transcription factors and protein kinases under physiological conditions [129,130]. Hsp90 is thought not to be involved in protein folding or unfolding in a manner similar to chaperones such as GroEL/Hsp60 or DnaK/Hsp70, but rather believed to promote subtle structural changes in substantially folded substrate proteins [129]. However, Hsp90 can recognize and bind chemically or heat-denatured substrates, thereby preventing their nonspecific aggregation [139]. This general protective chaperone function may be especially important under stress conditions. Chaperone function of Hsp90 under stress conditions has been largely ignored.

In eukaryotes, Hsp90 collaborates with co-chaperones/cofactors such as Hop/Sti1, Sgt1, FKBP51/52, Cdc37, Aha1, and p23/Sba1 to mediate the conformational regulation of substrate proteins [129,130]. Co-chaperones/cofactors bind to Hsp90 and may regulate the Hsp90's ATPase activity. At least some of them recruit specific substrate proteins for Hsp90.

5.6.2 Cyanobacterial HtpG

5.6.2.1 Genes

It appears that cyanobacterial genome contains a single copy of the *htpG* gene as the *S. elongatus* PCC 7942 genome from which the *htpG* gene was cloned for the first time [140]. The *Synechococcus* HtpG polypeptide has a predicted molecular weight of 72,602. The size of HtpG is shorter than that of Hsp90 since HtpG lacks a C-terminal extension and a linker sequence/highly charged region that connects the N-terminal domain with the middle domain.

5.6.2.2 Gene Expression under Various Stresses

Expression of the *Synechococcus htpG* gene is low under normal growth conditions, but greatly increases during thermal stress [140]. The *htpG* gene is also induced by low temperature, high light intensity, or methylviologen stress in *Synechococcus* with concomitant increase in the level of the HtpG protein [89,90].

DNA microarray analysis has shown that expression of *Synechocystis htpG* is enhanced by high light intensity, salt, hyperosmotic, heat, UV-B, peroxide, or alkali (pH 10) stress [8,27–31,33,141]. Northern blot analysis has shown that *Synechocystis htpG* is induced upon heat shock [34,35]. Proteome analysis has revealed that the level of *Synechocystis* HtpG protein increases upon heat shock [37] and during prolonged UV-B irradiation [85].

5.6.2.3 Regulatory Mechanism for Gene Expression

The induction of the *Synechocystis htpG* gene expression is very rapid, intense, and transient upon heat shock as shown in Figure 5.4 [35]. Expression of the *htpG* gene is greatly modulated by light and photosynthetic electron transport. The heat shock induction of *htpG* is abolished almost altogether in the dark [35]. DCMU inhibits the light-induced accumulation of the *htpG* mRNA, whereas DBMIB does not. Similar results were obtained with the *Synechococcus htpG* gene (M. Hossain and H. Nakamoto, unpublished data, 2003). Thus, the redox state of plastoquinone may be involved in the light-dependent heat induction/expression of cyanobacterial *htpG* genes.

Nothing is known about the regulatory mechanism for the expression of the *htpG* gene. However, one interesting observation is that the enhancement of *Synechocystis groEL* expression due to the absence of the transcriptional repressor HrcA under normal conditions represses the heat induction of *htpG* [34], while the level of *Synechococcus* GroEL protein is elevated in an *htpG* mutant [90]. *Synechocystis htpG* as well as *groEL1* and *groEL2* are up-regulated in an *hspA* deletion mutant [36]. Thus, the cellular level of molecular chaperones may influence expression of other chaperone genes by unknown mechanisms. These results indicate that a regulatory cross talk exists in the expression of these chaperone genes.

5.6.2.4 Biochemical Properties

Synechococcus HtpG consists of three major domains as do other members of the Hsp90 protein family [135]. Like its *E. coli* homolog [142], it forms a dimer (E. Kojima and H. Nakamoto, unpublished data, 2001). However, it oligomerizes at a high temperature like the eukaryotic Hsp90 [143]. The ATPase activity of the cyanobacterial HtpG is inhibited by the antibiotic radicicol (S. Menagawa and H. Nakamoto, unpublished data, 2008), a specific inhibitor of Hsp90 [144].

Cyanobacterial HtpG binds heat-denatured proteins including the 30 kDa rod linker polypeptide of the phycobilisome (see later) and model substrates such as malate dehydrogenase and citrate synthase, and suppresses their nonspecific aggregation [135,145]. The N-terminal domain as well as the middle domain binds a heat-denatured protein substrate to suppress its aggregation [135]. The substrate binds with higher affinity to the middle domain than to the N-terminal domain, indicating that the middle domain contains the major substrate binding site. Our study indicates that the substrate-binding stoichiometry is fixed, one substrate molecule per HtpG dimer (two HtpG protomers). Our stoichiometry is consistent with the structure of an Hsp90-substrate complex determined by single-particle electron microscopy [146]. In this complex, the substrate kinase (Cdk4) is bound in an asymmetric fashion to the N-terminal and middle domains of one subunit with a stoichiometry of one substrate molecule per one Hsp90 dimer.

A number of small molecules are known to bind eukaryotic Hsp90 and inhibit its chaperone function. The natural product geldanamycin was the first reported inhibitor of Hsp90 [147]. Small molecules that control Hsp90's function may have clinical value since we know that Hsp90 is implicated in various diseases including cancer. Cancer cells use Hsp90 to facilitate the function of numerous oncoproteins, and thus much effort has been devoted to the development of Hsp90 inhibitors to treat cancer.

We employed chemical arrays to screen ligands for *Synechococcus* HtpG. Cyclic lipopeptide antibiotics were found to specifically bind to the N-terminal domain of HtpG [145]. They inhibit the chaperone function of HtpG that suppresses thermal aggregation of substrate proteins without affecting its ATPase activity. One of these antibiotics, colistin sulfate salt, was shown to increase surface hydrophobicity of the N-terminal domain of HtpG and induce oligomerization of HtpG via interaction of the N-terminal domains.

Eukaryotic Hsp90 is involved in the folding, maturation, and stabilization of a numerous, but specific set of substrate proteins [130]. In contrast, there had been no report about substrate proteins for HtpG until we identified uroporphyrinogen decarboxylase as a target for *Synechococcus* HtpG by the yeast two-hybrid screening method [148]. We detected complex formation between the enzyme and HtpG via various *in vivo* and *in vitro* analyses [149]. The interaction suppressed the enzyme activity, suggesting a regulatory role of HtpG in tetrapyrrole biosynthesis since uroporphyrinogen decarboxylase plays an important role due to its location at the first branch-point in the tetrapyrrole biosynthesis pathway. De-regulation or perturbations of the pathway in the absence of HtpG may result in susceptibility to heat and/or oxidative stress (see later) since misregulation of tetrapyrrole metabolism can lead to severe photo-oxidative stress [150].

Other substrates that we have found are linker polypeptides of the phycobilisome [135]. In an *htpG* mutant cell grown under non-stress conditions, the cellular level of the 30 kDa rod linker polypeptide was reduced. It is known that cellular levels of Hsp90 substrates are reduced when eukaryotic Hsp90 function is inhibited. *In vitro* studies with purified HtpG and phycobilisome showed that HtpG interacts with the 30 kDa linker polypeptide as well as other linker polypeptides. The interaction resulted in suppression of their aggregation.

In eukaryotes, Hsp90 collaborates with Hsp70/Hsp40 by forming a complex [151,152]. In the assembly of this complex, Hop/Sti1, a cochaperone/cofactor plays a role as an adaptor protein. Unlike eukaryotic Hsp90, no cochaperones/cofactors for HtpG have been detected. However, there are indications that cyanobacterial HtpG is involved in complex formation with DnaK and GroEL. We previously reported that a novel, high-molecular-weight complex containing GroEL and DnaK accumulates at 16°C, although the accumulation is abolished in the *htpG* mutant [89]. The results prompted us to postulate that HtpG forms a multi-chaperone system with DnaK and/or GroEL. Recently, it has been shown by immunoprecipitation assays that HtpG from psychrophilic bacteria can interact with DnaK and/or GroEL [153].

5.6.2.5 Cellular Localization

We have observed that native and recombinant *Synechococcus* HtpG are mainly present in the soluble protein fraction of cell extracts. HtpG is also detected in *Synechococcus* membrane fractions [148] and the thylakoid membrane in *S. platensis* [103].

5.6.2.6 Cellular Function

Although eukaryotes require functional cytoplasmic Hsp90 for viability under all conditions [154], the bacterial HtpG proteins are dispensable under normal growth conditions and even under heat stress in heterotrophic bacteria although the *htpG* genes are induced under heat stress in these bacteria [155,156]. Thus, unlike for GroEL and DnaK, studies of *E. coli* have not provided the model for subsequent investigations of Hsp90 in other organisms. However, we have shown an indispensable role of HtpG in cyanobacterial growth and survival under heat stress [140,157]. An *htpG* mutant was unable to grow at the moderately high temperature (45°C), while the wild-type *S. elongatus* PCC 7942 was able to grow. The disruptant lost both basal and acquired thermotolerances. Similar results were obtained with an *htpG* null mutant of *Synechocystis* [158]. Among several *Synechocystis* strains whose stress related genes were mutated, an *htpG* mutant showed the most severe defect at a high temperature [159]. Thus, HtpG plays an essential role for the thermal stress management in cyanobacteria. This raises an important question as why cyanobacterial *htpG* mutants are so sensitive to heat stress? It is reasonable to assume that the protein quality control is disturbed in the absence of a molecular chaperone. Regarding the protein quality control in cyanobacterial cells, it must be important to take care of phycobilisomes since phycobilisome proteins occupy 50% of total cellular proteins. Consistent with an important role of HtpG in the quality control of phycobilisomes under heat stress, phycobilisome proteins were greatly denatured/degraded when the *Synechococcus htpG* mutant was heat-treated (Figure 5.3B).

Like heterotrophic bacteria, the mutation in the *htpG* gene does not affect the photoautotrophic growth of cyanobacteria under normal growth conditions [140,158]. However, this does not mean the absence of cellular function of HtpG under normal growth conditions. Although at a low level, the *htpG* mRNA and the HtpG protein can be detected under non-stress conditions [89,140]. We observed that the phycocyanin level in the mutant cell was reduced to 70% of that in wild-type *Synechococcus* on a per cell basis [135]. This phenotype may be due to the reduced level of the 30 kDa rod linker polypeptide in the mutant cells as described earlier. Alternatively, the phenotype may be caused by the mutant's susceptibility to iron stress. It is known that iron stress in cyanobacteria causes the loss of chlorophyll and phycocyanin [160]. The *htpG* mutant exhibits an iron-starved phenotype under the standard growth conditions [161].

HtpG is also induced under low temperature and plays an important role regarding cold stress [89]. The inactivation of the *Synechococcus htpG* gene resulted in slower than wild-type growth at 20°C. The *htpG* mutant stopped growing when the temperature dropped from 30°C to 16°C. In contrast, the wild type continued to grow at the low temperature [89]. Furthermore, complete inactivation of photosynthetic oxygen evolution occurred in the mutant when it was shifted down to 16°C for 3 h, whereas a significant level of photosynthesis was still retained in the wild type. HtpG also plays a role in cold adaptation by γ-proteobacteria from mild (*Shewanella oneidensis*) and cold environments (*Shewanella frigidimarina*), as indicated by the fact that their viability became less at 12°C than at 30°C in the presence of the specific inhibitor of Hsp90, 17-allylamino-geldanamycin [153].

HtpG is induced and plays a role under high light and/or oxidative stresses [90]. The *htpG* mutation caused a decrease in the survival rate and an increase in photoinhibition when cells were incubated under high light irradiance. Photoinhibition is a process whose extent is determined by the balance between the rate of photodamage to PSII and the rate of its repair. HtpG may either suppress the photodamage or facilitate the repair directly or indirectly under/after the light stress. The low temperature sensitivity of the *htpG* mutant may also be due to increase in photoinhibition in the absence of HtpG. The *htpG* mutant was highly sensitive to methyl viologen when it was grown on an agar plate [90]. Thus, HtpG plays a role in the acclimation to oxidative stress. Induction/cellular accumulation of HtpG under oxidative stress has also been observed in pathogenic bacteria [162,163], and mutation of *htpG* renders the cells more sensitive to stress induced by reactive oxygen species [163].

As described earlier, a disruptant of *hspA* encoding a small Hsp homologue, shows decreased growth rates at moderately high temperatures, and loss of both basal and acquired thermotolerances, which resembles the phenotype of an *htpG* disruptant. *In vitro* studies have shown that both small Hsp and Hsp90 can bind nonnative proteins and inhibit their aggregation. Thus, we tested whether constitutive expression of HspA can functionally replace HtpG in *Synechococcus*. HspA did not improve the basal and acquired thermotolerance of the *htpG* disruptant at a lethal temperature, although it did improve thermotolerances of the wild type [161], suggesting that cellular function or chaperone mechanisms of HtpG differ significantly from that of HspA.

In *S. elongatus* PCC 7942, *htpG*-overexpression causes remarkable growth inhibition [148]. Thus, the cellular level of HtpG may have to be strictly regulated.

5.7 ClpB

5.7.1 Introduction

ClpB/Hsp104 belongs to the class 1 family of Clp/Hsp100 AAA+ (ATPases associated with various cellular activities) proteins [164]. The family is defined by the presence of two AAA+ modules of ~200–250 amino acids that comprise an α-helical domain and a Walker-type nucleotide-binding domain [164]. Among members of this family, ClpB/Hsp104 is unique in containing a relatively long intervening region (linker, spacer or a middle domain) between the two ATP-binding domains. *Saccharomyces cerevisiae* Hsp104 and *Thermus thermophilus* ClpB form a homo-hexameric (two-tiered hexameric) ring in the presence of a nucleotide [165,166]. On the other hand, the *E. coli* ClpB is reported to form hexameric or heptameric ring structures [167]. Nucleotides such as ATPγS and ADP stabilize the hexameric ClpB. ClpB/Hsp104 neither promotes protein folding nor suppresses protein aggregation [168], although an anti-aggregation effect of HSP104 was found recently [169]. ClpB/Hsp104 is an ATP dependent unfoldase/disaggregase that functions with the DnaK/Hsp70 chaperone system, to dissolve/unfold protein aggregates [165,170,171].

A probable mechanism for protein disaggregation by ClpB/Hsp104 may be as follows [164,172]. ClpB/Hsp104 binds a substrate protein(s) that is unfolded by forcible translocation through the central channel of ClpB/Hsp104 using the energy of ATP hydrolysis. Upon ATP-hydrolysis dependent release from ClpB/Hsp104, the substrate protein may refold spontaneously or with the help of the DnaK/Hsp70 chaperone system. The chaperone system may also assist in the extraction of polypeptides from aggregates by presenting unstructured regions of the aggregate to ClpB/Hsp104.

5.7.2 Cyanobacterial ClpB

5.7.2.1 Genes

The first gene encoding a ClpB homolog, ClpB1, was cloned and characterized from *S. elongatus* PCC 7942 [173]. Subsequently, a second *Synechococcus clpB* gene (*clpB2*) was identified [174]. Each cyanobacterial genome generally contains multiple copies of the *clpB* gene (CyanoBase). For example, in the genomes of *Synechocystis* sp. PCC 6803, there are also two *clpB* genes. *Synechococcus* ClpBII shares 71% similarity with ClpBI in *Synechococcus* and *Synechocystis*.

5.7.2.2 Gene Expression under Various Stresses

DNA microarray analysis has shown that expression of *Synechocystis* slr1641, which encodes ClpB1, is enhanced by high light intensity, salt, hyperosmotic, heat, or UV-B stress [27–30]. Expression of *Synechocystis* slr0156, which encodes ClpB2, is enhanced by heat shock [8,30]. Northern blot analysis showed that transcripts of *Synechocystis* slr1641 accumulates greatly, but transiently upon heat shock, while transcription of slr0156 is constitutive and is not affected by heat shock [34].

The level of *Synechococcus* ClpB1 protein is very low under normal growth conditions, but increases under heat and cold stress [88,173]. On the other hand, *Synechococcus* ClpB2 is a constitutive protein, and is not induced significantly by heat shock or other stresses [174].

Interestingly, in a *Synechocystis hrcA* mutant in which both *groEL1* and *groEL2* genes are overexpressed, expression of *clpB1* is suppressed, while *clpB2* is overexpressed [34], indicating the presence of a regulatory cross talk in the expression of these genes.

5.7.2.3 Cellular Function

ClpB1 and ClpB2 from *S. elongatus* PCC7942 have been well-characterized. A Δ*clpB1* strain shows no significant alteration in phenotype when compared with the wild type, suggesting that ClpB1 is dispensable for *Synechococcus* cells under non-stress conditions [173]. However, the absence of ClpB1 causes a great reduction in the ability of cells to acquire (develop) thermotolerance with respect to photosynthetic parameters such as oxygen evolution [173], whereas it does not increase the cell's susceptibility to sudden and extreme heat shocks. The situation is reversed for *E. coli*, in which loss of ClpB increases the cell's susceptibility to sudden heat shock but does not affect its ability to develop thermotolerance [175]. In *E. coli*, other heat-induced functions may contribute more to induced thermotolerance than ClpB does. Interestingly, the *clpB1* gene has a second internal translational start site within the *clpB* transcript, resulting in a truncated 79-kDa form in addition to the full-length 93-kDa protein [173]. The truncated form is functional and has the same capacity as the 93-kDa one to confer thermotolerance [176]. In *E. coli*, the truncated form ClpB-79 may have a regulatory effect on the full-length ClpB-93 [177]. The ATPase activity of the ClpB-79 protein is not stimulated by proteins, while that of the ClpB-93 protein is, suggesting that the truncated form lacks a region responsible for protein binding. ClpB-79 inhibits the protein-activated ATPase activity of ClpB-93. It is not known whether the truncated form of the cyanobacterial ClpB1 has similar biochemical functions.

ClpB2, a 97-kDa protein, is an essential protein for cell viability as it is impossible to disrupt the *clpB2* gene [174]. ClpB2 cannot complement ClpB1 for acquired thermotolerance [174], indicating that ClpB2 has an essential role that is not related to the acquisition of thermotolerance. Unlike ClpB1, no truncated form of ClpB2 is produced. In contrast to *Synechococcus* ClpB2, the *E. coli* ClpB, which is 56% identical to the *Synechococcus* ClpB1, complements the *Synechococcus clpB1* mutant and restores its ability to develop thermotolerance [178].

A Δ*clpB1* strain shows higher susceptibility to sudden and extreme cold shocks than the wild type, but it can develop an equivalent degree of cold-tolerance as the wild type after being preconditioned at a permissive low temperature [88].

5.8 CONCLUDING REMARKS AND FUTURE DIRECTIONS

Studies have revealed that regulation of chaperone gene expression and function of chaperones in cyanobacteria are unique compared with those in *E. coli* or other heterotrophic bacteria whose studies have provided the model for subsequent investigations of molecular chaperones in other organisms.

In terms of regulation of cyanobacterial chaperone gene expression, the CIRCE/HrcA system is the only mechanism in which the operator and its corresponding repressor have been clarified. A potential *cis*-acting element to which a protein binds has been detected upstream of the *hspA* gene that encodes a small Hsp [42]. The element has never been reported, and thus the system may be involved in a novel regulatory mechanism. Further biochemical and genetic studies have to be conducted to clarify the system. Unique regulatory elements such as K-box and MARS have been shown to be essential for high temperature stress and/or light induction of the *groE(S)L* and *dnaK* genes [80,94]. It is important to clarify the heat/light signal transduction pathway resulting in the stress induction of the chaperone genes.

Cyanobacteria seem to depend more strongly on molecular chaperones under stress than *E. coli* or other heterotrophic bacteria. For example, a *Synechococcus htpG* mutant is highly sensitive to high temperature, while *htpG* mutants of various heterotrophic bacteria do not show a clear phenotype. Thus, cyanobacteria are useful organisms to elucidate function of molecular chaperones under stress. It is important to elucidate how a molecular chaperone plays this important role.

In the DnaJ, GroEL, DnaK, and ClpB families, there are multiple members. It is reasonable to surmise that the individual members have specific functions and distinct substrate specificity, but little is known about this important issue. The DnaK/DnaJ/GrpE chaperone system plays roles in many cellular processes. In order to understand its chaperone mechanism at a molecular level, it is necessary to reconstitute the chaperone system *in vitro*. In eukaryotes, Hsp90 forms multiprotein complexes several of which require essential collaboration with the Hsp70 chaperone system. Information about multi-chaperone complexes in bacteria including cyanobacteria is scarce. It is important to analyze interactions among various molecular chaperones in order to understand the cellular chaperone network for the maintenance of proteome homeostasis.

ACKNOWLEDGMENTS

This work was supported in part by Grant-in-aids for Scientific Research (C) (No. 21580083) to H.N. from the Ministry of Education, Science, Sports and Culture of Japan. The author thanks Dr. Kouji Kojima for many productive discrissions and for help in preparing for the figures.

REFERENCES

1. Richter, K., Haslbeck, M., and Buchner, J., The heat shock response: Life on the verge of death, *Mol. Cell*, 40, 253, 2010.
2. Hartl, F.U., Bracher, A., and Hayer-Hartl, M., Molecular chaperones in protein folding and proteostasis, *Nature*, 475, 324, 2011.
3. Horváth, I. et al., Heat shock response in photosynthetic organisms: Membrane and lipid connections, *Prog. Lipid Res.*, 51, 208, 2012.
4. Ellis, R.J., Macromolecular crowding: An important but neglected aspect of the intracellular environment, *Curr. Opin. Struct. Biol.*, 11, 114, 2001.
5. Suzuki, I. et al., The histidine kinase Hik34 is involved in thermotolerance by regulating the expression of heat shock genes in *Synechocystis*, *Plant Physiol.*, 138, 1409, 2005.
6. Imamura, S. et al., Purification, characterization, and gene expression of all sigma factors of RNA polymerase in a cyanobacterium, *J. Mol. Biol.*, 325, 857, 2003.
7. Tuominen, I. et al., The SigB sigma factor mediates high-temperature responses in the cyanobacterium *Synechocystis* sp. PCC6803, *FEBS Lett.*, 580, 319, 2006.
8. Singh, A.K. et al., The heat shock response in the cyanobacterium *Synechocystis* sp. strain PCC 6803 and regulation of gene expression by HrcA and SigB, *Arch. Microbiol.*, 186, 273, 2006.
9. Horváth, I. et al., Membrane physical state controls the signaling mechanism of the heat shock response in *Synechocystis* PCC 6803: Identification of *hsp17* as a 'fluidity gene', *Proc. Natl. Acad. Sci. U.S.A.*, 95, 3513, 1998.
10. Nakamoto, H., Tanaka, N., and Ishikawa, N., A novel heat shock protein plays an important role in thermal stress management in cyanobacteria, *J. Biol. Chem.*, 276, 25088, 2001.
11. Kojima, K. et al., Roles of the cyanobacterial *isiABC* operon in protection from oxidative and heat stresses, *Physiol. Plant.*, 128, 507, 2006.
12. Nakamoto, H. and Vígh, L., The small heat shock proteins and their clients, *Cell. Mol. Life Sci.*, 64, 294, 2007.
13. Haslbeck, M. et al., Some like it hot: The structure and function of small heat-shock proteins, *Nat. Struct. Mol. Biol.*, 12, 842, 2005.
14. Kim, K.K., Kim, R., and Kim, S.H., Crystal structure of a small heat-shock protein, *Nature*, 394, 595, 1998.
15. van Montfort, R.L. et al., Crystal structure and assembly of a eukaryotic small heat shock protein, *Nat. Struct. Biol.*, 8, 1025, 2001.
16. Ehrnsperger, M. et al., Binding of non-native protein to Hsp25 during heat shock creates a reservoir of folding intermediates for reactivation, *EMBO J.*, 16, 221, 1997.
17. Lee, G.J. et al., A small heat shock protein stably binds heat-denatured model substrates and can maintain a substrate in a folding-competent state, *EMBO J.*, 16, 659, 1997.
18. Mogk, A. et al., Refolding of substrates bound to small Hsps relies on a disaggregation reaction mediated most efficiently by ClpB/DnaK, *J. Biol. Chem.*, 278, 31033, 2003.
19. Cashikar, A.G., Duennwald, M., and Lindquist, S.L., A chaperone pathway in protein disaggregation. Hsp26 alters the nature of protein aggregates to facilitate reactivation by Hsp104, *J. Biol. Chem.*, 280, 23869, 2005.
20. Haslbeck, M. et al., Disassembling protein aggregates in the yeast cytosol. The cooperation of Hsp26 with Ssa1 and Hsp104, *J. Biol. Chem.*, 280, 23861, 2005.
21. Jaya, N., Garcia, V., and Vierling, E., Substrate binding site flexibility of the small heat shock protein molecular chaperones, *Proc. Natl. Acad. Sci. U.S.A.*, 106, 15604, 2009.
22. Stengel, F. et al., Quaternary dynamics and plasticity underlie small heat shock protein chaperone function, *Proc. Natl. Acad. Sci. U.S.A.*, 107, 2007, 2010.
23. Franzmann, T.M. et al., The activation mechanism of Hsp26 does not require dissociation of the oligomer, *J. Mol. Biol.*, 350, 1083, 2005.

24. Roy, S.K. and Nakamoto, H., Cloning, characterization, and transcriptional analysis of a gene encoding an alpha-crystallin-related, small heat shock protein from the thermophilic cyanobacterium *Synechococcus vulcanus*, *J. Bacteriol.*, 180, 3997, 1998.
25. Kaneko, T. et al., Sequence analysis of the genome of the unicellular cyanobacterium *Synechocystis* sp. strain PCC6803. II. Sequence determination of the entire genome and assignment of potential protein-coding regions, *DNA Res.*, 3, 109, 1996.
26. Liu, X. et al., Purification and characterization of two small heat shock proteins from *Anabaena* sp. PCC 7120, *IUBMB Life*, 57, 449, 2005.
27. Hihara, Y. et al., DNA microarray analysis of cyanobacterial gene expression during acclimation to high light, *Plant Cell*, 13, 793, 2001.
28. Kanesaki, Y. et al., Salt stress and hyperosmotic stress regulate the expression of different sets of genes in *Synechocystis* sp. PCC 6803, *Biochem. Biophys. Res. Commun.*, 290, 339, 2002.
29. Huang, L. et al., Global gene expression profiles of the cyanobacterium *Synechocystis* sp. strain PCC 6803 in response to irradiation with UV-B and white light, *J. Bacteriol.*, 184, 6845, 2002.
30. Inaba, M. et al., Gene-engineered rigidification of membrane lipids enhances the cold inducibility of gene expression in *Synechocystis*, *J. Biol. Chem.*, 278, 12191, 2003.
31. Li, H. et al., Differential gene expression in response to hydrogen peroxide and the putative PerR regulon of *Synechocystis* sp. strain PCC 6803, *J. Bacteriol.*, 186, 3331, 2004.
32. Ohta, H. et al., Identification of genes expressed in response to acid stress in *Synechocystis* sp. PCC 6803 using DNA microarrays, *Photosynth. Res.*, 84, 225, 2005.
33. Summerfield, T.C. and Sherman, L.A., Global transcriptional response of the alkali-tolerant cyanobacterium *Synechocystis* sp. strain PCC 6803 to a pH 10 environment, *Appl. Environ. Microbiol.*, 74, 5276, 2008.
34. Nakamoto, H., Suzuki, M., and Kojima, K., Targeted inactivation of the *hrcA* repressor gene in cyanobacteria, *FEBS Lett.*, 549, 57, 2003.
35. Asadulghani, Suzuki, Y., and Nakamoto, H., Light plays a key role in the modulation of heat shock response in the cyanobacterium *Synechocystis* sp PCC 6803, *Biochem. Biophys. Res. Commun.*, 306, 872, 2003.
36. Asadulghani et al., Comparative analysis of the *hspA* mutant and wild-type *Synechocystis* sp. strain PCC 6803 under salt stress: Evaluation of the role of *hspA* in salt-stress management, *Arch. Microbiol.*, 182, 487, 2004.
37. Suzuki, I., Simon, W.J., and Slabas, A.R., The heat shock response of *Synechocystis* sp. PCC 6803 analysed by transcriptomics and proteomics, *J. Exp. Bot.*, 57, 1573, 2006.
38. Fulda, S. et al., Proteome analysis of salt stress response in the cyanobacterium *Synechocystis* sp. strain PCC 6803, *Proteomics*, 6, 2733, 2006.
39. Webb, R., Reddy, K.J., and Sherman, L.A., Regulation and sequence of the *Synechococcus* sp. strain PCC 7942 *groESL* operon, encoding a cyanobacterial chaperonin, *J. Bacteriol.*, 172, 5079, 1990.
40. Roy, S.K., Hiyama, T., and Nakamoto, H., Purification and characterization of the 16-kDa heat-shock-responsive protein from the thermophilic cyanobacterium *Synechococcus vulcanus*, which is an alpha-crystallin-related, small heat shock protein, *Eur. J. Biochem.*, 262, 406, 1999.
41. Kojima, K. and Nakamoto, H., Post-transcriptional control of the cyanobacterial hspA heat-shock induction, *Biochem. Biophys. Res. Commun.*, 331, 583, 2005.
42. Kojima, K. and Nakamoto, H., Specific binding of a protein to a novel DNA element in the cyanobacterial small heat-shock protein gene, *Biochem. Biophys. Res. Commun.*, 297, 616, 2002.
43. Kortmann, J. et al., Translation on demand by a simple RNA-based thermosensor, *Nucleic Acids Res.*, 39, 2855, 2011.
44. Narberhaus, F., Translational control of bacterial heat shock and virulence genes by temperature-sensing mRNAs, *RNA Biol.*, 7, 84, 2010.
45. Sundby, C. et al., Conserved methionines in chloroplasts, *Biochim. Biophys. Acta*, 1703, 191, 2005.
46. Giese, K.C. and Vierling, E., Changes in oligomerization are essential for the chaperone activity of a small heat shock protein in vivo and *in vitro*, *J. Biol. Chem.*, 277, 46310, 2002.
47. Török, Z. et al., *Synechocystis* HSP17 is an amphitropic protein that stabilizes heat-stressed membranes and binds denatured proteins for subsequent chaperone-mediated refolding, *Proc. Natl Acad. Sci. U.S.A.*, 98, 3098, 2001.
48. Basha, E. et al., The identity of proteins associated with a small heat shock protein during heat stress in vivo indicates that these chaperones protect a wide range of cellular functions, *J. Biol. Chem.*, 279, 7566, 2004.

49. Giese, K.C. and Vierling, E., Mutants in a small heat shock protein that affect the oligomeric state. Analysis and allele-specific suppression, *J. Biol. Chem.*, 279, 32674, 2004.
50. Giese, K.C. et al., Evidence for an essential function of the N terminus of a small heat shock protein in vivo, independent of in vitro chaperone activity, *Proc. Natl. Acad. Sci. U.S.A.*, 102, 18896, 2005.
51. Nakamoto, H., Suzuki, N., and Roy, S.K., Constitutive expression of a small heat-shock protein confers cellular thermotolerance and thermal protection to the photosynthetic apparatus in cyanobacteria, *FEBS Lett.*, 483, 169, 2000.
52. Nakamoto, H. and Honma, D., Interaction of a small heat shock protein with light-harvesting cyanobacterial phycocyanins under stress conditions, *FEBS Lett.*, 580, 3029, 2006.
53. Lee, S. et al., HSP16.6 is involved in the development of thermotolerance and thylakoid stability in the unicellular cyanobacterium, *Synechocystis* sp. PCC 6803, *Curr. Microbiol.*, 40, 283, 2000.
54. Nitta, K. et al., Ultrastructural stability under high temperature or intensive light stress conferred by a small heat shock protein in cyanobacteria, *FEBS Lett.*, 579, 1235, 2005.
55. Lee, S. et al., A 16.6-kilodalton protein in the cyanobacterium *Synechocystis* sp. PCC 6803 plays a role in the heat shock response, *Curr. Microbiol.*, 37, 403, 1998.
56. Grossman, A.R. et al., The phycobilisome, a light-harvesting complex responsive to environmental conditions, *Microbiol. Rev.*, 57, 725, 1993.
57. Sakthivel, K., Watanabe, T., and Nakamoto, H., A small heat-shock protein confers stress tolerance and stabilizes thylakoid membrane proteins in cyanobacteria under oxidative stress, *Arch. Microbiol.*, 191, 319, 2009.
58. Genevaux, P., Georgopoulos, C., and Kelley, W.L., The Hsp70 chaperone machines of *Escherichia coli*: A paradigm for the repartition of chaperone functions, *Mol. Microbiol.*, 66, 840, 2007.
59. Vos, M.J. et al., Structural and functional diversities between members of the human HSPB, HSPH, HSPA, and DNAJ chaperone families, *Biochemistry*, 47, 7001, 2008.
60. Hill, R.B., Flanagan, J.M., and Prestegard, J.H., ^{1}H and ^{15}N magnetic resonance assignments, secondary structure, and tertiary fold of *Escherichia coli* DnaJ(1–78), *Biochemistry*, 34, 5587, 1995.
61. Cajo, G.C. et al., The role of the DIF motif of the DnaJ (Hsp40) co-chaperone in the regulation of the DnaK (Hsp70) chaperone cycle, *J. Biol. Chem.*, 281, 12436, 2006.
62. Szabo, A. et al., A zinc finger-like domain of the molecular chaperone DnaJ is involved in binding to denatured protein substrates, *EMBO J.*, 15, 408, 1996.
63. Linke, K. et al., The roles of the two zinc binding sites in DnaJ, *J. Biol. Chem.*, 278, 44457, 2003.
64. Shi, Y.Y., Hong, X.G., and Wang, C.C., The C-terminal (331–376) sequence of *Escherichia coli* DnaJ is essential for dimerization and chaperone activity: A small angle X-ray scattering study in solution, *J. Biol. Chem.*, 280, 22761, 2005.
65. Langer, T. et al., Successive action of DnaK, DnaJ and GroEL along the pathway of chaperone-mediated protein folding, *Nature*, 356, 683, 1992.
66. Oguchi, K. et al., Sequence and analysis of a *dnaJ* homologue gene in cyanobacterium *Synechococcus* sp. PCC7942, *Biochem. Biophys. Res. Commun.*, 236, 461, 1997.
67. Düppre, E., Rupprecht, E., and Schneider, D., Specific and promiscuous functions of multiple DnaJ proteins in *Synechocystis* sp. PCC 6803, *Microbiology*, 157, 1269, 2011.
68. Sato, M. et al., Characterization of *dnaJ* multigene family in the cyanobacterium *Synechococcus elongatus* PCC 7942, *Biosci. Biotechnol. Biochem.*, 71, 1021, 2007.
69. Koksharova, O.A. and Wolk, C.P., A novel gene that bears a DnaJ motif influences cyanobacterial cell division, *J. Bacteriol.*, 184, 5524, 2002.
70. Watanabe, S. et al., Protection of psbAII transcript from ribonuclease degradation in vitro by DnaK2 and DnaJ2 chaperones of the cyanobacterium *Synechococcus elongatus* PCC 7942, *Biosci. Biotechnol. Biochem.*, 71, 279, 2007.
71. Carpousis, A.J., The RNA degradosome of *Escherichia coli*: An mRNA-degrading machine assembled on RNase E, *Annu. Rev. Microbiol.*, 61, 71, 2007.
72. Braig, K. et al., The crystal structure of the bacterial chaperonin GroEL at 2.8 Å, *Nature*, 371, 578, 1994.
73. Xu, Z., Horwich, A.L., and Sigler, P.B., The crystal structure of the asymmetric GroEL-GroES-(ADP)7 chaperonin complex, *Nature*, 388, 741, 1997.
74. Hunt, J.F. et al., The crystal structure of the GroES co-chaperonin at 2.8 Å resolution, *Nature*, 379, 37, 1996.
75. Horwich, A.L. and Fenton, W.A., Chaperonin-mediated protein folding: Using a central cavity to kinetically assist polypeptide chain folding, *Q. Rev. Biophys.*, 42, 83, 2009.
76. Chitnis, P.R. and Nelson, N., Molecular cloning of the genes encoding two chaperone proteins of the cyanobacterium *Synechocystis* sp. PCC 6803, *J. Biol. Chem.*, 266, 58, 1991.

77. Lehel, C. et al., A second *groEL*-like gene, organized in a *groESL* operon is present in the genome of *Synechocystis* sp. PCC 6803, *J. Biol. Chem.*, 268, 1799, 1993.
78. Furuki, M. et al., Cloning, characterization and functional analysis of *groEL*-like gene from thermophilic cyanobacterium *Synechococcus vulcanus*, which does not form an operon with *groES*, *Biochim. Biophys. Acta*, 1294, 106, 1996.
79. Tanaka, N., Hiyama, T., and Nakamoto, H., Cloning, characterization and functional analysis of *groESL* operon from thermophilic cyanobacterium *Synechococcus vulcanus*, *Biochim. Biophys. Acta*, 1343, 335, 1997.
80. Kojima, K. and Nakamoto H., A novel light- and heat-responsive regulation of the *groE* transcription in the absence of HrcA or CIRCE in cyanobacteria, *FEBS Lett.*, 581, 1871, 2007.
81. Lund, P.A., Multiple chaperonins in bacteria—Why so many?, *FEMS Microbiol. Rev.*, 33, 785, 2009.
82. Glatz, A. et al., Chaperonin genes of the *Synechocystis* PCC 6803 are differentially regulated under light-dark transition during heat stress, *Biochem. Biophys. Res. Commun.*, 239, 291, 1997.
83. Prakash, J.S. et al., An RNA helicase, CrhR, regulates the low-temperature-inducible expression of heat-shock genes *groES*, *groEL1* and *groEL2* in *Synechocystis* sp. PCC 6803, *Microbiology*, 156, 442, 2010.
84. Kovács, E. et al., The chaperonins of *Synechocystis* PCC 6803 differ in heat inducibility and chaperone activity, *Biochem. Biophys. Res. Commun.*, 289, 908, 2001.
85. Gao, Y. et al., Identification of the proteomic changes in *Synechocystis* sp. PCC 6803 following prolonged UV-B irradiation, *J. Exp. Bot.*, 60, 1141, 2009.
86. Sato, S., Ikeuchi, M., and Nakamoto, H., Expression and function of a *groEL* paralog in the thermophilic cyanobacterium *Thermosynechococcus elongatus* under heat and cold stress, *FEBS Lett.*, 582, 3389, 2008.
87. Rajaram, H. and Apte, S.K., Nitrogen status and heat-stress-dependent differential expression of the cpn60 chaperonin gene influences thermotolerance in the cyanobacterium *Anabaena*, *Microbiology*, 154, 317, 2008.
88. Porankiewicz, J. and Clarke, A.K., Induction of the heat shock protein ClpB affects cold acclimation in the cyanobacterium *Synechococcus* sp. strain PCC 7942, *J. Bacteriol.*, 179, 5111, 1997
89. Hossain, M.M. and Nakamoto, H., HtpG plays a role in cold acclimation in cyanobacteria, *Curr. Microbiol.*, 44, 291, 2002.
90. Hossain, M.M. and Nakamoto, H., Role for the cyanobacterial HtpG in protection from oxidative stress, *Curr. Microbiol.*, 46, 70, 2003.
91. Yamazawa, A. et al., UV-A-induced expression of GroEL in the UV-A-resistant marine cyanobacterium *Oscillatoria* sp. NKBG 091600, *Microbiology*, 145, 949, 1999.
92. Schumann, W., The *Bacillus subtilis* heat shock stimulon, *Cell Stress Chaperones*, 8, 207, 2003.
93. Rajaram, H. and Apte, S.K., Differential regulation of *groESL* operon expression in response to heat and light in *Anabaena*, *Arch. Microbiol.*, 192, 729, 2010.
94. Sato, M. et al., Expression analysis of multiple *dnaK* genes in the cyanobacterium *Synechococcus elongatus* PCC 7942, *J. Bacteriol.*, 189, 3751, 2007.
95. Huq, S. et al., Comparative biochemical characterization of two GroEL homologs from the cyanobacterium *Synechococcus elongatus* PCC 7942, *Biosci. Biotechnol. Biochem.*, 74, 2273, 2010.
96. Ybarra, J. and Horowitz, P.M., Inactive GroEL monomers can be isolated and reassembled to functional tetradecamers that contain few bound peptides, *J. Biol. Chem.*, 270, 22962, 1995.
97. Rospert, S. et al., Identification and functional analysis of chaperonin 10, the *groES* homolog from yeast mitochondria, *Proc. Natl. Acad. Sci. U.S.A.*, 90, 10967, 1993.
98. Kovács, E. et al., Characterisation of a GroEL single-ring mutant that supports growth of *Escherichia coli* and has GroES-dependent ATPase activity, *J. Mol. Biol.*, 396, 1271, 2010.
99. Qamra, R., Srinivas, V., and Mande, S.C., *Mycobacterium tuberculosis* GroEL homologues unusually exist as lower oligomers and retain the ability to suppress aggregation of substrate proteins, *J. Mol. Biol.*, 342, 605, 2004.
100. Kovács, E. et al., Heat-stress induces association of the GroEL-analog chaperonin with thylakoid membranes in cyanobacterium *Synechocystis* PCC 6803, *Plant Physiol. Biochem.*, 32, 285, 1994.
101. Srivastava, R., Pisareva, T., and Norling, B., Proteomic studies of the thylakoid membrane of *Synechocystis* sp. PCC 6803, *Proteomics*, 5, 4905, 2005.
102. Huang, F. et al., Proteomics of *Synechocystis* sp. strain PCC 6803: Identification of plasma membrane proteins, *Mol. Cell Proteomics*, 1, 956, 2002.
103. Hongsthong, A. et al., Proteome analysis at the subcellular level of the cyanobacterium *Spirulina platensis* in response to low-temperature stress conditions, *FEMS Microbiol. Lett.*, 288, 92, 2008.

104. Török, Z. et al., Evidence for a lipochaperonin: Association of active protein-folding GroESL oligomers with lipids can stabilize membranes under heat shock conditions, *Proc. Natl. Acad. Sci. U.S.A.*, 94, 2192, 1997.
105. Fayet, O., Ziegelhoffer, T., and Georgopoulos, C., The *groES* and *groEL* heat shock gene products of *Escherichia coli* are essential for bacterial growth at all temperatures, *J. Bacteriol.*, 171, 1379, 1989.
106. Chaurasia, A.K and Apte, S.K., Overexpression of the *groESL* operon enhances the heat and salinity stress tolerance of the nitrogen-fixing cyanobacterium *Anabaena* sp. strain PCC7120, *Appl. Environ. Microbiol.*, 75, 6008, 2009.
107. Flaherty, K.M., DeLuca-Flaherty, C., and McKay, D.B., Three-dimensional structure of the ATPase fragment of a 70 K heat-shock cognate protein, *Nature*, 346, 623, 1990.
108. Mayer, M.P. and Bukau, B., Hsp70 chaperones: Cellular functions and molecular mechanism, *Cell. Mol. Life Sci.*, 62, 670, 2005.
109. Zhu, X. et al., Structural analysis of substrate binding by the molecular chaperone DnaK, *Science*, 272, 1606, 1996.
110. Harrison, C.J. et al., Crystal structure of the nucleotide exchange factor GrpE bound to the ATPase domain of the molecular chaperone DnaK, *Science*, 276, 431, 1997.
111. Rüdiger, S., Buchberger, A., and Bukau, B., Interaction of Hsp70 chaperones with substrates, *Nat. Struct. Biol.*, 4, 342, 1997.
112. Nimura, K., Yoshikawa, H., and Takahashi, H., Identification of *dnaK* multigene family in *Synechococcus* sp. PCC7942, *Biochem. Biophys. Res. Commun.*, 201, 466, 1994.
113. Nimura, K., Yoshikawa, H., and Takahashi, H., Sequence analysis of the third *dnaK* homolog gene in *Synechococcus* sp. PCC7942, *Biochem. Biophys. Res. Commun.*, 201, 848, 1994.
114. Blanco-Rivero, M.C., Takabe, T., and Viale, A.M., Functional differences between cyanobacterial DnaK1 chaperones from the halophyte *Aphanothece halophytica* and the freshwater species *Synechococcus elongatus* expressed in *Escherichia coli*, *Curr. Microbiol.*, 51, 164, 2005.
115. Segal, R. and Ron, E.Z., Regulation and organization of the *groE* and *dnaK* operons in Eubacteria, *FEMS Microbiol. Lett.*, 138, 1, 1996.
116. Saito, H. and Uchida, H., Organization and expression of the *dnaJ* and *dnaK* genes of *Escherichia coli* K12, *Mol. Gen. Genet.*, 164, 1, 1978.
117. Lee, B.H. et al., Isolation and characterization of a *dnaK* genomic locus in a halotolerant cyanobacterium *Aphanothece halophytica*, *Plant. Mol. Biol.*, 35, 763, 1997.
118. Homuth, G. et al., The *dnaK* operon of *Bacillus subtilis* is heptacistronic, *J. Bacteriol.*, 179, 1153, 1997.
119. Rupprecht, E., Düppre, E., and Schneider, D., Similarities and singularities of three DnaK proteins from the cyanobacterium *Synechocystis* sp. PCC 6803, *Plant Cell Physiol.*, 51, 1210, 2010.
120. Mary, I. et al., Effects of high light on transcripts of stress-associated genes for the cyanobacteria *Synechocystis* sp. PCC 6803 and *Prochlorococcus* MED4 and MIT9313, *Microbiology*, 150, 1271, 2004.
121. Nimura, K., Takahashi, H., and Yoshikawa, H., Characterization of the *dnaK* multigene family in the cyanobacterium *Synechococcus* sp. strain PCC7942, *J. Bacteriol.*, 183, 1320, 2001.
122. Hibino, T. et al., Molecular characterization of DnaK from the halotolerant cyanobacterium *Aphanothece halophytica* for ATPase, protein folding, and copper binding under various salinity conditions, *Plant Mol. Biol.*, 40, 409, 1999.
123. Nimura, K., Yoshikawa, H., and Takahashi, H., DnaK3, one of the three DnaK proteins of cyanobacterium *Synechococcus* sp. PCC7942, is quantitatively detected in the thylakoid membrane, *Biochem. Biophys. Res. Commun.*, 229, 334, 1996.
124. Rupprecht, E. et al., Three different DnaK proteins are functionally expressed in the cyanobacterium *Synechocystis* sp. PCC 6803, *Microbiology*, 153, 1828, 2007.
125. Katano, Y., Nimura-Matsune, K., and Yoshikawa, H., Involvement of DnaK3, one of the three DnaK proteins of cyanobacterium *Synechococcus* sp. PCC7942, in translational process on the surface of the thylakoid membrane, *Biosci. Biotechnol. Biochem.*, 70, 1592, 2006.
126. Hongsthong, A. et al., Subcellular proteomic characterization of the high-temperature stress response of the cyanobacterium *Spirulina platensis*, *Proteome Sci.*, 7, 33, 2009.
127. Paek, K.H. and Walker, G.C., *Escherichia coli dnaK* null mutants are inviable at high temperature, *J. Bacteriol.*, 169, 283, 1987.
128. Blum, P. et al., Physiological consequences of DnaK and DnaJ overproduction in *Escherichia coli*, *J. Bacteriol.*, 174, 7436, 1992.
129. Pearl, L.H. and Prodromou, C., Structure and mechanism of the Hsp90 molecular chaperone machinery, *Annu. Rev. Biochem.*, 75, 271, 2006.

130. Wandinger, S.K., Richter, K., and Buchner, J., The Hsp90 chaperone machinery, *J. Biol. Chem.*, 283, 18473, 2008.
131. Prodromou, C. et al., Identification and structural characterization of the ATP/ADP-binding site in the Hsp90 molecular chaperone, *Cell*, 90, 65, 1997.
132. Panaretou, B. et al., ATP binding and hydrolysis are essential to the function of the Hsp90 molecular chaperone *in vivo*, *EMBO J.*, 17, 4829, 1998.
133. Nemoto, T. et al., Mechanism of dimer formation of the 90-kDa heat-shock protein, *Eur. J. Biochem.*, 233, 1, 1995.
134. Harris, S.F., Shiau, A.K., and Agard, D.A., The crystal structure of the carboxy-terminal dimerization domain of *htpG*, the *Escherichia coli* Hsp90, reveals a potential substrate binding site, *Structure*, 12, 1087, 2004.
135. Sato, T. et al., HtpG, the prokaryotic homologue of Hsp90, stabilizes a phycobilisome protein in the cyanobacterium *Synechococcus elongatus* PCC 7942, *Mol. Microbiol.*, 76, 576, 2010.
136. Shiau, A.K. et al., Structural Analysis of *E. coli* hsp90 reveals dramatic nucleotide-dependent conformational rearrangements, *Cell*, 127, 329, 2006.
137. Ali, M.M. et al., Crystal structure of an Hsp90-nucleotide-p23/Sba1 closed chaperone complex, *Nature*, 440, 1013, 2006.
138. Dollins, D.E. et al., Structures of GRP94-nucleotide complexes reveal mechanistic differences between the hsp90 chaperones, *Mol. Cell*, 28, 41, 2007.
139. Wiech, H. et al., Hsp90 chaperones protein folding *in vitro*, *Nature*, 358, 169, 1992.
140. Tanaka, N. and Nakamoto, H., HtpG is essential for the thermal stress management in cyanobacteria, *FEBS Lett.*, 458, 117, 1999.
141. Marin, K. et al., Gene expression profiling reflects physiological processes in salt acclimation of *Synechocystis* sp. strain PCC 6803, *Plant Physiol.*, 136, 3290, 2004.
142. Spence, J. and Georgopoulos, C., Purification and properties of the *Escherichia coli* heat shock protein, HtpG, *J. Biol. Chem.*, 264, 4398, 1989.
143. Yonehara, M. et al., Heat-induced chaperone activity of HSP90, *J. Biol. Chem.*, 271, 2641, 1996.
144. Roe, S.M. et al., Structural basis for inhibition of the Hsp90 molecular chaperone by the antitumor antibiotics radicicol and geldanamycin, *J. Med. Chem.*, 42, 260, 1999.
145. Minagawa, S. et al., Cyclic lipopeptide antibiotics bind to the N-terminal domain of the prokaryotic Hsp90 to inhibit the chaperone activity, *Biochem. J.*, 435, 237, 2011.
146. Vaughan, C.K. et al., Structure of an Hsp90-Cdc37-Cdk4 complex, *Mol. Cell*, 23, 697, 2006.
147. Whitesell, L. et al., Inhibition of heat shock protein HSP90-pp60v-src heteroprotein complex formation by benzoquinone ansamycins: Essential role for stress proteins in oncogenic transformation, *Proc. Natl. Acad. Sci. U.S.A.*, 91, 8324, 1994.
148. Watanabe, S. et al., Studies on the role of HtpG in the tetrapyrrole biosynthesis pathway of the cyanobacterium *Synechococcus elongatus* PCC 7942, *Biochem. Biophys. Res. Commun.*, 352, 36, 2007.
149. Saito, M. et al., Interaction of the molecular chaperone HtpG with uroporphyrinogen decarboxylase in the cyanobacterium *Synechococcus elongatus* PCC 7942, *Biosci. Biotechnol. Biochem.*, 72, 1394, 2008.
150. Mochizuki, N. et al., The cell biology of tetrapyrroles: A life and death struggle, *Trends Plant Sci.*, 15, 488, 2010.
151. Smith, D.F. et al., Identification of a 60-kilodalton stress-related protein, p60, which interacts with hsp90 and hsp70, *Mol. Cell. Biol.*, 13, 869, 1993.
152. Wegele, H., Müller, L., and Buchner, J., Hsp70 and Hsp90—A relay team for protein folding, *Rev. Physiol. Biochem. Pharmacol.*, 151, 1, 2004.
153. García-Descalzo, L. et al., Identification of in vivo HSP90-interacting proteins reveals modularity of HSP90 complexes is dependent on the environment in psychrophilic bacteria, *Cell Stress Chaperones*, 16, 203, 2011.
154. Borkovich, K.A. et al., hsp82 is an essential protein that is required in higher concentrations for growth of cells at higher temperatures, *Mol. Cell. Biol.*, 9, 3919, 1989.
155. Bardwell, J.C. and Craig, E.A., Ancient heat shock gene is dispensable, *J. Bacteriol.*, 170, 2977, 1988.
156. Versteeg, S., Mogk, A., and Schumann, W., The *Bacillus subtilis htpG* gene is not involved in thermal stress management, *Mol. Gen. Genet.*, 261, 582, 1999.
157. Jackson, S.E., Queitsch, C., and Toft, D., Hsp90: From structure to phenotype, *Nat. Struct. Mol. Biol.*, 11, 1152, 2004.
158. Fang, F. and Barnum, S.R., The heat shock gene, *htpG*, and thermotolerance in the cyanobacterium, *Synechocystis* sp. PCC 6803, *Curr. Microbiol.*, 47, 341, 2003.

159. Rowland, J.G. et al., Identification of components associated with thermal acclimation of photosystem II in *Synechocystis* sp. PCC6803, *PLoS One*, 5, e10511, 2010.
160. Guikema, J.A. and Sherman, L.A., Organization and function of chlorophyll in membranes of cyanobacteria during iron starvation, *Plant Physiol*., 73, 250, 1983.
161. Kojima, K. and Nakamoto, H., Constitutive expression of small heat shock protein in an *htpG* disruptant of the cyanobacterium *Synechococcus* sp. PCC 7942, *Curr. Microbiol*., 50, 272, 2005.
162. Steeves, C.H. et al., Oxidative stress response in the opportunistic oral pathogen *Fusobacterium nucleatum*, *Proteomics*, 11, 2027, 2011.
163. Dang, W., Hu, Y.H., and Sun, L., HtpG is involved in the pathogenesis of *Edwardsiella tarda*, *Vet. Microbiol*., 152, 394, 2011.
164. Doyle, S.M. and Wickner, S., Hsp104 and ClpB: Protein disaggregating machines, *Trends Biochem. Sci*., 34, 40, 2009.
165. Parsell, D.A., Kowal, A.S., and Lindquist, S., *Saccharomyces cerevisiae* Hsp104 protein. Purification and characterization of ATP-induced structural changes, *J. Biol. Chem*., 269, 4480, 1994.
166. Lee, S. et al., The structure of ClpB: A molecular chaperone that rescues proteins from an aggregated state, *Cell*, 115, 229, 2003.
167. Akoev, V. et al., Nucleotide-induced switch in oligomerization of the AAA+ ATPase ClpB, *Protein Sci*., 13, 567, 2004.
168. Glover, J.R. and Lindquist, S., Hsp104, Hsp70, and Hsp40: A novel chaperone system that rescues previously aggregated proteins, *Cell*, 94, 73, 1998.
169. Arimon, M. et al., Hsp104 targets multiple intermediates on the amyloid pathway and suppresses the seeding capacity of Abeta fibrils and protofibrils, *J. Mol. Biol*., 384, 1157, 2008.
170. Mogk, A. et al., Identification of thermolabile *Escherichia coli* proteins: Prevention and reversion of aggregation by DnaK and ClpB, *EMBO J*., 18, 6934, 1999.
171. Goloubinoff, P. et al., Sequential mechanism of solubilization and refolding of stable protein aggregates by a bichaperone network, *Proc. Natl. Acad. Sci. U.S.A*., 96, 13732, 1999.
172. Haslberger, T., Bukau, B., and Mogk, A., Towards a unifying mechanism for ClpB/Hsp104-mediated protein disaggregation and prion propagation, *Biochem. Cell Biol*., 88, 63, 2010.
173. Eriksson, M.J. and Clarke, A.K., The heat shock protein ClpB mediates the development of thermotolerance in the cyanobacterium *Synechococcus* sp. strain PCC 7942, *J. Bacteriol*., 178, 4839, 1996.
174. Eriksson, M.J. et al., Novel form of ClpB/HSP100 protein in the cyanobacterium *Synechococcus*, *J. Bacteriol*., 183, 7392, 2001.
175. Squires, C.L. et al., ClpB is the *Escherichia coli* heat shock protein F84.1, *J. Bacteriol*., 173, 4254, 1991.
176. Clarke, A.K. and Eriksson, M.J., The truncated form of the bacterial heat shock protein ClpB/HSP100 contributes to development of thermotolerance in the cyanobacterium *Synechococcus* sp. strain PCC 7942, *J. Bacteriol*., 182, 7092, 2000.
177. Park, S.K. et al., Site-directed mutagenesis of the dual translational initiation sites of the *clpB* gene of *Escherichia coli* and characterization of its gene products, *J. Biol. Chem*., 268, 20170, 1993.
178. Eriksson, M.J. and Clarke, A.K., The *Escherichia coli* heat shock protein ClpB restores acquired thermotolerance to a cyanobacterial *clpB* deletion mutant, *Cell Stress Chaperones*, 5, 255, 2000.

6 Heat Stress Management in *Synechocystis* PCC 6803

The Interplay between Membranes and Stress Protein Molecular Chaperones

Attila Glatz, Zsolt Török, László Vígh, and Ibolya Horváth*

CONTENTS

6.1 INTRODUCTION

Exposure of organisms to a sudden increase in temperature triggers a ubiquitous and homeostatic cellular stress response. This heat shock response is characterized by rapid induction of genes encoding heat shock proteins (HSPs), which are divided into seven functional classes: metabolism, DNA/RNA repair, protein degradation, regulation, transport, cell organization, and molecular chaperones [1], and references therein. An almost identical functional classification of stress proteins was suggested by Kültz [2]. The predominant class of HSPs, the molecular chaperones [3], comprises five major and evolutionarily conserved families—ClpB (in eukaryotes HSP100), HtpG (HSP90), *dnaK* (HSP70), GroESL also called as chaperonins (HSP60), and small heat shock proteins (sHSPs). Under physiological conditions, they assist in several important cellular processes including the correct folding of *de novo* synthesized proteins, the assembly of oligomeric structures, the membrane transport of proteins, etc. [4]. Upon heat stress, polypeptides tend to lose their native conformation, which leads to the increased exposure of hydrophobic amino acid sequences normally buried inside the molecule. Molecular chaperones are able to recognize these patches and

* Corresponding author: attila@brc.hu

help to prevent their aggregation and facilitate the refolding of the damaged proteins [1,5]. Based on their activity, chaperones can be divided into three classes. "Foldases" (*dnaK*, GroEL) assist in the correct (re)-folding of polypeptides together with their co-chaperones. The "holdases" (sHSPs and HtpGs) are able to bind nonnative proteins and prevent their aggregation, but do not directly facilitate their folding. The third class of chaperones ("unfoldases/disaggregases," ClpB family) is capable to dissolve aggregates [1,5,6].

The photosynthetic cyanobacterium *Synechocystis* sp. PCC 6803 has several features that make this strain particularly suitable for studying stress response at the molecular level. It has previously been shown that in higher plant cells exposed to heat stress the photosynthetic apparatus (especially photosystem II) is irreversibly damaged prior to impairment of other cellular functions [7,8]. The origin of the higher plant chloroplast is believed to be the result of endosymbiosis of a cyanobacterial progenitor [9]. Early, ^{14}C-labeling experiments with *Synechocystis* 6803 have demonstrated that the cyanobacterium responds to heat shock by elevated synthesis of at least four HSPs (*dnaK*, chaperonins, and the sHSP HSP17) [10]. It has also been shown that growth temperature is able to modulate the acquired thermotolerance of *Synechocystis* 6803 [11]. Therefore, detailed characterization of the heat shock response of this cyanobacterium might contribute to understanding the stress–defense mechanisms of higher plants [12].

6.2 BIG FIVE: FUNCTION AND REGULATION OF MOLECULAR CHAPERONES IN *SYNECHOCYSTIS* 6803

6.2.1 ClpB Family

ClpBs belongs to the class 1 family of AAA (ATPases associated with various cellular functions) proteins [13,14]. They form dynamic hexameric structures containing two different nucleotide binding sites in each monomer [15]. The mechanism of ClpB action is obscure, but they are thought to pull misfolded proteins through the central channel of the hexamer in an unfolded state enabling proteins to become refolded after releasing them [16]. The details of the folding are still under debate, but ClpB's "disaggregase" might act together with the *dnaK* chaperone machinery [1,5].

According to Cyanobase, *Synechocystis* 6803 contains two open reading frames (ORFs) encoding potential ClpB chaperones (*clpB1 and clpB2*) [17].

As regards their cellular functions, clpB1-deficient mutant grows normally at 30°C but shows weak heat sensitivity when incubated at high temperature (44°C). This temperature sensitivity is greatly increased in *hsp17/clpB1* double mutant, suggesting the genetic interaction of the two chaperone families. Interestingly, attempts to delete *clpB2* gene have failed, indicating that this gene is essential in *Synechocystis* 6803 [18]. Regarding their transcriptional regulation, based on microarray experiments—similar to other main chaperone members—ClpB1 is upregulated when cells are subjected to different kinds of stresses (heat, high light, salt, etc.) (see [19]). The mechanism of the heat activation as well as the exact cellular function of ClpBs in *Synechocystis* 6803 remains to be established.

6.2.2 HtpG Family

The members of the HSP90 families form dimer at normal growth temperatures. It does not seem as promiscuous in its substrate spectrum as the other chaperone members belonging to "holdases" [1,20]. In eukaryotes, they collaborate with several cofactors to mediate the conformational changes of substrates including transcriptional factors and protein kinesis under physiological conditions [21]. It is still an open question whether HSP90 changes its substrate spectrum upon stress conditions [1]. However, by keeping mutant proteins in wild-type conformation, Hsp90 is suggested to have an evolutionary significance [22].

Synechocystis 6803 contains only a single *htpG* gene [17]. The null mutant shows no growth defect at 30°C but fails to grow at higher (40°C) temperature. It has also been demonstrated that the *htpG*-deficient cells are unable to acquire thermotolerance [23]. Interestingly, the loss of *htpG* function varies phenotypically in prokaryotes: while *Escherichia coli* [24] and *Synechococcus* PCC 7942 [25] showed the same phenotype as *Synechocystis* 6803, the removal of htpG did not affect the thermotolerance of *Bacillus subtilis* [26]. This might indicate that—despite their close relationship—the *htpG* family members might have different protective role in the prokaryotic world.

Little is known about the transcriptional regulation of *htpG* gene in *Synechocystis* 6803. As might be expected from a heat shock gene, its transcript level is increased upon heat shock in wild-type strain. The induction level was lower in a sigma factor mutant *sigB*, as demonstrated by microarray data [27], but the details of the regulation are still obscure.

6.2.3 DnaK Family

DnaK-type proteins belong to the most conserved "holdase/foldase"-type chaperones in living organisms. Under normal conditions—together with their co-chaperones DnaJ (HSP40) and GrpE (nucleotide exchange factor [NEF])—they assist in the folding of the *de novo* synthesized proteins and the aggregation and refolding of damaged proteins [28]. HSP70 consists of two domains: an ATPase at the N-terminus and a C-terminal substrate binding domain. HSP70 interacts with hydrophobic segments of the proteins in an ATP-dependent manner: the "posthydrolysis ADP state" protein has high affinity to the substrates. J-domain-containing proteins bind and deliver the nonnative proteins to Hsp70 [29]. The J domains interact with the ATPase domain of Hsp70 and stimulate ATP hydrolysis; finally, the nucleotide and substrate release is enhanced by the NEF [1].

The first *Synechocystis* 6803 *dnaK (dnaK2)* gene has been isolated in the early 1990s [30]. The genome sequencing project uncovered the existence of two other *dnaK* homologues, together with four *dnaJ*s and a single *grpE* gene [17]. *DnaK2* [31] and *dnaK3* proved to be essential, while *dnaK1* is dispensable [32]. All three *dnaK* proteins are able to interact with the lonely NEF GrpE [33]. Their interaction with the DnaJ co-chaperones remains to be established. The *dnaK2* partial mutant showed temperature-sensitive phenotype but was able to acquire thermotolerance [31]. Notably, most of the *dnaK3* proteins have been found in the membrane fraction of the *Synechocystis* 6803 under normal conditions. Interestingly, the presence of the two other *dnaK*s has also been detected in the same fraction [32]. This indicates that this family might be able to protect thylakoid membranes under stress conditions. Therefore, examination of (re)-distribution of the *dnaK* family members and the mechanism of protection of the photosynthetic apparatus under thermal stress would be useful to explore the membrane-defensive function of this important chaperone family.

Surprisingly, among the three *dnaK* genes, only *dnaK2* is heat inducible at mRNA [30,31] and protein levels [33]. All four *dnaJs* and the *grpE* genes were uninduced upon heat shock as tested by reverse transcriptase polymerase chain reaction (RT-PCR) [31]. Recently, although under different heat shock conditions (42°C vs. 45°C), an increase of two *dnaJ* transcripts was reported together with *dnaK2*, which seems to be Sigma B (or σB) dependent [27]. Importantly, the temperature threshold of the induction of the *dnaK2*—together with other chaperones—showed an evident correlation with the membrane's physical state of the *Synechocystis* 6803 [34], supporting the notion, that membrane acts as stress sensors in this organism [35,36].

6.2.4 Chaperonins

The GroESL-type chaperonins are the oldest known "foldase"-type chaperones, which widely distributed in the living organisms. They were originally described as proteins assisting in the folding of the overexpressed foreign ribulose biphosphate carboxilase (Rubisco) in *E. coli* [37]. The active complex consists of the tetradecameric ring of GroEL and a 7-mer GroES ring. The unfolded substrate enters into the cavity of GroEL ring, then the GroES oligomer closes the channel. After ATP-binding and hydrolysis, the substrate is released by opening the cavity or the cycle restarts again [1,5]. It should be noted that not only one but two GroES rings can bind to GroEL complex [38]. It has also been demonstrated that the GroESL oligomer can bind and stabilize membranes upon heat shock, while retaining its chaperone function [39].

The first chaperonin gene (*cpn60, groEL2*) of *Synechocystis* has been described by Chitnis and Nelson [30]. Since the N-terminal amino acid deduced from the published sequence did not agree exactly with the purified chaperonin protein of the *Synechocystis* 6803 [10], the existence of the second *groEL* gene—forming an operon with *groES*—has been uncovered [40]. They are able to bind thylakoid membranes upon sublethal heat stress and this association might provide protection against the hyperfluidizing and non-bilayer phase forming effects of high temperature [41,42]. Interestingly, when overexpressed in *groEL*-mutant *E. coli*—in contrast to GroEL1—GroEL2 failed to show chaperone activity as measured by Rubisco and T4 phage assembly. This finding might indicate that the two chaperonins have different chaperone functions [43].

Both chaperonin genes are heat inducible [30,40]. The upstream region of groESL1 and groEL genes contains a highly conservative inverted repeat element (CIRCE, for a review see Ref. [44]). The existence of the potential CIRCE-binding repressor, HrcA, has been predicted [45]. Deletion of the hrcA gene resulted in higher basal chaperonin mRNA level supporting the idea that HrcA functions as repressor [46]. In spite of the presence of the common transcriptional regulatory elements, the heat inducibility of the chaperonins differs when light conditions are changed or the photosynthetic electron transport is inhibited with diuron [45]. Using different mutants of the *Synechocystis* 6803 strain, it has been shown that novel regulatory sequences including K-box might direct the light-mediated transcription of the chaperonin genes [46].

6.2.5 Small Heat Shock Proteins

The sHSPs are the most diverse family of "holdase"-type chaperones. Their common structural feature is the α-crystallin domain (for recent reviews see [36,47]). Upon heat stress they form larger or smaller oligomeric complexes and bind nonnative proteins preventing their aggregation in an ATP-independent manner. The release and the folding of the clients require the cooperation of other ATP-dependent chaperones (e.g., *dnaK*; see [1]).

The *Synechocystis* 6803 sHSP (HSP17, also known as Hsp16.6 or HspA) has many interesting features besides its holdase activity. Being an amphitropic protein, it can bind to thylakoid membranes upon heat shock *in vivo* [34]. Binding to lipid membranes results in an upshifting in the heat-induced phase transition temperatures of the membranes; therefore, HSP17 might have membrane-protecting role as well [48]. Detailed analysis shows that the nature of sHSP/membrane interaction depends on the lipid composition and the extent of lipid unsaturation, suggesting that sHSP can modulate the lipid polymorphism and the fluidity of the membranes [49]. Mutants in the N-terminal arm, namely L9P and Q16R, showed altered interaction with thylakoid and lipid membranes and modified thylakoid functions. These mutants are unusual in that they retain their oligomeric state and chaperone activity *in vitro* but fail to confer thermotolerance *in vivo*. We found that both mutant proteins had dramatically altered membrane/lipid interaction properties. Whereas L9P showed strongly reduced binding to thylakoid and model membranes, Q16R was almost exclusively membrane-associated, properties that may be the cause of reduced heat tolerance of cells carrying these mutations [50].

Synechocystis 6803 contains a single *hsp17* gene [17]. Northern analysis revealed that the gene is greatly induced when cells were subjected to higher temperatures [34]. Supporting aforementioned membrane-protecting function, the *hsp17* gene is highly upregulated when the physical state of *Synechocystis* 6803 membrane was manipulated either by growth temperature or by treating the cells with the membrane fluidizer, benzyl alcohol. This sensitivity led us to give the "fluidity" nickname to the gene [34]. The detailed analysis of *hsp17* regulation remains to be established.

6.3 EVOLUTION OF MEMBRANE SENSOR HYPOTHESIS: STRESS SENSORS AND SIGNALING IN *SYNECHOCYSTIS* 6803

According to the classic model, during thermal stress, proteins tend to lose their native conformation and serve as a signal to induce the heat shock response [51]. It has long been known that during rapid temperature fluctuations membranes are among the most thermally sensitive macromolecular structures within cells [52]. Not surprisingly, therefore, membranes are one of the main cellular targets of temperature adaptation. Many organisms, including cyanobacteria, are known to adapt mostly via changing the extent of unsaturation of the alkyl chains of their constituent membrane lipids. The readily transformable unicellular cyanobacteria have proved to be particularly suitable for studies of temperature stress response at the molecular level. The general features of their plasma and thylakoid membranes resemble those of the chloroplasts of higher plants in terms of both lipid composition and the assembly of membranes [12].

Being totally nonspecific in its effects on the cellular metabolism, temperature is a "blunt instrument" with which to dissect the mechanisms involved in this response. Nevertheless, application of the technique of homogeneous catalytic hydrogenation allowed the well-controlled *in vivo in situ* modulation of unsaturated membrane lipids with apparently no effects other than ordering of the membrane interior, under isothermal conditions. Catalytic hydrogenation with novel, water-soluble ruthenium and rhodium complexes [53,54] has been demonstrated to be a useful method for investigation of the molecular basis of cellular chilling sensitivity in various *in vivo* unicellular model systems, such as cyanobacteria, the green alga *Dunaliella salina,* or the ciliated protozoan *Tetrahymena mimbres*, and also *in vitro*, using isolated membranes of very different origins (see refs in [55]). Noteworthy, that with appropriate selection of the reaction conditions, the *in vivo* lipid hydrogenation can be confined to the cell surface membranes. Based on results obtained by using this method, it has been shown that modification of the membrane's physical state is able to induce the cold-responsive Δ12 desaturase (*desA*) gene in *Synechocystis* 6803 under isothermal conditions [56]. The surface membrane-selective hydrogenation of *Synechocystis* stimulated the transcription of *desA* gene in the same way as cooling. These and other results led firstly to the hypothesis that cellular membranes might be involved in temperature sensing in *Synechocystis* 6803 (for reviews, see [12,35,36]. In *Synechocystis* 6803, it has been demonstrated that changes in thylakoid membrane's physical properties alter the temperature threshold of the induction of heat shock genes (*groESL, groEL2, dnaK2*, and especially *hsp17*) [34]. In subsequent and analogous studies, changes in membrane composition and/or fluidity in yeast, *E. coli*, plant and mammalian cells uniformly altered the "set point" for heat shock protein expression, with expression initiated at lower temperatures in cells with more fluid membranes [57–62]. A general strategy to improve the native folding of recombinant proteins is to increase the cellular concentration of viscous organic compounds, termed osmolytes, or of molecular chaperones per se that can prevent aggregation and/or actively scavenge and convert aggregates into natively refoldable species. Addition of the membrane fluidizer heat-shock inducer benzyl alcohol to *E. coli* resulted in similar high yields of refolding as that of obtained with plasmid-encoded co-expression of molecular chaperones *dnaK*-DnaJ-GrpE, ClpB, GroEL-GroES, and IbpA-IbpB [63].

Moreover, we suggest that membranes as thermal sensors and HSPs form a regulatory network (for reviews, see Refs [36,64,65]). A lipid selective association of a subpopulation of cyanobacterial HSP (first shown for the HSP60/GroEL and the HSP17) with membranes can cause an increased molecular order and reduced propensity to form membrane disrupting non-bilayer lipid phase [39,42,48,49,66]. Most recent studies of Guzzo and Colleagues identified the amino acids essential for damaged protein protection and also those responsible for membrane stabilization (rigidification) within the alfa-crystallin domain of Lo18, the small HSP of the lactic acid bacteria *Oenococcus oeni* [67]. Similar studies are in progress in our laboratory with *Synechocystis* HSPs.

Thus, we propose that HSP–membrane interaction may lead to downregulation of the heat shock gene expression [36,64,65,68]. The membrane–lipid interaction of certain HSPs proved to be an important regulator of membrane function also in mammalian cells [36,69]. We reasoned that since the association of certain amphitropic HSP and lipid molecular species remodel the preexisting hyperstructure of membranes, the specific HSP–lipid interactions may be an unrecognized mean for the spatial separation and distinct compartmentalization of specific signaling molecules to lipid microdomains. It is tempting to speculate that being either "residents" or "visitors" (i.e., located within or transiently on the surface of membranes) [70], a specific subset of these HSPs may participate in those homeostatic controlling mechanisms, which ultimately retailor the lipid composition and consequently basic functions of HSP-hosting membranes. Indeed, we have shown that the rapidly formed monoglucosyldiacylglycerol (MGlcDG) may act as "heat shock lipid" that can stabilize membranes at the early phase of thermal stress via its extreme high level of saturated fatty acyl chains and by its unique capability to interact selectively with and thereby to "anchor" the HSP17 in *Synechocystis* thylakoids [66]. Recently, Shimojima et al. further examined the underlying mechanism of such an increase of MGlcDG following heat shock in *Synechocystis* by measuring MGlcDG synthase (Sll1377) activity [71]. Whereas the Sll1377 at protein level remained unchanged, a high-temperature-dependent activation of Sll1377 was observed exclusively in the membrane fraction of *Synechocystis*. Surprisingly, if expressed in *E. coli*, the recombinant MGlcDG synthase was also found primarily in the membrane fraction and its activity increased upon temperature elevation. Obviously, a mechanism common to the membranes of *Synechocystis* and *E. coli* upregulates post-translationally the MGLcDG synthase activity and it is suggested that at least in part this leads to the change of the membrane hyperstructure (fluidity, microdomain organization) caused by temperature elevation. In this context, Sll1377 could be a "heat-sensing" protein playing a key role in regulating cell signaling at high temperature and its membrane-controlled activation may be a more rapid and efficient response than transcriptional/translational upregulation or covalent modification. Taken together, these results suggest that elevated temperature first activates the synthase of the "heat shock lipid" MGlcDG, followed by an increased level of MGlcDG that ultimately stabilizes further the membranes via binding to HSP17, thereby promoting cell survival. It is noted that high-temperature-dependent hyperactivation of the heat shock lipid synthase has also been observed in the membrane fraction of another cyanobacterium, *Anabaena variabilis* [72]. An additional role of the high-temperature-dependent increase in MGlcDG could be to preserve membrane integrity via its highly saturated fatty acids, in a way as we had evidenced by *in vitro* catalytic hydrogenation more than 20 years ago [73–75].

In conclusion, the primary composition and physical state of the lipid phase of membranes can control both the expression and membrane association of preexisting HSPs at the early phase of heat stress. Interaction of preexisting or newly formed HSPs with membranes (together with a hitherto unexplored cross-talk of stress protein and membrane lipid homeostasis) may represent an additional layer of this control.

6.4 ADDITIONAL FACTORS INVOLVED IN THE REGULATION OF THE CHAPERONE GENES OF *SYNECHOCYSTIS*

As regards other factors involved in the regulation of the chaperone genes of *Synechocystis*, the importance of light conditions and the photosynthetic electron transport has been shown in the case of the chaperonin genes (*groESL operon, groEL2*) and the existence of *HrcA*-type repressing mechanism was suggested [45] and proved also by Nakamoto et al. [76]. Recently, a novel, HrcA-independent-positive regulatory mechanism was also demonstrated, together with the increased survival rate of the hrcA mutant compared to wild-type cells upon applying lethal temperature stress [46]. Although under different growth and stress conditions, the histidine kinase *hik34* mutant showed similar resistance to lethal heat treatment [77]. Based on microarray data, they also have shown that *hik34* mutation had no significant effect on heat-induced gene expression, but when overexpressed the HSP induction is repressed. These results might indicate that *hik34* negatively regulates the heat shock genes [77]. It should be noted that hik34 (similarly to hik33) is also involved in the regulation of salt-, hyperosmotic-, and hydrogen peroxide stress (reviewed in Ref. [19]).

Synechocystis 6803 contains nine RNA polymerase sigma factors of which sigB proved to be heat inducible at protein level [78]. Inactivation of sigB was shown to decrease the steady-state mRNA level of different heat shock genes after shifting the cells from 32°C to 43°C for 60 min, but whether the rate of induction or the stability of the corresponding mRNAs changes remains to be established [79]. Recently, another sigma factor, sigC, was suggested to be involved in the heat acclimation of *Synechocystis* 6803 in a carbon dioxide concentration-dependent manner [80]. Microarray analysis has also shown that the topology of the DNA is also involved in the regulation of genes induced by cold, heat, and salt stress [81].

However, since the structure and function of all cellular macromolecules (proteins, lipids, RNA, DNA, etc.) are sensitive to the temperature (changes), the emergence of new sensor candidates might be expected (for review, see Ref. [82]). The question is the rate of sensitivity, the sequence of events, and the place of the sensor. Cyanobacterial membranes are very sensitive, highly organized structures, usually therefore they might be put at the first place of the external stress signaling pathways. Recent results have demonstrated that the "membrane-sensor" theory might be extended to higher eukaryotes (for plants see Ref. [83], for mammals see Ref. [61] as well [33].

ACKNOWLEDGMENT

This work was supported by the Hungarian National Scientific Research Foundation (OTKA NN 76716 and K68804).

REFERENCES

1. Richter, K., Haslbeck, M., and Buchner, J., The heat shock response: Life on the verge of death, *Mol. Cell*, 40, 253, 2010.
2. Kültz, D., Molecular and evolutionary basis of the cellular stress response, *Annu. Rev. Physiol.*, 67, 225, 2005.
3. Ellis, R.J., van der Vies, S.M., and Hemmingsen, S.M. The molecular chaperone concept, *Biochem. Soc. Symp.*, 55, 145, 1989.
4. Hartl, F.U. and Hayer-Hartl, M., Converging concepts of protein folding in vitro and in vivo, *Nat. Struct. Mol. Biol.*, 16, 574, 2008.
5. Finka, A., Mattoo, R.U.H., and Goloubinoff, P., Meta-analysis of heat- and chemically upregulated chaperone genes in plant and human cells, *Cell Stress Chaperones*, 16, 15, 2011.
6. Horváth, I. et al., Heat shock response in photosynthetic organisms: The membrane connections, *Prog. Lipid Res.*, 51, 208, 2012.
7. Berry, J.A. and Bjorkman, O., Photosynthetic response and adaptation to temperature in higher plants, *Annu. Rev. Plant Physiol.*, 31, 491, 1980.
8. Rowland, J.G. et al., Identification of components associated with thermal acclimation of photosystem II in *Synechocystis* sp. PCC6803, *PLoS One*, 5 (5), e10511, 2010, doi:10.1371/journal.pone.0010511.
9. Giovannoni, S.J. et al., Evolutionary relationships among cyanobacteria and green chloroplasts, *J. Bacteriol.*, 170, 3584, 1988.
10. Lehel, Cs. et al., Heat shock protein synthesis of the cyanobacterium Synechocystis PCC 6803: Purification of the GroEL-related chaperonin, *Plant Mol. Biol.*, 18, 327, 1992.
11. Lehel, Cs. et al., Growth temperature modulates thermotolerance and heat shock response of cyanobacterium *Synechocystis* PCC 6803, *Plant Physiol. Biochem.*, 31, 81, 1993.
12. Glatz, A. et al., The *Synechocystis* model of stress: From molecular chaperones to membranes, *Plant Physiol. Biochem.*, 37, 1, 1999.
13. Sauer, R.T. et al., Sculpting the proteome with AAA(+) proteases and disassembly machines, *Cell*, 119, 9, 2004.
14. Bukau, B., Weissman, J., and Horwich, A., Molecular chaperones and protein quality control, *Cell*, 125, 443, 2006.

15. Martin, A., Baker, T.A., and Sauer, R.T., Rebuilt AAA + motors reveal operating principles for ATP-fuelled machines, *Nature*, 437, 1115, 2005.
16. Doyle, S.M. et al., Asymmetric deceleration of ClpB or Hsp104 ATPase activity unleashes protein-remodeling activity, *Nat. Struct. Mol. Biol.*, 14, 114, 2007.
17. Kaneko, T. et al., Sequence analysis of the genome of the unicellular cyanobacterium *Synechocystis* sp. strain PCC 6803, II. Sequence determination of the entire genome and assignment of potential protein-coding regions, *DNA Res.*, 3, 109, 1996.
18. Giese, K.C. and Vierling, E., Changes in oligomerization are essential for the chaperone activity of a small heat shock protein in vivo and in vitro, *J. Biol. Chem.*, 277, 46310, 2002.
19. Los, D.A. et al., Stress sensors and signal transducers in cyanobacteria, *Sensors*, 10, 2386, 2010.
20. Picard, D., Heat-shock protein 90, a chaperone for folding and regulation, *Cell. Mol. Life Sci.*, 59, 1640, 2002.
21. Wandinger, S.K., Richter, K., and Buchner, J., The Hsp90 chaperone machinery, *J. Biol. Chem.*, 283, 18473, 2008.
22. Queitsch, C., Sangster, T.A., and Lindquist, S., Hsp90 as a capacitor of phenotypic variation, *Nature*, 417, 618, 2002.
23. Fang, F. and Barnum, S.R., The heat shock gene, htpG, and thermotolerance in the cyanobacterium, *Synechocystis* sp. PCC 6803, *Curr. Microbiol.*, 47, 341, 2003.
24. Bardwell, J.C. and Craig, E.A., Ancient heat shock gene is dispensable, *J. Bacteriol.*, 170, 2977, 1998.
25. Tanaka, N. and Nakamoto, H., HtpG is essential for the thermal stress management in cyanobacteria, *FEBS Lett.*, 458, 117, 1999.
26. Versteeg, S., Mogk, A., and Schumann, W., The *Bacillus subtilis* htpG gene is not involved in thermal stress management, *Mol. Gen. Genet.*, 261, 582, 1999.
27. Singh, A.K. et al., The heat shock response in the cyanobacterium *Synechocystis* sp. strain PCC 6803 and regulation of gene expression by HrcA and SigB, *Arch. Microbiol.*, 186, 273, 2006.
28. Mayer, M.P. and Bukau, B., Hsp70 chaperones: Cellular functions and molecular mechanism, *Cell. Mol. Life Sci.*, 62, 670, 2005.
29. Kampinga, H.H. and Craig, E.A., The HSP70 chaperone machinery: J proteins as drivers of functional specificity, *Nat. Rev. Mol. Cell Biol.*, 11, 579, 2010.
30. Chitnis P.R. and Nelson, N., Molecular cloning of the genes encoding two molecular chaperones of the cyanobacterium *Synechocystis* sp. PCC 6803, *J. Biol. Chem.*, 266, 58, 1991.
31. Varvasovszki, V. et al., Only one dnaK homolog, dnaK2 is transcriptionally active and essential in *Synechocystis*, *Biochem. Biophys. Res. Commun.*, 305, 641, 2003.
32. Rupprecht, E. et al., Three different DnaK proteins are functionally expressed in the cyanobacterium *Synechocystis* sp. PCC 6803, *Microbiology*, 153, 1828, 2007.
33. Rupprecht, E., Düppre, E., and Schneider, D., Similarities and singularities of three DnaK proteins from the cyanobacterium *Synechocystis* sp. PCC 6803, *Plant Cell Physiol.*, 51, 1210, 2010.
34. Horváth, I. et al., Membrane physical state controls the signaling mechanism of the heat shock response in *Synechocystis* PCC 6803: Identification of hsp17 as a novel "fluidity gene", *Proc. Natl. Acad. Sci. USA*, 95, 3513, 1998.
35. Vígh, L., Maresca, B., and Harwood, J., Does membrane physical state control the expression of heat shock and other genes? *Trends Biochem. Sci.*, 23, 369, 1998.
36. Horváth, I. et al., Membrane-associated stress proteins: More than simply chaperones, *BBA—Biomembranes*, 1778, 1653, 2008.
37. Goloubinoff, P., Gatenby, A.A., and Lorimer, G.H., GroE heat-shock proteins promote assembly of foreign prokaryotic ribulose bisphosphate carboxylase oligomers in *Escherichia coli*, *Nature*, 337, 44, 1989.
38. Török, Zs., Vígh, L., and Goloubinoff, P., Fluorescence detection of symmetric GroEL14(GroES7)2 heterooligomers involved in protein release during the chaperonin cycle, *J. Biol. Chem.*, 271, 16180, 1996.
39. Török, Zs. et al., Evidence for a lipochaperonin, association of active protein-folding GroESL oligomers with lipids can stabilize membranes under heat-shock conditions, *Proc. Natl. Acad. Sci. USA*, 94, 2192, 1997.
40. Lehel, Cs. et al., A second groEL-gene organized in a groESL operon is present in the genome of *Synechocystis* PCC 6803, *J. Biol. Chem.*, 268, 1799, 1993.
41. Kovács, E. et al., Molecular characterization, assembly and membrane association of the GroEL-type chaperonins in *Synechocystis* PCC 6803, in *Structure, Biogenesis and Dynamics of Biological Membranes*, Op den Kamp, J.A.F. Ed., NATO ASI Series, Springer-Verlag, Berlin, Germany, Series H: Cell Biology, H82, p. 253, 1994.

42. Kovács, E. et al., Heat stress induces association of the GroEL-analog chaperonin with thylakoid membranes in cyanobacterium, *Synechocystis* PCC 6803, *Plant Phys. Biochem.*, 32, 285, 1994.
43. Kovács, E. et al., The chaperonins of *Synechocystis* PCC 6803 differ in heat inducibility and chaperone activity, *Biochem. Biophys. Res. Commun.*, 289, 908, 2001.
44. Segal, G. and Ron, E.Z., Regulation of the groE and dnaK operons in Eubacteria, *FEMS Microbiol. Lett.*, 138, 1, 1996.
45. Glatz, A. et al., Chaperonin genes of the *Synechocystis* PCC 6803 are differently regulated under light-dark transition during heat stress, *Biochem. Biophys. Res. Commun.*, 239, 291, 1997.
46. Kojima, K. and Nakamoto, H., A novel light- and heat responsive regulation of the groE transcription in the absence of hrcA or CIRCE in cyanobacteria, *FEBS Lett.*, 581, 1871, 2007.
47. Nakamoto, H. and Vígh, L., The small heat shock proteins and their clients, *Cell. Mol. Life Sci.*, 64, 294, 2007.
48. Török, Z. et al., *Synechocystis* HSP17 is an amphitropic protein that stabilizes heat-stressed membranes and binds denatured proteins for subsequent chaperone-mediated refolding, *Proc. Natl. Acad. Sci. USA*, 98, 3098, 2001.
49. Tsvetkova, N.M. et al., Small heat-shock proteins regulate membrane lipid polymorphism, *Proc. Natl. Acad. Sci. USA*, 99, 13504, 2002.
50. Balogi, Z. et al., A mutant small heat shock protein with increased thylakoid association provides an elevated resistance against UV-B damage in *Synechocystis* 6803, *J. Biol. Chem.*, 283, 22983, 2008.
51. Morimoto, R.I., Regulation of the heat shock transcriptional response: Cross talk between a family of heat shock factors, molecular chaperones, and negative regulators, *Genes Dev.*, 12, 3788, 1998.
52. Vígh, L. et al., The significance of lipid composition for membrane activity: New concepts and ways of assessing function, *Prog. Lipid Res.*, 44, 303, 2005.
53. Vígh, L. et al., Hydrogenation of model and biomembranes using a water-soluble ruthenium phosphin-catalyst, *J. Mol. Catal.*, 22, 15, 1983.
54. Vígh, L. and Joó, F., Modulation of membrane fluidity of catalytic hydrogenation affects the chilling susceptibility of the blue-green alga, *Anacystis nidulans*, *FEBS Lett.*, 162, 423 1983.
55. Joó, F. et al., Complex hydrogenation/oxidation reactions of the water-soluble hydrogenation catalyst, palladium-di(sodium alizarine-monosulfonate) and the know-how of homogeneous hydrogenation of lipids in isolated biomembranes and living cells, *Anal. Biochem.*, 194, 34, 1991.
56. Vígh, L. et al., The primary signal in the biological perception of temperature: Pd-catalyzed hydrogenation of membrane lipids stimulated the expression of the *desA* gene in *Synechocystis* PCC 6803, *Proc. Natl. Acad. Sci. USA*, 90, 9090, 1993.
57. Carratu, L. et al., Membrane lipid perturbation sets the temperature of heat shock response in yeast, *Proc. Natl. Acad. Sci. USA*, 93, 3870, 1996.
58. Chatterjee, M.T., Khalawan, S.A., and Curran, B.P.G., Alterations in cellular lipids may be responsible for the transient nature of the yeast heat shock response, *Microbiology*, 146, 877, 2000.
59. Shigapova, N. et al., Membrane fluidization triggers membrane remodeling which affects the thermotolerance in *Escherichia coli*, *Biochem. Biophys. Res. Commun.*, 328, 1216, 2005.
60. Balogh, G. et al., The hyperfluidization of mammalian cell membranes acts as a signal to initiate the heat shock protein response, *FEBS J.*, 272, 6077, 2005.
61. Nagy, E. et al., Hyperfluidization-coupled membrane microdomain reorganization is linked to activation of the heat shock response in a melanoma cell line, *Proc. Natl. Acad. Sci. USA*, 104, 7945, 2007.
62. Saidi, Y. et al., The heat shock response in moss plants is regulated by specific calcium-permeable channels in the plasma membrane, *Plant Cell*, 21, 1, 2009.
63. de Marco, A. et al., Native folding of aggregation-prone recombinant proteins in *Escherichia coli* by osmolytes, plasmid- or benzyl alcohol over-expressed molecular chaperones, *Cell Stress Chaperones*, 10, 329, 2005.
64. Vigh, L. et al., Membrane-regulated stress response: A theoretical and practical approach, *Adv. Exp. Med. Biol.*, 594, 114, 2007.
65. Vigh, L. et al., Membrane regulation of the stress response from prokaryotic models to mammalian cells, *Ann. N. Y. Acad. Sci.*, 1113, 40, 2007.
66. Balogi Z. et al., "Heat shock lipid" in cyanobacteria during heat/light-acclimation, *Arch. Biochem. Biophys.*, 436, 346, 2005.
67. Weidmann, S. et al., Distinct amino acids of the *Oenococcus oeni* small heat shock protein L018 are essential for damaged protein protection and membrane stabilization, *FEMS Microbiol. Lett.*, 309, 8, 2010.

68. Vígh, L. and Maresca, B., Dual role of membranes in heat stress: As thermosensors modulate the expression of stress genes and, by interacting with stress proteins, re-organize their own lipid order and functionality, in *Cell and Molecular Responses to Stress*, Storey K.B. and Storey J.M. Eds., Elsevier, Amsterdam, the Netherlands, p. 173, 2002.
69. Horváth, I. and Vígh, L., Stability in times of stress, *Nature*, 463, 436, 2010.
70. Escriba, P.V. et al., Membranes: A meeting point for lipids, proteins and therapies, *J. Cell. Mol. Med.*, 12, 829, 2008.
71. Shimojima, M., Tsuchiya, M., and Ohta, H., Temperature-dependent hyper-activation of monoglucosyldiacylglycerol synthase is post-translationally regulated in *Synechocystis* sp. PCC 6803, *FEBS Lett.*, 583, 2372, 2009.
72. Sato, N. and Murata, N., Lipid biosynthesis in the blue-green alga (cyanobacterium), Anabaena variabilis. III. UDPglucose: Diacylglycerol glucosyltransferase activity in vitro, *Plant Cell. Physiol.*, 23, 1115, 1982.
73. Thomas, P.G. et al., Increased thermal stability of pigment-protein complexes of pea thylakoids following catalytic hydrogenation of membrane lipids, *Biochim. Biophys. Acta*, 849, 131, 1986.
74. Horváth, I. et al., Homogenous catalytic hydrogenation of the polar lipids of pea chloroplast in situ and the effects on lipid polymorphism, *Chem. Phys. Lipids*, 39, 251, 1986.
75. Horváth, G. et al., Role of lipids in the organization and function of Photosystem II studied by homogeneous catalytic hydrogenation of thylakoid membrane in situ, *BBA—Biomembranes*, 891, 68, 1987.
76. Nakamoto, H., Suzuki, K., and Kojima, K., Targeted inactivation of the hrcA repressor gene in cyanobacteria, *FEBS Lett.*, 549, 57, 2003.
77. Suzuki, I. et al., The histidine kinase hik34 is involved in thermotolerance by regulating expression of heat shock genes in *Synechocystis*, *Plant Physiol.*, 138, 1409, 2005.
78. Imamura, S. et al., Purification, characterization, and gene expression of all sigma factors of RNA polymerase in a cyanobacterium, *J. Mol. Biol.*, 325, 857, 2003.
79. Tuominen, I. et al., The SigB σ factor mediates high-temperature responses in the cyanobacterium *Synechocystis* PCC 6803, *FEBS Lett.*, 580, 319, 2006.
80. Tuominen, I. et al., Sigma factor sigC is required for heat acclimation of the cyanobacterium *Synechocystis* strain PCC 6803, *FEBS Lett.*, 582, 346, 2008.
81. Los, D.A., DNA supercoling regulates the stress inducible expression of genes in the cyanobacterium *Synechocystis*, *Mol. Biosyst.*, 5, 1904, 2009.
82. Klinkert, B. and Narberhaus, F., Microbial thermosensors, *Cell. Mol. Life Sci.*, 66, 2661, 2009.
83. Saidi,Y. et al., Membrane lipid composition affects plant heat sensing and modulates Ca(2+)-dependent heat shock response, *Plant Signal Behav.,* 12, 1530, 2010.

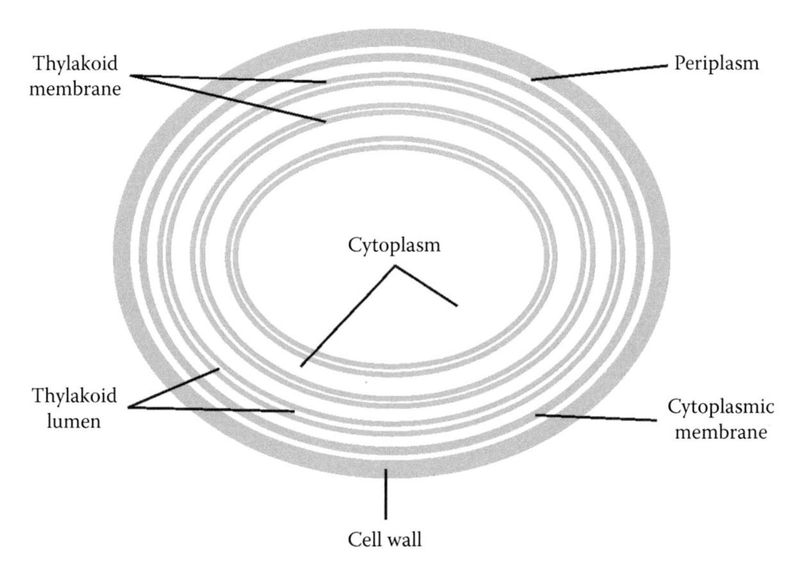

FIGURE 1.1 A schematic outline of the intracellular membranes and compartments in a cyanobacterial cell. The thylakoid membranes (green) contain chlorophyll *a* and perform both photosynthetic and respiratory electron transport, while the cytoplasmic membrane system (yellowish), which contains carotenoids, is involved only in respiration. As a consequence of photosynthetic and respiratory electron transport in thylakoid membranes, protons are brought into the thylakoid lumen, the space between a pair of thylakoid membranes. The resulting proton gradient across the thylakoid membrane is utilized for the synthesis of ATP. (Modified and adapted from Vermaas, W.F.J., *Encyclopedia of Life Sciences (ELS)*, John Wiley & Sons, Ltd, London, U.K., 2001. Copyright Wiley-VCH Verlag GmbH & Co. KGaA.)

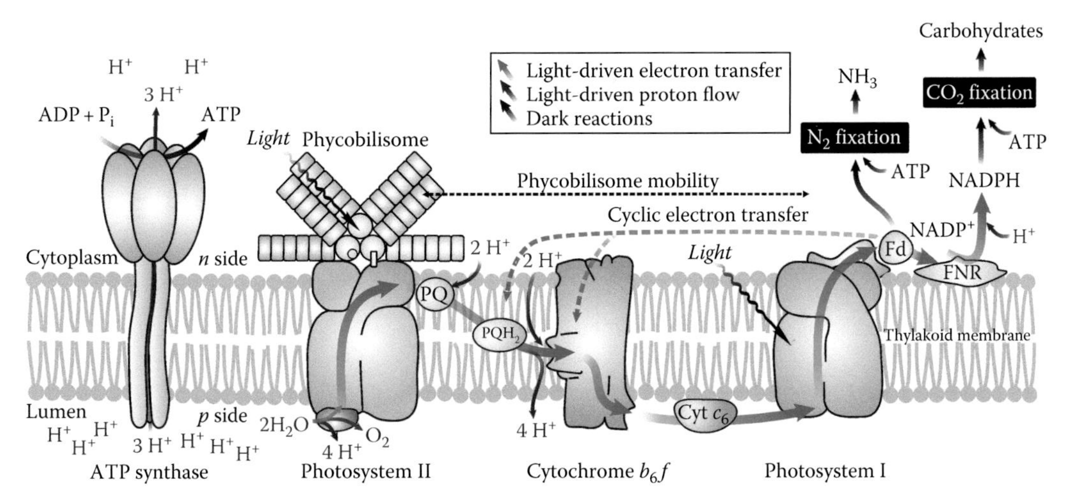

FIGURE 1.2 A schematic representation of the protein complexes involved in light-induced electron and proton transfer reactions of oxygenic photosynthesis in cyanobacteria. The arrows indicating the light-driven electron transfer and proton flow as well as some dark reactions are colored individually, as shown in the figure. Looking at the abbreviated components from the left of the diagram: PQ, plastoquinone; PQH_2, plastoquinol; Cyt c_6, cytochrome c_6 (also known as Cyt c_{553}); Fd, ferredoxin; FNR, ferredoxin-NADP reductase. Note that although the diagram does not show PBSs that are attached to PSI, PBSs can be redistributed to PSI due to their mobility (indicated by black dashed arrow). Also note that cyanobacteria use Cyt c_6 (as shown) or plastocyanin (not shown) to transfer electrons from Cyt b_6f to PSI. For the sake of simplicity, respiratory complexes (type 1 *NADPH dehydrogenase* that oxidizes NADPH to $NADP^+$; *Succinate dehydrogenase* that oxidizes succinate to fumarate and reduces PQ to PQH_2; and a *(terminal) oxidase* that reduces O_2 to water) are not shown. For further details, see text.

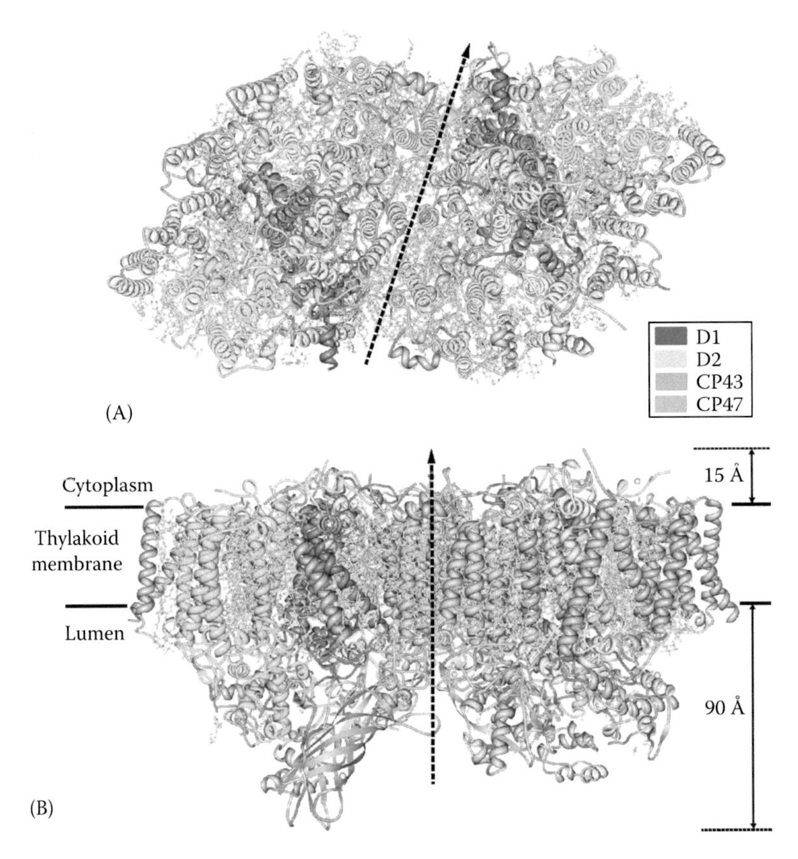

FIGURE 1.5 Overall structure of the cyanobacterial PSII dimer from *T. vulcanus*. (A) View of the dimer from the cytoplasmic side. (B) Side-on view (perpendicular to the membrane normal). The PSII core proteins D1, D2, CP43, and CP47 are colored individually, as shown in the figure, whereas others are colored in light gray. PSII extends to ~15 Å on the cytoplasmic side but to a much larger distance, ~90 Å, on the lumen side. Approximate boundary between the monomeric subunits of the homodimer is indicated by dashed arrows. See text for further details. The PSII model was generated with *RCSB Protein Workshop Viewer* using x-ray crystallographic coordinates deposited at Protein Data Bank (PDB) with ID 3ARC. (By permission from Macmillan Publishers Ltd. *Nature*, Umena, Y. et al., Crystal structure of oxygen-evolving photosystem II at a resolution of 1.9 Å, 473, 55, 2011. Copyright 2011.)

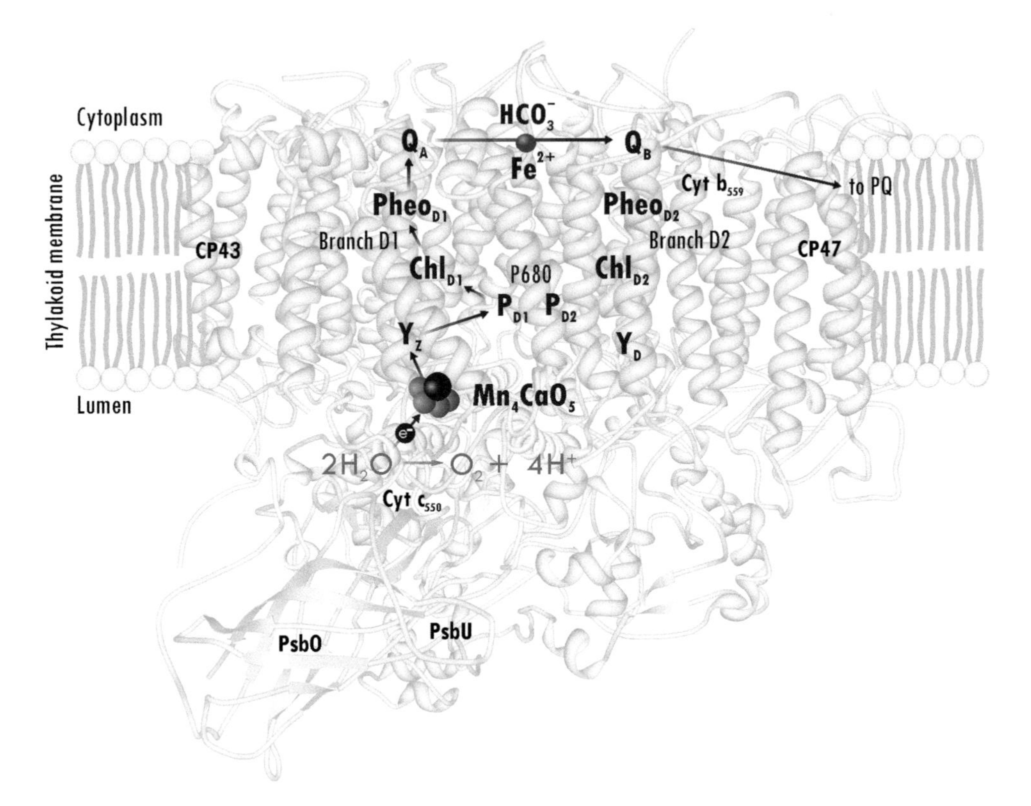

FIGURE 1.6 Schematic arrangement of electron transfer cofactors in cyanobacterial PSII monomer. Figure shows side-on view from the direction parallel to the membrane plane. All cofactors (dark blue labels) of the monomer are arranged in two branches on the D1 and D2 protein subunits (see Figure 1.5). The light-induced single electron transfer occurs mainly on the D1 protein of the PSII RC (so-called active branch). The direction of electron transfer is indicated by dark blue arrows. The location of some protein subunits is shown (black labels). Note that primary electron donor P680 (traditional definition) refers to a pair of Chl *a* molecules (P_{D1} and P_{D2}), and two accessory Chls (Chl_{D1} and Chl_{D2}). For further details and abbreviations, see text. The protein background of PSII monomer was generated with the *RCSB Protein Workshop viewer* using coordinates deposited at PDB under ID 3KZI.

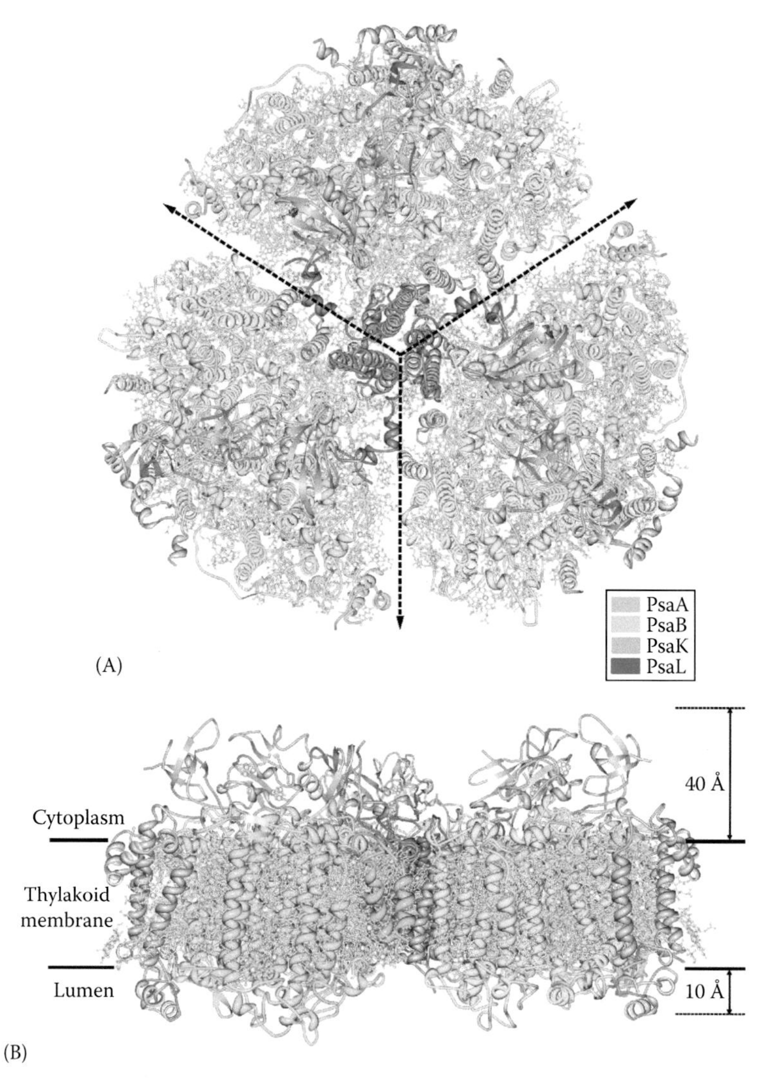

FIGURE 1.7 Overall structure of the cyanobacterial PSI trimer from *T. elongatus*. (A) View of the trimer from the cytoplasmic side. (B) Side-on view (from the direction perpendicular to the membrane normal). PSI core proteins PsaA, PsaB, PsaK, and PsaL are colored individually, as shown in the figure, whereas the others are colored in light gray. PSI extends to ~40 Å on the cytoplasmic side but to a lesser distance, ~10 Å, on the lumen side. Approximate boundary between monomeric RCs is indicated by dashed arrows. For details, see text. The PSI model was generated with the *RCSB Protein Workshop viewer* using coordinates from PDB with ID 3PCQ. (By permission from Macmillan Publishers Ltd. *Nature*, Jordan, P. et al., Three-dimensional structure of cyanobacterial photosystem I at 2.5 A resolution, 411, 909, 2001. Copyright 2001; *Nature*, Chapman, H.N. et al., Femtosecond X-ray protein nanocrystallography, 470, 73, 2011. Copyright 2011.)

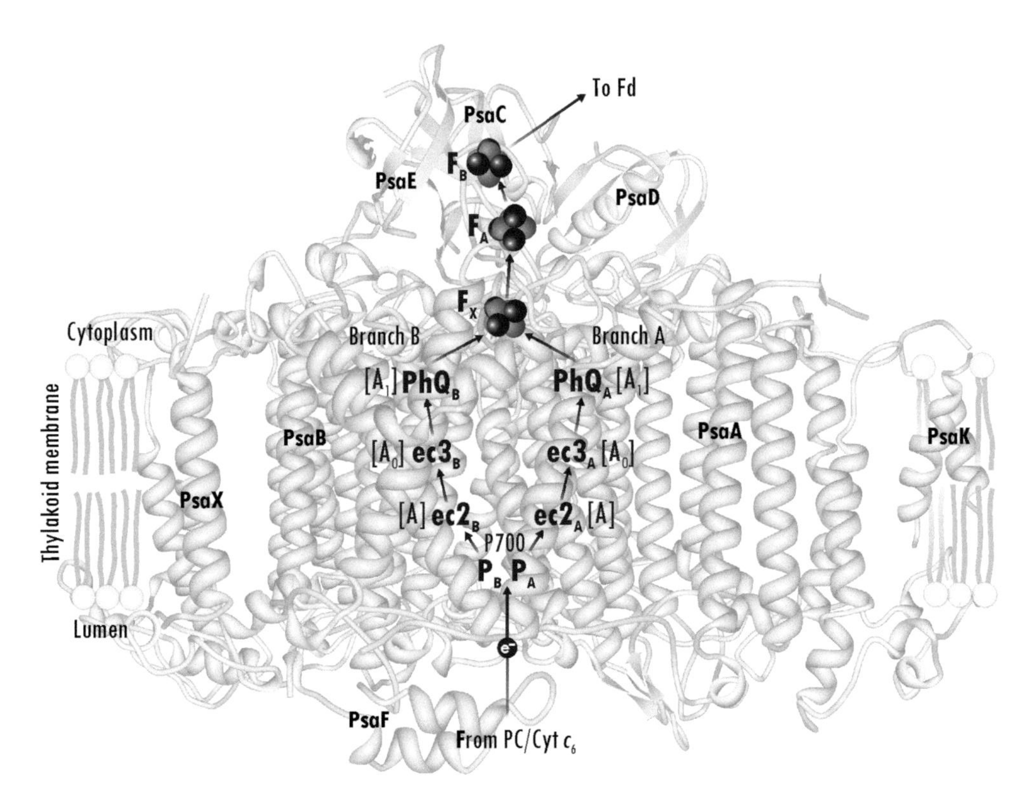

FIGURE 1.8 Schematic arrangement of electron transfer cofactors in cyanobacterial PSI monomer. This figure is a side view of the monomer from the direction of the membrane exposed periphery of PSI along the membrane plane into the center of the trimer. As in PSII, the electron transfer cofactors (dark blue labels) in PSI are organized in two parallel branches denoted as A and B (named after the protein subunits PsaA and PsaB, respectively; see Figure 1.7). The arrows indicate the direction of the electron transfer and the two possible electron transfer pathways with almost equal probability (for overviews of bidirectional electron transfer in PSI and its possible advantages for photosynthetic organism, see [210,214]). The electron transfer cofactors of two branches are indicated according to the nomenclature suggested by Redding and van der Est [210], while in brackets we show the commonly used (traditional) names that refer to the spectroscopic signatures of these cofactors. Note that the primary electron donor P700 (P_A and P_B) is used for a "special pair" of Chl *a* and Chl *a'* heterodimer. The location of some protein subunits is indicated by black labels. For further details and cofactor abbreviations, see text. The protein background of PSI monomer was modeled by the *RCSB Protein Workshop viewer* using coordinates deposited at PDB with ID 1JB0. (By permission from Macmillan Publishers Ltd. *Nature*, Jordan, P. et al., Three-dimensional structure of cyanobacterial photosystem I at 2.5 A resolution, 411, 909, 2001. Copyright 2001.)

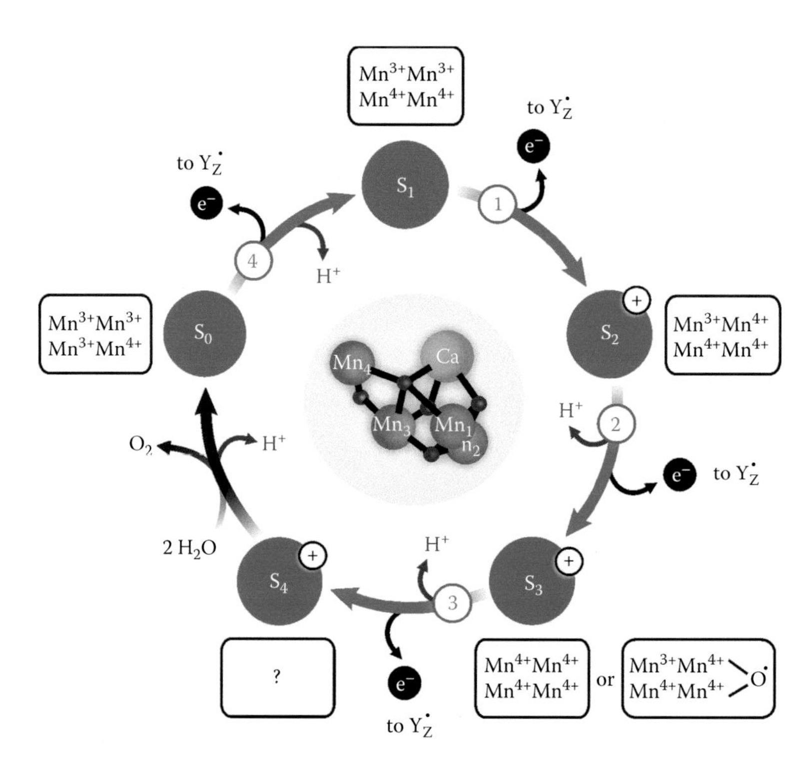

FIGURE 1.10 The Kok cycle (also known as the "oxygen clock") that illustrates the stepwise process of photosynthetic water oxidation by the Mn_4CaO_5 cluster of oxygenic organisms. Blue arrows indicate light-induced S-state transitions and the numbers in circles on the arrows indicate the number of light flashes required for that transition, assuming that in the dark the Mn_4CaO_5 cluster is mostly in the S_1 state. The $S_4 \rightarrow S_0$ transition does not require light and is shown in black. Note that depending on the S-state transition, either the electron or the proton is thought to be removed first. Currently discussed oxidation states of the four Mn ions of the Mn_4CaO_5 cluster in various S states are shown. For further details on photosynthetic water splitting, see text (this chapter), Chapter 2 (this book), and Refs. [266,267,320]. For the original version of the Kok cycle and other models, see [256–258]. The structural model of the Mn_4CaO_5 cluster in the center is as derived from the recent x-ray crystallographic PSII structure.

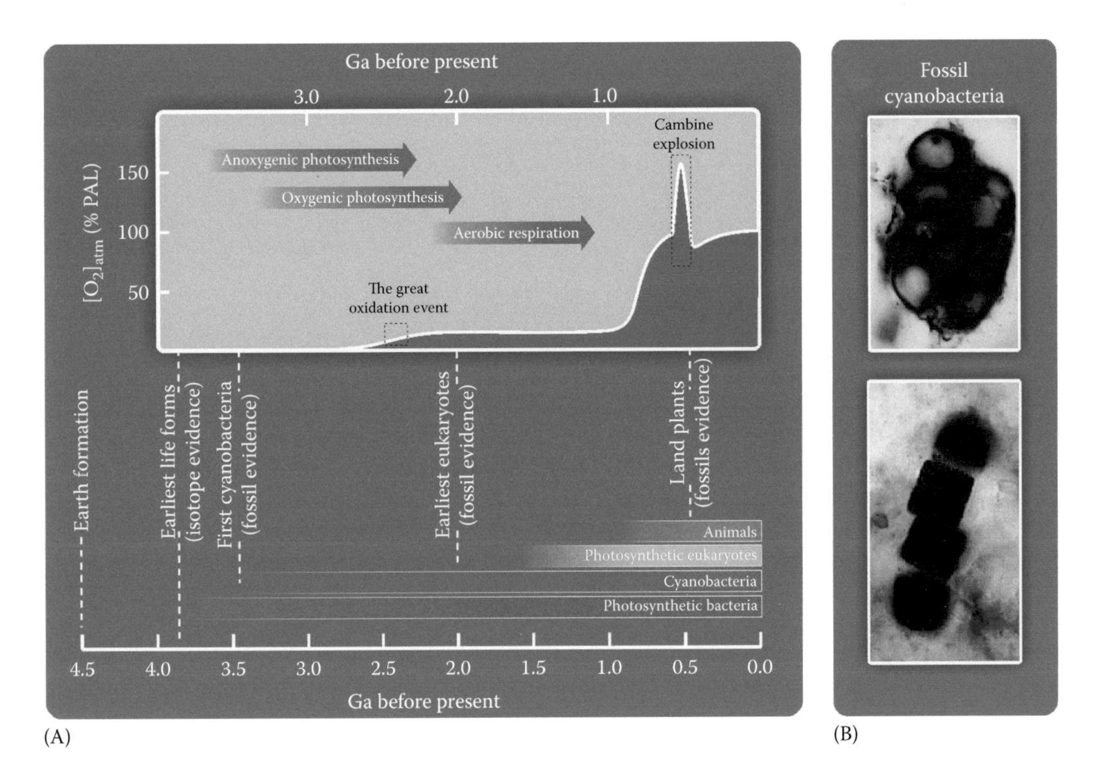

FIGURE 1.12 The role of cyanobacteria in the evolution of life and evolution of metabolic pathways during Earth's history. (A) Schematic view on the relationships between important events in the evolution of life and photosynthesis, origin of cyanobacteria (highlighted in blue), and atmospheric O_2 concentration (in % of present atmospheric level (PAL) of O_2) through geological times in billions of years (Ga). These relationships and the dates in the figure are approximations based on numerous literature data [1,7,9,157,289,290,292,296,298,302,303]. Only selected events of evolutionary diversification and the emergence of some organisms are shown. There is no firm conclusion as to when oxygenic photosynthesis was invented by primitive cyanobacteria. The earliest known fossil record of cyanobacteria occurring at 3.45 Ga [291,293] has been questioned [286,287]. Although there are other ample indications for the presence of primitive morphological forms of cyanobacteria in stomatolites about 3.5 Ga ago [294,295], they do not rule out the possibility that some of the earliest cyanobacteria behaved like anoxygenic photosynthetic bacteria. Concentration curve of atmospheric O_2 over time is displayed between a lower and an upper limits of PAL values provided in [290]. (B) Two representatives of ancient fossil cyanobacteria from the ~0.85-Ga-old Bitter Springs Chert of central Australia. On the top is a nonmobile colonial chroococcacean cyanobacterium (Coccoidal cyanobacteria), and on the bottom is the filamentous cyanobacterium *Palaeolyngbya* (Oscillatoriaceae). The Oscillatoriaceaen cyanobacteria have changed little or not at all over the last thousands of millions years [292]. The pictures of fossil cyanobacteria were kindly provided to the authors by J. William Schopf. (Based on Blankenship, R.E., *Plant Physiol.*, 154, 434, 2010; Hohmann-Marriott, M.F. and Blankenship, R.E., *Annu. Rev. Plant Biol.*, 62, 515, 2011; Falkowski, P.G., *Science*, 311, 1724, 2006; Raymond, J. and Blankenship, R.E., *Coord. Chem. Rev.*, 252, 377, 2008; Jagendorf, A.T. and Uribe, E., *Proc. Natl. Acad. Sci. USA*, 55, 170, 1966; Dunn, S.D. et al., *Biochemistry*, 40, 187, 2000; Van Walraven, H.S. et al., *FEBS Lett.*, 379, 309, 1996; Olson, J.M., *Photosynth. Res.*, 88, 109, 2006; Brasier, M.D. et al., *Nature*, 416, 76, 2002; Schopf, J.W., *Science*, 260, 640, 1993; Schopf, J.W., *Photosynth. Res.* 107, 87, 2011.)

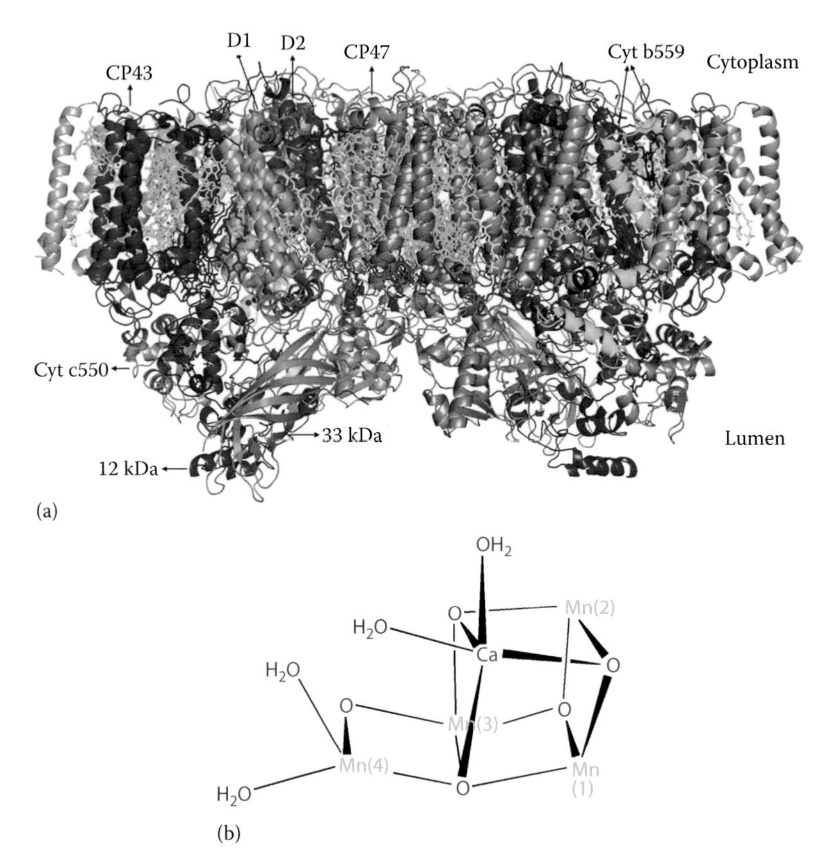

FIGURE 2.2 (a) Structure of a cyanobacterial PSII dimer [29]. View from a direction perpendicular to the membrane normal. Molecules in green, yellow, and blue represent chlorophylls, β-carotenes, lipids, and detergent molecules, respectively. Red and yellow balls at the lumenal surface represent Mn and Ca ions, respectively. Protein subunits are labeled in the figure. For clarity, water molecules are omitted. (b) The entire structure of the Mn_4CaO_5 cluster resembles a distorted chair, with the asymmetric cubane. (From Umena, Y. et al., *Nature*, 473, 55–60, 2011; Kawakami, K. et al., *J. Photochem. Photobiol. B*, 104, 9–18, 2011.)

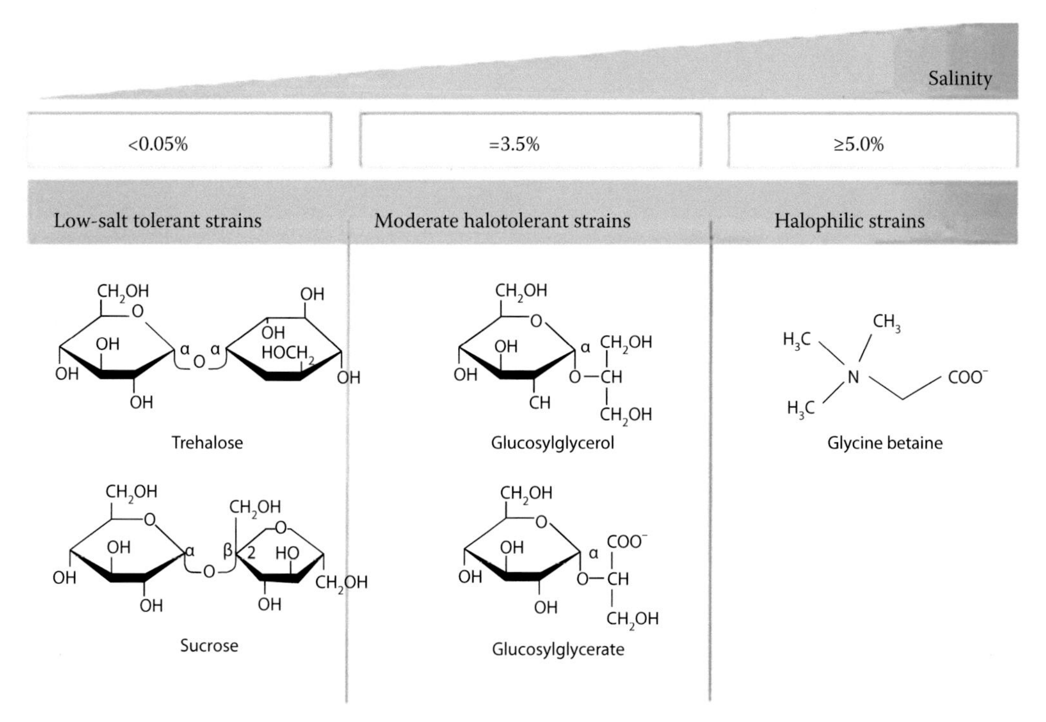

FIGURE 9.1 Structure of major compatible solutes of cyanobacteria in correlation with their overall salt resistance.

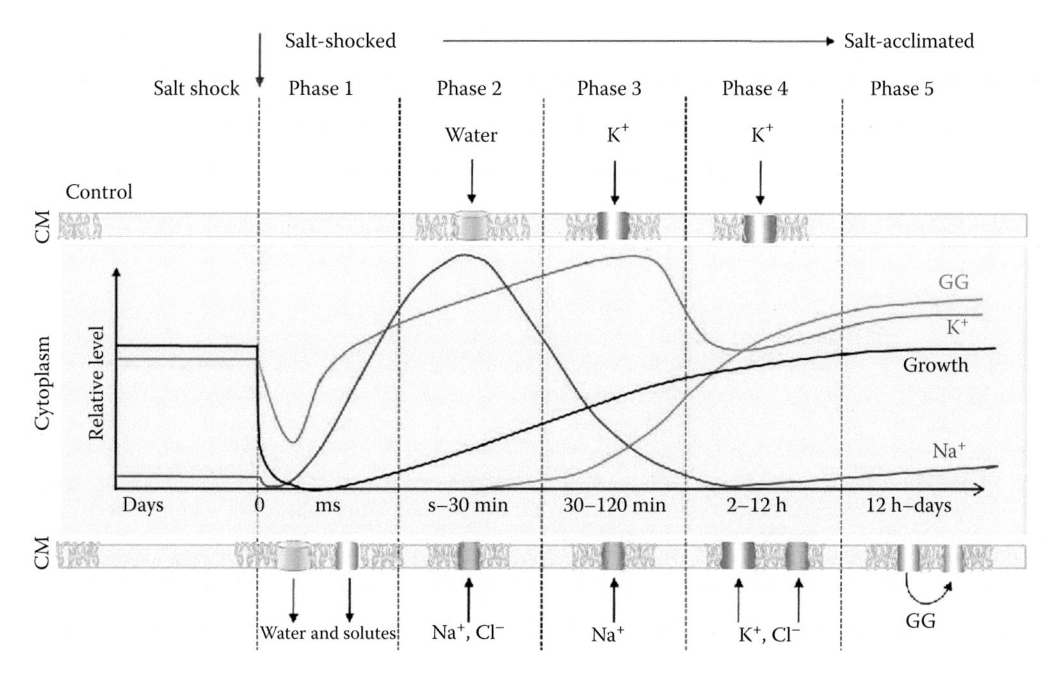

FIGURE 9.2 Schematic view of the kinetics of salt acclimation in the moderate halotolerant cyanobacterium *Synechocystis* sp. PCC 6803. Salt acclimation can be divided into five different phases. The changes in ion and water concentrations and the activation of compatible solute biosynthesis, for example, glucosylglycerol (GG), were used to distinguish different cellular states. The diagram shows the relative changes in parameters like growth, Na^+, K^+, and GG contents (CM—cytoplasmic membrane).

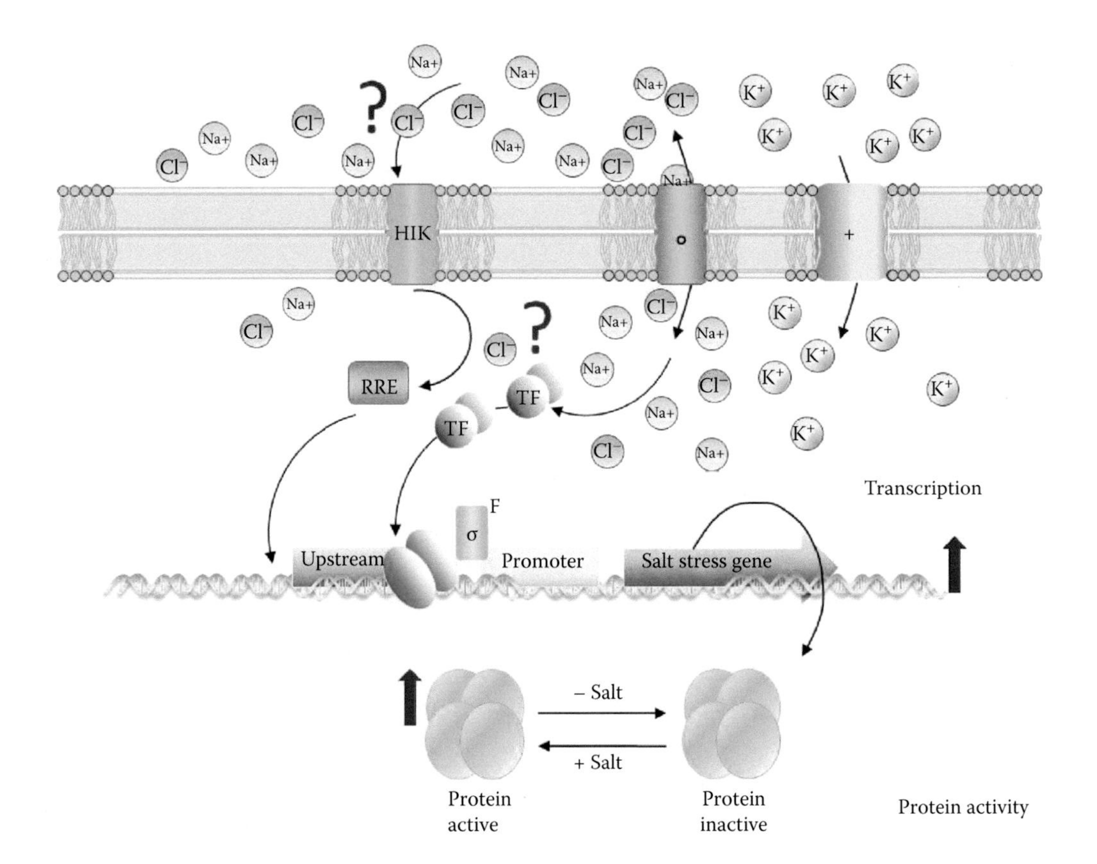

FIGURE 9.3 Schematic view of possible regulatory processes involved in the sensing of salt stress among cyanobacteria. The figure displays possible sensory molecules like two-component systems composed by histidine kinases (HIK) and the corresponding response regulators (RRE), transcriptional factors (TF) binding together with RRE proteins to upstream promoter elements, and alternative sigma factors (σ) directing the RNA-polymerase to specific basal promoter sequences. Transporters or pores involved in ion influx or efflux (e.g., transport of Na^+, Cl^-, and K^+ over the cell membrane) are included in the model explaining the possible (red?) action of external or internal ions in the direct sensing of salt stress situations. Red arrows indicate activation of transcription and/or enzyme activities by salt stress.

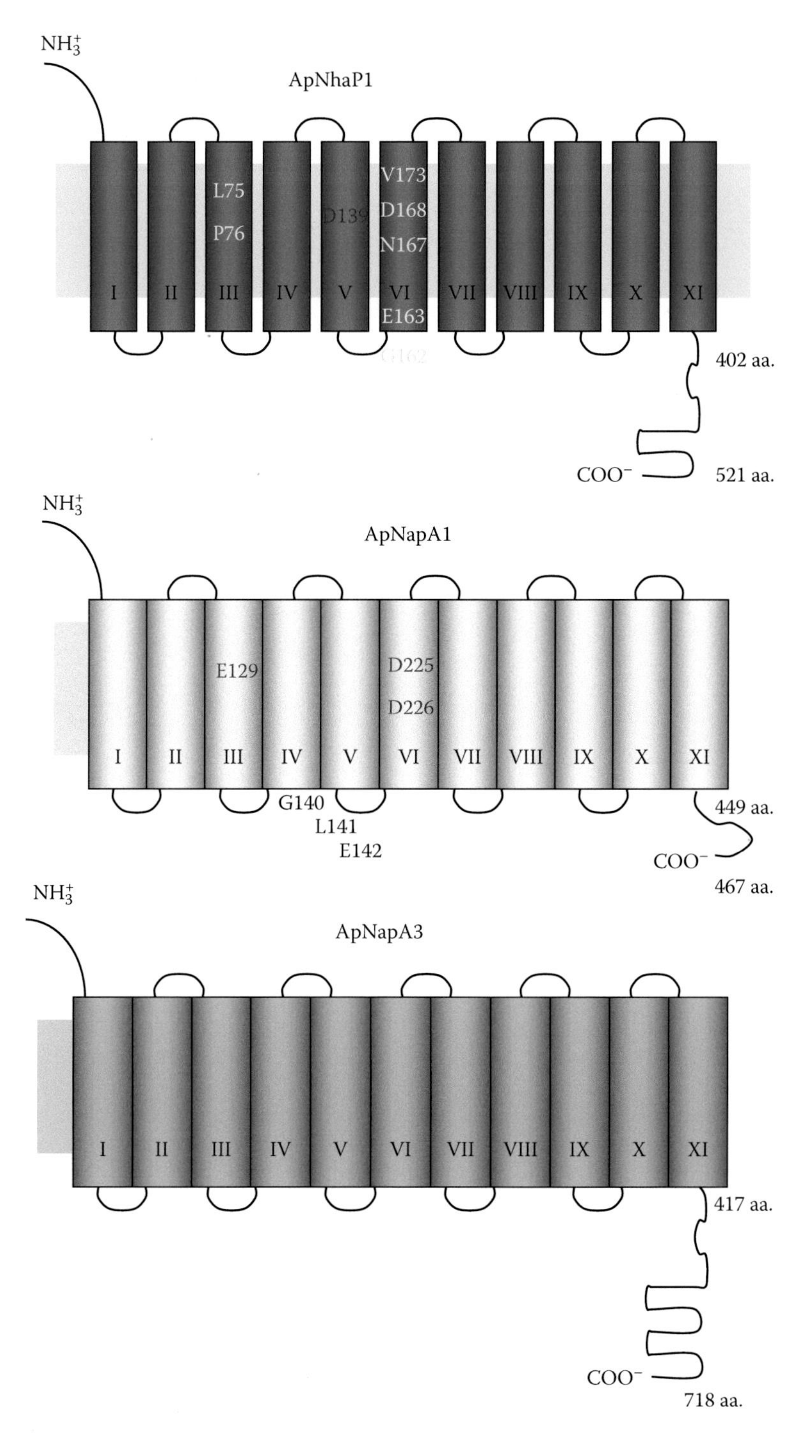

FIGURE 10.1 Predicted membrane topologies for cyanobacterial Na^+/H^+ antiporters. The possible trans-membrane of the ApNhaP1 (accession number: AB059562), ApNapA1 (accession number: AB193603), and ApNapA3 (unpublished data) sequences were deduced by a computer program TopPredII. Charged amino acids are indicated by red and yellow.

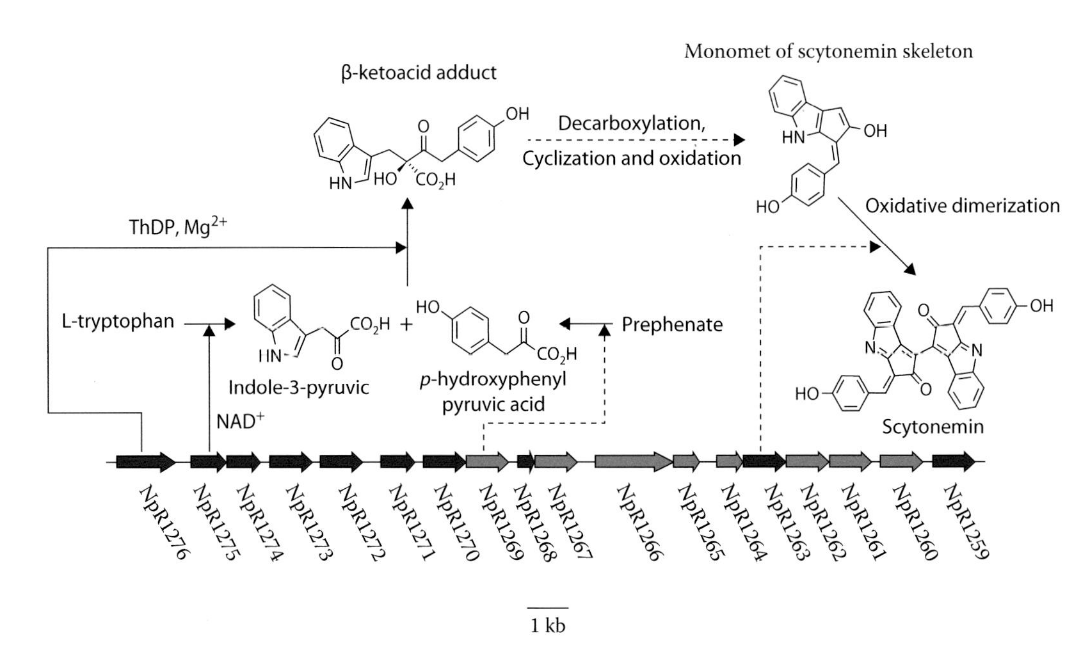

FIGURE 11.6 Biosynthetic pathway for scytonemin and corresponding gene products involved in each step. Continuous arrows represent gene products which are functionally characterized while gene products indicated by broken arrows are still to be functionally characterized for their involvement in corresponding steps. Gray arrows represent ORFs involved in the biosynthesis of aromatic amino acids while brown arrows represent ORFs of unknown function. (The model is based on the information from Soule, T. et al., *J. Bacteriol.*, 189, 4465, 2007; Balskus, E.P. and Walsh, C.T., *J. Am. Chem. Soc.*, 130, 15260, 2008; Adapted from Singh, S.P. et al, *Ageing Res. Rev.*, 9, 79, 2010.)

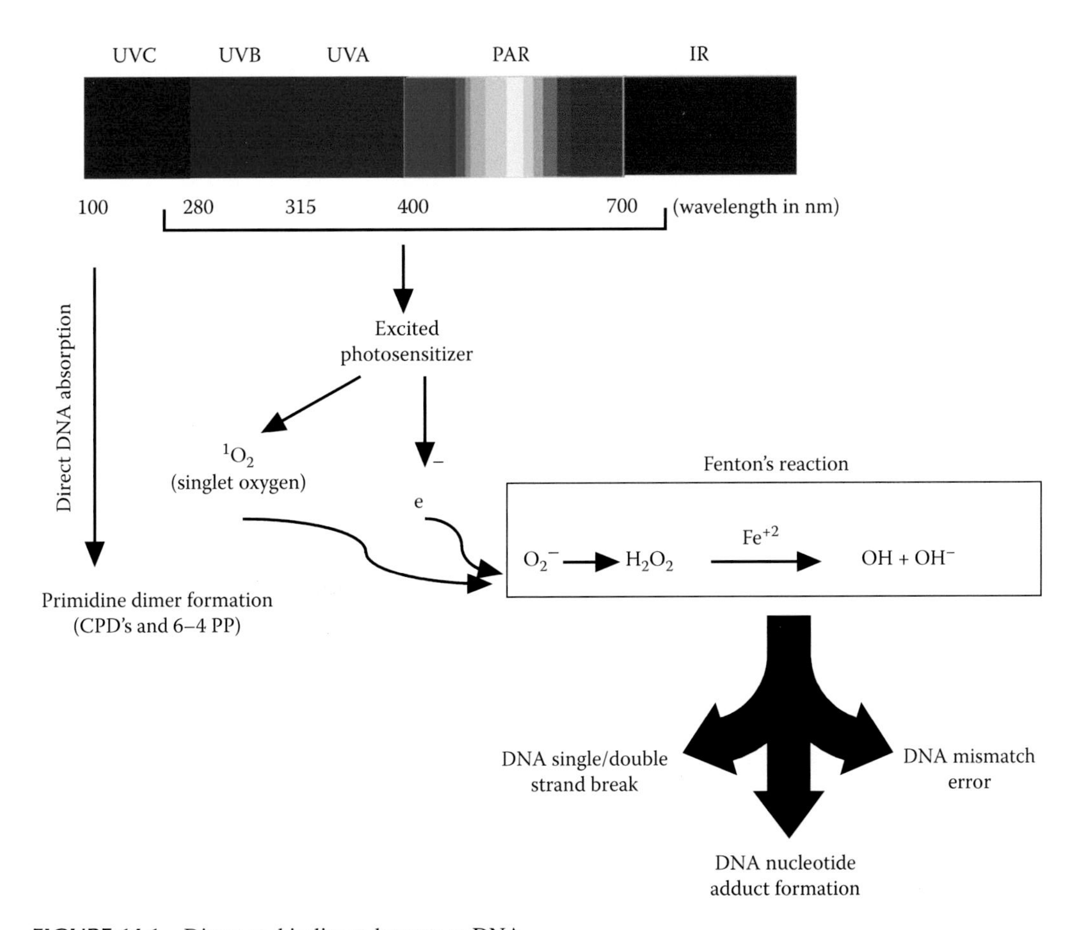

FIGURE 14.1 Direct and indirect damage to DNA.

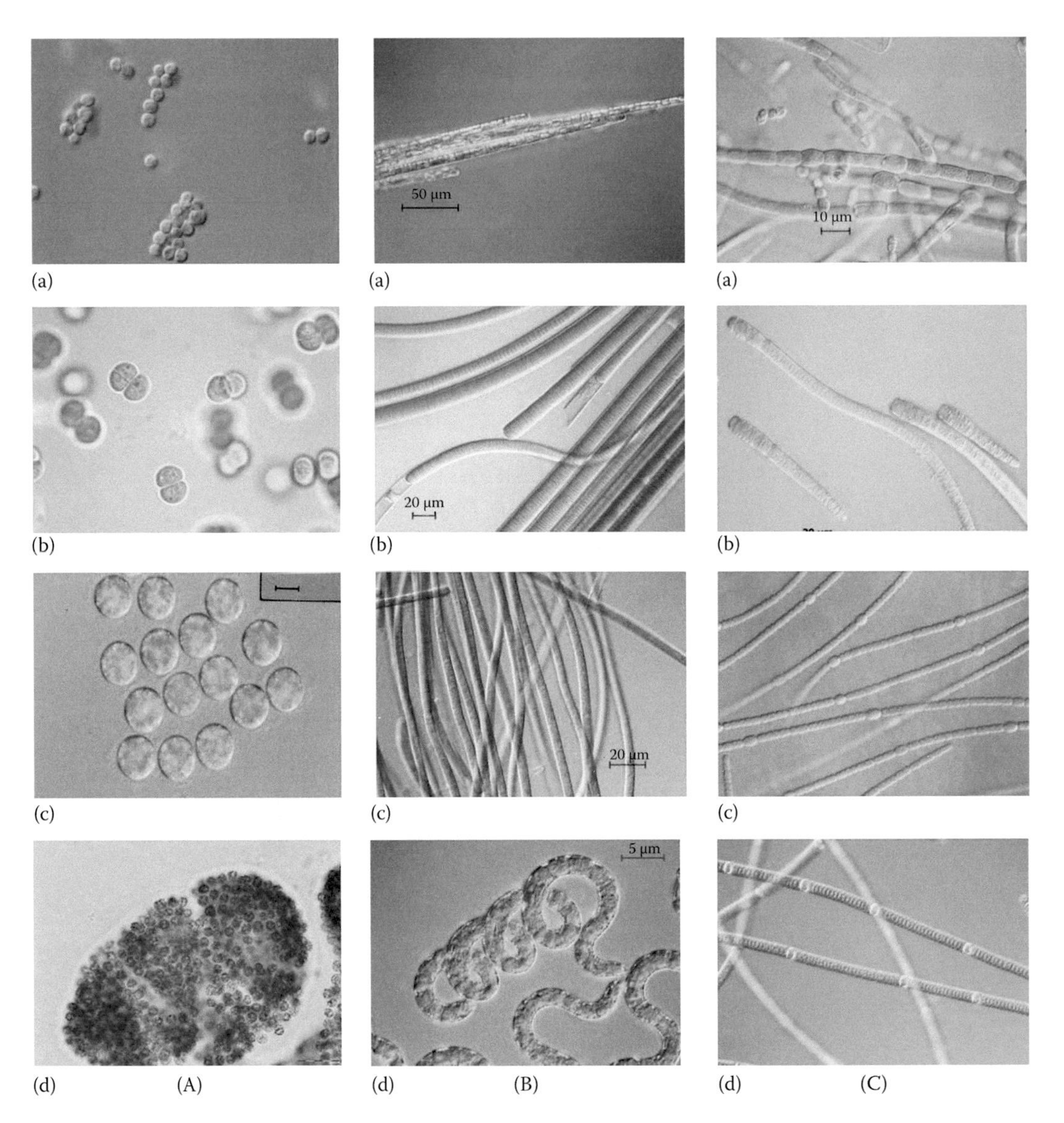

FIGURE 15.1 Morphological diversity of cyanobacteria. (A) Unicellular cyanobacteria. (a–d): *Crocosphaera* sp., *Chroococcus* sp., *Dermocarpa* sp. (division 2), and the colony-forming *Microcystis* sp. (B) non-heterocystous filamentous cyanobacteria (division 3). (a–d): *Trichodesmium* sp., *Lyngbya* sp., *Phormidium* sp., and *Spirulina* sp. (C) Heterocystous cyanobacteria. (a): *Fischerella* sp. with true branching (division 5); the other three belong to division 4, (b–d): *Calothrix* sp. with terminal heterocysts, *Anabaena* sp. and *Nodularia* sp. with intercalary heterocysts.

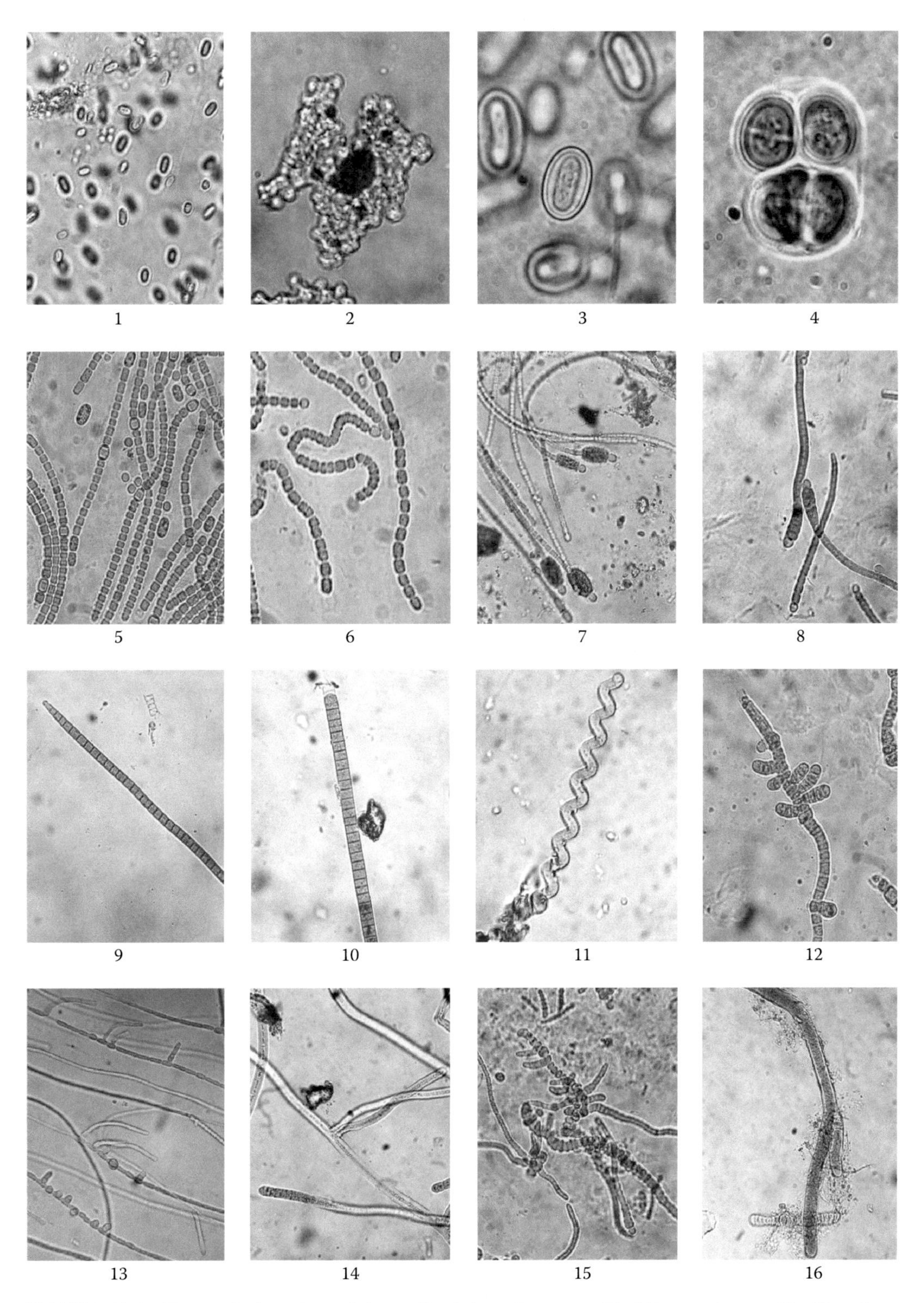

FIGURE 19.1 Name of Cyanobacterial sp. (1) *Aphanothece* sp., (2) *Microcystis* sp., (3) *Gloeothece* sp., (4) *Chrococcus* sp., (5) *Anabaena* sp., (6) *Nostoc* sp., (7) *Cylindrospermum* sp., (8) *Calothrix* sp., (9) *Oscillatoria* sp., (10) *Lyngbya* sp., (11) *Spirulina* sp., (12) *Westiella* sp., (13) *Nostochopsis* sp., (14) *Scytonema* sp., (15) *Fischerella* sp., and (16) *Tolypothrix* sp.

FIGURE 19.2 Terrestrial and freshwater cyanobacterial mats growing in their natural habitats in Mizoram, India.

7 Sensing and Molecular Responses to Low Temperature in Cyanobacteria

Jogadhenu S.S. Prakash, Pilla Sankara Krishna, and Sisinthy Shivaji*

CONTENTS

7.1 INTRODUCTION

Abiotic stress on living cells can be defined as environmental conditions capable of causing potentially injurious effects. Clearly if the magnitude and duration of the change in environmental factor is sufficient, then a certain degree of stress is perceived by the organism. Living organisms may not be able to acclimatize upon sudden exposure to severe stress thus leading to their death. But, they can acclimatize to gradual changes in environmental conditions by sensing them, followed by induction and/or optimization of various biochemical and cellular processes. Therefore, survival under abiotic stress depends primarily on the organism's ability to "sense" the change in the environment and transduce the signal to target genes in the genome, ultimately bringing about altered expression of genes. Thus, upon sensing the changes in environmental conditions, the proteins that are synthesized (stress-specific proteins) are involved in acclimation of the organism to the changed environment (Figure 7.1). Cyanobacteria use a

* Corresponding author: jsspsl@uohyd.ernet.in

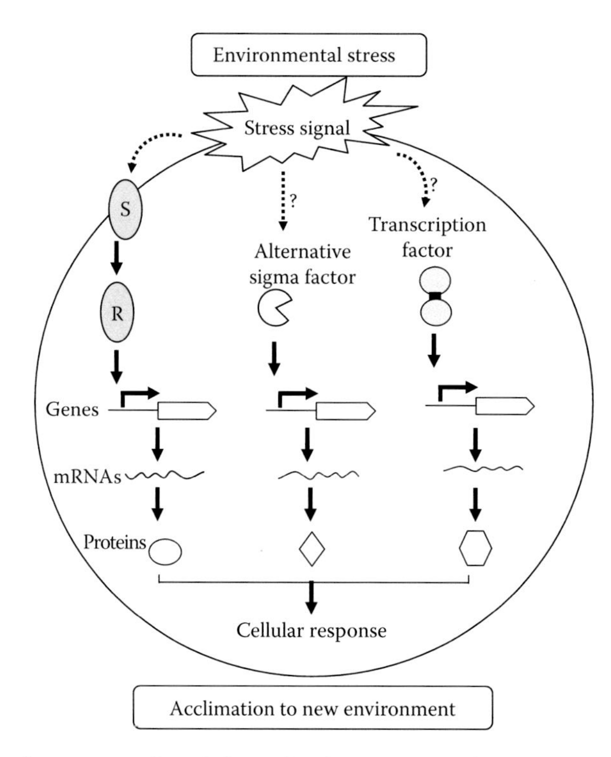

FIGURE 7.1 Schematic representation of the molecular responses of a cyanobacterial cell to changes in environmental conditions. In cyanobacteria, two-component signal transduction pathways involving histidine kinase and response regulator are predominant systems of sensing and perception of environmental stress. Alternative sigma factors and transcription factors are also known to participate in stress-regulated expression of genes in cyanobacteria. Downregulation of genes by two-component systems and transcriptional repressors are not shown in the figure. S, Sensory kinase; R, Response regulator.

two-component signal transduction system, involving a sensory kinase and a response regulator, as a means to perceive stress signal and to transduce the signal to the target genes for regulated expression. In addition different transcription factors and alternative sigma factors are proposed to be involved in regulation of stress-inducible gene expression (Figure 7.1).

7.1.1 Molecular Adaptation to Extreme Temperatures

Temperature stress is an inevitable abiotic stress and exhibits both diurnal and seasonal changes. Denaturation of macromolecules at high-temperature is the major challenge faced by microorganisms. Temperatures approaching 100°C normally denature DNA and proteins and increase the fluidity of membranes to lethal levels. In photosynthetic microorganisms, chlorophyll degrades above 75°C, affecting the process of photosynthesis. Low-temperature also has devastating effects on cells due to ice crystal formation at chilling temperatures. The ice crystals thus formed can rip cell membranes, and solution chemistry stops in the absence of liquid water. Thus freezing of intracellular water is almost invariably lethal to the living cells. Yet, in nature microorganisms inhabit extreme environmental niches, and microorganisms are known to survive temperatures ranging from below 0°C to 100°C. Hyperthermophiles can survive above 80°C and psychrophiles have been adapted to live below 15°C. Hyperthermophile enzymes can have higher temperature optimum, for example, activity up to 142°C for amylopullulanase [1]. Thermophiles exist among the phototrophic bacteria (cyanobacteria, purple and green bacteria), eubacteria (*Clostridium*, *Bacillus*, *Thiobacillus*, *Desulfotomaculum*, *Thermus*, *Actinomycetes*, *Spirochetes*, and numerous other genera), and the archaea (*Pyrococcus*, *Thermococcus*, *Thermoplasma*, *Sulfolobus*, and the *Methanogens*) [2].

Representatives of all major taxa inhabit temperatures just below 0°C. Some microorganisms are reported to survive and exist even at −18°C [3]. Capability of synthesizing heat stable or cold active enzymes is a possible way; organisms adapt to extremely high- and low-temperature conditions, respectively. Reports are available on the heat stability, thereby activity at high-temperatures for thermophilic enzymes and molecular adaptation, thereby optimized enzyme activity for few psychrophilic enzymes at cold temperatures [4–7]. Several cold active enzymes have been characterized from psychrophilic bacteria [8–11]. Optimum temperature for refolding activity of GroESL from *Oleispira*, a psychrophilic bacterium, is between 4°C and 12°C *in vitro* [12]. This suggests that functions of most proteins are optimized to a range of temperature in which the microorganism has been adapted to survive.

7.1.2 *Synechocystis* as a Model to Study Cold Acclimation

The cyanobacterium *Synechocystis* sp. PCC6803 (hereafter, *Synechocystis*) has been extensively used as a model system to study the molecular response of life forms to low-temperature stress. *Synechocystis* is a unicellular cyanobacterium specifically used for studies on abiotic stress responses at molecular level since it is believed to be the progenitor of the green plant chloroplast. It has a simple genome and proteome than higher plants, thus allowing for rapid analyses. The general features of the plasma membrane and thylakoid membranes of *Synechocystis* cells are similar to those of the chloroplasts of higher plants in terms of lipid composition and the assembly of membranes. Therefore, *Synechocystis* can be expected to serve as powerful model systems for studying the molecular mechanisms of the responses and acclimation to low-temperature stress [13–15], also these mechanisms may provide models that are applicable to plants as well. Some strains of cyanobacteria, such as *Synechocystis* sp. PCC 6803, *Synechococcus elongatus* PCC 7942, and *Synechococcus* sp. PCC 7002, are naturally competent and, thus, foreign DNA is incorporated into cells and is integrated into their genomes by homologous recombination at high frequency [16]. As a result, cyanobacteria are widely used by researchers for the production of mutants with disrupted genes of interest. The entire *Synechocystis* genome was sequenced by Kaneko et al. [17]. Genome sequences provide vast amounts of basic information, which can be exploited for genome-wide studies of gene expression. In 1999, Takara Bio Co. (Ohtu, Japan) initiated the production of a genome-wide cDNA microarray for the analysis of gene expression in *Synechocystis.* Their DNA microarray covers 3079 of the 3165 (97%) genes on the chromosome of *Synechocystis.*

7.1.3 Responses to Low-temperature at the Transcriptional Level

When microorganisms experience decrease in surrounding temperature, activity of various enzymes is drastically reduced, ultimately leading to decrease in the rate of metabolic pathways and physiological activities. Despite these adverse effects caused by low-temperature, microorganisms have innate strategies by virtue of which they are capable of acclimatizing to low-temperature. Crucial to this acclimatization process is the ability of microorganisms to sense changes, transduce signals to target genes to activate and/or enhance their expression for low-temperature acclimation. The northern blotting method can be used to analyze the gene expression at the transcriptional level of a limited number of genes at a time, but the method is time consuming. The development of DNA microarray technology allowed researchers to analyze expression of all the genes in an organism at the transcriptional level in one go by performing a single experiment. The first successful report on cyanobacterial global gene expression changes in response to cold stress was reported by Suzuki et al. [18]. In continuation of their first report on cold-regulated genes in *Synechocystis*, detailed studies on regulation of cold-stress-inducible expression changes were investigated using DNA microarray analysis by Norio Murata's group in Japan [18–21]. In these studies, *Synechocystis* cells, which had been grown at 34°C, were exposed to 24°C for 20, 60, and 180 min and then the expression of genes was monitored [21]. More than 100 genes were upregulated due to low-temperature stress [18,21]. Table 7.1 shows the

TABLE 7.1
Low-Temperature-Inducible Genes in *Synechocystis* sp. PCC6803

ORF	Gene	Encoded Protein	Ratio of Transcript Levels (24°C/34°C)		
			20 (min)	60 (min)	180 (min)
Desaturases					
sll1441	*desB*	Acyl-lipid desaturase (omega-3)	5.5 ± 0.4	4.8 ± 0.3	3.4 ± 0.3
slr1350	*desA*	Acyl-lipid desaturase (delta 12)	2.9 ± 0.2	2.5 ± 0.0	1.7 ± 0.2
sll0262	*desD*	Acyl-lipid desaturase (delta 6)	2.7 ± 0.2	2.4 ± 0.0	2.4 ± 0.2
RNA and DNA related proteins					
slr0083	*crhR*	ATP-dependent RNA helicase	13.2 ± 0.2	3.8 ± 0.1	1.7 ± 0.1
sll0517	*rbpA*	Putative RNA binding protein	5.3 ± 0.8	4.5 ± 1.0	3.8 ± 0.3
sll0825	*pcnB*	PolyA polymerase	3.2 ± 0.7	3.3 ± 0.3	2.8 ± 0.0
sll1742	*nusG*	Transcription antitermination protein	2.9 ± 0.1	2.9 ± 0.1	2.9 ± 0.1
sll1772	*mutS*	DNA mismatch repair protein MutS	3.0 ± 0.2	2.6 ± 0.1	1.8 ± 0.1
ssl0426	*ISY100t*	Putative transposase gene	3.0 ± 0.2	3.5 ± 0.3	2.1 ± 0.1
Metabolic enzymes					
slr1291	*ndhD2*	NADH dehydrogenase I chain M	16.5 ± 0.7	3.0 ± 0.0	2.8 ± 0.1
slr0423	*rlpA*	Rare lipoprotein A	4.5 ± 0.4	5.4 ± 0.1	5.5 ± 0.3
sll1541		Lignostilbene-alpha,beta-dioxygenase	4.4 ± 0.1	2.1± 0.0	1.9 ± 0.0
slr0549	*asd*	Aspartate beta-semialdehyde dehydrogenase	4.2 ± 0.3	3.1 ± 0.1	2.4 ± 0.3
slr0239	*cbiF*	Precorrin-4 C11-methyltransferase	4.0 ± 0.4	3.3 ± 0.2	2.0 ± 0.1
slr0399	*ycf39*	Quinone binding in photosystem II	4.0 ± 0.3	1.7 ± 0.0	1.9 ± 0.3
sll0784	*merR*	Nitrilase	3.7 ± 0.2	1.6 ± 0.0	2.2 ± 0.4
sll1454	*narB*	Ferredoxin-nitrate reductase	3.7 ± 0.1	1.6 ± 0.0	1.4 ± 0.0
slr0901	*moaA*	Molybdopterin biosynthesis protein A	3.3 ± 0.2	2.2 ± 0.1	1.7 ± 0.0
slr1238	*gshB*	Glutathione synthetase	3.2 ± 0.6	1.9 ± 0.1	1.6 ± 0.1
sll1245	*cytM*	Cytochrome cM	3.2 ± 0.2	2.7 ± 0.0	2.0 ± 0.1
slr0550	*dapA*	Dihydrodipicolinate synthase	3.2 ± 0.2	2.6 ± 0.1	1.7 ± 0.0
slr0553		Dephospho-CoA kinase	3.2 ± 0.0	2.9 ± 0.1	2.7 ± 0.1
slr1992	*gpx2*	Glutathione peroxidase	3.1 ± 0.3	2.1 ± 0.1	2.6 ± 0.2
slr1254	*crtP*	Phytoene dehydrogenase	3.0 ± 0.0	1.3 ± 0.0	1.7 ± 0.0
slr0646	*pbp*	Penicillin-binding protein 4	3.0 ± 0.4	1.9 ± 0.0	1.5 ± 0.1
sll1408	*pcrR*	Regulatory protein PcrR	2.9 ± 0.4	3.1 ± 0.4	1.8 ± 0.0
sll1258	*dcd*	Deoxycytidine triphosphate deaminase	2.8 ± 0.1	3.3 ± 0.3	3.8 ± 0.2
slr0743	*nusA*	N utilization substance protein	2.7 ± 0.1	3.3 ± 0.1	2.5 ± 0.1
slr1963		Water-soluble carotenoid protein	2.6 ± 0.0	3.0 ± 0.1	2.8 ± 0.0
sll1709	*gdh*	Glucose dehydrogenase	2.5 ± 0.0	2.9 ± 0.1	2.7 ± 0.0
sll1566	*ggpS*	Glucosyl-glycerol-phosphate synthase	2.5 ± 0.2	2.4 ± 0.2	2.4 ± 0.3
slr0252		Precorrin-6X reductase	2.3 ± 0.4	2.6 ± 0.2	2.1 ± 0.5
slr0676	*cysC*	Adenylylsulfate 3-phosphotransferase	1.5 ± 0.1	2.5 ± 0.1	2.9 ± 0.3
GTP binding proteins					
slr1974		Putative GTP-binding protein	4.6 ± 0.1	2.6 ± 0.0	1.8 ± 0.1
slr1521	*hflX*	GTP-binding protein HflX	4.5 ± 0.1	3.6 ± 0.1	1.8 ± 0.0
slr0426	*folE*	GTP cyclohydrolase I	3.6 ± 0.2	2.1 ± 0.1	2.2 ± 0.1
slr0321	*era*	GTP-binding protein Era	3.0 ± 0.9	2.5 ± 0.3	1.8 ± 0.1

TABLE 7.1 (continued)
Low-Temperature-Inducible Genes in *Synechocystis* sp. PCC6803

ORF	Gene	Encoded Protein	Ratio of Transcript Levels (24°C/34°C) 20 (min)	60 (min)	180 (min)
High-light-inducible proteins					
ssl2542	*hliA*	High-light-inducible protein	8.1 ± 0.6	2.1 ± 0.0	2.4 ± 0.4
ssl1633	*hliC*	CAB/ELIP/HLIP superfamily	4.2 ± 0.4	2.1 ± 0.1	2.2 ± 0.0
ssr2595	*hliB*	High-light-inducible protein	4.1 ± 0.1	1.1 ± 0.1	1.1 ± 0.1
ABC type transporters					
sll0086		Arsenical pump-driving ATPase	4.6 ± 0.4	2.7 ± 0.1	2.6 ± 0.0
slr1881	*natE*	ABC-type neutral amino acid transport	3.5 ± 0.4	2.4 ± 0.2	2.1 ± 0.2
sll0384	*cbiQ*	ABC-type cobalt transport system	2.9 ± 0.3	2.3 ± 0.1	2.4 ± 0.1
sll0374	*urtE*	ABC-type urea transport system	2.0 ± 0.1	3.0 ± 0.1	2.2 ± 0.1
Protein synthesis and folding					
sll1802	*rplB*	50S ribosomal protein L2	2.9 ± 0.0	2.3 ± 0.0	3.0 ± 0.1
sll1801	*rplW*	50S ribosomal protein L23	3.2 ± 0.0	2.1 ± 0.1	2.9 ± 0.1
sll1813	*rplO*	50S ribosomal protein L15	1.1 ± 0.0	2.7 ± 0.2	2.9 ± 0.1
sll1743	*rplK*	50S ribosomal protein L11	2.2 ± 0.0	2.3 ± 0.0	2.8 ± 0.2
sll1746	*rplL*	50S ribosomal protein L7/L12	1.7 ± 0.0	2.2 ± 0.0	2.8 ± 0.1
sll1800	*rplD*	50S ribosomal protein L4	2.9 ± 0.0	1.8 ± 0.0	2.7 ± 0.0
sll1810	*rplF*	50S ribosomal protein L6	1.3 ± 0.0	2.4 ± 0.0	2.6 ± 0.0
sll1799	*rplC*	50S ribosomal protein L3	3.6 ± 0.1	1.9 ± 0.0	2.6 ± 0.1
sll1744	*rplA*	50S ribosomal protein L1	1.9 ± 0.1	2.3 ± 0.0	2.6 ± 0.0
sll1812	*rpsE*	30S ribosomal protein S5	1.1 ± 0.0	2.4 ± 0.0	2.5 ± 0.0
ssl3437	*rpsQ*	30S ribosomal protein S17	1.7 ± 0.1	2.0 ± 0.1	2.5 ± 0.2
sll1808	*rplE*	50S ribosomal protein L5	1.5 ± 0.0	2.2 ± 0.1	2.5 ± 0.0
sll1803	*rplV*	50S ribosomal protein L22	2.3 ± 0.1	2.0 ± 0.0	2.4 ± 0.2
sll1809	*rpsH*	30S ribosomal protein S8	1.3 ± 0.1	1.9 ± 0.1	2.4 ± 0.1
ssl3436	*rpmC*	50S ribosomal protein L29	1.8 ± 0.1	1.9 ± 0.0	2.4 ± 0.1
sll1811	*rplR*	50S ribosomal protein L18	1.1 ± 0.0	2.1 ± 0.1	2.3 ± 0.0
ssl3441	*infA*	Translation initiation factor IF-1	1.8 ± 0.2	3.0 ± 0.1	2.6 ± 0.1
slr0744	*infB*	Translation initiation factor IF-2	2.9 ± 0.1	3.4 ± 0.0	2.2 ± 0.1
slr0974	*infC*	Translation initiation factor IF-3	2.1 ± 0.0	2.9 ± 0.2	1.6 ± 0.0
slr1105	*fus*	Elongation factor EF-G	3.4 ± 0.2	2.1 ± 0.0	2.3 ± 0.2
sll1261	*tsf*	Elongation factor EF-Ts	1.3 ± 0.0	2.1 ± 0.1	2.7 ± 0.1
sll0533	*tig*	Trigger factor	3.7 ± 0.1	4.1 ± 0.1	3.8 ± 0.2
slr0713	*tgt*	tRNA-guanine transglycosylase	3.0 ± 0.1	2.2 ± 0.2	2.2 ± 0.1
slr0955		tRNA/rRNA methyltransferase	8.0 ± 0.0	6.3 ± 0.1	4.6 ± 0.4
slr0922	*pth*	Peptidyl-tRNA hydrolase	2.3 ± 0.1	3.0 ± 0.3	1.4 ± 0.1
slr2076	*groEL1*	60 kDa chaperonin 1	0.6 ± 0.0	0.7 ± 0.0	4.4 ± 0.0
slr2075	*groES*	10 kD chaperonin	0.4 ± 0.0	0.5 ± 0.0	3.7 ± 0.0
sll0416	*groEL2*	60 kDa chaperonin 2	0.5 ± 0.0	0.7 ± 0.0	3.1 ± 0.0
Hypothetical proteins					
slr0082		Hypothetical protein	10.0 ± 0.6	2.3 ± 0.1	1.0 ± 0.0
slr1927		Hypothetical protein	6.1 ± 0.2	6.2 ± 0.4	3.2 ± 0.1

(*continued*)

TABLE 7.1 (continued)
Low-Temperature-Inducible Genes in *Synechocystis* sp. PCC6803

ORF	Gene	Encoded Protein	Ratio of Transcript Levels (24°C/34°C)		
			20 (min)	60 (min)	180 (min)
slr0959		Hypothetical protein	5.8 ± 1.5	4.6 ± 0.3	2.3 ± 0.2
sll1911		Hypothetical protein	5.3 ± 0.4	2.7 ± 0.2	3.3 ± 0.2
slr1544		Hypothetical protein	5.2 ± 0.2	1.3 ± 0.1	1.2 ± 0.0
sll1611		Hypothetical protein	4.4 ± 0.4	8.4 ± 0.5	6.7 ± 2.0
sll1483		Hypothetical protein	4.3 ± 0.4	3.2 ± 0.1	2.6 ± 0.4
slr0236		Hypothetical protein	4.1 ± 0.2	4.5 ± 0.1	3.0 ± 0.0
slr0755		Hypothetical protein	3.9 ± 0.9	2.7 ± 0.2	2.0 ± 0.4
sll0528		Hypothetical protein	3.7 ± 0.3	1.5 ± 0.1	0.9 ± 0.0
sll0494		Hypothetical protein	3.6 ± 0.1	2.4 ± 0.1	1.7 ± 0.1
sll0815		Hypothetical protein	3.6 ± 0.4	2.8 ± 0.0	2.3 ± 0.2
sll0462		Hypothetical protein	3.5 ± 0.4	3.7 ± 0.2	2.9 ± 0.0
slr0209		Hypothetical protein	3.5 ± 0.3	3.1 ± 0.1	2.6 ± 0.1
sll0185		Hypothetical protein	3.5 ± 0.1	5.0 ± 0.2	3.6 ± 0.0
slr1050		Hypothetical protein	3.4 ± 0.5	2.4 ± 0.3	1.6 ± 0.2
slr1747		Hypothetical protein	3.4 ± 1.1	1.5 ± 0.2	1.4 ± 0.1
slr1495		Hypothetical protein	3.3 ± 0.0	2.9 ± 0.0	2.0 ± 0.2
sll0263		Hypothetical protein	3.3 ± 0.2	2.8 ± 0.0	2.4 ± 0.1
sll0783		Hypothetical protein	3.3 ± 0.0	1.7 ± 0.0	3.5 ± 0.2
sll1769		Hypothetical protein	3.2 ± 0.1	2.3 ± 0.0	2.0 ± 0.0
sll0722		Hypothetical protein	3.2 ± 0.1	2.9 ± 0.1	2.4 ± 0.3
sll0005		Hypothetical protein	3.2 ± 0.2	2.7 ± 0.2	1.4 ± 0.1
slr2120		Hypothetical protein	3.1 ± 0.1	3.1 ± 0.0	1.5 ± 0.0
slr2126		Hypothetical protein	3.1± 0.1	3.0 ± 0.5	1.8 ± 0.2
sll1770		Hypothetical protein	3.0 ± 0.1	2.2 ± 0.1	1.8 ± 0.0
slr1677		Hypothetical protein	2.9 ± 0.0	2.8 ± 0.1	2.5 ± 0.1
slr1436		Hypothetical protein	2.8 ± 0.6	3.3 ± 0.5	3.6 ± 0.2
slr0742		Hypothetical protein	2.8 ± 0.0	3.4 ± 0.2	2.4 ± 0.0
slr1865		Hypothetical protein	2.3 ± 0.2	2.8 ± 0.0	2.2 ± 0.1

Source: Adapted from Prakash, J.S.S. et al., *Microbiology*, 156, 442, 2010; The complete list of genes with expression changes can be accessed at http://www.genome.jp/kegg-bin/get_htext?htext=Exp_DB&hier=1.

Cells that had been grown at 34°C were incubated at 24°C for 20, 60, and 180 min. Genes with ratios of transcript levels higher than 2.5 under any given time point were considered for making the final list of cold-inducible genes.

list of low-temperature-inducible genes along with fold change in their transcript levels during the course of low-temperature treatment. The upregulated genes could be categorized into several groups: (1) genes for acyl lipid desaturases; (2) proteins involved in RNA and DNA metabolism, such as CrhR, RbpA1, and MutS; (3) genes that encode for various metabolic enzymes; (4) genes for GTP binding proteins; (5) high-light-inducible genes; (6) ABC-type transporter genes; (7) genes that encode proteins involved in translation and protein folding; and (8) hypothetical genes that encode for proteins, whose functions are yet to be characterized.

Out of the low-temperature-inducible genes, some are strongly upregulated at the initial phase of low-temperature stress followed by a gradual decrease to the control levels. Low-temperature-inducible genes, such as *ndhD2*, *crhR*, *hliA*, *ycf39*, *tig*, *narB*, *gshB*, and genes for GTP binding proteins, *slr0082*, *slr1544,* and *sll0528*, belong to this category. Expression of some other genes was upregulated throughout the period of low-temperature treatment, suggesting their importance in acclimation to low-temperature stress. For example, *desB, desD, rbpA, pcnB, nusG, rlpA, gpx2, dcd, nusA, ggpS, fus, sll1611,* and several hypothetical genes are constantly upregulated upon low-temperature treatment. Perhaps, desaturases need to be constantly synthesized under low-temperatures in order to maintain the unsaturated fatty acid content at higher level to keep the membrane optimally fluidized. Genes that encode ribosomal proteins and several other genes that code for proteins involved in translation were upregulated after prolonged incubation of cells at low-temperature. Importantly, the genes for molecular chaperonins GroES and GroEL were upregulated only at prolonged incubation of cells at low-temperature, suggesting shift of cells to low-temperature gradually inhibits translation. To resume protein synthesis, which is inhibited under low-temperature, cells trigger the expression of genes that encode ribosomal proteins, initiation factors, elongation factors, and several other proteins required to reinitiate the translation. Once the translation is resumed, newly synthesized polypeptides could be the signal for upregulation of genes that encode chaperonin-refolding machinery [22].

7.2 LOW-TEMPERATURE-INDUCIBLE PROTEINS AND MECHANISMS OF COLD ACCLIMATION

7.2.1 Desaturases and Membrane Fluidity Optimization

Biological membranes are normally composed of saturated and unsaturated fatty acids, and by altering the proportion of these fatty acids the fluidity of the membrane could be effectively altered. Membranes are more fluid, when there is an increase in the unsaturated fatty acid content over the saturated fatty acid content, because the unsaturated fatty acid acyl chains exhibit poor packing due to the kink caused by the *cis* double bond [23,24]. The psychrotrophic cyanobacteria have been adapted to survive at chilling temperatures with a greater proportion of unsaturated fatty acids in their membrane, as compared with mesophilic bacteria [25,26]. Mesophilic bacteria when exposed to low-temperatures respond by increasing the level of unsaturated fatty acids in the membrane lipids [13,27]. This phenomenon of thermal regulation of membrane fluidity has been studied in detail in *Escherichia coli, Bacillus subtilis,* and *Synechocystis*. Cyanobacterial cells optimize their membrane fluidity in response to low-temperature by altering the extent of unsaturation of fatty acids, such as a decrease in the levels of C18:1(9) and C18:2(9,12) and increase in the levels of C18:3(9,12,15) [28–30]. In cyanobacteria, the fatty acid unsaturation is catalyzed by acyl-lipid desaturases: DesA, DesB, DesC, and DesD, which introduce a double bond in the Δ12, Δ15, Δ9, and Δ6 position of the fatty acid, respectively [29]. It has been a well-established fact that polyunsaturated fatty acids and acyl-lipid desaturases are essential for the cold acclimation of cyanobacteria [13,31]. Studies on *Synechocystis* sp. PCC 6803 revealed that the *desA*, *desB*, and *desD* genes but not *desC* are upregulated at the transcriptional level, when cells are transferred to a lower growth temperature; that is, from 34°C to 24°C (Table 7.1), and the activities of the enzymes, DesA, DesB, and DesD, increased, while that of DesC remained unchanged [32,33]. Targeted inactivation of the gene for the Δ12 desaturase (*desA*) in *Synechocystis* sp. PCC 6803 and *Synechococcus* sp. PCC 7002 exhibited a considerable increase in the level of C18:1(9) at the expense of the polyunsaturated fatty acid C18:3(9,12,15) in the membrane lipids and the strains were cold sensitive at 25°C compared to the wild-type strain [13,34,35]. A double mutant of *Synechocystis*, in which both Δ12 and Δ6 desaturases (*desA* and *desD*) were mutagenized, also showed a cold-sensitive phenotype, and in this mutant, an increase in the level of C18:1(9) at the expense of the polyunsaturated fatty acid C18:3 was observed in the membrane lipids. These observations confirmed that increase in the polyunsaturated fatty acids is essentially required for survival at low-temperatures. Among the four desaturase genes, *desB* was strongly upregulated in *Synechocystis* (Table 7.1). This suggests that the significance of enhancement in polyunsaturated fatty acid 18:3(9,12,15) content is for low-temperature acclimation. Further, when the *desC* gene for the acyl-lipid Δ9 desaturase of the thermophilic cyanobacterium *Synechococcus vulcanus* was introduced into *Nicotiana tobacum*, the chilling tolerance of these plants increased, indicating that introduction of unsaturated fatty acids increases cold tolerance [36].

7.2.2 Cold-Inducible RNA Helicases

RNA helicases are ubiquitously distributed in all the biological kingdoms [37]. Extensive studies have demonstrated that they are active in modulating the secondary structure of RNAs by unwinding RNAs in an ATP-dependent manner [38]. They participate in various cellular processes in which RNAs are involved. In model experimental organisms, such as *E. coli*, a cold-inducible RNA helicase, CsdA, has been suggested to participate in the assembly of ribosomes [39]. Another RNA helicase, RhlB, is a component of "cold shock degradosome complex" and enhances the degradation of mRNAs by unwinding the secondary structures so as to facilitate cleavage of mRNA by RNase E [40,41]. The cyanobacterium *Anabaena* sp. PCC 7120 contains two genes for RNA helicases; the *crhC* gene is induced only under low-temperature stress whereas the *crhB* gene is induced by salt, low-temperature, and light stress, as well as by nitrogen limitation [42]. Biochemical characterization and cellular localization of CrhC suggested that it is a membrane-bound protein and may be involved in translocation of proteins across the plasma membrane under low-temperature conditions [43].

One of the earliest and largest transcriptional responses that occur during exposure of *Synechocystis* to low-temperature is the induction of the *crhR* transcript (Table 7.1). CrhR is a redox regulated and cold-inducible RNA helicase that catalyzes both ATP-dependent unwinding of secondary structures of RNA and annealing of complementary RNA strands [44]. It is suggested that CrhR destabilizes the secondary structures of mRNAs, thereby allowing cells to overcome the cold-induced blockage of translation that occurs at low-temperature. We demonstrated that *Synechocystis* fails to cold acclimate on *crhR* gene deletion [45]. Subsequently, we reported that CrhR is essentially required for executing state transitions and for the regulation of photosystem stoichiometry at low-temperature [46].

7.2.3 Cold-Inducible RNA Binding Proteins

RNA-binding proteins contain RNA recognition motif (RRM) and could be involved in many post-transcriptional regulation processes [47]. They are divided into two classes: Class I RNA binding proteins contain only RRM, but do not contain long conserved C-terminal domain, which is seen in Class II Rbps (RbpG) [48]. Class I Rbp proteins are further divided into two types: type I Rbps are cold-inducible and contain a C-terminal glycine-rich domain and type II Rbps do not possess C-terminal glycine-rich domain and show no or a slight response to cold induction [49,50]. In the filamentous species *Anabaena variabilis*, a type I RNA-binding protein, RbpA1, affects the gene expression and the *rbpA1* mutant showed abnormal regulation of heterocyst differentiation at low-temperature [51]. In *Synechocystis*, a type II RNA-binding protein, *Rbp3*, is specifically required for maintaining the mRNA levels of *desA*, *desB*, *desD*, and *ccrl* [52]. In a unicellular species, *Synechococcus* sp. strain PCC 7942, an *rbpl* mutant showed greatly reduced growth at 20°C [53]. An *rbp3* mutant of *Synechocystis* did not exhibit any slow growth phenotype at low-temperature but showed reduced fatty acid desaturation of membrane lipids [52].

7.2.4 Ribosomal Proteins and Translation

An immediate effect on cyanobacterial cells upon shift to low-temperature is inhibition of protein synthesis, presumably due to blockage at the initiation phase of translation, due to formation of mRNA secondary structures. Perhaps to resume the cold-induced inhibition of translation, a number of genes that encode ribosomal proteins, initiation factors, and elongation factors are upregulated upon exposure of *Synechocystis* cells to low-temperature (Table 7.1). Induced expression and thus excess amount of translation-initation factors, elongation factors, ribosomal proteins and proteins involved in protein-folding are necessary for acclimation of the translational apparatus to cold, since translational activity always appears to decrease in bacteria upon exposure to low-temperatures. RbfA, a 30S ribosomal-subunit associated protein, is thought to play a crucial role in relieving the translational block at low-temperature in *E. coli* [54].

7.2.5 Folding and Quality Control of Proteins during Cold Acclimation

Molecular chaperones that facilitate protein folding and the proteases that degrade mis-folded proteins are important for maintaining protein quality control. Low-temperature causes denaturation and aggregation of proteins. Studies suggest that folding of proteins, and/or maintenance of quaternary structure of proteins at low-temperature, might be essential for cold adaptation process [55]. Hsc66, a member of the Hsp70 class of molecular chaperone, is cold inducible and has been proposed to act as cold shock molecular chaperone in *E. coli* [56]. The importance of chaperone function at low-temperature was demonstrated in *E. coli* by heterologous expression of the GroES and GroEL of the Antarctic bacterium *Oleispira antarctica* strain RB-8T, which supported the growth of *E. coli* at 4°C [12]. The cold-inducible trigger factor (TF) binds to nascent polypeptides on ribosomes and enhances GroEL's affinity to unfolded proteins, thus promoting degradation of certain polypeptides at low-temperature. Inactivation of the gene for TF resulted in reduced viability of *E. coli* to cold [57]. Two caseinolytic proteases (ClpB and ClpP) of *Synechococcus* sp. 7942 were also found essential for growth at low-temperature [58]. Thus, protein folding and protein degradation also play a major role in the cold adaptation process.

7.2.6 Other Low-Temperature-Inducible Genes

In *Synechocystis*, several genes that code for metabolic enzymes are upregulated during low-temperature stress. The gene expression of *ndhD2*, which encodes a subunit of NADH dehydrogenase, is strongly upregulated upon low-temperature treatment [59]. In addition, genes that code for glutathione synthetase, glutathione peroxidase, phytoene dehydrogenase, glucosyl-glycerol-phosphate synthase, and glucose dehydrogenase are upregulated during exposure of *Synechocystis* cells to low-temperatures. It is believed that shift of cells to low-temperature induces cellular dehydration that is similar to the effect caused by hyperosmotic stress. Therefore, upregulation of genes involved in the synthesis of a compatible solute, glucosylglycerol supports a common effect caused by both these stresses on cell. Upregulation of genes that encode high-light-inducible proteins reflects probable oxidative damage at low-temperatures [60]. The expression of the *cytM* gene for cytochrome C_M is known to be upregulated upon shift of cells to low-temperature (Table 7.1) [61]. Cyt C_M replaces plastocyanin and/or cytochrome C_6, which donate electrons to photosystem I under optimal growth conditions [62]. The genes for plastocyanin and cytochrome C_6 are strongly downregulated at low-temperatures. The substitution of cytochrome C_M in place of regular electron donors might be advantageous under low-temperature stress.

Apart from all the aforementioned cellular responses to low-temperature, a number of unknown molecular mechanisms might be occurring in the cells, in response to low-temperature, as the DNA microarray analyses revealed upregulation of several genes that code for hypothetical proteins (Table 7.1).

7.3 LOW-TEMPERATURE SENSING AND SIGNAL TRANSDUCTION

7.3.1 Membrane Fluidity as a Low-Temperature Signal

A biological membrane that acts as an interface between the external and internal environments of the cell could thus be one of the primary sensors of low-temperature [63]. When exposed to low-temperatures, membranes exhibit a decrease in fluidity and become rigid. Such a rigidification of the membrane activates a sensor located in the membrane [15]. Norio Murata's group performed a series of experiments using *Synechocystis* to demonstrate membrane rigidification as a low-temperature signal. Under isothermal conditions, rigidity of the membrane by palladium-catalyzed hydrogenation of the unsaturated fatty acids activates *desA* gene, which in turn is under the control of a membrane-bound sensory kinase, Hik33 [64,65]. A *Synechocystis* double mutant, in which *desA* and *desD* genes were inactivated, showed rigidified plasma membrane at physiological temperatures [66]. DNA microarray analysis of this mutant was reported to enhance the inducibility of expression of cold-inducible genes [20]. Thus, rigidification of membrane lipids apparently enhanced the response of gene expression to low-temperature in *Synechocystis*. These results suggest that decrease in membrane fluidity serves as a primary signal for low-temperature perception. Later, a two-component system consisting of a membrane-bound sensory kinase and a cytoplasmic response regulator was discovered to be involved in cold-signal transduction [18,65,67].

7.3.2 Two-Component Systems for Low-Temperature Signal Perception

In cyanobacteria, the two-component signal transduction system (composed of a histidine kinase and a response regulator) transduces the stress signal to target genes for altered expression. The two protein components utilize phosphorylation as a means of information transfer between each other [68]. The sensor is normally an integral membrane protein consisting of membrane-spanning domains, a signal-recognition domain, and a kinase domain. Upon recognition of signal, the histidine kinase gets activated due to autophosphorylation of a histidine residue by the kinase domain. Thus activated kinase transfers the phosphate to an aspartate on the response regulator [69]. However, in contrast to the aforementioned simple two-component signal transduction systems, multistep phosphorelay systems have been discovered in both prokaryotes and eukaryotes [68,70]. These multiple-step-type signal transduction pathways have a hybrid-type histidine kinase in which both the histidine kinase and the response regulator receiver domains are present in the same protein. Studies revealed that a two-component signal transduction system is involved in perception of cold signal to target genes in *Synechocystis* [65]. Suzuki et al. systematically inactivated each putative histidine kinase (*hik*) genes by inserting a spectinomycin-resistance gene cassette to identify sensors responsible for signal perception of various stimuli and identified a membrane-bound Hik33 as a potential sensor of cold signal [71]. Hik33 has a highly conserved histidine kinase domain at its carboxyl terminus, two membrane-spanning domains at its amino terminus, and a type-P linker and a leucine zipper in the middle region, characteristic of several membrane-bound histidine kinases from various organisms [65,72,73]. Low-temperature-induced rigidification of membrane lipids around Hik33 might change the conformation of the trans-membrane domains of Hik33, leading to dimerization and activation by autophosphorylation [20]. Further, DNA microarray analysis of Δ*hik33* cells indicated that Hik33 regulates the expression of about 50% of the cold-inducible genes (Figure 7.2) [18]. Cold inducibility of remaining genes was not regulated by Hik33, suggesting at least one other pathway for cold-signal transduction. Subsequently, it was demonstrated that the response regulator, Rre26, is a cognate response regulator of Hik33 in cold-signal transduction (Figure 7.2) [65].

7.3.3 Low-Temperature Sensing through Changes in DNA Supercoiling

Change in DNA supercoiling is suggested to play important roles in the regulation of global gene expression in response to environmental stress [74–76]. Environmental stress dependent alterations in the DNA supercoiling might be a mechanism employed by the cell for the regulated expression of stress responsive genes [77]. The expression of many genes is dependent on DNA conformation, which in turn depends on temperature-dependent changes in DNA supercoiling [78,79]. We have recently reported that at low-temperature inhibition of DNA gyrase, which induces negative supercoils in DNA, prevented the upregulation of majority of cold-inducible genes, thus suggesting negative supercoiling as a regulator of cold-inducible gene expression [76] (Figure 7.2). Salt stress also affects the negative supercoiling of DNA and regulates gene expression [76,80,81]. It is suggested that the stress-induced changes in superhelicity of genomic DNA provide an important permissive background for successful acclimatization of cyanobacterial cells to stress conditions.

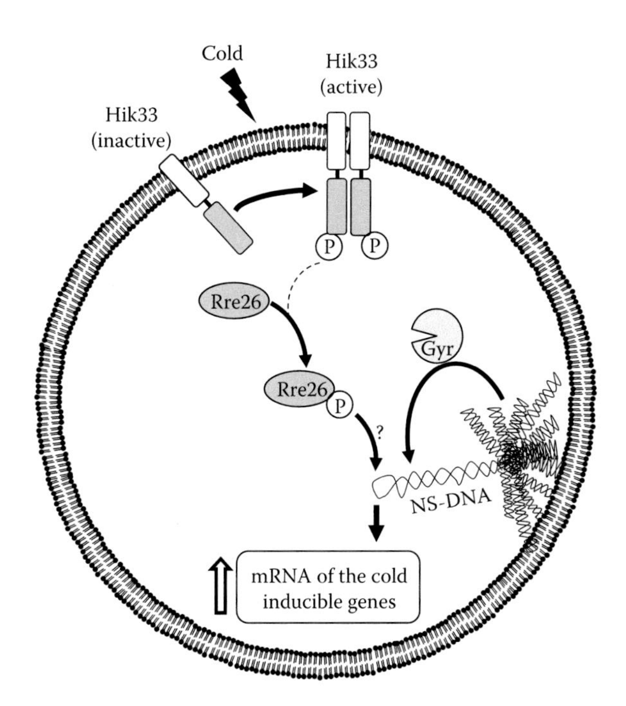

FIGURE 7.2 Diagrammatic representation of signal transduction for the sensing and transduction of low-temperature signal in cyanobacterium *Synechocystis* sp. PCC6803. Upon cold shock, the membrane becomes rigid. Cold-induced rigidity of the membrane activates an inactive sensory kinase (Hik33). Hik33 dimer gets autophosphorylated and transfers the phosphate to the response regulator (Rre26). Activated-Rre26 probably binds to the negatively supercoiled upstream DNA regions of target cold-inducible genes and activates their expression. The model also demonstrates a link between expression of cold-inducible genes and DNA supercoiling. Upon cold shock, DNA gyrase gets activated by an unknown signal transduction pathway and introduces negative supercoils in DNA. It is proposed that the response regulator can activate transcription of several cold-inducible genes, only when DNA is in negatively supercoiled conformation. Open arrow corresponds to the enhanced mRNA levels of 21 cold-inducible genes listed in Murata and Los [65]. (Adapted and modified from Shivaji, S. and Prakash, J.S.S., *Arch. Microbiol.*, 192, 85, 2010.)

7.3.4 Low-Temperature Sensing through Changes in RNA Conformation

RNA molecules act as temperature sensors, due to the ability to modulate their secondary structure and form intermolecular RNA/RNA hybrids due to changes in temperature [82–84]. In *E. coli*, a temperature regulated *dsrA* gene codes for a small RNA, DsrA. The DsrA in turn regulates translation of RpoS message by an anti-antisense mechanism [85,86].

7.4 CONCLUSIONS

Studies with DNA microarrays have identified a number of genes with changes in their expression in response to cold stress. Membrane fluidity optimization is one of the key responses of cyanobacteria to cold stress. Studies proved beyond doubt that the cold-induced membrane rigidification is the signal for cold-regulated gene expression in *Synechocystis*. Functional roles played by cold-inducible RNA helicases and RNA binding proteins need a further detailed research. Functions of a number of cold-inducible proteins are yet to be worked out with respect to cold acclimation. A two-component signal transduction system consisting of Hik33 and Rre26 regulates majority of cold-inducible genes. Yet, we are far from understanding many key aspects of cyanobacterial signal transduction in response to low-temperature. It should be emphasized that an important contribution to further progress can be made by studies on small-RNA-mediated, transcription-factor-mediated, and sigma-factor-mediated regulation of cold-inducible genes. The domain in the sensory kinase that perceives the membrane rigidity is yet not known. Therefore, studies directed toward finding answers to the earlier questions would ultimately unravel the molecular basis of cold perception, signal transduction, and survival at low-temperature. A combination of physiological, biochemical, and genetic approaches should lead to a deeper understanding of the mechanisms of low-temperature stress responses in cyanobacteria.

ACKNOWLEDGMENTS

This work was supported by a grant from the Department of Science and Technology (DST), Project No: SR/SO/BB-39/2004 (grant to J.S.S.P.), by a grant from the India–Japan Cooperative Sciences Programme of the Department of Science and Technology, Government of India, and the Japanese Society for the Promotion of Science, Government of Japan (Project No: DST/INT/JSPS/P-125/2011).

REFERENCES

1. Schuliger, J.W. et al., Purification and characterization of a novel amylolytic enzyme from ES4, a marine hyperthermophilic archaeum, *Mol. Marine Biol. Biotech.*, 2, 76, 1993.
2. Rothschild, L.J. and Mancinelli, R.L., Life in extreme environments, *Nature*, 409, 1092, 2001.
3. Clarke, A., Evolution and low temperatures, in *The Impact of the Physical Environment*, Rothschild, L. and Lister, A., Eds., Academic Press, London, U.K., 2003, p. 187.
4. Ladenstein, R. and Antranikian, G., Proteins from hyperthermophiles: Stability and enzymatic catalysis close to the boiling point of water, in *Biotechnology of Extremophiles*. Antranikian, G., Ed., Springer, Berlin, Germany, 1998, p. 37.
5. Demirjian, D.C., Morri's-Varas, F., and Cassidy, C.S., Enzymes from extremophiles, *Curr. Opin. Chem. Biol.*, 5, 144, 2001.
6. Arnott, M.A. et al., Thermostability and thermoactivity of citrate synthases from the thermophilic and hyperthermophilic archaea, *Thermoplasma acidophilum* and *Pyrococcus furiosus*, *J. Mol. Biol.*, 304, 657, 2000.
7. Feller, G. and Gerday, C., Psychrophilic enzymes: Molecular basis of cold adaptation, *Cell. Mol. Life Sci.*, 53, 830, 1997.
8. Alam, S.I. et al., Purification and characterisation of extracellular protease produced by *Clostridium* sp. from Schirmacher oasis, Antarctica, *Enzyme Microb. Technol.*, 36, 824, 2005.
9. Ray, M.K. et al., Extracellular protease from the Antarctic yeast *Candida humicola*, *Appl. Environ. Microbiol.*, 58, 1918, 1992.
10. Singh, A.K. and Shivaji, S., A cold-active and a heat-labile t-RNA modification GTPase from a psychrophilic bacterium *Pseudomonas syringae* (Lz4W), *Res. Microbiol.*, 161, 46, 2010.
11. Reddy, G.S.N., Rajagopalan, G., and Shivaji, S., Thermolabile ribonucleases from Antarctic psychrotrophic bacteria: Detection of the enzyme in various bacteria and purification from *Pseudomonas fluorescens*, *FEMS Microbiol. Lett.*, 122,122, 1994.
12. Ferrer, M. et al., Chaperonins govern growth of *Escherichia coli* at low temperatures, *Nat. Biotech.*, 21, 1266, 2003.
13. Murata, N. and Wada, H., Acyl-lipid desaturases and their importance in the tolerance and acclimatization to cold of cyanobacteria, *Biochem. J.*, 308, 1, 1995.
14. Glatz, A. et al., The *Synechocystis* model of stress: From molecular chaperons to membranes, *Plant Physiol. Biochem.*, 37, 1, 1999.
15. Los, D.A. and Murata, N., Membrane fluidity and its roles in the perception of environmental signals, *Biochim. Biophys. Acta*, 1666, 142, 2004.
16. Williams, J.G.K., Construction of specific mutations in photosystem II reaction center by genetic engineering methods in *Synechocystis* 6803, *Methods Enzymol.*, 167, 1988, 766.
17. Kaneko, T. et al., Sequence analysis of the genome of the unicellular cyanobacterium *Synechocystis* sp. strain PCC 6803. II. Sequence determination of the entire genome and assignment of potential protein-coding regions, *DNA Res.*, 3, 109, 1996.
18. Suzuki, I. et al., Low temperature-regulated genes under control of the low temperature sensor Hik33 in *Synechocystis*, *Mol. Microbiol.*, 40, 235, 2001.

19. Mikami, K. et al., The histidine kinase Hik33 perceives osmotic stress and low temperature stress in *Synechocystis* sp. PCC 6803, *Mol. Microbiol.*, 46, 905, 2002.
20. Inaba, M. et al., Gene-engineered rigidification of membrane lipids enhances the low-temperature inducibility of gene expression in *Synechocystis*, *J. Biol. Chem.*, 278, 12191, 2003.
21. Prakash, J.S.S. et al., An RNA helicase, CrhR, regulates the low temperature-inducible expression of heat shock genes *groES*, *groEL1* and *groEL2* in *Synechocystis* sp. PCC6803, *Microbiology*, 156, 442, 2010.
22. Thieringer, H.A., Jones, P.G., and Inouye, M., Cold shock and adaptation, *BioEssays*, 20, 49, 1998.
23. McElhaney, R.N., Effects of membrane lipids on transport and enzymic activities, *Current Topics in Membranes and Transport*, Vol. 17. S. Razin, S., and Rottem S., Eds., Academic Press, New York, 1982, p. 317.
24. Mansilla, M.C. et al., Control of membrane lipid fluidity by molecular thermosensors, *J. Bacteriol.*, 186, 6681, 2004.
25. Chintalapati, S. et al., A novel Δ9 acyl-lipid desaturase, DesC2, from cyanobacteria acts on fatty acids esterified to the sn-2 position of glycerolipids, *Biochem. J.*, 398, 207, 2006.
26. Chintalapati, S. et al., Desaturase genes in a psychrotolerant Nostoc sp. Are constitutively expressed at low temperature, *Biochem. Biophys. Res. Commun.*, 362, 81, 2007.
27. Russell, N.J., Mechanisms of thermal adaptation in bacteria: Blueprints for survival, *Trends Biochem. Sci.*, 9, 108, 1984.
28. Sato, N. and Murata, N., Temperature shift-induced responses in lipids in the blue-green alga, *Anabaena variabilis*. The central role of diacylmonogalactosylglycerol in thermoadaptation, *Biochim. Biophys. Acta*, 619, 353, 1980.
29. Murata, N., Wada, H., and Gombos, Z., Modes of fatty-acid desaturation in cyanobacteria, *Plant Cell Physiol.*, 33, 933, 1992.
30. Wada, H. and Murata, N., Temperature-induced changes in the fatty acid composition of the cyanobacterium, *Synechocystis PCC6803*, *Plant Physiol.*, 92, 1062, 1990.
31. Nishida, I. and Murata, N., Chilling sensitivity in plants and cyanobacteria: The crucial contribution of membrane lipids, *Annu. Rev. Plant Physiol. Plant Mol. Biol.*, 47, 541, 1996.
32. Los, D.A. et al., The temperature dependent expression of the desaturase gene desA in *Synechocystis* PCC6803, *FEBS Lett.*, 318, 57, 1993.
33. Los D.A., Ray, M.K., and Murata, N., Differences in the control of temperature-dependent expression of four genes for desaturases in *Synechocystis* sp. PCC 6803, *Mol. Microbiol.*, 25, 1167, 1997.
34. Tasaka, Y. et al., Targeted mutagenesis of acyl-lipid desaturases in *Synechocystis*: Evidence for the important roles of polyunsaturated membrane lipids in growth, respiration and photosynthesis, *EMBO J.*, 15, 6416, 1996.
35. Sakamoto, T. et al., Low-temperature-induced desaturation of fatty acids and expression of desaturase genes in the cyanobacterium *Synechococcus* sp. *PCC 7002*, *FEMS Microbiol. Lett.*, 152, 313, 1997.
36. Orlova, I.V. et al., Transformation of tobacco with a gene for the thermophilic acyl-lipid desaturase enhances the chilling tolerance of plants, *Plant Cell Physiol.*, 44, 447, 2003.
37. Rocak, S. and Linder, P., DEAD-box proteins: The driving forces behind RNA metabolism, *Nat. Rev. Mol. Cell. Biol.*, 5, 232, 2004.
38. Tanner, N.K. and Linder, P., DExD/H box RNA helicases: From generic motors to specific dissociation functions, *Mol. Cell*, 8, 251, 2001.
39. Lauri, P., Virumae, K., and Remme, J., Ribosome assembly in *Escherichia coli* strains lacking the RNA helicase DeaD/CsdA or DbpA, *FEBS J.*, 275, 3772, 2008.
40. Carpousis, A.J., Vanzo, N.F., and Raynal, L.C., mRNA degradation, a tale of poly(A) and multiprotein machines, *Trends Genet.*, 15, 24, 1999.
41. Prud'homme-Genereux, A. et al., Physical and functional interactions among RNase E, polynucleotide phosphorylase and the cold-shock protein, CsdA: Evidence for a 'cold shock degradosome', *Mol. Microbiol.*, 54, 1409, 2004.
42. Chamot, D. et al., A low temperature shock induced cyanobacterial RNA helicase, *J. Bacteriol.*, 181, 1728, 1999.
43. El-Fahmawi, B. and Owttrim, G.W., Polar-biased localization of the low temperature stress-induced RNA helicase, CrhC, in the cyanobacterium *Anabaena* sp. strain PCC 7120, *Mol. Microbiol.*, 50, 1439, 2003.
44. Chamot, D. et al., RNA structural rearrangement via unwinding and annealing by the cyanobacterial RNA helicase, CrhR, *J. Biol. Chem.*, 280, 2036, 2005.
45. Rowland, J.G. et al., Proteomics reveals a role for the RNA helicase CrhR in the modulation of multiple metabolic pathways during cold acclimation of *Synechocystis* sp. PCC6803, *J. Prot. Res.*, 10, 3674, 2011.

46. Sireesha, K. et al., RNA helicase, CrhR is indispensable for the energy redistribution and the regulation of photosystem stoichiometry at low temperature in Synechocystis sp. PCC6803, *Biochim. Biophys. Acta.*, 1817, 1525, 2012
47. Albà, M. M. and Pagès, M., Plant proteins containing the RNA-recognition motif, *Trends Plant Sci.*, 3, 15, 1998.
48. Hamano, T. et al., Characterization of RNA-binding properties of three types of RNA-binding proteins in *Anabaena* sp. PCC 7120, *Cell. Mol. Biol.*, 50, 613, 2004.
49. Maruyama, K., Sato, N., and Ohta, N., Conservation of structure and cold-regulation of RNA-binding proteins in cyanobacteria: Probable convergent evolution with eukaryotic glycine-rich RNA-binding proteins, *Nucleic Acids Res.*, 27, 2029, 1999.
50. Sato, N., A family of cold-regulated RNA-binding protein genes in the cyanobacterium *Anabaena variabilis* M3, *Nucleic Acids Res.*, 23, 2161, 1995.
51. Sato, N. and Wada, A., Disruption analysis of the gene for a cold-regulated RNA-binding protein, rbpA1, in Anabaena: Cold-induced initiation of the heterocyst differentiation pathway, *Plant Cell Physiol.*, 37, 1150, 1996.
52. Tang, Q., Tan, X., and Xu, X., Effects of a type-II RNA-binding protein on fatty acid composition in *Synechocystis* sp. PCC 6803, *Chin. Sci. Bull.*, 55, 2416, 2010.
53. Sugita, C. et al., Targeted deletion of genes for eukaryotic RNA-binding proteins, Rbp1 and Rbp2, in the cyanobacterium *Synechococcus* sp. strain PCC 7942: Rbp1 is indispensable for cell growth at low temperatures, *FEMS Microbiol. Lett.*, 176, 155, 1999.
54. Jones, P.G. and Inouye, M., RbfA, a 305 ribosomal binding factor, is a cold-shock protein whose absence triggers the cold-shock response, *Mol. Microbiol.,* 21, 1207, 1996.
55. Strocchi, M. et al., Low temperature-induced systems failure in *Escherichia coli*: Insights from rescue by cold-adapted chaperones, *Proteomics*, 6, 193, 2006.
56. Lelivelt, M.J. and Kawula, T.H., Hsc66, an Hsp70 homolog in Escherichia coli, is induced by cold shock but not by heat shock, *J. Bacteriol.*, 177, 4900, 1995.
57. Kandror, O. and Goldberg, A.L., Trigger factor is induced upon cold shock and enhances viability of *Escherichia coli* at low temperatures, *Proc. Natl Acad. Sci. U S A*, 94, 4978, 1997.
58. Porankiewiz, J., Schelin, J., and Clarke, A.K., The ATP-dependent Clp protease is essential for acclimation to UV-Band low temperature in the cyanobacterium *Synechococcus. Mol. Microbiol.,* 29, 275, 1998.
59. Ohkawa, H., Pakrasi, H. B., and Ogawa, T., Two types of functionally distinct NAD(P)H dehydrogenases in *Synechocystis* sp. Strain PCC6803, *J. Biol. Chem.*, 275, 31630, 2000.
60. He, Q. et al., The high light inducible polypeptides in *Synechocystis* PCC6803. Expression and function in high light, *J. Biol. Chem.*, 276, 306, 2001.
61. Malakhov, M.P., Wada, H., Los, D.A., Semenenko, V.E., and Murata, N., A new type of cytochrome c from *Synechocystis* PCC6803, *J. Plant Physiol.*, 144, 259, 1994.
62. Malakhov, M.P., Malakhova, O.A., and Murata, N., Balanced regulation of expression of the gene for cytochrome cM and that of genes for plastocyanin and cytochrome c6 in *Synechocystis*, *FEBS Lett.*, 444, 281, 1999.
63. Rowbury, R.J., Temperature effects on biological systems: Introduction, *Sci. Prog.*, 86, 1, 2003.
64. Vigh. L. et al., The primary signal in the biological perception of temperature: Pd-catalyzed hydrogenation of membrane lipids stimulated the expression of the *desA* gene in *Synechocystis* PCC 6803, *Proc. Natl Acad. Sci. U S A,* 90, 9090, 1993.
65. Murata, N. and Los, D.A., Genome-wide analysis of gene expression characterizes histidine kinase Hik33 as an important component of the cold-signal transduction in cyanobacteria, *Physiol. Plant.*, 57, 235, 2006.
66. Szalontai, B. et al., Membrane dynamics as seen by Fourier transform infrared spectroscopy in a cyanobacterium, *Synechocystis* PCC 6803: The effects of lipid unsaturation and the protein-to-lipid ratio, *Biochim. Biophys. Acta*, 1509, 409, 2000.
67. Los, D.A. et al., Stress responses in *Synechocystis*: Regulated genes and regulatory systems, in *The Cyanobacteria: Molecular Biology, Genomics and Evolution*, Herrero, A., and Flores, E., Eds., Caister Academic Press, Norfolk, VA, 2008, p. 117.
68. Morici, L.A., Frisk, A., and Schurr, M.J., Two component regulatory systems, in *Molecular Paradigms of Infectious Disease Emerging Infectious Diseases of the 21st Century,* Nickerson, C.A., and Schurr, M.J., Eds., Springer, New York, 2006, p. 502.
69. Novikova, I., Moshkov, E., and Los D.A., Protein sensors and transducers of cold and osmotic stress in cyanobacteria and plants, *Mol. Biol.*, 41, 427, 2007.

70. Zhang, W. and Shi, L., Distribution and evolution of multiple-step phosphorelay in prokaryotes: Lateral domain recruitment involved in the formation of hybrid-type histidine kinases, *Microbiology*, 151, 2159, 2005.
71. Suzuki, I. et al., The pathway for perception and transduction of low-temperature signals in *Synechocystis*, *EMBO J.*, 19, 1327, 2000.
72. Park, H., Saha, S.K., and Inouye, M., Two-domain reconstitution of a functional protein histidine kinase, *Proc. Natl Acad. Sci. USA*, 95, 6728, 1998.
73. Sakamoto, T. and Murata, N., Regulation of the desaturation of fatty acids and its role in tolerance to cold and salt stress, *Curr. Opin. Microbiol.*, 5, 206, 2002.
74. Wang, J.C. and Lynch, A.S., Transcription and DNA supercoiling, *Curr. Opin. Genet. Dev.*, 3, 764, 1993.
75. Weinstein-Fischer, D., Elgrably-Weiss, M., and Altuvia, S., Escherichia coli response to hydrogen peroxide: A role for DNA supercoiling, topoisomerase I and Fis, *Mol. Microbiol.*, 35, 1413, 2000.
76. Prakash, J.S.S. et al., DNA supercoiling regulates the stress-inducible expression of genes in *Synechocystis* sp. PCC 6803, *Mol. Biosyst.*, 5, 1904, 2009.
77. Los, D.A., The effect of low-temperature-induced DNA supercoiling on the expression of the desaturase genes in *Synechocystis*, *Cell Mol. Biol.*, 50, 605, 2004.
78. Grau, R. et al., DNA supercoiling and thermal regulation of unsaturated fatty acid synthesis in *Bacillus subtilis*, *Mol. Microbiol.*, 11, 933, 1994.
79. Hurme, R. and Rhen, M., Temperature sensing in bacterial gene regulation—What it all boils down to, *Mol. Microbiol.*, 30, 1, 1998.
80. Cheung, K.J. et al., A microarray-based antibiotic screen identifies a regulatory role for supercoiling in the osmotic stress response of *Escherichia coli*, *Genome Res.*, 13206, 2003.
81. Conter, A., Menchon, C., and Gutierrez, C., Role of DNA supercoiling and *rpoS* sigma factor in the osmotic and growth phase-dependent induction of the gene *osmE* of *Escherichia coli* K12, *J. Mol. Biol.*, 273, 75, 1997.
82. Andersen, J. and Delihas, N., micF RNA binds to the 5 end of ompF mRNA and to a protein from *Escherichia coli*, *Biochemistry*, 29, 9249, 1990.
83. Lease, R.A. and Belfort, M., A trans-acting RNA as a control switch in Escherichia coli: DsrA modulates function by forming alternative structures, *Proc. Natl. Acad. Sci. USA*, 97, 9919, 2000.
84. Narberhaus, F., Waldminghaus, T., and Chowdhury, S., RNA thermometers, *FEMS Microbiol. Rev.*, 20, 1, 2005.
85. Sledjeski, D.D., Gupta, A., and Gottesman, S., The small RNA, DsrA, is essential for the low temperature expression of RpoS during exponential growth in *Escherichia coli*, *EMBO J.*, 15, 3993, 1996.
86. Majdalani, N. et al., DsrA RNA regulates translation of RpoS message by an anti-antisense mechanism independent of its action as an antisilencer of transcription, *Proc. Natl. Acad. Sci. USA*, 95, 12462, 1998.
87. Shivaji, S. and Prakash, J.S.S., How do bacteria sense and respond to low temperature? *Arch. Microbiol.*, 192, 85, 2010.

8 Salt Toxicity and Survival Strategies of Cyanobacteria

*Poonam Bhargava and Ashish Kumar Srivastava**

CONTENTS

8.1 INTRODUCTION

Ever-increasing population in developing countries is exerting negative effects on ecological balance in general and agriculture in particular because of the requirement of more food for sustainability. Use of different fertilizers and ill-managed irrigation practices year after year results in decrease in soil fertility and crop productivity [1]. Of the various abiotic stresses, salinity is of prime concern due to its injurious effects and wide-spreading nature. Salinity may be defined as increase in the total salt concentration, mainly chlorides and sulfates of sodium, calcium, and magnesium in the soil. However, sodium constitutes the major component of soil salinity [2]. About half of the total irrigated land world over is affected by salinity, but this problem is more confined to the tropical part of the world [3]. This is due to the fact that the major part of agricultural lands is located in tropical countries that are already overburdened with population.

Approximately half of the world population consumes rice as staple food and, therefore, greater portion of total agricultural land is currently used for rice production. Frequent canal irrigation and water logging make the conditions suitable for the spreading of soil salinity in India. The saline soil is mainly characterized by high electrical conductivity and sodium absorption ratio [4]. Decrease in soil fertility as a consequence of salinity adversely affects the crops and soil microflora including cyanobacteria.

Evolution of cyanobacteria around 3.5 billion years ago created major transition in the history of life due to initiation of oxygenic photosynthesis [5–7]. Their enormous adaptability to diverse environment made them potential candidates to occupy almost every ecological niche ranging from fresh to marine environment and cold to hot springs [8]. Further, vast diversity of cyanobacteria may be due to their genome plasticity [9]. It is worth stating that cyanobacterial morphology varies from unicellular to filamentous, unbranched to branched, and nonheterocystous to heterocystous [10,11]. Among the various habitat, cyanobacteria are the dominant inhabitant of the paddy fields where

* Corresponding author: ashish.mzu@gmail.com

they contribute significantly to the nitrogen content of soil [12]. Thus, cyanobacteria are the only group of organisms in the entire living system that has both oxygen-evolving photosynthetic and biological nitrogen fixing capability, though both are spatially separated. Their cosmopolitan distribution, prokaryotic nature, and close resemblance with higher plants make cyanobacteria excellent candidate for understanding stress management [13].

Salinity is known to inhibit almost every process of cyanobacteria including growth, photosynthesis, cellular homeostasis, and osmotic pressure of cell [14–16], which results in the production of reactive oxygen species (ROS) [17,18], change in gene expression, and protein synthesis [19–21], leading to programmed cell death (PCD) [22]. Zhu [3] reported that salinity causes ionic and osmotic stress as a primary effect and oxidative stress as a secondary effect. Osmotic stress is more of a physical stress and occurs when the concentration of molecules in solution outside of the cell is different than that inside the cell. When this happens, water flows either into or out of the cell by osmosis, thereby altering the intracellular environment. Thus, osmotic stress is one of the major components of salt stress.

Critical perusal of literature suggests that a number of reviews related to cyanobacterial response to salt stress are available [23–31]. Unfortunately most of them are focused on one or the other aspect related to physiology, biochemistry, and molecular biology of salt tolerance in cyanobacteria. The present review will focus on the different aspects of salinity toxicity and tolerance offered by cyanobacteria at physiological, biochemical, and molecular level.

8.2 SODIUM EXTRUSION AND EFFECTS OF SALINITY ON PLASMA MEMBRANE AND CYTOPLASM

The disturbance in intracellular Na^+ concentration leads to alteration in metabolism due to its vital role in different physiological processes such as growth, uptake system, water photolysis by PS II, activation of nitrogenase, and internal pH regulation [32]. An increase in intracellular Na^+ occurs just after onset of salinity, mainly NaCl stress [17]. This increase results in loss of intracellular K^+ from the cell, which causes various pleiotropic effects because it has a major role in cellular homeostasis, maintenance of cell turgor, and regulation of expression of certain osmoresponsive genes [33]. The main route for Na^+ influx is diffusion [34]. However, probably an H^+/Na^+ and Na^+/K^+ antiporter is involved in Na^+ efflux from the cell using the proton gradient across the plasma membrane formed by cytochrome oxidase and/or by hydrolysis of ATP [35]. The required ATP pool is constantly maintained by the activity of F_0/F_1-ATPase during photo- and oxidative phosphorylation. The rate of ATP hydrolysis is maintained constant in thylakoid, but under high salt concentration it increases by 5- to 20-fold (depending on the strains) in the plasma membrane of the cells [36].

The role of P-ATPase in active extrusion of Na^+ from cyanobacterial cells is also established [37]. Further, the aa3-type cytochrome oxidase in plasma membrane and in thylakoids of cyanobacteria is also involved in Na^+ extrusion from the cell to maintain cellular homeostasis. In spite of the remarkable differences in the salt tolerance in cyanobacteria, the basic mechanism of Na^+ transport appears to be identical [32]. However, they differ in the rates and magnitude of influx and efflux, affinity of their carrier for Na^+ and Na^+ accumulation capacity.

Divalent calcium (Ca^{2+}) also plays a very important role in salt adaptation. It is now a well-established fact that the onset of salinity enhances the intracellular Ca^{2+} level. Torrecilla et al. [38] reported the salt and osmotic stress-dependent transient increase in intracellular Ca^{2+} in *Anabaena* using aequorin. Liu and Zhu [39] observed that extracellular Ca^{2+} supply reduces the ill effects of salinity probably by facilitating K^+/Na^+ selectivity. This finding provides evidence that Ca^{2+} signaling may be involved in early sensing of stresses including salinity. Further, primary and secondary transporters as well as ion channels (Ca^{2+}, K^+, Na^+) are involved in calcium transport mechanism in several genera of bacteria [40,41]. Like eukaryotes, cyanobacteria also have Ca^{2+}-ATPases and ion channels [42,43], which may be involved in transient Ca^{2+} accumulation.

Along with the ion channels and ATPases, changes in membrane fluidity vis-à-vis lipid composition also help in salt adaptation. Salt stress is known to cause desaturation of the fatty acids of membrane lipids and thus alteration in membrane fluidity [44]. Cyanobacteria have significant amount of lipids (fats and oils), which are rich in essential fatty acids such as C_{18} linoleic (18:2ω6) and γ-linolenic acid (18:3ω3) acids and their C_{20} derivatives, eicosapentaenoic acid (20:5ω3) and arachidonic acid (20:4ω6) [44].

Nonetheless, cyanobacteria contain lipids and fatty acids as membrane components, storage products, metabolites, and sources of energy [44]. The lipid content is significantly increased under different stressful conditions [44]. Salinity causes lipid peroxidation [17], which can be divided into three steps: (1) initiation, (2) propagation, and (3) termination. The initiation step is considered as rate limiting and involves the activation of molecular oxygen to produce ROS, which can be accelerated under different stresses [17,45–47]. Once the process is initiated, the other two steps of propagation and termination take place readily, thereby damaging the membrane [48]. The most deleterious effect of lipid peroxidation is disturbance in ion flow across the membrane and therefore cellular homeostasis [49]. However, it is not clear if peroxidation is considered as a cause of membrane damage and metabolic disorder or a secondary effect of these processes. Nevertheless the lipid peroxidation is also considered to be an indirect parameter to measure the oxidative stress by several workers [50,51].

Many reports have suggested that lipids may be involved in the protection against salt stress [44,52–54]. When photosynthetic organisms are exposed to salt stress, the fatty acids of membrane lipids are desaturated. The enhanced tolerance could be due both to the increased resistance of the photosynthetic machinery to the salt-induced damage and to the increased ability of *desA*$^+$ cells to repair the photosynthetic and Na^+/H^+ antiport systems [55]. Allakhverdiev et al. [55] reported that the increase in unsaturation of the fatty acids of membrane lipids enhances the salt tolerance of the photosynthetic apparatus of *Synechococcus*. Not only this, K^+/Na^+ channels and water channels located in the plasma membrane may be depressed due to unsaturation of the membrane lipids. Likewise, the activity of the Na^+/H^+ antiporter system consisting of the Na^+/H^+ antiporter(s) and H^+-ATPase(s) is also enhanced by increased fluidity of the membrane due to salt stress resulting in the protection of PS I and PS II activities as reported by Padan and Schuldiner [56]. The activities of several membrane bound enzymes are also affected by altered membrane fluidity [57,58].

Changes in lipid composition include increase in the proportion of unsaturated fatty acids and phosphatydyl glycerides in the mildly halotolerant *Synechococcus* PCC 6311 [52]. Likewise, change in the proportion of mono/digalactosyl diacyl glycerols, the main fatty acid of plasma membrane of cyanobacteria, was observed by Ritter and Yopp [54]. This change in polar lipids is responsible for the change in the ion exchange properties of plasma membrane [14]. Further, decrease in protein to lipid ratio leads to increased cytochrome oxidase and H^+/Na^+ antiport activities in *Synechococcus* PCC 6301, in which salt stress results in preferential increase of plasma membrane rather than thylakoid located respiratory activities [36].

Alterations in cytoplasm depend on the concentration and duration of salt stress as studied in freshwater cyanobacterium *Synechococcus* PCC6311 [59]. The cyanobacterium depicted a significant change in internal cell organization such as decrease in cytoplasm density, complete loss of cellular granules (glycogen and carboxysomes), and modification of the appearance of DNA. The rapid decomposition of cellular granules is necessary to cope with the salt-induced reduction in nutrient uptake to supply carbon and nitrogen molecule for the metabolic substrate and for osmoregulation. Nonetheless, salinity is also found to impair the cell division in cyanobacteria; as a consequence the cell size increases [60].

8.3 SALINITY SELECTIVELY INHIBITS PHOTOSYNTHESIS

Cyanobacterial photosynthesis is mainly confined to the four multi-protein complexes embedded in thylakoid membrane present in cytoplasm, namely, photosystem I (PS I), photosystem II (PS II), cytochrome b/f complex, and ATP synthase, which are responsible for photosynthetic electron transport. Further, plastoquinone (PQ) and plastocyanin (PC) function as mobile electron carriers.

The PS II involved in the noncyclic electron transfer is responsible for the production of NADPH, whereas PS I mediates cyclic electron transfer and is involved in ATP generation [61]. The electron is generated by photolysis of water through water oxidizing cycle, present in close proximity to PS II.

The importance of sodium in the photosynthetic process cannot be denied [62] but excess is highly deleterious [63–65]. The beneficial effects can be evident from Na^+-deficient *Anabaena torulosa* and deleterious effects from hyper-accumulation of Na+, both of which showed loss of photosynthetic pigment and severe inhibition of oxygen evolution [66]. Change in chlorophyll content depends on the nature, habitat, and morphology of cyanobacteria. Salinity-induced decrease in chlorophyll content of *Anabaena doliolum* was reported by Singh and Kshatriya [16] and Srivastava et al. [17]. Contrary to this, chlorophyll content of *Spirulina platensis* enhanced after salt stress [67].

Phycobiliproteins, responsible for light harvesting in cyanobacteria, are also sensitive to salt stress. In contrast to other photosynthetic pigments, carotenoid content enhances under salt stress and provides protection to chlorophyll due to absorption of photon as well as its antioxidative properties, which is further discussed in Section 8.4.

The prime target of salt stress appears to be the PS II reaction center in general and its 34 kDa Dl protein in particular [65]. The decrease in PS II activity may be attributed to salinity-induced increase in the number of Q_B nonreducing reaction centers as reported by Lu and Vonshak [67] in *S. platensis*. However, this effect was found to be more pronounced if the salt treated cells were exposed to PAR (photosynthetically active radiation) but no significant effect was observed if kept in dark [64]. The photochemical response of PS II in *S. platensis* to salt stress exhibits two distinct phases: the first phase is independent of light and characterized by a rapid decline in PS II activity followed by a subsequent recovery, and the second phase is denoted by a progressive decrease in PS II activity, which is exposure time dependent. The first phase response may be due to the downregulation of the light harvesting capacity mainly because of significant loss of the major light harvesting pigment phycocyanin. In addition to this, the mechanism of excess excitation energy dissipation by the state transition in salt stressed cells does not appear to be enough to prevent the PS II damage during the second phase.

Notwithstanding the aforementioned, salt stress induces the PS I activity specifically [66] and causes increase in fluorescence emission profile, indicating an increase in the PS I/PS II ratio. This may be either due to salinity-induced downregulation of PS II synthesis or upregulation of PS I reaction center. However, the change in PS I electron transfer capacity is reported to be faster than P_{700} content adjustment under salt stress [63]. This indicates an activated cyclic electron flow through pre-existing PS I centers during the adaptation period in *Synechocystis* PCC6803, resulting in delay in synthesis of new PS I. Therefore, PS I should not be considered a limiting factor for the electron flow in cells. Not only the PS I activity enhances under salt stress, but the activity of complex IV (ATP synthase) also increases, which suggest the involvement of the PS I reaction center and components other than cytochrome c oxidase. The enhanced ATP synthase activity can be attributed to the over demand of ATP under salt stress to increase the active extrusion of Na^+ and restoration of cellular homeostasis. In view of the sensitivity of PS II and enhanced PS I under salt stress, the alteration of normal photosynthetic electron transfer is well anticipated. An alternate electron transfer route from PQ to cytochrome C_{553} via PC is reported in cyanobacteria [68]. Decreased PS II vis-à-vis water oxidizing cycle and enhanced PS I activities suggest the possible involvement of an alternate electron source for PS I activity. Garlick et al. [69] have reported that cyanobacteria may take an electron from hydrogen, hydrogen sulfide, sodium sulfide, and NADH to meet the requirement of PS I to operate the cyclic electron flow.

The dark reaction of photosynthesis is also affected by salt stress. Moisander et al. [70] suggested that carbon assimilation is suppressed by salt stress. It has been demonstrated that the uptake of C^{14} highly reduces under salt stress [21]. Since cyanobacteria are oxygen-evolving photosynthetic prokaryotes, their type I RuBisCO plays a key role in depicting stress-specific responses due to its both carboxylase and oxygenase activities [71]. Cyanobacteria have evolved a mechanism of carbon concentration for improving the carboxylation by their relatively insufficient RuBisCO [72]. However, under conditions of carbon limitation as in salt stress conditions, the oxygenase activity prevails over the normal carboxylase. Sivakumar et al. [73] have reported salinity-induced oxygenase activity of RuBisCO in *Sesbania*. This leads to the operation of photorespiratory glycolate pathway (C2 cycle), which has not only been experimentally demonstrated in different cyanobacteria [74,75] but characterized in *Synechocystis* PCC6803 at the molecular level [76]. Salinity-induced glycolate metabolism has been studied by Srivastava et al. [21] in *Anabaena doliolum*. Details of this pathway are discussed later in the review.

8.4 SALINITY-INDUCED OXIDATIVE STRESS AND ANTIOXIDATIVE DEFENSE SYSTEM

Being photosynthetic prokaryotes, cyanobacterial metabolism exploits the potential of interaction with oxygen; photosynthesis in general and electron transfer chain in particular. Therefore, generation of ROS during photosynthesis is a normal phenomenon but any stress condition provokes increased production of toxic oxygen derivatives [77]. Therefore, photo-oxidative damage should not be considered independently but as a part of the overall response of the cells to stress. Under moderate concentration, these ROS have several important functions such as lignification of the cell wall of higher plants, defense against pathogen, sensing and adaptation to stress conditions, and induction of apoptosis by modulating gene expression [78]. The oxidative burst is mainly characterized by hyper-accumulation of various ROS such as superoxide anion $\left(O_2^{-1}\right)$, hydroxyl (OH^-), and oxygen free radical ($O^\cdot$). Among the various ROS, superoxide is first produced at chlorophyll pigments associated with electron transport system [79]. The lifetime of excited chlorophyll singlet state is short but depends on the physiological conditions of the cell. The excited chlorophyll is used for energy transfer in the form of electron to other acceptor of the light reaction. Despite the aforementioned, other two possible routes of decay are radiative decay and conversion to the triplet chlorophyll state [80,81]. However, the latter can transfer its energy to the ground state triplet molecular oxygen to produce superoxide anion. Further, there are other possible sites for generations of superoxide anion at the reducing site of PS I [82]. The reduced ferredoxin, which reduces molecular oxygen to the superoxide radical and H_2O_2, is then formed through dismutation of O_2^{-1} [83,84]. Salt stress is known to enhance the generation of ROS by deviating the normal process of photosynthesis and increasing the aforementioned reactions [17,85]. Further, salinity-induced inhibition of the rate of carbon assimilation [86] results in decrease in the regulation capacity of the photosynthetic electron transport system [87]. However, the production of ROS will increase if the thermal energy dissipation energy is lower than energy intercepted by the photosynthetic pigments [80,81].

Further, the alteration in photosynthetic electron flow and decreased PS II activity result in enhanced production of ROS, therefore, oxidative burst in the cells. The oxidative burst occurs when the production of ROS exceeds to the scavenging by the antioxidative defense system. This complex array of antioxidative defense system has both enzymatic and nonenzymatic counterparts. The enzymatic antioxidative part consists of superoxide dismutase (SOD), catalase (CAT), ascorbate peroxidase (APX), and glutathione reductase (GR), whereas nonenzymatic one has carotenoid (CAR), ascorbate (ASA), glutathione reduced (GSH), and α-tocopherol (α-TOC). Both the counterparts work in a cooperative manner, and the detoxification cycle is known as the Halliwell–Asada pathway. This system not only controls the steady state level of ROS but also allows them to perform important functions at specific sites under a variety of environmental conditions and at different developmental stages of the organism [17]. These enzymes not only involve in scavenging of ROS but also participate in repair mechanism. SOD is the first enzyme that dismutates the O_2^{-1} into H_2O_2, which detoxifies into water and oxygen molecule by the activities of CAT and APX. However, APX activity is coordinated with GR activity in the Halliwell–Asada pathway. Three types of SOD reported in cyanobacteria *Tolypothrix, Anabaena,* and *Lyngbya* are Mn-SOD, Fe-SOD, and Hy-SOD (Fe-Mn-SOD), respectively. It is worth stating that CAT is not a robust enzyme in *Anabaena doliolum* because it is susceptible to photo-inactivation and degradation [88]. Likewise, in *Tolypothrix*, it showed sensitivity to salt stress but was found higher under osmotic stress. Therefore, under such situation APX emerges as an H_2O_2 scavenging enzyme in cyanobacteria. In contrast to the aforementioned, there is a division in the function of CAT and APX in higher plants, where the CAT is essential for the destruction of the photorespiratory H_2O_2 in C_3 plants, and APX scavenges H_2O_2 in chloroplast (which do not contain CAT) [80,81].

Unfortunately, no enzyme has been reported so far for the scavenging of the singlet oxygen and triplet chlorophyll [89]. Therefore, apart from the enzymatic antioxidants; nonenzymatic antioxidants also have a significant role. Among the various nonenzymatic antioxidants, CAR has been vividly studied, which is universally present in all organisms from bacteria to human [90]. One of the CARs, β-carotene, is known to protect the light harvesting pigments in the antenna complexes against photochemical damage caused by excited triplet state via triplet–triplet energy transfer from chlorophyll to carotenoid [91] and other ROS [17,18]. The efficiency of the CAR depends on the number of the conjugated double bonds [92]. On the other hand, α-TOC is a potent scavenger of H_2O_2 and OH^{-1} radicals and/or prevents lipid peroxidation [93,94]. Further, the low molecular weight ASA has the ability to scavenge a wide range of ROS, e.g., superoxide anion, singlet oxygen, and peroxide. It is also a chain breaking antioxidant [95] and capable of regenerating α-TOC from the tocopheroxyl radical bound to the cell membrane [96]. On the other hand, GSH not only helps in maintaining the redox state, but is also a precursor for biosynthesis of proline, a potent osmoprotectant and antioxidant under salt stress [97].

8.5 SALINITY SELECTIVELY AFFECTS RESPIRATION AND METABOLISM

The site of respiration in general and respiratory electron transfer in particular are the cytoplasmic and thylakoid membranes. The relative distribution of respiratory electron transport capacity in these two membranes depends on the growth conditions. Salt stress may induce respiration activity in either one or both sites. Further, under salt stress *Synechocystis* PCC 6803 showed an increase in dark respiration without any change in the electron transport through complex I and II, but increased complex IV activity. Further, in view of the location of the sites of photosynthetic and respiratory electron transport, the P700 content and respiratory enzyme consumption are jointly regulated. The photosynthetic and reparatory electron flow share numerous electron transport intermediates. This can also be explained in the light of the involvement of a common NADH dehydrogenase as studied in mutated *Synechocystis* PCC6803 defective in this enzyme. Electron from ferredoxin can also enter the transport system possibly passing via the same complex(es) that catalyze NADPH oxidation [98]. Sodium is one of the essential nutrients for the metabolism of cyanobacteria. It plays an important role in nitrogen metabolism [63]. The cyanobacterial nitrogenase activity depends on Na^+ concentration, as it has a stabilizing effect. This can be due to the influence of Na^+ on the transport of certain cations such as Ca^{2+}, which is essential for heterocyst differentiation, and anions like phosphate, amino acid, and sugars. However, the synthesis of nitrogenase is free from Na^+ requirement [99,100]. Effect of high NaCl (~2%) on nitrogenase activity is not universal. It has been demonstrated that nitrogenase activity enhances under salt stress in *Anabaena doliolum* to meet the requirement of the alternate electron source in the form of H_2, produced due to the activity of nitrogenase–hydrogenase complex for PS I activity. In contrast to this, high salinity was found to inhibit the nitrogenase activity of *Anabaena azollae* [101] and *Nostoc muscorum* [102]. Further, Fernandes et al. [103] reported the NaCl-induced inhibition of nitrogenase activity of *Anabaena torulosa* and *Anabaena* strain L-31. They further reported that the deleterious effects of salinity stress on nitrogen fixation are, therefore, a consequence of the ionic component rather than the osmotic component. It has been demonstrated that nitrogenase activity is more sensitive to salt stress than photosynthesis [62,104–106], because fixed carbon is generally utilized in the synthesis of osmoprotectants leading to the less availability of carbon as energy source for nitrogenase. Intracellular phosphate and phosphate uptake both were found to be reduced significantly with decrease in the assimilatory enzyme alkaline phosphatase under salt stressed *Anabaena doliolum* [107–108]. Thus we can see that almost all the major biochemical processes including photosynthesis and respiration are affected but salinity.

8.6 ACCUMULATION OF COMPATIBLE SOLUTES IN SALT STRESS

The accumulation of nontoxic organic compounds also known as osmoprotectant or compatible solutes is one of the foremost tolerance mechanisms adopted by cyanobacteria [109–112]. These solutes help to prevent the water loss and plasmolysis and therefore maintain the osmoticum of the cell. Sucrose [113], trehalose [14], glucosylglycerol (GG) [114], glutamate, and glycine betaine [111,115] are the common osmoprotectants present in cyanobacteria. Unlike eubacteria, cyanobacteria synthesize all their osmotica from the environment. The photosynthetic carbon fixation, primary source of carbon, and endogenous glycogen [116] are used for the synthesis of compatible solutes. Sometimes, exogenous glucose is also used as an alternative carbon source under dark conditions, but allows only a limited synthesis [14]. The accumulation of osmoprotectant substances is immediate just after the exposure to salt stress. A hyper-saline condition causes reversible and rapid extrusion of these solutes into the medium [117]; of these the carbon thus metabolized can be re-incorporated into glycogen. Among the various compatible solutes, GG is extensively studied in cyanobacteria. Mikkat et al. [118] have described a system able to transport (GG) actively in *Synechocystis* PCC 6803. In addition to the aforesaid, Csonka and Hanson [119] reported the synthetic pathway of glycine betaine in *Aphanothece halophytica*.

There is no universal osmoprotectant accumulation in all cyanobacteria. GG is the main solute in *Synechocystis* PCC 6803 [120] and PCC 6714, while sucrose and trehalose in *Synechococcus* PCC 6301 and *Anabaena*, and glycine/glutamatebetaine in *Spirulina* and *Aphanothece*. Likewise, trehalose forms a major intracellular osmoticum in cells of *Rivularia atra* and *R. bullata* and sucrose in *Nostoc muscorum*. The enzymes involved in GG biosynthesis in *Synechocystis* PCC 6803 are GG-phosphate-synthase and GG-phosphatase [121]. However, the control point of synthesis is not well characterized, but it is assumed that these are synthesized inactively and salt stress reversibly activate then *in vitro*. However, *in vivo* activation does not require *de novo* protein synthesis [14]. Further, two enzyme activities involved in glycine betaine synthesis in plants have been demonstrated in cyanobacteria. These enzymes are betaine-aldehyde dehydrogenase responsible for choline oxidation and *S*-adenosylhomocysteine hydrolase involved in *S*-adenosyl-cysteine conversion in *Spirulina subsalsa* [122] and *Aphanothece halophytica* [123], respectively. Page-Sharp et al. [124] have reported the accumulation of trehalose and sucrose in *Scytonema* in response to salt stress. However, for the first time the reported trehalose phosphorylase in cyanobacteria was the only catabolic enzyme detected for trehalose; neither trehalase nor phosphotrehalase activities could be detected. Interestingly, the release of fixed carbon in response to salinity shock may be involved in the development of lichen symbioses, where carbohydrates are transferred from photosynthetically active algal cells to fungal counterpart [104]. Figure 8.1 summarizes the physiological and biochemical changes in cyanobacteria trying to adapt to salt stress.

8.7 SALINITY-INDUCED GENES AND SIGNAL TRANSDUCTION

Stress-induced genes can be defined as those genes whose expression increases or is induced only under unfavorable environmental conditions [121,125]. Out of a large number of salt responsive genes, some of them are induced very early and remain induced only for a short time. These genes have been thought to trigger the actual response to salt stress. Induction of salt responsive genes is one of the very complex mechanisms. One of the reasons for this complexity is the number of genes involved. In fact, transcription of almost 10% of the entire genome of *Anabaena torulosa* is reported to be regulated by changes in salinity [126].

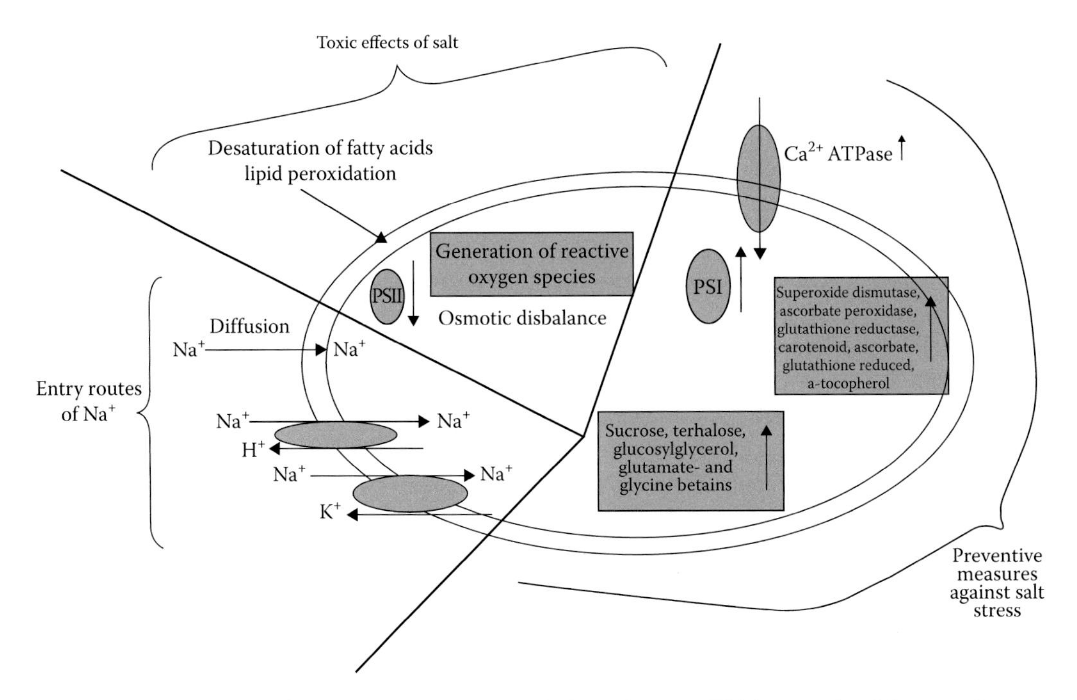

FIGURE 8.1 Physiological and biochemical changes during salt stress. Thick upward and downward arrows indicate increase and decrease, respectively.

Microarrays have been the method of choice for deciphering stress-induced gene expression. Salt is known to increase expression of certain ribosomal protein genes such as *prs* and *ftsh* [125]. As mentioned earlier, salt stress induces the antioxidative defense system and thus the genes for heat shock proteins (*hspA*, *dnaK*, *dnaJ*, *htrA*, *groEL2*, and *clpB*) and superoxide dismutase (*sodB*) are also expressed. Similarly the genes to maintain the osmoticum such as those involved in biosynthesis of the osmoprotector GG (*glpD* and *ggpS*), and the σ70 factor (two *rpoD* genes) are also induced. Further activation of the genes that code for proteases (*HtrA* and *ClpB*) and chaperone proteins (*DnaK*, *DnaJ*, and *GroEL*) suggests that negative effects of salt and hyperosmotic stresses destabilize proteins, as in heat shock.

Vinnemeier and Hagemann [127] used the subtractive RNA hybridization procedure to identify genes transcribed preferentially under salt stress in the cyanobacterium *Synechocystis* sp. PCC 6803. Apart from the aforementioned genes they identified *cpn60* (encoding GroEL; a molecular chaperone) and *isiA* (encoding a chlorophyll-binding protein), genes encoding a protein of unknown function (*slr0082*) and a putative RNA helicase (*slr0083*) as salt-regulated. Similarly expression of several other genes including the *ggpS* gene for glucosylglycerolphosphate synthase [128], the *crh* gene for RNA helicase [127], the *isiA* gene for iron-stress-inducible protein A [129], the *isiB* gene for flavodoxin and the *petH* gene for ferredoxin: NADP1 reductase [98] has been reported. Apart from the known involvement of unknowns cannot be underestimated. Kanesaki et al. [130] have reported regulation of more than 30 unknowns (genes with unknown function) by salt stress.

A collection of 17 salt-sensitive mutants of the cyanobacterium *Synechocystis* sp. strain PCC6803 obtained by random cartridge mutagenesis by Karandashova et al. [131] revealed the essential genes involved in salt tolerance. They found that two genes coding for enzymes involved in biosynthesis of the osmolyte GG were affected in nine mutants, and two mutants defective in a glycoprotease encoding gene *gcp* showed a reduced salt resistance. Four genes (*slr*1799, *slr*1087, *sll*1061, and *sll*1062) were found to be essential for salt tolerance in cyanobacteria. These genes code for proteins not yet functionally characterized.

So we find that among the salt-induced genes there are (1) ribosomal protein genes (*rpl2, rpl3, rpl4,* and *rpl23)* suggesting destability of ribosomes and the requirement of the corresponding proteins L2, L3, and K$ for the maintenance of the activities of ribosomes [132,133]; (2) *FtsH*, a gene responsible for the degradation of photodamaged D1 protein of the photosynthetic pathway [134] and compensating for this degradation the enhanced expression of the *prc* and *ctpA* genes responsible for the synthesis of D1 protein; (3) *glpD* and *ggpS* genes for proteins involved in the synthesis of GG [134,135] to overcome the osmotic stress induced by salt; (4) expression of heat shock proteins and chaperones (*hspA*, *dnaK*, *dnaJ*, *htrA*, *groEL2*, and *clpB*) to avoid accumulation of denatured proteins; and (5) induction of *rpoD* genes (*sll2012* and *sll0306*) that encode σ 70 factors to compensate for the salt-induced depression of transcription, and of course there are many others along with those genes for which the function is not known.

Thus we see that induction of each and every gene has a role and is highly regulated. There are early responsive genes and late responsive genes. So the question that automatically arises is: What induces these genes or what senses the stress and sends signals for the particular gene to be expressed? In cyanobacteria this signaling pathway involved is similar to the *E. coli* two-component system. Serine/threonine protein kinases (Spks) and sigma factors of RNA polymerase are conserved as potential candidates of sensors and transducers of environmental signals. The study of this two-component system generally utilizes the technique of targeted mutagenesis. This along with microarray analysis has given way to in-depth knowledge of the signaling systems in cyanobacteria [136–138]. Synechocystis has been the cyanobacterium of choice to study the signaling pathway because of the availability of its complete genome sequence and it being naturally competent. It has been found that the *Synechocystis* includes 44 putative genes for *Hiks* [139] in its chromosome and three on its plasmids [140]. The second component that is the *Rres* gene is also present in similar numbers with 42 putative genes on the chromosome and three on the plasmids. The association of each *Hik* with its corresponding *Rres* has been a challenge for scientists. This is because these genes are scattered and not present in an orderly pair as in the case of *E. coli.* So to find out the partner for each *Hik* gene, several knockouts and the gene expression have to be tested.

In general, in all organisms there are two kinds of signals: (1) positive and (2) negative. In case of positive regulation, *Hik* via-à-vis *Rres* remain inactive under non-stress conditions. Under stress conditions, *Hik* becomes phosphorylated and activates *Rres*, which activates the genes responsible to combat stress that are otherwise silent. However, in the case of negative regulation, the Hik is active under non-stress conditions, and when the appropriate environmental stress is applied, the Hik and Rre become inactive. Genes that are expressed or repressed under non-stress conditions become silent or are released from repression, respectively. As a result, the levels of expression of genes decrease in the former case and appear to increase in the latter case [141].

Comprehensive analysis of the Synechocystis genome along with several knockout studies has demonstrated five two-component systems that participate in the perception and transduction of salt stress. DNA microarray of the *hik* mutant library upon exposure of the cells to salt (NaCl) stress revealed that the inducibility of gene expression by elevated levels of NaCl was significantly affected in *hik16* (*slr1805*), *hik33* (*sll0698*), *hik34* (*slr1285*), and *hik4* (*sll1229*) mutant cells [142]. Five Hik-Rre systems, namely, Hik33-Rre31, Hik10-Rre3, Hik16-Hik41-Rre17, Hik2-Rre1, and Hik34-Rre1, have been found that are involved in the perception of salt stress and transduction of the signal [138]. These sets of genes generally regulate different sets of genes. Typical examples are the Hik33-Rre31 and Hik34-Rre1 pairs mentioned earlier. The Hik33-Rre31 two-component system regulates the expression of *fabG* and *gloA* as hyperosmotic-stress-specific genes and that of *pgr5*, *nblA1,* and *nblA2* as oxidative-stress-specific genes. The Hik34-Rre1 system regulates the expression of *sll1107* as a salt-stress-specific gene and that of *htpG* as a hyperosmotic-stress-specific gene.

Recently Liang et al. [143] revealed a serine protein kinase G (*SpkG*) to be directly involved in salt stress signaling. They examined differential expression of four serine/threonine kinases, *SpkC*, *SpkD*, *SpkF*, and *SpkG*, at seven different stress conditions. They found that the transcriptional level was upregulated of *spkG* and downregulated of *spkC* under high-salt-stress conditions. The spk deletion mutants, *DspkC* and *DspkG*, were constructed and their growth characteristic showed that growth of *spkG* mutant was completely impaired under salt stress. They concluded that *SpkD* senses salt signal directly rather than mediating signals among other kinases.

The current concept of two-component systems does not explain how the Hik33-Rre31 system controls the differential expression of the two different sets of genes under different types of stress. It seems reasonable to assume the existence of some factor(s) that provide each two-component system with the strict specificity that is related to the specific nature of the stress. An example of such a factor might be a small polypeptide SipA (Ssl3451), which enhances the activity of *Hik33 in vivo* and *in vitro* [144].

8.8 SALT-INDUCED CHANGES IN PROTEOME

All the aforementioned salinity-induced changes in physiological and biochemical processes ultimately lead to changes at the molecular level, which includes alteration in protein profile and gene expression. Proteins are one of the most important targets of salinity [20] and therefore their degradation synergies are the deleterious effects of salt stress. The onset of salt stress results in the disturbance of cellular homeostasis, which damages the housekeeping proteins. The NaCl-induced oxidative damage is the foremost reason for the misfolding of proteins [145]. The prime site for ROS attack to proteins is the side chains of amino acids [146]. The oxidation of cysteine (SH) to cysteine (S-S), methionine to methionine sulfoxide, and formation of 3,4-dihydroxyphenylalanine (DOPA) and various hydroxyleucines from tyrosine and leucine, respectively, are some of the examples of protein damage by ROS. The small changes in primary structure of proteins may result in major changes in higher-order structures [147]. Covalent intermolecular cross-linking of proteins takes place due to the formation of disulfide bridges or dityrosine cross-links under severe stress conditions. The fragmentation of proteins because of direct disintegration (main chain scission and side-chain scission) mechanism except peptide bond hydrolysis is another pathway of protein damage.

However, the final line of salt tolerance is change in protein profile because all the physiological and biochemical changes result in alteration at the molecular level, which involves gene expression and protein profile. The physiological basis of cellular response to salt stress includes the alteration in protein profile [148]. A number of workers have selected the study of proteome for better understanding of stress response [20,149–154] because the fold upregulation of genes is not always in tune with the respective gene products, i.e., proteins, which is actually involved in stress tolerance [154]. Therefore, proteomics become a promising and complementary tool in post-genomic era. In addition to the aforementioned, functional gene expression can only be achieved by proteomic analysis, which also shows the stress-induced posttranslational modifications [153]. Genome sequencing followed by proteome analysis using two-dimensional gel electrophoresis combined with micro-sequencing to characterize expressed protein is the basis for the proteomics [155]. Our understanding of cellular events in different cyanobacteria including *Synechocystis* PCC 6803 and *Anabaena* PCC 7120 at the molecular level is greatly improved after the sequencing of their entire genome [142,156]. Several advances over the 2DE have emerged in the recent past, and worth mentioning is the iTRAQ. It is usually based on a peptide labeling strategy, where four or eight amine reactive isobaric reagents are used, enabling the comparison of several phenotypes in one experiment (including technical and biological replicates). The quantification occurs at the MS/MS stage, and thus requires a tandem mass spectrometer [157]. MS/MS leads to fragmentation of peptides into individual amino acids allowing far more accurate sample analyses [158]. Recently a eurohaline cyanobacterium from a salt lake in southern Libya, in the heart of the Sahara, has been isolated and iTRAQ was used to identify proteins responsible for salt adaptation [159].

Salinity, like other abiotic stresses, causes three types of changes in protein profile, viz., (1) expression of a group of proteins, (2) disappearance or suppression of another group of proteins, and (3) expression of a new set of proteins, commonly known as salt responsive proteins [19]. The change in protein profile may be divided into two categories: (1) early shock response, which occurs just after the onset of salt stress, and (2) late adoptive response, which occurs when the organism is exposed to salt stress for a long time. A good correspondence between proteomic and transcriptomic analysis exists for short-term salt shock as assessed by DNA microarray analysis [160]. However, data of proteomic and transcriptomic analysis differ for almost half of the proteins or genes in the long-term salt acclimatization response [161]. The basic reason behind the change in protein profile is salinity-induced dehydration of cells. One of the important proteins, which induce in cyanobacteria by salinity stress, is ubiquitin. Earlier it was thought that the ubiquitin is the characteristic feature of eukaryotes, but Durner and Böger [162] for the first time identified this protein in cyanobacterium *Anabaena* PCC 7120. This protein has an important role in cellular housekeeping because it provides an ATP-dependent proteolytic mechanism that specifically degrades abnormal proteins. Therefore, ubiquitin helps in degrading those proteins that lose the active conformation due to salinity. Not only this, other proteases are also involved in degradation of abnormal proteins to release amino acids for reutilization of these for protein synthesis. This concept is also termed "Proteolysis secondary antioxidant defense" [163,164]. Although not unique to the salt stress response, proteolytic enzymes are considered the secondary line of defense for the antioxidant defense system.

Further, the proteins, which respond to salinity stress, are of four types: (1) general stress proteins, which can respond to any stress condition, and (2) salt-stress-specific proteins such as enzymes involved in synthesis of the compatible solutes, (3) enzymes of general metabolism, and (4) hypothetical proteins with unknown function. Some general stress proteins [129,165] such as Dna K or Hsp 70 [152] and Gro EL or Hsp 60 [161,166] have been reported to induce in response to salt stress. These stress proteins are also known as molecular chaperons because they assist housekeeping proteins to achieve their normal conformation. These proteins not only work under stress conditions but also have function to assist protein targeting in cells under normal conditions. Further, these proteins are much conserved in nature so they also have some evolutionary importance [167,168]. It is worth stating that knowledge of the salt stress proteins is still far less than expected. This may be due to the fact that about 59% of the upregulated genes encode hypothetical proteins, which are not functionally characterized [161]. Two proteins of molecular weight 40 and 67 kDa have been reported in *Anabaena* sp. [169]. Further, two periplasmic proteins of unknown function encoded by gene *slr0924* and *slr1485* were accumulated under salt stress in *Synechocystis* PCC 6803 [152]. Recently Huang et al. [169] reported the salt-induced proteome of plasma membranes of Synechocystis PCC 6803. The ABC-transporter proteins with specific reference to glucosylglycerol-binding protein (GgtB), regulatory subunit of ATP-dependent Clp protease, the HhoA-protease, a HlyD protein, a protein belonging to the CheY superfamily and subunits of ATP synthase were found to increase by two-fold in salt acclimated cells. However, all the ABC-transporter proteins carry a putative lipoprotein anchor to the plasma membrane at their N-terminal end; however, in other gram-negative bacteria, substrate binding proteins are believed to be soluble proteins in the periplasmic space. The other proteins that enhanced under salt stress include, FutA2 and Vipp1, which are involved in the protection of PS I and II [170].

Fulda et al. [161] reported the hyper-accumulation of elongation factor-Tu (EF-Tu) in *Synechocystis* PCC6803 under salt stress, but the truncated product of EF-Tu may be its dominant existence in different variants in a cell due to the role in prokaryotic cytoskeleton [171]. However, whether these smaller proteins represent simple breakdown or have any biological significance is not yet clear. The DNA- and RNA-binding proteins are also known to induce under salt stress due to their role in stabilization of nucleic acid structure, which are influenced by salinity-induced change in water potential and ionic balance.

Figure 8.2 summarizes some of the important proteins up- and downregulated in response to salinity.

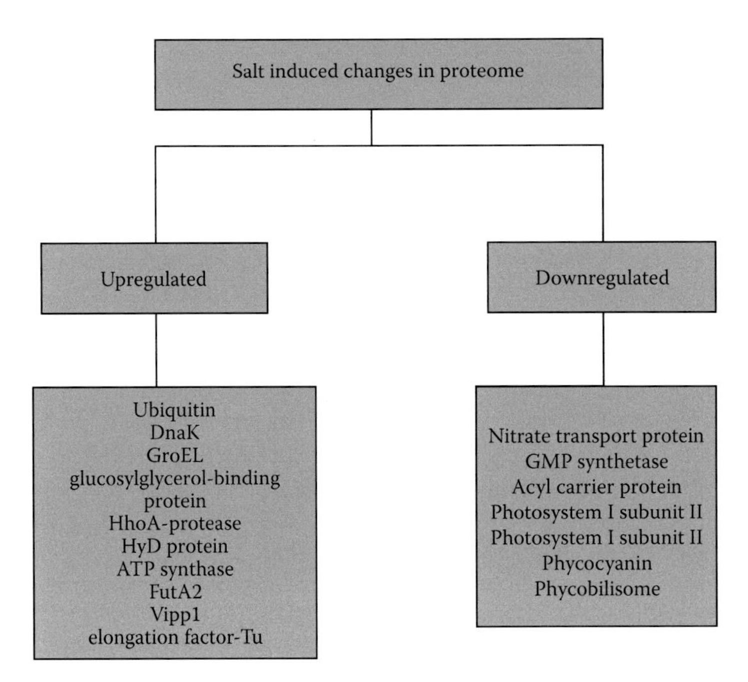

FIGURE 8.2 Some of the prominent proteins up- and downregulated during salt stress.

8.9 CONCLUSION

Salinity represents a major abiotic factor which hampers the growth of cyanobacteria. Apart from reducing the amount of available water, it increases the ionicity of the medium. Salt stress typically affects all the biochemical processes of cyanobacteria including photosynthesis, respiration, and metabolism. The basic mechanism of salt acclimation involves the active extrusion of toxic inorganic ions and the accumulation of compatible solutes, including sucrose, trehalose, GG, and glycine betaine. Apart from the physiological processes salt acclimation also involves changes in the transcriptional and translational levels. Recent "-omics" technologies using the advanced model *Synechocystis* 6803 have revealed a comprehensive picture of the dynamic process of salt acclimation involving the differential expression of hundreds of genes. However, the mechanisms involved in sensing specific salt stress signals are not well resolved. Further several genes reported in salt acclimation are hypothetical or unknown. In future these genes will need to be characterized for detailed and in-depth analysis of salinity-induced salt stress in cyanobacteria.

ACKNOWLEDGMENT

Poonam Bhargava is thankful to DST for research grant and also to the Director, Birla Institute of Scientific Research for facilities. Ashish Kumar Srivastava is thankful to District Magistrate, Tehri Garhwal for providing official facility.

REFERENCES

1. Garcia-Moya, E. and Castro, P.M. Saline grassland near Mexico City, in *Primary Productivity of Grass Ecosystems of the Tropics and Sub-Tropics*, Long, S.P., Jones, M.B., and Roberts, M.J., Eds., Chapman & Hall, London, U.K., 1992, p. 70.
2. Onkware, A.O. Effect of soil salinity on plant distribution and production at Lobura delta lake, Bogoria National Reserves, Kenya, *Austral Ecol.*, 25, 140, 2000.
3. Zhu, J.-K. Plant salt tolerance, *Trends Plant Sci.*, 6, 66, 2001.

4. Tanji, K.K. Nature and extent of agricultural salinity, in *Agricultural Salinity Assessment and Management*, Tanji, K.K. Ed., American Society of Civil Engineers, New York, 1990, p. 1.
5. Castenholz, R.W. Phylum BX: Cyanobacteria: Oxygenic photosynthetic bacteria, in *Bergey's Manual of Systematic Bacteriology*, 2nd edn. Vol. 1, Boone, D.R. and Castenholz, R.W., Eds., Springer, New York, 2001, p. 473.
6. Whitton, B.A. Diversity, ecology and taxonomy of the cyanobacteria, in *Photosynthetic Prokaryotes*, Mann, N.H. and Carr, N.G. Eds., Plenum Press, New York, 1992, p. 1.
7. Woese, C.R. Bacterial evolution, *Microbiol. Rev.*, 51, 221, 1987.
8. Schopf, J.W. The fossil record: Tracing the roots of the cyanobacterial lineage, in *The Ecology of Cyanobacteria*, Whitton, B.A. and Potts, M., Eds., Kluwer Academic Publishers, Dordrecht, the Netherlands, 2000, p. 13.
9. Roberts, T.M., Klotz, L.C., and Loeblich, A.R. Characterization of a blue-green algal genome, *J. Mol. Biol.*, 110, 341, 1977.
10. Desikachary, T.V. *Cyanophyta*. Indian Council of Agricultural Research, New Delhi, 1959, p. 686.
11. Rippka, R. et al. Generic assignments, strain histories and properties of pure cultures of cyanobacteria, *J. Gen. Microbiol.*, 111, 1, 1979.
12. Singh, R.N. *The Role of Blue Green Algae in Nitrogen Economy of Indian Agriculture*, Indian Council of Agricultural Research, New Delhi, 1961, p. 1.
13. Gray, M.W. and Doolittle, W.F. Has the endosymbiont hypothesis been proven? *Microbiol. Rev.*, 46, 1, 1982.
14. Joset, F., Jeanjean, R., and Hagemann, M. Dynamics of the response of cyanobacteria to salt stress: Deciphering the molecular events, *Physiol. Plant.*, 96, 738, 1996.
15. Oren, A. Salts and brines. *Ecology of Cyanobacteria: Their Diversity in Time and Space*, Whitton, B.A. and Potts, M. Eds., Kluwer Academic Publishers, Dordrecht, the Netherlands, 2000, p. 281.
16. Singh, D.P. and Kshatriya, K. NaCl-induced oxidative damage in the cyanobacterium *Anabaena doliolum*, *Curr. Microbiol.*, 44, 411, 2002.
17. Srivastava, A.K., Bhargava, P., and Rai, L.C. Salinity and copper-induced oxidative damage and changes in antioxidative defense system of *Anabaena doliolum*, *W. J. Microbiol. Biotechnol.*, 22, 1291, 2005.
18. Srivastava, A.K. et al. Effect of pretreatment of salt, copper and temperature on ultraviolet-B-induced antioxidants in diazotrophic cyanobacterium *Anabaena doliolum*, *J. Basic Microbiol.*, 46, 135, 2006.
19. Apte, S.K. and Bhagwat, A.A. Salinity stress-induced proteins in two nitrogen-fixing *Anabaena* strains differentially tolerant to salt, *J. Bacteriol.*, 171, 909, 1989.
20. Fulda, S. et al. Proteomics of *Synechocystis* sp. strain PCC 6803 identification of periplasmic proteins in cells grown at low and high salt concentrations, *Eur. J. Biochem.*, 267, 5900, 2000.
21. Srivastava, A.K. et al. A physiological and proteomic analysis of salinity induced changes in *Anabaena doliolum*, *Environ. Exp. Bot.*, 64, 49, 2008.
22. Ning, S.B. et al. Salt stress induced programmed cell death in prokaryotic organism, *Anabaena*, *J. Appl. Microbiol.*, 93, 15, 2002.
23. Gimmler, H. Primary sodium plasma membrane ATPases in salt-tolerant algae: Facts and fictions, *J. Exp. Bot.*, 51,1171, 2000.
24. Allakhverdiev, S.I. and Murata, N. Salt stress inhibits photosystems II and I in cyanobacteria, *Photosynth. Res.*, 98, 529, 2008.
25. Klähn, S. and Hagemann, M. Compatible solute biosynthesis in cyanobacteria, *Environ. Microbiol.*, 13, 51, 2011.
26. Hagemann, M. Molecular biology of cyanobacterial salt acclimation, *FEMS Microbiol. Rev.*, 35, 87, 2011.
27. Novikova, G.V., Moshkov, I.E., and Los' D.A. Protein sensors and transducers of cold, hyperosmotic and salt stresses in cyanobacteria and plants, *Mol. Biol. (Mosk.)*, 41, 478, 2007.
28. Oren, A. and Gunde-Cimerman, N. Mycosporines and mycosporine-like amino acids: UV protectants or multipurpose secondary metabolites? *FEMS Microbiol. Lett.*, 269, 1–10, 2007.
29. Karandashova, I.V. and Elanskaia, I.V. Genetic control and mechanisms of salt and hyperosmotic shock resistance in cyanobacteria, *Genetika*, 41,1589, 2005.
30. Murata, N. and Suzuki, I. Exploitation of genomic sequences in a systematic analysis to access how cyanobacteria sense environmental stress, *J. Exp. Bot.*, 57, 235, 2006.
31. Sakamoto, T. and Murata, N. Regulation of the desaturation of fatty acids and its role in tolerance to cold and salt stress, *Curr. Opin. Microbiol.*, 5, 208, 2002.
32. Apte, S.K. and Thomas, J. Membrane electrogenesis and sodium transport in filamentous nitrogen-fixing cyanobacteria, *Eur. J. Biochem.*, 154, 395, 1986.
33. Alahari, A. and Apte, S.K. Pleiotropic effects of potassium deficiency in heterocystous, nitrogen-fixing cyanobacterium, *Anabaena torulosa*, *Microbiol.*, 144, 1557, 1998.

34. Pashinger, H. DCCD-induced sodium uptake by *Anacystis nidulans*, *Arch. Microbiol.*, 113, 285, 1977.
35. Iwano, M. Selective effect of salt stress on the activity of two ATPases in the cell membrane of *Nostoc muscorum*, *Plant Cell Physiol.*, 36, 1297, 1995.
36. Peschek, G.A. et al. Correlation between immuno-gold labels and activities of the cytochrome-c oxidase (aa_3-type) in membranes of salt stressed cyanobacteria, *FEMS Microbiol. Lett.*, 124, 431, 1994.
37. Kanamaru, K., Kashiwagi, S., and Mizuno, T. A copper-transporting P-type ATPase found in the thylacoid membrane of the cyanobacterium *Synechococcus* sp. PCC7942, *Mol. Microbiol.*, 13, 369, 1994.
38. Torrecilla, I. et al. A calcium signal is involved in heterocyst differentiation in the cyanobacterium *Anabaena* sp. PCC 7120, *Microbiology*, 150, 3731, 2001.
39. Liu, J. and Zhu, J.-K. Proline accumulation and salt induced gene expression in a salt-hypersensitive mutant of *Arabidopsis*, *Plant Physiol.*, 114, 591, 1997.
40. Norris, V. et al. Calcium signalling in bacteria, *J. Bacteriol.,* 178, 3677, 1996.
41. Paulsen, R., Bahner, M., and Huber, A. The PDZ assembled "transducisome" of microvillar photoreceptors: The TRP/TRPL problem. *Pflugers Arch.*, 439, R181, 2000.
42. Ren, D. et al. A prokaryotic voltage-gated sodium channel, *Science*, 294, 2372, 2001.
43. Raeymaekers, L. et al. Expression of a P-type Ca^{2+}-transport ATPase in *Bacillus subtilis* during sporulation, *Cell Calcium*, 32, 93, 2002.
44. Singh, S.C., Sinha, R.P., and Häder, D.-P. Role of lipids and fatty acids in stress tolerance in cyanobacteria, *Acta Protozool.*, 41, 297, 2002.
45. Bhargava, P. et al. Phytochelatin plays a role in UV-B tolerance in N_2-fixing cyanobacterium *Anabaena doliolum*, *J. Plant Physiol.*, 162, 1220, 2005.
46. Mishra, Y., Bhargava, P., and Rai, L.C. Differential induction of enzymes and antioxidants of antioxidative defense system in *Anabaena doliolum* exposed to heat stress. *J. Thermal Biol.,* 30, 524, 2005.
47. Mallick, N. and Rai, L.C. Response of the antioxidant systems of the nitrogen fixing cyanobacterium *Anabaena doliolum* to copper, *J. Plant Physiol.*, 155, 146, 1999.
48. Frankel, E.N. Chemistry of autoxidation: Mechanism, products and flavor significance, in *Flavor Chemistry of Fats and Oils*, Min, D.B., and Smouse, T.H., Eds., Champaign, IL: AOCS Press, 1985, p. 1.
49. Medina, J.H. et al. Flavonoids: A new family of benzodiazepine receptor ligands, *Neurochem. Res.*, 22, 419, 1997.
50. Fridovich, I. (1986) Superoxide dismutases, *Adv. Enzymol.*, 58, 62, 1986.
51. Elstner, E.F., Wagner, G.A., and Schutz, W. Activated oxygen in green plants in relation to stress situations, *Curr. Top. Plant Biochem. Physiol.*, 7, 159, 1988.
52. Huflejt, M. et al. Changes in membrane lipid composition during saline growth of the freshwater cyanobacterium *Synechococcus* 6311, *Plant Physiol.*, 94, 1512, 1990.
53. Khamutov, G. et al. Membrane lipid composition, fluidity, and surface charge changes in response to growth of the freshwater cyanobacterium *Synechococcus* 6311 under high salinity, *Arch. Biochem. Biophys.*, 277, 263, 1990.
54. Ritter, D. and Yopp, J.H. Plasma membrane lipid composition of the halophilic cyanobacterium *Aphanothece halophytica*. *Arch. Microbiol.*, 159, 435, 1993.
55. Allakhverdiev, S.I. et al. Unsaturated fatty acids in membrane lipids protect the photosynthetic machinery against salt-induced damage in *Synechococcus*, *Plant Physiol.*, 125, 1842, 2001.
56. Padan, E. and Schuldiner, S. Molecular physiology of the Na^+/H^+ antiporter in *Escherichia coli*, *J. Exp. Biol.*, 196, 443, 1994.
57. Kamada, Y. et al. The protein kinase C-activated MAP kinase pathway of *Saccharomyces cerevisiae* mediates a novel aspect of the heat shock response, *Genes Dev.*, 9, 1559, 1995.
58. Kates, M., Pugh, E.L., and Ferrante, G. Regulation of membrane fluidity by lipid desaturases, *Biomembranes*, 12, 379, 1984.
59. Lefort-Tran, M., Pouphile, M., and Spath, S. Cytoplasmic membrane changes during adaptation of the fresh water cyanobacterium *Synechococcus* 6311 to salinity, *Plant Physiol.*, 87, 767, 1988.
60. Ferjani, A. et al. Glucosylglycerol, a compatible solute, sustains cell division under salt stress, *Plant Physiol.*, 131, 1628, 2003.
61. Thomas, D.J. et al. Photoinhibition and light-induced cyclic electron transport in *ndhB*$^-$ and *psaE*$^-$ mutants of *Synechocystis* sp. PCC 6803, *Plant Cell Physiol.*, 42, 803, 2001.
62. Thomas, J. and Apte, S.K. Sodium requirement and metabolism in nitrogen-fixing cyanobacteria, *J. Biosci.*, 6, 771, 1984.
63. Jeanjean, R. et al. Exposure of cyanobacterium *Synechocystis* PCC6803 to salt stress induces concerted changes in respiration and photosynthesis, *Plant Cell Physiol.*, 34, 1073, 1993.

64. Lu, C. and Zhang, J. Role of light in the response of PSII photochemistry to salt stress in the cyanobacterium *Spirulina platensis*, *J. Exp. Bot.*, 51, 911, 2000.
65. Putty-Reddy, S. et al. The effects of salt stress on photosynthetic electron transport and thylakoid membrane proteins in the cyanobacterium *Spirulina platensis, J. Biochem. Mol. Biol.*, 38, 481, 2005.
66. Bhadauria, P. et al. Physiological and biochemical characterization of a diazotrophic cyanobacterium *Anabaena cylindrica* under NaCl stress, *Curr. Microbiol.*, 55, 334, 2007.
67. Lu, C.M. and Vonshak, A. Photoinhibition in outdoor *Spirulina platensis* cultures assessed by polyphasic chlorophyll fluorescence transients, *J. Appl. Phycol.*, 11, 355, 1999.
68. Bhargava, P. et al. Excess copper induces anoxygenic photosynthesis in *Anabaena doliolum*: A homology based proteomic assessment of its survival strategy, *Photosynth. Res.*, 96, 61, 2008.
69. Garlick, S., Oren, A., and Padan, E. Occurrence of facultative anoxygenic photosynthesis among filamentous and unicellular cyanobacteria, *J. Bacteriol.*, 129, 623, 1977.
70. Moisander, P.H., McClinton, E., and Paerl, H.W. Salinity effects on growth, photosynthetic parameters, and nitrogenase activity in estuarine planktonic cyanobacteria, *Microb. Ecol.*, 43, 432, 2002.
71. Kellogg, E.A. and Juliano, N.D. Structure and function of RuBisCO and their implications for phylogenetic studies, *Am. J. Bot.*, 84, 413, 1997.
72. Badger, M.R. and Price, G.D. CO_2 concentrating mechanisms in cyanobacteria: Molecular components, their diversity and evolution, *J. Exp. Bot.*, 54, 609, 2003.
73. Sivakumar, P., Sharmila, P., and Saradhi, P.P. Proline alleviates salt-stress-induced enhancement in ribulose-1,5-bisphosphate oxygenase activity, *Biochem. Biophys. Res. Commun.*, 279, 512, 2000.
74. Norman, E.G. and Colman, B. Formation and metabolism of glycolate in the cyanobacterium *Coccochloris peniocystis*, *Arch. Microbiol.*, 157, 375, 1992.
75. Renström, E., Rai, A.N., and Bergman, B. Glycolate metabolism in cyanobacteria, *Physiol. Plant.*, 75, 144, 1989.
76. Eisenhut, M. et al. The plant-like C2 glycolate cycle and the bacterial-like glycerate pathway cooperate in phosphoglycolate metabolism in cyanobacteria, *Plant Physiol.*, 142, 333, 2006.
77. Halliwell, B. and Gutteridge, J.M. C. *Free Radical in Biology and Medicine*, 3rd edn. Oxford University Press, London, U.K., 1999.
78. Neill, S.J. et al. Nitric oxide is a novel component of abscisic acid signaling in stomatal guard cells, *Plant Physiol.*, 128, 13, 2002.
79. Cadenas, E. Biochemistry of oxygen toxicity, *Annu. Rev. Biochem.*, 58, 79, 1989.
80. Foyer, C.H. et al. Adaptations of photosynthetic electron transport, carbon assimilation, and carbon partitioning in transgenic *Nicotiana plumbaginifolia* plants to changes in nitrate reductase activity, *Plant Physiol.*, 104, 171, 1994.
81. Foyer, C.H. et al. Short term effects of nitrate, nitrite and ammonium assimilation on photosynthesis, carbon partitioning and protein phosphorylation in maize, *Planta*, 192, 211, 1994.
82. Asada, K. Production and action of oxygen species in photosynthetic tissues, in *Causes of Photooxidative Stress and Amelioration of Defense Systems in Plants*, Foyer, C.H. and Mullineaux, P.M., Eds., CRC Press, Boca Raton, FL, 1994, p. 77.
83. Foyer, C.H. and Harbinson, J. Oxygen metabolism and the regulation of photosynthetic electron transport, in *Causes of Photooxidative Stress and Amelioration of Defense Systems in Plants*, Foyer, C.H. and Mullineaux, P.M., Eds., CRC Press, Boca Raton, FL, 1994, p. 1.
84. Furbank, R.T. and Badger, M.R. Oxygen exchange associated with electron transport and photophosphorylation in spinach thylakoids, *Biochim. Biophys. Acta*, 723, 400, 1983.
85. Bohnert, H.J. and Jensen, R.G. Strategies for engineering water-stress tolerance in plants, *Trends Biotechnol.*, 14, 89, 1996.
86. Agastian, P., Kingsley, S.J., and Vivekanandan, M. Effect of salinity on photosynthesis and biochemical characteristics in mulberry genotypes. *Photosynthetica*, 38, 287, 2000.
87. Reddy, P.G. et al. Interleukin 2 treatment of *Staphylococcus aureus* mastitis, *Cytokine*, 4, 227, 1992.
88. Feierabend, J. and Engel, S. Photoinactivation of catalase in vitro and in leaves, *Arch. Biochem. Biophys.*, 251, 567, 1986.
89. Frank, H.A. and Cogdell, R.J. Carotenoids in photosynthesis, *Photochem. Photobiol.*, 63, 257, 1996.
90. Britton, G., Liaaen-Jensen, S., and Pfander, H.P. Eds., *Carotenoids*, Vols. I and II. Birkheuser Verlag, Basel, Switzerland, 1995.
91. Kim, S.Y. et al. Establishment of the vernalization-responsive, winter-annual habit in *Arabidopsis* requires a putative histone H3 methyl transferase, *Plant Cell*, 17, 3301, 2005.
92. Foote, C.S., Chang, Y.C., and Denny, R.W. Chemistry of singlet oxygen X. Carotenoid quenching parallels biological protection, *J. Am. Chem. Soc.*, 92, 5216, 1970.

93. Munné-Bosch, S. and Alegre, L. The function of tocopherols and tocotrienols in plants, *Crit. Rev. Plant Sci.*, 21, 31, 2002.
94. Trebst, A., Depka, B., and Holländer-Czytko, H. A specific role for tocopherol and of chemical singlet oxygen quenchers in the maintenance of photosystem II structure and function in *Chlamydomonas reinhardtii. FEBS Lett.*, 43, 2157, 2002,
95. Foyer, C.H. Ascorbic acid, in *Antioxidants in Plants*, Alsher, R.G. and Hess, J.L. Eds., CRC Press, Boca Raton, FL, 1993, p. 31.
96. Beyer, R.E. The role of ascorbate in antioxidant protection of biomembranes: Interaction with vitamin E and coenzyme Q, *J. Bioenerg. Biomembr.*, 26, 349, 1994.
97. Khedr, A.H.A. et al. Proline induces the expression of salt-stress-responsive proteins and may improve the adaptation of *Pancratium maritmum* L. to salt stress, *J. Exp. Bot.*, 54, 2553, 2003.
98. van Thor, J.J. et al. Salt shock-inducible photosystem I cyclic electron transfer in *Synechocystis* PCC6803 relies on binding of ferredoxin:NAD^{P+} reductase to the thylakoid membranes via its CpcD phycobilisome-linker homologous N-terminal domain, *Biochim. Biophys. Acta.*, 1457, 129, 2000.
99. Allen, M.B. and Arnon, D.I. Studies on nitrogen-fixing blue-green algae. I. Growth and nitrogen fixation by *Anabaena cylindrica* Lemm, *Plant Physiol.*, 30, 366, 1955.
100. Peterson, R.B. and Wolk, C.P. High recovery of nitrogenase activity and of ^{55}Fe-labeled nitrogenase in heterocysts isolated from *Anabaena variabilis*, *Proc. Natl. Acad. Sci. U S A,* 75, 6271, 1978.
101. Rai, V., Tiwari, S.P., and Rai, A.K. Effect of NaCl on nitrogen fixation of unadapted and NaCl-adapted *Azolla pinnata-Anabaena azollae*, *Aquat. Bot.*, 71, 109, 2001.
102. Bhargava, S. and Singh, K. Differential response of NaCl resistant mutants of the cyanobacterium *Nostoc muscorum* to salinity and osmotic stress, *W. J. Microb. Biotech.*, 22, 783, 2006.
103. Fernandes, T.A., Iyer, V., and Apte, S.K. Differential responses of nitrogen-fixing cyanobacteria to salinity and osmotic stresses, *Appl. Environ. Microbiol.*, 59, 899, 1993.
104. Reed, R.H. and Stewart, W.D.P. Physiological responses of *Rivularia atra* to salinity: Osmotic adjustment in hyposaline media, *New Phytologist,* 95, 595, 1983.
105. Tel-Or, E. Adaptation to salt to the photosynthetic apparatus in cyanobacteria, *FEBS Lett.*, 110, 253, 1980.
106. Tel-Or, E. Response of N_2-fixing cyanobacteria to salt, *Appl. Environ. Microbiol.*, 40, 689, 1980.
107. Rai, A.K. and Sharma, N.K. Phosphate metabolism in the cyanobacterium *Anabaena doliolum* under salt stress, *Curr. Microbiol.*, 52, 6, 2006.
108. Singh, S.K. et al., Factors modulating alkaline phosphatase activity in the diazotrophic rice-field cyanobacterium, *Anabaena oryzae*, *World J. Microbiol. Biotech.*, 22, 927, 2006.
109. Hagemann, M. et al. Biochemical characterization of glucosylglycerol-phosphate synthase of *Synechocystis* sp. strain PCC 6803: Purification and overexpression change its salt dependence, *Curr. Microbiol.*, 43, 278, 2001.
110. Hagemann, M., Erdmann, N., and Wittenburg, E. Synthesis of glucosylglycerol in salt-stressed cells of the cyanobacterium *Microcystis firma*, *Arch. Microbiol.*, 148, 275, 1987.
111. Reed, R.H. et al. Organic solute accumulation in osmotically stressed cyanobacteria, *FEMS Microbiol. Rev.*, 39, 51, 1986.
112. Nakamura, N. et al. The vesicle docking protein p115 binds GM130, a *cis*-Golgi matrix protein, in a mitotically regulated manner, *Cell*, 89, 445, 1997.
113. Hagemann, M. and Erdmann, N. Environmental stresses, in *Cyanobacterial Nitrogen Metabolism and Environmental Biotechnology*, Rai, A.K. Ed. Narosa Publishing House, New Delhi, India, 1997, p. 156.
114. Erdmann, N., Fulda, S., and Hagemann, M. Glucosylglycerol accumulation during salt acclimation of two unicellular cyanobacteria, *J. Gen. Microbiol.*, 138, 363, 1992.
115. Mackay, M.A., Norton, R.S., and Borowitzka, L.J. Organic osmoregulatory solutes in cyanobacteria, *J. Gen. Microbiol.*, 130, 177, 1984.
116. Erdmann, N., Berg, C., and Hagemann, M. Missing salt adaptation of *Microcystis firma* (cyanobacterium) in the dark, *Arch. Hydrobiol.*, 114, 521, 1989.
117. Fulda, S., Hagemann, M., and Libbert, E. Release of glucosylglycerol from the cyanobacterium *Synechocystis* spec. SAG 92.79 by hypoosmotic shock, *Arch. Microbiol.*, 153, 405, 1990.
118. Mikkat, S., Hagemann, M., and Schoor, A. Active transport of glucosylglycerol is involved in salt adaptation of the cyanobacterium *Synechocystis* sp. strain PCC 6803, *Microbiology*, 142, 1725, 1996.
119. Csonka, L.N. and Hanson, A.D. Prokaryotic osmoregulation: Genetics and physiology, *Annu. Rev. Microbiol.*, 45, 569, 1991.
120. Marin, K. et al., Osmotic stress in *Synechocystis* sp. PCC 6803: Low tolerance towards nonionic osmotic stress results from lacking activation of glucosylglycerol accumulation, *Microbiology*, 152: 2023, 2006.

121. Hagemann, M. and Erdmann, N. Activation and pathway of glucosylglycerol synthesis in the cyanobacterium *Synechocystis* sp. PCC 6803, *Microbiology*, 140, 1427, 1994.
122. Gabbay-Azaria, R. and Tel-Or, E. Mechanisms of salt tolerance in cyanobacteria, in *Plant Responses to the Environment*, Gresshoff, P.M., Ed., CRC Press, Boca Raton, FL, 1993, p. 692.
123. Sibley, M.H. and Yopp, J.H. Regulation of S-adenosylhomocysteine hydrolase in the halophilic cyanobacterium *Aphanothece halophytica*: A possible role in glycinebetaine biosynthesis, *Arch. Microbiol.*, 149, 43, 1987.
124. Page-Sharp, M., Behm, C.A., and Smith, G.D. Involvement of the compatible solutes trehalose and sucrose in the response to salt stress of a cyanobacterial *Scytonema* species isolated from desert soils, *Biochim. Biophys. Acta.*, 1472, 519, 1999.
125. Karandashova, I.V. and Elanskaya, I.V. Genetic control and mechanisms of salt and hyperosmotic stress resistance in cyanobacteria, *Russian J. Gen.*, 41, 1311, 2005.
126. Apte, S.K. and Haselkorn, R. Cloning of salinity stress-induced genes from the salt-tolerant nitrogen-fixing cyanobacterium *Anabaena torulosa*, *Plant Mol. Biol.*, 15, 723, 1990.
127. Vinnemeier, J. and Hagemann, M. Identification of salt-regulated genes in the genome of the cyanobacterium *Synechocystis* sp. strain PCC 6803 by subtractive RNA hybridization, *Arch. Microbiol.*, 172, 377, 1999.
128. Engelbrecht, F., Marin, K., and Hagemann, M. Expression of the ggpS gene, involved in osmolyte synthesis in the marine cyanobacterium *Synechococcus* sp strain PCC 7002, revealed regulatory differences between this strain and the freshwater strain *Synechocystis* sp strain PCC 6803, *Appl. Environ. Microbiol.*, 65, 4822, 1999.
129. Vinnemeier, J., Kunert, A., and Hagemann, M. Transcriptional analysis of the isiAB operon in salt-stressed cells of the cyanobacterium *Synechocystis* sp. PCC 6803, *FEMS Microbiol. Lett.*, 169, 323, 1998.
130. Kanesaki, Y. et al. Salt stress and hyperosmotic stress regulate the expression of different sets of genes in *Synechocystis* sp. PCC 6803, *Biochem. Biophys. Res. Commun.*, 290, 339, 2002.
131. Karandashova, I. et al. Identification of genes essential for growth at high salt concentrations using salt-sensitive mutants of the cyanobacterium *Synechocystis* sp. PCC 6803, *Curr. Microbiol.*, 44, 184, 2002.
132. Ban, N. et al. The complete atomic structure of the large ribosomal subunit at 2.4 Å resolution, *Science*, 289, 905, 2000.
133. Nissen, P. et al. The structural basis of ribosome activity in peptide bond synthesis, *Science*, 289, 920, 2000.
134. Kaneko, T. et al. Sequence analysis of the genome of the unicellular cyanobacterium *Synechocystis* sp. strain PCC 6803. I. Sequence features in the 1 Mb region from map positions 64% to 92% of the genome, *DNA Res.*, 2, 153, 1995.
135. Hagemann, M. et al. Characterization of a glucosylglycerol-phosphate accumulating, salt-sensitive mutant of the cyanobacterium *Synechocystis* sp. PCC 6803, *Arch. Microbiol.*, 166, 83, 1996.
136. Suzuki, I. et al. The pathway for perception and transduction of low-temperature signals in *Synechocystis*, *EMBO J.*, 19, 1327, 2000.
137. Paithoonrangsarid, K. et al. Five histidine kinases perceive osmotic stress and regulate distinct sets of genes in *Synechocystis*, *J. Biol. Chem.*, 279, 53078, 2004.
138. Shoumskaya, M.A. et al. Identical Hik-Rre systems are involved in perception and transduction of salt signals and hyperosmotic signals but regulate the expression of individual genes to different extents in *Synechocystis*, *J. Biol. Chem.*, 280, 21531, 2005.
139. Mizuno, T., Kaneko, T., and Tabata, S. Compilation of all genes encoding bacterial two-component signal transducers in the genome of the cyanobacterium, *Synechocystis* sp. strain PCC 6803, *DNA Res.*, 3, 407, 1996.
140. Kaneko, T. et al. Structural analysis of four large plasmids harboring in a unicellular cyanobacterium, *Synechocystis* sp. PCC 6803, *DNA Res.*, 10, 221, 2003.
141. Los, D.A. et al. Stress sensors and signal transducers in cyanobacteria, *Sensors*, 10, 2386, 2010.
142. Marin, K. et al. Identification of histidine kinases that act as sensors in the perception of salt stress in *Synechocystis* sp. PCC 6803, *Proc. Natl. Acad. Sci. USA*, 100, 9061, 2003.
143. Liang, C. et al. Serine/threonine protein kinase SpkG is a candidate for high salt resistance in the unicellular cyanobacterium *Synechocystis* sp. PCC 6803, *PLoS One* 6(5): e18718, 2011.
144. Davies, K.J.A. Protein damage and degradation by oxygen radicals, *J. Biol. Chem.*, 262, 9895, 1987.
145. Davies, K.J.A. and Delsingnore, M.E. Protein damage and degradation by oxygen radicals III. Modification of secondary and tertiary structure, *J. Biol. Chem.*, 262, 9902, 1987.
146. Stadtman, E.R. Oxidation of free amino acids and amino acid residues in proteins by radiolysis and by metal-catalyzed reactions, *Annu. Rev. Biochem.*, 62, 797, 1993.

147. Hagemann, M. et al. Flavodoxin accumulation contributes to enhanced cyclic electron flow around photosystem I in salt-stressed cells of *Synechocystis* sp. strain PCC 6803, *Physiol. Plant*, 105, 670, 1999.
148. Duché, O. et al. Salt stress proteins induced in *Listeria monocytogenes*, *Appl. Environ. Microbiol.*, 68, 1491, 2002.
149. Ehling-Schulz, M. et al. The UV-B stimulon of the terrestrial cyanobacterium *Nostoc commune* comprises early shock proteins and late acclimation proteins, *Mol. Microbiol.*, 46, 827, 2002.
150. Ferianc, P., Farewell, A., and Nyström, T. The cadmium-stress stimulon of *Escherichia coli* K-12, *Microbiology*, 144, 1045, 1998.
151. Fulda, S. et al. MycN sensitizes neuroblastoma cells for drug-induced apoptosis, *Oncogene*, 18, 1479, 1999.
152. Parker, R. et al. An accurate and reproducible method for proteome profiling of the effects of salt stress in the rice leaf lamina, *J. Exp. Bot.*, 57, 1109, 2006.
153. Yan, S. et al. Proteomic analysis of salt stress responsive proteins in rice root, *Proteomics*, 5, 235, 2005.
154. Sazuka, T., Yamaguchi, M., and Ohara, O. Cyano2Dbase updated: linkage of 234 protein spots to corresponding genes through N-terminal microsequencing, *Electrophoresis*, 20, 2160, 1999.
155. Kaneko, T. et al. Complete genomic sequence of the filamentous nitrogen-fixing cyanobacterium *Anabaena* sp. strain PCC 7120, *DNA Res.,* 8, 227, 2001.
156. Kaneko, T. et al. Sequence analysis of the genome of the unicellular cyanobacterium *Synechocystis* sp. strain PCC6803. II. Sequence determination of the entire genome and assignment of potential protein-coding regions (supplement), *DNA Res.*, 3, 185, 1996.
157. Oda, Y. et al. Accurate quantitation of protein expression and site-specific phosphorylation, *Proc. Natl. Acad. Sci. USA*, 96, 6591, 1999.
158. Pandhal, J. et al. A cross-species quantitative proteomic study of salt adaptation in a halotolerant environmental isolate using ^{15}N metabolic labelling, *Proteomics*, 8, 2266, 2008.
159. Pandhal, J., Wright, P.C., and Biggs, C.A. Proteomics with a pinch of salt: A cyanobacterial perspective, *Saline Systems*, 4,1, 2008.
160. Fulda, S. et al. Proteome analysis of salt stress response in the cyanobacterium *Synechocystis* sp. strain PCC 6803, *Proteomics*, 6, 2733, 2006.
161. Durner, J. and Böger, P. Ubiquitin in the prokaryote Anabaena *variabilis*, *J. Biol. Chem.*, 220, 3270, 1995.
162. Grune, T., Reinheckel, T., and Davies, K.J.A. Degradation of oxidized proteins in K562 human hematopoietic cells by proteasome, *J. Biol. Chem.*, 271, 15504, 1996.
163. Grune, T. et al. Peroxynitrite increases the degradation of aconitase and other cellular proteins by proteasome, *J. Biol. Chem.*, 273, 10857, 1998.
164. Hihara, Y. et al. DNA microarray analysis of cyanobacterial gene expression during acclimation to high light, *Plant Cell*, 13, 793, 2001.
165. Suzuki, I. et al. Cold-regulated genes under control of the cold sensor Hik33 in *Synechocystis*, *Mol. Microbiol.*, 40, 235, 2001.
166. Gupta, R.S. et al. Molecular signatures in protein sequences that are distinctive of cyanobacterial and plastid homologs, *Int. J. Syst. Evol. Microbiol.*, 53, 1833, 2003.
167. Gupta, R.S. et al. Sequencing of heat shock protein 70 (DnaK) homologs from *Deinococcus proteolyticus* and *Thermomicrobium roseum* and their integration in a protein based phylogeny of prokaryotes, *J. Bacteriol.*, 179, 345, 1997.
168. Schwartz, S.H. et al. Regulation of an osmoticum responsive gene in *Anabaena* sp strain PCC7120, *J. Bacteriol.*, 180, 6332, 1998.
169. Huang, F. et al. Proteomic screening of salt-stress-induced changes in plasma membranes of *Synechocystis* sp. strain PCC 6803, *Proteomics*, 6, 910, 2006.
170. Sakayori, T., Shiraiwa, Y., and Suzuki, I. A *Synechocystis* homolog of SipA protein, Ssl3451, enhances the activity of the histidine kinase Hik33, *Plant Cell Physiol.*, 50, 1439, 2009.
171. Mayer, F. Cytoskeletons in prokaryotes, *Cell. Biol. Int.*, 27, 429, 2003.

9 Cyanobacterial Salt Stress Acclimation

Genetic Manipulation and Regulation

*Nadin Pade and Martin Hagemann**

CONTENTS

9.1 INTRODUCTION

The capacity of organisms to respond to fluctuations in salt and/or osmotic concentrations is important to survive in a great variety of habitats, particularly in aquatic environments. Generally, cells try to keep their internal osmotic composition and turgor pressure constant; therefore, an increase or a decrease in external salt concentration and/or water availability poses a challenge for the cellular metabolism and for the survival of the cells. To avoid negative effects, microorganisms have developed two effective strategies for the acclimation to changing salt concentrations: the "salt-in-strategy" and the "salt-out-strategy" [1,2].

The "salt-in-strategy" is restricted to a small number of true halophiles, for example, aerobic, halophilic archaea of the order Halobacteriales [1], anaerobic, halophilic bacteria of the order Halanaerobiales [3], and the extreme halophilic representative *Salinibacterruber* [4]. All these prokaryotes accumulate inorganic salts (mainly KCl) in molar concentrations in the cytoplasm, which are at least as high as the external salt concentration in their surrounding medium to ensure turgor pressure necessary for cell growth. Hence, it becomes apparent that the presence of high concentrations of KCl in the cytoplasm requires an adaptation of all proteins to this salt-rich reaction medium [5], restricting these specialized organisms to a few niches.

* Corresponding author: martin.hagemann@uni-rostock.de

The "salt-out-strategy" is used by the majority of prokaryotic and all eukaryotic organisms to acclimate to high or changing salt concentrations. This strategy involves two main processes. First, it includes the accumulation of small organic molecules called compatible solutes, which do not interfere with the metabolism of the cell [6]. Second, it depends on the active export of inorganic ions that diffuse into the cytoplasmic, metabolically active space along the electrochemical gradients. Basically, the internal ion concentration is kept at low levels and the accumulation of compatible solute warrants the osmotic balance. Thus, the combined accumulation of mainly compatible solutes and some ions makes the cell hyperosmotic compared to the external medium, which allows water uptake, turgor maintenance, and growth. There exists a great chemical variety of compatible compounds, ranging from amino acids and their derivatives to sugars and polyols. In principle, they could be taken up from the medium or synthesized *de novo* [1,7]. The latter strategy is preferred by photoautotrophic organisms such as cyanobacteria. Adaptation to a high-salt tolerance level using the "salt out" strategy requires only the acquisition of a few specialized proteins, such as ion transporters and compatible solute synthesis enzymes, while the majority of proteins remain unchanged; that is, they show optimal performance at comparatively low salt concentrations.

The screening of almost 200 cyanobacterial strains from different habitats revealed an interesting correlation between the strain-specific salt resistance level and the nature of its dominating compatible solutes [2,8]. Low-salt tolerant strains (resisting freshwater up to full seawater conditions) accumulate the sugars trehalose and/or sucrose, while moderate halotolerant strains (resisting freshwater to threefold seawater salt levels) are characterized by the accumulation of glucosylglycerol and glucosylglycerate, with the latter restricted to true marine isolates of rather stenohaline behavior [9]. Halophilic cyanobacteria, which tolerate very high up to saturated salt concentrations, synthesize the compatible solute glycine betaine (Figure 9.1). The cellular level of the corresponding compatible solute is determined by the salt concentration of the surrounding medium [10]. Besides these main groups, some other compatible solutes were found only in a few strains or under special environmental conditions; for example, some hypersaline strains have been reported to accumulate glutamate betaine as a second compatible solute besides glycine betaine [11]. Singh et al. [12] showed that proline overaccumulation,

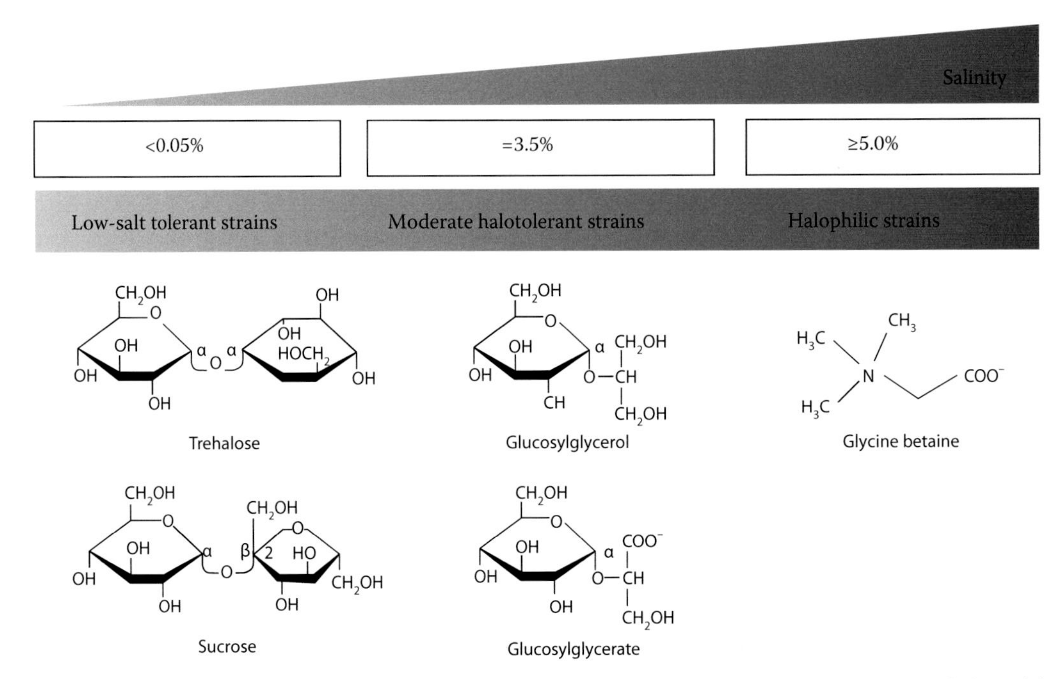

FIGURE 9.1 **(See color insert.)** Structure of major compatible solutes of cyanobacteria in correlation with their overall salt resistance.

observed after proline supplementation of the growth medium, conferred an enhanced salinity tolerance to *Nostoc muscorum*. Proline, as a minor compatible solute, has also been found transiently in salt-shocked cyanobacteria, for example, in cells of *Synechococcus* sp. PCC 7418 [10].

9.2 GENETIC MANIPULATION OF CYANOBACTERIAL SALT ACCLIMATION

In the pregenomic era, the accumulation of distinct compatible solutes and a certain degree of salt tolerance were found to be closely correlated. To find a direct verification that these two traits are linked, genetic approaches were performed in the last 20 years (Table 9.1).

The directed mutation of potential salt stress genes is applicable to many freshwater and moderate halotolerant strains such as *Synechococcus* sp. PCC 7942 (6301), *Nostoc* (*Anabaena*) sp. PCC 7120, *Synechococcus* sp. PCC 7002, and especially *Synechocystis* sp. PCC 6803. Unfortunately, none of the hypersaline strains can be genetically manipulated; therefore, genes and proteins from such strains, for example, *Aphanothece halophytica* (*Synechococcus* sp. PCC 7418), were functionally verified by their overexpression in freshwater strains, mostly *Synechococcus* sp. PCC 7492, and in defined mutants of *Escherichia coli*.

9.2.1 Mutant Generation

The mutation strategy revealed that sucrose and glucosylglycerol are synthesized by similar pathways. First a sugar-phosphate synthase generates a phosphorylated intermediate, then specific sugar phosphatases release the compatible solute. Sucrose is made by the combined action of a UDP-glucose-dependent sucrose-phosphate synthase (SpsA) and sucrose-phosphate phosphatase (Spp). The two proteins are encoded by separate genes or by one large gene for a fusion protein [2,13–15]. Glucosylglycerol (GG) is synthesized by an ADP-glucose-dependent glucosylglycerol-phosphate synthase (GgpS) and a glucosylglycerol-phosphate phosphatase (GgpP) [16,17]. The mutation of *spsA* and *ggpS* or *ggpP* resulted not only in the inability to synthesize sucrose and glucosylglycerol,

TABLE 9.1
Selected Salt-Sensitive Mutants of Cyanobacteria

Strain	Gene	Phenotype	References
Synechocystis sp. PCC 6803			
Mutants 143; 406; 549		Salt-sensitive	[33]
*gcp::aph*II	*gcp*	Nitrogen starvation phenotype, changes in ultrastructure, reduced salt tolerance	[35]
*ggpS::aph*II	*ggpS*	Salt-sensitive	[17]
*stpA::aph*II	*stpA, ggpP*	Salt-sensitive	[16]
*slr*1564::*aph*II	*sigF*	Slightly salt-sensitive	[73]
*slr*1799::*aph*II	*slr*1799	Salt-sensitive	[34]
*slr*1087::*aph*II	*slr*1089	Salt-sensitive	[34]
Δ*sll*1061–*sll*1062::*aph*II	*slr*1061 and/or *sll*1062	Salt-sensitive	[34]
Δ*isiA::aph*II	*isiA*	Slightly salt-sensitive	[34]
*hspA*mutant	*hspA*	Salt-sensitive	[36]
Synechococcus sp. PCC 7002			
Δ*ggpS::aph*II(GK02–1)	*ggpS*	Salt-sensitive	[18]
Synechococcus elongates PCC 6301			
Δ*yfr1*	RNA *yfr1*	Reduced growth under oxidative stress and high-salt-stress conditions	[37]

respectively, but also in salt-sensitive phenotypes of corresponding mutants of *Anabaena* sp. PCC 7119/*Synechococcus* sp. PCC 7492 and *Synechocystis* sp. PCC 6803/*Synechococcus* sp. PCC 7002, respectively [15–18]. In enterobacteria, yeast and plants, the disaccharide trehalose is synthesized by a similar, so-called OtsAB pathway [2]; however, homologous proteins are not present in the genomes of trehalose-synthesizing cyanobacteria. Instead, the TreYZ pathway was identified to be responsible for trehalose synthesis in *Nostoc* (*Anabaena*) sp. PCC 7120 or *N. muscorum*, because mutants defective in maltooligosyl trehalose synthase (EC 5.4.99.15, TreY) and maltooligosyl trehalose trehalohydrolase (EC 3.2.1.141, TreZ) were unable to synthesize trehalose [19,20]. It should be mentioned that the *treYZ* mutations did not significantly change the salt tolerance of these strains.

Compared to compatible solute biosynthesis, the molecular basis for active ion transport is less clearly supported by genetic approaches. Sodium ion export by salt-exposed cyanobacteria seems to be mainly driven by a specific isoform of Na^+/H^+ antiporters, named NhaS3 [21–23]. However, the corresponding genes in *Synechococcus* sp. PCC 7942 and *Synechocystis* sp. PCC 6803 turned out to be essential for cell viability; therefore, only partly segregated mutants (these mutant cells contain both wild-type and mutated gene copies) could be analyzed that showed slightly decreased salt resistance. Additionally, an Mrp-like ion transport system (most likely exhibiting Na^+/H^+ antiport activity) is probably involved in ion homeostasis, because corresponding mutants of *Nostoc* (*Anabaena*) sp. PCC 7120 or *Synechococcus* sp. PCC 7942 showed a reduced tolerance toward high salt concentrations [24,25].

Active uptake of K^+ is part of the salt acclimation in many organisms including cyanobacteria. K^+ is believed to play a signaling role (discussed in the following) and contributes as an ionic compatible solute to the overall osmotic balance [26,27]. It should also be mentioned that low-salt-grown cells contain relatively high amounts of about 150 mM K^+ in the cytoplasm, which plays a crucial role in turgor generation and in the maintenance of protein structures. While in enterobacteria such as *E. coli* the ATP-dependent K^+ uptake system KdpABC is mainly responsible for the salt-stimulated K^+ import, cyanobacteria seem to employ mostly a Ktr-like system. This notion is based on the findings that a *kdpA* mutant of *Synechocystis* sp. PCC 6803 did not display a salt-sensitive phenotype but the *ktr* mutants of this strain showed a clearly diminished salt tolerance level [28,29]. Probably, the cyanobacterial Kdp system plays a role in K^+ uptake at very low external conditions [30] and seems to be also involved in desiccation tolerance [31].

The mutant studies mentioned earlier support the notion that compatible solute accumulation and ion transporters are essential mechanisms for a successful salt acclimation among cyanobacteria. Initially, genes coding for proteins with essential functions in salt acclimation were searched by the random generation of salt-sensitive mutants (mostly done with *Synechocystis* sp. PCC 6803 [32–34]). The subsequent characterization of these mutants showed that the majority of them were affected in genes for compatible solute biosynthesis. Analyses of the genetic lesions in the other salt-sensitive mutants led to the identification of additional genes/proteins, such as a periplasmic glycoprotease [35] and some hypothetical proteins [34] that are obviously needed to support growth of *Synechocystis* sp. PCC 6803 at high salinities (Table 9.1). However, in most of these cases, it is still not clear if the defect in these genes/proteins is directly affecting salt tolerance or indirectly affecting it by decreasing general cellular fitness. It has often been observed that a decreased overall stress tolerance also diminished salt resistance levels [2]. Other such examples are mutations in the small heat shock protein HspA [36] or in the noncoding RNA Yfr1 [37], which decreased the salt tolerance but also the resistance to other stress factors.

In the last few years, many "omics" approaches were undertaken to identify salt-regulated genes and proteins in *Synechocystis* sp. PCC 6803 [38–42] as well as *Synechococcus* sp. PCC 7002 [43]. In the case of *Synechocystis* sp. PCC 6803, mRNAs for many so-called hypothetical proteins were found to be strongly increased after salt shock treatments. Their salt-dependent regulation can be taken as a hint that additional, so far unknown processes might also be involved in the cyanobacterial salt acclimation. Approaches to mutate these genes (e.g., the top-salt-induced genes *sll*1862/1863 encoding for unknown proteins [40]) have been undertaken in *Synechocystis* sp. PCC 6803. To date, these attempts have not resulted in the observation of a salt-sensitive phenotype of these mutants, making an essential function of these highly salt-regulated genes/proteins in the cyanobacterial salt acclimation questionable [44].

9.2.2 Gene Overexpression

Many cyanobacterial strains can be genetically manipulated, but this is not possible with the well-characterized halophilic strain *A. halophytica*. Therefore, many potential salt stress genes/proteins of this strain have been analyzed through overexpression in defined *E. coli* mutants (biochemical verification) and/or in the low halotolerant, freshwater cyanobacterium *Synechococcus* sp. PCC 7942 (functional verification by salt tolerance improvement). These strategies were used to identify the genes for the glycine betaine synthesis enzymes from *A. halophytica*. In contrast to *E. coli* and plants, which usually produce glycine betaine by an oxidative pathway (BetAB), glycine betaine synthesis in *Aphanothece* is performed by three methylations using glycine as a precursor [45]. Moreover, the functional expression of these genes in *Synechococcus* sp. PCC 7942 led to glycine betaine synthesis and improved salt tolerance of this low tolerant strain [45,46]. The observed salt tolerance improvement after expression of the glycine methylation pathway clearly exceeded the protective effect reported earlier, when glycine betaine synthesis genes from heterotrophic bacteria (oxidative pathways starting from choline) were expressed in *Synechococcus* sp. PCC 7942 [47,48].

A salt-regulated Na^+/H^+ antiporter gene from *A. halophytica* was also expressed in the freshwater strain *Synechococcus* sp. PCC 7942, which again resulted in an improved salt resistance. Notably, the salt resistance level of the Na^+/H^+-antiporter-expressing *Synechococcus* sp. PCC 7942 strain was higher compared to the strain expressing glycine betaine synthesis genes [45]. Similarly, the accumulation of glucosylglycerol inside the sucrose-accumulating, freshwater strain *Synechococcus* sp. PCC 6301 had also only minor effects on the salt resistance level [2]. These findings seem to indicate that despite the correlation between the type of compatible solute and salt resistance limit among cyanobacteria, the final salt resistance level is merely dependent on the capacity of ion export than the kind of compatible solute. Recently, a primary Na^+-transporting ATPase was identified in *A. halophytica* (similar genes occur also in *Synechococcus* sp. PCC 7002), which might also play a role in Na^+ export from cells at high-salt conditions. Again, the functional expression of the Na^+-ATPase gene cluster in *Synechococcus* sp. PCC 7942 increased the salt resistance significantly [49], supporting the notion that ion export might be the limiting factor for salt resistance in this freshwater strain.

9.2.3 Application of Cyanobacterial Salt Resistance Genes

Improvements in salt tolerance are not only interesting from the academic point of view. An increasing amount of arable land is lost by increasing soil salinity; therefore, plant breeders want to increase the low-salt tolerance of most crop plants, for example, by the expression of microbial salt resistance genes. A few attempts were reported in which cyanobacterial salt tolerance genes have been successfully used for such purposes. The glycine methylation pathway for glycine betaine biosynthesis from *A. halophytica* was expressed in Arabidopsis, leading to glycine betaine accumulation and increased salt tolerance [46]. Glucosylglycerol biosynthesis was introduced into Arabidopsis, which resulted also in a slightly improved salt resistance of the transgenic plant lines [50]. There is also commercial interest to use purified compatible solutes as additives to cosmetics as moistures and to pharmaceuticals as stabilizers (http://www.bitop.de//).

9.3 DYNAMICS OF CYANOBACTERIAL SALT ACCLIMATION

Studies using the moderately halotolerant strains *Synechocystis* sp. PCC 6803 and PCC 6147 as well as the freshwater strains *Synechococcus* sp. PCC 7942 and PCC 6301 showed that the salt acclimation of cyanobacteria is a dynamic process that includes the activation and inactivation of transporters, the synthesis of compatible solutes, and the global reprogramming of gene expression patterns [40,51]. Five different phases for the salt acclimation can be distinguished (Figure 9.2; [2]).

If cells are exposed to a salt shock exceeding the threshold salt difference of 300 mM NaCl, channels and pores in the cytoplasmic membrane open immediately and the cell releases ultimately water and many solutes [52]. Consequently, the cells rapidly shrink [53]. In the second phase, which persists several minutes, the cell volume recovers by an influx of external salts, like Na^+ and Cl^- [51]. The ion influx causes reuptake of water, because the water potential decreases and therefore the cell volume becomes restored. However, the high salt concentration in the cytoplasm inhibits the cellular metabolism, including photosynthesis, transcription, and translation [40,54]. Subsequently, a third phase takes place persisting for several hours that is characterized by an exchange of the toxic Na^+ to the less toxic K^+. To accomplish this exchange, ion transport systems must be activated, such as Na^+/H^+ antiporters and the Mrp and Ktr systems. It became apparent that cyanobacteria also contain ionic solutes (especially K^+) in osmotically significant amounts [55] upon addition of NaCl, sucrose, or sorbitol to the medium [56]. The compatible solute biosynthesis becomes also activated during this early acclimation phase.

Phase 4 is characterized by an increased synthesis of compatible solutes and the efflux of K^+ and Cl^-. During this phase, the ion concentrations return to the initial, prestress levels, whereas the compatible solute, such as glucosylglycerol, concentration is increased and reaches the salt-strength-specific steady-state level. With the decrease in the internal ion concentration, the toxic ionic effects decline and, therefore, metabolic activities such as photosynthesis or the gene expression can be

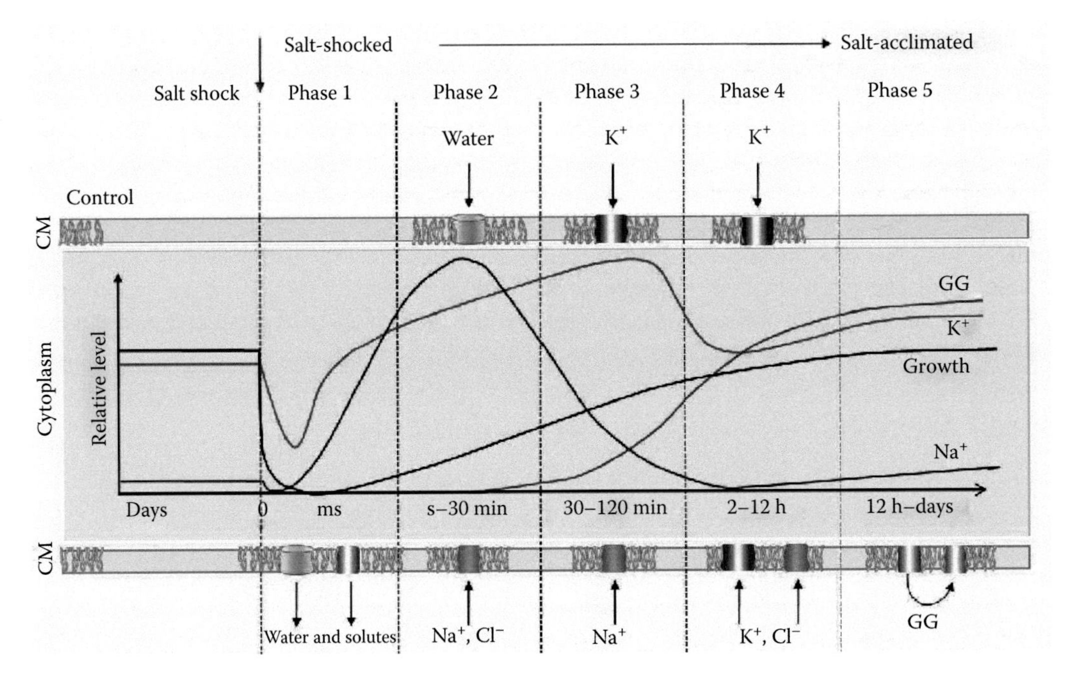

FIGURE 9.2 (See color insert.) Schematic view of the kinetics of salt acclimation in the moderate halotolerant cyanobacterium *Synechocystis* sp. PCC 6803. Salt acclimation can be divided into five different phases. The changes in ion and water concentrations and the activation of compatible solute biosynthesis, for example, glucosylglycerol (GG), were used to distinguish different cellular states. The diagram shows the relative changes in parameters like growth, Na^+, K^+, and GG contents (CM—cytoplasmic membrane).

restored [40,54]. After 1–2 days, growth and cell division are resumed and phase 5 is reached. In the salt-acclimated, new steady state, the cells are characterized by a high concentration of the compatible solute glucosylglycerol (GG) and a low ion concentration, which equals the steady state level of cells grown in freshwater medium (Figure 9.2).

9.4 SALT STRESS SENSING

9.4.1 Two-Component Systems

If living organisms are exposed to changing environmental conditions, they must adjust their metabolism to the new environment. To accomplish this purpose, the kind and strength of the stress must be perceived. Generally, the salt stress signaling is rather poorly understood.

In many prokaryotic cells, signal transduction systems involve a two-component system that consists of a histidine protein kinase (HIK) and a response regulator protein (RRE). The HIK perceives the environmental signal by its sensory domain (usually membrane bound or in the periplasm). Subsequently, the HIK forms a homodimer and becomes auto-phosphorylated at a histidine residue in its kinase domain (usually in the cytoplasmic space), which has a high potential to transfer the phosphoryl group to an aspartate residue in the RRE. Consequently, the conformation of RRE is changed and eventually the active RRE can bind to upstream promoter sequences (Figure 9.3). Two-component sensor-transducer systems are well studied in *E. coli* [57] and *Bacillus subtilis* [58]. In *E. coli*, the two-component system consisting of EnvZ as HIK and OmpR as RRE regulates the outer membrane porins OmpC and OmpF in a salt-dependent manner. However, even in this excellently studied model system, the signal that is primarily sensed upon salt stress is not clearly defined [59].

The role of HIK/RRE pairs in salt stress recognition was analyzed using corresponding mutants of cyanobacteria, particularly of *Synechocystis* sp. PCC 6803. Initially, a library of HIK mutants (about 45 single mutants) was generated [60] and analyzed regarding genome-wide changes in the salt-induced gene expression pattern by the DNA microarray technique [39]. Later, a similar strategy was applied to a library of RREs [61]. In contrast to *E. coli* and *B. subtilis*, where the genes for HIKs and the corresponding RREs are often found in operons or closely linked, only 14 HIKs are located in the vicinity of genes for potentially cognate RREs in *Synechocystis* sp. PCC 6803. The other 30 HIKs are located separately from genes coding for RREs. Therefore, it is difficult to predict the pairs of HIKs and RREs [62]. Together with many additional experiments, the screening of the two-component system mutants of *Synechocystis* sp. PCC 6803 by the microarray technique allowed the identification of putative salt sensing and more general stress-sensing mechanisms (reviewed by Los et al. [63]).

All the attempts regarding salt stress resulted in the identification of five two-component systems in *Synechocystis* sp. PCC 6803, which are involved in the activation of certain groups of salt-stress-induced genes, namely, HIK33-RRE31, HIK10-RRE3, HIK16-HIK41-RRE17, HIK2-RRE1, and HIK34-RRE1 [39,61]. Screening of the *hik* mutant library upon exposure to hyperosmotic stress revealed that the same five two-component systems are involved [64]. This finding corresponds to the assumption that these two-component systems are not only responsible for the expression for certain genes under specific stress conditions, but each separately contributes to a specific part of the general stress response. For example, the HIK33-RRE31 two-component system regulates the expression of *fabG* and *gloA* as hyperosmotic stress-specific genes and also that of *pgr5*, *nblA1*, and *nblA2* as oxidative stress-specific genes [65]. Recently, it has been reported that the binding of a small protein to the HIK33 alters its behavior [66], which could explain how similar HIK/RRE systems can regulate similar but also different gene groups in response to different stresses. It has to be mentioned that the salt-regulated expression of none of the essential genes for salt acclimation, that is, genes for proteins involved in ion export or compatible solute biosynthesis, was affected in any of the single HIK or RRE mutants. Correspondingly, none of these mutants showed a salt-sensitive phenotype. Therefore, currently, we have to assume that other sensory mechanisms might be responsible for the regulation of the salt-specific acclimation processes.

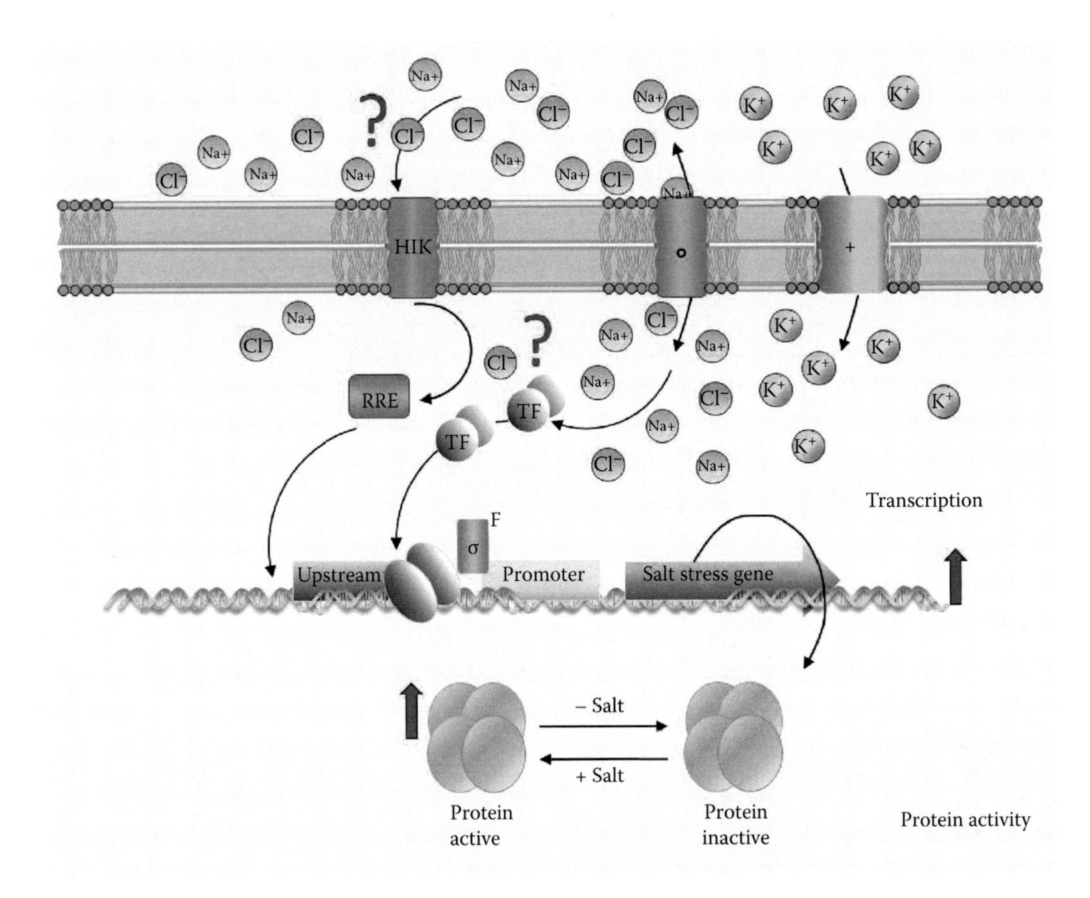

FIGURE 9.3 **(See color insert.)** Schematic view of possible regulatory processes involved in the sensing of salt stress among cyanobacteria. The figure displays possible sensory molecules like two-component systems composed by histidine kinases (HIK) and the corresponding response regulators (RRE), transcriptional factors (TF) binding together with RRE proteins to upstream promoter elements, and alternative sigma factors (σ) directing the RNA-polymerase to specific basal promoter sequences. Transporters or pores involved in ion influx or efflux (e.g., transport of Na^+, Cl^-, and K^+ over the cell membrane) are included in the model explaining the possible (red?) action of external or internal ions in the direct sensing of salt stress situations. Red arrows indicate activation of transcription and/or enzyme activities by salt stress.

9.4.2 Promoters, Sigma, and Transcriptional Factors

For the global acclimation to different environmental conditions, the changed transcription pattern of genes is mediated by the activation of different sigma factors in many bacteria. A sigma factor is a prokaryotic transcription initiation factor, which enables specific binding of the RNA polymerase to the basal gene promoter [67]. With a multiplicity of sigma factors (σ), bacteria such as *E. coli* have the possibility to alter its transcriptional program in response to stress [68]. The sigma factors of cyanobacteria can be divided into three groups. The principle sigma factor called σ^{70} represents the first group. It is essential for cell viability and responsible for the expression of housekeeping genes. The second group, σ^{70}-like factors, is not essential for the cell viability and growth. The third group comprises alternative sigma factors, which are often involved in stress acclimation or reactions toward extra-cytoplasmic signals [67,68].

Compared to cyanobacteria, the transcriptional regulation has been much better analyzed in heterotrophic bacteria. For example, in *B. subtilis*, stress exposure including high salt led to the induction of σ^B-dependent genes [69], while in *E. coli*, osmotic and salt stress is part of the σ^S

regulon [70]. Analyses of salt-stress-exposed cyanobacteria using microarrays showed that mRNAs for certain group 2 σ factors, for example, σ^B, showed increased levels [39,71]. Due to unclear nomenclature, it should be mentioned that the σ^B among cyanobacteria is not homologous to σ^B of *B. subtilis*. The screening of group 2 sigma mutants of *Synechocystis* sp. PCC 6803 revealed that only double or triple mutants of σ^B and other σ factors showed not only reduced salt but also general stress tolerance [72]. Instead, the σ^F, belonging to group 3, appeared to be more specifically involved in the transcriptional initiation of salt stress genes (Figure 9.3). A *Synechocystis* sp. PCC 6803 mutant defective in σ^F showed a reduced salt tolerance and also a decreased expression of genes, encoding for enzymes involved in the synthesis of compatible solutes like glucosylglycerol [73,74]. Phylogenetic analyses indicate that the σ^F factor of *Synechocystis* sp. PCC 6803 is related to *B. subtilis* σ^B [68,73]. Furthermore, it was shown that the promoter sequence of σ^F-dependent genes in *Synechocystis* sp. PCC 6803 is similar to that of σ^B-dependent genes in *B. subtilis* [75].

Transcriptomic and proteomic phenotyping [38,42] led to the identification of many salt-regulated genes and proteins in *Synechocystis* sp. PCC 6803. These data could be used for a systematic analysis of salt-regulated promoters in this cyanobacterial model based on the recently published promoter-mapping data [76]. Such an analysis has been only performed with a few mapped, salt-regulated promoters [72,77], which did not allow defining a salt-stress box, that is, a DNA sequence responsible for specific salt induction of genes among cyanobacteria. A detailed analysis of the *ggpS* promoter in *Synechocystis* sp. PCC 6803 revealed some insights into the salt-stress-dependent expression of this gene for the synthesis of glucosylglycerol. The *ggpS* promoter overlaps with an ORF, encoding a small, alkaline protein. Mutation of this overlapping ORF led to full induction of the promoter under low-salt conditions, showing that this small protein acts as a repressor for the *ggpS* gene under noninducing, low-salt conditions [78]. Interestingly, a similar genomic organization, that is, a small ORF upstream of *ggpS*, was found in two other, glucosylglycerol-accumulating cyanobacterial strains, *Synechococcus* sp. PCC 7002 and *Acaryochloris marina* [78].

Additionally, to regulatory proteins such as sigma and transcriptional factors, an important role of small RNAs in cyanobacterial stress acclimation is emerging [79]. It can be expected that the complex response toward salt stress will also employ the action of these interesting, new regulatory modules that seem to be more frequent than initially thought [80].

9.4.3 Direct Ionic Effects on DNA and Protein

Despite the possible roles of HIK/RRE systems, sigma factors, and other regulatory molecules, we still lack a complete understanding of mechanisms leading to salt-stress-regulated gene expression and/or enzyme activation. Therefore, many researchers have considered the involvement of alternative or additional mechanisms such as a direct role of ions and/or water amount in the sensing and signaling of salt stress [59]. Such signals could directly influence the structure of macromolecules such as proteins, for example, by interaction with the surface layer of water and/or surface charge, and DNA, for example, by influencing the supercoiling of this molecule.

Salt shock treatments on cyanobacteria are known to induce rapid fluxes of water and ions in and out of the cytoplasmic space [51,80]. These changes can easily serve as direct signals to initiate salt-stress-specific acclimation processes (Figure 9.3). Accordingly, a mutation of AqpZ, the major aquaporin in *Synechocystis* sp. PCC 6803, led not only to significantly altered shrinking behavior but also clear changes in the gene expression pattern after salt and osmotic shocks [81]. It is also known that key processes of salt acclimation such as GgpS activity for glucosylglycerol biosynthesis [82] can be directly activated by adding high amounts of salts, that is, NaCl or KCl, while the extraction of ions by dialysis leads to inactivation (reviewed by Hagemann [2]). The mechanism underlying the reversible activation/inactivation of GgpS was recently elucidated [83]. In a series of elegant experiments, it was shown that nucleic acids or polyphosphates, that is, DNA but also RNA, bind in a presumably nonsequence-specific manner to the GgpS protein, leading to its inactivation under low-salt conditions. The release of the nucleic acids from the GgpS protein can be simply induced by adding increasing amounts of inorganic ions, NaCl or KCl. This treatment resembles the ion influx into intact cells during phase 2 or phase 3 of salt shock acclimation (see Figure 9.2). These findings suggest that *in vivo* the GgpS protein is mostly bound to DNA under low-salt conditions, making glucosylglycerol biosynthesis impossible. After salt shock, the influx of high ion amounts releases the entire enzyme molecules from the inhibiting nucleic acids, leading to fully activated glucosylglycerol biosynthesis (Figure 9.3). In salt-acclimated cells, the ion concentration is only minimally increased and only a small proportion of nucleic acids is released; however, combined with the increased amount of GgpS protein [72,78], this small active fraction of GgpS is sufficient to produce enough glucosylglycerol to balance the demand by cell growth [83]. A crucial role of the cellular K^+ amount, which increases transiently after the salt shock, was also found in *E. coli.* It was shown that K^+, together with glutamate, changes the promoter affinity of the RNA polymerase, leading to a preferred binding to promoters of salt-stress-induced genes without the coaction of any salt-specific transcriptional factor [84].

DNA supercoiling also seems to play an important role in the salt-stress-specific gene expression pattern, making different gene and promoter regions more or less accessible for the recognition by RNA polymerase. It has been shown that supercoiling of DNA responds to environmental perturbations [85–87], including salt and hyperosmotic stress. In one example, cells of *Synechocystis* sp. PCC 6803 were exposed to salt, cold, and heat stress, in the presence or absence of novobiocin, and the genome-wide expression of genes was analyzed [88]. Novobiocin is a potent inhibitor of DNA gyrase and functions by inhibiting the GyrB subunit of the enzyme [89,90]. Novobiocin treatment induced the expression of many genes in *Synechocystis* sp. PCC 6803 that were found to be induced after salt or osmotic stresses as well [88].

It can be assumed that the combined action of changed cellular ion/water contents on DNA supercoiling, RNA polymerase promoter recognition, and activities of enzymes for compatible solute biosynthesis, as well as ion transporters might be sufficient to adjust the cellular metabolism to the high- or low-salt environment.

9.5 CONCLUSIONS

Different levels of salt resistance enabled cyanobacteria to settle in habitats from low up to saturating salt concentrations. For some model cyanobacteria, the basic salt acclimation strategy is well characterized on physiological and molecular levels, that is, genes for compatible solute biosynthesis and uptake as well as ion transporters are well studied. This knowledge allows the annotation of salt resistance strategies of nonstudied cyanobacterial strains based on their genome sequences. However, the regulation of salt acclimation, particularly specific salt stress signal recognition and transfer, is not well understood. In this regard, it is necessary to mention that addition of salt to freshwater media induces a very complex stress situation, that is, ionic, osmotic, oxidative, and other specific sub-stresses act at the same time on the cell. Therefore, one of the main problems is to distinguish between acclimation toward general stress or to specific salt stress, that is, increased Na^+ or Cl^- ions. The emerging new technologies such as high throughput expression analysis applied to defined mutants promise progress to solve this urgent problem. The comprehensive understanding of cyanobacterial salt stress acclimation will be useful for the application of cyanobacterial mass cultivation to produce bioenergy or valuable chemical feedstock and can be further used as a gene resource to manipulate the salt resistance of salt-sensitive crop plants.

ACKNOWLEDGMENTS

The work on cyanobacterial salt acclimation at Rostock University was generously supported by grants of the DFG (Deutsche Forschungsgemeinschaft).

REFERENCES

1. Galinski, E.A. and Trüper, H.G., Microbial behaviour in salt-stressed ecosystems, *FEMS Microbiol. Rev.*, 15, 95, 1994.
2. Hagemann, M., Molecular biology of cyanobacterial salt acclimation, *FEMS Microbiol. Rev.*, 35, 87, 2011.
3. Oren, A., The genera *Haloanaerobium*, *Halobacteroides* and *Sporohalobacter*, in *The Prokaryotes. A Handbook on the Biology of Bacteria: Ecophysiology, Isolation, Identification, Applications*, Balows, A., Trüper, H.G., Dworkin, M., Harder, W., and Schleifer, K.-H., Eds., Springer-Verlag, New York, 1992, p. 1893.
4. Oren, A. and Mana, L., Amino acid composition of bulk protein and salt relationships of selected enzymes of *Salinibacter ruber*, an extremely halophilic bacterium, *Extremophiles*, 6, 217, 2002.
5. Oren, A., Bioenergetic aspects of halophilism, *Microbiol. Mol. Biol. Rev.*, 63, 334, 1999.
6. Brown, A.D., Microbial water stress, *Bacteriol. Rev.*, 40, 803, 1976.
7. Poolman, B. and Glaasker, E., Regulation of compatible solute accumulation in bacteria, *Mol. Microbiol.*, 29, 397, 1998.
8. Reed, R.H. et al., Organic solute accumulation in osmotically stressed cyanobacteria, *FEMS Microbiol. Lett.*, 39, 51, 1986.
9. Klähn, S. et al., Glucosylglycerate: A secondary compatible solute common to marine cyanobacteria from nitrogen-poor environments, *Environ. Microbiol.*, 12, 83, 2010.
10. Fulda, S. et al., Analysis of stress responses in the cyanobacterial strains *Synechococcus* sp. PCC 7942, *Synechocystis* sp. PCC 6803, and *Synechococcus* sp. PCC 7418: Osmolyte accumulation and stress proteins synthesis, *J. Plant Physiol.*, 154, 240, 1999.
11. MacKay, M.A., Norton, R., and Borowitzka, L.J., Organic osmoregulatory solutes in cyanobacteria, *J. Gen. Microbiol.*, 130, 2177, 1984.
12. Singh, A.K. et al., Evidence for a role of L-proline as a salinity protectant in the cyanobacterium *Nostoc muscorum*, *Plant Cell Environ.*, 19, 490, 1996.
13. Hagemann, M. and Marin, K., Salt-induced sucrose accumulation is mediated by sucrose-phosphate-synthase in cyanobacteria, *J. Plant Physiol.*, 155, 424, 1999.
14. Curatti, L. et al., Sucrose-phosphatesynthase from *Synechocystis* sp. Strain PCC 6803: Identification of the *spsA* gene and characterization of the enzyme expressed in *Escherichia coli*, *J. Bacteriol.*, 180, 6776, 1998.
15. Porchia, A.C., Curatti, L., and Salerno, G.L., Sucrose metabolism in cyanobacteria: Sucrose synthase from *Anabaena* sp. strain PCC 7119 is remarkably different from the plant enzymes with respect to substrate affinity and amino-terminal sequence, *Planta*, 210, 234, 1999.
16. Hagemann, M. et al., The *stpA* gene form *Synechocystis* sp. strain PCC 6803 encodes the glucosylglycerol-phosphate phosphatase involved in cyanobacterial osmotic response to salt shock, *J. Bacteriol.*, 179, 1727, 1997.
17. Marin, K. et al., The *ggpS* gene from *Synechocystis* sp. strain PCC 6803 encoding glucosyl-glycerol-phosphate synthase is involved in osmolyte synthesis, *J. Bacteriol.*, 180, 4843, 1998.
18. Engelbrecht, F., Marin, K., and Hagemann, M., Expression of the *ggpS* gene, involved in osmolyte synthesis in the marine cyanobacterium *Synechococcus* sp. strain PCC 7002, revealed regulatory differences between this strain and the freshwater strain *Synechocystis* sp. strain PCC 6803, *Appl. Environ. Microbiol.*, 65, 4822, 1999.
19. Higo, A. et al., The role of a gene cluster for trehalose metabolism in dehydration tolerance of the filamentous cyanobacterium *Anabaena* sp. PCC 7120, *Microbiology*, 152, 979, 2006.
20. Sakamoto, T. et al., Accumulation of trehalose in response to desiccation and salt stress in the terrestrial cyanobacterium *Nostoc commune*, *Phycol. Res.*, 57, 66, 2009.
21. Wang, H.L., Postier, B.L., and Burnap, R.L., Polymerase chain reaction-based mutageneses identify key transporters belonging to multigene families involved in Na^+ and pH homeostasis of *Synechocystis* sp. PCC 6803, *Mol. Microbiol.*, 44, 1493, 2002.
22. Elanskaya, I.V. et al., Functional analysis of the Na^+/H^+ antiporter encoding genes of the cyanobacterium *Synechocystis* PCC 6803, *Biochemistry (Moscow)*, 67432, 2002.
23. Billini, M., Stamatakis, K., and Sophianopoulou, V., Two members of a network of putative Na^+/H^+ antiporters are involved in salt and pH tolerance of the freshwater cyanobacterium *Synechococcus elongates*, *J. Bacteriol.*, 190, 6318, 2008.
24. Blanco-Rivero, A. et al., *mrpA*, a gene with roles in resistance to Na+ and adaptation to alkaline pH in the cyanobacterium *Anabaena* sp. PCC 7120, *Microbiology*, 151, 1671, 2005.

25. Fukaya, F. et al., An Mrp-like cluster in the halotolerant cyanobacterium *Aphanothece halophytica* functions as a Na+/H+ antiporter, *Appl. Environ. Microbiol.*, 75, 6626, 2009.
26. Sutherland, D. et al., Osmotic regulation of transcription: Induction of the *proU* betaine transport system is dependent on accumulation of intracellular potassium, *J. Bacteriol.*, 168, 805, 1986.
27. Booth, I.R. and Higgins, C.F., Enteric bacteria and osmotic stress: Intracellular potassium glutamate as a secondary signal of osmotic stress, *FEMS Microbiol. Rev.*, 75, 239, 1990.
28. Shibata, M. et al., Genes essential to sodium-dependent bicarbonate transport in cyanobacteria: Function and phylogenetic analysis, *J. Biol. Chem.*, 277, 18658, 2002.
29. Berry, S. et al., Potassium uptake in the unicellular cyanobacterium *Synechocystis* sp. strain PCC 6803 mainly depends on a Ktr-like system encoded by slr1509 (ntpJ), *FEBS Lett.*, 548, 53, 2003.
30. Ballal, A., Basu, B., and Apte, S.K., The Kdp-ATPase system and its regulation, *J. Biosci.*, 32, 559, 2007.
31. Katoh, H., Asthana, R.K., and Ohmori, M., Gene expression in the cyanobacterium *Anabaena* sp. PCC7120 under desiccation, *Microb. Ecol.*, 47, 164, 2004.
32. Jeanjean, R. et al., Mutants of the cyanobacterium *Synechocystis* PCC 6803 impaired in respiration and unable to tolerate high salt concentrations, *FEMS Microbiol. Lett.*, 68, 125, 1990.
33. Hagemann, M. and Zuther, E., Selection and characterization of mutants of the cyanobacterium *Synechocystis* sp. PCC 6803 unable to tolerate high salt concentrations, *Arch. Microbiol.*, 158, 429, 1992.
34. Karandashova, I. et al., Identification of genes essential for growth at high salt concentrations using salt-sensitive mutants of the cyanobacterium *Synechocystis* sp. PCC 6803, *Curr. Microbiol.*, 44, 184, 2002.
35. Zuther, E., Schubert, H., and Hagemann, M., Mutation of a gene encoding a putative glycoprotease leads to reduced salt tolerance, altered pigmentation, and cyanophycin accumulation in the cyanobacterium *Synechocystis* sp. strain PCC 6803, *J. Bacteriol.*, 180, 1715, 1998.
36. Asadulghani et al., Comparative analysis of the *hspA* mutant and wild-type *Synechocystis* sp. strain PCC 6803 under salt stress: Evaluation of the role of *hspA* in salt-stress management, *Arch. Microbiol.*, 182, 487, 2004.
37. Nakamura, T. et al., A cyanobacterial non-coding RNA, Yfr1, is required for growth under multiple stress conditions, *Plant Cell Physiol.*, 48, 1309, 2007.
38. Kanesaki, Y. et al., Salt stress and hyperosmotic stress regulate the expression of different sets of genes in *Synechocystis* sp. PCC 6803, *Biochem. Biophys. Res. Commun.*, 290, 339, 2007.
39. Marin, K. et al., Identification of histidine kinases that act as sensors in the perception of salt stress in *Synechocystis* sp. PCC 6803, *Proc. Natl. Acad. Sci. U.S.A.*, 100, 9061, 2003.
40. Marin, K. et al., Gene expression profiling reflects physiological processes in salt acclimation of *Synechocystis* sp. strain PCC 6803, *Plant Physiol.*, 136, 3290, 2004.
41. Fulda, S. et al., Proteomics of *Synechocystis* sp. strain PCC 6803. Identification of periplasmic proteins in cells, grown at low and high salt concentrations, *Eur. J. Biochem.*, 267, 5900, 2000.
42. Fulda, S. et al., Proteome analysis of salt stress response in the cyanobacterium *Synechocystis* sp. strain PCC 6803, *Proteomics*, 6, 2733, 2006.
43. Ludwig, M. and Bryant, D.A., Transcription profiling of the model cyanobacterium *Synechococcus* sp. strain PCC 7002 by Next-Gen (SOLiD™) sequencing of cDNA, *Front. Microbiol.*, 2, e41, 2011
44. Hagemann, M. et al., unpublished data.
45. Waditee, R. et al., Overexpression of a Na^+/H^+ antiporter confers salt tolerance on a freshwater cyanobacterium, making it capable of growth in sea water, *Proc. Natl. Acad. Sci. U.S.A.*, 99, 4109, 2002.
46. Waditee, R. et al., Genes for direct methylation of glycine provide high levels of glycinebetaine and abiotic-stress tolerance in *Synechococcus* and *Arabidopsis*, *Proc. Natl. Acad. Sci. U.S.A.*, 102, 1318, 2005.
47. Deshnium, P. et al., Transformation of *Synechococcus* with a gene for choline oxidase enhances tolerance to salt stress, *Plant Mol. Biol.*, 29, 897, 1995.
48. Nomura, M. et al., *Synechococcus* sp. PCC 7942 transformed with *Escherichia coli bet* genes produces glycine betaine from choline and acquires resistance to salt stress, *Plant Physiol.*, 107, 703, 1995.
49. Soontharapirakkul, K. et al., Halotolerant cyanobacterium *Aphanothece halophytica* contains an Na+-dependent F_1F_0-ATPsynthase with a potential role in salt-stress tolerance, *J. Biol. Chem.*, 286, 10169, 2011.
50. Klähn, S. et al., Expression of the *ggpPS* gene for glucosylglycerol biosynthesis from *Azotobacter vinelandii* improves the salt tolerance of *Arabidopsis thaliana*, *J. Exp. Bot.*, 60, 1679, 2009.
51. Reed, R.H. et al., Multiphasic osmotic adjustment in a euryhaline cyanobacterium, *FEMS Microbiol. Lett.*, 28, 225, 1985.
52. Matsuda, N. et al., Na^+-dependent K^+ uptake Ktr system from the cyanobacterium *Synechocystis* sp. PCC 6803 and its role in the early phases of cell adaptation to hyperosmotic shock, *J. Biol. Chem.*, 279, 54952, 2004.

53. Blumwald, E., Mehlhorn, R.J., and Packer, L., Studies of osmoregulation in salt adaptation of cyanobacteria with ESR spin-probe techniques, *Proc. Natl. Acad. Sci. U.S.A.*, 80, 2599, 1983.
54. Hagemann, M., Fulda, S., and Schubert, H., DNA, RNA, and protein synthesis in the cyanobacterium *Synechocystis* sp. PCC 6803 adapted to different salt concentrations, *Curr. Microbiol.*, 28, 201, 1994.
55. Miller, D.M. et al., Ion metabolism in a halophilic blue green alga *Aphanothece halophytica*, *Arch. Microbiol.*, 111, 145, 1976.
56. Reed, R.H. and Stewart, W.D.P., Osmotic adjustment and organic solute accumulation in unicellular cyanobacteria from freshwater and marine habitats, *Mar. Biol.*, 88, 1, 1985.
57. Stock, A.M., Robinson, V.L., and Goudreau, P.N., Two-component signal transduction, *Annu. Rev. Biochem.*, 69, 183, 2000.
58. Aguilar, P.S. et al., Molecular basis of thermosensing: A two-component signal transduction thermometer in *Bacillus subtilis*, *EMBO J.*, 20, 1681, 2001.
59. Wood, J.M., Osmosensing by bacteria: Signals and membrane-based sensors, *Microbiol. Mol. Biol. Rev.*, 63, 230, 1999.
60. Suzuki, I. et al., The pathway for perception and transduction of low-temperature signals in *Synechocystis*, *EMBO J.*, 19, 1327, 2000.
61. Shoumskaya, M.A. et al., Identical Hik-Rre systems are involved in perception and transduction of salt signals and hyperosmotic signals but regulate the expression of individual genes to different extents in *Synechocystis*, *J. Biol. Chem.*, 280, 21531, 2005.
62. Murata, N. and Suzuki, I., Exploitation of genomic sequences in a systematic analysis to access how cyanobacteria sense environmental stress, *J. Exp. Bot.*, 57, 235, 2006.
63. Los, D.A. et al., Stress responses in *Synechocystis*: Regulated genes and regulatory systems, in *The Cyanobacteria-Molecular Biology, Genomics and Evolution*, Herrero, A. and Flores E., Eds., Caister Academic Press., Norfolk, U.K., 2008, p. 117.
64. Mikami, K. et al., The histidine kinase Hik33 perceives osmotic stress and cold stress in *Synechocystis* sp. PCC 6803, *Mol. Microbiol.*, 46, 905, 2002.
65. Los, D.A. et al., Stress sensors and signal transducers in cyanobacteria, *Sensors*, 10, 2386, 2010.
66. Sakayori, T., Shiraiwa, Y., and Suzuki, I., A *Synechocystis* homolog of SipA protein, Ssl3451, enhances the activity of the histidine kinase Hik33, *Plant Cell Physiol.*, 50, 1439, 2009.
67. Imamura, S. and Asayama, M., Sigma factors for cyanobacterial transcription, *Gene Regul. Syst. Biol.*, 2009, 65, 2009.
68. Gruber, T.M. and Gross, C.A., Multiple sigma subunits and the partitioning of bacterial transcription space, *Annu. Rev. Microbiol.*, 57, 441, 2003.
69. Hecker, M., Pané-Farré, J., and Völker, U., SigB-dependent general stress response in *Bacillus subtilis* and related Gram-positive bacteria, *Annu. Rev. Microbiol.*, 61, 215, 2007.
70. Hengge-Aronis, R., Back to log phase: Sigma S as a global regulator in the osmotic control of gene expression in *Escherichia coli*, *Mol. Microbiol.*, 21, 887, 1996.
71. Kanesaki, Y. et al., Histidine kinases play important roles in the perception and signal transduction of hydrogen peroxide in the cyanobacterium, *Synechocystis* sp. PCC6803, *Plant J.*, 49, 313, 2007.
72. Pollari, M. et al., Characterization of single and double in activation strains reveals new physiological roles for group 2 sigma factors in the cyanobacterium *Synechocystis* sp. PCC6803, *Plant Physiol.*, 147, 1994, 2008.
73. Huckauf, J. et al., Stress responses of *Synechocystis* sp. strain PCC 6803 mutants impaired in genes encoding putative alternative sigma factors, *Microbiology*, 146, 2877, 2000.
74. Marin, K. et al., Salt-dependent expression of glucosylglycerol-phosphate synthase, involved in osmolyte synthesis in the cyanobacterium *Synechocystis* sp. strain PCC 6803, *J. Bacteriol.*, 184, 2870, 2002.
75. Imamura, S. et al., Purification, characterization, and gene expression of all sigma factors of RNA polymerase in a cyanobacterium, *J. Mol. Biol.*, 325, 857, 2003.
76. Mitschke, J. et al., An experimentally anchored map of transcriptional start sites in the model cyanobacterium *Synechocystis* sp. PCC 6803, *Proc. Natl. Acad. Sci. U.S.A.*, 108, 2124, 2011.
77. Vinnemeier, J. and Hagemann, M., Identification of salt-regulated genes in the genome of the cyanobacterium *Synechocystis* sp. strain PCC 6803 by subtractive RNA hybridization, *Arch. Microbiol.*, 172, 377, 1999.
78. Klähn, S. et al., The gene ssl3076 encodes a protein mediating the salt-induced expression of ggpS for the biosynthesis of the compatible solute glucosylglycerol in *Synechocystis* sp. strain PCC 6803, *J. Bacteriol.*, 192, 4403, 2010.
79. Georg, J. et al., Evidence for a major role of antisense RNAs in cyanobacterial gene regulation, *Mol. Syst. Biol.*, 5, 305, 2009.

80. Blumwald, E., Mehlhorn, R.J., and Packer, L., Ionic osmoregulation during salt adaptation of the cyanobacterium *Synechococcus* 6311, *Plant Physiol.*, 73, 377, 1983.
81. Shapiguzov, A. et al., Osmotic shrinkage of cells of *Synechocystis* sp. PCC 6803 by water efflux via aquaporins regulates osmostress-inducible gene expression, *Microbiology*, 151, 447, 2005.
82. Hagemann, M. et al., Biochemical characterization of glucosylglycerol-phosphate synthase of *Synechocystis* sp. strain PCC 6803: Comparison of crude, purified, and recombinant enzymes, *Curr. Microbiol.*, 43, 278, 2002.
83. Novak, J.F., Stirnberg, M., and Marin, K., A novel mechanism of osmosensing: Salt dependent protein-DNA interaction in the cyanobacterium *Synechocystis* sp. PCC 6803, *J. Biol. Chem.*, 286, 3235, 2011.
84. Gralla, J.D. and Vargas, D.R. Potassium glutamate as transcriptional inhibitor during bacterial osmoregulation, *EMBO J.*, 25, 1515, 2005.
85. Wang, J.C. and Lynch, A.S., Transcription and DNA supercoiling, *Curr. Opin. Genet. Dev.*, 3, 764, 1993.
86. Higgins, C.F. et al., A physiological role for DNA supercoiling in the osmotic regulation of gene expression in *S. typhimurium* and *E. coli*, *Cell*, 52, 569, 1988.
87. Dorman, C.J., DNA supercoiling and bacterial gene expression, *Sci. Prog.*, 89, 151, 2006.
88. Prakash, J.S. et al., DNA supercoiling regulates the stress inducible expression of genes in the cyanobacterium *Synechocystis*, *Mol. Biosyst.*, 5, 1904, 2009.
89. Maxwell, A., DNA gyrase as a drug target, *Biochem. Soc. Trans.*, 27, 48, 1999.
90. Lewis, R.J., Tsai, F.T.F., and Wigley, D.B., Molecular mechanisms of drug inhibition of DNA gyrase, *Bioassays*, 18, 661, 1996.

10 Regulatory Mechanisms of Cyanobacteria in Response to Osmotic Stress

Aran Incharoensakdi and Rungaroon Waditee-Sirisattha*

CONTENTS

10.1 INTRODUCTION

Cyanobacteria are pioneering and ubiquitous oxygenic phototrophs with diverse habitats on the planet. Existing in different morphologies ranging from the unicellular to the filamentous, cyanobacteria perform different modes of metabolism, easily switching from one mode to another [1,2]. For example, under adverse environmental conditions, such as a hypersaline (i.e., condition in which the salinity of the water is above $35\,g\,L^{-1}$) environment, cyanobacteria employ a variety of mechanisms to maintain and adjust their internal osmotic status. One such mechanism is the exclusion of sodium (Na^+) ions from cells to maintain ion homeostasis [3–5]. Another mechanism is the ability to accumulate low-molecular-weight organic-compatible solutes [6–8]. In addition, strategies for cyanobacteria to survive during acclimation to osmotic stress also involve the synthesis of

* Corresponding author: aran.i@chula.ac.th

stress-related proteins as well as the appropriate adjustments of various metabolic activities related to photosynthesis and nitrogen metabolism.

In this review, we describe various regulatory mechanisms of cyanobacteria in response to osmotic stress with NaCl stress in particular. The perception and transduction of salt-stress signals by cyanobacteria is not included in the present review as this has been recently described [9]. We place special emphasis on (1) regulation of ion homeostasis as the mechanism mediated by Na^+/H^+ antiporter, cation/H^+ transporter, Mrp-like protein, and Na^+-ATPase; (2) accumulation of compatible solutes; (3) synthesis of salt-stress proteins; (4) regulation of photosynthesis; and (5) nitrogen metabolism. Thus, this review will impart knowledge on cyanobacterial responses to osmotic stress via activation of diverse regulatory and physiological cellular mechanisms.

10.2 REGULATION OF ION HOMEOSTASIS

Our world is faced with adverse environmental conditions, such as pollution and water stress (including drought, flooding, and salinity), affecting our biota on the land and in the water. Among these, salinity can be considered as one of the most significant environmental problems facing the world [10]. It is well established that dramatic changes in the salt levels cause osmotic and ionic stresses among plants including cyanobacteria. When faced with a change in salt concentrations in their natural habitats, cyanobacteria employ strategies to maintain ion homeostasis within the cell in order to maintain normal growth and survival under high salinity. The best characterized of these strategies is the increased expression and activity of Na^+/H^+ and additional monovalent cation/H^+ antiporters [11–14]. These transmembrane proteins not only maintain the intracellular pH by uptake of protons but also by utilizing outward monovalent cation gradients. Multiple cation/proton antiporters have been identified in cyanobacteria [3,5,15,16].

The Na^+/H^+ antiporters are integral membrane proteins that catalyze the exchange of Na^+ for H^+ across membranes and play a major role in pH and sodium homeostasis [3,14]. The primary energy source for this system in most organisms is the proton electrochemical gradient ($\Delta\mu_H{}^+$) across the cytoplasmic membrane. The proton electrochemical gradient is derived either from respiratory electron transport or at the expense of ATP formed during substrate-level phosphorylation by activity of the membrane ATPase [17]. In addition, Na^+/H^+ antiporters have been shown to be involved in diverse physiological processes such as maintenance of cell volume [18], morphogenesis [19], extrusion of the extra H^+ generated during metabolism. Moreover, these transporters are believed to be crucial for salt tolerance, especially when the organism lacks primary Na^+ pumps or when the Na^+ pumps are not operative, which is often observed in higher plants [3].

10.2.1 Multiplicity and Physiological Functions of Na^+/H^+ and Cation/H^+ Antiporters in Cyanobacteria

To date, genomic sequence projects of several species from bacteria [20] to higher plants [21] including rice [22,23] and to mammals [24] have been completed, and many more are in the pipeline. Looking at the Na^+/H^+ antiporters among the sequenced genomes, we can find the Na^+/H^+ antiporters cluster in several families. These are CPA (cation:proton antiporter) families (i.e., CPA1, CPA2, and CPA3); and Nha families (i.e., NhaA, NhaB, NhaC, and NhaD). Among them, physiological and molecular genetic studies of the Na^+/H^+ antiporters system in *Escherichia coli* are the most comprehensive and contribute extensively to our understanding of the physiology of Na^+/H^+ antiporters [14,25]. Three antiporters in *E. coli*, NhaA, NhaB, and ChaA, are known and their functional characteristics have been well described [25,26]. In the dicot model, *Arabidopsis*, the Na^+/H^+ antiporters have been explored to a great extent, including cloning and screening of mutants defective in salt tolerance. For example, the *AtNHX1* gene encoding a tonoplast Na^+/H^+ antiporter has been cloned and identified by similarity to the yeast vacuolar antiporter *NHX1* gene [27]. On the other hand, the *SOS1* gene encoding a plasma membrane Na^+/H^+ antiporter has been identified by screening of mutants defective in salt tolerance. This was the first isolation of a Na^+ efflux transporter across plasma membrane in *Arabidopsis*. *SOS1* putative gene plays a major role for Na^+ extrusion across plasma membrane and controls long distance Na^+ transport from root to shoot [28,29].

In cyanobacteria, entire genome sequences of several cyanobacterial species are available and will be most useful for assigning individual Na^+/H^+ antiporter members to gene families. Interestingly, a variety of Na^+/H^+ antiporters behave cooperatively in the cyanobacteria. For instance, in a moderately salt-tolerant cyanobacterium, *Synechocystis* sp. PCC 6803, six genes have been annotated as genes encoding putative Na^+/H^+ antiporters. These are *slr1727* (NhaS1), *sll0273* (NhaS2), *sll0689* (NhaS3), *slr1595* (NhaS4), *slr0415* (NhaS5), and *sll0556* (NhaS6) [3,16,30,31]. These genes show very high homology to those found in higher plants. Phylogenetic relationship of these Na^+/H^+ antiporters suggests that the NhaS1 and NhaS2 closely resemble the CPA1 family, while the NhaS3, NhaS4, and NhaS5 closely resemble the CPA2 family. The NhaS3 has a high affinity for both sodium and lithium ions [3]. A recent study using immunogold labeling revealed that the NhaS3 was localized in the thylakoid membrane of *Synechocystis* sp. PCC 6803 [16].

The multiplicity of Na^+/H^+ antiporters is also found in a freshwater-type cyanobacterium, *Synechococcus elongatus* PCC 7942. Analysis of the recently completed genomic sequence of *S. elongatus* revealed the presence of seven putative Na^+/H^+ antiporters (Nha1, Nha2, Nha3, Nha4, Nha5, Nha6, and Nha7) [15]. Among these putative proteins, the Nha3 protein showed Na^+/H^+ antiporter activity in everted membrane vesicles and successfully complemented the salt-sensitive phenotype of recipient TO114 cells. In contrast, the Nha1, Nha4, Nha6, and Nha7 proteins showed low Na^+/H^+ antiporter activity and were not able to complement the salt-sensitive phenotype of TO114 cells. Additionally, inactivation of six of the seven *nha* genes in *S. elongatus* revealed that these genes have essential roles in growth at different salt concentrations and pH.

In a halotolerant cyanobacterium, *Aphanothece halophytica* isolated from the Dead Sea in Israel, a shot-gun genome sequencing revealed at least eight putative genes encoding Na^+/H^+ antiporters [5,32]. These putative antiporters belong to CPA1 (ApNhaP1, ApNhaP2) and CPA2 (ApNapA1-1, ApNapA1-2), respectively. The ApNhaP1 exhibits the exchange activity with novel ion specificity as well as its high activity over a wide pH range [5]. In contrast, ApNapA1-1 and ApNapA1-2 exhibit strongly pH-dependent Na^+/H^+ antiporter activities with higher activities at alkaline pH [32]. Moreover, the ability of ApNapA1-2 to take up K^+ was also demonstrated suggesting its function as a Na^+/K^+ antiporter. This property of ApNapA1-2 is beneficial to the cells under salt stress due to the fact that the ratio of Na^+/K^+ inside the cells can be maintained at a low level.

10.2.1.1 Crucial Role of Conserved Amino Acids

Functional characterization of the conserved amino acids of NhaP- and NapA-type Na^+/H^+ antiporters was conducted in *Synechocystis* sp. PCC 6803 [30] and a halotolerant cyanobacterium *A. halophytica* [5,32]. Putative genes encoding the Na^+/H^+ antiporter from *Synechocystis* sp. PCC 6803 have been functionally characterized and expressed in the salt-sensitive *E. coli* mutant in which three putative Na^+/H^+ antiporters are disrupted [33]. Among them, only NhaS1 (CPA1 family) and NhaS3 (CPA2 family) complemented the salt-sensitive *E. coli* mutant and could exhibit the exchange activity of Na^+, Li^+ for H^+.

NhaS1 (or SynNhaP) can exchange $Na^+(Li^+)$ for H^+ over a wide range of pH ranging from 5 to 9, and their activities are insensitive to amiloride [30]. Kinetic analysis revealed that NhaS1 is a low affinity Na^+/H^+ antiporter [3]. Mutagenesis study revealed that the replacement of Asp-138 with Glu-138 or Tyr-138 in SynNhaP could abolish the Na^+/H^+ and Li^+/H^+ exchange activities of SynNhaP, indicating the importance of Asp-138 for the antiporter activity of SynNhaP [30]. The importance of aspartic acid (Asp-139) in transmembrane segment for the exchange activity was also identified in ApNhaP1 (Waditee et al., unpublished). In the case of ApNapA (CPA2 family), the importance of Glu129, Asp225, and Asp226 in the transmembrane segment and Glu142 in the loop region for the exchange activity has been demonstrated [32].

10.2.1.2 Regulation of Na^+/H^+ Antiporter Selectivity by C-Terminus

Prediction of membrane topology of the *Synechocystis*- and the *Aphanothece*-SynNhaP1 and ApNhaP1, respectively, revealed the presence of 11 transmembrane segments with a long hydrophilic cytoplasmic tail in the C-terminal part (123 amino acid residues in the case of SynNhaP1 and 124 amino acid residues in the case of ApNhaP1). Functional studies of the chimera proteins by exchanging the C-terminus between SynNhaP1 and ApNhaP1 and vice versa have revealed the importance of the C-terminal tail for ion specificity [5,30]. The putative SynNhaP1 exhibited the exchange of Na^+, Li^+ for H^+ whereas ApNhaP1 exhibited the exchange of Na^+, Ca^{2+} for H^+. Although both of them contain a similar number of amino acids at the C-terminal tail, their charges are different. The C-terminus of ApNhaP1 has more positive charges (22^+, 14^-) while that of SynNhaP1 has more negative charges (15^+, 24^-). The chimera ASNhaP, in which a long C-terminus of ApNhaP1 was replaced with that of SynNhaP1, could exhibit the exchange of Li^+ for H^+ that has never been observed in the wild-type ApNhaP1. The exchange activities observed in the chimera between ApNhaP1 and SynNhaP1 demonstrated the importance of the charge on C-terminal tail, which could have the bearing on their ion specificity. The truncated versions of SynNhaP1 by deletion of C-terminus not only diminished the exchange of antiporter activity but also increased the affinity for Na^+ and Li^+ ion [34]. Based on these results, it was hypothesized that the C-terminal tail of NhaP played a role for ion specificity and could affect the Km of the exchange activity. In higher plant, the C-terminus of AtNHX1 appears to be involved in the determination of the ion selectivity of the transporter [35]. There is evidence that the Nha1p C-terminus is involved in the cell response to sudden changes in environmental osmolarity [36]. Recently, it has been shown that a conserved domain in the tail region of Nha1p plays an important role in localization and salinity-resistant cell growth [37]. The topology of various types of Na^+/H^+ antiporters from *A. halophytica* is shown in Figure 10.1. The ApNapA3 contains a very long C-terminal tail compared to those of ApNhaP1 and ApNapA1.

10.2.2 Mrp-Like Protein and Other Cation Transporters

The majority of the Na^+/H^+ antiporters are classified as members of the CPA1 and CPA2 families. However, a very recent study has shown that membrane proteins in CPA3 family could exhibit the exchange of Na^+ for H^+. Genes encoding the Mrp-like proteins (multiple resistance and pH adaptation), belonging to CPA3 family, were found in *A. halophytica* [38] and *Anabaena* sp. PCC 7120 [39,40]. Mrp systems are widespread among bacteria and archaea. Their structure consists of highly unusual multi-subunits with a unique complexity [41].

In *A. halophytica*, the *mrp* homolog gene cluster *mrpCD1D2EFGAB* (Ap-*mrp*) was identified from the shot-gun genome sequencing. The Ap-*mrp* complemented the salt-sensitive phenotype of TO114 and exhibited Na^+/H^+ and Li^+/H^+ exchange activities, indicating that Ap-Mrp functions as a Na^+/H^+ antiporter.

In addition to the CPA1, CPA2, and CPA3, a number of transporters may contribute to osmotic stress tolerance. Global transcriptional response of *Synechocystis* sp. PCC 6803 revealed up-regulation of the *slr2057* transcript encoding a water channel protein (ApqZ). This channel protein, which is responsive to osmotic shock, had elevated transcript levels at pH 10. The cells exhibited altered cell shrinkage and gene expression under hyperosmotic stress conditions [42]. The *slr0875* gene encoding a high-conductance mechanosensitive channel has recently been reported [43].

10.2.3 Role of Na^+-ATPase

Cells thriving in a high Na^+ ion concentration usually have a mechanism to maintain low intracellular Na^+ levels. The removal of Na^+ from cytoplasm out of the cells via plasma membrane or via tonoplast membrane into the vacuoles in higher plants has been established. In most plant cells,

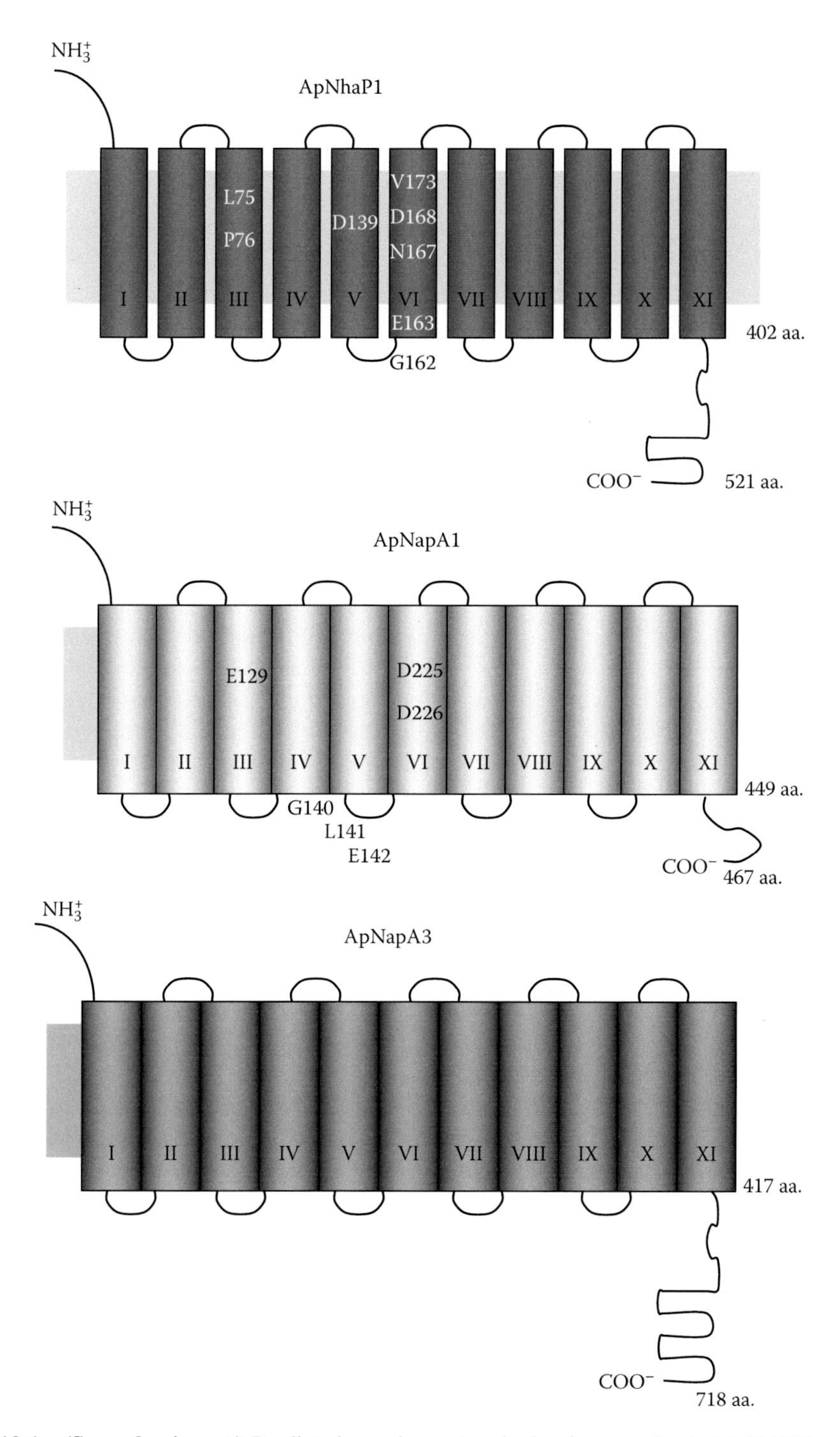

FIGURE 10.1 **(See color insert.)** Predicted membrane topologies for cyanobacterial Na^+/H^+ antiporters. The possible transmembrane of the ApNhaP1 (accession number: AB059562), ApNapA1 (accession number: AB193603), and ApNapA3 (unpublished data) sequences were deduced by a computer program TopPredII. Charged amino acids are indicated by red and yellow.

the removal of Na^+ from cytoplasm is mediated through Na^+/H^+ antiporters located at both plasma membrane and tonoplast membrane. The operation of the Na^+/H^+ antiporter is made possible by utilizing proton motive force or a proton gradient generated by a primary proton pump (H^+-ATPase in the plasma membrane and the tonoplast membrane; and H^+-pyrophosphatase in the tonoplast membrane) as a driving force. A role for H^+-ATPase in response to salt stress has been reported in several higher plants. A greater amount of H^+-ATPase transcript was detected in a halophyte *Atriplex mummularia* than in a glycophyte salt-sensitive tobacco [44]. Increases in the vacuolar H^+-ATPase activities were found in roots and shoots of a halophyte *Sueada salsa* under salt stress [45].

Many studies in cyanobacteria have implicated H^+-pumping involvement in Na^+ extrusion of salt-treated cyanobacteria including that of membrane vesicles. The H^+-pumping activity is attributed to either the respiratory activity or the activity of P-type H^+-ATPase [44]. Recently, a study in the halotolerant *A. halophytica* has demonstrated the presence of H^+-ATPase in *A. halophytica* which was assigned to a P-type ATPase based on its inhibition by vanadate [46]. The P-type H^+-ATPase in the cytoplasmic membrane of *A. halophytica* showed an enhanced activity in cells grown under high-salt conditions. The increase in H^+-ATPase activity can provide a higher electrochemical gradient necessary for a better operation of a Na^+/H^+ antiporter in *A. halophytica* under salt stress.

Another primary pump involved in Na^+ extrusion, the so-called Na^+-ATPase, has been found in bacteria, fungi, yeast, algae, and some higher plants [47–52]. On the contrary, the existence of Na^+-ATPase in cyanobacteria is still controversial. An early study in *Synechococcus* sp. PCC 7942 by Ritchie (1992) [53] concluded that a primary active transport for Na^+ is necessary in cells thriving under alkaline conditions. In the marine cyanobacterium *Oscillatoria brevis*, it has been shown that NaCl could stimulate light-supported generation of membrane potential, thus suggesting a light-dependent primary Na^+ pump in the cytoplasmic membrane of *O. brevis* [54]. In *Synechocystis* sp. PCC 6803, it was suggested that Na^+-ATPase might play a role in salt resistance [12]. This was based on the findings that the mutant disrupted in a putative Na^+-ATPase subunit showed high NaCl sensitivity. Very recently, a study in the halotolerant cyanobacterium *A. halophytica* has shown the existence of Na^+-ATPase which can translocate Na^+ into proteoliposomes [55]. This enzyme was assigned to an F-type ATPase based on the band pattern of SDS–PAGE analysis as well as the inhibition of activity by dicyclohexyl carbodiimide (DCCD) together with the protection of DCCD inhibition by Na^+.

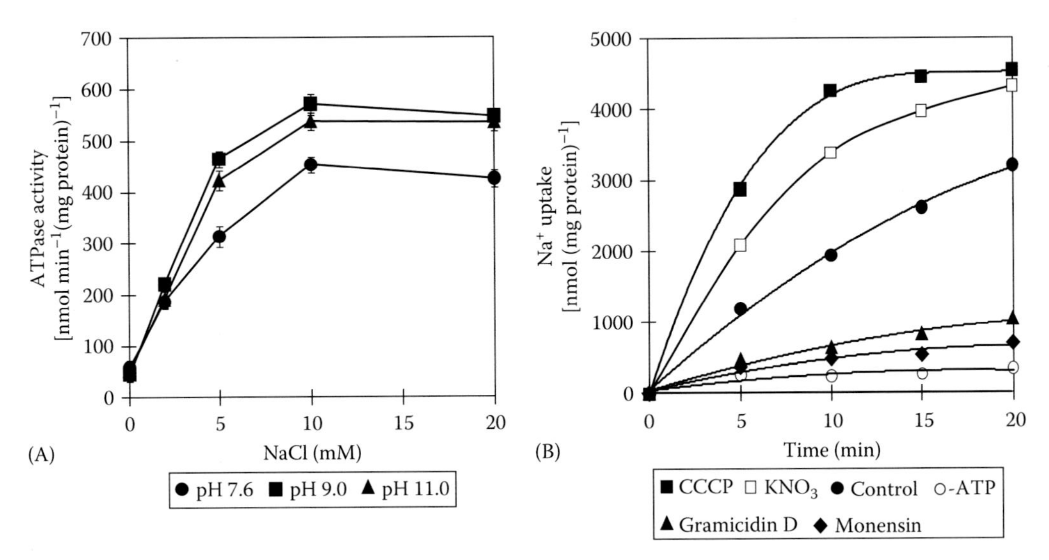

FIGURE 10.2 Activity of the purified Na^+-ATPase at various pHs (A) and Na^+ uptake activity of the proteoliposomes reconstituted with the purified Na^+-ATPase (B). (From Soontharapirakkul, K. and Incharoensakdi, A., *BMC Biochem.*, 11, 30, 2010. With permission.)

The activity of Na^+-ATPase increased with increasing concentration of NaCl as shown in Figure 10.2A. This Na^+-ATPase was more active at high pH of 9.0 and 11.0 than at low pH of 7.6. The presence of 10 mM NaCl could stimulate the enzyme activity by 9-fold at pH 7.6, and 11-fold at pH 9.0 and 11.0. In addition, the increase of both NaCl concentration and pH of the growth medium led to the increase of Na^+-ATPase activity. This suggests that Na^+-ATPase in *A. halophytica* might play a role in Na^+ homeostasis as well as pH regulation inside the cells.

The proteoliposomes reconstituted with Na^+-ATPase were able to take up Na^+ upon supplementation with ATP (Figure 10.2B). The hydrolysis of ATP by Na^+-ATPase provided the driving force needed for Na^+ uptake. The uptake of Na^+ was stimulated by protonophore and the permeant anion nitrate, whereas the uptake was abolished by monensin and gramicidin. This suggested the operation of electrogenic Na^+ transport mediated by Na^+-ATPase in *A. halophytica*. Most recently, the operon of the F-type Na^+-ATPase (*ApNa$^+$-atp*) of *A. halophytica* has been characterized [56]. This operon is composed of nine genes with homologous operon found in some cyanobacteria such as *Synechococcus* sp. PCC 7002 and *Acaryochloris marina*. An *E. coli* mutant DK8 (deficient in ATP synthase) transformed with *ApNa$^+$-atp* exhibited Na^+-dependent ATP hydrolysis activity. In addition, a freshwater *Synechococcus* sp. PCC 7942 transformed with *ApNa$^+$-atp* also had an increased tolerance to salt stress, which was attributed to the presence of Na^+-ATPase in the cytoplasmic membrane of *Synechococcus* sp. PCC 7942. Taken together, the role of ATPase in *A. halophytica* for salt tolerance has been established. The H^+-ATPase provides the proton motive force to facilitate Na^+/H^+ antiporter to extrude Na^+ from the cytoplasm. On the other hand, Na^+-ATPase functions as a primary Na^+-pump to exclude Na^+ from the cells and this occurs more favorably at an alkaline pH where the proton electrochemical gradient is low or insufficient.

10.3 ACCUMULATION OF COMPATIBLE SOLUTES

For cells thriving under high-salinity conditions, accumulation of low-molecular-weight organic-compatible solutes (osmoprotectants) is one of the important strategies to cope with high osmotic stress. As in other living organisms, cyanobacteria possess an uptake system and/or biosynthetic pathway for accumulating compatible solutes under osmotic-stress conditions. Compatible solutes are low-molecular mass organic compounds such as sugars, some amino acids, and quaternary ammonium compounds [6–8] with no net charge. They can be accumulated in high amounts without negatively interfering with cellular metabolism [57]. The accumulation of compatible solutes in stressed cells can compensate for the difference in water potential and enable cells to take up water, leading to the maintenance of turgor pressure.

It has been demonstrated long time ago that different salt-tolerant cyanobacteria can accumulate different types of compatible solutes. Furthermore, the types of compatible solutes accumulated inside the cells can determine the degree of salt tolerance among various cyanobacteria. For example, three groups of salt-tolerant cyanobacteria are classified according to the types of compatible solutes the cells accumulate. The low salt-tolerant cyanobacteria with maximum tolerance up to 0.7 M NaCl accumulate sucrose and/or trehalose [58]. The moderately salt-tolerant (maximum tolerance up to 1.25 M NaCl) accumulate glucosylglycerol [59,60], whereas the highly salt-tolerant (maximum tolerance up to 2.5–3.5 M NaCl) accumulate glycine betaine or glutamate betaine [7,61]. The structures of compatible solutes accumulated inside cyanobacteria are shown in Figure 10.3.

10.3.1 TREHALOSE AND SUCROSE

The non-reducing disaccharide, trehalose, is the compatible solute accumulated in cyanobacteria in response to a number of environmental stresses such as heat, water, or osmotic stress [62]. Trehalose is composed of two glucose molecules with a 1α–1α bond linkage. It has other functions apart from being a compatible solute; for example, it has been shown to be a carbon source [63–65]. In certain bacteria, trehalose can be a constituent of cell wall glycolipids [66,67]. The use of trehalose as a

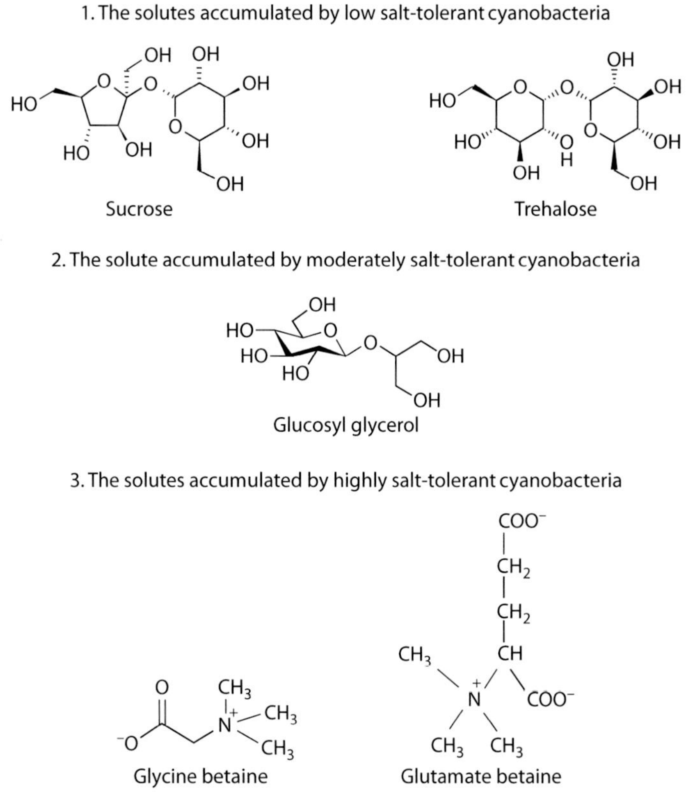

1. The solutes accumulated by low salt-tolerant cyanobacteria

Sucrose

Trehalose

2. The solute accumulated by moderately salt-tolerant cyanobacteria

Glucosyl glycerol

3. The solutes accumulated by highly salt-tolerant cyanobacteria

Glycine betaine

Glutamate betaine

FIGURE 10.3 The structure of compatible solutes accumulated in salt-stressed cyanobacteria.

reserve compound is found in yeast [68,69]. The most common pathway for the synthesis of trehalose is via the reactions catalyzed by trehalose phosphate synthase (TPS) and trehalose phosphate phosphatase (TPP). In this pathway, glucose-6-phosphate accepts one glucose molecule donated by UDP-glucose to form trehalose-6-phosphate, which is catalyzed by TPS. Trehalose-6-phosphate is further dephosphorylated by TPP to produce trehalose. This pathway of trehalose synthesis is found in bacteria and yeast as well as in higher plants and insects [70].

In cyanobacteria, the activities of both TPS and TPP are present in the crude extracts of *Scytonema* sp. [71]. ADP-glucose is a better substrate for TPS than UDP-glucose. The enhancement of TPS and TPP activities occurs in salt-shocked *Scytonema* sp. together with the increase in trehalose contents. Another pathway existing in *Anabaena* sp. PCC 7120 is the TreY/TreZ pathway [72]. In this pathway, trehalose is obtained from the conversion of maltooligosaccharides by two enzymatic steps catalyzed by maltooligosyltrehalose synthase (TreY) and maltooligosyltrehalose trehalohydrolase (TreZ). These two enzymes are encoded by a gene cluster of *treY* and *treZ*. Recently, another gene, *treH* encoding a protein trehalase (TreH) has also been found in *Anabaena* sp. PCC 7120 [58].

One important property of trehalose is to stabilize and protect the membranes and proteins of the cells under stress, especially those that can survive cycles of dehydration and rehydration [73,74]. The metabolism of trehalose with regard to the accumulation in response to desiccation and salt stress was well studied in the cyanobacterium *Nostoc punctiforme* IAM M-15 m [75]. Trehalose was undetectable in fully hydrated cell, which was in contrast to its accumulation under desiccation stress. During desiccation mRNAs for the three enzymes (TreH, TreY, and TreZ) involved in trehalose metabolism were detected at similar levels. The stress by NaCl treatment could also induce the accumulation of trehalose in *N. punctiforme*. The degradation of trehalose by trehalase (TreH) was strongly inhibited in the presence of 10 mM NaCl as opposed to the uninhibited activity of trehalose synthetic enzymes, namely maltooligosyltrehalose synthase (TreY) and hydrolase (TreZ). In a freshwater cyanobacterium *Anabaena* sp. PCC 7120, the three genes (*treH*, *treY*, *treZ*) for trehalose metabolism are markedly upregulated upon dehydration of the cells [58]. Nevertheless, only a small amount of trehalose is detected while a large amount of sucrose is accumulated, thus suggesting that the function of trehalase (TreH) may negatively regulate dehydration tolerance. This contention is supported by the evidence that the gene disruption of TreH leads to an increase in both the amount of accumulated trehalose and the dehydration tolerance [58]. Furthermore, the induction of TreH synthesis can be demonstrated in *Nostoc commune* DRH1 following rehydration [76]. However, since the accumulation of trehalose is low and therefore unlikely to stabilize and protect membranes and proteins during dehydration, it has been further demonstrated that a chaperone Dnak induced by trehalose might be important for the dehydration tolerance of *Anabaena* sp. PCC 7120 [58]. In a terrestrial cyanobacterium *N. commune*, trehalose accumulation is an important feature in response to water loss during desiccation [75]. Similar levels of trehalose accumulation are detected in *N. commune* under salt stress by treatment with 0.2 M NaCl. In another strain of *N. commune*, namely, *N. commune* DRH1, sucrose is also present in abundance (0.8 mg g^{-1} dry weight), second only to trehalose (1 mg g^{-1} dry weight) during desiccation [76]. The levels of both sucrose and trehalose decreased to less than 0.3 mg g^{-1} dry weight after 30 min rehydration. The ability of sucrose to confer desiccation tolerance on bacteria has been demonstrated in recombinant *E. coli*. The expression of sucrose phosphate synthase gene of *Synechocystis* sp. PCC 6803 in *E. coli* can increase desiccation tolerance of the recombinant cells [77]. At present it remains unclear why freshwater *Anabaena* sp. PCC 7120 accumulates sucrose as the sole compatible solute whereas *N. commune* and *Nostoc punctiforme* IAM M-15, both of terrestrial habitat, accumulate trehalose. A likely explanation may be due to the higher capacity of trehalose degrading enzyme than synthetic enzyme in *Anabaena* sp. PCC 7120, all of which are upregulated during dehydration [58]. The mat-forming cyanobacterium *Phormidium autumnale* isolated from sewage plant in Israel was reported to accumulate 6.2 μg trehalose μg^{-1} chlorophyll in response to matric water stress [78]. Upon salt stress, a similar maximal trehalose concentration was detected in *P. autumnale*. The intracellular content of trehalose corresponded with the stress condition. After 30 min release of water stress, about 70% of trehalose remained and only 20% was detected after 8 h of release. This reduced content of trehalose was likely due to its breakdown into phosphorylated glucose after the release of water stress. Other mat-forming cyanobacteria grown in high NaCl concentration and contain high levels of trehalose (3–6 μg μg^{-1} chlorophyll) are *Microcoleus chthonoplastes* [79], *Microchaete grisea*, and *Rivularia atra* [80,81]. Exceptionally high levels of trehalose (9–182 μg μg^{-1} chlorophyll) have been reported for cyanobacterial mats consisting mainly of *Oscillatoria* sp. [79]. Not only does the mat-forming cyanobacterium *M. chthonoplastes* accumulate trehalose under osmotic stress, but glucosylglycerol is found to coexist. However, under hypoosmotic stress the relative proportion of trehalose with respect to glucosylglycerol is elevated [82]. Most mat-forming cyanobacteria respond to a decrease in salinity by excretion of the osmolytes. This suggests that glucosylglycerol is predominantly excreted in *M. chthonoplates* under hypoosmotic stress. In addition, the ability of *M. chthonoplates* to degrade and partially ferment glucosylglycerol has been reported [83], thus enabling this organism to thrive under a wide range of salinities.

In freshwater cyanobacteria, a unicellular *Synechococcus* sp. PCC 6301 and a filamentous *Anabaena* sp. PCC 7120 both accumulate sucrose during adaptation to salt stress [84]. The increased sucrose contents are attributed to an enhancement of sucrose phosphate synthase (SPS) activity which synthesizes sucrose phosphate from UDP-glucose and fructose. Sucrose is finally obtained after the hydrolysis of sucrose phosphate by the action of sucrose phosphate phosphatase (SPP). Sucrose in cyanobacteria can also be synthesized by another enzyme, namely sucrose synthase (SuS). This enzyme catalyzes the reversible reaction that converts UDP-glucose plus fructose to sucrose plus UDP. The enzyme activity has been reported in *Anabaena* sp. PCC 7119 [85]. However, SuS does not appear to have a role in salt-acclimated cyanobacteria. Sucrose could not be detected in SusA-overexpressing *Anabaena* sp. PCC 7119 cells, whereas the *susA* mutant showed an increased level of sucrose [86]. This suggests that SusA is involved in the breakdown of sucrose *in vivo* and therefore plays a role in diazotrophic growth of *Anabaena* sp. PCC 7119. Recently, another compatible solute apart from sucrose, namely fructose-oligosaccharide, has been reported to accumulate inside *Anabaena* strains exposed to salt stress [87]. An increase of five- to eightfold in fructose-oligosaccharide was observed in *Anabaena* strains after 24 h exposure to 80 mM NaCl. However, under this condition the major compatible solute in *Anabaena* strains was sucrose whose contents were twice as high as fructose-oligosaccharide.

In *Synechocystis* sp. PCC 6803, only a small amount of sucrose can be detected in salt-loaded cells. The absence of the osmoprotective effect of sucrose in *Synechocystis* cells is probably due to its being degraded inside the cells adapted to high-salt condition. Higher accumulation rate of sucrose than glycosylglycerol is detected in *Synechocystis* sp. PCC 6803 during the initial phase of salt stress. However, when cells become salt-acclimated, glycosylglycerol represents the predominant compatible solute [88]. In this respect it is noted that the *agp* mutant disrupted in ADP-glucose synthesis of *Synechocystis* sp. PCC 6803 could resist salt stress with high levels of accumulated sucrose as a compatible solute [89]. The high levels of sucrose were due to the increased UDP-glucose pool that could serve as a substrate for sucrose synthesis. The *agp* mutant was unable to synthesize not only glucosylglycerol but also glycogen. This led to the accumulation of UDP-glucose, which is a precursor for glycogen synthesis.

10.3.2 Glucosylglycerol

Moderately salt-tolerant cyanobacteria usually accumulate glucosylglycerol as a compatible solute in response to osmotic stress. However, the accumulation of disaccharide, either sucrose or trehalose or both, can be observed during initial phase of osmotic upshock. This arises from the fact that the biosynthesis of disaccharide is much faster than glucosylglycerol. The pool of disaccharide represents 90% of its maximum after 8 h of an osmotic upshock, whereas 90% of maximum pool of glucosylglycerol is available after 24–28 h. The accumulation of glucosylglycerol has been extensively studied in *Synechocystis* sp. PCC 6803. The molecule of glucosylglycerol is synthesized by a two-step reaction catalyzed by glucosylglycerol phosphate synthase (GGPS) and glucosylglycerol phosphate phosphatase (GGPP). The changes in these enzyme activities are regulated by chemical modulation of enzyme activities as well as the up-regulation/down-regulation of the gene expressing the two enzymes. Under low-salt conditions, GGPS is inactive and constitutively expressed, whereas under high-salt conditions, GGPS is active and highly expressed [90]. *Synechocystis* sp. PCC 6803 not only synthesizes glucosylglycerol as an adaptation to salt stress but can also uptake exogenous glucosylglycerol. The uptake rate is rather low in wild-type cells under salt stress. However, in mutants disrupted in glucosylglycerol synthesis, the uptake of exogenous glucosylglycerol is sufficient for adaptation to salt [59]. Trehalose as well as sucrose can also be taken up by *Synechocystis* cells with trehalose being able to exert nearly the same osmoprotective effect as glucosylglycerol.

Another interesting finding in regard to the response of *Synechocystis* sp. PCC 6803 toward nonionic osmotic stress has recently been reported [91]. The osmotic stress exerted by exogenously added sorbitol rendered the cells tolerant to the osmotic stress with similar growth rate to the NaCl-stressed cells. The differences between sorbitol-stressed and NaCl-stressed cells were the different types of compatible solutes accumulated inside the cells. Cells stressed with 200 and 500 mM NaCl up to 4 h showed an increase of NaCl concentration-dependent glucosylglycerol contents, which were ascribed to the activation by NaCl of both GGPS and GGPP. In contrast, intracellular glucosylglycerol was present in trace amounts after 4 h stress with sorbitol. Under this condition sorbitol was accumulated inside the cells with similar level to that of the accumulation of glucosylglycerol under NaCl-stress condition. The absence of glucosylglycerol accumulation in sorbitol-treated cells was attributed to the inactivation of the GGPS, despite the increased expression of its gene.

The accumulation of glucosylglycerol as a major compatible solute has also been reported in a marine cyanobacterium *Synechococcus* sp. PCC 7002. However, the regulatory mechanism for the synthesis of glucosylglycerol in *Synechococcus* sp. PCC 7002 is different from that of the freshwater cyanobacterium *Synechocystis* sp. PCC 6803 [92]. The modes of glucosylglycerol accumulation in these two cyanobacteria suggest that the synthesis of glucosylglycerol is regulated mainly by transcriptional control in the marine strain and by posttranslational control in the freshwater strain. This is based on the observation that the salt-induced increase in GGPS activity was inhibited by treatment of *Synechococcus* cells with chloramphenicol. In contrast, this treatment in *Synechocystis* cells had no effect on the constitutive high level of GGPS activity. Northern blot hybridization analysis revealed that the transcript of *ggpS* gene increased rapidly and reached maximum after 40 min salt-shock treatment in *Synechococcus* cells. The activity of GGPS in extracts obtained from low-salt grown *Synechococcus* cells was not detected. However, significant enzyme activity was observed in extracts obtained from 2 h salt-shocked cells. This clearly demonstrates the transcriptional regulation of GGPS by salt shock in *Synechococcus* sp. PCC 7002. In contrast, salt-shocked *Synechocystis* sp. PCC 6803 showed little changes in enzyme activity compared to the low-salt grown cells. Nevertheless, the enzyme activities from extracts of cells grown under both conditions showed the activation by NaCl added during the assay. Overall, this indicates that the regulation of GGPS in *Synechocystis* is not under transcriptional control but rather the post-transcriptional modification of the pre-existing enzyme.

The important role of glucosylglycerol in protection of cells against salt stress has been shown to be due to the sustainability of cells division [93]. The analysis by electron microscopy revealed an incomplete separation of daughter cells disrupted in *ggpS* gene. Normal cells division could resume in this mutant under salt stress when glucosylglycerol was supplemented to the culture medium. This suggests that glucosylglycerol taken up by the mutant can counteract the effect of salt stress probably by releasing the arrest of the separation of the daughter cells caused by NaCl.

10.3.3 Glycine Betaine

Glycine betaine (*N,N,N*-trimethylglycine, GB) is a major osmoprotectant to protect organisms from high salinity [6–8]. It is synthesized by a choline oxidation route and by glycine methylation. The well-known biosynthetic pathways of GB include a classical two-step oxidation of choline. The first step is catalyzed by choline monooxygenase in plants [94], choline dehydrogenase in animals [95] and bacteria [96], and choline oxidase in some bacteria [97,98]. The second step is catalyzed by betaine aldehyde dehydrogenase in all organisms [95,99,100]. Unlike bacteria and higher plants, the halotolerant cyanobacterium *A. halophytica* possesses a novel biosynthetic pathway for GB. Waditee et al. [101] have successfully identified a novel biosynthetic pathway for GB in *A. halophytica*, which is different from other living organisms, via a three-step methylation of glycine catalyzed by two new methyltransferases. Glycine/sarcosine-*N*-methyltransferase catalyzes the methylation reactions of glycine and sarcosine, yielding sarcosine and dimethylglycine, respectively, with *S*-adenosylmethionine acting as the methyl donor. The other enzyme, dimethylglycine-*N*-methyltransferase, specifically catalyzes the methylation of dimethylglycine to GB [101].

The conversion of glycine to GB by methyltransferase activities supports an earlier study in *A. halophytica* where [$_{14}$C]-labeled GB could be detected in cells incubated with [$_{14}$C]-labeled glycine [102]. It is noteworthy that the incubation of salt-stressed *A. halophytica* with [$_{14}$C]-labeled choline resulted in an increase of [$_{14}$C]-labeled GB as well as the detectable [$_{14}$C]-labeled betaine aldehyde intermediate. In line with this finding, *A. halophytica* could uptake exogenous choline at an enhanced rate under salt-stress conditions [103]. Taken together, this suggests that *A. halophytica* might also utilize the choline oxidation pathway as a minor route to synthesize GB especially under salt-stress conditions.

The accumulation of GB in response to salt stress in *A. halophytica* was first reported by Ishitani et al. [104]. The authors also showed that the photosynthesis in the cells is markedly reduced upon salt upshock from 0.5 to 1.5 and 2.0 M NaCl. The synthesis of GB can overcome the inhibition of photosynthesis due to salt stress. The protection by GB against the deleterious effect of salts has been demonstrated for RuBisCO in *A. halophytica* [105,106]. The accumulated GB can also occur as a result of the transport from the environment. The uptake activity has been shown in

A. halophytica, *Synechocystis* DUN 52, and *Dactylococcopsis salina* [107]. Recently, the betaine transporter (BetT) from *A. halophytica* with functional GB transport activity has been isolated and characterized [108]. BetT belongs to the betaine-choline-carnitine transporter (BCCT) family. However, BetT transporter specifically transported GB with little or no inhibition by choline, γ-aminobutyric acid, betaine aldehyde, sarcosine, dimethylglycine, and amino acids such as proline. In addition, this BetT has high activity under alkaline pH conditions. BetT can also facilitate the growth of a freshwater *Synechococcus* sp. PCC 7942 in seawater supplemented with GB, thus suggesting the role of BetT for salt tolerance.

The fate of the accumulated GB has been studied in *A. halophytica* [107,109]. It was found that GB could be degraded inside the cells by betaine-homocysteine methyltransferase upon hypoosmotic stress as well as starvation [109]. In contrast, previous studies in *A. halophytica* showed the excretion of GB into the medium under salt downshock [107]. It is likely that GB can be both degraded and excreted in order to reduce the intracellular content of GB, thus avoiding the bursting of the cells due to water influx.

The transfer of genes encoding enzymes involved in the two-step oxidation pathway for GB resulted in only very low amounts of GB synthesized in the recombinant strains. After supplementing the medium with choline, a slightly higher GB accumulation was observed [110,111]. In contrast, the transfer of genes for the three-step methylation of glycine resulted in much higher GB biosynthesis and conferred greater salt-tolerance in the freshwater *Synechococcus* sp. PCC 7942 as well as in the *Arabidopsis*, which was correlated by the accumulation of elevated levels of GB [112].

GB accumulates as an osmoprotectant in stressed cells. So far, there are only a few different pathways for the biosynthesis of GB in living cells: (1) by a choline oxidation and (2) by glycine methylation. Choline is synthesized from ethanolamine, which is derived from serine. Serine and glycine are interconvertible through the activity of serine hydroxymethyltransferase. Therefore, two routes for the biosynthesis of GB can utilize serine as an upstream precursor.

It has been shown that providing substrate for GB synthesis via the choline oxidation pathway could enhance GB to some extent [111]. Similarly, supplementation of glycine could enhance the GB level significantly via glycine methylation (a series of three *N*-methylations of glycine), which is catalyzed by the two new methyltransferases (see earlier). These results suggest that provision of substrate is crucial for the enhancement of the GB levels. Based on these observations it is desirable to increase the levels of glycine as a means for boosting betaine accumulation in those plants overexpressing both the cyanobacterial methyltransferases and choline oxidation. Waditee et al. [113] have identified a 3-phosphoglycerate dehydrogenase (PGDH) gene by partially sequencing the *A. halophytica* genome, and then identifying the gene based on homology searches. Interestingly, this gene exhibits salt-inducible expression in *A. halophytica*.

Expression of the *A. halophytica* PGDH gene in *E. coli* increases GB accumulation, despite the fact that *E. coli* synthesizes GB via a serine route [113]. Taken together, these data suggest the importance of the *ApPGDH* gene in serine or glycine engineering for the enhancement of GB accumulation through the choline oxidation or the glycine methylation pathway in both microbes and plants.

10.4 SYNTHESIS OF SALT-STRESS PROTEINS

Various environmental stresses including salt stress have been shown to induce modifications of protein synthesis in several organisms [114–118]. An early study in cyanobacteria was done using two *Anabaena* strains differing in their tolerance capability. The study reported that the synthesis of several proteins in salt-sensitive strain was inhibited, whereas the synthesis of certain proteins was highly enhanced in salt-tolerant strain during salt stress [119]. The conclusion drawn from this study indicates that some proteins preferentially synthesized during salt stress may be important for the adaptation of cyanobacteria to osmotic stress. The drought-resistant cyanobacteria, namely *P. autumnale*, LPP4, and *Chroococcidiopsis* sp., synthesize at least two to three new proteins having a molecular mass of 30 and 40–45 kDa during drought stress [120]. Salt stress also induces the synthesis of similar-size proteins in these three cyanobacteria. Acidic proteins of molecular masses between 30 and 39 kDa are accumulated in high concentration in a desiccation-tolerant *N. commune* [121]. These proteins were suggested to be water stress proteins with a protective function against desiccation stress. Furthermore, these proteins occurred in microheterogeneous forms in high amounts, which were probably due to *in vivo* proteolytic cleavage. Further studies in *N. commune* also revealed that Fe-superoxide dismutase was the third most abundant soluble protein in cells of *N. commune* CHEN/1986 after prolonged storage in the desiccated state [122]. This protein was released from dried cells upon rehydration. The high levels of Fe-superoxide dismutase are important to counteract the damaging effects of oxidative stress during multiple cycles of desiccation and rehydration, especially when UV radiation effects are also involved.

Osmotic stress imposed by sucrose can also evoke the synthesis of at least 12 proteins in a freshwater *Anabaena* sp. strain L-31 [123]. These newly synthesized [$_{35}$S] methionine-labeled proteins, ranging from 11.5 to 85 kDa, were detected in cells exposed to either 100 or 350 mM sucrose for 30 min. A role for these proteins in osmotic stress tolerance is proposed based on two findings; firstly these proteins were present with the observed osmotolerance, and secondly some proteins, especially the high molecular weight, were absent with the reduction in osmotolerance after prolonged osmotic stress for 5 days.

Alterations in protein synthesis after salt shock have been reported for *Synechocystis* sp. PCC 6803 by Hagemann et al. [124]. By employing two-dimensional PAGE analysis of [^{35}S] methionine-labeled proteins, the authors showed that three classes of proteins are detected after prolonged salt shock with 684 mM NaCl, namely, proteins synthesized at the same rate, at a reduced rate, and at an enhanced rate compared to the control. From these results, they suggested that the proteins induced by salt stress are unlikely to be responsible for short-term salt adaptation of *Synechocystis* sp. PCC 6803. Short-term adaptation may involve the biochemical activation of the pre-existing proteins. Furthermore, the proteins synthesized at an enhanced rate during salt shock may function in the long-term adaptation phase.

During the last decade, there have been a lot of investigations employing proteomics as well as transcriptomic techniques. These approaches provide powerful tools to study responses of organisms to various environmental stresses. A large number of so-called stress proteins can be identified through the use of such techniques. An early study in *Synechocystis* sp. PCC 6803 adapted to salt stress was done by combining two-dimensional PAGE for enhanced resolution with MALDI-TOF MS. The MALDI-TOF MS enables the identification of proteins by soft ionization methods with a TOF mass spectrometer. Initially, protein fragments after digestion with protease are immobilized on a solid phase matrix and then ionized by laser. TOF analysis can reveal multiple peptide masses, which can be matched to theoretical masses in a database. By this way, Fulda et al. [125] identified 57 different periplasmic proteins in which some were enhanced, induced, or reduced by salt stress. Most of these 57 proteins were hypothetical proteins. Several salt-enhanced proteins were involved in the synthesis or the modification of extracellular layers of *Synechocystis* cells. Similar approach by Fulda et al. [126] could also identify 55 soluble proteins with some proteins detected after 2 h stress but most proteins were detected after 5 days acclimation to 684 mM NaCl. It is noteworthy that only a small number of proteins such as enzymes involved in the synthesis of the compatible solute glucosylglycerol were specifically induced by salt stress. Most of the salt-induced proteins were involved in general stress responses. Some examples of these proteins are heat shock proteins, such as HspA, DnaK2, GroEL1, GroEL2, and GroES, as well as proteins counteracting the effect of oxidative stress such as superoxide dismutase and peroxiredoxin. Similar studies in plasma membrane proteins of *Synechocystis* sp. PCC 6803 could identify 106 salt-stress proteins [127]. The levels of 25 proteins are either increased or decreased. Most salt-induced proteins are substrate-binding proteins of ABC transporter such as the glucosylglycerol-binding protein, the iron-binding protein, the phosphate-binding protein, and the nitrate/nitrite-binding protein. Increases in the last three substrate binding proteins suggest the needed strategy for cells to overcome the problem of nutrient deficiency arising from salt stress.

Recently, a system biology approach to study the response of *Synechocystis* sp. PCC 6803 to high-salt environment has been performed by Biggs's group in Sheffield. Using proteomics combined with targeted analysis of transcripts they were able to quantitate and identify 378 proteins in *Synechocystis* cells and about 40% of these proteins were differentially expressed in cells stressed with 6% NaCl for 9 days [128]. Most proteins that are induced by salt stress correspond to those previously reported, which were analyzed by peptide-mass-fingerprinting using MALDI-TOF MS [125–127], for example, GGPS, superoxide dismutase, etc. However, a number of interesting proteins not previously reported for *Synechocystis* have emerged. For example, an approximately two-fold increase in nitrate transporter is observed by salt stress. This protein is shown to confer the protection against salt stress by inhibiting the influx of Na^+ in two *Anabaena* strains [129,130]. In addition, an anti-sigma β factor was identified and shown to be reduced during salt stress. Another protein that was decreased after salt stress is an ATP-dependent CLP protease subunit. This protein is involved in the degradation of protein aggregates during stress and can function as a molecular chaperon. This may suggest that the aggregation of some particular proteins is important for salt-stress tolerance in *Synechocystis* cells.

10.5 REGULATION IN PHOTOSYNTHESIS

Salt stress constitutes the ionic effect due to the excess of Na^+ and the osmotic effect due to a deficit of water. Changes in pigments contents of cyanobacteria under salt stress have been reported. Moderate salt stress imposed by NaCl (342 mM) increases chlorophyll *a* content in *Synechocystis* sp. PCC 6803 [131]. Reduction of chlorophyll *a* content occurs in *Synechocystis* cells under severe salt stress at 684 or 1026 mM NaCl [131]. However, based on pigment contents, cells adapted to 1026 mM NaCl showed maximal photosynthesis. This suggested that cells with reduced chlorophyll *a* could better utilize the absorbed light for photosynthesis than cells with excess chlorophyll *a*. In a halotolerant cyanobacterium *A. halophytica*, salt-stressed cells (grown in 2.0 M NaCl) contain higher levels of chlorophyll *a* than control cells (grown in 0.5 M NaCl) [132]. The increase in chlorophyll *a* by salt stress is more apparent at a later than at an early stage of growth.

Another important pigment in cyanobacteria is phycobilin, which normally associates with proteins to form phycobiliproteins. Three major phycobiliproteins widely found in cyanobacteria are phycocyanin, allophycocyanin, and phycoerythrin. These phycobiliproteins are attached to the stromal surface of thylakoid membranes and serve as accessory light-harvesting antenna for PSII and PSI. Salt stress decreases phycocyanin contents in *Synechocystis* cells, which leads to the reduction of energy transfer from phycocyanin to PSII reaction centers, resulting in a decrease in photosynthesis. In *Spirulina platensis* exposed to 0.8 M NaCl for 12 h, chlorophyll contents remain unchanged whereas phycocyanin contents significantly decrease by about 50% compared to the control [133].

Photosynthetic oxygen evolution is inhibited in most cyanobacteria under salt stress. The decrease of evolved oxygen is associated with the decrease in the activity of PSII in *Synechocystis* sp. PCC 6803 [134]. The inactivation of *Synechocystis* cells is ascribed to the inhibition of D1 protein of PS II [135]. In higher plants, there have been reports of both salt-inhibited and salt-insensitive PSII activities [107,136]. Various salt-stress conditions could activate PSI activity in *Synechocystis* sp. PCC 6803 [137], *A. halophytica* [138], and *S. platensis* [139].

Studies in *S. platensis* exposed to 0.8 M NaCl for 9 h have shown a 40% decrease in oxygen evolution mediated by PSII with a concomitant decrease of electron transport in PSII to a similar extent [139]. In contrast, the activity of PSI electron transport was increased by about 220% with concomitant increase in P700 amount to a similar extent. The inhibition of PSII activity was attributed to a 40% loss of D1 protein, located in the PSII reaction center, after 9 h salt treatment. On the other hand, the increase in PSI activity and P700 reflects the need of the cells to build up ATP pool via cyclic photophosphorylation for the extrusion of excess Na^+ from the cytoplasm. Very recently, a study on the bioenergetic aspect of *S. platensis* in response to salt stress has been reported. It was found that the energy transfer from phycobilisomes to PSII and PSI either decreased or increased, respectively, in cells exposed to 0.8 M NaCl for 12 h [140]. The activity of PSI was also enhanced in salt-stressed *Spirulina* cells. In this respect it is interesting to note that an increase in cyclic electron flow has been reported in cyanobacteria under salt stress [137,138]. In *A. halophytica*, the contents of PSI as well as NAD(P)H-dehydrogenase also increase with significant electron donation to PSI reaction centers via the NAD(P)H-dehydrogenase complex. The increase in PSI activity in salt-stressed cells may be necessary to provide energy for the synthesis and/or transport of compatible solutes as well as for the extrusion of Na^+ during acclimation to salt-stress conditions. The target for salt stress can be at the water-oxidizing complex or at the PSII reaction center. In *S. platensis*, both the water-oxidizing complex and PSII are the target for salt stress [133,139].

The osmotic stress on *Synechococcus* cells imposed by 1.0 M sorbitol depresses oxygen-evolving activity of PSII and electron transport activity of PSI by 60% and 30%, respectively, after 2 h [135]. On the other hand, a similar study in *Synechocystis* sp. PCC 6803 showed that long-term osmotic stress by 0.7 M sorbitol for 10 days completely inhibited oxygen-evolving activity of PSII [141]. In addition, both the *psbA* transcript and D1 protein levels were drastically reduced, suggesting D1 protein as a target for osmotic stress in *Synechocystis* cells. It is also of interest to note that ionic stress by NaCl had only little effect on PSII activity as well as on *psbA* transcript and D1 protein. Taken together, this suggests that long-term osmotic stress is more detrimental to photosynthesis than ionic stress in *Synechocystis* cells.

Cyanobacteria with different salt-stress tolerance capacity show differences in photosynthetic carbon fixation under salt stress. An increase in NaCl concentration resulted in an increase in carbon fixation in a halotolerant *A. halophytica* [132] but no increase was observed in a freshwater *Aphanothece stagnina* [142]. The increase in carbon fixation corresponded with the increase in RuBisCO activity as well as an increase in RuBisCO content [132].

In contrast to *A. halophytica*, the expression of RuBisCO gene (rbcL/S) has been reported to be repressed by osmotic and salt stress in *Anabaena* sp. PCC 7120 [143]. However, it is noted that the decrease in the expression of RuBisCO gene in *Anabaena* cells was observed after 24 h stress, whereas the increase in carbon fixation in *A. halophytica* was observed after a long-term stress of 7 days. In this respect, a study in the halotolerant green alga, *Dunaliella*, has also shown an up-regulation of RuBisCO gene upon salt stress with 3.0 M NaCl using cells grown for several weeks [144]. This suggests that *Dunaliella* responds to salt stress by increasing carbon assimilation to provide energy sources needed for the synthesis of a compatible solute glycerol.

10.6 REGULATION IN NITROGEN METABOLISM

10.6.1 Nitrate Uptake and Its Reduction

An early study on nitrate uptake was reported in *Anacystis nidulans* R2 [145]. This study showed that nitrate uptake is accomplished by a Na^+/NO_3^- cotransport system, which differs from a H^+/NO_3^- cotransport observed in most higher plants. Subsequently, a Na^+-dependent nitrate uptake was reported in both non-stressed and salt-stressed cells of a halotolerant cyanobacterium *A. halophytica* [146]. The salt-stressed *A. halophytica* has a reduction in nitrate uptake (V_{max}) by 50%. Different observations were reported in a nitrogen-fixing salt tolerant cyanobacterium *Anabaena torulosa* [147]. The cells grown in salt-stress medium containing 170 mM NaCl showed an enhanced nitrate uptake rate. This enhancement of nitrate uptake was suggested to play a role in salt tolerance in *A. torulosa* via the inhibition of Na^+ influx. The small increase in nitrate taken up by the salt-stressed cells rules out the possibility that nitrate may act as an effective osmolyte. Similar mechanism of salt-stress tolerance has been reported in a freshwater nitrogen-fixing cyanobacterium *Anabaena doliolum* [130]. It was shown that the influx of Na^+ into the cells is minimum when nitrate is simultaneously available to the cells. This suggests that salt tolerance depends on the control of Na^+ influx, which can be reduced or inhibited when nitrate is taken up by the cells. The uptake of nitrate is also stimulated by an increase in osmolality of the assay medium generated by either NaCl or sorbitol in *Synechocystis* sp. PCC 6803 [148]. The stimulation of nitrate uptake by Na^+ is abolished in mutants disrupted in nitrate assimilation activities such as nitrate and nitrite reductases.

The studies on the effect of salinity on nitrate reduction have mostly been done in higher plants and bacteria [149–151]. In general, nitrate reductase is suppressed under salt stress. The study on the changes of nitrate reductase in response to salt stress in cyanobacteria is very limited. The enzyme in cyanobacteria is a ferredoxin-linked nitrate reductase, which is different from a nicotinamide nucleotide-dependent counterpart found in higher plants [149]. The activity of nitrate reductase is influenced by nitrogen sources in the growth medium. Nitrate induced a fivefold increase in nitrate reductase activity in *A. halophytica* after 8 days [152]). A small increase (~1.5-fold) in activity was observed with glutamine but not with ammonium after 4 days. The enhancement of nitrate reductase by nitrate is not unusual. Nitrate has been shown to activate the transcription of the nitrate assimilation operons of *Synechococcus* PCC 7942 and *Plectonema boryanum* [153]. Nitrate reductase of *A. halophytica* is activated by increasing concentration of NaCl up to 200 mM. At 300 mM NaCl or higher the enzyme is severely inhibited. It is therefore necessary that intracellular NaCl concentration of *A. halophytica* is kept not higher than 200 mM to allow for continued cell growth under salt-stress condition. The low level of intracellular Na^+ is achieved by extruding Na^+ in exchange for H^+, which is mediated by a Na^+/H^+ antiporter.

Recent experiments in *A. doliolum* have been conducted to identify the target of salt tolerance in regard to nitrate and ammonium assimilation pathways. By comparing the activities associated with these two nitrogen assimilation pathways using salt-tolerant mutant and wild-type cells, it was found that nitrate reductase activities were similarly decreased in both the mutant and the wild type upon 100 mM NaCl stress [154]. In contrast, the glutamine synthase (GS) activity was increased 156% in the mutant compared to only 35% in the wild type. Furthermore, about 1.5-fold higher ammonium uptake rate was observed in the mutant under salt stress. Together, this study indicates that the salt tolerance of *A. doliolum* is dependent on ammonium rather than nitrate assimilation.

The enzymes for ammonium assimilation are GS and glutamate synthase (GOGAT). The function of these two enzymes is important for maintaining C–N balance in the cells. A third enzyme involved in ammonium assimilation, glutaminase, has recently been reported in *Synechocystis* sp. PCC 6803. This enzyme catalyzes the conversion of glutamine to glutamate and ammonia. The glutaminase encoded by *slr2079* in *Synechocystis* was cloned and expressed as histidine-tagged fusion protein in *E. coli* [155]. The purified enzyme was stimulated (twofold) by 1 M NaCl with concomitant twofold increased affinity for glutamine. However, this glutaminase is not essential for growth or photosynthetic oxygen evolution activity. The Δ*slr2079* mutant cells could grow slightly faster than the wild-type cells under high-salt-stress condition imposed by 700 mM NaCl but not under osmotic stress imposed by 700 mM sorbitol. The involvement of *slr2079* in salt tolerance was demonstrated by the results showing higher oxygen evolution rate in the Δ*slr2079* mutant than in the wild type. This suggests the negative regulatory role of *slr2079* in salt tolerance. The *slr2079* gene also plays a role in the expression of some salt-stress-related genes in *Synechocystis* cells. Disruption of *slr2079* led to an increased expression of two salt-enhanced genes, *gdhB* and *prc*, encoding glucose dehydrogenase B and carboxy-terminal protease, respectively. On the other hand, a decreased expression of two salt-repressed genes, *desD* and *guaA*, encoding delta-6 desaturase and GMP synthetase, respectively, was observed in Δ*slr2079* mutant cells. This suggests that *slr2079* negatively regulates the expression of salt tolerance genes in *Synechocystis* cells.

10.6.2 Nitrogen Fixation and Hydrogen Production

Nitrogen-fixing organisms can reduce dinitrogen to ammonia, which is catalyzed by nitrogenase. This enzyme is sensitive to oxygen produced by photosynthetic organisms such as cyanobacteria. Several diazotrophic filamentous cyanobacteria contain specialized cells called heterocysts which protect the nitrogenase from the oxygen [156]. However, aerobic nitrogen fixation can occur in non-heterocystous filamentous cyanobacteria such as *Oscillatoria* spp. and unicellular cyanobacteria such as *Gloeothece* spp. and *Synechococcus* sp. SF1.

The responses of nitrogenase to salt and osmotic stress have been reported in two *Anabaena* strains, a salt-sensitive *Anabaena* sp. strain L-31 and a salt-tolerant *A. torulosa*. Both strains showed inhibition of nitrogenase by NaCl, whereas remarkable enhancement of acetylene reduction activity was observed in sucrose-treated cells [157,158]. Protection against salt stress could be achieved by supplementation of ammonium to the medium. However, such treatment afforded no protection against osmotic stress. The osmotic stress induced by non-permeable solutes such as mannitol and polyethyleneglycol ($M_r \geq 6000$) had no positive effect on nitrogenase, unlike the effect caused by a permeable sucrose. The increased nitrogenase activity is attributed to the increased synthesis of dinitrogenase reductase. This increased dinitrogenase reductase level was not detected in cells stressed with NaCl. Therefore, it can be concluded that ionic and osmotic stresses differentially regulate cyanobacterial nitrogenase activity.

Nitrogenase activity of *Anabaena* sp. PCC 7120 is inhibited by increasing concentration of NaCl up to 200 mM [159]. However, nitrogenase activity seems to require the presence of Na^+ because complete removal of NaCl from the nutrient solution results in a marked decrease in nitrogenase activity. In *Nostoc muscorum*, treatment with 60 mM NaCl leads to 50% reduction in nitrogenase activity with the complete loss of activity at 90 mM NaCl [160].

The nitrogenase activity in nitrogen-fixing cyanobacteria also produces hydrogen in addition to ammonia. Recent investigations on hydrogen production by *Synechocystis* sp. PCC 6803, a non-nitrogen-fixing cyanobacterium, have reported the activation of Hox-hydrogenase activity by ionic and osmotic stress. A nearly twofold increase in hydrogen production is induced by the presence of either NaCl or sorbitol at 20 mOsm kg^{-1} [161]. A further increase in osmolality by NaCl decreased hydrogen production to the level where no NaCl or sorbitol was present. On the other hand, an increase of sorbitol osmolality higher than 20 mOsm kg^{-1} could still maintain high production of hydrogen. This suggests that high concentration of NaCl may result in the inhibition of Hox-hydrogenase of *Synechocystis* cells. In a nitrogen-fixing cyanobacterium *Anabaena siamensis* strain TISTR 8012, an osmotic stress induced by 0.5% fructose (which also acts as a carbon source) can stimulate hydrogen production by twofold [162]. This increased hydrogen production by fructose occurred despite the increase in hydrogen uptake activity. Hydrogen production using 0.5% sucrose was less efficient than that using 0.5% fructose, partly due to a reduced uptake of sucrose into *Synechocystis* cells.

10.6.3 Regulation of Polyamine Metabolism

Another group of nitrogen-containing compounds, polyamines, has recently received considerable attention owing to its possible role in abiotic stress resistance [163]. Polyamines are polycationic molecules that are ubiquitously found in all prokaryotes and eukaryotes. The most common polyamines in most living cells are spermidine (a triamine), putrescine (a diamine), and spermine (a tetraamine) in order of their abundance in the cells. Early studies implicating polyamines as osmolytes have been done in higher plants. An increase in intracellular polyamine contents in response to salt stress has been reported in several higher plants including wheat, sorghum, maize, and tomato [164–166]. In rice, the salt sensitivity has been shown to correlate with the reduction of spermidine and spermine levels, although a large increase in putrescine level [167].

The manipulation of polyamine biosynthesis in plants can create plants with drought-tolerant phenotype. Arginine decarboxylase (ADC) is an enzyme responsible for converting arginine to putrescine via agmatine and *N*-carbamoylputrescine intermediates. The transgenic rice expressing *adc* gene from *Datura stramonium* can tolerate drought stress induced by 20% polyethyleneglycol (M_r 8000) [168]. The wild-type rice contains increasing endogenous putrescine levels after the onset of drought stress. The transgenic rice, on the other hand, contains much higher putrescine levels together with the increase in spermidine and spermine synthesis, which ultimately leads to the protection of transgenic rice from drought stress. The increased contents of spermidine and spermine are attributed to the enhanced levels of *samdc* transcript in the transgenic rice. SAMDC generates decarboxylated *S*-adenosylmethionine, which further donates an aminopropyl group to putrescine and spermidine to produce spermidine and spermine, respectively. The wild-type contains much less abundance of *samdc* transcript with the consequent reduction of spermidine and spermine leading to a wilting and rolling of leaves under drought stress.

The first study on the changes of polyamine contents under the influence of salt stress and osmotic stress has been reported for *Synechocystis* sp. PCC 6803 [169]. A long-term osmotic stress by sorbitol led to a threefold increase in total polyamine contents while only a small increase was observed under NaCl salt stress. This suggests that the machinery involved in the synthesis of polyamines is better activated by the osmotic effect than the ionic effect. In this respect it is noteworthy that different sets of genes in *Synechocystis* cells are induced by osmotic stress (0.5 M sorbitol) and salt stress (0.5 M NaCl) for 20 min [170]. An in-depth analysis revealed that spermine was specifically induced by salt stress, whereas osmotic stress preferentially induced spermidine. For putrescine, an increase in the content occurred by osmotic stress induced by sorbitol at a concentration as low as 4 mM. No further increase in putrescine was observed by raising sorbitol concentration to 700 mM. The maintenance of moderate levels of putrescine is beneficial for the survival of *Synechocystis*. It has been suggested that high level of putrescine is toxic due to the oxidation products arising from the activity of diamine oxidase causing damage to plasma membrane [171].

The underlying molecular mechanism for the increase of polyamines content in *Synechocystis* is attributed to the activation of *adc* genes, namely *adc1* (*slr1312*) and *adc2* (*slr0662*), by salt and osmotic stress. The *adc* transcript level was maximally observed in cells treated with 550 mM NaCl, accounting for two- and fourfold of those in 700 mM sorbitol treated and untreated cells, respectively [169]. The high level of *adc* transcript induced by osmotic and salt stress is partially attributed to the increased stability of the transcript in the stressed cells. By incubating the cells in the presence of rifampicin, an inhibitor of transcriptional initiation, control cells showed no detectable *adc* transcript after 2 h, whereas significant level of the transcript was detected in salt and osmotic stressed cells [169]. The high contents of spermidine and spermine in the stressed cell could facilitate the formation of the polyamine-*adc* transcript complex, leading to the increase in the stability of the transcript.

The osmotic stress-induced accumulation of polyamines in *Synechocystis* cells seems to contribute no significant role with regard to the osmoprotectant ability due to low level of the accumulation. The increase in polyamines contents may represent the acclimatization of the cells in response to the stress. The main role of polyamines is to maintain a cation–anion balance during a long-term osmotic and salt stress as shown in tomato leaves [172]. *Synechocystis* cells contain Na^+/H^+ antiporters which export Na^+ in exchange for H^+ during salt-stress acclimation. The excess Cl^- inside the cells would be balanced by the positively charged polyamine molecules.

The intracellular contents of polyamines in *Synechocystis* can be increased by de novo synthesis, which represents the major source of intracellular polyamines. However, cells may also rely on another source of cellular polyamines via the activity of the polyamine transport system. Studies on polyamine transport in cyanobacteria have been scarce. The earliest study in *A. nidulans* reported the transport of putrescine, which occurs by passive diffusion and ion-trapping mechanisms [173]. Recent studies in *Synechocystis* sp. PCC 6803 have shown a rapid-energy-dependent uptake of polyamines driven by proton gradient and membrane potential [174,175]. The transport of polyamines has been extensively investigated in *E. coli* where the uptake system belongs to ATP binding cassette (ABC) polyamine transporters [176–178]. The ABC-type polyamine transport system in *E. coli* consists of the substrate-binding protein (PotD), two transmembrane proteins (PotB, PotC), and a membrane-bound ATPase (PotA). PotD-like protein is annotated in *Synechocystis* based on sequence similarity to *E. coli* PotD [176]. The uptake of both putrescine and spermidine in *Synechocystis* is stimulated by the presence of low concentration of NaCl or sorbitol [174,175]. Similar pattern of the activation of the uptake by either NaCl or sorbitol was obtained, suggesting the osmotic effect rather than the ionic effect as the activator.

The protection of *Synechocystis* growth against salt stress by polyamine has been reported [175]. The investigation involved the determination of cells growth under normal and salt-stress conditions in the presence of different concentrations of spermidine. Under normal growth condition in BG11 medium, the presence of spermidine at 0.5, 1.0, and 2 mM slowed down the growth of *Synechocystis* in proportion to its concentration. The growth was completely inhibited after 3 days. The inhibition of growth was attributed to the toxicity of intracellular high contents of spermidine observed at days 6 and 9 during the cultivation. Cells growth was different under salt stress in the presence of spermidine. The presence of 0.5 and 1.0 mM spermidine alleviated the growth inhibition by salt stress showing a similar (at 1.0 mM spermidine) or a better growth rate (at 0.5 mM spermidine) during the first 3 days of growth. However, cell growth was completely inhibited after 3 days in the presence of 1.0 or 2.0 mM spermidine. Cells still retained high growth rate under salt stress in the presence of low concentration (0.5 mM) of spermidine. The high toxic levels of intracellular spermidine in cells grown under salt stress in the presence of 1 mM spermidine may be the contributing factor leading to the cessation of growth after 3 days. It is interesting to note that the presence of low concentration of spermidine (0.5 mM) under salt-stress condition resulted in a continuous growth after 3 days. In contrast, under the same condition and without salt stress, no growth was recorded. It is unclear as to the mechanism underlying this observation. It may be that the excess Cl^- inside the cells under salt stress is optimally balanced by a polycationic spermidine accumulated inside the cells under salt stress.

The uptake of polyamines into *Synechocystis* sp. PCC 6803 has recently been shown to involve PotD protein [179]. PotD is a polyamine-binding protein and *Synechocystis* PotD shows 24% identity to *E. coli* PotD [31]. *Synechocystis* PotD can bind all three classes of polyamines; however, preferential binding toward spermidine was observed. The transcript levels of *potD* gene in *Synechocystis* exposed to 3 days salt (550 mM NaCl) or osmotic (300 mM sorbitol) stress increased by about 1.5- and 2-fold, respectively [179]. On the other hand, PotD protein levels were decreased and unchanged, respectively, in salt- and osmotic-stressed cells. It is speculated that some PotD proteins might be degraded during long-term stress for 3 days. However, after a longer duration of stress, that is, 5 days, the cells showed an increase in PotD content by 30%–40% [180]. The increase of PotD under osmotic stress is beneficial to the cells because more polyamines available in the environment can be taken up by the cells to serve as an additional osmolyte.

10.7 CONCLUSIONS

Cyanobacteria can evolve mechanisms that allow their survival under various adverse environmental conditions including high salinity. The immediate regulatory response of cyanobacteria to salt stress involves the exclusion of Na^+ from the cytoplasm, which is accomplished by Na^+/H^+ antiporters and to a lesser extent by a Na^+-ATPase, which has recently been reported in the alkaliphilic halotolerant cyanobacterium *A. halophytica*. The next phase of the response is the accumulation of compatible solutes achieved by de novo synthesis or the uptake from the environment or both. During the acclimation to salt stress, cyanobacteria are able to synthesize a complex array of stress-related proteins including those involved in the synthesis of compatible solutes and various transporters as well as some hypothetical proteins with unknown functions.

The present review is designed to address several important aspects of regulatory mechanisms of cyanobacteria in response to osmotic stress with NaCl stress in particular. The scheme in Figure 10.4 illustrates the generalized salt-stress responses by cyanobacteria. We highlight areas that have seen substantial progress in recent years and discuss subjects that are still a matter of controversy. There are emerging fields like those of salt-sensing and signal transduction, which have not been covered here but are likely to attract considerable attention in the future. We hope that the present review will be useful for students and established researchers alike, in the context of cyanobacteria response to environmental stresses. In the near future, we are likely to witness the discovery of many more networks intricately involved in cyanobacterial stress response. Last but not least, it is

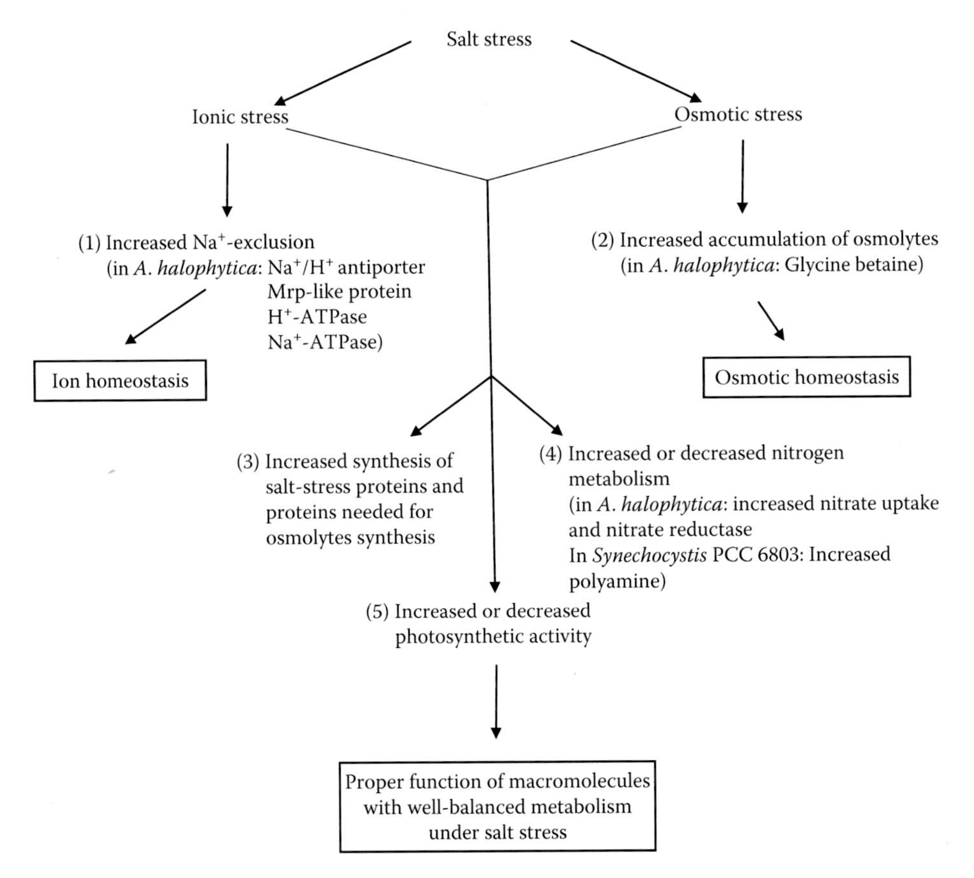

FIGURE 10.4 Predominant salt-stress responses operating in cyanobacteria. (1)–(3) represent the immediate and short-term response, whereas (4) and (5) denote variable changes depending on cyanobacterial species and account for minor contribution toward salt-stress resistance.

anticipated that a better understanding on the molecular mechanism of *A. halophytica* in response to salt and osmotic stress will be forthcoming after its complete genome sequence is available in the near future.

ACKNOWLEDGMENTS

The work in A.I. laboratory was supported by the Commission on Higher Education, CHE (University Staff Development Consortium), and the Thai Government Stimulus Package 2 (TKK 2555). Most of the research on polyamine received financial support from the Thailand Research Fund through the Royal Golden Jubilee Ph.D. Program, and Thailand National Research University Project of CHE (FW0659A). R.W.S. appreciates the support from Faculty of Science, Chulalongkorn University (A1B1-MICO (TRF)-12 and RES-A1B1-30).

REFERENCES

1. Hagemann, M. 2011. Molecular biology of cyanobacterial salt acclimation. *FEMS Microbiol. Rev.* 35: 87–123.
2. Kumar, K., Mella-Herrera, R. A., and Golden, J. W. 2010. Cyanobacterial heterocysts. *Cold Spring Harb. Perspect. Biol.* 2(4): a000315. doi: 10.1101/cshperspect.a000315.
3. Inaba, M., Sakamoto, A., and Murata, N. 2001. Functional expression in *Escherichia coli* of low-affinity and high-affinity $Na^+(Li^+)/H^+$ antiporters of *Synechocystis*. *J. Bacteriol.* 183: 1376–1384.

4. Padan, E., Venturi, M., Gerchman, Y., and Dover, N. 2001. Na(+)/H(+) antiporters. *Biochim. Biophys. Acta* 1505: 144–157.
5. Waditee, R., Hibino, T., Tanaka, Y., Nakamura, T., Incharoensakdi, A., and Takabe, T. 2001. Halotolerant cyanobacterium *Aphanothece halophytica* contains an Na(+)/H(+) antiporter, homologous to eukaryotic ones, with novel ion specificity affected by C-terminal tail. *J. Biol. Chem.* 276: 36931–36938.
6. Rhodes, D. and Hanson, A. D. 1993. Quaternary ammonium and tertiary sulfonium compounds in higher plants. *Annu. Rev. Plant Physiol. Mol. Biol.* 44: 357–384.
7. Takabe, T., Nakamura, T., Nomura, M., Hayashi, Y., Ishitani, M., Muramoto, Y., Tanaka, A., and Takabe, T. 1998. Glycinebetaine and the genetic engineering of salinity tolerance in plants. In *Stress Responses of Photosynthetic Organisms*, Satoh, K. and Murata, N. (Eds.). Elsevier Science, Amsterdam, the Netherlands, pp. 115–132.
8. Kempf, B. and Bremer, E. 1998. Uptake and synthesis of compatible solutes as microbial stress response to high-osmolality environments. *Arch. Microbiol.* 170: 319–330.
9. Los, D. A., Zorina, A., Sinetova, M., Kryazhov, S., Mironov, K., and Zinchenko, V. V. 2010. Stress sensors and signal transducers in cyanobacteria. *Sensors* 10: 2386–2415.
10. Zhu, J.-K. 2001. Plant salt tolerance. *Trends Plant Sci.* 6: 66–71.
11. Elanskaya, I. V., Karandashova, I. V., Bogachev, A. V., and Hagemann, M. 2002. Functional analysis of the Na^+/H^+ antiporter encoding genes of the cyanobacterium *Synechocystis* PCC 6803. *Biochemistry (Mosc.)* 67: 432–440.
12. Wang, H. L., Postier, B. L., and Burnap, R. L. 2002. Polymerase chain reaction-based mutageneses identify key transporters belonging to multigene families involved in Na^+ and pH homeostasis of *Synechocystis* sp. PCC 6803. *Mol. Microbiol.* 44: 1493–1506.
13. Waditee, R., Hossain, G. S., Tanaka, Y., Nakamura, T., Shikata, M., Takano, J., Takabe, T., and Takabe, T. 2004. Isolation and functional characterization of Ca^{2+}/H^+ antiporters from cyanobacteria. *J. Biol. Chem.* 279: 4330–4338.
14. Padan, E., Bibi, E., Ito, M., and Krulwich, T. A. 2005. Alkaline pH homeostasis in bacteria: New insights. *Biochim. Biophys. Acta* 1717: 67–68.
15. Billini, M., Stamatakis, K., and Sophianopoulou, V. 2008. Two members of a network of putative Na^+/H^+ antiporters are involved in salt and pH tolerance of the freshwater cyanobacterium *Synechococcus elongatus. J. Bacteriol.* 190: 6318–6329.
16. Tsunekawa, K., Shijuku, T., Hayashimoto, M., Kojima et al. 2009. Identification and characterization of the Na^+/H^+ antiporter Nhas3 from the thylakoid membrane of *Synechocystis* sp. PCC 6803. *J. Biol. Chem.* 284: 16513–16521.
17. Oren, A. 1999. Bioenergetic aspects of halophilism. *Microbiol. Mol. Biol. Rev.* 63: 334–348.
18. Blumwald, E., Aharon, G. S., and Apse, M. P. 2000. Sodium transport in plant cells. *Biochim. Biophys. Acta* 1465: 140–151.
19. Soong, T. W., Yong, T. F., Ramanan, N., and Wang, Y. 2000. The *Candida albicans* antiporter gene CNH1 has a role in Na^+ and H^+ transport, salt tolerance, and morphogenesis. *Microbiology* 146: 1035–1044.
20. Moszer, I. 1989. The complete genome of *Bacillus subtilis*: From sequence annotation to data management and analysis. *FEBS Lett.* 430: 28–36.
21. Arabidopsis Genome Initiative. 2000. Analysis of the genome sequence of the flowering plant *Arabidopsis thaliana. Nature* 408: 796–815.
22. Goff, S. A., Ricke, D., Lan, T. H., Presting, G. et al. 2002. A draft sequence of the rice genome (*Oryza sativa* L. ssp. *japonica*). *Science* 296: 92–100.
23. Yu, J., Hu, S., Wang, J., Wong, G. K. et al. 2002. A draft sequence of the rice genome (*Oryza sativa* L. ssp. *indica*). *Science* 296: 79–92.
24. Brown, S. D. and Hancock, J. M. 2006. The mouse genome. *Genome Dyn.* 2: 33–45.
25. Padan, E. and Schuldiner, S. 1996. Bacterail Na^+/H^+ antiporters: Molecular biology, biochemistry, and physiology. In *Handbook of Biological Physics*, Konings, W. N., Kaback, H. R., and Lolkema, J. S. (Eds.). Elsevier Science, Amsterdam, the Netherlands, pp. 501–531.
26. Rosen, B. P. 1986. Ion extrusion systems in *Escherichia coli. Methods Enzymol.* 125: 328–336.
27. Gaxiola, R. A., Rao, R., Sherman, A., Grisafi, P., Alper, S. L., and Fink, G. R. 1999. The *Arabidopsis thaliana* proton transporters, *AtNhx1* and *Avp1*, can function in cation detoxification in yeast. *Proc. Natl. Acad. Sci. U.S.A.* 96: 1480–1485.
28. Shi, H., Ishitani, M., Kim, C., and Zhu, J. K. 2000. The *Arabidopsis thaliana* salt tolerance gene SOS1 encodes a putative Na^+/H^+ antiporter. *Proc. Natl. Acad. Sci. U.S.A.* 97: 6896–6901.

29. Shi, H., Lee, B. H., Wu, S. J., and Zhu, J. K. 2003. Overexpression of a plasma membrane Na^+/H^+ antiporter gene improves salt tolerance in *Arabidopsis thaliana. Nat. Biotechnol.* 21: 81–85.
30. Hamada, A., Hibino, T., Nakamura, T., and Takabe, T. 2001. Na^+/H^+ antiporter from *Synechocystis* species PCC 6803, homologous to SOS1, contains an aspartic residue and long C-terminal tail important for the carrier activity. *Plant Physiol.* 125: 437–446.
31. Kaneko, T., Sato, S., Kotani, H., Tanaka, A. et al. 1996. Sequence analysis of the genome of the unicellular cyanobacterium *Synechocystis* sp. strain PCC6803. II. Sequence determination of the entire genome and assignment of potential protein-coding regions (supplement). *DNA Res.* 30: 109–136.
32. Wutipraditkul, N., Waditee, R., Incharoensakdi, A., Hibino, T., Tanaka, Y., Nakamura, T., Shikata, M., Takabe, T., and Takabe, T. 2005. Halotolerant cyanobacterium *Aphanothece halophytica* contains NapA-type Na^+/H^+ antiporters with novel ion specificity that are involved in salt tolerance at alkaline pH. *Appl. Environ. Microbiol.* 71: 4176–4184.
33. Ohyama, T., Igarashi, K., and Kobayashi, H. 1995. Physiological role of the *chaA* gene in sodium and calcium circulations at a high pH in *Escherichia coli. J. Bacteriol.* 176: 4311–4315.
34. Waditee, R., Buaboocha, T., Kato, M., Hibino, T., Suzuki, S., Nakamura, T., and Takabe, T. 2006. Carboxyl-terminal hydrophilic tail of a *NhaP* type Na^+/H^+ antiporter from cyanobacteria is involved in the apparent affinity for Na^+ and pH sensitivity. *Arch. Biochem. Biophys.* 450: 113–121.
35. Yamaguchi, T., Apse, M. P., Shi, H., and Blumwald, E. 2003. Topological analysis of a plant vacuolar Na^+/H^+ antiporter reveals a luminal C terminus that regulates antiporter cation selectivity. *Proc. Natl. Acad. Sci. U.S.A.* 100: 12510–12515.
36. Kinclova, O., Potier, S., and Sychrova, H. 2002. Difference in substrate specificity divides the yeast alkali-metal-cation/H^+ antiporters into two subfamilies. *Microbiology* 148: 1225–1232.
37. Mitsui, K., Kamauchi, S., Nakamura, N., Inoue, H., and Kanazawa, H. 2004. A conserved domain in the tail region of the *Saccharomyces cerevisiae* Na^+/H^+ antiporter (Nha1p) plays important roles in localization and salinity-resistant cell-growth. *J. Biochem.* 135: 139–148.
38. Fukaya, F., Promden, W., Hibino, T., Tanaka, Y., Nakamura, T., and Takabe, T. 2009. An Mrp-like cluster in the halotolerant cyanobacterium *Aphanothece halophytica* functions as a Na^+/H^+ antiporter. *Appl. Environ. Microbiol.* 75: 6626–6629.
39. Blanco-Rivero, A., Leganés, F., Fernández-Valiente, E., Calle, P., and Fernández-Piñas, F. 2005. *mrpA*, a gene with roles in resistance to Na^+ and adaptation to alkaline pH in the cyanobacterium *Anabaena* sp. PCC7120. *Microbiology* 151: 1671–1682.
40. Blanco-Rivero, A., Leganés, F., Fernández-Valiente, E., and Fernández-Piñas, F. 2009. *mrpA* (all1838), a gene involved in alkali and Na(+) sensitivity, may also have a role in energy metabolism in the cyanobacterium *Anabaena* sp. strain PCC 7120. *J. Plant Physiol.* 166: 1488–1496.
41. Saier, Jr., M. H., Eng, B. H., Fard, S., Garg, J. et al. 1999. Phylogenetic characterization of novel transport protein families revealed by genome analyses. *Biochim. Biophys. Acta* 1422: 1–56.
42. Shapiguzov, A., Lyukevich, A. A., Allakhverdiev, S. I., Sergeyenko, T. V., Suzuki, I., Murata, N., and Los, D. A. 2005. Osmotic shrinkage of cells of *Synechocystis* sp. PCC 6803 by water efflux via aquaporins regulates osmostress-inducible gene expression. *Microbiology* 151: 447–455.
43. Summerfield, T. C. and Sherman, L. A. 2008. Global transcriptional response of the alkali-tolerant cyanobacterium *Synechocystis* sp. strain PCC 6803 to a pH 10 environment. *Appl. Environ. Microbiol.* 74: 5276–5284.
44. Niu, X., Narasinshan, M. L., Salzman, R. A., Bressan, R. A., and Hasegawa, P. M. 1993. NaCl regulation of plasma membrane H^+-ATPase gene expression in a glycophyte and a halophyte. *Plant Physiol.* 103: 713–718.
45. Yang, M. F., Song, J., and Wang, B. S. 2010. Organ-specific responses of vacuolar H^+-ATPase in the shoots and roots of C halophyte *Sueada salsa* to NaCl. *J. Integr. Biol.* 52: 308–314.
46. Wiangnon, K., Raksajit, W., and Incharoensakdi, A. 2007. Presence of a Na^+-stimulated P-Type ATPase in the plasma membrane of the alkaliphilic halotolerant cyanobacterium *Aphanothece halophytica. FEMS Microbiol. Lett.* 270: 139–145.
47. Glynn, I. M. and Karlish, S. 1975. The sodium pump. *Annu. Rev. Physiol.* 37: 13–55.
48. Heefner, D. I. and Harold, F. M. 1982. ATP-driven sodium pump in *Streptococcus faecalis. Proc. Natl. Acad. Sci. U.S.A.* 79: 2798–2802.
49. Haro, R., Garciadeblas, B., and Rodriguez-Navarro, A. 1991. A novel P-type ATPase from yeast involved in sodium transport. *FEBS Lett.* 291: 189–191.
50. Balnokin, Y. V. and Popova, L. G. 1994. The ATP-driven Na^+-pump in the plasma membrane of the marine unicellular alga, *Platymonas viridis. FEBS Lett.* 343: 61–64.
51. Shono, M., Hara, Y., Wada, M., and Fujii, T. 1996. A sodium pump in the plasma membrane of the marine alga *Heterosigma akashiwo. Plant Cell Physiol.* 37: 365–388.

52. Kaieda, N., Wakagi, T., and Koyama, N. 1998. Presence of Na^+-stimulated V-type ATPase in the membrane of a facultatively anaerobic and halophilic alkaliphile. *FEMS Microbiol. Lett.* 167: 57–61.
53. Ritchie, R. J. 1992. Sodium transport and the origin of the membrane potential in the cyanobacterium *Synechococcus* R-2 (*Anacystis nidulands*) PCC 7942. *J. Plant Physiol.* 139: 320–330.
54. Brown, I. I., Fadeyer, S. I., Kirik, I. I., Severina, I. I., and Skulacher, V. P. 1990. High-dependent $\Delta\mu$ Na^+-generation and utilization in the marine cyanobacterium *Oscillatoria brevis. FEBS Lett.* 270: 203–206.
55. Soontharapirakkul, K. and Incharoensakdi, A. 2010. Na^+-stimulated ATPase of alkaliphilic halotolerant cyanobacterium *Aphanothece halophytica* translocates Na^+ into progteoliposomes via Na^+ uniport mechanism. *BMC Biochem.* 11: 30.
56. Soontharapirakkul, K., Promden, W., Yamada, N., Kageyama, H., Incharoensakdi, A., Iwamoto-Kihara, A., and Takabe, T. 2011. Halotolerant cyanobacterium *Aphanothece halophytica* contains a Na^+-dependent F_1F_0-ATP synthase with potential role in salt-stress tolerance. *J. Biol. Chem.* 286: 10169–10176.
57. Brown, A. D. 1976. Microbial water stress. *Bacteriol. Rev.* 40: 803–846.
58. Higo, A., Katoh, H., Ohmori, K., Ikeuchi, M., and Ohmori, M. 2006. The role of a gene cluster for trehalose metabolism in dehydration tolerance of the filamentous cyanobacterium *Anabaena* sp. PCC 7120. *Microbiology* 152: 979–987.
59. Mikkat, S., Hagermann, M., and Schoor, A. 1996. Active transport of glucosylglycerol is involved in salt adaptation of the cyanobacterium *Synechocystis* sp. strain PCC 6803. *Microbiology* 142: 1725–1732.
60. Hagemann, M., Richter, S., and Mikkat, S. 1997. The *ggtA* gene encodes a subunit of the transport system for the osmoprotective compound glucosylglycerol in *Synechocystis* sp. strain PCC 6803. *J. Bacteriol.* 179: 714–720.
61. Reed, R. H. and Stewart, W. D. P. 1988. The response of cyanobacteria to salt stress. In *Biochemistry of the Algae and Cyanobacteria*, Roger, L. J. and Gallon, J. R. (Eds.). Oxford University Press, Oxford, U.K., pp. 217–231.
62. Potts, M. 1999. Mechanisms of desiccation tolerance in cyanobacteria. *Eur. J. Phycol.* 34: 319–328.
63. Strom, A. R. and Kaasen, I. 1993. Trehalose metabolism in *Escherichia coli*: Stress protection and stress regulation of gene expression. *Mol. Microbiol.* 8: 205–210.
64. Horlacher, R. and Boss, W. 1997. Characterization of TreR, the major regulator of the *Escherichia coli* trehalose system. *J. Biol. Chem.* 272: 13026–13032.
65. Arguelles, J. C. 2000. Physiological role of trehalose in bacteria and yeast: A comparative analysis. *Arch. Microbiol.* 174: 217–224.
66. De Smet, K. A., Weston, A., Brown, I. N., Young, D. B., and Robertson, B. D. Three pathways for trehalose biosynthesis in mycobacteria. *Microbiology* 146: 199–208.
67. Richards, A. B., Krakowka, S., Dexter, L. B., Schmid, H., Wolterbeek, A. P. M., Waalkens-Berendsen, D. H., Shigoyuki, A., and Kurimoto, M. 2002. Trehalose: A review of properties, history to use and human tolerance, and results of multiple safety studies. *Food Chem. Toxicol.* 40: 871–898.
68. Elbein, A. D. 1974. The metabolism of α,α-trehalose. *Adv. Carbohydr. Chem. Biochem.* 30: 227–256.
69. Thevelein, J. M. 1984. Regulation of trehalose mobilization in fungi. *Microbiol. Rev.* 48: 42–59.
70. Avonce, N., Mendoza-Vargas, A., Morett, E., and Iturriaga, G. 2006. Insights on the evolution of trehalose biosynthesis. *BMC Evol. Biol.* 6: 109.
71. Page-Sharp, M., Behm, C. A., and Smith, G. D. 1999. Involvement of the compatible solutes trehalose and sucrose in the response to salt stress of a cyanobacterial *Scytonema* species isolated from desert soils. *Biochim. Biophys. Acta* 1472: 519–528.
72. Katoh, H., Asthana, R. K., and Ohmori, M. 2004. Gene expression in the cyanobacterium *Anabaena* sp. PCC 7120 under desiccation. *Microb. Ecol.* 47: 164–174.
73. Weisburd, S. 1988. Death-defying dehydration. *Sci. News* 13: 107–110.
74. Colaco, C., Sen, S., Thangavelu, M., pinder, S., and Roser, B. 1992. Extraordinary stability of enzyme dried in trahalose: Simplified molecular biology. *Biotechnology (NY)* 10: 1007–1011.
75. Yoshida, T. and Sakamoto, T. 2009. Water-stress induced trehalose accumulation and control of trehalose in the cyanobacterium *Nostos punctiforme* IAM M-15. *J. Gen. Appl. Microbiol.* 55: 135–145.
76. Hill, D. R., Peat, A., and Potts, M. 1994. Biochemistry and structure of the glycan secreted by desiccation-tolerant *Nostoc commune* (cyanobacteria). *Protoplasma* 182: 126–148.
77. Billi, D., Wright, D. J., Helm, R. F. Prickett, T., Potts, M., and Crowe, J. H. 2000. Engineering desiccation tolerance in *Escherichia coli. Appl. Environ. Microbiol.* 66: 1680–1684.
78. Hershkovitz, N., Oren, A., and Cohen, Y. 1991. Accumulation of trehalose and sucrose in cyanobacteria exposed to matric water stress. *Appl. Environ. Microbiol.* 57: 645–648.
79. Stal, L. and Reed, R. H. 1987. Low-molecular mass carbohydrate accumulated in cyanobacteria from a marine microbial mat in response to salt. *FEMS Microbiol. Ecol.* 45: 305–312.

80. Reed, R. H. and Stewart, W. D. P. 1983. Physiological responses of *Rivularia atra* to salinity: Osmotic adjustment in hyposaline media. *New Phytol.* 95: 595–603.
81. Warr, S. R. C., Reed, R. H., and Stewart, W. D. P. 1987. Low-molecular-weight carbohydrate biosynthesis and the distribution of cyanobacteria (blue-green-algae) in marine environments. *Br. Phycol. J.* 22: 175–180.
82. Karsten, U. 1996. Growth and organic osmolytes of geographically different isolates of *Microcoleus chthonoplastes* (cyanobacteria) from benthic microbial mats: Response to salinity change. *J. Phycol.* 32: 501–506.
83. Moezelaar, R., Bijvank, S. M., and Stal, L. J. 1996. Fermentation and sulfur reduction in the mat-building cyanobacterium *Microcoleus chthonoplastes. Appl. Environ. Microbiol.* 65: 4659–4665.
84. Hagemann, M. and Marin, K. 1999. Salt-induced sucrose accumulation is mediated by sucrose-phosphate-synthase in cyanobacteria. *J. Plant Physiol.* 155: 424–430.
85. Porchia, A. C., Curatti, L., and Salerno, G. L. 1999. Sucrose metabolism in cyanobacteria: Sucrose synthase from the plant enzymes with respect to substrate affinity and amino-terminal sequence. *Planta* 210: 34–40.
86. Curatti, L., Flores, E., and Salerna, G. 2002. Sucrose is involved in the diazotrophic metabolism of the heterocyst-forming cyanobacterium *Anabaena* sp. *FEBS Lett.* 513: 175–178.
87. Salerno, G. L., Porchia, A. C., Vargas, W. A., and Abdian, P. L. 2004. Fructose-containing oligosaccharides: Novel compatible solutes in *Anabaena* cells exposed to salt stress. *Plant Sci.* 167: 1003–1008.
88. Marin, K., Kanesaki, Y., Los, D. A., Murata, N., Suzuki, I., and Hagemann, M. 2004. Gene expression profiling reflects physiological processes in salt acclimation of *Synechocystis* sp. strain PCC 6803. *Plant Physiol.* 136: 3290–3300.
89. Miao, X., Wu, Q., Wu, G., and Zhao, N. 2003. Sucrose accumulation in salt-stress cells of *agp* gene deletion-mutant in cyanobacterium *Synechocystis* sp. PCC 6803. *FEMS Microbiol. Lett.* 218: 71–77.
90. Hagemann, M. and Erdmann, N. 1994. Activation and pathway of GG synthesis in the cyanobacterium *Synechocystis* sp. PCC 6803. *Microbiology* 140: 1427–1431.
91. Marin, K., Stirnberg, M., Eisenhut, M., Kramer, R., and Hagemann, M. 2006. Osmotic stress in *Synechocystis* sp. PCC 6803: Low tolerance towards nonionic osmotic stress results from lacking activation of glucosylglycerol accumulation. *Microbiology* 152: 2023–2030.
92. Engelbrecht, F., Marin, K., and Hagemann, M. 1999. Expression of the *ggpS* gene, involved in osmolyte synthesis in the marine cyanobacterium *Synechococcus* sp. strain 7002, revealed regulatory difference between this strain and the fresh water strain *Synechocystis* sp. strain PCC 6803. *Appl. Environ. Microbiol.* 65: 4822–4829.
93. Ferjani, A., Mustardy, L., Sulpice, R., Marin, K., Suzuki, I., Hagemann, M., and Murata, N. 2003. Glucosylglycerol, a compatible solute, sustains cell division under salt stress. *Plant Physiol.* 131: 1628–1637.
94. Rathinasabapathi, B., Burnet, M., Russell, B. L., Gage, D. A., Liao, P. C., Nye, G. J., Scott, P., Golbeck, J. H., and Hanson, A. D. 1997. Choline monooxygenase, an unusual iron-sulfur enzyme catalyzing the first step of glycine betaine synthesis in plants: Prosthetic group characterization and cDNA cloning. *Proc. Natl. Acad. Sci. U.S.A.* 94: 3454–3458.
95. Lamark, T., Kaasen, I., Eshoo, M. W., McDougall, J., and Strom, A. R. 1991. DNA sequence and analysis of the bet genes encoding the osmoregulatory choline-glycine betaine pathway of *Escherichia coli. Mol. Microbiol.* 5: 1049–1064.
96. Boch, J., Kempf, B., Schmid, R., and Bremer, E. 1996. Synthesis of the osmoprotectant glycine betaine in *Bacillus subtilis*: Characterization of the gbsAB genes. *J. Bacteriol.* 178: 5121–5129.
97. Yamada, H., Mori, N., and Tani, Y. 1979. Properties of choline oxidase of *Cylindrocarpon didymum* M-1. *Agric. Biol. Chem.* 43: 2173–2177.
98. Hayashi, H., Alia, Mustardy, L., Deshnium, P., Ida, M., and Murata, N. 1997. Transformation of *Arabidopsis thaliana* with the *codA* gene for choline oxidase; accumulation of glycinebetaine and enhanced tolerance to salt and cold stress. *Plant J.* 12, 133–142.
99. Weretilnyk, E. A. and Hanson, A. D. 1990. Molecular cloning of a plant betaine-aldehyde dehydrogenase, an enzyme implicated in adaptation to salinity and drought. *Proc. Natl. Acad. Sci. U.S.A.* 87: 2745–2749.
100. Chern, M. K. and Pietruszko, T. 1995. Human aldehyde dehydrogenase E3 isozyme is a betaine aldehyde dehydrogenase. *Biochem. Biophys. Res. Commun.* 213: 561–568.
101. Waditee, R., Tanaka, Y., Aoki, K., Hibino, T., Jikuya, H., Takano, J., Takabe, T., and Takabe, T. 2003. Isolation and functional characterization of N-methyltransferases that catalyze betaine synthesis from glycine in a halotolerant photosynthetic organism *Aphanothece halophytica. J. Biol. Chem.* 278: 4932–4942.

102. Incharoensakdi, A. and Wutipraditkul, N. 1999. Accumulation of glycinebetaine and its synthesis from radioactive precursors under salt-stress in the cyanobacterium *Aphanothece halophytica*. *J. Appl. Phycol.* 11: 515–523.
103. Incharoensakdi, A. and Karnchanatat, A. 2003. Salt stress enhances choline uptake in the halotolerant cyanobacterium *Aphanothece halophytica*. *Biochim. Biophys. Acta* 1621: 102–109.
104. Ishitani, M., Takabe, T., Kojima, K., and Takabe, T. 1993. Regulation of glycinebetaine accumulation in the halotolerant cyanobacterium *Aphanothece halophytica*. *Aust. J. Plant Physiol.* 20: 693–703.
105. Incharoensakdi, A., Takabe, T., and Akazawa, T. 1986. Effect of betaine on enzyme activity and subunit interaction of ribulose-1,5-bisphosphate carboxylase/oxygenase from *Aphanothece halophytica*. *Plant Physiol.* 81: 1044–1049.
106. Incharoensakdi, A. and Takabe, T. 1988. Determination of intracellular chloride ion concentration in a halotolerant cyanobacterium *Aphanothece halophytica*. *Plant Cell Physiol.* 29: 1073–1075.
107. Moore, D. J., Reed, R. H., and Stewart, W. D. P., 1987. A glycine betaine transport system in *Aphanothece halophytica* and other glycine betaine-synthesizing cyanobacteria. *Arch. Microbiol.* 147: 399–405.
108. Laloknam, S., Tanaka, K., Buaboocha, T., Waditee, R., Incharoensakdi, A., Hibino, T., Tanaka, Y., and Takabe, T. 2006. Halotolerant cyanobacterium *Aphanothece halophytica* contains a betaine transporter active at alkaline pH and high salinity. *Appl. Environ. Microbiol.* 72: 6018–6026.
109. Incharoensakdi, A. and Waditee, R. 2000. Degradation of glycinebetaine by betaine-homocysteine methyltransferase in *Aphanothece halophytica*: Effect of salt downshock and starvation. *Curr. Microbiol.* 41: 227–231.
110. Deshnium, P., Los, D. A., Hayashi, H., Mustardy, L., and Murata, N. 1995. Transformation of *Synechococcus* with a gene for choline oxidase enhances tolerance to salt stress. *Plant Mol. Biol.* 29: 897–907.
111. Nomura, M., Ishitani, M., Takabe, T., Rai, A. K., and Takabe, T. 1995. *Synechococcus* sp. PCC7942 transformed with *Escherichia coli* bet genes produces glycine betaine from choline and acquires resistance to salt stress. *Plant Physiol.* 107: 703–708.
112. Waditee, R., Bhuiyan, M. N., Rai, V., Aoki, K., Tanaka, Y., Hibino, T., Suzuki, S., Takano, J., Jagendorf, A. T., Takabe, T., and Takabe, T. 2005. Genes for direct methylation of glycine provide high levels of glycinebetaine and abiotic-stress tolerance in *Synechococcus* and *Arabidopsis*. *Proc Natl. Acad. Sci. U.S.A.* 102: 1318–1323.
113. Waditee, R., Bhuiyan, N. H., Hirata, E., Hibino, T., Tanaka, Y., Shikata, M., and Takabe, T. 2007. Metabolic engineering for betaine accumulation in microbes and plants. *J. Biol. Chem.* 282: 34185–34193.
114. Berg, G. R., Innis, W. E., and Heikkila, J. 1987. Stress proteins and thermotolerance in psychrotrophic yeast from arctic environments. *Can. J. Microbiol.* 33: 383–389.
115. Ramagopal, S. 1987. Salinity stress induced tissue specific proteins in barley seedlings. *Plant Physiol.* 84: 324–331.
116. Edeman, L., Czarencka, E., and Key, J. L. 1988. Induction and accumulation of heat shock specific poly (A^+) RNAs and proteins in soybean seedlings during arsenite and cadmium treatments. *Plant Physiol.* 86: 1048–1056.
117. Parida, A. K., Das, A. B., Mitra, B., and Mohanty, P. 2004. Salt-stress induced alterations in protein profile and protease activity in the mangrove *Bruguiera parviflora*. *Z. Natureforsch.* 59c: 408–414.
118. Yan, S., Tang, Z., Zu, W., and Sun, W. 2005. Proteomic analysis of salt stress-responsive proteins in rice root. *Proteomics* 5: 235–244.
119. Apte, S. K. and Bhagwat, A. A. 1989. Salinity-stress-induced proteins in two nitrogen-fixing *Anabaena* strains differently tolerant to salt. *J. Bacteriol.* 171: 909–915.
120. Hershkovitz, N., Oren, A., Post, A., and Cohen, Y. 1991. Induction of water-stress proteins in cyanobacteria exposed to matric-or osmotic-water stress. *FEMS Microbiol. Lett.* 83: 169–172.
121. Scherer, S. and Potts, M. 1989. Novel water stress protein from a desiccation-tolerant cyanobacterium. *J. Biol. Chem.* 264: 12546–12553.
122. Shirkey, B., Kovarcik, D. P., Wright, D. J., Wilmoth, G., Prickett, T. F., Helm, R. F., Gregory, E. M., and Potts, M. 2000. Active Fe-containing superoxide dismutase and abundant *sodF* mRNA in *Nostoc commune* (Cyanobacteria) after years of desiccation. *J. Bacteriol.* 182: 189–197.
123. Iyer, V., Ferbandes, T., and Apte, S. K. 1994. A role for osmotic stress-induced proteins in the osmotolerance of a nitrogen-fixing cyanobacterium, *Anabaena* sp. strain L-31. *J. Bacteriol.* 176: 5868–5870.
124. Hagemann, M., Wolfel, L., and Kruger, B. 1990. Alteration of protein synthesis in a cyanobacterium *Synechocystis* sp. PCC 6803 after a salt shock. *J. Microbiol.* 136: 1393–1399.
125. Fulda, S., Huang, F., Nilsson, F., Hagemann, M., and Norling, B. 2000. Proteomics of *Synechocystis* sp. strain PCC 6803. Identification of periplasmic proteins in cells grown at low and high salt concentrations. *Eur. J. Biochem.* 267: 5900–5907.

126. Fulda, S., Mikkat, S., Huang, F., Huckauf, J., Marin, K., Norling, B., and Hagemann, M. 2006. Proteome analysis of salt stress response in the cyanobacterium *Synechocystis* sp. strain PCC 6803. *Proteomics* 6: 2733–2745.
127. Huang, F., Fulda, S., Hagemann, M., and Norling, B. 2006. Proteomic screening of salt-stress-induced changes in plasma membranes of *Synechocystis* sp. strain PCC 6803. *Proteomics* 6: 910–920.
128. Pandhal, J., Noirel, J., Wright, P. C., and Biggs, C. A. 2009. A systems biology approach to investigate the response of *Synechocystis* sp. PCC 6803 to a high salt environment. *Saline Systems* 5: 8.
129. Reddy, B. R., Apte, S. K., and Thomas, J. 1987. Relationship between sodium influx and salt tolerance of the nitrogen-fixing cyanobacteria. *Appl. Environ. Microbiol.* 53: 1934–1939.
130. Rai, A. K. and Abraham, G. 1995. Relationship of combined nitrogen-sources to salt tolerance in freshwater cyanobacterium *Anabaena doliolum. J. Appl. Bacteriol.* 78: 501–506.
131. Schubert, H. and Hagemann, M. 1990. Salt effect on 77 K fluorescence and photosynthesis in the cyanobacterium *Synechocystis* spec. PCC 6803. *FEMS Microbiol. Lett.* 71: 169–172.
132. Takabe, T., Incharoensakdi, A., Arakawa, K., and Yokota, S. 1988. Co_2 fixation rate and RuBisCO content increase in the halotolerant cyanobacterium, *Aphanothece halophytica*, frown in high salinities. *Plant Physiol.* 88: 1120–1124.
133. Lu, C. and Vonshak, A. 2002. Effects of salinity on photosystem II function in cyanobacterial *Spirulina platensis* cells. *Physiol. Plant* 114: 405–413.
134. Schubert, H., Fulda, S., and Hagemann, M. 1993. Effects of adaptation to different salt concentrations on photosynthesis and pigmentation of the cyanobacterium *Synechocystis* sp. PCC 6803. *J. Plant Physiol.* 142: 291–295.
135. Allakhverdiev, S. I., Sakamoto, A., Nishiyama, Y., Inaba, M., and Murata, N. 2000. Ionic and osmotic effects of NaCl-induced inactivation of photosystem I and II in *Synechococcus* sp. *Plant Physiol.* 123: 1047–1056.
136. Tiwari, B. S., Bose, A., and Ghosh, B. 1997. Photosynthesis in rice under a salt stress. *Photosynthetica* 34: 303–306.
137. Jeanjean, R., Matthijs, H. C. P., Onana, B., Havaux, M., and Joset, F. 1993. Exposure of the cyanobacterium *Synechocystis* PCC 6803 to salt stress induces concerted changes in respiration and photosynthesis. *Plant Cell Physiol.* 34: 1073–1079.
138. Hibino, T., Lee, B. H., Rai, A. K., Ishikawa, H., Kojima, H., Tawada, M., Shimoyama, H., and Takabe, T. 1996. Salt enhances photosystem I content and cyclic electron flow via NAD(P)H dehydrogenase in the halotolerant cyanobacterium *Aphanothece halophytica. Aust. J. Plant Physiol.* 23: 321–330.
139. Lu, C. and Vonshak, A. 1999. Characterization of PSII photochemistry in salt-adapted cells of cyanobacterium *Spirulina platensis. New Phytol.* 141: 231–239.
140. Zhang, T., Gong, H., Wen, X., and Lu, C. 2010. Salt stress induces a decrease in excitation energy transfer from phycobilisomes to photosystem II but an increase in photosystem I in the cyanobacterium *Spirulina platensis. J. Plant Physiol.* 167: 951–958.
141. Jantaro, S., Mulo, P., Jansen, T., Incharoensakdi, A., and Mäenpää, P. 2005. Effects of long-term ionic and osmotic stress conditions on photosynthesis in the cyanobacterium *Synechocystis* sp. PCC 6803. *Funct. Plant Biol.* 32: 807–815.
142. Rai, A. K. 1990. Biochemical characteristics of photosynthetic response to various external salinities in halotolerant and freshwater cyanobacteria. *FEMS Microbiol. Lett.* 69: 177–180.
143. Mori, S., Castoreno, A., and Lammers, P. J. 2002. Transcript levels of *rbcR1*, *ntcA* and *rbcL/S* genes in cyanobacterium *Anabaena* sp. PCC7120 are down regulated in response to cold and osmotic stress. *FEMS Microbiol. Lett.* 213: 167–173.
144. Liska, A. J., Shevchenko, A., Pick, U., and Katz, A. 2004. Enhanced photosynthesis and redox energy production contribute to salinity tolerance in *Dunaliella* as revealed by homology-based proteomics. *Plant Physiol.* 136: 2806–2817.
145. Rodriguez, R., Lara, C., and Guerrero, M. G. 1992. Nitrate transport in the cyanobacterium *Anacystis nidulans* R2: Kinetic and energetic aspects. *Biochem. J.* 282: 639–643.
146. Incharoensakdi, A. and Wangsupa, I. 2003. Nitrate uptake by the halotolerant cyanobacterium *Aphanothece halophytica* grown under non-stress and salt-stress conditions. *Curr. Microbiol.* 47: 255–259.
147. Reddy, B. R., Apte, S. K., and Thomas, J. 1989. Enhancement of cyanobacterial salt tolerance by combined nitrogen. *Plant Physiol.* 89: 204–210.
148. Baebprasert, W., Karnchanatat, A., Lindblad, P., and Incharoensakdi, A. 2011. Na^+-stimulated nitrate uptake with increased activity under osmotic upshift in *Synechocystis* sp. strain PCC 6803. *World J. Microbiol. Biotechnol.* 27: 2467–2473.

149. Losada, M. and Guerrero, M. G. 1979. The photosynthetic reduction of nitrate and its regulation. In *Photosynthesis in Relation to Model Systems*, Barber, J. (Ed.). Elsevier, Amsterdam, the Netherlands, pp. 365–408.
150. Abdelbaki, G. K., Siefritz, F., Man, H.-M., Weiner, H., Kaldenhoff, R., and Kaiser, W. M. 2000. Nitrate reductase in *Zea mays* L. under salinity. *Plant Cell Environ.* 23: 515–521.
151. Parida, A. K. and Das, A. B. 2004. Effects of NaCl stress on nitrogen and phosphorous metabolism in a true mangrove *Bruguiera parviflora* grown under hydroponic culture. *J. Plant Physiol.* 161: 921–928.
152. Thaivanich, S. and Incharoensakdi, A. 2007. Purification and characterization of nitrate reductase from the halotolerant cyanobacterium *Aphanothece halophytica. World J. Microbiol. Biotechnol.* 23: 85–92.
153. Andriesse, X., Bakker, H., and Weisbeek, P. 1990. Analysis of nitrate reductase genes in cyanobacteria. In *Inorganic Nitrogen in Plants and Microorganisms*, Ullrich, W. R., Rigano, C., Fuggi, A., and Aparicio, P. J. (Eds.)—. Springer-Verlag, Berlin, Germany, pp. 323–326.
154. Kshatriya, K., Singh, J. S., and Singh, D. P. 2009. Salt tolerant mutant of *Anabaena doliolum* exhibiting efficient ammonium uptake and assimilation. *Physiol. Mol. Biol. Plants* 15: 377–381.
155. Zhou, J., Zhou, J., Young, H., Yan, C., and Huang, F. 2008. Characterization of a sodium-regulated glutaminase from cyanobacterium *Synechocystis* sp. PCC 6803. *Sci. China Ser. C* 51: 1066–1075.
156. Walsby, A. E. 2007. Cyanobacterial heterocysts: Terminal pores proposed as sites of gas exchange. *Trends Microbiol.* 15: 340–349.
157. Fernandes, T. A., Iyer, V., and Apte, S. K. 1993. Differential responses of nitrogen fixing cyanobacteria to salinity and osmotic stresses. *Appl. Environ. Microbiol.* 59: 899–904.
158. Fernandes, T. A. and Apte, S. K. 2000. Differential regulation of nitrogenase activity by ionic and osmotic stresses and permeable sugars in the cyanobacterium *Anabaena* sp. strain L-31. *Plant Sci.* 150: 181–189.
159. Rai, A. K. and Tiwari, S. P. 1999. Response to NaCl of nitrate assimilation and nitrogenase activity in the cyanobacterium *Anabaena* sp. PCC 7120 and its mutants. *J. Appl. Microbiol.* 87: 877–883.
160. Bhargava, S., Saxena, R. K., Pandey, P. K., and Bisen, P. S. 2003. Mutational engineering of the cyanobacterium *Nostoc muscorum* for resistance to growth-inhibitory action of LiCl and NaCl. *Curr. Microbiol.* 47: 5–11.
161. Baebprasert, W., Lindblad, P., and Incharoensakdi, A. 2010. Response of H_2 production and Hox-hydrogenase activity to external factors in the unicellular cyanobacterium *Synechocystis* sp. PCC 6803. *Int. J. Hydrogen Energy* 35: 6611–6616.
162. Khetkorn, W., Lindblad, P., and Incharoensakdi, A. 2010. Enhanced biohydrogen production by the N_2-fixing cyanobacterium *Anabaena siamensis* strain TISTR 8012. *Int. J. Hydrogen Energy* 35: 12767–12776.
163. Bouchereau, A., Aziz, A., Larher, F., and Martin-Tanguy, J. 1999. Polyamines and environmental challenges: Recent development. *Plant Sci.* 140: 103–125.
164. Erdie, L., Trived, S., Takeda, K., and Matsumoto, H. 1990. Effects of osmotic and salt stresses on the accumulation of polyamines in leaf segments from wheat varieties differing in salt and drought tolerance. *J. Plant Physiol.* 137: 165–168.
165. Erdie, L., Szegletes, Z., Barabas, K., and Pestenacz, M. 1996. Responses in polyamine titer under osmotic and salt stress in sorghum and maize seedlings. *J. Plant Physiol.* 147: 599–603.
166. Aziz, A., Martin-Tanguy, J., and Larher, J. 1999. Salt stress-induced proline accumulation and changes in tyramine and polyamine levels are linked to ionic adjustment in tomato leaf discs. *Plant Sci.* 145: 83–91.
167. Krishnamurthy, R. and Bhagwat, K. A. 1989. Polyamines as modulators of salt tolerance in rice. *Plant Physiol.* 91: 500–504.
168. Capell, T., Bassie, L., and Christou, P. 2004. Modulation of the polyamine biosynthetic pathway in transgenic rice confers tolerance to drought stress. *Proc. Natl. Acad. Sci. U.S.A.* 101: 9909–9914.
169. Jantaro, S., Mäenpää, P., Mulo, P., and Incharoensakdi, A. 2003. Contents and biosynthesis of polyamines in salt and osmotically stressed cells of *Synechocystis* sp. PCC 6803. *FEMS Microbiol. Lett.* 228: 129–135.
170. Paithoonrangsarid, K., Shoumskaya, M. K., Kanesaki, Y., Satoh, S., Tabata, S., Los, D. A., Zinchenko, V. V., Hayashi, H., Tanticharoen, M., Suzuki, I., and Murata, N. 2004. Five histidine kinase perceive osmotic stress and regulate distinct sets of genes in *Synechocystis. J. Biol. Chem.* 279: 53078–53086.
171. Ditomasso, J. M., Shaff, J. E., and Kochian, L. V. 1989. Putrescine-induced wounding and its effects on membrane integrity and ion transport processes in roots of intact corn seedings. *Plant Physiol.* 90: 988–995.

172. Santa-Cruz, A., Acosta, M., Perez-Alfocea, F., and Bolarin, M. C. 1997. Changes in free polyamine levels induced by salt stress in leaves of cultivated and wild tomato species. *Physiol. Plant.* 101: 341–346.
173. Guarino, L. A. and Cohen, S. S. 1979. Uptake and accumulation of putrescine and its lethality in *Anacystis nidulands. Proc. Natl. Acad. Sci. U.S.A.* 76: 3184–3188.
174. Raksajit, W., Maenpaa, P., and Incharoensakdi, A. 2006. Putrescine transport in a cyanobacterium *Synechocystis* sp. PCC 6803. *J. Biochem. Mol. Biol.* 39: 394–399.
175. Raksajit, W., Yodsang, P., Maenpaa, P., and Incharoensakdi, A. 2009. Characterization of spermidine transport system in a cyanobacterium, *Synechocystis* sp. PCC 6803. *J. Microbiol. Biotechnol.* 19: 447–454.
176. Igarashi, K. and Kashiwagi, K. 1999. Polyamine transport in bacteria and yeast. *Biochem. J.* 344: 633–642.
177. Igarashi, K., Ito, K., and Kashiwagi, K. 2001. Polyamine uptake systems in *Escherichia coli. Res. Microbiol.* 152: 271–278.
178. Igarashi, K. and Kashiwagi, K. 2010. Characteristics of cellular polyamine transport in prokaryotes and eukaryotes. *Plant Physiol. Biochem.* 48: 506–512.
179. Brandt, A. M., Raksajit, W., Yodsang, P. Mulo, P., Incharoensakdi, A., Salminen, T., and Maenpaa, P. 2010. Characterization of the substrate-binding PotD subunit in *Synechocystis* sp. strain PCC 6803. *Arch. Microbiol.* 192: 791–801.
180. Yodsang, P., Raksajit, W., Brandt, A.-M., Salminen, T. A., Mäenpää, P., and Incharoensakdi, A. 2011. Recombinant polyamine-binding protein of *Synechocystis* sp. PCC 6803 specifically binds to and is induced by polyamines. *Biochemistry (Mosc.)* 76: 713–719.

11 Molecular Mechanisms of UV-B Stress Tolerance in Cyanobacteria

Richa, Rajeshwar P. Sinha, and Donat P. Häder*

CONTENTS

11.1 INTRODUCTION

Cyanobacteria are a primitive group of Gram-negative photoautotrophic prokaryotes having a cosmopolitan distribution thermally adapted from hot springs to the cold Arctic and Antarctic regions [1]. They were probably the first organisms to release oxygen by splitting water during photosynthesis into the then anoxygenic environment that appeared during the Precambrian era (between 2.8 and 3.5×10^9 years ago) and thus provided a favorable condition for the evolution of aerobic life [2]. Cyanobacteria are major biomass producers both in aquatic as well as terrestrial ecosystems and are a valuable source of various natural products of medicinal and industrial importance [3,4]. In addition, their inherent capacity to fix atmospheric nitrogen makes them important for rice-growing regions where they add to fertility of the rice paddy fields as natural biofertilizer [5].

The increase in solar UV radiation (UVR) reaching the Earth's surface due to anthropogenically released chemicals such as chlorofluorocarbons (CFCs), chlorocarbons (CCBs), and organobromides (OBMs) has become a subject of concern in the last few decades [6]. Although very small proportions of solar UVR contribute to the total irradiance on the Earth's surface (UV-C; 0%, UV-B; <1% and UV-A; <7%), this part of the solar spectrum is highly energetic. The UV-B radiation has the greatest potential for cell damage by directly influencing the structure of proteins and DNA which have absorption maxima in this region and indirectly via the production of reactive oxygen species (ROS). In contrast, UV-A irradiation has indirect effects via energy transfer from UV-A-stimulated chromophores to the DNA target or via photosensitized (chlorophylls and phycobilins can act as photosensitizers) production of ROS.

* Corresponding author: r.p.sinha@gmx.net

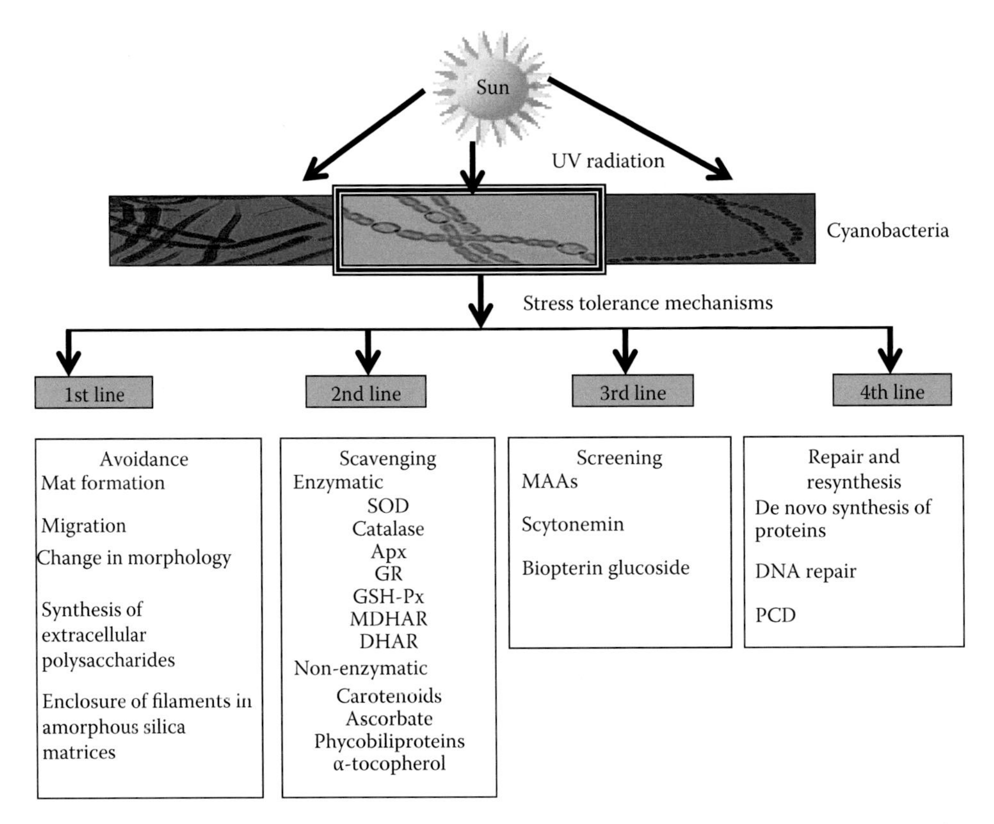

FIGURE 11.1 Tolerance mechanisms employed by cyanobacteria to counteract the damaging effects of ultraviolet radiation.

Cyanobacteria which are simultaneously exposed to photosynthetically active radiation (PAR; 400–700 nm) and UVR have evolved a number of mitigation strategies (tolerance mechanisms) to counteract the direct and indirect damaging effects of UVR. These strategies include avoidance, scavenging of ROS by non-enzymatic and enzymatic antioxidant molecules, synthesis of UV-absorbing/screening compounds such as mycosporine-like amino acids (MAAs) and scytonemin, repair of UV-induced damage of DNA, and resynthesis of proteins (Figure 11.1). This chapter deals with the molecular mechanisms of the mitigation strategies (tolerance mechanisms) employed by cyanobacteria to cope with harmful solar radiation.

11.2 AVOIDANCE AS A FIRST LINE OF DEFENSE

Several cyanobacteria which are exposed to high solar UVR in their natural habitats have developed avoidance as a first line of defense mechanisms. This includes migration from high to low UVR levels in the water column, formation of mats containing different cyanobacterial species or filaments enclosed in amorphous silica matrices, changes in morphology to increase self-shading, and synthesis of extracellular polysaccharides. Motile cyanobacteria can escape from intensive solar radiation by downward migration into the mat communities [7] or by sinking deeper into the water column [8]. The monitoring of vertical gliding motility patterns of *Oscillatoria cf. laetevirens* and *Spirulina cf. subsalsa* from hypersaline ponds near Guerrero Negro, Mexico, revealed that these cyanobacteria protect their photosynthetic machinery from UV-A and UV-B radiation damage by downward migration [9]. The highest incidence of migration in response to UV-B compared to UV-A and PAR was reported in *Microcoleus chthonoplastes* [10]. It was proposed that the deposits of amorphous silica matrices would have strongly attenuated Archean and early Precambrian levels of UVR, depicting their protective nature [11]. An effective protective mechanism against photoinhibition in *Arthrospira platensis* was found to be due to self-shading as a result of decreased helix pitch in the presence of UV-B to obtain a more compact spiral structure [12]. The synthesis of extracellular glycan in *Nostoc commune* was stimulated by UV-B and was proposed to provide UV resistance by increasing the effective path length for the absorption of radiation [13].

11.3 SCAVENGING AS A SECOND LINE OF DEFENSE

Once UVR reaches inside the cell, it interacts with oxygen and other organic compounds to produce toxic ROS such as superoxide (O_2^-), hydroxyl radical (OH^-), or hydrogen peroxide (H_2O_2), resulting in oxidative stress. To overcome the oxidative stress, cyanobacteria have developed antioxidant systems as a second line of defense mechanisms against UVR. This system includes non-enzymatic and enzymatic antioxidants. Non-enzymatic antioxidants comprise ascorbate (vitamin C), α-tocopherol (vitamin E), carotenoids, and reduced glutathione (GSH). The enzymatic antioxidants include superoxide dismutase (SOD), catalase, glutathione peroxidase (GSH-Px), and the enzymes involved in the ascorbate–glutathione cycle to detoxify ROS, such as ascorbate peroxidase (APX), monodehydroascorbate reductase (MDHAR), dehydroascorbate reductase (DHAR), and glutathione reductase (GR) [14]. Carotenoids protect cells against photooxidative damage by absorbing energy from excited chlorophyll molecules and quenching singlet state oxygen while α-tocopherol prevents lipid peroxidation by scavenging ROS [15]. However, ascorbate plays an important role in direct quenching of ROS, regenerating α-tocopherol and acts as a substrate in both violaxanthin de-epoxidase and APX reactions. Glutathione protects thiol groups in various enzymes and is also involved in α-tocopherol and ascorbate regeneration through the glutathione-ascorbate cycle [15]. SOD scavenges superoxide radicals and converts them to hydrogen peroxide which is further converted to water and O_2 via a combined catalase-peroxide system [16]. To date the presence of three groups of catalases (monofunctional haem-containing catalases, bifunctional haem-containing catalase–peroxidases, and nonhaem-manganese catalases) has been reported in 20 cyanobacterial genomes [17]. Depending upon the nature of the catalytic metals, four forms of SODs (FeSOD, MnSOD, Cu/ZnSOD, and NiSOD) are present. Comparative genomic analysis by Priya et al. [18] revealed that NiSOD is the only SOD found in primitive cyanobacteria, Fe and Mn occupy the higher orders of cyanobacteria, and Cu/ZnSOD is rare in cyanobacteria. These workers also revealed that the primitive unicellular *Prochlorococcus* possesses only NiSOD while the more evolved middle order forms of cyanobacteria possess a combination of Fe and Ni or Fe and Mn SODs. The most evolved filamentous, heterotrichous, and heterocystous forms predominantly have only Fe and Mn metalloforms [18].

Ehling-Schulz et al. [13] reported the changes in the carotenoids pattern of cyanobacterium *N. commune* in response to UV-B irradiation, and myxoxanthophyll and echinenone were suggested to act as outer membrane-bound UV-B photoprotectors. An increase in the activity of SOD and APX was observed in *Nostoc spongiaeforme* and *Phormidium corium* as a result of high-light (PAR; 400–700 nm) treatment [19]. The exogenous addition of antioxidants such as ascorbic acid and *N*-acetylcysteine (NAC) was found resulting in higher survival rate of *Anabaena* sp. due to reduction in chlorophyll bleaching, damage to the photosynthetic apparatus, lipid peroxidation and DNA strand breaks under UVR-induced oxidative damage [20]. Shirkey et al. [21] reported that an accumulation of active iron superoxide dismutase (FeSOD) in desiccated field cyanobacterium *N. commune* reverses the effects of oxidative stress imposed by multiple cycles of desiccation and rehydration during the UV-A or UV-B irradiation *in situ*. Two NADPH-dependent glutathione peroxidase-like proteins have been characterized in *Synechocystis* PCC 6803 and were found essential for the protection of membranes against lipid peroxidation [22]. Recently, another type of peroxidase called rubrerythrin-homolog (RbrA) was identified in *Anabaena* PCC 7120 and was found to protect the nitrogenase enzyme against oxidative stress [23]. In addition to known antioxidants, MAA mycosporine-glycine can also function as a biological antioxidant as it was found to effectively suppress various detrimental effects of the type-II photosensitization and decrease the levels of singlet oxygen generated by eosin Y or methylene blue [24].

11.4 SCREENING AS A THIRD LINE OF DEFENSE

Screening of damaging UVR by UV-absorbing compounds has been developed by several cyanobacteria as a third line of defense mechanism under prolonged UVR exposure. MAAs (Figure 11.2) and scytonemin are well known UV-absorbing/screening compounds that provide photoprotection against UV-B and/or UV-A radiation. MAAs are small (<400 Da), colorless, water-soluble compounds composed of a cyclohexenone or cyclohexenimine chromophore conjugated with the nitrogen substituent of an amino acid or its imino alcohol. Generally, the ring system contains a glycine subunit at the third carbon atom. Some MAAs also contain sulfate esters or glycosidic linkages through the imine substituent. Differences between the absorption spectra of MAAs are due to variations in the attached side groups and nitrogen substituents. Several characteristics of MAAs such as strong UV absorption maxima between 310 and 362 nm (Figure 11.3) and high molar extinction coefficients (ε = 28,100–50,000 M^{-1} cm^{-1}) suggest their role in UV-screening. In addition, photostability in both distilled and sea water in the presence of photosensitizers and high resistance against physico-chemical stressors like temperature, strong UVR, various solvents as well as pH make them successful photoprotectants in various habitats and organisms.

MAAs protect the cells by absorbing highly energetic UVR and then dissipating excess energy in the form of heat to their surroundings [25]. These compounds prevent 3 out of 10 photons from hitting cytoplasmic targets in cyanobacteria [26] and there is clear evidence that the presence of MAAs protects vital functions in phytoplankton from deleterious short wavelength radiation [27,28]. They can also act as antioxidants to prevent damage by ROS resulting from UVR [29]. The synthesis of MAAs in cyanobacteria is dependent on available nitrogen [30], and growth media with a nitrogen source support the highest MAA synthesis [31]. The biosynthesis of MAA in cyanobacteria is dependent on photosynthesis for the carbon source, as an externally added carbon source was found to overcome the negative effect of DCMU on MAA biosynthesis. Normally, *Anabaena*

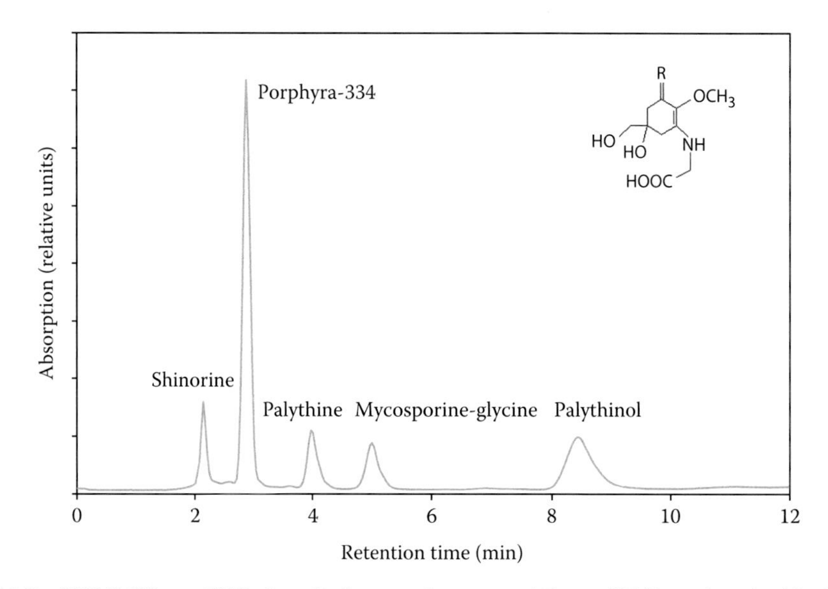

FIGURE 11.2 HPLC (Waters 2998 photodiode array detector and Pump 600 E, equipped with a Licrospher RP 18 column and guard; 5 mm packing) chromatogram of some mycosporine-like amino acids (MAAs) showing their retention times. Wavelength for detection 330 nm, mobile phase 0.2% (v/v) acetic acid in double-distilled water, run isocratically at a flow rate of 1.0 mL min^{-1}. Inset showing the general structure of MAAs.

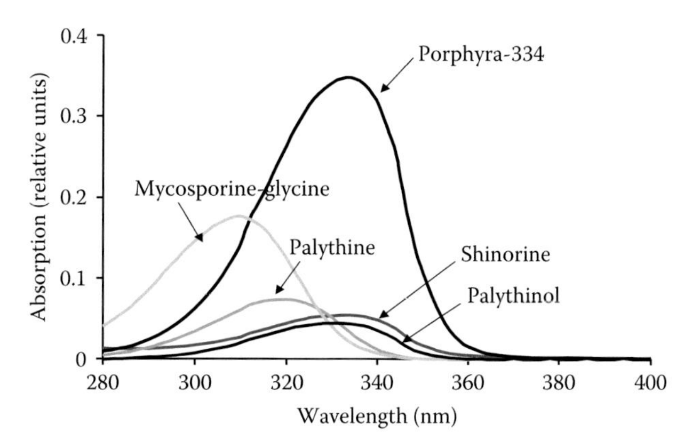

FIGURE 11.3 Absorption spectra of the corresponding MAAs as shown in Figure 11.2.

strains studied so far have been found to produce a single MAA, shinorine, in response to UVR [32,33]. However, recently, the biosynthesis of three MAAs (mycosporine-glycine, porphyra-334, and shinorine) in the rice-field cyanobacterium (*Anabaena doliolum*) was reported. Mycosporine-glycine was found to be under constitutive control and other two MAAs were specifically induced by UV-B radiation [33]. Recently, Oyamada et al. [34] reported the photoprotection of human fibroblast cells against UVR and their proliferation by three MAAs (shinorine, porphyra-334, and mycosporine glycine). Misonou et al. [35] have found that MAAs effectively block thymine dimer formation by UVR *in vitro*. Thus, MAAs may also have a pharmaceutical and industrial importance.

There is still a great deal of controversy concerning the precise mechanisms of MAA biosynthesis. However, it is evident that 3-dehydroquinate formed during the early stages of the shikimate pathway serves as a precursor for the synthesis of fungal mycosporines and MAAs via gadusols [33,36]. The synthesis of MAAs has been reported to occur in bacteria, cyanobacteria, phytoplankton, and macroalgae but not in animals, where these compounds are supposed to be accumulated either via the food chain or synthesized by their symbiotic algal partner due to the lack of the shikimate pathway, the predicted pathway for their biosynthesis [32,36]. Thus, MAAs provide protection from UV radiation not only in their producers, but also to primary and secondary consumers. However, recently, the gene for the shikimate pathway in the metazoan *Nematostella vectensis* has been disclosed by Starcevic et al. [37]. Genome mining of certain strains identified a combination of genes, YP_324358 (predicted DHQ synthase) and YP_324357 (O-methyltransferase), which were present only in *Anabaena variabilis* PCC 7937 and missing in the other studied cyanobacteria [38]. Phylogenetic analysis revealed that these two genes are transferred from a cyanobacterial donor to dinoflagellates and finally to metazoa by a lateral gene transfer event [38]. All other cyanobacteria, which have these two genes, also had another copy of the DHQ synthase gene. The study provides first insight into the genes of cyanobacteria involved in MAA biosynthesis (Figure 11.4) and thus widens the field of research for molecular, bioinformatics, and phylogenetic analysis of these evolutionary and industrially important compounds. Based on the results it was proposed that YP_324358 and YP_324357 gene products are involved in the biosynthesis of the common precursor (deoxygadusol) of all MAAs [38,39]. Mycosporine-glycine is considered as a prime MAA synthesized in the shikimate pathway that further goes through chemical and/or biochemical conversions to produce secondary MAAs [40–42].

Another UVR-screening compound, scytonemin (Figure 11.5), is exclusively produced by cyanobacteria. It is a yellow-brown lipid-soluble and inducible pigment located in the extracellular polysaccharide sheath of some cyanobacterial species [43]. Scytonemin is a dimer composed of indolic and phenolic subunits having a molecular mass of 544 Da and has an *in vivo* absorption maximum at 370 nm. A purified scytonemin shows maximum absorption at 386 nm but it also absorbs significantly at 252, 278, and 300 nm. Scytonemin synthesis is primarily induced by

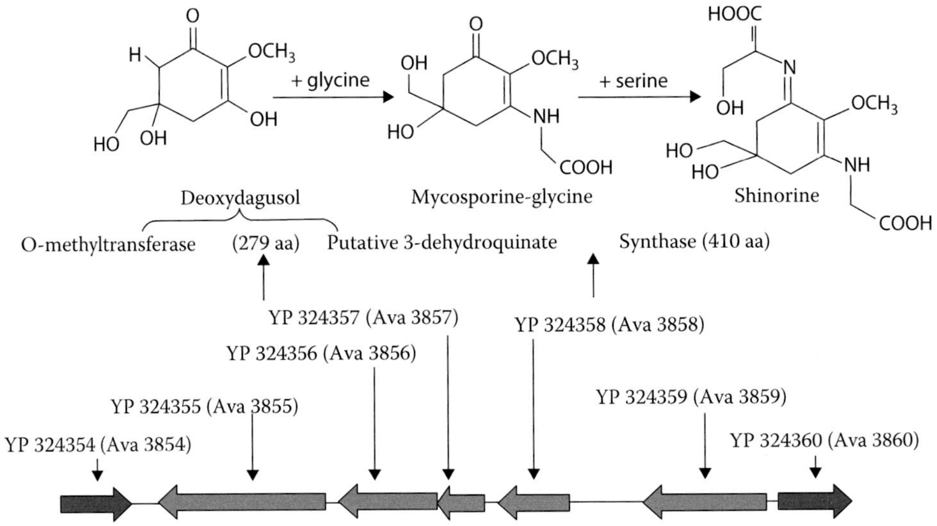

Deoxydagusol + glycine → Mycosporine-glycine + serine → Shinorine

O-methyltransferase (279 aa) Putative 3-dehydroquinate Synthase (410 aa)

YP 324354 (Ava 3854)
YP 324355 (Ava 3855)
YP 324356 (Ava 3856)
YP 324357 (Ava 3857)
YP 324358 (Ava 3858)
YP 324359 (Ava 3859)
YP 324360 (Ava 3860)

FIGURE 11.4 Genes involved in MAAs biosynthesis. (Modified from Singh, S.P. et al., *Genomics*, 95, 120, 2010.)

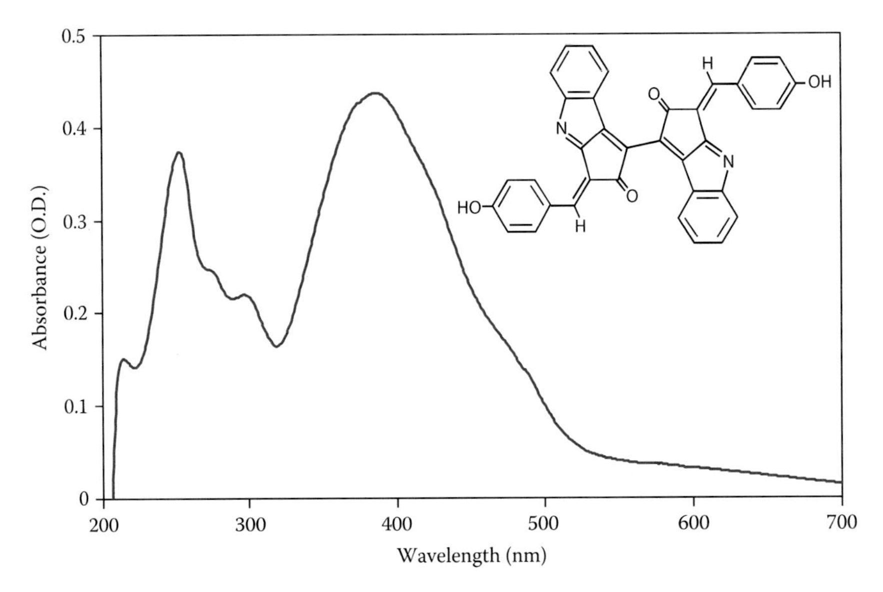

FIGURE 11.5 Absorption spectrum and molecular structure of scytonemin.

UV-A radiation whereas blue, green, or red light at the same fluence rates (intensity) does not have a significant influence [44]. Dillon et al. [45] showed that both increase in temperature and oxidative stress in combination with UV-A have a synergistic effect on the synthesis of scytonemin. The presence of scytonemin in the cyanobacterial sheath has been reported to reduce the entry of UV-A radiation into the cell by 90% [44,46]. Scytonemin is highly stable and performs its screening activity without any further metabolic investment even under prolonged physiological inactivity (e.g., desiccation). Thus, scytonemin can also protect cyanobacteria when other ultraviolet-protective mechanisms such as active repair of damaged cellular components are ineffective [47]. Three new pigments such as tetramethoxyscytonemin, dimethoxyscytonemin, and scytonine from the organic extracts of *Scytonema* sp. which are derived from the scytoneman skeleton of scytonemin have been reported [48].

Scytonemin was proposed to be synthesized from secondary metabolites of aromatic amino acid biosynthesis by condensation of tryptophan and phenylpropanoid-derived subunits [49]. Based on genetic, biochemical, and sequence similarity evidence, six putative genes in the scytonemin gene cluster (NpR1276 to NpR1271) with no previously known protein functions are shown likely to be involved in the assembly of scytonemin from central metabolites in the genome of *Nostoc punctiforme* ATCC 29133 [50]. The assessment of these gene clusters disclosed the occurrence of redundant copies of genes responsible for aromatic amino acid biosynthesis. The evolutionary phylogenetic analysis revealed that the scytonemin gene cluster is distributed across several cyanobacterial lineages, which led to the proposal that the distribution of this gene cluster is best explained by assuming an ancient evolutionary origin. Recently, Balskus and Walsh [51] have presented the possible biosynthetic route for scytonemin biosynthesis and identified the acyloin reaction as a key step in constructing the carbon framework of this ecologically and evolutionarily important pigment. They have also functionally characterized two enzymes involved in the initial step of scytonemin biosynthesis encoded by ORF NpR 1275 and NpR 1276 from the gene cluster identified by Soule et al. [50]. However, the products of NpR 1263 and NpR 1269 ORFs are yet to be functionally characterized. NpR1275, which resembles a leucine dehydrogenase, was shown to catalyze the oxidative deamination of tryptophan to form indole-3-pyruvic acid (IPA). IPA, along with the tyrosine precursor *p*-hydroxyphenylpyruvic acid, acts as the substrate for an acyloin reaction catalyzed by NpR1276 (a homolog to thiamine diphosphate [ThDP]-dependent acetolactate synthase). From a comparative study of four other cyanobacterial strains, Soule et al. [52] on the basis of conservation and location suggested that two supplementary conserved clusters (NpF5232 to NpF5236) and a putative two-component regulatory system (NpF1277 and NpF1278) are possibly involved in scytonemin biosynthesis and regulation, respectively (Figure 11.6). A UV-A-absorbing pigment, biopterin glucoside (BG; a compound

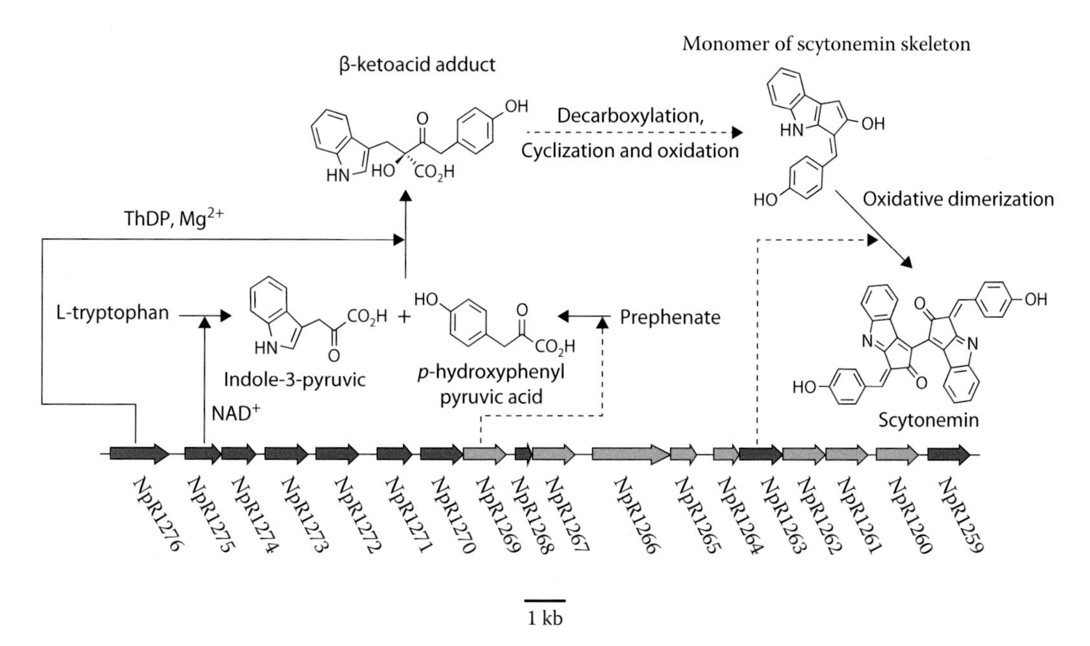

FIGURE 11.6 **(See color insert.)** Biosynthetic pathway for scytonemin and corresponding gene products involved in each step. Continuous arrows represent gene products which are functionally characterized while gene products indicated by broken arrows are still to be functionally characterized for their involvement in corresponding steps. Gray arrows represent ORFs involved in the biosynthesis of aromatic amino acids while brown arrows represent ORFs of unknown function. (The model is based on the information from Soule, T. et al., *J. Bacteriol.*, 189, 4465, 2007; Balskus, E.P. and Walsh, C.T., *J. Am. Chem. Soc.*, 130, 15260, 2008; Adapted from Singh, S.P. et al, *Ageing Res. Rev.*, 9, 79, 2010.)

chemically related to the pteridine pigments found in butterfly wings), with absorption maxima at 256 and 362 nm, has been purified from a marine planktonic cyanobacterium *Oscillatoria* sp. collected from the coastal areas of Japan [53]. Additionally, *Anacystis nidulans*, *A. variabilis*, *Nostoc muscorum,* and *Nostoc maseaum* were also found to produce pteridines in relatively large concentrations.

11.5 REPAIR AND RESYNTHESIS AS A FOURTH LINE OF DEFENSE

The repair and resynthesis of sensitive targets in the cyanobacteria becomes critically important as a fourth line of defense mechanism when UVR escapes from the first, second, and third line of defense mechanisms and damage of bio-molecules such as DNA and proteins occurs. Although the presence of multiple copies of genomic DNA in the cyanobacteria may nullify the effect of single mutations by UVR, the existence of several DNA repair mechanisms strengthen them additionally to cope with the radiation. These mechanisms include photoreactivation by photolyase (Figure 11.7) which converts UV-induced dimers into monomers, dark or excision repair and the recombinational repair. The presence of a UV-inducible photoreactivation system has been reported in strains of *Anabaena* sp. such as *Anabaena* sp. PCC 7120, *A. variabilis* PCC 7937, *Anabaena* sp. M-131 and *A. variabilis* sp. PCC 7118 [54]. The genes for photolyase homologs have been identified in *Synechocystis* sp. PCC 6803 and were functionally characterized for their roles in photoreactivation [55,56]. Han et al. [57] have also reported photoreactivation of UV-B-induced inhibition of photosynthesis in *Anabaena* sp. An increase in the transcript level of recA and the concomitant increase in the abundance of the corresponding 37–38-kDa polypeptide were reported in *A. variabilis* after UV exposure [58]. The gene for the DNA repair enzyme Fpg (formamidopyrimidine-DNA glycosylase) has been reported from *Synechococcus elongatus* and was suggested to be involved in photoprotection against oxidative damage [59]. Thus, cyanobacteria are capable of repairing DNA lesions by both photoreactivation and excision repair. Recently, photorepair of UV-B-induced damaged DNA has been reported in the cyanobacterium *A. variabilis* PCC 7937 [60].

In addition to DNA repair, cyanobacteria are also capable of synthesizing new proteins to replace damaged copies. The UV-B shock and acclimation response of UV-B stress was found to be a completely different and complex phenomenon influencing a total of 493 proteins [61] comprising of an early shock response influencing 214 proteins and a late acclimation response influencing 279 proteins. An increased turnover of D1 and D2 proteins of the photosystem II reaction center was found to be responsible for UV-B resistance in cyanobacteria *Synechocystis* sp. [62]. Instant turnover of photosystem II reaction center proteins helps in acclimatization and provides resistance against UVR in cyanobacteria [63]. In addition, recent studies show that cyanobacteria may also undergo apoptosis or programmed cell death (PCD) when a cell is damaged

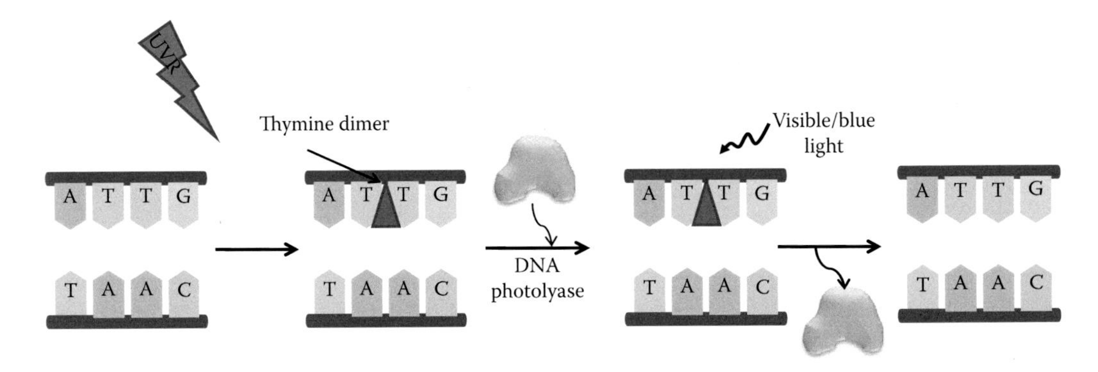

FIGURE 11.7 UV-induced DNA damage and mechanism of photoreactivation in cyanobacteria.

beyond repair. An autocatalytic PCD has been shown to operate in a nitrogen-fixing cyanobacterium (*Trichodesmium* sp.) and was found to be induced by high irradiance, iron starvation, and oxidative stress [64]. The caspase activity as well as proteins reacting to human caspase-3 antibodies was reported in *Trichodesmium* sp. [64]. The freshwater cyanobacterium *Anabaena* has also been found to activate PCD and increases general protease activity after exposure to univalent-cation salts [65].

11.6 SIGNAL TRANSDUCTION

It is clear that cyanobacteria have developed a number of mitigation strategies to counteract the detrimental effects of UV-B radiation; however, to adapt to new conditions, cells must first perceive the environmental signals and then transduce them to the corresponding apparatus to mediate the responses accordingly. In cyanobacteria, the underlying molecular mechanisms of UV-B signal perception and transduction are not well understood. UV-B radiation also results in the generation of ROS in cyanobacteria which may participate in transducing the signals under UV stress. UVR sensitive L-type calcium channels were detected in *Anabaena* sp. and *N. commune* [66], and UV-B exposure was found to increase the intracellular calcium concentration in the cyanobacterium *Anabaena* sp. [67]. This indicates that cyanobacteria also use calcium in transducing the UV signals and calmodulin-like protein; an important component of calcium-dependent signaling has been also characterized in the cyanobacterium *Nostoc* sp. PCC 6720 [68]. In *Anacystis nidulans* R-2, the UV-shock proteins were also induced by methyl viologen, indicating the role of ROS in transducing UV signals [69]. Thus, it is most likely that cyanobacteria transduce UV-B signals via ROS and calcium and might regulate a number of adaptive processes under UVR exposure. Other signaling molecules such as cAMP and cGMP are also found in cyanobacteria, and a change in cellular homeostasis of cGMP was found to be involved in the signal transduction under UV stress and permit the repair of the UV-B-damaged photosystem II in *Synechocystis* PCC 6803 [70]. However, not enough information is available about the signal transduction pathway regarding the induction of photoprotective compounds. Portwich and Garcia-Pichel [71] have reported the presence of a UV-sensitive photoreceptor, reduced pterin in *Chlorogloeopsis* PCC 6912. The prediction of a reduced pterin as UV-B specific photoreceptor was based on an action spectrum obtained for MAAs induction and inhibition of MAAs synthesis in the presence of an inhibitor of pterin biosynthesis and antagonist of excited states of pterin. From results of Sinha et al. [72] it was difficult to confirm that pterin is the only photoreceptor for MAA induction since the MAA content decreased in the presence of all inhibitors in comparison to the cells treated with UV-B only. Shick [73] has proposed that ROS transduce UV signals in zooxanthellate corals and suggested an involvement of lipid peroxides or related products in up-regulating the biosynthesis of primary MAAs. The synthesis of MAAs has also been found to be induced by several abiotic factors in addition to UVR, suggesting the presence of a common signal transduction pathway for their induction rather than the more specific one [31].

11.7 CONCLUSIONS

Cyanobacteria have several targets susceptible to UVR stress; however, these organisms have developed several lines of defense mechanisms that sustain their successful growth and survival in various habitats receiving high solar UVR. The balance between damage and defense mechanisms also has ecological importance as it maintains the productivity and nitrogen economy of an ecosystem with concomitant regulation of climate change. UVR causes damage to cyanobacteria either directly or indirectly through oxidative damage. ROS are known to induce the antioxidant system to remove toxic species. There are several classes of SODs and catalases found in cyanobacteria but characterization of specific SODs and catalases expressed under UVR stress should be the aim of future research. The fragmentation of cyanobacterial filaments under UVR may possibly occur due to PCD of severely damaged cells in a filament; however, degradation of proteins that maintains the structure, has been proposed for the filament breakage. The process of PCD and related proteins is found in cyanobacteria to play an important role under oxidative stress; however, the role of PCD under UVR stress may be a new field for future research. Identification of DNA and protein repair mechanisms as well as signal pathways for these mechanisms will also be a challenge for future research work. The molecular structure and reaction mechanism of DNA repair enzymes (photolyase and glycosylase) from cyanobacteria is also not well understood. The molecular biology of UV-absorbing/screening MAAs as well as scytonemin is in its infancy and some of the gene products involved in the initial step of scytonemin biosynthesis have been functionally characterized. Moreover, these photoprotective compounds have several other potential functions with therapeutic properties that may be exploited in commercial applications. Further development of biotechnology for human photoprotection and research must be specially focused on the analysis, biosynthesis, and mode of action of several unknown photoprotective compounds.

REFERENCES

1. Sinha, R.P. and Häder, D.-P., Photobiology and ecophysiology of rice field cyanobacteria, *Photochem. Photobiol.*, 64, 887–896, 1996.
2. Fischer, W.F., Life before the rise of oxygen, *Nature*, 455, 1051–1052, 2008.
3. Rastogi, R.P. and Sinha, R.P., Biotechnological and industrial significance of cyanobacterial secondary metabolites, *Biotech. Adv.*, 27, 521–539, 2009.
4. Rastogi, R.P., Richa, and Sinha, R.P., Biotechnological exploitations of cyanobacteria. In: *Plant Genome Diversity, Conservation and Manipulation* (Eds. Roy, B.K., Chaudhary, B.R., and Sinha, R.P.) Narosa Publishing House Pvt. Ltd., New Delhi, India, 2011, pp. 43–57.
5. Vaishampayan, A. et al., Cyanobacterial biofertilizers in rice agriculture, *Bot. Rev.*, 67, 453–516, 2001.
6. Crutzen, P.J., Ultraviolet on the increase, *Nature*, 356, 104–105, 1992.
7. Quesada, A. and Vincent, W.F., Strategies of adaptation by Antarctic cyanobacteria to ultraviolet radiation, *Eur. J. Phycol.*, 32, 335–342, 1997.
8. Reynolds, C.S., Oliver, R.L., and Walsby, A.E., Cyanobacterial dominance: The role of buoyancy regulation in dynamic lake environments, *N. Z. J. Mar. Freshwater Res.*, 21, 379–390, 1987.
9. Kruschel, C. and Castenholz, R.W., The effect of solar UV and visible irradiance on the vertical movements of cyanobacteria in microbial mats of hypersaline waters, *FEMS Microbiol. Ecol.*, 27, 53–72, 1998.
10. Bebout, B.M. and Garcia-Pichel, F., UV B-induced vertical migrations of cyanobacteria in a microbial mat, *Appl. Environ. Microbiol.*, 61, 4215–4222, 1995.
11. Phoenix, V.R. et al., Chilean high altitude hot-spring sinters: A model system for UV screening mechanisms by early Precambrian cyanobacteria, *Geobiology*, 4, 15–28, 2006.
12. Wu, H. et al., Effects of solar UV radiation on morphology and photosynthesis of filamentous cyanobacterium *Arthrospira platensis*, *Appl. Environ. Microbiol.*, 71, 5004–5013, 2005.
13. Ehling-Schulz, M., Bilger, W., and Scherer, S., UV-B-induced synthesis of photoprotective pigments and extracellular polysaccharides in the terrestrial cyanobacterium *Nostoc commune*, *J. Bacteriol.*, 179, 1940–1945, 1997.
14. He, Y.-Y. and Häder, D.-P., Reactive oxygen species and UV-B: Effect on cyanobacteria, *Photochem. Photobiol. Sci.*, 1, 729–736, 2002.
15. Niyogi, K.K., Photoprotection revisited: Genetics and molecular approaches, *Annu. Rev. Plant Physiol.*, 50, 333–359, 1999.
16. Tel-Or, E., Huflejt, M.E., and Packer, L., Hydroperoxide metabolism in cyanobacteria, *Arch. Biochem. Biophys.*, 246, 396–402, 1986.
17. Latifi, A., Ruiz, M., and Zhang, C.-C., Oxidative stress in cyanobacteria, *FEMS Microbiol. Rev.*, 33, 258–278, 2009.
18. Priya, B. et al., Comparative analysis of cyanobacterial superoxide dismutases to discriminate canonical forms, *BMC Genomics*, 8, 435, 2007.

19. Bhandari, R. and Sharma, P.K., High-light-induced changes on photosynthesis, pigments, sugars, lipids and antioxidant enzymes in freshwater (*Nostoc spongiaeforme*) and marine (*Phormidium corium*) cyanobacteria, *Photochem. Photobiol.*, 82, 702–710, 2006.
20. He, Y.-Y. and Häder, D.-P., UV-B-induced formation of reactive oxygen species and oxidative damage of the cyanobacterium *Anabaena* sp.: Protective effects of ascorbic acid and *N*-acetyl-L-cysteine, *J. Photochem. Photobiol. B: Biol.*, 66, 115–124, 2002.
21. Shirkey, B. et al., Active Fe-containing superoxide dismutase and abundant sodF mRNA in *Nostoc commune* (Cyanobacteria) after years of desiccation, *J. Bacteriol.*, 182, 189–197, 2000.
22. Gaber, A. et al., Induction and functional analysis of two reduced nicotinamide adenine dinucleotide phosphate-dependent glutathione peroxidase-like proteins in *Synechocystis* PCC 6803 during the progression of oxidative stress, *Plant Physiol.*, 136, 2855–2861, 2004.
23. Zhao, W., Ye, Z., and Zhao, J., RbrA, a cyanobacterial rubrerythrin, functions as a FNR-dependent peroxidase in heterocysts in protection of nitrogenase from damage by hydrogen peroxide in *Anabaena* sp. PCC 7120, *Mol. Microbiol.*, 66, 1219–1230, 2007.
24. Suh, H.-J., Lee, H.-W., and Jung, J., Mycosporine glycine protects biological systems against photodynamic damage by quenching singlet oxygen with a high efficiency, *Photochem. Photobiol.*, 78, 109–113, 2003.
25. Conde, F.R., Churio, M.S., and Previtali, C.M., The deactivation pathways of the excited-states of the mycosporine-like amino acids shinorine and porphyra-334 in aqueous solution, *Photochem. Photobiol. Sci.*, 3, 960–967, 2004.
26. Garcia-Pichel, F., Wingard, C.E., and Castenholz, R.W., Evidence regarding the UV sunscreen role of a mycosporine-like compound in the cyanobacterium *Gloeocapsa* sp., *Appl. Environ. Microbiol.*, 59, 170–176, 1993.
27. Klisch, M. et al., Mycosporine-like amino acids (MAAs) protect against UV-B-induced damage in *Gyrodinium dorsum Kofoid*, *J. Plant Physiol.*, 158, 1449–1454, 2001.
28. Singh, K.L. and Sinha, R.P., UV-absorbing compounds in algae. In: *Advances in Life Sciences* (Eds. Sinha, R.P., Sharma, N.K., and Rai, A.K.) I.K. International Publishing House Pvt. Ltd., New Delhi, India, 2011, pp. 213–239.
29. Dunlap, W.C. and Yamamoto, Y., Small-molecule antioxidants in marine organisms: Antioxidant activity of mycosporine-glycine, *Comp. Biochem. Physiol. B Biochem. Mol. Biol.*, 112, 105–114, 1995.
30. Singh, S.P. et al., Effects of abiotic stressors on synthesis of the mycosporine-like amino acid shinorine in the cyanobacterium *Anabaena variabilis* PCC 7937, *Photochem. Photobiol.*, 84, 1500–1505, 2008.
31. Singh, S.P. et al., Role of various growth media on shinorine (mycosporine-like amino acid) concentration and photosynthetic yield in *Anabaena variabilis* PCC 7937, *World J. Microbiol. Biotechnol.*, 24, 3111–3115, 2008.
32. Sinha, R.P., Singh, S.P., and Häder, D.-P., Database on mycosporines and mycosporine-like amino acids (MAAs) in fungi, cyanobacteria, macroalgae, phytoplankton and animals, *J. Photochem. Photobiol. B, Biol.*, 89, 29–35, 2007.
33. Singh, S.P. et al., Mycosporine-like amino acids (MAAs) profile of a rice-field cyanobacterium *Anabaena doliolum* as influenced by PAR and UVR, *Planta*, 229, 225–233, 2008.
34. Oyamada, C. et al., Mycosporine-like amino acids extracted from scallop (*Patinopecten yessoensis*) ovaries: UV protection and growth stimulation activities on human cells, *Mar. Biotechnol.*, 10, 141–150, 2008.
35. Misonou, T. et al., UV-absorbing substance in the red alga *Porphyra yezoensis* (Bangiales, Rhodophyta) block thymine photodimer production, *Mar. Biotechnol.*, 5, 194–200, 2003.
36. Shick, J.M. and Dunlap, W.C., Mycosporine-like amino acids and related gadusols: Biosynthesis, accumulation, and UV-protective functions in aquatic organisms, *Annu. Rev. Physiol.*, 64, 223–262, 2002.
37. Starcevic, A. et al., Enzymes of the shikimic acid pathway encoded in the genome of a basal metazoan, *Nematostella vectensis*, have microbial origins, *Proc. Natl. Acad. Sci. U.S.A.*, 105, 2533–2537, 2008.
38. Singh S.P. et al., Genome mining of mycosporine-like amino acid (MAA) synthesizing and non-synthesizing cyanobacteria: A bioinformatics study, *Genomics*, 95, 120–128, 2010.
39. Singh S.P., Häder D.-P., and Sinha, R.P., Cyanobacteria and ultraviolet radiation (UVR) stress: Mitigation strategies, *Ageing Res. Rev.*, 9, 79–90, 2010.
40. Carreto, J.I., Carignan, M.O., and Montoya, N.G., A high resolution reverse-phase liquid chromatography method for the analysis of mycosporine-like amino acids (MAAs) in marine organisms, *Mar. Biol.*, 146, 237–252, 2005.

41. Callone, A.I. et al., Biotransformation of mycosporine like amino acids (MAAs) in the toxic dinoflagellate *Alexandrium tamarense*, *J. Photochem. Photobiol. B: Biol.*, 84, 204–212, 2006.
42. Conde, F.R., Churio, M.S., and Previtali, C.M., Experimental study of the excited-state properties and photostability of the mycosporine-like amino acid palythine in aqueous solution, *Photochem. Photobiol. Sci.*, 6, 669–674, 2007.
43. Sinha, R.P. and Häder, D.-P., UV-protectants in cyanobacteria, *Plant Sci.*, 174, 278–289, 2008.
44. Garcia-Pichel, F. and Castenholz, R.W., Characterization and biological implications of scytonemin, a cyanobacterial sheath pigment, *J. Phycol.*, 27, 395–409, 1991.
45. Dillon, J.G. et al., Effect of environmental factors on the synthesis of scytonemin, a UV-screening pigment, in a cyanobacterium (*Chroococcidiopsis* sp.), *Arch. Microbiol.*, 177, 322–331, 2002.
46. Garcia-Pichel, F., Sherry, N.D., and Castenholz, R.W., Evidence for an ultraviolet sunscreen role of the extracellular pigment scytonemin in the terrestrial cyanobacterium *Chlorogloeopsis* sp, *Photochem. Photobiol.*, 56, 17–23, 1992.
47. Fleming, E.D. and Castenholz, R.W., Effects of periodic desiccation on the synthesis of the UV-screening compound, scytonemin, in cyanobacteria, *Environ. Microbiol.*, 9, 1448–1455, 2007.
48. Bultel-Poncé, V. et al., New pigments from the terrestrial cyanobacterium *Scytonema* sp. collected on the Mitaraka Inselberg, French Guyana, *J. Nat. Prod.*, 67, 678–681, 2004.
49. Proteau, P.J. et al., The structure of scytonemin, an ultraviolet sunscreen pigment from the sheaths of cyanobacteria, *Experientia*, 49, 825–829, 1993.
50. Soule, T. et al., Molecular genetics and genomic analysis of scytonemin biosynthesis in *Nostoc punctiforme* ATCC 29133, *J. Bacteriol.*, 189, 4465–4472, 2007.
51. Balskus, E.P. and Walsh, C.T., Investigating the initial steps in the biosynthesis of cyanobacterial sunscreen scytonemin, *J. Am. Chem. Soc.*, 130, 15260–15261, 2008.
52. Soule, T. et al., A comparative genomics approach to understanding the biosynthesis of the sunscreen scytonemin in cyanobacteria, *BMC Genomics*, 10, 336–345, 2009.
53. Matsunaga, T. et al., An ultraviolet (UV-A) absorbing biopterin glucoside from the marine planktonic cyanobacterium *Oscillatoria* sp., *Appl. Microbiol. Biotechnol.*, 39, 250–253, 1993.
54. Levine, E. and Thiel, T., UV-inducible DNA repair in the cyanobacteria *Anabaena* spp., *J. Bacteriol.*, 169, 3988–3993, 1987.
55. Hitomi, K. et al., Bacterial cryptochrome and photolyase: Characterization of two photolyase-like genes of *Synechocystis* sp. PCC6803, *Nucl. Acids Res.*, 28, 2353–2362, 2000.
56. Ng, W.-O. and Pakrasi, H.B., DNA photolyase homologs are the major UV resistance factors in the cyanobacterium *Synechocystis* sp. PCC 6803, *Mol. Gen. Genet.*, 264, 924–930, 2001.
57. Han, T., Sinha, R.P., and Häder, D.-P., UV-A/blue light-induced reactivation of photosynthesis in UV-B irradiated cyanobacterium, *Anabaena* sp., *J. Plant Physiol.*, 158, 1403–1413, 2001.
58. Owttrim, G.W. and Coleman, J.R., Regulation of expression and nucleotide sequence of the *Anabaena variabilis recA* gene, *J. Bacteriol.*, 171, 5713–5719, 1989.
59. Mühlenhoff, U., The FAPY-DNA glycosylase (Fpg) is required for survival of the cyanobacterium *Synechococcus elongatus* under high light irradiance, *FEMS Microbiol. Lett.*, 187, 127–132, 2000.
60. Rastogi, R.P. et al., Ultraviolet-B-induced DNA damage and photorepair in the cyanobacterium *Anabaena variabilis* PCC 7937, *Env. Exp. Bot.*, 74, 280–288, 2011.
61. Ehling-Schulz, M. et al., The UV-B stimulon of the terrestrial cyanobacterium *Nostoc commune* comprises early shock proteins and late acclimation proteins, *Mol. Microbiol.*, 46, 827–843, 2002.
62. Sass, L. et al., Repair of UV-B induced damage of Photosystem II via de novo synthesis of the D1 and D2 reaction centre subunits in *Synechocystis* sp. PCC 6803, *Photosynth. Res.*, 54, 55–62, 1997.
63. Campbell, D. et al., The cyanobacterium *Synechococcus* resists UV-B by exchanging photosystem II reaction-center D1 proteins, *Proc. Natl. Acad. Sci. U.S.A.*, 95, 364–369, 1998.
64. Berman-Frank, I. et al., The demise of the marine cyanobacterium, *Trichodesmium* spp., via an autocatalyzed cell death pathway, *Limnol. Oceanogr.*, 49, 997–1005, 2004.
65. Ning, S.B. et al., Salt stress induces programmed cell death in prokaryotic organism *Anabaena*, *J. Appl. Microbiol.*, 93, 15–28, 2002.
66. Sinha, R.P. et al., Effects of UV and visible light on cyanobacteria at the cellular level, *Photochem. Photobiol. Sci.*, 1, 553–559, 2002.
67. Richter, P. et al., Calcium signals from heterocysts of *Anabaena* sp. after UV irradiation, *J. Plant Physiol.*, 154, 137–139, 1999.
68. Onek, L.A., Lea, P.J., and Smith, R.J., Isolation and characterization of a calmodulinlike protein from the cyanobacterium *Nostoc* sp. PCC 6720, *Arch. Microbiol.*, 161, 352–358, 1994.

69. Shibata, H., Baba, K., and Ochiai, H., Near-UV irradiation induces shock proteins in *Anacystis nidulans* R-2; possible role of active oxygen, *Plant Cell Physiol.*, 32, 771–776, 1991.
70. Cadoret, J.-C. et al., Cyclic nucleotides, the photosynthetic apparatus and response to UV-B stress in the cyanobacterium *Synechocystis* sp. PCC 6803, *J. Biol. Chem.*, 280, 33935–33944, 2005.
71. Portwich, A. and Garcia-Pichel, F., A novel prokaryotic UVB photoreceptor in the cyanobacterium *Chlorogloeopsis* PCC 6912, *Photochem. Photobiol.*, 71, 493–498, 2000.
72. Sinha, R.P. et al., Wavelength-dependent induction of a mycosporine-like amino acid in a rice-field cyanobacterium, *Nostoc commune*: Role of inhibitors and salt stress, *Photochem. Photobiol. Sci.*, 2, 171–176, 2003.
73. Shick, J.M., The continuity and intensity of ultraviolet irradiation affect the kinetics of biosynthesis, accumulation, and conversion of mycosporine-like amino acids (MAAs) in the coral *Stylophora pistillata*, *Limnol. Oceanogr.*, 49, 442–458, 2004.

12 Zinc Homeostasis in Cyanobacteria

*Lee Hudek, Agnes Michalczyk, Brett A. Neilan, and M. Leigh Ackland**

CONTENTS

12.1 INTRODUCTION

The transition metal ions including cobalt, copper, iron, manganese, nickel, and zinc are essential for the viability of cyanobacteria as well as many other organisms. Cellular trace elements are required for the structure and function of many molecules essential for life including proteins and nucleic acids [1]. Metal ions are involved in a range of essential processes in particular enzyme functions where they are required for catalysis or may act as catalytic cofactors, for example, in reversible oxidation–reduction and hydrolytic reactions [2,3]. Individual metal ions may perform multiple biochemical roles or may be limited to just one role.

Higher levels of iron, manganese, magnesium, and copper are found in cyanobacteria in contrast to non-photosynthesizing organisms, as they are essential cofactors for oxygenic electron transfer during photosynthesis [4]. While metal ions are found in the environment, organisms must regulate cellular accumulation and maintain their metal homeostasis. Essential trace metals may pose a major oxidative risk factor through their capacity to generate free radicals [4]. The increased metal requirements of cyanobacteria in turn increase the need to tightly regulate intracellular levels to prevent oxidative and free-radical damage from occurring [4].

Heavy metal stress results in the inactivation of biomolecules by inhibiting essential functional groups or by displacing essential metal ions. Elevated levels of intracellular metals, including zinc and copper, may also induce the production of reactive oxygen species (ROS) [5]. Cells are able to protect against ROS through antioxidative enzymatic systems including superoxide dismutase (SOD), catalase (CAT), peroxidase (POD), and nonenzymatic systems including ascorbate, glutathione, and phenolic compounds [5].

Coordination chemistry investigates metal ion–protein complexes, focusing on the mechanisms for distinguishing and discriminating between metal ions [6]. This provides much insight for understanding the mechanisms responsible for metal ion transport and maintaining homeostatic levels.

* Corresponding author: leigh.ackland@deakin.edu.au

12.2 STRUCTURE AND FUNCTION OF BACTERIAL METAL TRANSPORT GENES

In prokaryotes, zinc confers protein structure and is involved in enzyme catalysis and signaling [1]. Zinc homeostasis in cyanobacteria is based on export and uptake systems, which are controlled by their own regulators [7,8]. The current understanding of bacterial zinc transporters has focused on the Adenosine Triphosphate (ATPase) Binding Cassette (ABC type), Zinc uptake (Znu) system [9–12]. While the ABC transporters are present in most bacteria, they are not exclusive to bacteria; they are also found in eukaryotes, where they play a variety of physiological roles. Mutations in ABC transport systems have been linked to numerous human genetic disorders relating to the transport of cytotoxic molecules and resistance to antibiotics, herbicides, and chemotherapeutic drugs [11].

In bacteria, the Znu system is classified as part of the cluster 9 family of metal transporters, within the ABC grouping [13]. The cluster 9 family classification is based on the architecture of the metal binding site, comprising conserved histidine residues. The most characterized ABC metal transport system in bacteria facilitates cation transport via two transmembrane domains. These transmembrane domains are the ZnuB component of the ZnuABC system [11]. While ZnuB indicates ligand specificity, the two hydrophobic nucleotide binding domains (NBDs), ZnuA and ZnuC, power the transport cycle through the well-described ATP switch model [11,14]. This operates through an NBD switch between a dimeric confirmation, closed around two ATP molecules, and a nucleotide-free dimeric open confirmation [11]. Studies of the Znu system in bacteria have identified the structural and functional domains including the metal binding domains. This information is useful in understanding zinc transport in cyanobacteria. Znu transporters have been studied to a lesser extent in cyanobacteria (*Synechocystis* and *Synechococcus* species); however, similarities in these transporters with bacterial (*Escherichia coli*, *Salmonella*, *Streptococcus*, and *Staphylococcus* species) Znu transporters have been established [9,11,12,15–18].

The high affinity Zn^{2+} uptake systems described for *E. coli* are linked to the activity of ABC transporters [19]. The zinc uptake genes, *znuA*, *znuB*, and *znuC*, belonging to the previously mentioned cluster 9 family, encode periplasmic binding proteins, integral membrane protein, and an ATPase component of the transporter, respectively [16]. Znu comprises a large family of binding proteins that may either have Zn^{2+}, Mn^{2+}, or Fe^{2+} as their substrate [16]. This is based on previously conducted sequence similarity searches.

Streptococcus pneumoniae AdcA is a specific zinc binding protein belonging to the cluster 9 protein family. AdcA is distinguished by a histidine-rich region that may constitute a metal (Zn) binding site. *S. pneumoniae* adc mutants have been used for identification and classification of cyanobacteria zinc transporters based on sequence analyses and functionality [14]. The histidine-rich binding site in AdcA was described as sharing similarities with ZRT1, which is responsible for Zn uptake in *Saccharomyces cerevisiae* [14,20]. Further analysis into the evolution and structural relationships of the cluster 9 family of ABC transporters in *E. coli*, *Haemophilus influenzae*, and *Synechocystis* sp. revealed that their respective proteins YebL, H10119, and Syn9 also shared these histidine-rich motifs at the same open reading frame as AdcA, leading to the notion that these proteins act as zinc permeases [14,15].

The Znu system and its regulators are well described in bacteria, and are understood to have originated from prokaryotes. The solute carrier (*slc*) family of genes has been well studied in eukaryotic systems with two families established as facilitating zinc uptake and efflux [21]. The *slc30* family, also known as the *c*ation *d*iffusion *f*acilitator (*cdf*) genes, is responsible for the efflux of zinc from cells. The *slc39* family, formally known as the *zip* (ZRT-, *I*RT-like *p*rotein) genes, is responsible for the uptake of zinc into the cell.

Both the *slc30* and *slc39* families encompass members with histidine-rich motifs, which are found in the cytoplasmic loop between the fourth and fifth transmembrane domains [21]. These fourth and fifth transmembrane domains are also believed to bind metal during transport [21]. Based on these described characteristics for the Slc30 and Slc39 transporters, the structures of putative zinc transporters in *Nostoc punctiforme* belonging to the Slc30 and Slc39 families have been determined through producing predicted protein transmembrane domain graphs and predicted protein structures [22].

Zip proteins transport zinc and other metal ion substrates from the extracellular space into the cytoplasm in both eukaryotes and prokaryotes [23]. Characteristically, Zip transporters have eight predicted transmembrane domains with similar predicted topologies [21]. The N and C termini of the proteins are located on the extracellular side of the membrane [21]. Zip proteins also contain a long loop region, which is located between the third and fourth transmembrane domains. A histidine-rich sequence is common in this loop; however, the function of this loop domain is unclear. The presence of multiple histidine residues suggests there may be a capacity to bind and transport metals including zinc or metal transport regulation [21,23].

The bacterial Zip, ZupT from *E. coli*, was the first Zip transporter identified and characterized in a prokaryote [24]. Most metal transport systems in bacteria and cyanobacteria demonstrate the capacity to transport multiple metals. The substrate specificity is dependent on the positioning of specific amino acid residues, such as histidine for zinc or cysteine for copper, exemplified by the presence of the classic CXXC copper binding motif [6]. The substrate specificity for *E. coli* ZupT is broad, encompassing cadmium, cobalt, iron, manganese, and zinc [24]. Zinc uptake by ZupT is inhibited by cadmium, suggesting that ZupT preferentially transports cadmium over zinc [6,24]. Cellular stress from cadmium uptake may be further exacerbated, as zinc deficiency is a possibility if zinc is out-competed by cadmium [6,24]. The saving measure for reducing cell stress from these types of events is the presence of unique transport systems and, in particular, efficient metal efflux systems, which typically complement the uptake systems, enabling metal homeostasis to be maintained for multiple metals. It is predictive that while these genes have only recently been identified in cyanobacteria, their function will be similar to the characterized *E. coli* homologues.

The *cdf* family is well known to transport zinc and other metal ions from the cytoplasm into intracellular organelles in eukaryotes, or to the outside of the cell in both eukaryotes and prokaryotes [21,25]. Cation diffusion facilitator proteins are characterized by having six predicted transmembrane domains [21].

The bacterial *cdf* family includes the zinc transport Yiip, Znt, Zit, and the Cobalt–zinc–cadmium (Czc) proteins. The CzcD protein from *Ralstonia metallidurans* was one of the first bacterial Czc efflux/resistance Cdf systems characterized [26,27]. The substrate range for Czc and the *E. coli* ZitB includes zinc, cobalt, cadmium, and nickel, but not magnesium, manganese, or cadmium [26]. The Cdf Yiip from *E. coli* may compensate for the inability of the Czc protein to transport magnesium, demonstrating affinity for both magnesium and zinc [27]. The ZntA from *E. coli* is best established as a zinc efflux protein, but has also shown to transport lead, nickel, cobalt, copper, cadmium, and zinc [28].

The ZntA of *E. coli* is a P-type ATPase, which is distinguished from other ATPase families by its enzymatic reactions, where the phospho-aspartate is an intermediate in the ATPase driven cation cycle [18]. The P-type ATPases in prokaryotes are membrane located, with multiple regions of conserved amino acid residues, that form a heptapeptide bond, with an aspartate residue phosphorylated and dephosphorylated during the ATPase transport cycle [18,29]. The P-type ATPases, which includes the ZntA, exclusively transport divalent cations including Zn^{2+} [18,29]. The specific cation and the direction of transport, either intracellular or extracellular, are specific to each different transporter [18]. The ZntA fits into the second group of two main groups of P-type ATPases [30]. The first group predominantly mediates transport of monovalent cations and the second mediates transport of divalent cations [30].

The current understanding for cyanobacterial Cdf proteins is limited to bioinformatic and transcript analyses [22,31]. Predictions from *E. coli* and *R. metallidurans* lead to the assumption that cyanobacteria Cdf proteins may have broad substrates including the divalent cations cadmium, cobalt, manganese, magnesium, iron, and zinc. The putative role for the cyanobacteria Cdfs will be to efflux excess metal ions from the cells as in *E. coli* and *R. metallidurans* to maintain homeostatic zinc levels.

12.3 REGULATORY ELEMENTS FOR ZINC UPTAKE AND EFFLUX

Transcriptional regulatory mechanisms are required to control the regulation of ABC transporters, such as the Znu transport complex. Transcriptional regulators include the ArsR–SmtB group of regulons and the Ferric uptake regulator (Fur), which have broad substrate ranges including divalent cations Cd^{2+}, Co^{2+}, and Zn^{2+}. A more specialized transcriptional repressor that interacts with the Znu system is the Zinc uptake regulator (Zur) [29]. Zur has been previously identified as belonging to the Fur family of transcriptional repressors based on structural and functional similarities [29].

Control of intracellular metal levels is crucial for cells to maintain metalloprotein function, while avoiding toxicosis. One such group of DNA binding proteins that regulates excess intracellular metals, via controlling the downstream production of proteins that in the case of bacteria export or in eukaryotes sequester zinc, includes the *ars*, *cad*, and *smt* operons [32–34]. There have been numerous representatives of the *ars*, *cad*, and *smt* operons that encode regulatory proteins, including metallothionein, for an array of metals including iron arsenic, cobalt, cadmium, and zinc regulators identified including SmtA and SmtB from *Synechococcus* PCC7942, CadC from *Bacillus firmus*, ZiaR from *Synechocystis* PCC6803, CzrA from *Staphylococcus aureus*, and BmxR from *Oscillatoria brevis* [17,33,35,36]. While these proteins respond to zinc, they may have broad substrate ranges that may also include iron, nickel, cobalt, and manganese.

The Furs have also been identified in bacteria as broad substrate uptake gene transcriptional repressors [37]. Furs are established as a member of the oxidative-stress response regulators in *E. coli* [37–39]. Fur initiates the intracellular uptake of iron and other divalent cations when cells are exposed to elevated levels [37]. Fur has also been identified as regulating multiple genes encoding metal-requiring proteins, through repressing small RNAs (sRNA) named RyhB [39]. The sRNAs play a regulatory role by modifying the activity of the metal-requiring proteins and the stability and translation of mRNA. The positive regulation of multiple genes by Fur through sRNAs, by the degradation of mRNA, can impact on the regulation of *zur* genes under oxidative stresses, resulting in increased sensitivity to elevated levels of zinc and other metals [37–39]. Other proteins positively regulated by Fur include superoxide dismutase (SodB), ferritins (Bfr and Ftn), aconitase (AcnA), and fumarase (FumC), and proteins of the acid stress response regulation [39].

The product of the *zur* gene is a cytoplasmic protein that is widespread among bacteria and cyanobacteria [12]. Zinc uptake regulation genes regulate several high affinity uptake systems, mainly the Znu system [12]. This is supported by investigations into *Streptomyces coelicolor*, where Zur was found to negatively regulate the *znuA* gene expression and in addition was shown to regulate its own expression [40].

Zur genes are described as having a very limited function in relation to Zn^{2+} uptake and metabolism in an environment poor in Zn^{2+} [12]. This could be indicative of the uptake systems being in a constant state of upregulation to ensure any available metals are taken up for survival. Investigations into the *S. coelicolor* Zur indicate it is a zinc-specific transcriptional regulator, which not only regulates a putative high-affinity zinc uptake system but also regulates some ribosomal proteins predicted to be involved in zinc mobilization [41]. Based on comparative genome analysis of zinc regulons, Zur is hypothesized to regulate the zinc mobilization of several paralogous ribosomal proteins [41,42]. Ribosomal proteins containing a zinc binding motif from the *S. coelicolor* Zur demonstrated a capacity to form zinc storage structures that can be replaced with Zur-regulated paralogs lacking the zinc binding motif to retain other metals or no metals [41]. This indicates the importance of Zur in maintaining intracellular zinc levels, preventing toxicosis, while ensuring essential trace levels are maintained and reserved for times of zinc deficiency.

Multiple sequence alignments analyses have been used in previous studies to show conserved residues, which are proposed to be the metal binding sites. Multiple sequence alignment of Fur-like regulatory proteins in *B. subtilis*, *E. coli*, and *S. epidermidis* (Fur: BsuFur, EcoFur, SepFur) (PerR: BsuPerR) (Zur: BsuZur and EcoZur) identified a carboxyl-terminal metal binding domain, containing a cluster of conserved histidine and cysteine residues [43]. A conserved recognition helix was also identified in an amino-terminal domain containing a proposed helix-turn-helix motif [43].

12.4 PHYSIOLOGICAL METAL UPTAKE ACROSS THE CELL WALL

Cyanobacteria can be distinguished by the composition of their cell surfaces. These surfaces are complex structures, containing distinct layers with unique molecular functional groups and metal binding properties [44]. The surface layers of cyanobacteria form protective coatings, adhere cells in filaments, and/or act as molecular sieves. Frequently, the external layer of the cell wall is covered by S layers formed by two-dimensional crystalline arrays of a single species of glycoprotein which covers the entire surface of a cell, and by carbohydrate structures beneath which lie the outer membrane, peptidoglycan layer, and cytoplasmic membrane [45]. The surface layer of cyanobacteria may also form an immobilizing external coating for metal adsorption [46].

The initial adsorption of free-metal ions to the surface of cyanobacterial cells is the first important step in the acquisition of metal ions from the environment. A major function of the cell wall is to allow sufficient transport of nutrients and metabolites into and out of the cell [45]. The cell wall of cyanobacteria is generally classified as gram negative based on peptidoglycan content and varies across cyanobacteria genera, with peptidoglycan thickness ranging from 10 nm in unicellular strains such as *Synechocystis* sp. to more than 700 nm in filamentous species such as *Oscillatoria princeps* [45]. In order for the metals to infiltrate the thick multilayered cell walls and become internalized, cyanobacteria have evolved modified metal transport systems in comparison to other bacteria [6]. Porins embedded in the outer membrane facilitate nonselective passive diffusion of metal ions across the outer membrane, while active transport systems operate both in the outer and inner membranes. The active transport systems still share homology with other bacterial transport systems particularly in the metal binding regions. The key role of metal uptake proteins is to bring the metals inside the cell and make them readily available for use in different parts of the cell [6]. The biochemical reactions that require metals may be localized to particular structures within the cells [22,47].

The intracellular structures of cyanobacteria play important roles in cellular metal distribution and detoxification [47]. Polyphosphate granules are commonly formed in cyanobacteria, mainly comprising phosphorus, magnesium, potassium, and calcium [47]. Polyphosphate granules may also contain other divalent cations including iron and zinc [47]. The trapping of excess intracellular metal ions within polyphosphate granules provides a nonspecific detoxification method for cyanobacterial cells in the presence of excess metal levels [47].

Carboxysomes are a polyhedral microcompartment found in cyanobacteria that contain an array of the carbon-fixing enzyme ribulose-biphosphate carboxylase/oxygenase (RuBisCO) [48]. Carboxysomes are carbon fixation sites that enhance carbon fixation by concentrating CO_2 at the active RuBisCO site, thus providing CO_2 for photosynthesis, nutrient production, and cell growth [48]. Zinc ions directly coordinate the amino acid sites of the carbonic anyhydrase in the carboxysome shell and are essential for carbonic anyhydrase activity [48].

The inner thylakoid membranes present in some cyanobacteria form an important site for the compartmentalization of toxic oxygen and by-products from photosynthesis. Thylakoids reduce the cellular damage caused by toxic oxygen molecules, promoting photosynthesis and enhancing cell growth [35]. Thylakoid membranes and carboxysomes may potentially provide a site for metal localization, particularly manganese, iron, copper, and magnesium, as these are cofactors for the activation and functioning of the photosynthetic electron transport machinery [35]. Metals associated with thylakoids and carboxysomes are likely to occur at picomolar to nanomolar levels, but may fluctuate with growth conditions and accumulate during trace element-deprived conditions, ensuring that a repository of elements is available for photosynthesis and cell growth.

12.5 PREDICTING STRUCTURE AND FUNCTION OF CYANOBACTERIAL METAL TRANSPORT GENES

The availability of cyanobacterial genomic sequence data has been expanding since 1996 [4,49]. At present, at least 40 cyanobacterial genome sequences have been completed, with this number ever increasing [31]. The availability of genomic sequences has enabled the mapping of genes and gene clusters within genomes with the ultimate outcome of identifying and characterizing the functionality of individual genes.

The modeling of cyanobacterial genomic and protein sequences, analysis of their function and expression, and quantifying local cellular binding sites for metals provides essential insights into metal uptake and efflux mechanisms that are highly conserved across all organisms. Metal transport genes in most organisms are distinguished based on the presence of metal binding motifs such as the MXCXXC, copper binding motif, or by their allocation to operons consisting of multiple subunits [50]. The classification of the transport systems is based on (a) putative membrane topology, (b) protein family, (c) bioenergetics, and (d) substrate specificity [51].

12.6 CELLULAR REQUIREMENTS AND EFFECTS OF ZINC

All organisms require zinc, with variability in cellular concentrations and sensitivity occurring between cells and species. Zinc plays an essential role in the structuring and functioning of a myriad of proteins, including regulatory, structural, and enzymatic proteins [52]. Bioavailable zinc may occur in an ionic form, or complexed to organic molecules where it is bound with different affinities [53]. As elevated levels of extracellular zinc are toxic while trace levels are required for survival, organisms are forced to tightly regulate intracellular zinc levels.

Zinc availability in the environment is governed by many factors. The presence of multiple ions may result in competition for uptake, where zinc may be substituted by another ion, particularly if zinc is at a lower concentration or has a lower affinity for the uptake system than the competing ion. The zinc transport systems of cyanobacteria may show a capacity to discriminate between metal ions, preferentially transporting zinc ions (Zn^{2+}), while excluding toxic metal ions such as Cd^{2+}. Different uptake systems vary in their specificity for different metals and the different metals have a range of affinities for most coordinating environments in the order $Mg^{2+}/Ca^{2+} < Mn^{2+} < Fe^{2+} < Co^{2+} < Ni^{2+} < Cu^{2+} \geq Zn^{2+}$, an equilibrium series known as the Irving–Williams Series [54]. Other factors that restrict zinc availability in the environment include pH and temperature.

It has been proposed that the symbiosis of cyanobacteria with plants may increase metal availability to the host [54,55]. In published examples to date, however, it appears that cyanobacteria have more of a metal resistance role rather than increasing metal availability to the host [54,56]. The most well-characterized symbiotic relationship where cyanobacteria are purported to be involved in heavy metal resistance is seen in cyanobacterial–fungal relationships [55,57]. It is also proposed that cyanobacteria may provide some metal resistance to cycads through symbiosis in the cycad coralloid roots [57].

Studies have identified that cyanobacteria are suitable candidates for the removal of heavy metals from contaminated waters based on cellular metal adsorption. This is primarily based on the adsorption of metals to the surface layers and the cell envelope [1,58,59]. While kinetic data of the uptake and efflux of metals, including zinc, have been widely described in cyanobacteria, the molecular mechanisms facilitating the regulation of intracellular levels is characterized to a lesser extent. Information describing the intracellular and molecular zinc toxicity responses is only just becoming available. The cyanobacterial species *N. punctiforme* has shown a significant sensitivity to zinc ($ZnCl_2$) with cell viability significantly reduced at concentrations over 22 μM [22]. Investigations into zinc transporter genes and their transcriptional regulatory elements in cyanobacteria may provide insight into how intracellular zinc levels are regulated, and how cyanobacteria modulate environmental zinc levels.

12.7 CONCLUSION

Metal homeostasis is essential for cyanobacterial cellular functioning. To obtain zinc and other metals from the environment, the first step is the adsorption of ions to the cell surface. Adsorbed metals move through the extracellular matrices and cell wall by passive transport including porins into the periplasmic space. Zinc in the periplasmic space is then acquired intracellularly through the activity of transport proteins including the ABC type Znu system and the Slc39 Zip proteins, which transport the zinc into the cytoplasm where it is a prerequisite for the structure and function of a myriad of proteins including enzymes. To avoid excessive accumulation of intracellular zinc, cyanobacteria have uptake regulatory proteins, including the Fur and Zur proteins that repress uptake proteins, while zinc efflux proteins such as the Cdf proteins export excess intracellular zinc from the cell. The combination of these active and passive mechanisms functions in a synchronized mode to regulate zinc homeostasis in cyanobacteria and is vital for avoiding the stress of both zinc over-accumulation and starvation.

REFERENCES

1. Baptista, M.S. and T.M. Vasconcelos, Cyanobacteria metal interactions: Requirements, toxicity, and ecological implications. *Crit. Rev. Microbiol.*, 32, 127–137, 2006.
2. Vallee, B.L. and D.S. Auld, Zinc coordination, function, and structure of zinc enzymes and other proteins. *J. Biol. Chem.*, 29, 5647–5659, 1990.
3. Vallee, B.L. and A. Glades, The metallobiochemistry of zinc enzymes. *Adv. Enzymol. Relat. Areas Mol. Biol.*, 56, 283–430, 1984.
4. Scholnick, S. and N. Keren, Metal homeostasis in cyanobacteria and chloroplasts. Balancing benefits and risks to the photosynthetic apparatus. *Plant Physiol.*, 141, 805–810, 2006.
5. Xu, J. et al., Cyanobacteria MT gene *SmtA* enhance zinc tolerance in *Arabidopsis*. *Mol. Biol. Rep.*, 37, 1105–1110, 2010.
6. Ma, Z., F.E. Jacobsen, and D.P. Giedroc, Coordination chemistry of bacterial metal transport and sensing. *Chem. Rev.*, 109, 4644–4681, 2009.
7. Tripathi, B.N., S.K. Mehta, and J.P. Gaur, Differential sensitivity of *Anabaena doliolum* to Cu and Zn in batch and semicontinuous cultures. *Ecotoxicol. Environ. Saf.*, 56, 311–318, 2003.
8. Yamamoto, K. and A. Ishihama, Transcriptional response of *Escherichia coli* to external zinc. *J. Bacteriol.*, 187, 6333–6340, 2005.
9. Blencowe, D.K. and A.P. Morby, Zn(II) metabolism in prokaryotes. *FEMS Microbiol. Rev.*, 27, 291–311, 2003.
10. Fath, M.J. and R. Kolter, ABC transporters: Bacterial exporters. *Microbiol. Rev.*, 57, 995–1017, 1993.
11. Linton, K.J. and C.F. Higgins, Structure and function of ABC transporters: The ATP switch provides flexible control. *Eur. J. Physiol.*, 453, 555–567, 2007.
12. Patzer, S.I. and K. Hantke, The ZnuABC high-affinity zinc uptake system and its regulator Zur in *Escherichia coli*. *Mol. Microbiol.*, 28, 1199–1210, 1998.
13. Banerjee, S. et al., Structural determinants of metal specificity in the zinc transport protein ZnuA from *Synechocystis* 6803. *J. Mol. Biol.*, 333, 1061–1069, 2003.
14. Dintilhac, A. et al., Competence and virulence of *Streptococcus pneumoniae*: Adc and PsaA mutants exhibit a requirement for Zn and Mn resulting from inactivation of putative ABC metal permeases. *Mol. Microbiol.*, 25, 727–739, 1997.
15. Dintilhac, A. and J.P. Claverys, The adc locus, which affects competence for genetic transformation in *Streptococcus pneumoniae*, encodes an ABC transporter with a putative lipoprotein homologous to a family of *Streptococcal adhesins*. *Res. Microbiol.*, 148, 119–131, 1997.
16. Chandra, B.R., M. Yogavel, and A. Sharma, Structural analysis of ABC-family periplasmic zinc binding protein provides new insights into mechanism of ligand uptake and release. *J. Mol. Biol.*, 367, 970–982, 2007.
17. Liu, T. et al., A novel cyanobacterial SmtB/ArsR family repressor regulates the expression of a CPx-ATPase and a metallothionein in response to both Cu(I)/Ag(I) and Zn(II)/Cd(II). *J. Biol. Chem.*, 279, 17810–17818, 2004.
18. Phung, L.T., G. Ajlani, and R. Haselkorn, P-type ATPase from the cyanobacterium *Synechococcus* 7942 related to the human Menkes and Wilson disease gene products. *Proc. Natl. Acad. Sci. U.S.A.*, 91, 9651–9654, 1994.

19. Li, H. and G. Jogl, Crystal structure of the zinc-binding transport protein ZnuA from *Escherichia coli* reveals an unexpected variation in metal coordination. *J. Mol. Biol.*, 368, 1358–1366, 2007.
20. Zhao, H. and D. Eide, The yeast ZRTJ gene encodes the zinc transporter protein of a high-affinity uptake system induced by zinc limitation. *Proc. Natl. Acad. Sci. U.S.A.*, 93, 2454–2458, 1996.
21. Eide, D.J., Zinc transporters and the cellular trafficking of zinc. *Biochim. Biophys. Acta*, 1763, 711–722, 2006.
22. Hudek, L. et al., Bioinformatic and expression analyses of genes mediating zinc homeostasis in *Nostoc punctiforme. Appl. Environ. Microbiol.*, 75, 784–791, 2009.
23. Eide, D.J., The SLC39 family of metal ion transporters. *Eur. J. Physiol.*, 447, 796–800, 2004.
24. Taudte, N. and G. Grass, Point mutations change specificity and kinetics of metal uptake by ZupT from *Escherichia coli. BioMetals*, 23, 643–656, 2010.
25. Palmiter, R.D. and L. Huang, Efflux and compartmentalization of zinc by members of the SLC30 family of solute carriers. *Eur. J. Physiol.*, 447, 744–751, 2004.
26. Anton, A. et al., Characteristics of zinc transport by two bacterial cation diffusion facilitators from *Ralstonia metallidurans* CH34 and *Escherichia coli. J. Bacteriol.*, 186, 7499–7507, 2004.
27. Neis, D.H., How cells control zinc homeostasis. *Science*, 317, 1695–1696, 2007.
28. Hou, Z. and B. Mitra, The metal specificity and selectivity of ZntA from *Escherichia coli* using the acyl-phosphate intermediate. *J. Biol. Chem.*, 278, 28455–28461, 2003.
29. Patzer, S.I. and K. Hantke, The zinc-responsive regulator Zur and its control of the znu gene cluster encoding the ZnuABC zinc uptake system in *Escherichia coli. J. Biol. Chem.*, 275, 24321–24332, 2000.
30. Gatti, D., B. Mitra, and B.P. Rosen, *Escherichia coli* soft metal Ion-translocating ATPases. *J. Biol. Chem.*, 275, 34009–34012, 2000.
31. Blindauer, C.A., Zinc-handling in cyanobacteria: An update. *Chem. Biodivers.*, 5, 1990–2013, 2008.
32. Blindauer, C.A. et al., Multiple bacteria encode metallothioneins and SmtA-like zinc fingers. *Mol. Microbiol.*, 45, 1421–1432, 2002.
33. Harvie, D.R. et al., Predicting metals sensed by ArsR-SmtB repressors: Allosteric interference by a non-effector metal. *Mol. Microbiol.*, 59, 1341–1356, 2006.
34. Nies, D.H., Efflux-mediated heavy metal resistance in prokaryotes. *FEMS Microbiol. Rev.*, 27, 313–339, 2003.
35. Cavet, J.S., G.P.M. Borrelly, and N.J. Robinson, Zn, Cu and Co in cyanobacteria: Selective control of metal availability. *FEMS Microbiol. Rev.*, 27, 165–181, 2003.
36. Mack Ivey, D. et al., The cadC gene product of alkaliphilic *Bacillus firmus* OF4 partially restores Na+ resistance to an *Escherichia coli* strain lacking an Na!/H+ antiporter (NhaA). *J. Bacteriol.*, 174, 4878–4884, 1992.
37. Masse', E. et al., Small RNAs controlling iron metabolism. *Curr. Opin. Microbiol.*, 10, 140–145, 2007.
38. Altuvia, S., Regulatory small RNAs: The key to coordinating global regulatory circuits. *J. Bacteriol.*, 186, 6679–6680, 2004.
39. Hantke, K., Iron and metal regulation in bacteria. *Curr. Opin. Microbiol.*, 4, 172–177, 2001.
40. Owen, G.A. et al., Zinc-responsive regulation of alternative ribosomal protein genes in *Streptomyces coelicolor* involves zur and sigmaR. *J. Bacteriol.*, 189, 4078–4086, 2007.
41. Shin, J.H. et al., The zinc-responsive regulator Zur controls a zinc uptake system and some ribosomal proteins in *Streptomyces coelicolor* A3(2). *J. Bacteriol.*, 189, 4070–4077, 2007.
42. Panina, E.M., A.A. Mironov, and M.S. Gelfand, Comparative genomics of bacterial zinc regulons: Enhanced ion transport, pathogenesis, and rearrangement of ribosomal proteins. *Proc. Natl. Acad. Sci. U.S.A.*, 100, 9912–9917, 2003.
43. Gaballa, A. and J.D. Helmann, Identification of a zinc-specific metalloregulatory protein, Zur, controlling zinc transport operons in *Bacillus subtilis. J. Bacteriol.*, 180, 5815–5821, 1998.
44. Yee, N. et al., Characterization of metal-cyanobacteria sorption reactions: A combined macroscopic investigation. *Environ. Sci. Technol.*, 38, 775–782, 2004.
45. Hoiczyk, E. and A. Hansel, Cyanobacterial cell walls: News from an unusual prokaryotic envelope. *J. Bacteriol.*, 182, 1191–1199, 2000.
46. Schuster, B. and U.B. Sleytr, Composite S-layer lipid structures. *J. Struct. Biol.*, 168, 207–216, 2009.
47. Andrade, L. et al., Zinc detoxification by a cyanobacterium from a metal contaminated bay in Brazil. *Braz. Arch. Biol. Technol.*, 47, 147–152, 2004.
48. Sawaya, M.R. et al., The structure of beta-carbonic anyhydrase from the carboxysomal shell reveals a distinct subclass with one active site for the price of two. *J. Biol. Chem.*, 281, 7546–7555, 2006.
49. Kaneko, T. and T. Satoshi, Complete genome structure of the unicellular cyanobacterium *Synechocystis* sp. PCC6803. *Plant Cell Physiol.*, 38, 1171–1176, 1997.

50. Banci, L. et al., Understanding copper trafficking in bacteria: Interaction between the copper transport protein CopZ and the N-terminal domain of the copper ATPase Copa from *Bacillus subtilis*. *J. Bacteriol.*, 42, 1939–1949, 2003.
51. Paulsen, I.T., M.K. Sliwinski, and M.H. Saier, Microbial genome analyses: Global comparisons of transport capabilities based on phylogenies, bioenergetics and substrate specificity. *J. Mol. Biol.*, 277, 573–592, 1998.
52. Frederickson, C.J. et al., Importance of zinc in the central nervous system: The zinc-containing neuron. *J. Nutr.*, 130, 1471S–1483S, 2000.
53. Allen, H.E., R.H. Hall, and T.D. Brisbin, Metal speciation: Effects on aquatic toxicity. *Environ. Sci. Technol.*, 14, 441–443, 1980.
54. Dupont, C.L. et al., History of biological metal utilization inferred through phylogenomic analysis of protein structures. *Proc. Natl. Acad. Sci. U.S.A.*, Early Edition, 1–6, 2010.
55. Brown, D.H. and R.P. Beckett, Differential sensitivity of lichens to heavy metals. *Ann. Bot.*, 52, 51–57, 1983.
56. Babich, H. and G. Stotzky, Toxicity of zinc to fungi, bacteria and coliphages: Influences of chloride ions. *Appl. Environ. Microbiol.*, 36, 906–914, 1978.
57. Nilsson, M., U. Rasmussen, and B. Bergman, Cyanobacterial chemotaxis to extracts of host and nonhost plants. *FEMS Microbiol. Ecol.*, 55, 382–390, 2006.
58. El-Enany, A.E. and A.A. Issa, Cyanobacteria. as a biosorbent of heavy metals in sewage water. *Environ. Toxicol. Pharmacol.*, 8, 95–101, 2000.
59. Mehta, S.K. and J.P. Gaur, Use of algae for removing heavy metal ions from wastewater: Progress and prospects. *Crit. Rev. Microbiol.*, 25, 113–152, 2005.

Part II

Cellular Responses and Ecophysiology

13 Cyanobacteria in Symbiosis
Cellular Responses

*Mayashree B. Syiem** and *Amar Nath Rai*

CONTENTS

13.1 INTRODUCTION

In nature, no organism exists in isolation and interactions among organisms are necessary for survival, ecological balance, and constant evolution. Such interactions could range from casual to extreme where two different species live on or in one another permanently or for a considerable period of their life involving exchange of metabolites between the partners [1,2]. Such close associations are described as symbiosis [3] and the associating partners are termed as "symbionts" with the larger partner commonly referred as the "host." Accordingly, symbiosis can be described as a life style in which originally independent entities form integrated units as a result of series of events directed toward intimate and orchestrated development for improved living and survival together in association. During the process of development, the partners often lose their independent identity. In the history of evolution of life on earth, such intermingling and permanent dependency of organisms may have contributed toward evolution and origin of species [2,4–8].

The depths of symbiotic relationship vary widely. In some cases, it could be parasitic, causing little harm to one of the partners. In others, it is pathogenic, where damage could be severe enough to cause disease symptoms. In these cases, the mode of extracting nutrients from the partner by the symbiont is necrotrophic that kills the host cells. Or such an interaction could be mutualistic where both partners benefit from the association and neither partner is harmed. Symbionts in mutualistic associations use "biotrophic" mode of nutrient exchange between the living cells of the partners. For such exchange, most biotrophs attain high structural–functional specialization

* Corresponding author: mayashreesyiem@yahoo.co.in

to ascertain success in the symbiotic endeavor [9–11]. Examples of such mutualistic symbioses are those where cyanobacteria are one of the partners.

As a group of ancient photosynthetic prokaryotes, cyanobacteria are widespread in a variety of terrestrial and aquatic biocenoses. They have immense environmental importance being accountable for large percentage of global CO_2 and N_2 fixation. They display large morphological diversity, metabolic flexibility, diazotrophy combined with a wide range of nutritional capabilities that range from obligate photoautotrophy to heterotrophy [2,12–14]. Their ability to (1) use light energy to generate reductants splitting water for photosynthesis and (2) reduce atmospheric nitrogen allows them to grow and colonize wide range of environments with minimal requirement of light, water, air, and a few inorganic nutrients [6]. However, the extreme sensitivity of the enzyme nitrogenase responsible for biological nitrogen fixation to molecular oxygen makes oxygenic photosynthesis incompatible to simultaneous nitrogen fixation [15]. Many multicellular cyanobacteria, for example, from the genera *Nostoc* and *Anabaena,* overcome this incompatibility by differentiating their vegetative cells at regular intervals into specialized cells called heterocysts dedicated for nitrogen fixation. The remaining vegetative cells perform photosynthesis, thereby spatially separating the two incompatible processes [16]. In unicellular cyanobacteria, these two processes are temporally separated with cycles of photosynthesis alternating with nitrogen fixation. Again, under conditions of low light intensity, phosphate limitation, sulfate starvation, cold, and desiccation, the vegetative cells of filamentous forms differentiate into spore-like structures called akinetes to survive through these adverse growth conditions. Additionally, positive or negative stress signals for growth trigger all vegetative cells of a filament to divide forming short, small celled, motile, heterocyst-lacking filaments called hormogonia that are used for short distance dispersion [17]. The process of hormogonia production is uncoupled from biomass increase and DNA replication. Completion of hormogonia formation is signaled by DNA replication, differentiation of heterocysts, and resumption of cellular growth in these filaments to establish new colonies [18]. All these genetic, structural, and functional plasticities bestow great diversity and versatility to the cyanobacteria, enabling them to inhabit extremes of environments and niches [2,19–21].

N_2-fixation and photosynthetic abilities of cyanobacteria fully enable them of an independent existence in nature; however, for reasons not clearly established, several of them form associations with a wide range of organisms. The symbiotic association of cyanobacteria with eukaryotes could be traced back to at least 2.1 billion years ago [7,22–24] and includes non-photosynthetic protists belonging to the group Glaucophyta, bacteria, animals (marine sponges and echiuroid worms), fungi, and plants [2,10,25–30]. Cyanobacteria have also been reported to be present in the hollow shafts of hairs of polar bears [31]. Among plants, cyanobacteria are seen in symbiotic associations with algae (diatoms), fungi (lichens), bryophytes (liverworts, hornworts, and mosses), pteridophytes (*Azolla*), gymnosperms (cycads), and angiosperms (*Gunnera*). All these plants, except fungi, are photosynthetic and are therefore autotrophic in nature using atmospheric CO_2 as a source of carbon nutrition. However, they are still dependent where nitrogen nutrition is concerned. The nitrogen-fixing ability of cyanobacteria is obviously valued by the non-nitrogen-fixing organisms, and by entering into symbiotic association with these organisms many plants have achieved N-autotrophy. Where fungus is the other symbiotic partner, cyanobacteria may at times provide both carbon and nitrogen nutrition to the fungal host.

In most plant symbioses, the cyanobiont is *Nostoc* (Table 13.1). However, cyanobacteria that enter into symbiotic association with fungi-forming lichens are much more diverse. The location of the cyanobiont could be extracellular or intracellular. The host tissue and organ that are involved in forming symbiosis (such as cells, bladders, cavities, sporophytes, fronds, root nodules, or stem nodules) also show great diversity.

Investigations of cyanobacterial symbioses at physiological, biochemical, and molecular levels are prerequisite to comprehend reasons and events leading to establishment of successful symbiosis. Such understanding can generate knowledge for manipulation of these organisms for their potential biotechnological use such as improving crop production, populating N-deficient soil by symbiotic plants to overcome desertification, bioremediation of polluted environment, etc.

TABLE 13.1
Cyanobacterial–Plant Symbioses

Host (Plants)	Type	Cyanobacteria	Comments
Heterotrophs: Fungi	Lichenized	*Nostoc*,[a] *Scytonema*,[a] *Fishcerella*,[a,b] *Calothrix*,[a] Unicellular cyanobacteria[c]	~20% of all known fungi form lichens (of 13,500 spp.: 98% Ascomycetes, 0.4% Basidiomyces, 1.6% Deuteromycetes). Of these, 10% of known lichens contain cyanobionts (bipartite) and 3%–4% contain both a cyanobiont and a green alga as phycobiont (tripartite).[d] The symbioses are intercellular in nature
	Other		Only one Phycomycete: *Geosiphon pyriforme*, a coenocytic soil fungus closely related to arbuscular mycorrhiza-forming fungi of the genus *Glomus*[e] is known till date. The cyanobiont is present in the bladder
Autotrophs: Algae (diatoms)	Marine (*Rhizosolenis*, *Hemiaulus*, *Bacterriastrum*, *Chaetoceros*)	*Richelia*, *Calothrix*[a]	*Rhizosolenia clevei* and other species[f]; *Hemiaulus indica*, *H. hauckii*, *H. membranaceus*, *H. sinensis*, *Bacteriastrum*, *Chaetoceros*, *Roperia tessellate*, *H. Steptotheca indica*, *Neostreptotheca subindica* enter into symbiotic associations. Cyanobacterium is located in the periplasmic space
	Freshwater (*Rhopalodia*, *Epithemia*, *Denticulata*)	Coccoid cyanobacteria	Diatoms involved are *Rhopalodia gibba*, *R. gibberula*; *Epithemia adnata*, *E. sorex*, *E. turgida*, *E. zebra*; *Denticulata vanheurcki*. Cyanobacterium is visible as bluish-green inclusions in the cytoplasm
Bryophytes	Liverworts: *Blasia*, *Cavicularia*	*Nostoc*[a]	Only 2 of the 330 known liverwort genera: *Blasia pusilla*, *Cavivularia densa* form symbiosis. Symbiosis is intercellular. Cyanobacterium resides in mucilage filled cavities on undersurface of hornwort/liverwort gametophytic thallus
	Hornworts: *Anthoceros*, *Notothylus*	*Nostoc*[a]	Four of the six extant hornwort genera are symbiotically competent: *Anthoceros laevis*, *A. husnotii*, *A. punctatus*; *Dendroceros* sp, *Notothylas* sp., *Phaeoceros laevis*
	Mosses: *Sphagnum*	*Nostoc*[a]	Cyanobacterium inhabits the hyaline cells of the moss
Pteridophytes: *Azolla*	Water fern	*Nostoc* sp.[g]	All seven extant species of the genus *Azolla* are competent:[h] *Azolla caroliniana*, *A. filiculoides*, *A. mexicana*, *A. microphylla*, *A. rubra* (*A. japonicum*) *A. pinnata* (var. *pinnata* and *imbricate*), *A. nilotica*. Cyanobacterium is located in the mucilage filled cavities on the ventral surface of the dorsal lobes of the leaves. Symbiosis is intercellular

(*continued*)

TABLE 13.1 (continued)
Cyanobacterial–Plant Symbioses

Host (Plants)	Type	Cyanobacteria	Comments
Gymnosperms	Cycads: *Cycas*, *Macrozamia*	*Nostoc* sp.[a]	All known cycads: 150 spp. of 10 genera belonging to 3 families form symbiosis. Cycadaceae: *Cycas*, *Stangeriaceae*: *Stangeria*, *Zamiaceae*: *Bowenia*, *Ceratozamia*, *Dioon*, *Encephalartos*, *Lepidozamia*, *Macrozamia*, *Microcycas*, *Zamia*.[i] Cyanobacterium is present in the cortical zone of the coralloid roots. Symbiosis is intercellular
Angiosperm: *Gunnera*	Haloragaceae	*Nostoc* sp.[a]	All 65 known species of *Gunnera*[j] are symbiotically competent. Cyanobacterium resides inside host cell in stem nodules. Symbiosis is intracellular

[a] Heterocystous form.
[b] *Fishcerella* includes *Hyphomorpha*, *Stigonema*, and *Mastigocladus*.
[c] *Gloeocapsa*, *Chroococcus*, *Synechocystis*, *Apanocapsa*, *Microcystis* [2,21,50,51].
[d] See Honegger [32], Schenk [33], and Hill [34]. Difficulties persist in the identification and nomenclature of mycobionts, requiring a combination of morphological, physiological and molecular approaches [35–39].
[e] See Schussler et al. [40]. and Gehrig et al. [41].
[f] See Villareal [42], Schenk [33], and Carpenter et al. [43].
[g] Usually called *Anabaena azollae*.
[h] Grouped into sections *Rhizosperma* (*A. pinnata* and *A. nilotica*) and *Azolla* (the other five spp.) [45,46].
[i] Stevenson [47] suggested a new genus, *Chigua*. Phylogenetic analysis with chloroplast DNA suggests that it is a sister group of the genus *Zamia* [48].
[j] See Bergman et al. [49].
[k] Two other hornworts, *Megaceros* and *Folioceros*, have no reported cyanobionts [44].

Since cyanobacteria are capable of independent existence and are prized symbiotic partners, they must be "enticed" into forming symbiosis. More so, because in symbiotic arrangements, the cyanobacteria are directed to migrate into pre-existing host structures (cavities, pea-coralloid roots, and stem glands) where the condition is microaerobic and/or non-photosynthetic and hence not ideal for cyanobacterial growth. Why then cyanobacteria enter into symbiosis and "agree" in principle to undergo modifications suitable for symbiotic living? The reasons are difficult to recognize. One major cause could be that in natural habitats in times of adverse environmental conditions nutrient scavenging becomes tricky as nutrients can be unevenly distributed. On the other hand, within the host tissues, the cyanobiont has accessibility to a stable environment and wider nutrient pool that the host accumulates due to its larger size and specialized appendages. In symbiosis they are offered a steady and continuous supply of metabolites in an environment that further offers protection against flooding, desiccation, exposure to extreme heat or cold as well as pollutants [2]. However, symbiosis as a habitat exerts influence, leading to changes appropriate for synchronized growth of the partners. Once committed to symbiosis, large changes initiated in the cyanobiont are almost certainly due to symbiotic conditions/partner. Undoubtedly, after the initial process of infection at the early stages of crafting the symbiosis, the host at all times controls the cyanobiont population, keeping it in steady relation to host biomass. In other words, the cyanobacterium is not under freedom to propagate unobstructed. In addition, host induces and the cyanobiont obliges with extensive morphological, physiological, and biochemical changes for efficient nutrient exchange. Transition from photoautotrophy to photo- or chemoheterotrophy, decrease in growth rate, and obvious increase in nitrogen fixation rate are the main functional modifications in the symbiotic cyanobacteria. There are other changes that include increase in cell size and heterocyst frequency; deficient peptidoglycan layer; disruption of intracellular contacts; trichome fragmentation; decrease in GS activity (except in cycad symbiosis);

and decreased ammonia assimilation [52]. Whether these changes from free-living life style could be explained as stress-responses toward living together with another organism is difficult to ascertain and is certainly not easy to answer in a straightforward manner. Are these changes always induced by the host or is it that these changes are brought about by the cyanobacteria themselves in response to their new environment in symbiosis or is it both? Again, such facts are difficult to establish. We shall discuss these aspects and try to analyze these changes as cellular responses toward the stress encountered by the cyanobacterium during the sequence of events leading to the formation of a mature and thriving association. The focus of this chapter will be on the structural and functional modifications in the cyanobacteria when in symbiosis vis-à-vis when they occur as free-living organisms in nature.

13.2 COMPATIBLE PARTNERS

13.2.1 Symbiotic Cyanobacteria

As seen from Table 13.1, not all N_2-fixing cyanobacteria enter into symbiotic associations. The most common are the heterocystous forms and it stands to reason that the main role of cyanobacteria in a symbiosis is to provide fixed nitrogen to the host which is incapable of using atmospheric nitrogen for its growth. Among the heterocystous cyanobacteria, *Nostoc* shows the widest range of hosts from algae to angiosperms. The members of the genera *Nostoc* are highly versatile and resilient and found across all ecological niches [53,54]. Their flexibility and adaptability to any environmental condition may be the reason behind their presence as symbiotic partner with such diverse hosts. Unicellular cyanobacteria are restricted to diatoms and lichens; however, their identities in some cases are yet to be established [2].

To be successful in developing a functional symbiosis, a cyanobacterium must have mechanism in place to send and to receive signals from host; suppress, avoid, or overcome any defenses mounted toward its advances by the plant as well as be capable of hormogonia formation, chemotaxis, heterocyst differentiation, and above all possess sufficient elasticity to live in host tissues that may present conditions of heterotrophy, low oxygen levels, and/or acidic mucilage. These criteria limit the number of cyanobacteria that enters into symbiosis. Thus, even though cyanobacterial symbioses cover a wide range of hosts, only a limited number of cyanobacteria and plants from each host group develop symbioses. That only a particular cyanobiont is approved entry into the host signifies the fact that there is a process to recognize and select the correct cyanobacteria at the level of entry. Host lectins specific to different sugars present on the cyanobacterial cell surface are implicated in such an early recognition process. Apart from these specific sugars, the presence of fimbriae and in some cases lectins has also been identified that may have roles in recognition and adherence to compatible hosts [1,55–57]. During initiation, cyanobionts have been shown to respond to "hormogonia inducing factors" (HIF), "hormogonia repressing factors" (HRF), and chemo-attractants released by a compatible host [58–61]. Symbiotic-specific genes or gene products have also been detected in cyanobionts. Specificity between host and cyanobiont, presence of HIF and chemoattractants during initiation of symbiosis, hormogonia formation by cyanobiont, coordinated growth and development of the symbioses, and various modifications in the cyanobiont indicate precise signaling-sensing between the competent partners, most likely via specific genes or gene products expressed only during symbiotic events. Investigations into the presence of these genes are in early stages of research; however, sequences similar to *nod* genes have been detected in the cyanobionts isolated from *Azolla* and *Gunnera* [62,63].

13.2.2 Eukaryotic Partners

In cyanobacteria–plant symbioses, plant partners commonly called hosts include members of the entire range of the plant kingdom. They vary from unicellular algae, fungi, bryophytes, pteridophytes, gymnosperm, and angiosperms (Table 13.2). However, as in the case of cyanobionts, the genera involved in symbiosis represent a small fraction of the entire plant population. Again, this may be because only these plants are competent in producing and responding to the complete range

TABLE 13.2
Heterocyst Frequency, Status of Photosynthetic Activity and Nutrient Exchange in Cyanobacterial–Plant Symbioses

Host Plant	Heterocyst Frequency	Photosynthetic Status	Nutrient Exchange: Fixed-Carbon	Nutrient Exchange: Fixed-Nitrogen
Diatoms	Unchanged	Active	Uncertain	Probably as NH_4^+ to the host
Fungi				
1. Bipartite	Unchanged	Active	As glucose to host from cyanobiont	As NH_4^+ to the host from the cyanobiont
2. Tripartite	15%–35%	Active	No fixed-C from cyanobiont to mycobiont	As NH_4^+ to the host from the cyanobiont
Bryophytes	40%–45%	Inactive	Probably sucrose to cyanobiont from host	As NH_4^+ to the host by the cyanobiont
Azolla	25%–30%	Inactive	Sucrose to cyanobiont from host	As NH_4^+ to the host by the cyanobiont
Cycads	~45%	Inactive	Fixed carbon from host to cyanobiont	Probably as glutamate/citrulline to the cyanobiont from host
Gunnera	60%–80%	Inactive	Fixed carbon from host to cyanobiont	As NH_4^+ to the host by the cyanobiont

of symbiotic-related compounds and factors. The host plays a significant role in starting and initiating developmental changes necessary for establishment of a fully functional symbiotic association. For example, a HIF produced in the mucilage filled *Gunnera* stem glands is capable of inducing hormogonia differentiation in the competent *Nostoc* strains [58]. Similarly, cycad [60] and bryophyte hosts [59] release chemoattractants in addition to HIF to facilitate hormogonia movement in the right direction. Following infection, hosts may release HRF to deter further differentiation of hormogonia in the cyanobiont and shift the cellular metabolism toward differentiation of heterocysts and N_2-fixation [30,64,65]. Flavonoids such as kaempferol and quercetin secreted by legumes that are implicated in symbiotic-specific genes in *Rhizobia* [66,67] are present in *Gunnera* [68]. Sequences similar to flavonoid responding genes in *Rhizobia* have been detected in cyanobionts [62,63]. Thus, a select group of plants that possess symbiotic-specific genes and/or gene products are capable of entering into symbiotic relationship with cyanobacteria.

The apparent advantage of entering into a symbiosis with N_2-fixing cyanobacteria is that the hosts receive fixed nitrogen produced from nitrogen fixation by the cyanobiont. Despite being a small percentage of the host biomass and being restricted in a localized area, the cyanobiont is able to meet the total fixed nitrogen demand of the host plant [42,69,70]. Such plants can thrive in N-deficient environment that are otherwise inhabitable. Non-photosynthetic fungi that form symbiosis with cyanobacteria receive both fixed-carbon and fixed-nitrogen from the cyanobiont and this way they acquire both C and N autotrophy [2]. These lichenized fungi are better equipped to thrive in harsh environments as they have access to source of metabolites from their cyanobacterial partner. However, biological N_2-fixation is an energy-demanding process and in free-living cyanobacteria, carbohydrate made during active photosynthesis is transported into heterocysts to meet the energy demand of N_2-fixation. In symbiosis, the requirement for fixed-N is much higher as the cyanobiont provides fixed nitrogen to both the partners. There are evidences that hosts other than fungi provide the fixed-C to the cyanobiont making it free to concentrate on nitrogen fixation. The cyanobionts may also provide UV protection to their symbiotic partners as they make N-containing sunscreen products: mycosporin-like amino acids [71] and scytonemin [8,72].

13.3 STRUCTURAL–FUNCTIONAL MODIFICATIONS IN CYANOBACTERIA DURING SYMBIOSIS

In symbiosis, various structural and functional alterations occur in the cyanobiont for efficient interactions and nutrient exchange. These modifications vary among different symbioses as well as from young to older regions within an association. In comparison to legume–rhizobia symbiosis, the cyanobacteria are the submissive partners in all associations with plants. In these symbioses, there is largely unidirectional flow of signals from plant partners to the cyanobiont. This observation is based on the facts that there are extensive morphological and physiological changes in the cyanobacteria compared to minor modifications (increase in size of the preexisting cavities and elaborate transfer cells for nutrient exchange) in the plant partners. The selective advantages to the plant host are obvious as it gets fixed N from the cyanobiont; however, such selective benefits are less apparent for the cyanobiont in these associations. Some suggest that these associations may be termed "commensal" since one partner (plant) is the clear beneficiary while the cyanobiont neither benefits nor harms [73–75]. None of these extensive morphological and physiological modifications in the cyanobiont however are outside their normal gamut of free-living growth state. Plant partners in these symbioses have acquired expertise to activate the regulatory pathways that function normally in free-living cyanobacteria. However, excepting for providing the plants with higher fixed N, the cyanobiont hardly affects the host [6]. Three physiological processes show reductions in symbiotic cyanobacteria. They are growth, CO_2 fixation, and NH_4^+ assimilation. In free-living state, the amount of CO_2 and atmospheric N fixed and ammonia assimilated is proportional to the organic nitrogen needed for the growth. Requirement to provide fixed N to the host during symbiosis uncouples these processes, indicating unbalanced growth in these conditions. In most cyanobacterial–plant symbioses, the cyanobiont population is maintained at a constant relation to the host biomass and there is consistent flow of nutrients per unit biomass. In the bryophyte *Anthoceros–Nostoc* symbiosis, *Anthoceros* regulates both the size of the symbiotic *Nostoc* colony and its nitrogenase activity [76].

The changes seen in various structural and functional aspects of a cyanobiont in true sense are wrong to compare with free-living forms maintained under optimized conditions in laboratory. More appropriate comparison would be if an isolate is characterized in the physical and metabolic environment prevailing in the symbiotic tissue of the host. However, such conditions are highly complex and difficult to mimic as metabolic status of a living entity changes on a moment to moment basis. In the following sections, we describe some of these changes compared to those we best understand, that is, against the free-living cyanobacteria in isolation under laboratory conditions.

13.3.1 Cell Size and Morphological Changes

During symbiosis, the host and the cyanobiont grow in synchrony with a firm control on the cyanobiont's biomass ratio to that of the host biomass. The cyanobiont is contained in specialized structures in the host that significantly limit its growth and cell division. The vegetative cells and heterocysts of *Nostoc* cyanobiont isolated from mature symbiotic colonies of *Azolla* [77] and other plants [73] are nearly fourfold larger than their free-living counterparts [6] and the cellular contacts between these cells are fragile. As a result the cyanobiont appears as aggregates of single cells [75]. The continued supply of nutrients with reduced rate of cell division might be the reason for bigger cell size in the cyanobiont. Other modifications include increased cell size, rounded cell shape, reduction in carboxysomes and glycogen granules, disappearance of polyphosphate bodies, thinner cell wall and reduced sheath material, altered cell surface antigens, and in some cases modified thylakoid arrangements [2,50,52,69,78–83]. These changes mature and stabilize as symbiotic events progress such that a gradient of these changes can be traced from young to old parts of a symbiotic tissue. A steadiness between the participating partners is maintained through the developing period of the symbiosis; however, this balance at times seems to break down in older parts [42,57,69]. New symbioses develop either by direct transmission of the cyanobiont from one generation to the next or by acquiring new cyanobiont from the environment. This may be the reason for low specificity of the cyanobiont for the plant host as seen in many cyanobacterial–plant symbioses [75]. In almost all symbioses, cyanobiont's growth is always synchronized with the host. For example, when growth ceases in *Azolla*, cell division in cyanobiont also stops. The cyanobacterial population and cell size increase from apical to older fronds only to the point when the control on the biomass ratio is applied. Then the cell division stops; however, the increase in cell size continues [45]. Similar pattern is seen in young to mature colonies in lichen, *Gunnera* plants, and in the thalli of liverworts and hornworts [1,44,49,84–88]. In *Anthoceros–Nostoc* symbiosis, a constant rate of nitrogen fixation is maintained through decrease in growth response and increase in specific nitrogenase activity. Though not yet established, it clearly indicates plant-mediated control over the cyanobiont growth and protein synthesis to maintain a constant growth rate in the plant partner [6].

13.3.2 Heterocyst Frequency

Heterocysts constitute 5%–10% of the total cell population in free-living cyanobacteria under aerobic conditions and develop at regular intervals in the absence of fixed nitrogen in the surrounding. Heterocysts host the nitrogen-fixing machinery and protect nitrogenase enzyme from oxygen toxicity [19]. When free-living cyanobacteria enter into symbiotic relationships, the entire responsibility of providing combined nitrogen falls on the cyanobionts. This is an enormous task as in many cases the host is sizably bigger than the cyanobiont biomass. Thus in symbiosis, cyanobiont fixes nitrogen at a higher rate than its free-living counterpart and does this by increasing its heterocyst frequency (Table 13.2) and altering the heterocyst spacing pattern. Several other changes are also seen during heterocyst differentiation in symbiosis. In free-living cyanobacteria, anaerobic conditions adversely affect heterocyst differentiation [89]; however, despite prevalence of microaerobic conditions in the host tissue, heterocysts develop in the cyanobiont. How the heterocyst differentiation is accelerated in symbiosis is unclear but heterocyst frequency could range from 25% to 80% of the entire cyanobacterial cells (Table 13.2), depending upon the association and maturity of the symbiotic colony [1]. Although heterocyst differentiation in free-living cyanobacteria occurs in response to N starvation, in symbiosis, heterocyst differentiation seems to occur even when the cyanobionts do not show any signs of N starvation [1,49]. Thus signal for initiation of heterocyst development cascade in symbiosis is independent of the N-status of the cyanobiont and might involve a symbiotic-specific signaling and sensing mechanism. Again, high concentration of combined N does not seem to affect heterocyst differentiation in symbiosis. In the symbiotic activities of *Anthoceros*, *Azolla*, and *Gunnera*, thicker heterocyst envelops suggest that fixed N is first transported to vegetative cells before being excreted. Hence, the vegetative cells are exposed to relatively high concentrations of combined N and despite that they continue to differentiate into heterocysts [6]. The increase in heterocyst frequency shows correlation with the age of the developing symbiosis. As the symbiotic tissue matures, so does the increase in the heterocyst number. For example, the heterocyst frequency in mature cyanobiont colonies in the cavities of the older parts of liverworts is higher than in the colonies initiated at the younger cavities [88]. Similarly, heterocyst number linearly grows as one moves away from apical leaves in *Azolla* [90]. Comparable trend is also seen in *Gunnera* and in lichens [1,86,87]. In free-living cyanobacteria, internal signaling is involved in controlling heterocyst differentiation and their spacing pattern. HetR is required for heterocyst differentiation. NtcA in turn controls transcription of *het*R. If this control is bypassed, heterocyst differentiation occurs even in the presence of combined-N sources, leading to increased heterocyst frequency [91]. The gene *pat*S is required to control the spacing of heterocyst in a filament. Differentiating heterocysts produce PatS (a 17-residue polypeptide) that has inhibitory effect on the adjacent vegetative cells, preventing them from differentiating into heterocysts [91,92]. In symbiosis, loss of filamentous habit may contribute toward the failure of the inhibitory effect of PatS. Alternately, all these gene functions are interrupted by secretion of effector molecules from the host or by endogenous regulation expressed in the cyanobiont by virtue of its presence in the symbiotic environment. This may explain the presence of double or multiple heterocysts seen in a row in symbiotic cyanobacteria (common in liverworts, *Azolla* and *Gunnera* cyanobionts where heterocyst frequency exceeds 40%). One other fact that comes to light is the correlation of heterocyst frequency with their C status. No increase in heterocyst frequency is seen in bipartite lichens where the cyanobiont is also responsible for provision of fixed-C to its fungal partner. In tripartite lichen, the cyanobiont is responsible only for its own C requirement and there is moderate increase in heterocyst frequency (15%). But in the cyanobionts of bryophytes, *Azolla*, cycads, and *Gunnera* where there is fixed-C flow from the hosts to the cyanobionts, heterocyst frequency increases much more (30%–80%).

13.3.3 Carbon Fixation

Free-living cyanobacteria are photosynthetically active and are independent toward meeting their carbon nutrition (Table 13.2). The C-fixation uses water as electron donor and thus produces molecular oxygen as byproduct of active photosynthesis. Since molecular oxygen is highly toxic to nitrogenase enzyme, C-fixation is restricted to vegetative cells. In symbioses involving cyanobacteria, at least one partner is photosynthetically active to achieve carbon autonomy for the whole symbiosis. In diatoms and lichens, the cyanobiont is involved in active photosynthesis [1,42,93]. The rate of C-fixation by the cyanobiont increases from young to mature sections of the lichen thallus [85]. However, in tripartite lichen symbiosis where a green alga is also one of the symbionts, the role of cyanobiont is more focused toward nitrogen fixation than C-fixation. In all other plant–cyanobacterial symbioses, photosynthesis is highly depressed in the cyanobiont. Photosynthetic activity is undetectable in cycad cyanobiont [94], while ~1% activity of that occurring in free-living state is recorded in *Gunnera* [95] and ~12% is documented in *Anthoceros* [96]. The cyanobiont present in the internal cephalodia of *Nephroma arcticum* is photosynthetically inactive. At closer investigation the cyanobiont was found to be covered by the fungal cortex and a layer of green alga limiting accessibility to light and showed severe reduction of carboxysomes that contains primary CO_2-fixing enzyme Rubisco [79]. This in turn might be the reason for lowered photosynthetic activity. In all other cyanobacteria–plant symbioses, the cyanobiont is photosynthetically inactive even when the photosynthetic pigments are present [57,69,97,98]. In bryophytes, the cyanobionts show normal levels of Rubisco but the photosynthetic O_2 evolution and CO_2-fixation is ~1%–12% of that of the level in the free-living cyanobacteria. Here, the lowered activity of the enzyme, and not the concentration (up to 90% of the usual level), accounts for the lowered CO_2-fixation in symbiosis [44,88,98]. The reasons for such repression or inactivation of Rubisco are uncertain but it could be due to microaerobic conditions. For example, heterocysts that lack Rubisco completely present microaerobic conditions suitable for nitrogen fixation [16]. Madan and Nierzwicki-Bauer [89] have also shown that anaerobiosis causes repression of *rbc*LS transcription in *Anabaena* 7120.

13.3.4 Nitrogen Fixation

In all cyanobacterial–plant symbioses, nitrogen fixation is carried out by the cyanobiont [50,93,99]. Increased heterocyst frequency ensures higher rate of nitrogen fixation aided by microaerobic condition prevalent in the symbiotic tissue [100]. Cost of high-nitrogen-fixation rate is supported by the fixed-C that moves from the hosts in all cyanobacterial–plant symbioses except in the cases of diatoms and lichens [2]. The increase in nitrogenase activity shows a correlation to the increased heterocyst frequency from young to mature colonies as the symbiosis develops [1,49,57,85–87,90,101]. However, the relationship between the higher heterocyst frequencies to increased nitrogenase activity seems to break down in old colonies, especially where there is occurrence of double or multiple heterocysts in sequence. Although all heterocysts contain nitrogenase enzyme [102], rate of nitrogen fixation correlates only with frequency of single heterocysts [86,87]. When heterocysts occur in a row, the middle ones are not in direct contact with vegetative cells and thus may not receive sufficient fixed-C for optimum nitrogen fixation [102].

13.3.5 NH_4^+ Assimilation

The primary ammonia assimilating enzyme in nitrogen fixing cyanobacteria is glutamine synthetase (GS) whose concentration and activity is double in the heterocysts in comparison to that in vegetative cells [103]. Expression of GS and nitrogenase is co-regulated and higher levels and activity of GS in heterocysts may be needed to assimilate ammonia derived from nitrogen fixation in the same cell. However, GS is regulated differently in symbiotic cyanobacteria. There is significant decline in GS activity in the cyanobionts of bryophytes (>60%), *Azolla* (70%), and in lichens (90%). This decrease varies from young to old colonies in a developing symbiosis with maximum decrease in the more mature colonies and coincides with amount of ammonia released by the cyanobiont to the host [1,57,84,85,99]. This decrease in GS activity has been shown to be due to repression of GS synthesis in lichens [1,80], to reduced transcription of *gln*A in *Azolla* [55] and to post-translational modifications in *Anthoceros* [44]. The effect of decreased levels of GS seems to manifest as partial loss of N control of nitrogenase in the cyanobiont. As against what is seen in free-living cyanobacteria, combined exogenous N sources such as nitrate or ammonia show little effect on symbiotic nitrogen fixation [104]. On the other hand, wherever the eukaryotic partner possesses a functional GS-glutamate synthase pathway, it can inhibit symbiotic nitrogen fixation probably because the exogenous combined N is metabolized by the host that reduces the need for combined N supply from the cyanobiont [105,106].

13.3.6 Metabolite Exchange

With the progression of a symbiosis, there is increased interaction between the partners leading to metabolic adjustments suitable for establishing biotrophic nutrient exchange—a process for which the symbiosis was established. In lichens, where the cyanobiont provides both fixed-C and fixed-N to the fungal partner, there is increasing fixation and transfer of both C and N as one moves from young to mature colonies. Similar trends are seen in other cyanobacterial–plant symbioses where increasing N transfer is seen from the cyanobiont to the host and that of fixed-C from the host to the cyanobiont. These exchanged molecules are normal products of the symbionts that are even produced in their free-living state; however, their rates of production and discharge are altered. The exchange of these metabolites happens at the cyanobacterial–host interface where the cyanobiont is in close contact with the host. Whether at these junctions the membrane structures are altered to facilitate better exchange and if the partners influence such alterations are not clear. In intracellular symbioses, the cyanobiont is always enveloped by a membrane of host origin. That there is the presence of H^+-ATPases affecting transport system in these membranes and membrane composition might be influenced by the endosymbionts suggests that they may have developed as a result of symbiotic interactions [107,108].

13.3.6.1 Fixed-C Exchange

As discussed earlier, the location of the cyanobiont in a symbiotic association is not always suitable for active photosynthesis. Moreover, the level of the primary C-fixing enzyme is also modulated in the symbiotic cyanobacteria. These two parameters ensure drastic reduction of photosynthesis in the cyanobiont. Therefore, to ensure a steady-state N_2-fixation, the cyanobiont probably has to rely on the plant host for the fixed-C supply [75]. Table 13.2 lists the form of fixed-C that is exchanged in symbiosis. There is no evidence of fixed-C movement in diatom symbiosis as both partners are photosynthetic [42,81]. The photosynthate moving from the cyanobiont to the mycobiont in bipartite lichen symbiosis is in the form glucose, which gets converted to mannitol in the fungal symbiont [10]. In tripartite lichens, there is little or no carbohydrate movement from the cyanobiont to the mycobiont as the third symbiotic partner—the green alga provides the necessary carbohydrate to the mycobiont while the cyanobiont is responsible for its own carbohydrate requirement [1]. Probably, modifications in cell wall synthesis lead to diversion of polysaccharides from cell wall synthesis to release [10]. In all other cyanobacterial–plant symbioses, there is photosynthate movement from the host to the cyanobiont and for all practical purposes the cyanobiont is functionally non-photosynthetic. The symbiotic organ of the host is also non-photosynthetic and fixed-C moves from the photosynthetically active locations to the symbiotic tissue. At least in liverworts, hornworts, and *Azolla* the translocated carbohydrate is in the form of sucrose [95,97,100]. In cycads, the host might directly provide reducing equivalents in place of photosynthate to the cyanobiont [86]. In free-living cyanobacteria, the carbohydrate translocated into heterocysts is catabolized through the oxidative pentose phosphate pathway to generate reductants for N_2-fixation and oxidative respiration. The oxidative pentose phosphate pathway can take either the cyclic or the lincar route in both heterocysts and vegetative cells to catabolize glucose made in photosynthesis. In cyclic pathway, glucose is completely oxidized to CO_2 while generating reductants, but for net assimilation of NH_4^+ produced in N_2-fixation, α-KG must be generated by diverting glucose through the linear pathway [109]. However, in symbiosis, the higher demand for reductants for N_2-fixation is unparallel to the low demand for carbon skeleton due to decreased growth rate, which indicates that in symbiosis the cyanobiont probably prefers the cyclic oxidative pentose phosphate pathway to avoid accumulation of the TCA cycle intermediate in toxic proportions [6].

13.3.6.2 Fixed-N Transfer

Atmospheric-N is fixed by the cyanobiont and transferred to the host in all cyanobacterial–plant symbioses [8,42,69,75,93,99,101,110]. In symbiosis, the growth rate of the cyanobiont is slower than its cultured isolates; therefore, the nitrogen fixed at higher rate during symbiosis becomes in excess of their growth requirement. This surplus combined-N is made available to the plant host (Table 13.2). In most cases including lichens, bryophytes, *Azolla*, cycads, and *Gunnera*, the translocated form of nitrogen from the cyanobionts has been determined as ammonia [69,81,111]. This translocated amount is almost as much as 50%–90% of the N_2 derived NH_3 in the lichens, 80%–90% [1] in bryophytes [112], 40%–50% in *Azolla* [113], and ~90% in *Gunnera* [111]. As pointed out earlier, reduction in GS quantity and possibly in GS activity in the heterocysts of the cyanobionts might be the reason for ammonia release from the cyanobionts to the hosts. The released ammonia is taken up into the host cells via an ammonium transporter of the host cells at the interface between the partners [114]. In cycads, however, the released form of nitrogen is probably glutamine or citrulline [86]. The cyanobiont in cycad symbiosis possesses unaltered GS levels and therefore the N assimilatory pathway of the cyanobiont probably is affected at a stage later than glutamine and citrulline synthesis. Ammonia obtained from the cyanobiont is assimilated via glutamate dehydrogenase in the mycobiont and via the GS-glutamate synthase (GOGAT) pathway in other symbiotic hosts [1,97,111]. The fixed N from the symbiotic tissue/organ to the other parts of the host is provided in the form of amino acids. In tripartite lichens, it is in the form of alanine; in *Azolla* glutamate, glutamine and ammonia moves from leaf cavities to stem apex; in cycads it is translocated as glutamine and citrulline [50,57]. The amino acid asparagine is the principal compound moving from the symbiotic gland to the other parts of the *Gunnera* plant via phloem [6,111,115].

13.4 CONCLUSION

Studies conducted by various researchers collectively indicate that a substantial number of cyanobacterial strains mostly representing the genus *Nostoc* form symbiotic associations with diverse plant partners. Symbiotic competence of the genus *Nostoc* is not a universal feature as not all strains are symbiotically competent. Likewise the majority of cyanobacterial strains entering into a symbiotic partnership appear to be non-specific for the plant partner. In nearly all cyanobacterial–plant symbioses, the cyanobacteria are the passive partners and the flow of signals for association is largely unidirectional from plants to the cyanobionts. Modifications in the cyanobacteria at structural and physiological levels are far more than the minor changes in the host during symbiosis. Symbiotically competent plant hosts have acquired the ability to produce metabolites that can induce mobility in the "desired" cyanobacteria, give direction to the motile cyanobacteria to pre-existing symbiotic-related tissues/organs, control their growth and modify their metabolism toward highly enhanced N_2-fixation. Cell wall synthesis at the junction of contact between the symbionts is also altered for better nutrient exchange. GS synthesis, expression, and/or activity are also modulated toward maximum release of N_2-derived ammonia to the host. This brings us back to the question of cyanobacteria being under stress during symbiosis. It certainly appears that way as the cyanobiont, otherwise capable of independent existence, allows tremendous modifications in itself to provide fixed-N to the host plants. The benefits to host are obvious; however, cyanobacterial contribution in the symbiotic partnership appears far more than the assumed benefits it receives in terms of protection from adverse weather conditions and/or continuous access to nutrients from the host plant. The initial signals from the host for symbiotic association must be compelling enough for the cyanobacteria to advance toward formation of the symbiosis. However, once in contact with the host, the host directs most of the association behavior of the cyanobacteria leading to events upon which the cyanobacteria do not appear to have much control.

Much still needs to be investigated to understand how host plants initiate and control the structural and physiological modifications in the respective cyanobionts. Since cyanobacteria are independent in their need for C and N nutrition, then why some of them are compatible for symbiotic associations while others are not? How the control on heterocyst differentiation is bypassed during symbiosis and what is the form of signal that plant sends to the cyanobiont to interfere with the heterocyst differentiation and spacing pattern? These are some of the questions that need investigations at a deeper level. Although the current understanding of these cyanobacterial–plant interactions is limited at the molecular level, Campbell et al. in 2007 have shown with microarray studies the complexity of hormogonia differentiation where almost 1800 genes are differentially transcribed. However, the processes of chemotaxis and motility of the cyanobacteria in establishing the association are still not well understood. Only a complete elucidation of the molecular mechanisms involved in cyanobacterial–plant symbiosis will be able to resolve whether the drastic modifications seen in the cyanobacteria in symbioses can be labeled as stress response or not.

REFERENCES

1. Rai, A. N., Cyanobacteria-fungal symbioses: The cyanolichens, in *Handbook of Symbiotic Cyanobacteria*, Rai, A. N., Ed., CRC Press, Boca Raton, FL, 1990.
2. Rai, A. N., Soderback, E., and Bergman, B., Cyanobacterium-plant symbioses, *New Phytol.*, 147, 449, 2000.
3. de Bary, A., Die erscheinung der symbiose, in *Vortrag auf der Versammlung der Naturforscher und Artz zu Cassel*, Verlag von K. J. Trubner, Ed., Strasburg, Germany, 1–30, 1879.
4. Lopez-Garcia, P. and Moreira, D., Metabolic symbiosis at the origin of eukaryotes, *Trends Biochem. Sci.*, 24, 88, 1999.
5. Tomitani, A. et al., Chlorophyll *b* and phycobilins in the common ancestor of cyanobacteria and chloroplasts, *Nature*, 400, 159, 1999.
6. Meeks, J. C. and Elhai, J., Regulation of cellular differentiation in filamentous cyanobacteria in free-living and plant-associated symbiotic growth states, *Microbiol. Mol. Biol. Rev.*, 66, 94, 2002.
7. Kopp, R. E. et al., The Palaeoproterozoic snowball Earth: A climate disaster triggered by the evolution of oxygenic photosynthesis, *Proc. Natl Acad. Sci. U.S.A.*, 102, 11131, 2005.
8. Usher, K. M., Bergman, B., and Raven, J. A., Exploring cyanobacterial mutualisms, *Annu. Rev. Ecol. Evol. Syst.*, 38, 255, 2007.
9. Ahmadjian, V. and Paracer, S., *Symbiosis*, New England University Press, Hanover, PA, 1986.
10. Smith, D. C. and Douglas, A. E., *The Biology of Symbiosis*, Edward Arnold, London, U.K., 1987.
11. Werner, D., *Symbiosis of Plants and Microbes*, Chapman & Hall, London, U.K., 1992.
12. Rippka, R., Photoheterotrophy and chemoheterotrophy among unicellular blue-green algae, *Arch. Microbiol.*, 87, 303, 1972.
13. Vasudevan, V. et al., Stimulation of pigment accumulation in *Anabaena* strains: Effect of light intensity and sugars, *Folia Microbiol.*, 51, 50, 2006.

14. Prasanna, R. et al., Influence of biofertilizers and organic amendments on nitrogenase activity and phototrophic biomass of soil under wheat, *Acta Agron. Hungarica*, 56, 149, 2008.
15. Fay, P., Oxygen relations of nitrogen fixation in cyanobacteria, *Microbiol. Rev.*, 56, 340, 1992.
16. Wolk, C. P., Ernst, A., and Elhai, J., Heterocyst metabolism and development, in *The Molecular Biology of Cyanobacteria*, Bryant, D. A., Ed., Kluwer Academic Publishers, Dordrecht, the Netherlands, 59–78, 1994.
17. Tandeau de Marsac, N., Differentiation of hormogonia and relationships with other biological processes, in *The Molecular Biology of Cyanobacteria*, Bryant, D. A., Ed., Kluwer Academic Publishers, Dordrecht, the Netherlands, 825–842, 1994.
18. Wong, F. C. and Meeks, J. C., Establishment of a functional symbiosis between the cyanobacterium *Nostoc punctiforme* and the bryophyte *Anthoceros punctatus* requires genes involved in nitrogen control and initiation of heterocyst differentiation, *Microbiology*, 148, 315, 2002.
19. Gallon, J. R., Reconciling the incompatible: N_2-fixation and O_2, *New Phytol.*, 122, 571, 1992.
20. Byrant, D. A., *The Molecular Biology of Cyanobacteria*, Kluwer Academic Publishers, Dordrecht, the Netherlands, 1994.
21. Bergman, B. et al., N_2-fixation by non-heterocystous cyanobacteria, *FEMS Microbiol. Rev.*, 19, 139, 1997.
22. Lockhart, P. J. et al., Sequence of *Prochloron didemni atp* BE and the inference of chloroplast origins, *Proc. Natl Acad. Sci. U.S.A.*, 89, 2742, 1992.
23. Raven, J. A., The evolution of cyanobacterial symbioses, *Proc. R. Irish. Acad. Biol. Environ.*, 102B, 3, 2002.
24. Raven, J. A., Evolution of cyanobacterial symbioses, *Proc. R. Irish. Acad. Biol. Environ.*, 102B, 19, 2002.
25. Rai, A. N. and Bergman, B., Cyanolichens, *Proc. R. Irish. Acad. Biol. Environ.*, 102B, 19, 2002.
26. Carpenter, E. J., Marine cyanobacterial symbioses, *Proc. R. Irish. Acad. Biol. Environ.*, 102B, 15, 2002.
27. Foster, R., Carpenter, E. J., and Bergman, B., Unicellular cyanobionts in open ocean dinoflagellates, radiolarians, and tintinnides: Ultrastructural characterization and immuno-localization of phycoerythrin and nitrogenase, *J. Phycol.*, 42, 453, 2006.
28. Adams, D. G. et al., Cyanobacterial-plant symbioses, in *The Prokaryotes: A Handbook on the Biology of Bacteria. Symbiotic Associations, Biotechnology, Applied Microbiology*, Dworkin, M., Falkow, S., Rosenberg, E., Schleifer, K. H., and Stackebrandt, E., Eds., Springer, New York, 331–363, 2006.
29. Bergman, B., Rasmussen, U., and Rai, A. N., Cyanobacterial associations, in *Associative and Endophytic Nitrogen-Fixing Bacteria and Cyanobacterial Associations*, Elmerich, C. and Newton, W. E., Eds., Kluwer Academic Publishers, Dordrecht, the Netherlands, 257–301, 2007.
30. Adams, D. G. and Duggan P. S., Cyanobacteria-bryophyte symbiosis, *J. Exp. Bot.*, 59(5), 1047, 2008.
31. Lewin, R. A. and Robinson, P., The greening of polar bears in zoos, *Nature (Lond)*, 278, 445, 1979.
32. Honegger, R., Functional aspects of the lichen symbiosis, *Annu. Rev. Plant. Physiol. Plant Mol. Biol.*, 42, 553, 1991.
33. Schenk, H. E. A., Cyanobacterial symbioses, in *The Prokaryotes*, Balows, A., Truper, H. G., Dworkin, M., Harder, W., and Schleifer, K. H., Eds., Springer-Verlag, Berlin, Germany, 3819–3854, 1992.
34. Hill, G. E., Trait elaboration via adaptive mate choice: Sexual conflict in the evolution of signals of male quality, *Ethol. Ecol. Evol.*, 6, 351, 1994.
35. Jorgensen, P. M., Difficulties in lichen nomenclature, *Mycotaxon*, 40, 497, 1991.
36. Gargas, A. and Taylor, J. W., Polymerase chain reaction (PCR) primers for amplifying and sequencing nuclear 18S rDNA from lichenized fungi, *Mycologia*, 84, 589, 1992.
37. Bridge, P. D. and Hawksworth, D. L., What molecular biology has to tell us at the species level in licehnizes fungi, *Lichenologist*, 30, 307, 1998.
38. Nimis, P. L., A critical appraisal of modern generic concepts in lichenology, *Lichenologist*, 30, 227, 1998.
39. Rambold, G., Friedl, T., and Beck, A., Photobionts in lichens: Possible indicators of phylogenetic relationships, *The Bryologist*, 101, 392, 1998.
40. Schussler, A. et al., *Geosiphon pyriforme*, an endosymbiotic association of fungus and cyanobacteria: The spore structure resembles that of arbuscular mycorrhizal (AM) fungi, *Botanica Acta*, 107, 36, 1994.
41. Gehrig, H., Schussler, A., and Kluge, M., *Geosiphon pyriforme*, a fungus forming endocytobiosis with *Nostoc* (Cyanobacteria), is an ancestral member of the Glomales: Evidence by SSU rRNA analysis, *J. Mol. Evol.*, 43, 71, 1996.
42. Villareal, T. A., Marine nitrogen-fixing diatom-cyanobacteria symbiosis, in *Marine Pelagic Cyanobacteria: Trichodesmium and Other Diazotrophs*, Carpenter, E. J., Capone, D. G., and Reuter, J. G., Eds., Kluwer Academic Publishers, Dordrecht, the Netherlands, 163–175, 1992.
43. Carpenter, E. J. et al., Extensive blooms of N_2 fixing diatom/cyanobacterial association in tropical Atlantic Ocean, *Mar. Ecol. Prog. Ser.*, 185, 273, 1999.

44. Meeks, J. C., Cyanobacterial–bryophyte associations, in *Handbook of Symbiotic Cyanobacteria*, Rai, A. N., Ed., CRC Press, Boca Raton, FL, 43–63, 1990.
45. Braun-Howland, E. B. and Nierzwicki-Bauer, S. A., *Azolla-Anabaena* symbiosis: Biochemistry, physiology, ultrastructure, and molecular biology, in *Handbook of Symbiotic Cyanobacteria*, Rai, A. N., Ed., CRC Press, Boca Raton, FL, 65–117, 1990.
46. Watanabe, I. and van Hove, C., Phylogenetic, molecular and breeding aspects of Azolla-Anabaena symbiosis, in *Pteridology in Perspective*, Camus, J. M., Gibby, M., and Johns, R. J., Eds., Royal Botanic Gardens, Kew, U.K., pp. 611–619, 1996.
47. Stevenson, D. W., Morphology and systematics of the Cycadales, *Mem. N. Y. Bot. Gard.*, 57, 8, 1990.
48. Caputo, P., Stevenson, D. W., and Wurtzel, E. T., A phylogenetic analysis of American *Zamiaceae* (Cycadales) using chloroplast DNA restriction fragment length polymorphisms, *Brittonia*, 43, 135, 1991.
49. Bergman, B., Johansson, C., and Soderback E., The *Nostoc–Gumnera* symbiosis, *New Phytol.*, 122, 379, 1992.
50. Stewart, W. D. P., Rowell, P., and Rai, A. N., Cyanobacteria-eukaryotic plant symbioses, *Annu. Microbiol. (Institut Pasteur)*, 134B, 205, 1983.
51. Galun, M. and Bubrick, P., Physiological interactions between the partners of lichen symbiosis, in *Encyclopedia of Plant Physiology New Series*, Linskens, H. F. and Heslop Harrison, J., Eds., Springer-Verlag, Berlin, Germany, 362–401, 1984.
52. Gorelova, O. A., Communication of cyanobacteria with plant partners during association formation, *Microbiology*, 75(4), 465, 2006.
53. Friedl, T. and Budel, B., Photobionts, in *Lichen Biology*, Nash, T. H., Ed., Cambridge University Press, Cambridge, U.K., 8–23, 1996.
54. Potts, M., The anhydrobiotic cyanobacterial cell, *Physiol. Plant.*, 97, 788, 1996.
55. Nierzwicki-Bauer, S. A., *Azolla-Anabaena* symbiosis: Use in agriculture, in *Handbook of Symbiotic Cyanobacteria*, Rai, A. N., Ed., CRC Press, Boca Raton, FL, 119–136, 1990.
56. Kardish, M. et al., Lectins from the lichen *Nephroma laevigatum* Arch: Localization and function, *Symbiosis*, 11, 47, 1991.
57. Bergman, B. et al., Cyanobacterial-plant symbiosis, *Symbiosis*, 14, 61, 1992.
58. Rasmussen, U., Johansson, C., and Bergman, B., Early communication in the *Gunnera-Nostoc* symbiosis: Plant-induced cell differentiation and protein synthesis in the cyanobacterium, *Mol. Plant Microbe Interact.*, 7, 696, 1994.
59. Campbell, E. L. and Meeks, J. C., Characteristics of hormogonia formation by symbiotic *Nostoc* sp. in response to the presence of *Anthoceros punctatus* or its extracellular products, *Appl. Environ. Microbiol.*, 55, 125, 1989.
60. Bergman, B., Matveyev, A., and Rasmussen, U., Chemical signaling in cyanobacterial-plant symbioses, *Trends Plant Sci.*, 1, 191, 1996.
61. Watts, D., Knight, C. D., and Adams, D. G., Characterization of plant exudates inducing chemotaxis in nitrogen-fixing cyanobacteria, in *The Photosynthetic Prokaryotes*, Peschek, G. A., Loffelhardt, W., and Schmetterer, G., Eds., Kluwer Academic/Plenum Publishers, New York, 679–684, 1999.
62. Plazinski, J. et al., Indigenous plasmids in *Anabaena azollae*: Their taxonomic distribution and existence of regions of homology with symbiotic genes of *Rhizobium*, *Can. J. Microbiol.*, 37, 171, 1991.
63. Rasmussen, U. et al., A molecular characterization of the *Gunnera-Nostoc* symbiosis: Comparison with *Rhizobium*- and *Agrobacterium*-plant interactions, *New Phytol.*, 133, 391, 1996.
64. Cohen, M. F. and Meeks, J. C., A hormogonium regulating locus, *hrmUA*, of the cyanobacterium *Nostoc punctiforme* strain ATCC 29133 and its response to an extract of a symbiotic plant partner *Anthoceros punctatus*, *Mol. Plant Microbe Interact.*, 10, 280, 1997.
65. Meeks, J. C. et al., Developmental alternatives of symbiotic *Nostoc punctiforme* in response to its plant partner *Anthoceros punctatus*, in *The Photosynthetic Prokaryotes*, Peschek, G. A., Loffelhardt, W., and Schmetterer, G., Eds., Kluwer Academic/Plenum Publishers, New York, 665–678, 1999.
66. Van Rhijn, P. and Vanderleyden, J., The *Rhizobium*-plant symbioses, *Microbiol. Rev.*, 59, 124, 1995.
67. Cohen, M. F. and Yamasaki, H., Flavonoids-induced expression of a symbiosis-related gene in the cyanobacterium *Nostoc punctiforme*, *J. Bacteriol.*, 182, 4644, 2000.
68. Patricia, P. et al., Flavonoids of *Guznnera* subgenera *Mlaisandra*, *Panke*, and *Perpensum* (*Gunnleraceae*), *Suppl. Am. J. Bot.*, 76, 264, 1989.
69. Rai, A. N., *Handbook of Symbiotic Cyanobacteria*, CRC Press, Boca Raton, FL, 1990.
70. Osborne, B. A. et al., Use of nitrogen by the *Nostoc-Gunnera tinctoria* (Molina) Mirbel symbiosis, *New Phytol.*, 120, 481, 1992.

71. Shick, J. M. and Dunlap, W. C., Mycosporine-like amino acids and related gadusols: Biosynthesis, accumulation and UV-protective functions in aquatic organisms, *Annu. Rev. Physiol.*, 64, 223, 2002.
72. Proteau, P. J. et al., The structure of scytonemin, an UV pigment from the sheaths of cyanobacteria, *Experientia*, 49, 825, 1993.
73. Meeks, J. C., Symbiosis between nitrogen-fixing cyanobacteria and plants, *Bioscience*, 48, 266, 1998.
74. Meeks, J. C., Molecular mechanisms in the nitrogen-fixing *Nostoc*-Bryophyte symbiosis, in *Molecular Basis of Symbiosis*, Overmann, J., Ed., Springer, Berlin, Germany, 165–196, 2005.
75. Meeks, J. C., Physiological adaptations in nitrogen-fixing *Nostoc*-plant symbiotic associations, *Microbiol. Monogr.*, 8, 181, 2009.
76. Enderlin, C. S. and Meeks, J. C., Pure culture and reconstitution of the *Anthoceros–Nostoc* symbiotic association, *Planta*, 158, 157, 1983.
77. Becking, T. H. and Donze, M., Pigment distribution and nitrogen-fixation in *Anabaena-Azollae*, *Plant Soil*, 61, 203, 1981.
78. Rai, A. N., Studies on the nitrogen-fixing lichen, *Peltigera aphthosa* Willd, PhD thesis, Dundee University, Dundee, U.K., 1980.
79. Bergman, B. and Rai, A. N., The *Nostoc-Nephroma* symbiosis: Localization, distribution pattern and levels of key proteins involved in nitrogen and carbon metabolism of the cyanobiont, *Physiol. Plant.*, 77, 216, 1989.
80. Janson, S., Rai, A. N., and Bergman, B., The marine lichen *Lichina confinis* (O.F. Mull) C.Ag.: Ultrastructure and localization of glutamine synthetase, phycoerythrin and ribulose 1,5-bisphosphate carboxylase/oxygenase in the cyanobiont, *New Phytol.*, 124, 149, 1993.
81. Janson, S., Rai, A. N., and Bergman, B., Intracellular cyanobiont *Richelia intracellularis*: Ultrastructure and immunolocalization of phycoerythrin, nitrogenase, Rubisco and glutamine synthetase, *Mar. Biol.*, 124, 1, 1995.
82. Schussler, A. et al., Characterization of the *Geosiphon pyriforme* symbiosome by affinity techniques: Confocal laser scanning microscopy (CSLM) and electron microscopy, *Protoplasma*, 190, 53, 1996.
83. Jager, K. M. et al., Suncellular element analysis of a cyanobacterium (*Nostoc* sp.) in symbiosis with *Gummera manicata* by ESI and EELS, *Bot. Acta*, 110, 151, 1997.
84. Rowell, P., Rai, A. N., and Stewart, W. D. P., Studies on the nitrogen metabolism of the lichen *Peltigera aphthosa* and *Peltigera canina*, in *Lichen Physiology and Cell Biology*, Brown, D. H., Ed., Plenum Press, New York, 145–160, 1985.
85. Hill, D. J., The control of cell cycle in microbial symbionts, *New Phytol.*, 112, 175, 1989.
86. Lindblad, P. and Bergman, B., The cycad-cyanobacterial symbiosin, in *Handbook of Symbiotic Cyanobacteria*, Rai, A. N., Ed., CRC Press, Boca Raton, FL, 137–159, 1990.
87. Soderback, E., Lindblad, P., and Bergman, B., Developmental patterns related to nitrogen fixation in the *Nostoc-Gunnera magellanica* Lam. symbiosis, *Planta*, 182, 355, 1990.
88. Rodgers, G. A. and Stewart, W. D. P., The cyanophyte-hepatic symbiosis I. Morphology and physiology, *New Phytol.*, 78, 441, 1977.
89. Madan, A. P. and Nierzwicki-Bauer, S. A., In-situ detection of transcripts for ribulose 1,5-bisphosphate carboxylase in cyanobacterial heterocysts, *J. Bacteriol.*, 175, 7301, 1993.
90. Hill, D. J., The role of *Anabaena* in *Azolla–Anabaena* symbiosis, *New Phytol.*, 78, 611, 1977.
91. Haselkorn, R., How cyanobacteria count to 10, *Science*, 282, 891, 1998.
92. Yoon, H.-S. and Golden, J. W., Heterocyst pattern formation controlled by a diffusible peptide, *Science*, 282, 935, 1998.
93. Kluge, M., Mollenhauer, R., and Kape, R., *Geosiphon pyriforme*, an endosymbiotic consortium of a fungus and a cyanobacteria (*Nostoc*) fixes nitrogen, *Bot. Acta*, 105, 343, 1992.
94. Lindblad, P., Rai, A. N., and Bergman, B., The *Cycas revoluta-Nostoc* symbiosis: Enzyme activities of nitrogen and carbon metabolism in the cyanobiont, *J. Gen. Microbiol.*, 133, 1695, 1987.
95. Soderback, E. and Bergman, B., The *Nostoc-Gunnera* symbiosis: Carbon fixation and translocation, *Physiol. Plant.*, 89, 125, 1993.
96. Steinberg, N. A. and Meeks, J. C., Physiological sources of reductant for nitrogen fixation activity in *Nostoc* sp. strain UCD 7801 in symbiotic association with *Anthoceros punctatus*, *J. Bacteriol.*, 173, 7324, 1991.
97. Peters, G. A. and Meeks, J. C., The *Azolla-Anabaena* symbiosis: Basic biology, *Annu. Rev. Plant Physiol. Mol. Biol.*, 40, 193, 1989.
98. Rai, A. N. et al., *Anthoceros–Nostoc* symbiosis: Immunoelectronmicroscopic localization of nitrogenase, glutamine synthetase, phycoerythrin and ribulose-1,5-bisphosphate carboxylase/oxygenase in the cyanobiont and the culture (free-living) isolate *Nostoc* 7801, *J. Gen. Microbiol.*, 135, 385, 1989.

99. Rowell, P. and Kerby, N. W., Cyanobacteria and their symbionts, in *Biology and Biochemistry of Nitrogen Fixation*, Dilworth, M. and Glen, A., Eds., Elsevier, Amsterdam, the Netherlands, 373–407, 1991.
100. Stewart, W. D. P. and Rodgers, G. A., The cyanophyte-hepatic symbiosis. 11. Nitrogen fixation and interchange of nitrogen and carbon, *New Phytol.*, 18, 459, 1977.
101. Rai, A. N., Nitrogen metabolism, in *Handbook of Lichenology*, Galun, M., Ed., CRC Press, Boca Raton, FL, 201–237, 1988.
102. Bergman, B., Lindbald, P., and Rai, A. N., Nitrogenase in free-living and symbiotic cyanobacteria: Immunoelectron microscopic localization, *FEMS Microbiol. Lett.*, 35, 75, 1986.
103. Bergman, B. et al., Immuno-gold localization of glutamine synthetase in a nitrogen-fixing cyanobacterium (*Anabaena cylindrica*), *Planta*, 166, 329, 1985.
104. Stewart, W. D. P., Rowell, P., and Rai, A. N., Symbiotic nitrogen fixing cyanobacteria, in *Nitrogen Fixation*, Stewart, W. D. P. and Gallon, J. R., Eds., Academic Press, London, U.K., 239–277, 1980.
105. Rai, A. N., Rowell, P., and Stewart, W. D. P., NH_4^+ assimilation and nitrogenase regulation in the lichen *Peltigera aphthosa* Willd, *New Phytol.*, 85, 545, 1980.
106. Campbell, E. L. and Meeks, J. C., Evidence for plant-mediated regulation of nitrogenase expression in the *Anthoceros–Nostoc* symbiotic association, *J. Gen. Microbiol.*, 138, 473, 1992.
107. Smith, S. E. and Smith, F. A., Structure and function of the interface in biotrophic symbioses and their relation to nutrient transport, *New Phytol.*, 114, 1, 1990.
108. Quispel, A., A search for signals in endophytic micro-organisms, in *Molecular Signals in Plant-Microbe Communications*, Verma, D. P. S., Ed., CRC Press, Boca Raton, FL, 471–491, 1992.
109. Summers, M. L. et al., Genetic evidence of a major role for glucose-6-phosphate dehydragenase in nitrogen fixation and dark growth of the cyanobacterium *Nostoc* sp. strain ATCC29133, *J. Bacteriol.*, 177, 6184, 1995.
110. Villareal, T. A., Nitrogen fixation by the cyanobacterial symbiont of the diatom genus *Hemiaulus*, *Mar. Ecol.*, 76, 201, 1991.
111. Silvester, W. B, Parsons, R., and Watt, P. W., Direct measurement of release and assimilation of ammonia in the *Gunnera-Nostoc* symbiosis, *New Phytol.*, 132, 617, 1996.
112. Meeks, J. C. et al., Fixation of [^{13}N] N_2 and transfer of fixed nitrogen in the *Anthoceros–Nostoc* symbiotic association, *Planta*, 164, 406, 1985.
113. Meeks, J. C. et al., *Azolla-Anabaena* relationship. XIII. Fixation of [^{13}N] N_2, *Plant Physiol.*, 84, 883, 1987.
114. Tyreman, S. D., Whitehead, L. F., and Day, D. A., A channel-like transporter for NH_4^+ on the symbiotic interface of N_2-fixing plants, *Nature*, 378, 629, 1995.
115. Stock, P. A. and Silvester, W. B., Phloem transport of recently-fixed nitrogen in the *Gunnera-Nostoc* symbiosis, *New Phytol.*, 126, 259, 1994.

14 A Global Understanding of Light Stress in Cyanobacteria

Environmental and Bioproducts Perspectives

*Nishikant V. Wase, Saw Ow Yen, and Phillip C. Wright**

CONTENTS

14.1 INTRODUCTION

Solar radiation is one of the most important factors required for the growth of cyanobacteria in their natural habitat. It is also the ultimate source of energy on Earth, as this energy is transformed into chemical energy generally by the photosynthetic organisms. Changes in the light intensity can be observed on broad timescale and over several orders of magnitude. For example, in terrestrial environments, short-term changes in the niche environment can occur because of shading due to cloud cover, whereas in the aquatic environment restriction of light can be caused by turbid water

* Corresponding author: p.c.wright@sheffield.ac.uk

or focusing of light by water surface causing rapid fluctuations in light intensity. Such variability in the light intensity is often rapid, can be deleterious, and presents a major challenge for survival of photosynthetic organisms. Due to massive industrialization and depletion of the stratospheric ozone layer, there has been a rapid increase in UV radiation, which has deleterious effects on the growth and survival of photosynthetic organisms [1]. Although light is obviously an essential factor for photosynthesis, it often causes destruction when it saturates the light harvesting complexes causing photoinhibition [2]. For additional information on photoinhibition of photosystem II (PS-II), the reader is directed to a recent review [2]. Oxygenic photosynthesis generates light-induced formation of reactive oxygen species (ROS) that cause damage to the PS-II [3–6]. Since light is essential for normal growth and survival, such damage is unavoidable in cyanobacteria. Apart from photodamage, there are various events that occur inside the cyanobacterial cell including critical roles of perception of light signals by the light sensors, damage of membrane proteins in the form of unsaturation of fatty acids, DNA damage, protein damage, redox homeostasis, induction of heat shock response, and apoptosis. All these events will be discussed in more detail in the following subsections.

14.2 PERCEPTION AND TRANSDUCTION OF LIGHT

After the completion of the genome sequence of the unicellular *Synechocystis* sp. PCC 6803, signal transduction studies in cyanobacteria received a tremendous boost. As many as 80 signal transduction genes were identified in *Synechocystis* sp. PCC 6803 [7], and more than 400 in filamentous cyanobacterium *Nostoc punctiforme* ATCC 29133 [8]. A list of sequenced cyanobacteria and number of (predicted) signal transduction genes is given in Table 14.1. It is difficult to provide a comprehensive account of cyanobacterial light signal transduction literature here, but the reader is directed to some interesting recent reviews on cyanobacterial light sensing mechanisms [9–12]. Generally, cyanobacteria sense light using specialized proteins called phytochromes. Cph1 was the first phytochrome identified (from *Synechocystis* sp. PCC 6803) [13] in cyanobacteria. Structurally, phytochromes consist of an N-terminal photosensory core and a C-terminal regulatory domain [9,14,15]. The chromophore molecule (recognizes a specific wavelength) resides in the pocket formed by the photosensory core. The photosensory core is generally made up of three structural domains: GAF (cGMP phosphodiesterase/adenylatecyclase), PAS (Per-Arnt-Sim), and PHY (phytochrome specific) domain. Generally the cyanobacterial phytochromes vary from the typical PAS-GAF-PHY architecture. Some might lack the PAS domain, as found in *Synechocystis* PCC 6803 Cph2, while others might possess the GAF domain only [16,17]. These cyanobacterial phytochromes (also called cyanobacteriochromes [CBCRs]) extend the photosensory range to shorter wavelengths of visible light [16]. During their study, Rockwell et al. [16] also showed that the dual-cysteine photosensor from *N. punctiforme* ATCC 29133 can sense the full spectrum from near infrared to near ultraviolet. It was also observed that most of the phytochromes contain histidine kinase as their output domain. A short list of validated phytochromes found in cyanobacteria is given in Table 14.2.

TABLE 14.1
List of Putative Two-Component Proteins Found in Sequenced Cyanobacteria

Organism	No. of Two-Components Genes
Microcystis aeruginosa NIES-843	37
Anabaena sp.	139
Nostoc punctiforme ATCC 29133	400
Cyanothece sp. ATCC 51142	53
Trichodesmium erythraeum IMS 101	9
Thermosynechococcus elongatus BP-1	41
Arthrospira platensis NIES-39	133

TABLE 14.2
List of Validated Signal Transduction Proteins in Cyanobacteria

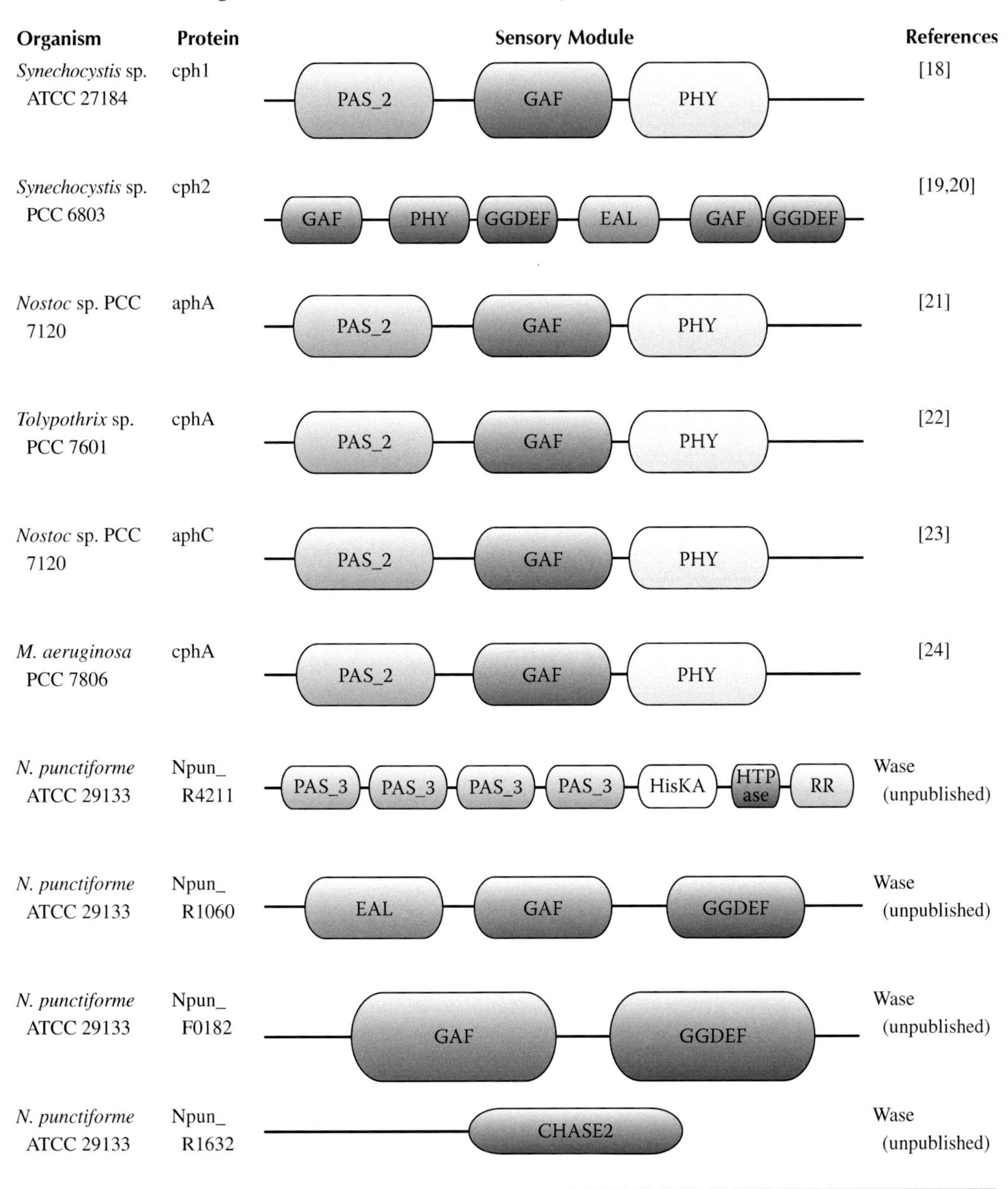

Organism	Protein	Sensory Module	References
Synechocystis sp. ATCC 27184	cph1	PAS_2 – GAF – PHY	[18]
Synechocystis sp. PCC 6803	cph2	GAF – PHY – GGDEF – EAL – GAF – GGDEF	[19,20]
Nostoc sp. PCC 7120	aphA	PAS_2 – GAF – PHY	[21]
Tolypothrix sp. PCC 7601	cphA	PAS_2 – GAF – PHY	[22]
Nostoc sp. PCC 7120	aphC	PAS_2 – GAF – PHY	[23]
M. aeruginosa PCC 7806	cphA	PAS_2 – GAF – PHY	[24]
N. punctiforme ATCC 29133	Npun_ R4211	PAS_3 – PAS_3 – PAS_3 – PAS_3 – HisKA – HTPase – RR	Wase (unpublished)
N. punctiforme ATCC 29133	Npun_ R1060	EAL – GAF – GGDEF	Wase (unpublished)
N. punctiforme ATCC 29133	Npun_ F0182	GAF – GGDEF	Wase (unpublished)
N. punctiforme ATCC 29133	Npun_ R1632	CHASE2	Wase (unpublished)

When light is received, as a response, the information is relayed via a two-component phosphorelay system and the phosphate is transferred to the response regulator (RR). This phosphorylation starts downstream signaling cascading events, which finally culminate into a physiological response in the cyanobacteria. Along with the phosphorylation of histidine, serine and threonine were also found to be phosphorylated in response to growth under different light conditions [25]. Despite significant progress made in the last few years, we are still far from understanding the key aspects of cyanobacterial signaling networks. The list of signaling domains is far from complete and new domains are defined with poorly defined function such as Chase2 sensor proteins. Also, there are many domains that are likely involved in signaling, but their functions are still unknown such as the tryptophan-rich sensory protein TspO [26].

14.3 UNSATURATION OF FATTY ACIDS IN LIGHT STRESS

The cell membrane is the barrier to the external environment, and therefore is the first organelle that senses stress. In all living organisms, membrane and lipid damage occurs in response to stress [27–29]. Additionally, in all photosynthetic organisms, most of the photosynthesis machinery is embedded in thylakoid membrane, which is primarily made up of proteins and lipids. Thus, changes in the composition of membrane lipids should affect the photosynthetic apparatus and consequently the photosynthesis efficiency. Cyanobacterial cells contain four major glycerolipids: monogalactosyldiacylglycerol (MGDG), digalactosyldiacylglycerol (DGDG), solfoquinovosyldiacylglycerol (SQDG), and phosphatidylglycerol (PG) [30]. The physiological role of DGDG was examined in *Synechocystis* sp. PCC 6803, and it was observed that the dgdA mutant showed normal growth under low light (20 μmoles photons m^{-2} s^{-1}). However, when the *Synechocystis* cells were transferred to high light (200 μmoles photons m^{-2} s^{-1}), growth was severely retarded [31]. Subsequently, this retardation of growth was recovered by an exogenous supply of DGDG in the growth medium. This study thus shows that a lack of DGDG has an adverse effect on the ability to repair the photosynthetic apparatus.

When cyanobacteria are exposed to strong light, photoinhibition occurs and the damaged D1 protein is replaced by a newly synthesized D1 protein for maintenance of PS-II homeostasis [32]. But, if the strong light continues, the reversible photoinhibition changes to irreversible photoinhibition and no more D1 protein is synthesized [2,33]. The photoinhibition of PS-II is further exacerbated by desaturation of fatty acids. It has been shown that during low temperature stress, unsaturation of fatty acids occurs and desA (Δ12 desaturase) gets induced in *Synechocystis* sp. PCC 6803 [28]. A similar increased expression of desA was found during high PAR (photosynthetically active radiation) or UV radiation exposure [29], indicating that unsaturation of fatty acids occurs during high light and UV radiation exposure. It has been observed that when *Synechococcus* PCC 7942 was transformed with the *desA* gene (Δ12 desaturase), unsaturation of membrane lipids increased, rendering the organism more tolerant against high light [34]. Thus, unsaturation of fatty acids is an important tool for cyanobacteria for high light stress tolerance and this unsaturation of the fatty acids also helps to increase the rate of de novo synthesis of D1 protein.

14.4 DNA DAMAGE AND REPAIR

Nucleic acids are important cellular components that are susceptible to light-induced damage. Damage to DNA occurs in a myriad of ways, ranging from normal base modification to stress-specific pyrimidine dimer formation during UV-R exposure. Normally, DNA damage can be classified into four types: DNA double strand breaks, DNA nucleotide adduct formation and base modification, DNA base pairing mismatch, and DNA single strand breaks [35]. The effect of strong visible light on *Synechocystis* sp. PCC 6803 on gene transcription was investigated using DNA microarrays [36]. During this study, the authors observed that expression of 160 genes was increased within 15 min when cells were transferred from low light (20 μmoles photons m^{-2} s^{-1}) to high light (300 μmoles photons m^{-2} s^{-1}), but no DNA repair genes were observed to be induced in this study. A similar large-scale global iTRAQ (isobaric tags for relative and absolute quantitation) based proteome profiling experiment was undertaken in the unicellular marine cyanobacterium *Prochlorococcus marinus* MED4 [37]. Cells were grown in low intensity (20 μE m^2 s^{-1}), medium intensity (60 μE m^2 s^{-1}) and high intensity (100 μE m^2 s^{-1}) light, but no protein related to DNA repair was found to be differentially regulated. But, when a similar transcriptomics experiment was performed in *P. marinus* PCC 9511, differential expression in genes associated with the DNA repair mechanism was observed [38]. During that study, the authors used 100 μmoles photons m^{-2} s^{-1} as the low light condition and 900 μmoles photons m^{-2} s^{-1} as the high light condition, supplemented by 7.59 Wm^{-2} of UV-A radiation and 0.57 Wm^{-2} UV-B radiations. The authors observed that surprisingly UV exposure did not result in significant upregulation of DNA repair genes, but instead these genes were significantly activated under high light. Further, the authors [38] observed that the *phrA* gene that is involved in the repair of DNA lesion by photoreactivation was strongly induced under high light.

In regard to the UV range of the light spectrum, because of the depletion of the ozone layer, there is an increased amount of harmful ultraviolet radiation incident on the Earth (UVR: 280–400 nm). The high energy UV-B sub-band (280–315 nm) has the largest potential to damage the biomolecules. UV-B can cause direct damage to DNA and proteins, and indirectly by the production of ROS [39–43] (refer to illustration in Figure 14.1). Although UV-A does not damage biomolecules directly, similar ROS generation was found in the filamentous cyanobacterium *Anabaena variabilis* sp. PCC 7937 when exposed to UV-A [6]. ROS are produced from oxygen when O_2^- and $\bullet OH^-$ react. Similarly ROS are also generated via the Fenton Reaction ($Fe^{2+} + H_2O_2 \rightarrow \bullet OH^- + OH^-$) and the generated $\bullet OH^-$ radicals attack the DNA directly [44]. Much of the cellular machinery that is involved in DNA damage sensing and repair is highly conserved in eukaryotes and prokaryotes [30,45]. The identification of DNA damage is a primary requirement in the cell, and the MutS protein generally performs the job of recognition and binding to the mismatch in base pairs. After binding to the site of DNA damage, the MutS protein recruits MutL to start the mismatch repair [46]. Thus, MutS and MutL proteins are involved in sensing DNA base mismatching in all organisms [47]. It has been shown that the *mutS* gene is induced by activation of stress protein RecA [48], which also helps in DNA repair and is a part of the bacterial SOS response. Along with this, endonuclease III which is another important component of the DNA repair pathway is found in the filamentous cyanobacterium *Anabaena* sp.

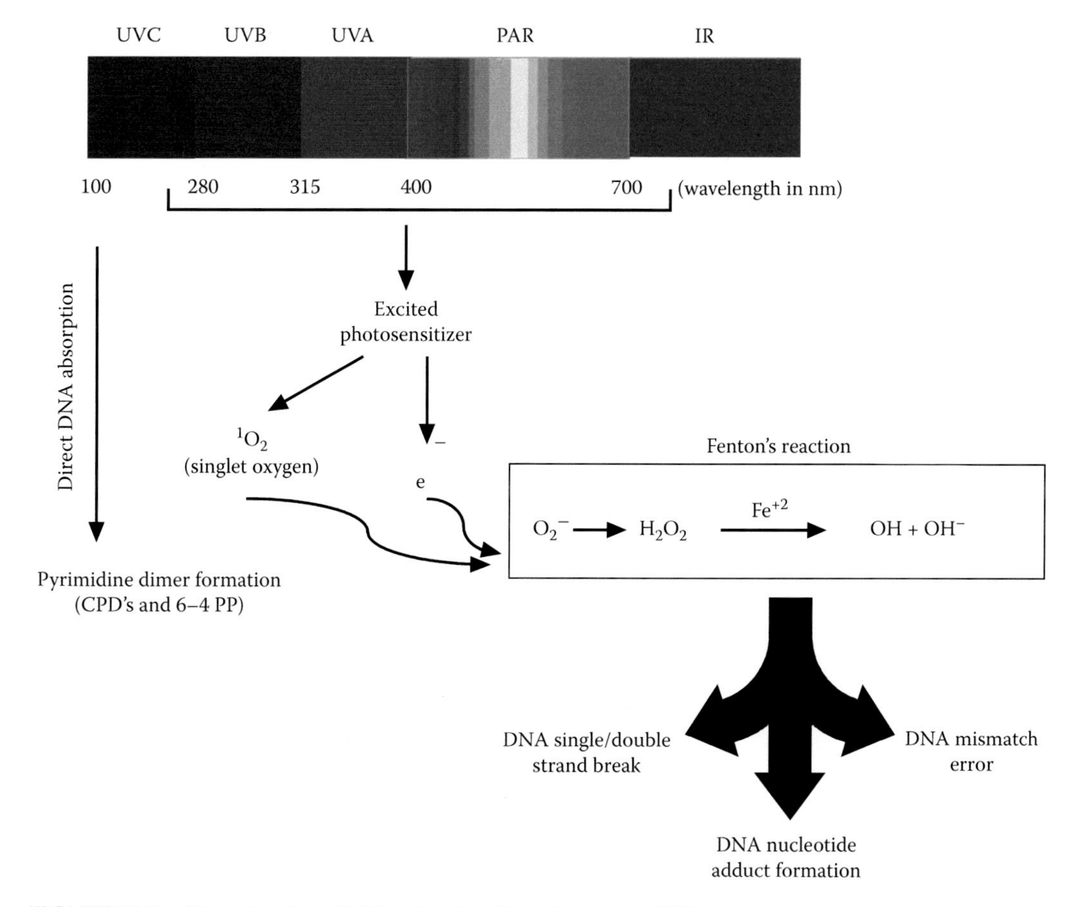

FIGURE 14.1 **(See color insert.)** Direct and indirect damage to DNA.

PCC 7120 [49]. It removes UV-light damaged nucleotides, recognizes pyrimidine dimers, and cleaves the phosphodiester bond immediately 5′ to the lesion [50]. Similarly, an enzyme complex called the UvrABC repair system is present in prokaryotes for recognition and procession of DNA lesions [51]. UvrA and UvrB subunits recognize the DNA damage and form a complex with the damaged site and an excision repair is performed at the 3′ end [52,53]. In a recent study, during which *P. marinus* sp. PCC 9511 cells were exposed to high light (HL)+UV-R (900 μmoles photons m^{-2} s^{-1} + 7.59 W m^{-2} of UV-A + 0.57 W m^{-2} UV-B) radiation, the *UvrA* genes were increased in expression. Similar findings were also observed by Zinser et al. [54] in another study, where uvrA transcript levels rapidly increased in the unicellular marine cyanobacterium *P. marinus* MED4 at moderate light (300 μmoles Q m^{-2} s^{-1}) intensity. During their study with *P. marinus* PCC 9511, Kolowrat et al. [38] also found an activation in the DNA mismatch repair (MMR) system. They observed that MutS was found lowest in abundance in the morning and it continuously increased during the high light period indicating that expression of this gene increases with accumulation of more UV and ROS induced mutations of DNA [38]. Along with the earlier pathway, there is a third pathway found activated called the homologous recombination pathway that acts on double strand break. Generally RuvA and RuvB form a complex and promote branch migration of Holliday junctions, and finally the endonuclease RuvC resolves the Holliday junction by nicking the DNA strand [55,56]. Kolowrat et al. [38] observed that RuvC was expressed during both HL and HL + UV conditions. Similar findings were observed in another study with the filamentous cyanobacterium *Anabaena* sp. PCC 7120 [57].

14.5 PROTEIN DAMAGE AND REPAIR

During various stresses, proteins were damaged mainly by oxidative or structural damage [23]. Although, enzymes did reverse the oxidative damage or protein refolding of some of the proteins but some of the proteins are irreversibly damaged. Thus, proteins that are damaged beyond repair need to be removed by proteolytic degradation and further regenerated by de novo synthesis. There are a number of proteins that perform the function of protein damage and repair in cyanobacteria such as amino peptidase (pepA) [58], ATP-dependent Clp protease adapter protein (clpS) [59], and ftsH protein [60]. The pepA are exopeptidase proteins that help in processing and turnover of intracellular proteins [58]. Recently, it was shown that methionine aminopeptidase is essential for N_2 starvation and high light acclimation in *Synechocystis* sp. PCC 6803 [61], and leucineaminopeptidase was differentially regulated during salt stress in *Synechocystis* sp. PCC 6803 [62]. Next, the Clp proteases are a class of proteins that are essential for protein degradation, dissolution, and processing of protein aggregates [59], but the mechanism of substrate recognition by these proteases is still unclear [63]. The presence of the Clp protease in cyanobacteria was first reported in the unicellular freshwater cyanobacterium *Synechococcus elongatus* PCC 7942, and it was shown that there are 10 distinct Clp proteins found in cyanobacteria, out of which four are HSP100 chaperones (ClpB1–2, ClpC, and ClpX), three ClpP proteins (ClpP1–3), a ClpP-like protein termed ClpR, and two adaptor proteins (ClpS1–2) [64]. Detailed information on the structural arrangement of the different Clp proteases with different HSP100 partners has been given elsewhere. The reader is referred to some recent studies where the heptamer arrangement of the clp protease and HSP has been elucidated [63–66]. Recently, a nitrogen starvation study in *Anabaena* sp. PCC 7120 revealed the strong role of Clp protease in the phycobilisomes degradation process [66]. Similarly, the third protein FtsH was also proposed to have strong photolytic properties. FtsH is a membrane-anchored ATP-dependent protease of 70 kDa in size [67] and has a strong role in the repair of the PS-II as shown in *Synechocystis* sp. PCC 6803 [68,69].

14.6 REDOX SENSORS AND REDOX HOMEOSTASIS

Normally, all the living cells have free radical scavenging systems to minimize and repair oxidative damage that include compounds such as ascorbate, glutathione, thioredoxin, and various antioxidant enzymes [70–73]. Cyanobacteria employ several proteins that have primary roles in cell redox homeostasis such as glutaredoxin [74], thioredoxin [75], peroxiredoxin [71,76], alkyl hydroperoxidereductase/thiol specific antioxidant/mal allergens family protein [77], and DSBA oxidoreductase [78].

In a recent study to uncover the anti-oxidative system in the cyanobacterium *Synechocystis* sp. PCC 6803, the authors showed the strong role of glutaredoxin, thioredoxin, and peroxiredoxin as antioxidative components of the metabolism [71]. Alkyl hydroperoxidereductase/thiol specific antioxidant/Mal allergens family protein is a protein complex found in cyanobacteria and shows a sequence similarity with peroxiredoxin [77]. A recent study proposed that alkyl hydroperoxidereductase (AhpC) plays an important role in combating multitude of stresses such as heat, salt, carbofuron, cadmium, copper, and UV-B [77]. The DSBA oxidoreductase possesses a thioredoxin-fold motif, and was shown to be vital in maintaining the redox balance in *Synechocystis* sp. PCC 6803 [78]. During this study, the authors showed that the loss of DsbA makes the cells hypersensitive to changes in redox potential that is induced by reducing DTT and glucose [78]. With regard to the light stress, previous studies performed by Hihara et al. in *Synechocystis* sp. PCC 6803 [36] showed that scavenging enzymes of ROS were not upregulated under high light conditions. They only observed induction of sodB and slr1992 (glutathione peroxidase). They also observed induction of four ftsH homologs that encodes for AAA-type protease. A similar stress response was observed by Huang et al. [79], where thioredoxin (trxM; slr0233) was observed to be induced by high UV-B light (2 h treatment) but not significantly changed when irradiated with 2 h of high light treatment. Similar findings were reported previously confirming the induction of slr1992 in both UV-B and high light [79].

14.7 HEAT SHOCK PROTEINS

During the cellular stress response, the macromolecular integrity is maintained by heat shock proteins, also known as molecular chaperones [80,81]. In conjunction with the DNA repair system, these chaperones make a formidable defense against damage to the macromolecular integrity during stress [82]. A schematic for putative functions of different chaperones in cyanobacteria is given in Figure 14.2. Chaperone proteins recognize unfolded and aggregated proteins and then try to disintegrate their aggregation, or refold them in their native functional state; or if both these tasks are not possible, they then help in their removal [83–85]. To date, as far as the authors are aware, seven types of chaperones have been described in cyanobacteria in various studies (groES, groEL, clpB, clpC/X, dnaJ, dnaK, and ATPase AAA-2), and these proteins could constitute an important part of the minimal stress proteome [63,83,84,86,87]. These proteins show strong evolutionary conservation in their function during stress. Briefly, GroEs is a 10 kDa protein and generally acts as a co-chaperone of GroEL, a 60 kDa chaperone [88] and functions in binding to partially folded/unfolded protein and helps to gain the native confirmation back. Another protein, DnaJ, is a 40 kDa protein and acts synergistically with DnaK (a 70 kDa protein) and both these proteins have strong roles in the suppression of protein aggregation [89]. Light stress dramatically increases the expression of heat shock genes as recorded by Hihara and coworkers in *Synechocystis* sp. PCC 6803 when cells were grown at a light intensity of 300 mmol photons m^{-2} s^{-1}. They observed increased levels of dnaK2 (sll0170), htpG (sll0430), groEL1 (slr2076), groEL2 (sll0416), and groES (slr2075) transcripts. Gao and coworkers [90] using a semi-quantitative densitometry-based proteomics experiment reported similar findings. They observed increased abundance of heat shock proteins in *Synechocystis* sp. PCC 6803 when the cells were exposed to PAR + UV-B radiation for 72, 84, and 96 h (170 μmoles photons m^{-2} s^{-1} + 1 W m^{-2}). sll0430 (HSP 90), slr2076 (ChaperoninGroEL), and sll0057 (HSP GrpE) were found increased in abundance. The authors of this chapter also observed increased abundance of chaperone proteins such as DnaK (Npun_F0567, Npun_F2986, and Npun_R5998) and ATpase AAA (Npun_F2157 and Npun_R5987) when *N. punctiforme* ATCC 29133 was exposed to HL + UV-A radiation for 20 days (Wase et al., unpublished data).

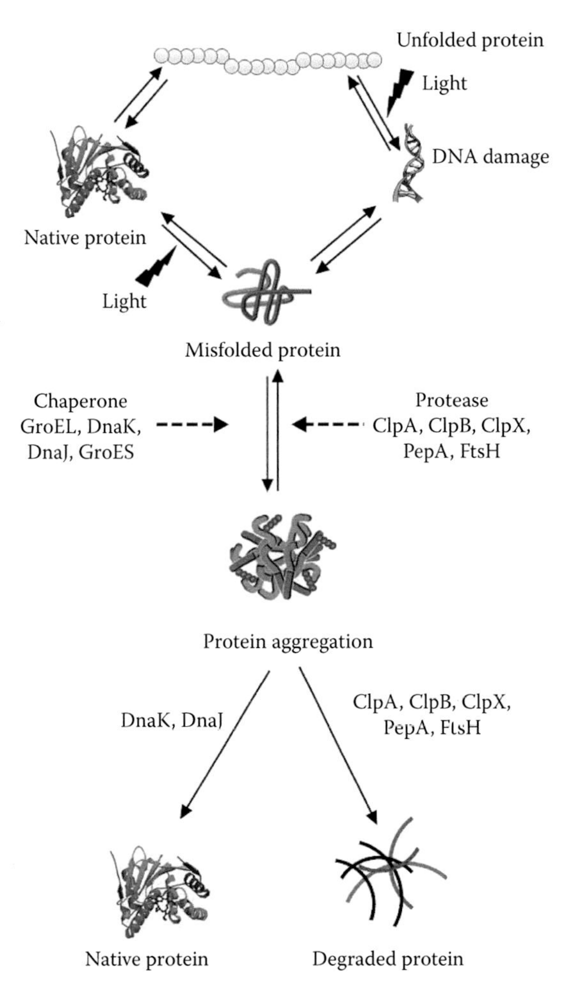

FIGURE 14.2 Schematic illustrating the putative function of chaperones and protease in bacteria. Proteins damaged/misfolded due to different stress or mutation and started to form aggregation. Formation of protein aggregates can be prevented by the proteolytic systems (ClpA, ClpX, ClpB, PepA, FtsH). High levels of misfolded proteins can overburden the capacity of chaperones and proteases and leads to protein aggregation which is rescued by DnaK, DnaJ, and ClpB. Protein aggregates can also be solubilized by the Clp system, FtsH, and peptidase, resulting in substrate hydrolysis. (Modified from Wickner, S. et al., *Science*, 286(5446), 1888, 1999.)

14.8 EFFECT OF HIGH LIGHT STRESS ON PHOTOSYNTHESIS

Modulation of major pathways of energy metabolism is one of the key aspects of cellular stress response [30]. Strong light reduces the expression of genes code for proteins that are involved in the photochemical reactions and light absorption. Huang et al. [79] observed that abundances for genes encoding vital photosynthetic apparatus proteins fell by almost 50% compared to control levels (*psbO*, *psbV*, *psbU*); however, other components of photosystem II (PS-II) such as *psbE* and *psbF* remained stable [36]. The transcript abundance of *psbA* encoding D1 was transiently increased threefold in 15 min, but later decreased. As regards the photosystem I (PS-I) components, almost all the transcripts were down-regulated 5- to 10-fold within 1 h when cultures were transferred from low light to high light. *psaA*, *psaB*, *psaC*, *psaD*, *psaF*, *psaJ*, and *psaL* were repressed continuously in both low light and high light but surprisingly *psaK* was induced within 1 h and remained so throughout the treatment [36]. Along with the downregulation of PS-I and PS-II, there was a coordinate downregulation of the light harvesting components as well. Most of the genes encoding the structural subunits of the light harvesting complex were repressed. Light harvesting complex components such as *apcA*, *apcB*, *apcC*, *apcE*, *apcF*, *cpcB*, *cpcC1* and *cpcC2*, *cpcD*, and *cpcG* were repressed in the transcript and in a 2DGE (2-dimensional gel electrophoresis) proteomics study as well.

14.9 EFFECT OF LIGHT STRESS ON CARBON CONCENTRATION MECHANISM

During the Phanerozoic period, CO_2 levels were probably 100-fold higher than present day conditions [91]. Because of this, cyanobacteria would not have needed a carbon concentration mechanism (CCM) to achieve effective photosynthesis [92]. Subsequently, there was a large decline in CO_2 levels and concomitant increase in O_2 (to near present day levels) and this may have prompted cyanobacteria to devise an alternative strategy for CCM so as to achieve effective carboxylation by the relatively inefficient Rubisco [93–95]. The basic concept of CCM is based on study on the model organisms *Synechococcus* sp. PCC 7942, *Synechococcus* sp. PCC 7002, and *Synechocystis* sp. PCC 6803 [96–99]. The main functional unit of the cyanobacterial CCM is the carboxysome, a proteinous microcompartment shell within the cell that contains Rubisco along with carbonic anhydrase (CA). The CA converts the accumulated cytosolic pool of HCO_3^- into CO_2 within the microcompartment [95]. Hihara et al. proposed a link between high light stress and low inorganic carbon to have an effect on the CCM as observed in *Synechocystis* sp. PCC 6803 [36]. They observed that ccmK, ccmM, and ccmN were strongly induced within 15 min when cells were transferred from low light (20 μmoles photons m^{-2} s^{-1}) to high light (300 μmoles photons m^{-2} s^{-1}). Similar findings were reported by Huang and coworkers along with strong induction of cmpA, cmpB, and cmpC when cells were transferred to high light (200 μE $m^{-2}s^{-1}$) [79]. In contrast, Mcginn and coworkers reported a discrepancy to this observation, since they reported that CCM related genes showed no significant stimulation when grown under high light (300μmoles photons m^{-2} s^{-1} supplemented with 3% (v/v) CO_2). Using RT-PCR, Woodger et al. [98] studied the transcriptional regulation of 30 CCM related genes in *Synechococcus* sp. PCC 7942 and observed that within 30–60 min, there was a rapid transient induction of carboxysome-associated proteins ccmK, ccmL, ccmM, ccmN, ccmO, RbcLS, and IcfA. The authors also observed a slow sustained induction of *psbA2* and rapid sustained induction of transporter genes *cmpA*, *cmpB*, *cmpC*, *cmpD*, *sbtA* and *ndhF3*, *ndhD3* and *chpY*. We feel that a low CO_2 supply and an additive effect of high light may be required for the CCM to activate but this assumption needs to be proven experimentally.

14.10 DOES HIGH LIGHT STRESS INDUCE APOPTOSIS?

In higher organisms, programmed cell death (PCD) or apoptosis is a common stress response when the stress level exceeds the cell's tolerance limits for maintenance of cellular integrity [100]. In higher plants and animals, this process is employed to avoid tumorigenesis and genetic instability of the organisms [101]. Some recent studies have shown that cyanobacteria may also undergo apoptosis or PCD [100,102]. This phenomenon was recently uncovered in a nitrogen fixing cyanobacterium *Trichodesmium* sp. in response to phosphorus and iron starvation in combination with high irradiance and oxidative stress [100,102]. Similar PCD events were reported during oxidative stress in *Microcystis aeruginosa* [103] and during salt stress in *Anabaena* [104] using a caspase protease assay. Jiang et al. [105] identified 58 putative metacaspases in 33 sequenced genomes of cyanobacteria. They observed that metacaspases are abundant in filamentous diazotrophic cyanobacteria, but are absent in all *Prochlorococcus* and marine *Synechococcus* except *Synechococcus* sp. PCC 7002 [105]. A list of metacaspase proteins in sequenced cyanobacteria is compiled and tabulated in Table 14.3.

Whether high light stress has any impact on apoptosis in cyanobacteria is an open question, but it is apparent that high light stress induces ROS formation that leads to oxidative stress and that might increase metacaspase abundance. However, more studies are needed to investigate the beneficial role of this mechanism under high light stress or high light stress combined with UV stress or photomediated oxidative stress.

TABLE 14.3
List of Putative Caspase Protein Found in All Sequenced Cyanobacteria

Organism	ORF Name
Acaryochloris marina (strain MBIC 11017)	AM1_G0162, AM1_4565
Anabaena variabilis (strain ATCC 29413/ PCC 7937)	Ava_0039, Ava_3246, Ava_4110, Ava_2183, Ava_3611, Ava_2184, Ava_3078, Ava_2381, Ava_3079
Arthrospira maxima CS-328	AmaxDRAFT_3444, AmaxDRAFT_0111, AmaxDRAFT_3443, AmaxDRAFT_0768, AmaxDRAFT_5231
Crocosphaera watsonii WH 8501	CwatDRAFT_1747
Cyanothece sp. (strain ATCC 51142)	cce_3645, cce_4657
Cyanothece sp. (strain PCC 7424) or *Synechococcus* sp. (strain ATCC 29155)	PCC7424_0759, PCC7424_4034, PCC7424_3482, PCC8801_4451, PCC8801_0557, PCC8801_2382
Cyanothece sp. (strain PCC 8802) or *Synechococcus* sp. (strain RF-2)	Cyan8802_0574, Cyan8802_4542, Cyan8802_2433
Cyanothece sp. CCY0110	CY0110_30191, CY0110_25216, CY0110_26068, CY0110_15465, CY0110_03184, CY0110_22127, CY0110_26807, CY0110_25566, CY0110_29229
Cyanothece sp. PCC 7822	Cyan7822DRAFT_4942, Cyan7822DRAFT_2079, Cyan7822DRAFT_3265
Cylindrospermopsis raciborskii CS-505	CRC_00461, CRC_00171
Lyngbya sp. PCC 8106	L8106_01040, L8106_16024, L8106_22561, L8106_11562, L8106_17212, L8106_16039, L8106_16014, L8106_14340
Microcoleus chthonoplastes PCC 7420	MC7420_7640, MC7420_7737, MC7420_2584, MC7420_5988, MC7420_4147, MC7420_5532, MC7420_1185, MC7420_945, MC7420_7152, MC7420_8059, MC7420_4148, MC7420_3523, MC7420_4082
M. aeruginosa (strain NIES-843)	MAE_17380, MAE_60910, MAE_24870
M. aeruginosa PCC 7806	IPF_5133
Nodularia spumigena CCY9414	N9414_06799, N9414_07866, N9414_23773, N9414_23778
Nostoc azollae' 0708	Aazo_1578, Aazo_2362
N. punctiforme (strain ATCC 29133/PCC 73102)	Npun_AR071, Npun_AF095, Npun_R3338, Npun_F0844, Npun_F5449
Raphidiopsis brookii D9	CRD_02671, CRD_00539
Synechococcus sp. PCC 7335	S7335_877, S7335_1133, S7335_5293, S7335_883, S7335_5146, S7335_4564, S7335_419
Trichodesmium erythraeum (strain IMS101)	Tery_2624, Tery_2012, Tery_1841, Tery_2471, Tery_2058, Tery_2760

14.11 COMPARATIVE ANALYSIS OF VARIOUS GLOBAL HIGH LIGHT AND UV STRESS STUDIES

In this section, a comparative assessment of large-scale proteomics and transcriptomics data obtained through different light stress studies in cyanobacteria has been made. We compare and contrast these so as to identify trends within the previously published dataset. This, in our opinion should allow us to pinpoint proteins/genes that are reliably expressed upon a different range of treatment.

14.11.1 RESPONSE TO HIGH LIGHT STRESS AT THE "OMICS" LEVEL IN THE MODEL CYANOBACTERIUM *SYNECHOCYSTIS* SP. PCC 6803

For performing this comparison, three studies were selected that consisted of two microarray studies by Hihara et al. [36] where gene expression during acclimation to high light was studied. The second microarray study selected was by Huang et al. [79] where the effect of high light and UV-B was studied on the transcriptome. The third study selected was a 2DGE study by Goa et al. [90] where *Synechocystis* sp. PCC 6803 cells were subjected to short- and long-term high light exposure supplemented with UV-B radiation. These studies vary in light intensity used and analytical method, and are performed by different researchers in different labs. This could likely lead to a high degree of variation (both biological and technical) in the reported changes in gene/protein expression. Nevertheless, we focus here only on the photosynthetic machinery, heat shock response, lipid metabolism, redox homeostasis CCM and DNA repair. A comparative summary of these three studies is given in Table 14.4.

14.11.2 Effect of High Light and High Light Supplemented with UV-B on Photosynthesis

The photosystem II Dl protein is highly susceptible to damage by the photogenerated radicals. *psbA2* and *psbA3* coding for the Dl polypeptide were induced and this was consistently found in both microarray studies, but this protein was not found induced in the 2DGE study [90]. The probable reason being would be its membrane bound nature and since membrane proteins are weakly (problematically) resolved by 2DGE, this protein was not found in the proteomics study and neither were other corresponding photosynthetic transcripts that were observed in the two microarray studies. The transcripts of genes involved in the oxygen-evolving complex *psbO* were also downregulated in both the transcriptome studies but again were not observed in the proteomics study. This is not so unusual, as the coverage of the proteome is often only a few tens of percent due to technical limitations.

In contrast to the PS-II, both the transcriptome studies revealed that almost all the components of PS-I were downregulated but no such observation was recorded in the 2DGE study. The light harvesting components were also found to be downregulated, including all the major allophycocyanin and phycocyanin components, but the proteomics study did not resolve any of the allophycoyanin components. Both the transcriptome studies showed that *apcA*, *apcB*, *apcC*, *apcE*, *apcF* were downregulated along with the phycocyanin *cpcB*, *cpcC1*, *cpcC2*, *cpcD*, and *cpcG* showing decreased expression as well. This suggests a probable slowdown of the phycobilisome synthesis and simultaneous degradation, thereby reducing the extent of damage through the saturation of light and ROS.

14.11.3 Heat Shock Response

Unfolded cytoplasmic proteins induce heat shock response [106]. The two transcriptomics studies and the 2DGE study showed that both high light and high light supplemented with UV-B elicited a heat shock response. In the two transcriptomics studies, heat shock regulon members such as *dnaK*, *dnaJ*, *htpG*, *hsp17*, *clpB*, *groES*, *groEL*, *groEL2*, and *grpE* were induced. Similar findings were reported by the 2DGE study as well. A strong heat shock response might be an indication of severe damage due to light induced stress.

14.11.4 Redox Homeostasis

Since high light and UV radiation induce the production of ROS, genes/proteins related scavenging enzymes for ROS were expected to be upregulated. However, Hihara et al. [36] observed induction of only two transcripts *sodB* and slr1992 (glutathione peroxidase). However, other ROS scavenging genes such as thioredoxin were not induced. The second study performed by Huang and coworkers [79] observed that the combined high light and UV radiation induce production of ROS at higher rate as compared to the HL alone; henceforth, more scavenging enzymes were found to be up-regulated in this study. They observed induction of glutaredoxin 3, thioredoxin M, slr1992, and slr1171. In fact slr1992 was found strongly induced in all the conditions (low UV-20 min; low UV-2 h; high UV-2 h and HL-2 h). The proteomics study by Gao et al. [90] did not decipher changes in any protein related to ROS scavenging.

TABLE 14.4
Comparative Overview of Three "Omics" Studies in Cyanobacteria

			Huang et al. [79] (UV-B + HL)		Gao et al. [90] (HL + UV-B)	
Function/Class	**Protein/Gene**	**Hihara (2001)**	**HL**	**UV-B + HL**	**8 h**	**3–4 days**
PS-II	PsbA2	↑	↑	↑	-	-
	PsbA3	↑	↑	↑	-	-
	psbV	↓	↓	-	-	-
	psbO	↓	↓	↓	-	-
	psbW	-	-	↑	-	-
PS-I	psaA	↓	↓	↓	-	-
	psaB	↓	↓	↓	-	-
	psaC	-	↓	↓	-	-
	psaD	↓	↓	↓	-	-
	psaE	-	↓	↓	-	-
	psaF	↓	↓	↓	-	-
	psaJ	↓	↓	↓	-	-
	psaK	-	↓	↓	-	-
	psaL	↓	↓	↓	-	-
Light harvesting complexes	apcA	↓	↓	↓	-	-
	apcB	↓	↓	↓	-	-
	apcC	↓	↓	↓	-	-
	apcE	↓	↓	↓	-	-
	apcF	↓	↓	↓	-	-
	cpcD	↓	↓	↓	-	-
	cpcG	↓	↓	↓	-	↓
	cpcB	-	-	-	↓	↓
	cpcC2	-	-	-	↓	↓
	cpcC1	-	-	-	-	↓
Heat shock proteins	dnaK	↑	-	↑	-	-
	dnaJ	-	-	↑	-	-
	htpG	↑	↑	-	-	↑
	hsp17	↑	↑	↑	-	-
	clpB	↑	-	↑	-	-
	groES	↑	↑	↑	-	-
	groEL	↑	↑		-	↑
	groEL2	-	↑	↑	-	-
	grpE	-	-	-	-	↑
Lipid metabolism	desA	↑	-	↓	-	-
	desB	↑	-	-	-	-
	CTP synthetase	-	↑	-	-	↓

High light and UV-R inducible genes in cyanobacterium *Synechocystis* sp. PCC 6803. Up arrow indicates that the corresponding gene/protein was found higher in abundance, down arrow indicates expression of gene/protein lower in abundance, and hyphen indicates that these genes/proteins were not detected.

14.11.5 Other Common Factors Observed in High Light and UV Stress in Cyanobacteria

The CCM is an extremely effective way to concentrate CO_2. The carboxysome, a protein microcompartment containing the Rubisco of the cell, is the central entity and is the main functional unit for CCM. The carboxysome also contains carbonic anhydrase (CA) that helps to convert the accumulated pool of $HCO3^-$ to CO_2. This leads to the localized increase of CO_2 to the tune of 1000-fold around the active site of Rubisco [93,95]. It has been proposed that the protein shell restricts the diffusion of CO_2 out of the carboxysome. The three large-scale studies compared here showed that during high light stress, CCM proteins such as ccmL, ccmM, ccmN, and ccmK were induced, but when the high light is supplemented with UV-B light surprisingly they were repressed, indicating that the CCM mechanism may have been shut down (see Table 14.4). The exact nature of UV-B effect on the CCM is not known to date and could be a potential area of work to pursue in future.

In conclusion, comparison of these three studies (high light and high light supplemented with UV-B radiation [both one proteomics and two transcriptomics studies] in the model cyanobacterium *Synechocystis* sp. PCC 6803) showed that there is differential regulation of key parts of the photosynthetic apparatus and other ancillary photosynthetic proteins along with genes/proteins involved in DNA damage and repair, cellular stress response, cellular homeostasis, and possible biomarkers for light sensing mechanism. To understand the further role of high light stress in CCM and apoptosis, further investigations are required.

14.12 CONCLUSIONS

Our knowledge about the stress response in bacteria in general has increased exponentially, but is still fragmentary as far as the response of cyanobacteria to light stress including UV-A/B & C is concerned. We feel that stress response involves an intricate interplay of several metabolic pathways and we feel that it is more complex than mere induction of heat shock response. Since many metabolic pathways contribute to the cell's response to stress. We feel that identification of the key players that are situated at strategic nodes of the metabolic network in cyanobacteria should provide better insight into the metabolic rearrangement of the cell. A feasible way of identifying such common regulators during the stress is the use of comparative proteomics. Technological advances during the last 10 years has enabled further studies in this field and using such approaches should increase our understanding of a highly functional conserved set of proteins and this should provide us with a powerful tool for manipulating stress tolerance in cyanobacterial cells.

Further questions arise when analyzing the cellular stress response in the context of an organism's temporary/permanent damage due to environmental change. Does the expression of a highly functional stress response make the organism more stress tolerant at the cost of decreased fitness in stable environments? What are the factors that favor survival during stressful period? What are the critical elements conserved across different stressors? Can we harness these conserved characters for commercial purposes? How do cells encountering stress decide between induction of PCD over repair. An insight into the mechanism of how cells assess the macromolecular damage may contribute significantly to this important decision for the cell whether or not to undergo PCD. These are some of the important unanswered questions that should drive future research on cyanobacteria physiology and metabolism under normal and stressed environments.

ACKNOWLEDGMENT

We thank the EPSRC for funding (EP/E036252/1) and the Indian Government for a PhD scholarship to NW.

REFERENCES

1. Singh, S.P. et al., Cyanobacteria and ultraviolet radiation (UVR) stress: Mitigation strategies. *Ageing Res Rev*, 2010. **9**(2): 79–90.
2. Murata, N. et al., Photoinhibition of photosystem II under environmental stress. *Biochim Biophys Acta*, 2007. **1767**(6): 414–421.
3. Allakhverdiev, S.I. et al., Salt stress inhibits the repair of photodamaged photosystem II by suppressing the transcription and translation of psbA genes in synechocystis. *Plant Physiol*, 2002. **130**(3): 1443–1453.
4. Gupta, S. and S.C. Agrawal, Survival of blue-green and green algae under stress conditions. *Folia Microbiol (Praha)*, 2006. **51**(2): 121–128.
5. Castielli, O. et al., Proteomic analyses of the response of cyanobacteria to different stress conditions. *FEBS Lett*, 2009. **583**(11): 1753–1758.
6. Rastogi, R.P. et al., Detection of reactive oxygen species (ROS) by the oxidant-sensing probe 2′,7′-dichlorodihydrofluorescein diacetate in the cyanobacterium *Anabaena variabilis* PCC 7937. *Biochem Biophys Res Commun*, 2010. **397**(3): 603–607.
7. Mizuno, T. et al., Compilation of all genes encoding bacterial two-component signal transducers in the genome of the cyanobacterium, *Synechocystis* sp. strain PCC 6803. *DNA Res*, 1996. **3**(6): 407–414.
8. Meeks, J.C. et al., An overview of the genome of *Nostoc punctiforme*, a multicellular, symbiotic cyanobacterium. *Photosynth Res*, 2001. **70**(1): 85–106.
9. Montgomery, B.L., Sensing the light: Photoreceptive systems and signal transduction in cyanobacteria. *Mol Microbiol*, 2007. **64**(1): 16–27.
10. van der Horst, M.A. et al., Photosensing in chemotrophic, non-phototrophic bacteria: Let there be light sensing too. *Trends Microbiol*, 2007. **15**(12): 554–562.
11. Rockwell, N.C. and J.C. Lagarias, A brief history of phytochromes. *ChemPhysChem*, 2010. **11**(6): 1172–1180.
12. Scheerer, P. et al., Light-induced conformational changes of the chromophore and the protein in phytochromes: Bacterial phytochromes as model systems. *ChemPhysChem*, 2010. **11**(6): 1090–1105.
13. Hughes, J. et al., A prokaryotic phytochrome. *Nature*, 1997. **386**(6626): 663.
14. Bordowitz, J.R. and B.L. Montgomery, Photoregulation of cellular morphology during complementary chromatic adaptation requires sensor-kinase-class protein RcaE in Fremyella diplosiphon. *J Bacteriol*, 2008. **190**(11): 4069–4074.
15. Montgomery, B.L. and J.C. Lagarias, Phytochrome ancestry: Sensors of bilins and light. *Trends Plant Sci*, 2002. **7**(8): 357–366.
16. Rockwell, N.C. et al., Diverse two-cysteine photocycles in phytochromes and cyanobacteriochromes. *Proc Natl Acad Sci U S A*, 2011. **108**(29): 11854–11859.
17. Ikeuchi, M. and T. Ishizuka, Cyanobacteriochromes: A new superfamily of tetrapyrrole-binding photoreceptors in cyanobacteria. *Photochem Photobiol Sci*, 2008. **7**(10): 1159–1167.
18. Yeh, K.C. et al., A cyanobacterial phytochrome two-component light sensory system. *Science*, 1997. **277**(5331): 1505–1508.
19. Wilde, A. et al., The cyanobacterial phytochrome Cph2 inhibits phototaxis towards blue light. *Mol Microbiol*, 2002. **44**(4): 981–988.
20. Wu, S.H. and J.C. Lagarias, Defining the bilin lyase domain: Lessons from the extended phytochrome superfamily. *Biochemistry*, 2000. **39**(44): 13487–13495.
21. Zhao, K.H. et al., Reconstitution of phycobilisome core-membrane linker, LCM, by autocatalytic chromophore binding to ApcE. *Biochim Biophys Acta*, 2005. **1706**(1–2): 81–87.
22. Sharda, S. et al., A non-hydrolyzable ATP derivative generates a stable complex in a light-inducible two-component system. *J Biol Chem*, 2009. **284**(49): 33999–34004.
23. Okamoto, S. et al., A phytochrome-like protein AphC triggers the cAMP signaling induced by far-red light in the cyanobacterium *Anabaena* sp. strain PCC7120. *Photochem Photobiol*, 2004. **80**(3): 429–433.
24. Straub, C. et al., A day in the life of *Microcystis aeruginosa* strain PCC 7806 as revealed by a transcriptomic analysis. *PLoS One*, 2011. **6**(1): e16208.
25. Zhang, C.C. et al., Protein phosphorylation on Ser, Thr and Tyr residues in cyanobacteria. *J Mol Microbiol Biotechnol*, 2005. **9**(3–4): 154–166.
26. Stowe-Evans, E.L. et al., Genomic DNA microarray analysis: Identification of new genes regulated by light color in the cyanobacterium *Fremyella diplosiphon. J Bacteriol*, 2004. **186**(13): 4338–4349.
27. Sakamoto, T. and N. Murata, Regulation of the desaturation of fatty acids and its role in tolerance to cold and salt stress. *Curr Opin Microbiol*, 2002. **5**(2): 206–210.

28. Wada, H. et al., The desA gene of the cyanobacterium *Synechocystis* sp. strain PCC6803 is the structural gene for delta 12 desaturase. *J Bacteriol*, 1993. **175**(18): 6056.
29. Kis, M. et al., Light-induced expression of fatty acid desaturase genes. *Proc Natl Acad Sci U S A*, 1998. **95**(8): 4209.
30. Kultz, D., Molecular and evolutionary basis of the cellular stress response. *Annu Rev Physiol*, 2004. **67**(1): 225–257.
31. Mizusawa, N. et al., Lack of digalactosyldiacylglycerol increases the sensitivity of *Synechocystis* sp. PCC 6803 to high light stress. *FEBS Lett*, 2009. **583**(4): 718–722.
32. Nixon, P.J. et al., Recent advances in understanding the assembly and repair of photosystem II. *Ann Bot*, 2010. **106**(1): 1–16.
33. Allakhverdiev, S.I. et al., Regulatory roles in photosynthesis of unsaturated fatty acids in membrane lipids, in *Lipids in Photosynthesis*, H. Wada and N. Murata, Eds., 2010, Springer, Dordrecht, the Netherlands, pp. 373–388.
34. Gombos, Z. et al., Genetic enhancement of the ability to tolerate photoinhibition by introduction of unsaturated bonds into membrane glycerolipids. *Plant Physiol*, 1997. **115**(2): 551–559.
35. Sinha, R.P. and D.P. Häder, UV-induced DNA damage and repair: A review. *Photochem Photobiol Sci*, 2002. **1**(4): 225–236.
36. Hihara, Y. et al., DNA microarray analysis of cyanobacterial gene expression during acclimation to high light. *Plant Cell*, 2001. **13**(4): 793–806.
37. Pandhal, J. et al., A quantitative proteomic analysis of light adaptation in a globally significant marine cyanobacterium *Prochlorococcus marinus* MED4. *J Proteome Res*, 2007. **6**(3): 996–1005.
38. Kolowrat, C. et al., Ultraviolet stress delays chromosome replication in light/dark synchronized cells of the marine cyanobacterium *Prochlorococcus marinus* PCC9511. *BMC Microbiol*, 2010. **10**: 204.
39. He, Y.Y. and D. Hader, Reactive oxygen species and UV-B: Effect on cyanobacteria. *Photochem Photobiol Sci*, 2002. **1**(10): 729–736.
40. He, Y.-Y. et al., Adaptation of cyanobacteria to UV-B stress correlated with oxidative stress and oxidative damage. *Photochem Photobiol*, 2002. **76**(2): 188–196.
41. Cadoret, J.C. et al., Cyclic nucleotides, the photosynthetic apparatus and response to a UV-B stress in the cyanobacterium *Synechocystis* sp. PCC 6803. *J Biol Chem*, 2005. **280**(40): 33935–33944.
42. Kumar, A. et al., Evidences showing ultraviolet-B radiation-induced damage of DNA in cyanobacteria and its detection by PCR assay. *Biochem Biophys Res Commun*, 2004. **318**(4): 1025–1030.
43. Lao, K. and A.N. Glazer, Ultraviolet-B photodestruction of a light-harvesting complex. *Proc Natl Acad Sci U S A*, 1996. **93**(11): 5258–5263.
44. Rai, P. et al., Localization of Fe_{2+} at an RTGR sequence within a DNA duplex explains preferential cleavage by Fe2+ and H_2O_2. *J Mol Biol*, 2001. **312**: 1089–1101.
45. Goosen, N. and G.F. Moolenaar, Repair of UV damage in bacteria. *DNA Repair*, 2008. **7**(3): 353–379.
46. Sixma, T.K., DNA mismatch repair: MutS structures bound to mismatches. *Curr Opin Struct Biol*, 2001. **11**(1): 47–52.
47. Fukui, K. et al., *Thermus thermophilus* MutS2, a MutS paralogue, possesses an endonuclease activity promoted by MutL. *J Biochem*, 2004. **135**(3): 375.
48. Khil, P.P. and R.D. Camerini-Otero, Over 1000 genes are involved in the DNA damage response of *Escherichia coli*. *Mol Microbiol*, 2002. **44**(1): 89–105.
49. Muro-Pastor, A.M. et al., Identification, genetic analysis and characterization of a sugar-non-specific nuclease from the cyanobacterium *Anabaena* sp. PCC 7120. *Mol Microbiol*, 1992. **6**(20): 3021–3030.
50. Kaur, B. and P.W. Doetsch, Ultraviolet damage endonuclease (Uve1p): A structure and strand-specific DNA endonuclease. *Biochemistry*, 2000. **39**(19): 5788–5796.
51. Truglio, J.J. et al., Prokaryotic nucleotide excision repair: The UvrABC system. *Chem Rev*, 2006. **106**(2): 233–252.
52. Truglio, J.J. et al., Interactions between UvrA and UvrB: The role of UvrB's domain 2 in nucleotide excision repair. *Embo J*, 2004. **23**(13): 2498–2509.
53. Machius, M. et al., Crystal structure of the DNA nucleotide excision repair enzyme UvrB from *Thermus thermophilus*. *Proc Natl Acad Sci U S A*, 1999. **96**(21): 11717–11722.
54. Zinser, E.R. et al., Choreography of the transcriptome, photophysiology, and cell cycle of a minimal photoautotroph, *Prochlorococcus*. *PLoS One*, 2009. **4**(4): e5135.
55. van Gool, A.J. et al., Assembly of the *Escherichia coli* RuvABC resolvasome directs the orientation of Holliday junction resolution. *Genes Dev*, 1999. **13**(14): 1861–1870.
56. West, S.C., Processing of recombination intermediates by the RuvABC proteins. *Annu Rev Genet*, 1997. **31**: 213–244.

57. Mazon, G. et al., LexA-binding sequences in Gram-positive and cyanobacteria are closely related. *Mol Genet Genomics*, 2004. **271**(1): 40–49.
58. Bartling, D. and E.W. Weiler, Leucine aminopeptidase from *Arabidopsis thaliana*. Molecular evidence for a phylogenetically conserved enzyme of protein turnover in higher plants. *Eur J Biochem*, 1992. **205**(1): 425–431.
59. Gottesman, S. et al., The ATP-dependent Clp protease of *Escherichia coli*. Sequence of clpA and identification of a Clp-specific substrate. *J Biol Chem*, 1990. **265**(14): 7886–7893.
60. Cheregi, O. et al., The role of the FtsH and Deg proteases in the repair of UV-B radiation-damaged photosystem II in the cyanobacterium *Synechocystis* PCC 6803. *Biochim Biophys Acta, Bioenerg*, 2007. **1767**(6): 820–828.
61. Drath, M. et al., An alternative methionine aminopeptidase, MAP-A, is required for nitrogen starvation and high-light acclimation in the cyanobacterium *Synechocystis* sp. PCC 6803. *Microbiology*, 2009. **155**(5): 1427.
62. Pandhal, J. et al., Comparative proteomics study of salt tolerance between a nonsequenced extremely halotolerant cyanobacterium and its mildly halotolerant relative using in vivo metabolic labeling and in vitro isobaric labeling. *J Proteome Res*, 2008. **8**(2): 818–828.
63. Dougan, D.A. et al., ClpS, a substrate modulator of the ClpAP machine. *Mol Cell*, 2002. **9**(3): 673–683.
64. Adrian, K.C. et al., The ATP-dependent Clp protease in chloroplasts of higher plants. *Physiol Plant*, 2005. **123**(4): 406–412.
65. Stanne, T.M. et al., Distinctive types of ATP-dependent Clp proteases in cyanobacteria. *J Biol Chem*, 2007. **282**(19): 14394–14402.
66. Karradt, A. et al., NblA, a key protein of phycobilisome degradation, interacts with ClpC, a HSP100 chaperone partner of a cyanobacterial Clp protease. *J Biol Chem*, 2008. **283**(47): 32394–32403.
67. Krzywda, S. et al., Crystallization of the AAA domain of the ATP-dependent protease FtsH of *Escherichia coli*. *Acta Crystallogr D Biol Crystallogr*, 2002. **58**(Pt 6 Pt 2): 1066–1067.
68. Nixon, P.J. et al., FtsH-mediated repair of the photosystem II complex in response to light stress. *J Exp Bot*, 2005. **56**(411): 357.
69. Silva, P. et al., FtsH is involved in the early stages of repair of photosystem II in *Synechocystis* sp PCC 6803. *Plant Cell*, 2003. **15**(9): 2152.
70. He, Y.Y. and D.P. Häder, UV-B-induced formation of reactive oxygen species and oxidative damage of the cyanobacterium *Anabaena* sp.: Protective effects of ascorbic acid and N-acetyl-cysteine. *J Photochem Photobiol B*, 2002. **66**(2): 115–124.
71. Hosoya-Matsuda, N. et al., Anti-oxidative stress system in cyanobacteria: Significance of type II peroxiredoxin and the role of 1-Cys peroxiredoxin in *Synechocystis* sp. strain PCC 6803. *J Biol Chem*, 2005. **280**(1): 840.
72. Taylor, B.L. and I.B. Zhulin, PAS domains: Internal sensors of oxygen, redox potential, and light. *Microbiol Mol Biol Rev*, 1999. **63**(2): 479–506.
73. Schottler, M.A. et al., The role of plastocyanin in the adjustment of the photosynthetic electron transport to the carbon metabolism in tobacco. *Plant Physiol*, 2004. **136**(4): 4265–4274.
74. Li, M. et al., Expression and oxidative stress tolerance studies of glutaredoxin from cyanobacterium *Synechocystis* sp. PCC 6803 in *Escherichia coli*. *Protein Expr Purif*, 2005. **42**(1): 85–91.
75. Zeller, T. and G. Klug, Thioredoxins in bacteria: Functions in oxidative stress response and regulation of thioredoxin genes. *Naturwissenschaften*, 2006. **93**(6): 259–266.
76. Latifi, A. et al., PrxQ-A, a member of the peroxiredoxin Q family, plays a major role in defense against oxidative stress in the cyanobacterium *Anabaena* sp. strain PCC7120. *Free Radic Biol Med*, 2007. **42**(3): 424–431.
77. Mishra, Y. et al., AhpC (alkyl hydroperoxide reductase) from *Anabaena* sp. PCC 7120 protects *Escherichia coli* from multiple abiotic stresses. *Biochem Biophys Res Commun*, 2009. **381**(4): 606–611.
78. Singh, A.K. et al., Identification of an atypical membrane protein involved in the formation of protein disulfide bonds in oxygenic photosynthetic organisms. *J Biol Chem*, 2008. **283**(23): 15762–15770.
79. Huang, L. et al., Global gene expression profiles of the cyanobacterium *Synechocystis* sp. strain PCC 6803 in response to irradiation with UV-B and white light. *J Bacteriol*, 2002. **184**(24): 6845–6858.
80. Lindquist, S., The heat-shock response. *Annu Rev Biochem*, 1986. **55**(1): 1151–1191.
81. Feder, M.E. and G.E. Hofmann, Heat-shock proteins, molecular chaperones, and the stress response: Evolutionary and ecological physiology. *Annu Rev Physiol*, 1999. **61**(1): 243–282.
82. Glatz, A. et al., The *Synechocystis* model of stress: From molecular chaperones to membranes. *Plant Physiol Biochem*, 1999. **37**(1): 1–12.
83. Bukau, B. and A.L. Horwich, The Hsp70 and Hsp60 chaperone machines. *Cell*, 1998. **92**(3): 351–366.

84. Langer, T. et al., Successive action of DnaK, DnaJ and GroEL along the pathway of chaperone-mediated protein folding. *Nature*, 1992. **356**: 683–689.
85. Gething, M.J. and J. Sambrook, Protein folding in the cell. *Nature*, 1992. **355**(6355): 33–45.
86. Neuwald, A.F. et al., AAA+: A class of chaperone-like ATPases associated with the assembly, operation, and disassembly of protein complexes. *Genome Res*, 1999. **9**(1): 27–43.
87. Wang, W. et al., Role of plant heat-shock proteins and molecular chaperones in the abiotic stress response. *Trends Plant Sci*, 2004. **9**(5): 244–252.
88. Horváth, I. et al., Membrane-associated stress proteins: More than simply chaperones. *Biochim Biophys Acta,– Biomembr*, 2008. **1778**(7–8): 1653–1664.
89. Trautinger, F. et al., Stress proteins in the cellular response to ultraviolet radiation. *J Photochem Photobiol, B*, 1996. **35**(3): 141–148.
90. Gao, Y. et al., Identification of the proteomic changes in *Synechocystis* sp. PCC 6803 following prolonged UV-B irradiation. *J Exp Bot*, 2009. **60**(4): 1141–1154.
91. Berner, R.A., Atmospheric oxygen over Phanerozoic time. *Proc Natl Acad Sci U S A*, 1999. **96**(20): 10955–10957.
92. Miyachi, S. et al., Historical perspective on microalgal and cyanobacterial acclimation to low- and extremely high-CO(2) conditions. *Photosynth Res*, 2003. **77**(2–3): 139–153.
93. Tchernov, D. et al., Massive light-dependent cycling of inorganic carbon between oxygenic photosynthetic microorganisms and their surroundings. *Photosynth Res*, 2003. **77**(2–3): 95–103.
94. McGinn, P.J. et al., Inorganic carbon limitation and light control the expression of transcripts related to the CO_2-concentrating mechanism in the cyanobacterium *Synechocystis* sp. strain PCC6803. *Plant Physiol*, 2003. **132**(1): 218–229.
95. Kaplan, A. and L. Reinhold, CO_2 concentrating mechanisms in photosynthetic microorganisms. *Annu Rev Plant Physiol Plant Mol Biol*, 1999. **50**: 539–570.
96. Benschop, J.J. et al., Characterisation of CO(2) and HCO(3) (-) uptake in the cyanobacterium *Synechocystis* sp. PCC6803. *Photosynth Res*, 2003. **77**(2–3): 117–126.
97. Ogawa, T. et al., A gene (ccmA) required for carboxysome formation in the cyanobacterium *Synechocystis* sp. strain PCC6803. *J Bacteriol*, 1994. **176**(8): 2374–2378.
98. Woodger, F.J. et al., Transcriptional regulation of the CO_2-concentrating mechanism in a euryhaline, coastal marine cyanobacterium, *Synechococcus* sp. strain PCC 7002: Role of NdhR/CcmR. *J Bacteriol*, 2007. **189**(9): 3335–3347.
99. Price, G.D. et al., Analysis of a genomic DNA region from the cyanobacterium *Synechococcus* sp. strain PCC7942 involved in carboxysome assembly and function. *J Bacteriol*, 1993. **175**(10): 2871–2879.
100. Berman-Frank, I. et al., Coupling between autocatalytic cell death and transparent exopolymeric particle production in the marine cyanobacterium *Trichodesmium. Environ Microbiol*, 2007. **9**(6): 1415–1422.
101. Jabs, T., Reactive oxygen intermediates as mediators of programmed cell death in plants and animals. *Biochem Pharmacol*, 1999. **57**(3): 231–245.
102. Berman-Frank, I. et al., The demise of the marine cyanobacterium, *Trichodesmium* spp., via an autocatalyzed cell death pathway. *Limnol Oceanogr*, 2004. **49**(4): 997–1005.
103. Ross, C. et al., Toxin release in response to oxidative stress and programmed cell death in the cyanobacterium *Microcystis aeruginosa. Aquat Toxicol*, 2006. **78**(1): 66–73.
104. Ning, S.B. et al., Salt stress induces programmed cell death in prokaryotic organism *Anabaena. J Appl Microbiol*, 2002. **93**(1): 15–28.
105. Jiang, Q. et al., Genome-wide comparative analysis of metacaspases in unicellular and filamentous cyanobacteria. *BMC Genomics*, 2010. **11**: 198.
106. Parsell, D. and S. Lindquist, The function of heat-shock proteins in stress tolerance: Degradation and reactivation of damaged proteins. *Annu Rev Genet*, 1993. **27**(1): 437–496.
107. Wickner, S. et al., Posttranslational quality control: Folding, refolding, and degrading proteins. *Science*, 1999. **286**(5446): 1888–1893.

15 Environmental Factors Regulating Nitrogen Fixation in Heterocystous and Non-Heterocystous Cyanobacteria

*Lucas J. Stal**

CONTENTS

15.1 INTRODUCTION

Nitrogen is one of the most important elements for organisms. It may account for up to 10% of the dry weight of biomass. Nitrogen occurs in amino acids—the building blocks of polypeptides, enzymes, and other proteins—but also in the nucleic acids (DNA and RNA), in the cell wall, and in chlorophyll. In all these compounds, nitrogen is present in its reduced (–3) state. Hence, organisms living at the expense of oxidized nitrogen such as dinitrogen, nitrate, or nitrite must reduce it before this nitrogen can be assimilated and used for the synthesis of structural cell material.

Nitrogen occurs in different redox states, varying from its most oxidized form (+5) as in nitrate (NO_3^-) to its most reduced form (–3) as in ammonia (NH_3), and it may therefore undergo transitions in its redox state catalyzed by bacteria and archaea. Reduced nitrogen may be oxidized and used as an energy source by chemotrophic bacteria, while oxidized nitrogen may be reduced and used as a terminal electron acceptor in anaerobic respiration by a variety of microorganisms. Together, these transitions are known as the microbial nitrogen cycle [1]. There are two important routes in this nitrogen cycle that lead to the conversion of "bound" nitrogen (e.g., nitrate, nitrite, ammonium, organic nitrogen) to gaseous nitrogen (mostly dinitrogen, N_2, and to some extent nitrous oxide, N_2O, or nitric oxide, NO). Denitrification converts nitrate eventually into N_2, while anaerobic

* Corresponding author: lucas.stal@nioz.nl

ammonium oxidation (ANAMMOX) converts ammonium and nitrite to N_2 [2]. Hence, all combined nitrogen will eventually be converted into N_2 and will escape into the atmosphere. The atmosphere is the largest source of nitrogen in the biosphere but it is unavailable to all but some bacteria and a few archaea. These diazotrophic ("feeding on dinitrogen") organisms possess an enzyme complex—nitrogenase—which reduces N_2 to ammonia which is subsequently assimilated [3]. Remarkably, not a single Eukarya is known that is capable of fixing dinitrogen, but there are many examples of Eukarya that live in symbiosis with diazotrophic Bacteria [4]. Diazotrophic microorganisms thus close the gap of the nitrogen cycle.

Nitrogenase catalyzes the reduction of N_2 to ammonium according to the following equation:

$$N_2 + 8[H] + 16\ ATP \rightarrow 2NH_4^+ + H_2 + 16\ ADP + 16\ P_i$$

Hence, the fixation of N_2 is an energetically expensive process [5]. Nitrogenase is an enzyme complex composed of two enzymes. Dinitrogenase reductase (the iron protein) is a homodimer encoded by *nifH* that reduces dinitrogenase (the iron-molybdenum protein) which is composed of two different subunits that are each present as dimers and encoded by *nifD* and *nifK* [6]. Dinitrogenase reductase reduces dinitrogenase at the expense of four ATP per two electrons using reduced ferredoxin as the electron donor [7,8]. As follows from the equation, the reduction of dinitrogen also results in the evolution of H_2. In most aerobic diazotrophs, this hydrogen is recycled by an uptake hydrogenase. When nitrogenase is not saturated with N_2, the amount of H_2 produced increases and when no N_2 is available, nitrogenase diverts all electrons to hydrogen. This has been used as a way to measure nitrogenase activity and has also been considered as a possibility for biotechnological H_2 production. Nitrogenase is not specific for N_2 but reduces many compounds with a triple bond [9]. For instance, it reduces acetylene to ethylene, which has become a popular way of determining nitrogenase activity [10]. Cyanide is also reduced by nitrogenase and it has been suggested that the detoxification of this compound might have been the ancient function of nitrogenase [11].

Nitrogenase evolved very early in the history of earth before the oxygenation of the atmosphere [12,13]. It may have had originally a different function such as the detoxification of cyanide while it evolved later to nitrogenase [11]. The original nitrogenase may have had only iron as cofactor. Molybdenum was probably in low supply due to its insolubility under low oxygen [8] and precipitation of molybdate with sulfide [13]. Iron-only nitrogenase can still be found in contemporaneous microorganisms [14,15]. Another alternative nitrogenase uses vanadium as a cofactor instead of molybdenum [16]. However, alternative nitrogenase always occurs besides the molybdenum enzyme and is synthesized when molybdenum is unavailable [17]. The origin in an anoxic earth is reflected in the extreme sensitivity of nitrogenase to oxygen. Nitrogenase requires an anoxic environment and together with its high demand for energy this restricts a diazotrophic mode of life for many microorganisms. Among the diazotrophic microorganisms, cyanobacteria claim a special place as they are the only oxygenic phototrophic organisms capable of fixing dinitrogen and because of their wide distribution and abundance [18].

15.2 DIVERSITY OF DIAZOTROPHIC CYANOBACTERIA

Cyanobacteria are a monophyletic group of oxygenic phototrophic bacteria that nevertheless exhibit a high diversity [19]. Their photosynthetic apparatus is similar to that of plants possessing two photosystems, using water as the electron donor and evolving oxygen.

Cyanobacteria evolved more than 2.5 billion years ago. As a result of their oxygenic photosynthesis, they enriched the atmosphere with oxygen [20]. There is a wealth of other evidence that is indicative of the presence of cyanobacteria 2.5 billion years ago, such as molecular proxies, microfossils, and phylogeny [21]. Microfossils as old as almost 3.5 billion years found in stromatolites have been interpreted as originating from cyanobacteria [22]. Certainly, the ancestors of cyanobacteria may have been anoxygenic and may have possessed one photosystem. Also, oxygenic photosynthesis may have taken place in isolated pockets in microbial ecosystems such as stromatolites and it may have taken a long time until the reduced earth became oxidized and free oxygen could accumulate in the atmosphere. The oldest microfossils had already a filamentous morphology. Fossil akinetes have been reported from 1.5-billion-year-old sediments and have been taken as evidence for the presence of heterocystous N_2-fixing cyanobacteria [23]. Through an endosymbiotic event, cyanobacteria gave rise to the evolution of the chloroplast and subsequently to algae and plants [24]. Hence, basically all O_2 is produced due to cyanobacteria.

Cyanobacteria are photoautotrophic and fix CO_2 through the Calvin–Benson–Bassham pathway. The photosynthetic reaction centers contain chlorophyll with chlorophyll *a* as the most common type. But chlorophyll *b* is found in *Prochloron* and *Prochlorothrix*, a group of unusual cyanobacteria [25], and *Prochlorococcus*, the most abundant phototroph in the ocean, contains divinyl derivatives of chlorophyll *a* and *b*, while chlorophyll *d* is found in *Acaryochloris* [26]. Most cyanobacteria contain phycobiliproteins that serve as light-harvesting pigments. The phycobiliproteins are organized in structures known as phycobilisomes which are situated on the thylakoid membranes and connected to the photosynthetic reaction centers [27]. The core of the phycobilisomes is composed of allophycocyanin from which stacks of phycocyanin stick out. These two types are always present in the phycobilisomes and have a blue chromophore. Some cyanobacteria contain in addition the red-colored phycoerythrin which comes in two types: phycoerythrobilin (PEB) and phycourobilin (PUB), which have different colors and light absorption characteristics [28]. Cyanobacteria with large amounts of phycoerythrin appear red and those with equal amounts of phycocyanin and phycoerythrin may appear brown to black. Some phycoerythrin-containing cyanobacteria are capable of adjusting the relative amounts of phycocyanin and phycoerythrin to acclimate optimally to the spectral light climate under which they thrive. This process is called complementary chromatic acclimation. Another type of chromatic acclimation adjusts the relative PEB and PUB amounts. This renders cyanobacteria a plethora of colors with which they can optimally occupy differently illuminated environments. *Prochloron*, *Prochlorothrix*, and *Prochlorococcus* do not contain phycobilisomes [29].

The cyanobacteria are morphologically diverse, and based on their morphology they have been divided into five divisions (Table 15.1) [30]. Division 1 comprises unicellular forms that multiply by binary fission in one or two planes, while division 2 are unicellular cyanobacteria that multiply by multiple fission. A mother cell divides into many small cells that are called baeocytes. The other three divisions comprise all filamentous cyanobacteria, which are composed of many cells organized in a trichome and therefore represent a simple form of multicellularity. Division 3 comprises all filamentous cyanobacteria composed of one cell type. Divisions 4 and 5 comprise cyanobacteria that differentiate heterocysts and in addition some may also differentiate akinetes. Heterocysts (also, and more correctly, called heterocytes) are cells that contain nitrogenase and are the site of dinitrogen fixation in these organisms. Akinetes are cells that are produced when the organism is exposed to conditions unsuitable for growth and they serve as survival stages. Akinetes germinate to hormogonia, short trichomes composed of one cell type, smaller than the vegetative

TABLE 15.1
Sections of Cyanobacteria and Their Main Characteristics

Section	Morphology	Cell Division	Cell Differentiation	N_2 Fixation
1	Unicellular	Two planes	Vegetative only	Anaerobic; temporal separation
2	Unicellular	Multiple fission	Baeocytes	Anaerobic
3	Filamentous	One plane	Vegetative only	Anaerobic; temporal separation; spatial separation (diazocytes)
4	Filamentous	One plane	Heterocysts and akinetes	Spatial separation (heterocysts)
5	Filamentous	Two planes	Heterocysts and akinetes	Spatial separation (heterocysts)

cells of the mature organism. Hormogonia are motile and can migrate away, thus serving the dispersion of the organism. Division 5 cyanobacteria differ by exhibiting cell division in two planes resulting in true branching. Division 4 cyanobacteria sometimes exhibit false branching, when two trichomes are enclosed by a common sheath from which one trichome partly breaks out. Examples of all types of cyanobacteria are shown in Figure 15.1.

Cyanobacteria occur in many size classes that comprise 2 orders of magnitude. Picocyanobacteria may be as small as 0.5 μm and some filamentous cyanobacteria may be as wide as 50 μm. Trichomes may be up to many tenths of millimeters long. Many cyanobacteria, unicellular as well as filamentous, form aggregates or colonies, sometimes surrounded by a common sheath, and may be visible

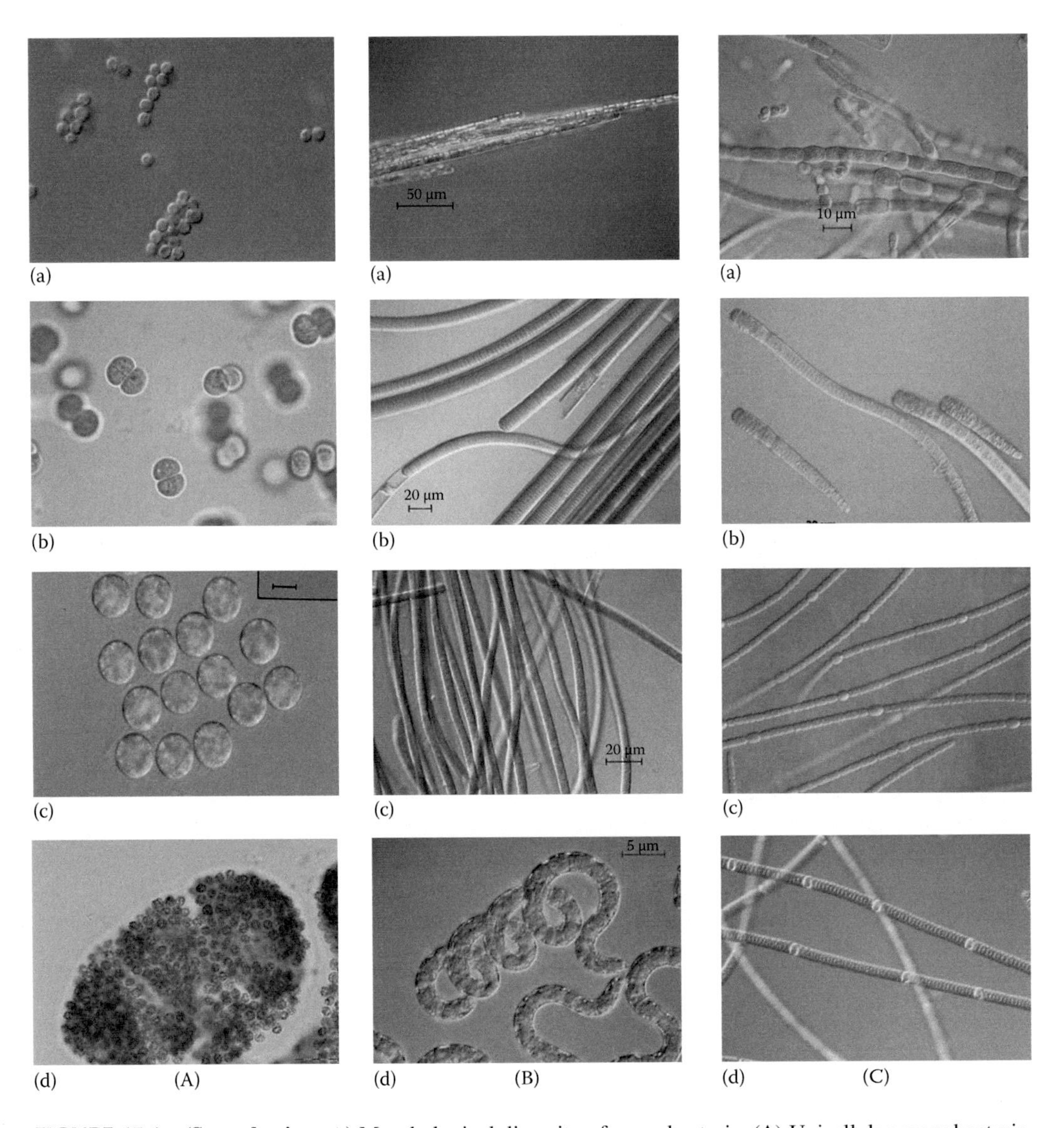

FIGURE 15.1 (See color insert.) Morphological diversity of cyanobacteria. (A) Unicellular cyanobacteria. (a–d): *Crocosphaera* sp., *Chroococcus* sp., *Dermocarpa* sp. (division 2), and the colony-forming *Microcystis* sp. (B) non-heterocystous filamentous cyanobacteria (division 3). (a–d): *Trichodesmium* sp., *Lyngbya* sp., *Phormidium* sp., and *Spirulina* sp. (C) Heterocystous cyanobacteria. (a): *Fischerella* sp. with true branching (division 5); the other three belong to division 4, (b–d): *Calothrix* sp. with terminal heterocysts, *Anabaena* sp. and *Nodularia* sp. with intercalary heterocysts.

macroscopically. Such aggregates and colonies exhibit multicellular behavior [31]. Cyanobacteria may also form macroscale structures such as microbial mats and biofilms or water blooms [32].

15.3 STRATEGIES OF DIAZOTROPHIC CYANOBACTERIA TO PROTECT NITROGENASE FROM OXYGEN

Many but not all cyanobacteria are capable of fixing N_2 [33]. As oxygenic phototrophic organisms, they can satisfy the demand of nitrogenase for energy and low-potential electrons from sun light and water. However, cyanobacteria had to evolve adaptations and strategies that provide a low-oxygen environment for nitrogenase.

15.3.1 ANAEROBIC NITROGEN-FIXING CYANOBACTERIA

A large number of unicellular and filamentous cyanobacteria are capable of expressing nitrogenase and fixing N_2 only under low-oxygen conditions and in the absence of oxygenic photosynthesis. This condition is fulfilled under anoxic conditions in the dark. However, the only way of cyanobacteria to generate energy under these conditions is by fermenting their intracellular storage compound (glycogen) [34]. This is a slow and inefficient process and allows only low nitrogenase activity, if at all. Some cyanobacteria have been shown to fix low amounts of N_2 under these conditions. Many anoxic environments are characterized by high levels of sulfide (sulfureta) which is a potent inhibitor of the oxygenic photosystem II, but not of the anoxygenic photosystem I. Some cyanobacteria living in these sulfureta are capable of switching to anoxygenic photosynthesis using sulfide as the electron donor and it has been shown that N_2 fixation is promoted under such conditions [35]. Since many cyanobacteria are motile, they might migrate between the oxygen-rich and anoxic compartments of an ecosystem and in this way satisfy their nitrogen demand and fix N_2 fixation in the anoxic ecosystem compartment [36]. Nevertheless, neither the ecology of the vast majority of anaerobic N_2-fixing cyanobacteria is known nor is known whether or not they live as diazotrophs.

15.3.2 SPATIAL SEPARATION OF NITROGEN FIXATION AND OXYGENIC PHOTOSYNTHESIS

The most advanced adaptation is the differentiation of a special cell type for N_2 fixation, the heterocyst, by cyanobacteria belonging to the *Nostocales* [37]. Well-known representatives include, among others, the genera *Anabaena*, *Nostoc*, *Calothrix*, and *Nodularia*. The heterocyst has lost the oxygenic photosystem II but retained photosystem I. Hence, it is capable of harvesting light energy but it cannot utilize water as an electron donor and does not evolve oxygen. The heterocyst is therefore dependent on the vegetative cells for the supply of fixed carbon and reducing equivalents that are supplied in the form of sucrose. The heterocyst has also an additional thick glycolipid cell wall layer that serves as a gas diffusion barrier and that limits the diffusion of oxygen into the cell [38]. Any oxygen that diffuses into the heterocyst is scavenged by respiration so that the cell interior is very low in oxygen. Nitrogenase is only expressed in the heterocyst where it is protected from oxygen inactivation. A gene rearrangement is necessary to transcribe the *nif* operon and this occurs simultaneously with the process of heterocyst differentiation [39]. The processes that lead to the differentiation of a vegetative cell into a heterocyst are complex and also not fully understood [40]. Heterocysts usually occur at regular distances along the trichome. They do not grow or divide. The divisions of the vegetative cells cause an ever increasing distance between two heterocysts and leads to the differentiation of a new heterocyst in the middle. The heterocyst frequency is usually around 5%. The N_2 fixed by the heterocyst is transported to the vegetative cell as amino acid. The nitrogen storage compound cyanophycin (composed of a poly-aspartate backbone and arginine bound to each aspartate) is present in some species in the pores connecting the heterocyst with the vegetative cell. Heterocystous cyanobacteria have spatially separated the incompatible processes of N_2 fixation and oxygenic photosynthesis in two different cell types. This allows these organisms to perform both processes simultaneously and utilize light energy directly for N_2 fixation. Heterocystous cyanobacteria also fix N_2 in the dark but usually at a much lower rate because the energy is generated by respiration and the oxygen diffusion into the heterocyst is limiting because of the glycolipid cell wall layer [41].

There is no fossil record of heterocysts but microfossils of akinetes, survival stages of some heterocystous cyanobacteria, suggest that these organisms may have evolved more than 1.5 billion years ago [42]. Also, the timing of the oxygenation of the atmosphere and the restrictions imposed by the oxygen-sensitivity of nitrogenase suggest such an early evolution of heterocysts.

There are also non-heterocystous cyanobacteria that are capable of fixing N_2 and oxygenic photosynthesis simultaneously [43]. The best-known example is the filamentous *Trichodesmium* (including *Katagmynene*), a group of cyanobacteria that is abundant in the tropical oceans. The strategy by which these cyanobacteria are able to fix N_2 is not precisely known [44]. Since it is unlikely that oxygenic photosynthesis and N_2 fixation occur simultaneously in the same cell, a spatial separation as in the case with heterocystous cyanobacteria is conceived. In some organisms, nitrogenase was detected in a selected number of adjacent cells in the trichome. These cells have been termed "diazocytes" [45]. In other strains, nitrogenase was located in all cells of the trichome [46]. Probably, these cells are capable of temporally switching off oxygenic photosynthesis in order to charge the cell with fixed N_2. Hence, this strategy has also been named as a combination of spatial and temporal separation of oxygenic photosynthesis and N_2 fixation. *Trichodesmium* is unable to fix N_2 in the dark because it is unable to maintain nitrogenase under these conditions and it differs in that with the heterocystous cyanobacteria. The enzyme is inactivated and synthesized de novo every day [47]. The putative unicellular N_2-fixing cyanobacterium UCYN-A expresses its *nif* genes in the light and in situ measurements have indicated that also N_2 fixation occurs during the daytime [48]. Since this organism is uncultured, little is known about it. Metagenomic sequencing of cells that were enriched through flow cytometric sorting revealed that this very tiny cyanobacterium lacks the genes for photosystem II and for CO_2 fixation [49]. In that respect, it shows similarity with the heterocyst. However, the small size of this organism would not allow it to maintain an intracellular environment sufficiently low in oxygen to maintain active nitrogenase. A possibility could be that this organism lives diazotrophically in symbiosis with another organism. Until now this enigma has not been solved.

15.3.3 Temporal Separation of Nitrogen Fixation and Oxygenic Photosynthesis

Several unicellular and filamentous diazotrophic cyanobacteria temporally separate the two incompatible processes by confining N_2 fixation to the dark (night), when photosynthesis obviously does not take place [43]. In the dark, the energy for N_2 fixation comes from respiration of endogenous storage compounds (glycogen) that have been accumulated during the light. These storage compounds also supply the reducing equivalents for N_2 fixation. Moreover, respiration must be sufficiently fast to generate intracellular conditions low enough in oxygen, which also imposes restrictions on the minimum cell size (because of the surface to volume ratio) or on the maximum O_2 concentration in the environment [41]. The temporal separation of N_2 fixation and photosynthesis in cyanobacteria is in most cases under control of a circadian clock [50]. When these cyanobacteria are transferred from a light–dark regime to continuous light, the cycling of both processes continues for a few days and oxygenic photosynthesis is suppressed in the light [51]. Usually, N_2 fixation is then much higher due to the availability of light energy. However, it must be emphasized that as far as known these cyanobacteria are all capable of diazotrophic growth under continuous light, although the marine planktonic *Crocosphaera* may be an exception to this rule [52]. It is possible that a population of cells in continuous light is composed of N_2-fixing cells, which have switched off oxygenic photosynthesis, and non-N_2-fixing cells, and that cells are able to switch between these two physiological states. The unicellular *Gloeothece* fixed N_2 predominantly in the light when grown under a light–dark regime in continuous culture, although this organism exhibits the typical temporal separation when grown in batch culture [53]. The N_2-fixing behavior of this organism differs in several aspects from that of, for example, *Crocosphaera* [52].

Although the strategies explained in this paragraph cover the main differences known among diazotrophic cyanobacteria, analyzing N_2 fixation in natural communities indicates that there is a plethora of different daily patterns emphasizing that various organisms exhibit many subtle variations [54]. Similarly, there is a range of O_2 tolerance for N_2 fixation in various cyanobacteria, of which the physiological basis is in most cases not well understood [19]. Cyanobacteria that are unable to fix N_2 under any condition as a rule do not contain *nif* genes. There is genomic evidence that the ancestral cyanobacterium was not diazotrophic [55], which would mean that some cyanobacteria acquired the *nif* genes while others did not. Evidence of horizontal gene transfer of a *nif* cluster in cyanobacteria has been reported [56].

15.4 ENVIRONMENTS OF DIAZOTROPHIC CYANOBACTERIA

Cyanobacteria occur in virtually any illuminated environment and sometimes also when there is very little light or even in complete darkness, for instance, as symbionts in the rhizosphere. An exception may be acidic environments for which hardly any report exists of cyanobacteria thriving in those places [57]. Diazotrophic cyanobacteria form no exception to that rule. Shortage of a source of combined nitrogen would select for diazotrophic organisms because these organisms have access to the abundant atmospheric dinitrogen. This shortage is often reflected in the ratio of combined nitrogen to phosphorus in the environment. When this N:P ratio is below the "Redfield" value of 16, it is usually considered as a sign of nitrogen shortage, while a higher ratio hints to nitrogen sufficiency [58]. However, a low N:P ratio may be due to the depletion of combined nitrogen when dinitrogen fixation is not possible, while under dinitrogen-fixing conditions the ratio will become high due to the depletion of the phosphate [59].

Heterocystous cyanobacteria may form blooms in fresh and brackish water bodies. Many freshwater lakes worldwide develop water blooms of cyanobacteria as a result of eutrophication [60]. When nitrogen becomes depleted while excess phosphate is present, heterocystous cyanobacteria develop. These organisms often possess gas vesicles with which they float to the surface. Such blooms are also known from brackish water bodies such as the Baltic Sea [61]. However, heterocystous cyanobacteria do not occur as free-living organisms in full-salinity sea water. In fact, the water column of the temperate and cold ocean seems to be devoid of cyanobacteria, including diazotrophic cyanobacteria. The reason for this is not known [41].

The warm (sub)tropical ocean is the habitat of non-heterocystous cyanobacteria [62]. Particularly, the filamentous *Trichodesmium* is an abundant diazotroph in the tropical ocean. This species possesses gas vesicles that make them buoyant bringing them to the photic surface waters. Also, the unicellular diazotrophic cyanobacteria *Crocosphaera* and, to a lesser extent, *Cyanothece* are common in the warmer oceanic waters. *Cyanothece* may be more abundant attached to other organisms or in biofilms on shallow, illuminated sediments. The uncultured putative unicellular diazotrophic cyanobacterium UCYN-A is also mostly confined to the warm waters and may contribute locally to N_2 fixation in the same order of magnitude as *Trichodesmium*.

Microbial mats are laminated microbial communities and mostly built by cyanobacteria [32]. Microbial mats occur worldwide in environments that are characterized by extreme environmental conditions such as high temperature, high salt whether or not combined with high alkalinity, and drought (hot and cold deserts). Microbial mats may also develop in intertidal sediments that, due to their steep physicochemical gradients and extreme and fluctuating conditions, are also considered to be extreme. As far as known, all cyanobacterial mats fix dinitrogen [63]. Microbial mats are complex and diverse microbial communities that also contain diazotrophic bacteria other than cyanobacteria [64]. Although these diazotrophs may be important under some conditions, in most cases cyanobacteria are the dominant N_2-fixers. Microbial mats are often formed by non-heterocystous filamentous cyanobacteria but unicellular diazotrophs may also occur, particularly in hot spring cyanobacterial mats [65]. Hot springs are also known for containing heterocystous cyanobacteria as is the case for some microbial mats in intertidal areas [66].

Heterocystous and unicellular diazotrophic cyanobacteria are further known from paddy fields where they provide fixed nitrogen to the rice plants, or from biofilms developing on illuminated cave walls. Extensive terrestrial microbial mats are known from the arctic tundra. Last but not least, diazotrophic cyanobacteria are found in numerous symbiotic relationships, both on land and in the sea [67]. While heterocystous cyanobacteria are largely excluded as free-living organisms from the plankton in the marine environment, they occur as endo and ectosymbionts in planktonic diatoms (such as the heterocystous *Richelia* in the diatom *Rhizosolenea*) [68]. Also, non-heterocystous cyanobacteria go into symbiotic relationships. Other well-known examples are the heterocystous cyanobacteria *Anabaena* in the water fern *Azolla*, *Nostoc* in *Cycas*, and in *Gunnera*, and tripartite lichens also contain diazotrophic cyanobacteria [67]. In all these examples, the cyanobacteria provide the host with combined nitrogen. This overview can only be limited and is by no means complete.

15.5 ROLE OF LIGHT FOR NITROGEN FIXATION BY CYANOBACTERIA

Cyanobacteria are phototrophs and they need light for growth and for the fixation of N_2. Heterocystous cyanobacteria, *Trichodesmium*, and UCYN-A fix N_2 in the light and utilize light energy directly. Heterocystous cyanobacteria can also fix N_2 in the dark at the expense of endogenous storage carbohydrate and respiratory metabolism. Non-heterocystous cyanobacteria that employ the temporal separation strategy use light energy indirectly. During the light phase, they store a large part of the harvested energy in the form of intracellular carbohydrate (glycogen), which is subsequently respired in the subsequent dark period to provide the energy and the reducing equivalents for N_2 fixation [50,51,69].

The development of blooms of heterocystous cyanobacteria has often been attributed to the increase in water temperature [70]. However, the effect of temperature may be indirect. In a bloom of heterocystous cyanobacteria in the Baltic Sea, the daily integrated light compensation point was 22.7 mol m^{-2} day^{-1}. This is the total amount of solar insolation that impinges the water surface for a 24 h period. Below this insolation, the total net production of the water column-integrated cyanobacterial biomass will be negative and this happened even in the middle of the summer in August, 1993, during 2 days [71]. Hence, the compensation point is critical for photosynthesis because below this point the losses of organic carbon due to respiration exceed the gains by photosynthetic CO_2 fixation. This was also seen from the depth-integrated photosynthesis. In this case, the euphotic depth was calculated as 8 m and is here defined as the depth with the compensation point light intensity. Hence, below that depth, the cyanobacteria showed a negative production. The total cumulative production of the cyanobacteria in the water column was at 27 m and is called the critical depth below which no net water column-integrated productivity is possible. Calculations demonstrated that neither the euphotic depth nor the critical depth varied with temperature, although the cyanobacterial biomass was larger due to the higher rate of photosynthesis at elevated temperatures. The euphotic and critical depths increase with the daily insolation during the year with the highest values in summer. When the mixing depth exceeds the critical depth, no net depth and daily integrated production can occur. In winter, the whole water column of the Baltic Sea is mixed until the halocline which is at 45 m. In summer, the higher temperature will cause thermal stratification of the water column occurs and the mixing depth may be at 17 m, well above the critical depth of 27 m thus allowing net production. Hence, while light determines the net production in this example, a higher temperature is required to decrease the mixing depth.

Interestingly, the daily water column-integrated N_2 fixation by these heterocystous cyanobacteria in the Baltic Sea was only marginally light dependent [71]. This can be understood when realizing that N_2 fixation occurs both in the light and in the dark and does not have a negative component such as respiration in the case of calculating the net photosynthetic integral. In fact, model calculations show that the daily depth-integrated N_2 fixation of a bloom of heterocystous cyanobacteria could be increased most by increasing dark N_2 fixation [41]. This can be achieved by increasing the rate of respiration and this can only be done by, counterintuitively, increasing the flux of O_2 into the heterocyst.

Similarly, it was shown that for microbial mats the daily integral of N_2 fixation was relatively independent of the daily integral of the insolation [72]. For many non-heterocystous cyanobacteria, N_2 fixation is fully confined to the dark and no direct relation of nitrogenase activity with light would be seen in these organisms, although one should keep in mind that indirectly light is responsible for the storage of glycogen, which is required to fuel N_2 fixation in the dark. For non-heterocystous cyanobacteria that are capable of fixing N_2 in the light, there may be a negative relationship with too high levels of irradiation causing oxygen super-saturation.

15.6 ROLE OF TEMPERATURE IN NITROGEN FIXATION BY CYANOBACTERIA

Temperature has a profound effect on N_2 fixation in heterocystous and non-heterocystous cyanobacteria [41]. Temperature has an inverse effect on the solubility of oxygen in water. Hence, at higher temperature, diazotrophic cyanobacteria are exposed to lower environmental oxygen concentrations. However, this is also the case with the solubility of N_2 and this may have a negative effect on the saturation of nitrogenase by N_2.

Temperature is positively related with metabolic rates and is usually expressed by Q_{10}, a factor that describes the increase or decrease in metabolic rate with a 10°C increase or decrease in temperature, respectively. For metabolic rates, this factor is ~2, hence a doubling with every 10°C rise in temperature (within the physiological range of the organism). Hence, a higher temperature will let the respiration increase and this will eventually allow the diazotrophic cell to become virtually anoxic (the remaining oxygen depends on the K_m for oxygen of the respiratory system). Similarly, it will increase the rate of N_2 fixation (and of course of all other metabolic processes in the cell). The Q_{10} of other non-metabolic processes may be very different. For instance, the Q_{10} of gas diffusion is a combination of the solubility of the gas in a liquid (which decreases with increasing temperature) and the diffusion rate (which increases with increasing temperature). The Q_{10} of O_2 diffusion into the cell is ~1.1, meaning that it hardly changes with temperature. Hence, we can calculate that at a certain temperature the rate of respiration equals that of the influx of O_2, which should scavenge theoretically all O_2 that enters the N_2-fixing cell.

The effects of temperature on the solubility and diffusion of N_2 and O_2 are similar. Hence, the flux of N_2 into the N_2-fixing cell will increase only marginally with increasing temperature. However, nitrogenase activity as a metabolic process will have a Q_{10} of ~2. Hence, with increasing temperature, it is expected that the reduction of N_2 by nitrogenase increases faster than the influx of N_2 into the N_2-fixing cell, eventually leading to an anazotic cell (cell that is effectively devoid of N_2, again the actual remaining N_2 concentration will depend on the K_m of nitrogenase), compared to anoxic for devoid of O_2 [41]. Under saturation of nitrogenase with dinitrogen may have disastrous consequences for the cell. In that case, nitrogenase will divert electrons to protons, producing H_2 instead of NH_3. Because this is at the expense of ATP, it is a futile process that will diminish the organism's ecological success.

In the ocean, diazotrophic cyanobacteria are predominantly present in the (sub)tropical regions with surface water temperatures above 25°C [73]. In the light of global change, ocean surface water temperatures are expected to increase. This would also increase the area of the ocean where diazotrophic cyanobacteria could thrive. However, it is not unlikely that in the tropical areas of the ocean, the surface water temperature becomes higher than beneficial for N_2 fixation for reasons explained earlier. Model calculations indicate that increases in the ocean surface water temperature hardly affect the global oceanic N_2 fixation, although the distribution of diazotrophic cyanobacteria will shift considerably [73].

15.7 ROLE OF SALINITY IN NITROGEN FIXATION BY CYANOBACTERIA

For a long time, the paradigm existed that N_2 fixation was unimportant in the marine environment [74]. Indeed, heterocystous cyanobacteria are often found in eutrophic freshwater lakes. Blooms of heterocystous cyanobacteria are also known from brackish environments with the Baltic Sea as one of the most prominent examples [70]. Since the discovery of the diazotrophy of the filamentous non-heterocystous cyanobacterium *Trichodesmium*, it also became clear that the ocean was a major contributor to global fixation of N_2 [75]. In fact, the ocean contributes approximately 50% of the non-anthropogenic global N_2 fixation. Subsequently, several other diazotrophs were found in the ocean, such as three different groups of unicellular cyanobacteria and cyanobacteria living in symbiosis with other algae [62]. Moreover, N_2-fixing cyanobacteria are common in marine microbial mats, coral reefs, and sea grass meadows. Hence, salinity is clearly not a factor that eliminates diazotrophic cyanobacteria. The factor temperature is probably essential and restricts non-heterocystous diazotrophic cyanobacteria to the warmer (sub)tropical ocean. It is unclear why heterocystous cyanobacteria are absent from the temperate and cold ocean. In the Baltic Sea, heterocystous cyanobacteria occur up to a salinity of 9‰ [76]. Most heterocystous cyanobacteria are (salt tolerant) freshwater species and a few can be characterized as brackish or marine. Cyanobacteria accumulate compatible solutes to equilibrate their cytoplasm to the environmental osmotic pressure. Compatible solutes are low-molecular, highly soluble organic compounds that do not interfere with the cellular metabolism (are compatible with it) [77]. For instance, freshwater cyanobacteria predominantly accumulate the disaccharides sucrose or trehalose, marine cyanobacteria accumulate the heteroside glucosyl-glycerol, and hypersaline cyanobacteria accumulate the amino acid derivative glycine betaine, although there are also exceptions to this classification. These osmolytes are eventually produced from CO_2 fixation but since heterocysts do not fix CO_2, osmolytes need to be imported from the vegetative cells or are synthesized by the heterocyst from photosynthetic products that are imported from the vegetative cell. Hence, it is possible that heterocysts are unable to adapt to higher salinity when they cannot accumulate osmolytes to the relevant concentration. Heterocysts import organic matter from the vegetative cells, which is needed as an electron donor for respiration to scavenge oxygen as well as for N_2 fixation. Sucrose has been shown to be transported from the vegetative cell to the heterocyst. Although sucrose is the osmolyte in many (freshwater/brackish) heterocystous cyanobacteria, it does not seem to be compatible with the metabolism in the heterocyst because it is catabolized [78].

Another possibility is that something else other than osmotic potential affects N_2 fixation in heterocystous cyanobacteria at full-salinity seawater. There is some evidence that the high concentration of sulfate in seawater is the cause of the absence of heterocystous cyanobacteria. It has been conceived that high sulfate concentrations would prevent the uptake of its structural analog molybdate, which provides an important cofactor for nitrogenase, and which may make use of the same transporter [79]. This hypothesis has several problems. Obviously, this mechanism would then not apply to non-heterocystous cyanobacteria but it is difficult to understand why. Also, one might expect that this would give a selective advantage for vanadium-containing alternative nitrogenase but this does not seem to be the case. Nitrate-reductase is also a molybdenum-containing enzyme and in seawater nitrate is used as a source of nitrogen; there is no evidence that this is limited by molybdenum availability. Moreover, there are also specific molybdate transporters that are not affected by sulfate and there is no experimental evidence that supports the hypothesis of the sulfate–molybdate antagonism. Nevertheless, lowering the sulfate concentration increased N_2 fixation at seawater salinity [80]. It is possible that a high concentration of sulfate draws on the pool of electrons through assimilatory sulfate reduction, which would be futile in the non-growing heterocyst. This idea is supported by the observation that molybdate, a competitive inhibitor of sulfate reduction, alleviated the effect of sulfate on nitrogenase activity [80].

15.8 ROLE OF NUTRIENTS IN NITROGEN FIXATION BY CYANOBACTERIA

The availability of combined nitrogen usually suppresses the expression of nitrogenase in diazotrophic cyanobacteria through a feedback mechanism on the transcription of the nitrogenase genes. Active nitrogenase will continue to fix N_2 until the enzyme activity ceases through the turnover of the enzyme. This may be quick under oxic conditions due to the irreversible inactivation of nitrogenase by oxygen. Under anoxic conditions, the turnover time may be much longer. Another factor is the nature of the combined nitrogen. Nitrate requires a specific transporter and in the cell it needs to be reduced to ammonia by nitrate and nitrite reductases, before it can be assimilated. Not all cyanobacteria are capable of utilizing nitrate due to the absence of the required enzymes. The majority of cyanobacteria can utilize nitrate but it requires eight electrons to reduce it to ammonia. Hence, the transport and reduction of nitrate is at a considerable cost of energy, albeit less than the energy demand of N_2 fixation.

Ammonia is ready to be assimilated. Ammonia is a gas and can diffuse freely through the cytoplasm membrane. Ammonia can be protonated to ammonium in a pH-dependent manner and the uptake of ammonium requires an active transport. In the cell, ammonium can be de-protonated to ammonia, which may diffuse out of the cell before it is assimilated. This process is called a futile cycling of ammonium that represents a loss of energy for the organism [81]. This may also occur when the organism fixes N_2. The ammonia resulting from the fixed N_2 may diffuse out of the cell before it is assimilated. This ammonia may then be used by other non-N_2-fixing organisms in the community. It has, for instance, been shown that N_2 fixed by heterocystous cyanobacteria was passed on in this way to the picocyanobacteria in the phytoplankton community [82].

Urea is a potential source of nitrogen for cyanobacteria that possess urease [83]. Organic nitrogen, for example, in the form of amino acids, can also be used by cyanobacteria that are capable of transporting it into the cell. For instance, the unicellular cyanobacterium *Gloeothece* fixes N_2 during the dark and exudes amino acids that are stored in the sheath of the organism to be used during the light period when photosynthetic CO_2 fixation accommodates the fixed nitrogen for the synthesis of structural cell material [84].

Phosphate is an essential nutrient for all organisms. It is necessary for cellular energy metabolism and it is part of the nucleic acids. It has been suggested that phosphate is the ultimate limiting factor in the ocean because nitrogen is abundantly available as atmospheric dinitrogen which is made available through the diazotrophs [85]. Phosphate concentrations are therefore often very low. Organic phosphate can be used by organisms that produce phosphatases [86]. These are extracellular enzymes that hydrolyze the phosphate group after which the organism can take it up. Other organisms that do not produce phosphatases may also benefit from the phosphate liberated in this way. Several cyanobacteria are capable of using phosphonates, phosphorus that is bonded to a carbon atom [87]. Cyanobacteria may take up phosphate in excess to their actual demand (luxury uptake) and store it intracellularly. This would allow the organism to continue to grow in the absence of extracellular phosphate, until the intracellular storage (granules of poly-phosphate) is exhausted. Whether or not phosphate is the ultimate limiting factor in the ocean is debated. N_2 fixation is limited to the (sub)tropical oceans and it is therefore unclear whether this paradigm is valid for the temporal and cold oceans. Moreover, iron has been suspected to limit growth of diazotrophic cyanobacteria [88].

Iron is an essential micronutrient in almost all organisms. In the biosphere, iron occurs basically in two oxidations states: reduced, ferrous, iron (Fe^{2+}), and oxidized, ferric, iron (Fe^{3+}). Except under acidic conditions, ferric iron is virtually insoluble. Ferrous iron is very soluble but in the presence of oxygen it readily oxidizes to ferric iron. This means that in oxic seawater, the concentrations of dissolved iron are normally very low. Many organisms produce siderophores that bind iron extracellularly and that are subsequently transported together with the iron into the cell. Nitrogenase contains iron as a cofactor and also the electron donor to nitrogenase, ferredoxin, requires iron. The iron demand is about 7–11 times higher in diazotrophs than in non-diazotrophic organisms [88]. An important source of iron is dust that deposits at sea [89]. This dust may, for instance, come from the Sahara desert and is transported by wind to the Atlantic Ocean. Iron-containing dust particles may be captured by, for instance, colonies of *Trichodesmium*. Under the influence of UV light, particles of ferric iron captured by *Trichodesmium* may be reduced to ferric iron and be taken up by the cyanobacterium. Also, blooms of heterocystous cyanobacteria in the Baltic Sea may be at least temporarily limited by iron, although fresh and brackish waters as well as coastal waters are not usually considered to be iron limited due to the coastal run-off and river discharge. However, when the standing stock of biomass is high, the demand of nutrients will be equally high and may exceed the supply, even when the supply is high in absolute terms.

15.9 CONCLUSIONS

The fixation of atmospheric dinitrogen (N_2) is an essential step in the nitrogen cycle that counteracts the loss of "bioavailable nitrogen" from the environment through denitrification and anaerobic ammonium oxidation. It is catalyzed by nitrogenase, an enzyme complex that is readily inactivated by oxygen. Moreover, the fixation of N_2 is at the expense of a high amount of biochemical energy. Hence, diazotrophic organisms require an anoxic environment or, when living in an oxic environment, need to provide nitrogenase with a low-oxygen environment and protect the enzyme from inactivation by oxygen. Diazotrophy occurs throughout the domain Bacteria and Archaea, but diazotrophic cyanobacteria are extraordinary because they combine oxygenic photosynthesis and autotrophy with N_2 fixation. Cyanobacteria satisfy the energy and electron demands by using the abundant solar energy and water, respectively. However, cyanobacteria are challenged to protect nitrogenase from oxygen inactivation. Diazotrophic cyanobacteria are morphologically diverse and this diversity determines both the niches and habitats they occupy and the strategy evolved for protecting nitrogenase. Diazotrophic cyanobacteria may be filamentous or unicellular. Filamentous cyanobacteria may or may not differentiate heterocysts for the fixation of N_2. Heterocystous cyanobacteria separate nitrogenase spatially from the oxygenic vegetative cells and they fix most of the N_2 during the day, simultaneous with photosynthesis. Most non-heterocystous cyanobacteria (unicellular and filamentous) separate both incompatible processes temporally and confine N_2 fixation to the night. The non-heterocystous, filamentous oceanic *Trichodesmium* fixes N_2 during the day, probably by a combination of spatial and temporal separation of oxygenic photosynthesis and N_2 fixation. A newly discovered group of uncultivated unicellular cyanobacteria (UCYN-A) lack photosystem 2 and therefore resemble a heterocyst in that respect. They fix N_2 also during the day and are probably living symbiotic.

Cyanobacteria occupy nearly every illuminated environment and so do their diazotrophic representatives. Free-living heterocystous cyanobacteria seem to be restricted to freshwater, brackish, and terrestrial environments and largely excluded from marine environments, but symbiotic species are quite important in the (tropical) marine environment. Heterocystous cyanobacteria are also found in benthic marine systems. Non-heterocystous diazotrophic cyanobacteria are particularly important in the phytoplankton of (sub)tropical oceans.

The fixation of dinitrogen by cyanobacteria responds to a variety of environmental factors. Light is obviously the ultimate source of energy but it also causes the evolution of oxygen. The total daily integrated fixation of dinitrogen by cyanobacteria is often only marginally light dependent. This is because the energy for dinitrogen fixation is delivered through the respiration of storage carbohydrate. Heterocystous cyanobacteria are most successful when they increase the dark nitrogenase activity by increasing the influx of oxygen and respiration. Temperature is a complex factor. Temperature affects the solubility and diffusion of gases, including oxygen and nitrogen, but temperature also affects the rates of physiological processes. In warm waters, the oxygen saturation is low while respiration is high. Together, this makes the heterocyst superfluous and gives way for non-heterocystous dinitrogen-fixing cyanobacteria that are typical for the warmer tropical and subtropical waters. It is unclear why heterocystous cyanobacteria are also absent from the colder temperate seas but it is possible that the heterocyst is unable to accumulate compatible solutes and therefore may have a limited salt tolerance. Heterocystous cyanobacteria are abundant in eutrophic freshwater lakes and brackish seas. Because diazotrophic cyanobacteria are not limited by nitrogen, it has been conceived that phosphate is the ultimate limiting nutrient. However, this assumes that the fixation of dinitrogen can take place and is not limited by other factors and this is not always the case.

REFERENCES

1. Canfield, D.E., Glazer, A.N., and Falkowski, P.G., The evolution and future of Earth's nitrogen cycle, *Science* 330, 192, 2010.
2. Jetten, M.S.M. et al., Biochemistry and molecular biology of anammox bacteria, *Crit. Rev. Biochem. Mol. Biol.*, 44, 65, 2009.
3. Peters, J.W., Fisher, K., and Dean, D.R., Nitrogenase structure and function: A biochemical-genetic perspective, *Annu. Rev. Microbiol.*, 49, 335, 1995.
4. Bergman, B. et al., On the origin of plants and relations to contemporary cyanobacterial-plant symbioses, *Plant Biotechnol.*, 25, 213, 2008.
5. Howard, J.B. and Rees, D.C., Nitrogenase: A nucleotide-dependent molecular switch, *Annu. Rev. Biochem.*, 63, 235, 1994.
6. Zehr, J. and Turner, P., Nitrogenfixation: Nitrogenase genes and gene expression, *Methods Microbiol.*, 30, 271, 2001.
7. Kim, J. and Rees, D.C., Nitrogenase and biological nitrogen fixation, *Biochemistry*, 33, 389, 1994.
8. Rees, D.C. and Howard, J.B., Nitrogenase: Standing at the crossroads, *Curr. Opin. Chem. Biol.*, 4, 559, 2000.
9. Burgess, B.K. and Lowe, D.J., Mechanism of molybdenum nitrogenase, *Chem. Rev.*, 96, 2983, 1996.
10. Hardy, R.W.F. et al., The acetylene-ethylene assay for N_2 fixation: Laboratory and field evaluation, *Plant Physiol.*, 43, 1185, 1968.
11. Fani, R., Gallo, R., and Liò, P., Molecular evolution of nitrogen fixation: The evolutionary history of the nifD, nifK, nifE, and nifN genes, *J. Mol. Evol.*, 51, 1, 2000.
12. Raymond, J. et al., The natural history of nitrogen fixation, *Mol. Biol. Evol.*, 21, 541, 2004.
13. Berman-Frank, I., Lundgren, P., and Falkowski, P., Nitrogen fixation and photosynthetic oxygen evolution in cyanobacteria, *Res. Microbiol.*, 154, 157, 2003.
14. Bishop, P.E. and Joerger, R.D., Genetics and molecular biology of alternative nitrogen fixation systems, *Annu. Rev. Plant Physiol. Plant Mol. Biol.*, 41, 109, 1990.
15. Loveless, T.M., Royden Saah, J., and Bishop, P.E., Isolation of nitrogen-fixing bacteria containing molybdenum-independent nitrogenases from natural environments, *Appl. Environ. Microbiol.*, 65, 4223, 1999.
16. Eady, R.R., Structure–function relationships of alternative nitrogenases, *Chem. Rev.*, 96, 3013, 1996.
17. Boison, G. et al., The rice field cyanobacteria *Anabaena azotica* and *Anabaena* sp. CH1 express vanadium-dependent nitrogenase, *Arch. Microbiol.*, 186, 367, 2006.
18. Fay, P., Oxygen relations of nitrogen fixation in cyanobacteria, *Microbiol. Rev.*, 56, 340, 1992.
19. Stanier, R.Y. and Cohen-Bazire, G., Phototrophic prokaryotes: The cyanobacteria, *Annu. Rev. Microbiol.*, 31, 225, 1977.
20. Drews, G., The evolution of cyanobacteria and photosynthesis, in: Peschek, G.A., ed., *Bioenergetic Processes of Cyanobacteria*, Springer-Verlag, New York, 2011, p. 265.
21. Sánchez-Baracaldo, P., Hayes, P.K., and Blank, C.E., Morphological and habitat evolution in the cyanobacteria using a compartmentalization approach, *Geobiology*, 3, 145, 2005.
22. Schopf, J.W., Fossil evidence of Archaean life, *Philos. Trans. R. Soc. Lond. B*, 361, 869, 2006.
23. Srivastava, P., Vindhyanakinetes: An indicator of mesoproterozoic biosphere evolution, *Orig. Life Evol. Biosph.*, 35, 175, 2005.
24. Moreira, D., Le Guyader, H., and Philippe, H., The origin of red algae and the evolution of chloroplasts, *Nature*, 405, 69, 2000.
25. Partensky, F., Hess, W.R., and Vaulot, D., *Prochlorococcus*, a marine photosynthetic prokaryote of global significance, *Microbiol. Mol. Biol. Rev.*, 63, 106, 1999.
26. Larkum, A.W.D. and Kühl, M., Chlorophyll *d*: The puzzle resolved, *Trends Plant Sci.*, 10, 355, 2005.
27. Glazer, A.N., Phycobilisomes: Structure and dynamics, *Annu. Rev. Microbiol.*, 36, 173, 1982.
28. Everroad, C. et al., Biochemical bases of Type IV chromatic adaptation in marine *Synechococcus* spp., *J. Bacteriol.*, 188, 3345, 2006.
29. Ting, C.S. et al., Phycobiliprotein genes of the marine photosynthetic prokaryote *Prochlorococcus*: Evidence for rapid evolution of genetic heterogeneity, *Microbiology*, 147, 3171, 2001.
30. Rippka, R. et al., Generic assignments strain histories and properties of pure cultures of cyanobacteria, *J. Gen. Microbiol.*, 111, 1, 1979.
31. Velicer, G.J., Social strife in the microbial world, *Trends Microbiol.*, 11, 330, 2003.
32. Stal, L.J., Physiological ecology of cyanobacteria in microbial mats and other communities, *New Phytol.*, 131, 1, 1995.
33. Stewart, W.D.P., Some aspects of structure and function in N_2-fixing cyanobacteria, *Annu. Rev. Microbiol.*, 34, 497, 1980.

34. Stal, L.J. and Moezelaar, R., Fermentation in cyanobacteria, *FEMS Microbiol. Rev.*, 21, 179, 1997.
35. Cohen, Y., Padan, E., and Shilo, M., Facultative anoxygenic photosynthesis in the cyanobacterium *Oscillatoria limnetica*, *J. Bacteriol.*, 123, 855, 1975.
36. Stal, L.J. and Heyer, H., Dark anaerobic nitrogen fixation (acetylene reduction) in the cyanobacterium *Oscillatoria* sp., *FEMS Microbiol. Ecol.*, 45, 227, 1987.
37. Adams, D.G., Heterocyst formation in cyanobacteria, *Curr. Opin. Microbiol.*, 3, 618, 2000.
38. Walsby, A.E., Cyanobacterial heterocysts: Terminal pores proposed as sites of gas exchange, *Trends Microbiol.*, 15, 340, 2007.
39. Golden, J.W., Robinson, S.J., and Haselkorn, R., Rearrangement of nitrogen fixation genes during heterocyst differentiation in the cyanobacterium *Anabaena*, *Nature*, 314, 419, 1985.
40. Zhang, C.-C. et al., Heterocyst differentiation and pattern formation in cyanobacteria: A chorus of signals, *Mol. Microbiol.*, 59, 367, 2006.
41. Stal, L.J., Is the distribution of nitrogen-fixing cyanobacteria in the oceans related to temperature? *Environ. Microbiol.*, 11, 1632, 2009.
42. Tomitani, A. et al., The evolutionary diversification of cyanobacteria: Molecular-phylogenetic and paleontological perspectives, *Proc. Natl Acad. Sci. U.S.A.*, 103, 5442, 2006.
43. Bergman, B. et al., N_2 fixation by non-heterocystous cyanobacteria, *FEMS Microbiol. Rev.*, 19, 139, 1997.
44. Berman-Frank, I. et al., Segregation of nitrogen fixation and oxygenic photosynthesis in the marine cyanobacterium *Trichodesmium*, *Science*, 294, 1534, 2001.
45. Fredriksson, C. and Bergman, B., Ultrastructural characterisation of cells specialised for nitrogen fixation in a non-heterocystous cyanobacterium, *Trichodesmium* spp., *Protoplasma*, 197, 76, 1997.
46. Ohki, K., Intercellular localization of nitrogenase in a non-heterocystous cyanobacterium (Cyanophyte) *Trichodesmium* sp. NIBB1067, *J. Oceanogr.*, 64, 211, 2008.
47. Capone, D.G. et al., Basis for diel variation in nitrogenase activity in the marine planktonic cyanobacterium *Trichodesmium thiebautii*, *Appl. Environ. Microbiol.*, 56, 3532, 1990.
48. Church, M.J. et al., Temporal patterns of nitrogenase (*nifH*) gene expression in the oligotrophic North Pacific Ocean, *Appl. Environ. Microbiol.*, 71, 5362, 2005.
49. Zehr, J.P. et al., Globally distributed uncultivated oceanic N_2-fixing cyanobacteria lack oxygenic photosystem II, *Science*, 322, 1110, 2008.
50. Sherman, L.A., Meunier, P., and Colón-López, M.S., Diurnal rhythms in metabolism: A day in the life of a unicellular, diazotrophic cyanobacterium, *Photosynth. Res.*, 58, 25, 1998.
51. Huang, T.C. et al., Circadian rhythm of the prokaryote *Synechococcus* sp. RF-1, *Plant Physiol.*, 92, 531, 1990.
52. Compaoré, J. and Stal, L.J., Oxygen and the light–dark cycle of nitrogenase activity in two unicellular cyanobacteria, *Environ. Microbiol.*, 12, 54, 2009.
53. Ortega-Calvo, J.J. and Stal, L.J., Diazotrophic growth of the unicellular cyanobacterium *Gloeothece* sp. PCC 6909 in continuous culture, *J. Gen. Microbiol.*, 137, 1789, 1991.
54. Villbrandt, M., Stal, L.J., and Krumbein, W.E., Interactions between nitrogen fixation and oxygenic photosynthesis in a marine cyanobacterial mat, *FEMS Microbiol. Ecol.*, 74, 59, 1990.
55. Shi, T. and Falkowski, P.G., Genome evolution in cyanobacteria: The stable core and the variable shell, *Proc. Natl Acad. Sci. U.S.A.*, 105, 2510, 2008.
56. Bolhuis, H. et al., Horizontal transfer of the nitrogen fixation gene cluster in the cyanobacterium *Microcoleus chthonoplastes*, *ISME J.*, 4, 121, 2010.
57. Dominic, T.K. and Madhusoodanan, P.V., Cyanobacteria from extreme acidic environments, *Curr. Sci.*, 77, 1024, 1999.
58. Karl, D.M. et al., Ecological nitrogen-to-phosphorus stoichiometry at station ALOHA, *Deep-Sea Res.*, 48, 1529, 2001.
59. Staal, M. et al., Nitrogen fixation along a north-south transect in the eastern Atlantic Ocean, *Limnol. Oceanogr.*, 52, 1305, 2007.
60. Paerl, H.W. and Huisman, J., Climate change: A catalyst for global expansion of harmful cyanobacterial blooms, *Environ. Microbiol. Rep.*, 1, 27, 2009.
61. Walsby, A.E., Hayes, P.K., and Boje, R., The gas vesicles, buoyancy and vertical distribution of cyanobacteria in the Baltic Sea, *Eur. J. Phycol.*, 30, 87, 1995.
62. Zehr, J.P., Nitrogen fixation by marine cyanobacteria, *Trends Microbiol.*, 19, 162, 2011.
63. Severin, I. and Stal, L.J., Diazotrophic microbial mats, in: Seckbach, J. and Oren, A., eds., *Microbial Mats. Modern and Ancient Microorganisms in Stratified Systems*, Springer Science, Heidelberg, Germany, 2010, p. 321.

64. Severin, I., Acinas, S.G., and Stal, L.J., Diversity of nitrogen-fixing bacteria in cyanobacterial mats, *FEMS Microbiol. Ecol.*, 73, 514, 2010.
65. Steunou, A.-S. et al., Regulation of nif gene expression and the energetics of N_2 fixation over the diel cycle in a hot spring microbial mat, *ISME J.*, 2, 364, 2008.
66. Roeselers, G. et al., Diversity of phototrophic bacteria in microbial mats from Arctic hot springs (Greenland), *Environ. Microbiol.*, 9, 26, 2007.
67. Rai, A.N., Söderbäck, E., and Bergman, B., Cyanobacterium-plant symbioses, *New Phytol.*, 147, 449, 2000.
68. Foster, R.A., Carpenter, E.J., and Bergman, B., Unicellular cyanobionts in open ocean dinoflagellates, radiolarians, and tintinnids: Ultrastructural characterization and immuno-localization of phycoerythrin and nitrogenase, *J. Phycol.*, 42, 453, 2006.
69. Stal, L.J. and Krumbein, W.E., Temporal separation of nitrogen fixation and photosynthesis in the filamentous, non-heterocystous cyanobacterium *Oscillatoria* sp., *Arch. Microbiol.*, 149, 76, 1987.
70. Laamanen, M. and Kuosa, H., Annual variability of biomass and heterocysts of the N_2-fixing cyanobacterium *Aphanizomenonflos-aquae* in the Baltic Sea with reference to *Anabaena* spp. and *Nodularia spumigena*, *Boreal. Environ. Res.*, 10, 19, 2005.
71. Stal, L.J. and Walsby, A.E., Photosynthesis and nitrogen fixation in a cyanobacterial bloom in the Baltic Sea, *Eur. J. Phycol.*, 35, 97, 2000.
72. Severin, I. and Stal, L.J., Light dependency of nitrogen fixation in a coastal cyanobacterial mat, *ISME J.*, 2, 1077, 2008.
73. Breitbarth, E., Oschlies, A., and LaRoche, J., Physiological constraints on the global distribution of *Trichodesmium*—Effect of temperature on diazotrophy, *Biogeosciences*, 4, 53, 2007.
74. Dugdale, R.C., Menzel, D.W., and Ryther, J.H., Nitrogen fixation in the Sargasso Sea, *Deep Sea Res.*, 7, 297, 1961.
75. Karl, D. et al., Dinitrogen fixation in the world's oceans, *Biogeochemistry*, 57/58, 47, 2002.
76. Stal, L.J. et al., BASIC: Baltic Sea cyanobacteria. An investigation of the structure and dynamics of water blooms of cyanobacteria in the Baltic Sea—Responses to a changing environment, *Cont. Shelf Res.*, 23, 1695, 2003.
77. Warr, S.R.C., Reed, R.H., and Stewart, W.D.P., The compatibility of osmotica in cyanobacteria, *Plant Cell Environ.*, 11, 137, 1988.
78. López-Igual, R., Flores, E., and Herrero, A., Inactivation of a heterocyst-specific invertase indicates a principal role of sucrose catabolism in heterocysts of *Anabaena* sp., *J. Bacteriol.*, 192, 5526, 2010.
79. Cole, J.J. et al., Sulfate inhibition of molybdate assimilation by planktonic algae and bacteria: Some implications for the aquatic nitrogen cycle, *Biogeochemistry*, 2, 179, 1986.
80. Stal, L.J., Staal, M., and Villbrandt, M., Nutrient control of cyanobacterial blooms in the Baltic Sea, *Aquat. Microb. Ecol.*, 18, 165, 1999.
81. Kleiner, D., Bacterial ammonium transport, *FEMS Microbiol. Rev.*, 32, 87, 1985.
82. Ohlendieck, U., Stuhr, A., and Siegmund, H., Nitrogen fixation by diazotrophic cyanobacteria in the Baltic Sea and transfer of the newly fixed nitrogen to picoplankton organisms, *J. Mar. Syst.*, 25, 213, 2000.
83. Collier, J.L., Brahamsha, B., and Palenik, B., The marine cyanobacterium *Synechococcus* sp. WH7805 requires urease (urea amidohydrolase, EC 3.5.1.5) to utilize urea as a nitrogen source: Molecular-genetic and biochemical analysis of the enzyme, *Microbiology*, 145, 447, 1999.
84. Flynn, K.J. and Gallon, J.R., Changes in intracellular and extracellular alpha-amino acids in *Gloeothece* during N_2-fixation and following addition of ammonium, *Arch. Microbiol.*, 153, 574, 1990.
85. Toggweiler, J.R., An ultimate limiting nutrient, *Nature*, 400, 511, 1999.
86. Yelloly, J.M. and Whitton, B.A., Seasonal changes in ambient phosphate and phosphatase activities of the cyanobacterium *Rivulariaatra* in intertidal pools at Tyne Sands, Scotland, *Hydrobiologia*, 325, 201, 1996.
87. White, A.K. and Metcalf, W.W., Microbial metabolism of reduced phosphorus compounds, *Annu. Rev. Microbiol.*, 61, 379, 2007.
88. Kustka, A. et al., A revised estimate of the iron use efficiency of nitrogen fixation, with special reference to the marine cyanobacterium *Trichodesmium* spp. (Cyanophyta), *J. Phycol.*, 39, 12, 2003.
89. Ridame, C. et al., Nutrient control of N_2 fixation in the oligotrophic Mediterranean Sea and the impact of Saharan dust events, *Biogeosci. Discus.*, 8, 2629, 2011.

16 Adaptation of Cyanobacteria to Anthropogenic and Natural Stress

The Role Played for Spontaneous Mutation

*Raquel Gonzalez, Camino García-Balboa, Eduardo Costas, and Victoria Lopez-Rodas**

CONTENTS

16.1 INTRODUCTION

Stromatolites of fossilized oxygen-producing cyanobacteria have been found from at least 2.8 billion years ago, possibly as far back as 3.5 billion years ago. In contrast to other organisms whose final destination was extinction, cyanobacteria can be found in almost every conceivable environment: from oceans to fresh water and bare rock to soil. Accordingly, these organisms have had to survive global extinction of many species as well as several environmental crises. Presently there are numerous extreme natural environments that support growth of vast community of cyanobacteria. However, some questions yet remain unanswered, such as: how have cyanobacteria achieved the adaptation mechanisms to survive and proliferate under complicated stressful conditions?

Cyanobacteria must defend themselves against diverse types of selective pressures resulting from natural phenomena. Nowadays, changes in environmental conditions are occurring at an unprecedented rate as a result of large-scale changes caused by human activities. The massive loss of diversity, homogenization of biotas, proliferation of opportunistic species, and unpredictable emergent novelties can be considered among the distinctive features of the future biosphere [1]. During the

* Corresponding author: vlrodas@vet.ucm.es

last century, the disappearance rate of species was estimated to be 500-fold higher than had been the case over the preceding centuries, giving rise to an annual extinction of 30,000 species out of the 11 million currently estimated [2]. The species' collapse can be mainly attributed to the incapacity of organisms to cope with drastic environmental changes occurring in their habitats [3].

Freshwater reservoirs, lakes, rivers, coastal, and ocean areas receive considerable amounts of anthropogenic pollutants that alter the chemical balance and biogeochemical cycles and appear to be a major cause of the biodiversity crisis [4,5]. The impact of these toxic compounds on biodiversity threatens all ecosystems, being particularly significant in those characterized by a slow response to change, such as aquatic systems.

The productions and emissions of pollutants are usually derived from human settlements, resource uses, and interventions, such as infrastructural development and construction, agricultural activities, industrial developments, urbanization, tourism, etc. Contaminants of major concern include: persistent organic pollutants, nutrients, oils, radionuclides, heavy metals, pathogens, antibiotics, sediments, litter, and debris, etc. [6]. For example, herbicides like DCMU and glyphosate are among the most significant anthropogenic pollutants in aquatic ecosystems [7]. Similarly, antibiotics like erythromycin are widely used in modern agriculture and aquaculture activities in many developed nations and released in various habitats [8]. Several algaecides including copper sulfate are used to control algal blooms of water reservoirs meant for human use [9].

Phytoplankton is the autotrophic component of the plankton community. This group includes prokaryotic (cyanobacteria) and eukaryotic (microalgae) photosynthetic organisms that grow near the surface of the water column where they are able to capture light for photosynthesis. Despite their microscopic size, phytoplankton is responsible for about half of the global primary production of oxygen. These organisms also drive essential biogeochemical cycles and export massive amounts of carbon to deep waters and sediments in the open ocean. In addition, they have a strong influence on the water–atmosphere gas exchanges [10], so the repercussive impact on phytoplankton populations will undoubtedly affect the rest of the components of the trophic web. These environmental changes affect the abundance of the aquatic biota, as they are exposed to unprecedented scenarios. However, phytoplankton can be highlighted as a likely target to experience this environmental pressure but little is known about the mechanism of functioning within these organisms under such extreme conditions.

Within limits, organisms may survive in stressed environments as a result of two distinct processes. First, the majority of cyanobacteria have an incredible phenotypic plasticity to acclimate to modifications in environmental parameters and are able to survive in adverse habitats as a result of physiological acclimation, usually resulting from modifications of gene expression [11,12]. However, when changes in environmental conditions exceed physiological limits, survival is determined exclusively on adaptation by natural selection if mutations provide the appropriate genetic variability and confer resistance [13]. The neo-Darwinian view that evolutionary adaptation occurs by selection of preexisting genetic variation was accepted early in multicellular organisms [14]. However, recent evolutionary studies in bacteria have suggested that hypothetical "adaptive mutation" could be a process resembling Lamarckism wherein the absence of lethal selection produces mutations that relieve selective pressure [15,16]. The key to resolving this postulation is to know the preadaptive or post-adaptive origin of new mutations. Surprisingly, there are almost no studies that have made a direct connection between the rates of origin of favored mutants and the process of adaptation [17]. The main reason for this lack of studies is the difficulty in measuring the rate of favored mutants directly in diploid, multicellular, sexual organisms living in well-defined populations. On the contrary, most microbes (including cyanobacteria and many microalgae) are unicellular, asexual organisms and have only one copy of genetic material and their populations comprise countless cells [18]. Therefore, the study of genetic adaptation of cyanobacteria is a clear approximation to the problem of the origin of favored mutants and the process of adaptation.

In this study, our framework was focused on evaluating, from an evolutionary point of view, the mechanisms involved in the adaptation of cyanobacteria to sudden environmental changes arising from novel anthropogenic pollutants or extreme natural environments. For this purpose, an experimental technique such as fluctuation analysis was rigorously performed [19]. In addition, we have assessed the mechanisms (fitness and mutation–selection balance) that allow cyanobacteria to withstand continuous exposure to contaminants. Finally, the ratchet protocol constitutes a novel way to explore, through a rigorous experimental model, the limits of genetic adaptation in cyanobacteria to adapt to increasing anthropogenic-induced changes in environmental conditions.

16.2 MATERIALS AND METHODS

16.2.1 Fluctuation Test

Fluctuation analysis is the best experimental model to demonstrate if the adaptation to lethal doses of toxic compounds or to natural forcing could take place in wild strains of cyanobacteria. Furthermore, this technique is particularly well suited to discriminate between cells that become resistant due to either: (a) acquired specific adaptation in response to the anthropogenic or natural stress (including both physiological adaptation or acclimation, and the remote possibility of mutations following the exposure), and (b) resistant cells arising from rare spontaneous mutations that occur randomly during growth of cyanobacteria prior to the exposure to the selective agent.

A modified Luria–Delbrück fluctuation analysis using liquid cultures was used to investigate the occurrence of resistance (Figure 16.1) [20,21]. The modification involves the use of liquid medium (BG11 medium for freshwater algae and cyanobacteria [Sigma Aldrich Chemie, Taufkirchen, Germany]) containing the selective agent instead of plating bacterial cultures on a solid medium, as done by Luria and Delbrück [19].

For correct interpretation of the results, two different sets of cultures were managed. In the first set (Figure 16.1, Set 1A and Set 1B), cultures flasks (ca 100), containing nonselective culture medium, were inoculated with $N_0 = 10^1$–10^2 wild-type cells, a number small enough to reasonably ensure that no preexisting mutants were present. The cultures were grown under nonselective conditions until they reached $N_t = 10^6$–10^8, and thereafter supplemented with a lethal dose of the selective agent used in this study. The dose was previously calculated from a dose–growth rate relationship, and a dose 2–4 times higher than those which showed 100% inhibition of growth was selected.

For set 2 (control, Figure 16.1, Set 2), 25–50 flasks were inoculated each with $N_t = 10^6$–10^8 cells originating from the same parental population used in set 1 experiment. In this case, cells were directly supplemented with selective medium containing the dose as used in set 1 experiment. Both sets of cultures were kept under selective conditions for a period of time long enough to allow resistant cells to grow (usually 75–90 days). Flasks were placed at 22°C under a continuous photon flux density of 60 μmol m^{-2} s^{-1} over the waveband 400–700 nm provided by cool white fluorescent tubes using the most suitable method depends on the specie used in the experiments a Beckman (Brea, Ca, USA) Z2 particle counter or a Uriglass settling chamber (Biosiga, Cona, Italy) in an inverted microscope (Axiovert 35, Zeiss Oberkóchen, Germany).

Analysis of data showed two different results in the set 1 experiment, each of them could be interpreted as the independent consequence of two different phenomena of adaptation [19,22]. In the first case, if resistant cells appeared by specific post-selective mutations or physiological adaptation, every cell is likely to have the same opportunity of developing resistance and cells per culture variance would be low (Figure 16.1, set 1A). Consequently, interculture (flask-to-flask) variation should be consistent with the Poisson model (and so, variance/mean ≈ 1). By contrast, if resistant cells arise before the exposure (i.e., genetic adaptation by rare spontaneous mutation occurring during the time in which the cultures grew to N_t from N_0 before selection), then the flask-to-flask variation would not be consistent with the Poisson model (i.e., variance/mean >1; Figure 16.1, Set 1B). Obviously, results may differ (0 resistant cells in each culture), indicating that neither selection by spontaneous mutations had occurred prior to exposure nor specific adaptation during the exposure had taken place.

The set 2 cultures were experimental controls of the fluctuation analysis. Anyway, resistance cells appear and interculture variance is expected to be low, because set 2 samples the variance of

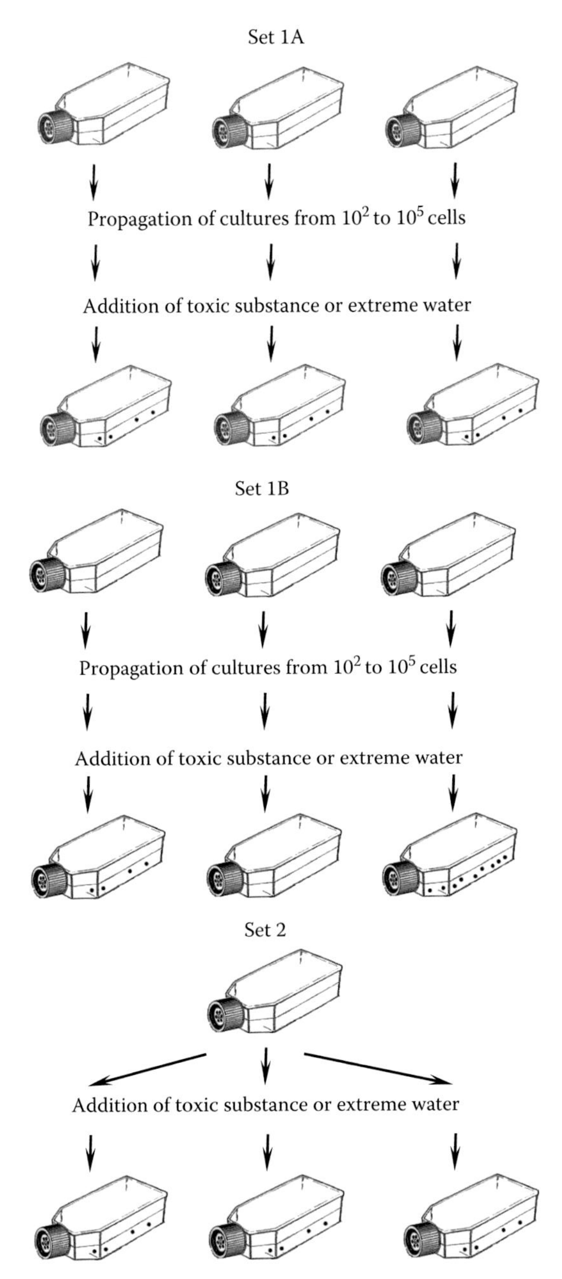

FIGURE 16.1 Schematic diagram of the modified Luria and Delbrück [19] fluctuation analysis. In the set 1, several cultures, each inoculated with small inoculum, were propagated until a high cell density was reached, and then a lethal dose of the toxic substance or extreme water was added. Set 1A: physiological adaptation (i.e., acclimation) or possible adaptive mutations. In this case, the number of resistant cells in all the cultures must be similar. Set 1B: adaptation by mutations occurring in the period of the propagation of cultures, that is, before exposure to the selective agent. One mutational event occurred late in the propagation of culture 2 (therefore, the density of resistant cells found is low) and early in the propagation of culture 3 (thus, density of resistant cells found is higher than in culture 2); no mutational events occurred in culture 1. Therefore, the variance/mean ratio of the number of resistant cells per replicate must be >1. Set 2 samples the variance of parental populations as an experimental control. In this case, the number of resistant cells in all the cultures must be similar (the distribution should be Poisson, with a variance approximately equal to the mean). (Modified from Lopez Rodas et al., *Eur. J. Phycol.*, 36, 179, 2001.)

the parental population. If the variance/mean ratio of set 1 was significantly greater than the variance/mean ratio of set 2 (fluctuation), this would confirm that resistant cells arose by rare mutations that occurred before the exposure to the pollutant or to the water samples from extremely stressful ecosystems. By contrast, if a similar variance/mean ratio between set 1 and set 2 was found, it would confirm that resistant cells arose during experimental exposure.

In addition, the fluctuation analysis also allows estimation of the rate of appearance of resistant cells. There are different approaches for accomplishing this estimation [23]. Due to methodological limitations imposed by fluctuation analysis using liquid cultures, the proportion of cultures from set 1 showing no resistant cells (P_0 estimator) was used to calculate the mutation rate (μ) using the following equation:

$$\mu = \frac{\text{Log}_e P_0}{(N_t - N_0)}$$

where N_0 and N_t were the initial and final cell population size, respectively.

16.2.2 Equilibrium State

If mutation from wild-type, sensitive allele to resistant cells is recurrent, new mutants may develop in each generation, and resistant alleles become detrimental to fitness when the selective agent is not present in the medium. Obviously, most of these mutants would be eliminated eventually by natural selection, if not by chance. At any given time, there will be a certain number of resistant cells that have not yet been eliminated. According to Kimura and Maruyama [24], the balance between μ and the rate of selective elimination (s) will determine the average number of such mutants, according to the equation

$$q = \frac{\mu}{(\mu + s)}$$

where

q is the frequency of the resistant allele

s is the coefficient of selection calculated, according to the equation

$$s = 1 - \left(\frac{m_T^r}{m_T^s}\right)$$

where m_T^r and m_T^s are the Malthusian fitness of resistant and sensitive cells measured in nonselective conditions, respectively.

16.2.3 Ratchet Protocol (Testing Limitations)

When genetic adaptation is found via fluctuation analysis, the adaptation is referred to by a given lethal dose of the toxic substance (usually, two to three times higher than that causing 100% inhibition of growth). However, the potential limit of adaptation to the highest concentration of the toxic substance is difficult to estimate via fluctuation analysis, since stronger selection pressures drastically reduce population size. This constraint can be overcome by performing experiments that include several values of selection pressure. To this end, Reboud et al. [25] developed an experimental model aimed at evaluating the maximal potential for herbicide resistance evolution in the green

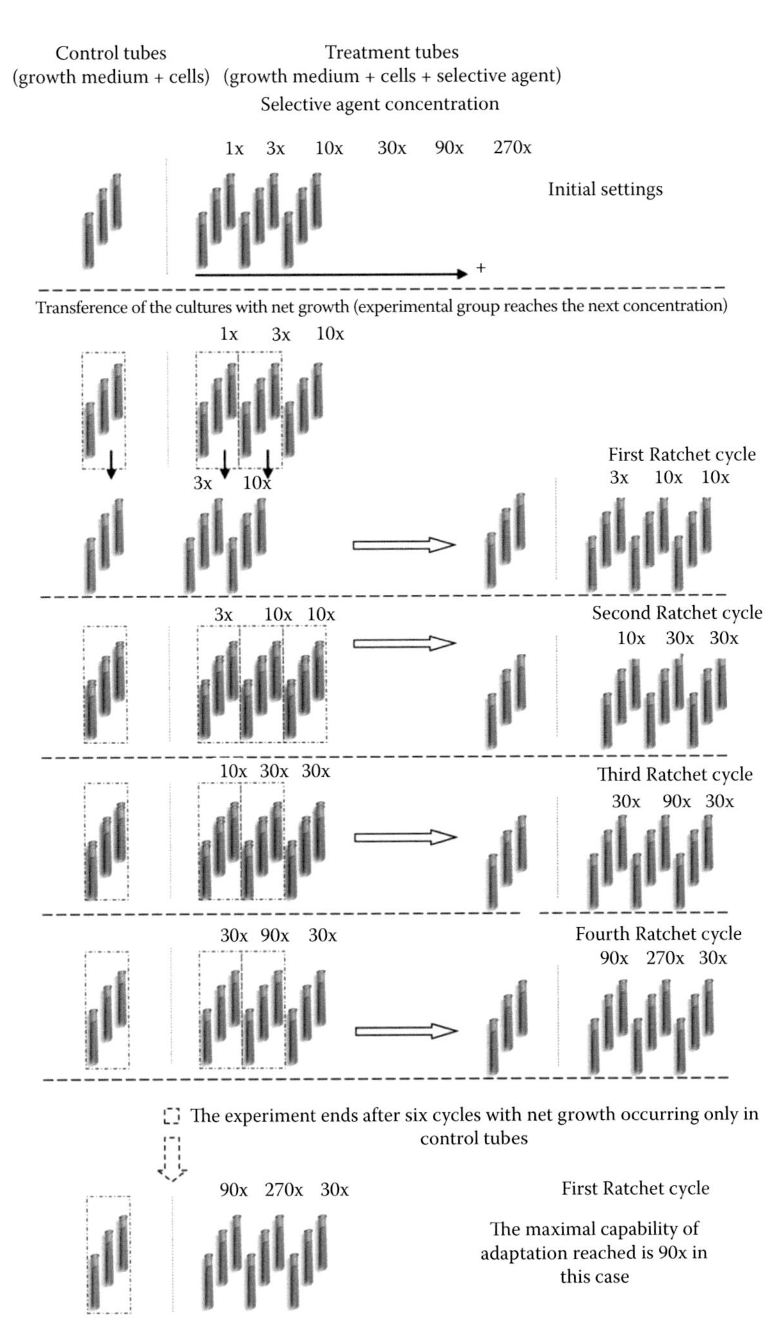

FIGURE 16.2 Schematic representation of the ratchet experimental design. Three replicates of the control cultures and three replicates of cultures for each of the three initial doses of the selected agent are present in each ratchet cycle. Each tube is transferred to the next concentration when the same net growth as the control tubes is reached (these tubes are represented as framed in the figure). Tubes that do not present net growth are maintained at the same concentration. A new ratchet cycle is considered each time the control tubes are transferred. The experiment ends after six cycles with net growth occurring only in the control tubes. At this point, the maximal capability of adaptation corresponds to the maximal concentration of the selective agent that presents net growth. (From Huertas et al., *New Phytol.*, 188, 478, 2010.)

microalga *Chlamydomonas*. This experiment was based on the use of different concentrations of herbicide, which, thereby, imposed different selection pressures. Furthermore, Orellana et al. [26] provided a modified procedure that allowed for maximizing the occurrence of mutants in microalgae and their selection by applying variable selection pressures. A significant refinement of this experimental procedure was achieved by using different replicates of each strain under each selection condition. This assures reproducibility and is referred to as ratchet assays [27]. The protocol aims at reaching equilibrium between strong selection intensity by means of ratcheting to increase doses of the selective agent, and at maintenance of a population size large enough to increase probability of rare spontaneous mutation that confers adaptation. Cultures must be ratcheted only up to a dose that supports population growth. This experimental procedure is then applied in several independent replicates (Figure 16.2).

During the initial step, replicates of the control cultures containing growth medium and replicates of cultures for each initial dose of the treatment are prepared. All cultures are counted by a Beckman Z2 particle counter and kept under the selective conditions for 20 days and then observed. Each culture is transferred to the next concentration when the same net growth of the control cultures has reached; cultures that do not present net growth are maintained at the same concentration. A new ratchet cycle is concluded each time the control cultures are transferred. The experiment ends after several cycles when net growth occurs only in the control cultures. At this point, the maximal capability of adaptation corresponds to the maximal concentration of the selective agent that presents net growth. Data are analyzed employing the classic equations of Novick and Szilard [28]. The following equations are used:

$$N_t = N_0\, 2^{t/T}$$

$$g = t/T$$

where

- N_t is the number of cells at time t
- N_0 is the number of cells at time 0
- t is the time
- T is the generation time
- g is the number of generations, which are used to estimate the number of generations during the ratchet experiments

16.3 RESULTS AND DISCUSSION

16.3.1 Example Cases of Genetic Adaptation of Cyanobacteria

Cyanobacteria are a very important segment as primary producers of aquatic ecosystems but are frequently exposed to diverse types of anthropogenic agents. Henceforth, the tolerance of these organisms to stressed environments is an important subject as it carries an ecological price tag. Numerous resistant mutants have been reported in cyanobacteria especially against herbicides [7]. However, the empirical evidence for the appearance of resistant mutants in cyanobacteria from wild types by spontaneous mutations has only been addressed recently by fluctuation analysis (Table 16.1).

During the last 10 years, we have analyzed the adaptation of cyanobacteria to numerous anthropogenic pollutants (herbicides, antibiotics, heavy metals, and others) [20,29,30] and to diverse types of extreme environments distributed throughout the world [31–35]. In particular, when cultures of cyanobacteria were exposed to the selective agent, their number drastically decreased in both experimental set 1 and set 2, most probably due to death of sensitive cells by the lethal effects of the exogenously added test substances. However, after prolonged incubation lasting for several

TABLE 16.1
Mutation Rate (μ), Coefficient of Selection, and Mutation–Selection Balance in the Genetic Adaptation by Different Microorganisms in the Presence of Different Substances

Anthropogenic Pollutants	Organisms	Mutation Rate (μ)	Coefficient of Selection (S)	Resistant Mutant Allele Frequency (q)
DCMU (30 μM)	*Pseudoanabaena planktonica*[a]	2.4×10^{-6}	0.84	1.6×10^{-3}
Glyphosate (120 mg L^{-1})	*Microcystis aeruginosa*[a]	3.6×10^{-7}	0.83	6.5×10^{-4}
Erythromycin (10 mg L^{-1})	*Pseudoanabaena planktonica*[a]	2.1×10^{-6}	0.82	1.6×10^{-3}
Copper sulfate (Cu^{2+} concentration of 10 μM)	*Microcystis aeruginosa*[a]	1.76×10^{-6}	0.76	1.5×10^{-3}

[a] Phytoplankton species obtained from the Algal Culture Collection, Genetics Laboratory, Veterinary Faculty, Complutense University, Madrid, Spain.

weeks, the fraction of cultures increased in density in the set 1, apparently due to growth of resistant variants. By contrast, all cultures of set 2 showed recovery, and resistant cells were prevalent in all culture flasks. Additionally, less resistant cell fluctuation was observed in set 2 (variance/mean < 1, consistent with Poisson variability), which indicates that the high fluctuation found in set 1 cultures may result from processes other than sampling error. As in set 1 cultures, the variance significance exceeded the mean (variance/mean ≫1); it may be inferred that resistant cells arose by rare, preselective spontaneous mutations occurring randomly during growth under nonselective conditions rather than by specific adaptation or post-selective mutations appearing in response to the selective agent.

Table 16.1 presents the data of the mutation rate from sensitive to resistances in two different species of cyanobacteria, namely, *Pseudoanabaena planktonica* and *Microcystis aeruginosa* against several anthropogenic substances. A representative group of pollutants have been chosen, which included DCMU (3-(3,4-dichlorophenyl)-1,1-dimethylurea) and glyphosate (both herbicides), antibiotics like erythromycin, and heavy metal like copper sulfate. Analysis of results showed that the mutation rates ranged from 0.36 to 1.76 mutants per 10^6 cells per generation. Since mutation can be recurrent in each generation, new mutant cells may be seen arising continuously. One of the main characteristics of resistant mutants in comparison to the wild type (parent) is that the former has a significantly lower growth rate than the latter, as is evident by q values of Table 16.1. It is worth mentioning here that mutations usually imply an energetic cost that may affect the survival of adapting populations [36,37]. Obviously, most of the resistant mutants are eventually eliminated by natural selection [38]. At any given time, the balance between continuous appearance of mutants and their selective elimination determine the number of remaining resistant mutants in algal populations growing in the absence of the toxic substance. This may be the case in wild-type populations growing in nonpolluted waters. As such, the population would be predominantly a clonal line of sensitive genotypes accompanied by a very small fraction of resistant mutant clones. In summary, rare spontaneous mutations conferring resistance seem to be a novel means of ensuring survival of cyanobacteria in polluted waters. The primary production supported by cyanobacteria and microalgae could be significantly lower than the present value as a consequence of the diminished growth rate of resistant mutants in comparison to wild-type cells.

Toxic waters from extreme aquatic environments could be considered natural labs with a natural toxic gradient. In a way, it facilitates the understanding of the adaptive process that allows the survival of mesophilic cyanobacteria in these environments. These habitats are characterized by extreme values of pH, toxic mineral concentrations, temperature, salinity, and other stress factors that limit the survival of mesophilic cyanobacteria. However, despite its toxicity, these waters are able to support a diverse phytoplankton community. If the stressful conditions do not exceed the limits of the physiological tolerance of cyanobacteria, survival is considered to be the result of physiological adaptation (i.e., acclimation) supported by alteration of gene expression [11]. On the other hand, in extreme environments as characterized by values of ecological factors exceeding the physiological limits of cyanobacteria, survival would depend exclusively on adaptive evolution, which results from the occurrence of new mutations that confer resistance [15,18,31–35]. However, if the toxicity is extremely high, adaptation cannot occur by any mechanism.

In general, cyanobacteria have an exceptional ability to withstand extreme temperatures, dryness, and high salinities [39]. Almost all the waters examined had a common feature—they have an extremely low pH—and cyanobacteria were not found in extreme acidic environments [40–45], suggesting a limit of 4.8 pH for growth and proliferation. This may be associated with the absence of ATPase transport and also with the peripheral location of the photosynthetic apparatus instead of chloroplast, as found in eukaryotic algae [41]. Nevertheless, several studies have reported the presence of cyanobacteria in some acid lakes having a pH of 2.9 from Germany [46]. Such organisms are called cyanellas, which are endosymbiotic cyanobacteria and live inside the host cell (eukaryotic algae). Host cells perform the same functions as the plastids [46]. The earlier assumption could be valid for unicellular cyanobacteria but not for endosymbiotic ones. To a certain extent, these findings correspond to our findings since *Microcystis aeruginosa* is also a unicellular organism.

In recent years, a number of studies of adaptation in several stressful environments around various parts of the world (i.e., Vulcano Island, southern Italy; Rio Tinto water, southwest Spain; diverse geothermal waters in Italy and Argentina; Agrio River-Caviahue Lake system Neuquén, Argentina) have been carried out. Table 16.2 shows the water characteristics and capacities of different organisms to adapt to certain extremophilic conditions of water based on the mutation rate (μ, mutants per cell per generation); the coefficient of selection against resistant mutant (s); and the frequency of the resistance allele (q), particularly in *Microcystis aeruginosa* during genetic adaptation to some geothermal waters, which have a neutral pH.

16.3.2 Limits of Genetic Adaptation to Contaminants

Recent works have shown that taxonomic group, ploidy level, growth rate, and habitat preference are involved in the ability of the different groups of phytoplankton to adapt to a contaminant [27,47]. With taxonomic group, chlorophyta showed the strongest capacity to adapt to the highest levels of selection, while cyanobacteria did not adapt to that extent. This suggests that the difference results from the fact that prokaryotic organisms are more adversely affected by anthropogenic pollutants (especially herbicides) than eukaryotic species [48]. Several studies pertaining to adaptation have demonstrated that members of cyanobacteria are usually the most sensitive to certain pollutants as compared to eukaryotic microalgae [27,49]. This finding has been attributed to the overlapping of respiratory and photosynthetic electron transports in cyanobacteria, as both pathways share numerous electron transport intermediaries that are the target of the triazine herbicides [50]. Bacillariophyta and haptophyta are known to show from moderate to scarce ability to adapt, respectively. Table 16.3 shows the data of differences in adaptation among different groups of phytoplankton against copper sulphate and simazine.

From an ecological point of view, it is interesting to highlight that the clear differences in adaptation to different agents depend on the types of habitat. This is evident from the fact that the greatest adaptation ability was found in phytoplankters from epicontinental freshwaters

TABLE 16.2
Water Characteristics, Mutation Rate (μ), Coefficient of Selection, and Mutation–Selection Balance in the Genetic Adaptation to Different Extreme Waters in *Microcystis aeruginosa*

Water Origin	Water Characteristics	Mutation Rate (μ)	Coefficient of Selection (q)	Resistant Mutant Allele Frequency (q)
Agrio River (Argentina)	pH: ≤ 4, 1 Extreme acidity from (SO_2, HCl and HF) Ca: 1250–100 mg L^{-1} Fe: 800–50 mg L^{-1} Si: 100–10 mg L^{-1} SO_4: 1200–100 mg L^{-1}	No adaptation		
Vulcano island (Italy)	pH: 3.1 Sulfhydric acid levels: 1.84 ± 0.10 g L^{-1} T: 30.3°C ± 0.5°C	No adaptation		
Tinto River (Spain)	pH: 1.7–2.5 Fe: 0.4–20.2 g L^{-1} Cu: 0.02–0.70 g L^{-1} Zn: 0.02–0.56 g L^{-1}	No adaptation		
Geothermal waters (Italy and Argentina)	pH: 2.5–7 [35]	1.1×10^{-5} (pH: 6.5) 7.9×10^{-6} (pH: 6.3) 2.7×10^{-6} (pH: 7.7) 1.1×10^{-6} (pH: 8.3)	0.93 0.86 0.92 0.78	1.2×10^{-5} 9.2×10^{-6} 2.9×10^{-6} 1.4×10^{-6}

comprising chlorophyta and cyanobacteria (usually the sink of arrays of several pollutants). In general, the recurrent presence of toxic substances in the environment has an effect on phytoplankton communities wherein sensitive organisms are excluded and resistant organisms are favored. Under such a condition, the tolerance of the population will be greater mainly due to the presence of resistant alleles in these populations [51]. In the same way but to a lesser magnitude, coastal marine microalgae are also exposed to anthropogenic agents as a result of river discharge and they exhibit intermediate adaptation abilities. Among all the groups, the most sensitive group is formed by oceanic microalgae species; these phytoplankton species live in habitats without pollutants and can therefore be more vulnerable to sudden exposures to pollutants than species inhabiting more contaminated regions. Thus, it can be hypothesized that the potential of phytoplankters to adapt to the toxic substances depends on previous evolutionary exposure history. Therefore, a sudden contamination episode could be relieved by freshwater phytoplankters but not by oceanic phytoplankters.

As a part of the differential ability of adaptation, the ratchet experiment allows investigation of whether the adaptation was by genetic mechanisms or by a physiological acclimation. In all the cases, it was observed that the adaptation to the highest selective pressure was by genetic mechanisms, as the time required for the acquisition of adaptation differed between replicates of the same species. Furthermore, the growth rate of the derived populations after the ratchet experiments was significantly lower than the growth rate in the ancestral populations prior to the ratchet experiments. Certain studies based on fluctuation analysis have also demonstrated the genetic origin of the resistant cells that appear when microalgae are exposed to simazine and copper sulfate, the same pollutants used in the ratchet experiments [30,52].

TABLE 16.3
Adaptation Increase (Times) and Number of Generations Required to Obtain Adaptation Comparing Cyanobacteria Species with Other Species of Phytoplankton

		Copper Sulphate		Simazine	
Cepa	**Characteristics (Taxonomic Group, Habitat, Cell Division)**	**Adaptation Increase (Times)**	**g**	**Adaptation Increase (Times)**	**g**
Dictyosphaerium	Chlrorophyta	270	230	90	300
chlorelloides[a]	Continental	270	253	90	300
	Rapid	90	184	90	255
Scenedesmus	Chlrorophyta	270	240		300
intermedius[a]	Continental	270	240		300
	Rapid	270	270		315
Microcystis	Cyanobacteria	30	84	9	75
aeruginosa (3D)[a]	Continental	10	48	9	120
	Moderate	10	48	9	93
Microcystis	Cyanobacteria	10	36	9	90
aeruginosa (6D)[a]	Continental	10	48	9	105
	Moderate	10	36	9	98
Microcystis	Cyanobacteria	10	48	9	76
aeruginosa (7D)[a]	Continental	10	48	9	76
	Moderate	10	36	9	75
Tetraselmis	Chlorophyta	—	—	10	150
suecica[a]	Coastal		—	1.0	165
	Rapid			10	150
Phaeodactylum	Bacillarophyta	—	—	4.5	150
tricornutum[a]	Coastal		—	4.5	165
	Rapid		—	4.5	150
Emiliania huxleyi[a]	Haptophyta	—	—	1.5	36
(CCMP371)	Oceanic		—	1.5	36
	Slow		—	1.5	32
Emiliania huxleyi[a]	Haptophyta	—	—	1.5	32
(CCMP372)	Oceanic		—	1.5	32
	Slow		—	1.5	36
Emiliania huxleyi[a]	Haptophyta	—	—	3	40
(CCMP373)	Oceanic		—	3	36
	Slow		—	3	X
Isochrysis	Haptophyta	—	—	1.5	40
galbana[a]	Oceanic		—	1.5	45
	Moderate		—	1.5	40
Monochrysis	Haptophyta	—	—	1.5	45
lutheri[a]	Oceanic		—	1.5	45
	Moderate		—	1.5	40

Cell division: rapid, one doubling every 2–3 days; moderate, one doubling every 3–4 days; slow, one doubling every 5–7 days.

[a] Phytoplankton species obtained from the Algal Culture Collection, Genetics Laboratory, Veterinary Faculty, Complutense University, Madrid, Spain.

16.4 CONCLUSIONS

The present study demonstrates that it is possible to predict the response of cyanobacterial populations to global change by designing suitable experimental setups. By setting such experiments, it is possible to know the capacity of genetic adaptation in these microorganisms. Although there still remain many uncertainties concerning the impact of anthropogenic contaminants on cyanobacteria, rare spontaneous mutations conferring resistance against extreme natural waters or toxic compounds seem enough to ensure survival of cyanobacterial populations in polluted or toxic waters. Additionally, Ratchet protocol, despite being an oversimplification of reality, can be considered a novel way to explore the prevalence of genetic adaptation in cyanobacteria and other phytoplankton groups. It is also pertinent to mention that if the organisms are unable to adapt to an increasing pollution, this will undoubtedly cause shifts in the composition of the phytoplankton community and may replace the impaired individuals by resistant ones. An absolute scenario cannot be envisaged at this stage; nevertheless, genetics will surely determine the survival of the organisms to the environmentally driven selection.

ACKNOWLEDGMENTS

This work has been financially supported by the Spanish Ministry of Sciences and Innovation through the grants CTM2008- 05680 C02-01/MAR and CGL2008-00652/BOS. The authors thank the Ministerio de Educación y Ciencia for the financial support through the FPU program.

REFERENCES

1. Myers, N. and Knoll, A.H., The biotic crisis and the future of evolution, *Proc. Natl Acad. Sci. U.S.A.*, 98, 5389, 2001.
2. Woodruff, D.S., Declines of biomes and biotas and the future of evolution, *Proc. Natl Acad. Sci. U.S.A.*, 98, 5471, 2001.
3. Ramakrishnan, B. et al., The impacts of environmental pollutants on microalgae and cyanobacteria, *Crit. Rev. Environ. Sci. Technol.*, 40, 699, 2010.
4. Tilman, D., Global environmental impacts of agricultural expansion: The need for sustainable and efficient practices, *Proc. Natl Acad. Sci. U.S.A.*, 96, 5995, 1999.
5. Malato, S. et al., Degradación de plaguicidas, in *Eliminación de Contaminantes por Fotocatálisis Heterogénea*, Blesa, M.A., Ed., Programa Iberoamericano de Ciencia y Tecnología para el Desarrollo (CYTED), La Plata, Argentina, p. 269, 2001.
6. Williams, C., Combating marine pollution from land-based activities: Australian initiatives, *Ocean Coastal Manag.*, 33, 87, 1996.
7. Koenig, F., Eukaryotic algae, cyanobacteria and pesticides, in *Algal Adaptation to Environmental Stresses, Physiological, Biochemical and Molecular Mechanisms*, Rai, L.C. and Gaur, J.P., Eds., Springer, Berlin, Germany, 2001, p. 389.
8. Sarmah, A.K., Meyer, M.T., and Boxall A.B.A., A global perspective on the use, sales, exposure pathways, occurrence, fate and effects of veterinary antibiotics (VAs) in the environment, *Chemosphere*, 65, 725, 2006.
9. Hrudey, S. et al., Remedial measures, in *Toxic Cyanobacteria in Water. A Guide to Their Public Health Consequences, Monitoring and Management*, Chorus, I. and Bartram, J., Eds., Routledge, London, U.K., 1999, 275.
10. Rost, B., Zondervan, I., and Wolf-Gladrow, D., Sensitivity of phytoplankton to future changes in ocean carbonate chemistry: Current knowledge, contradictions and research directions, *Marine Ecol. Prog. Series*, 373, 227, 2008.
11. Bradshaw, A.D. and Hardwick, K., Evolution and stress—Genotype and phenotype components, *Biol. J. Linn. Soc.*, 37,137, 1989.
12. Fogg, G.E., Algal adaptation to stress, some general remarks, in *Algal Adaptation to Environmental Stresses. Physiological, Biochemical and Molecular Mechanisms*, Rai, L.C. and Gaur, J.P., Eds., Springer, Berlin, Germany, 2001, p. 1.

13. Belfiore, N.M. and Anderson, S.L., Effects of contaminants on genetic patterns in aquatic organisms: A review, *Mutat. Res.*, 489, 97, 2001.
14. Sniegowski, P.D. and Lenski, R.E., Mutation and adaptation: The directed mutation controversy in evolutionary perspective, *Annu. Rev. Ecol. Syst.*, 26, 553, 1995.
15. Cairns, J., Overbaugh, J., and Miller, S., The origin of mutants, *Nature*, 335, 142, 1988.
16. Foster, P.L., Adaptive mutation: Implications for evolution, *BioEssays*, 22, 1067, 2000.
17. Sniegowski, P.D., Linking mutation to adaptation: Overcoming stress at the spa, *New Phytol.*, 166, 360, 2005.
18. Margulis, L. and Schwartz K.V., *Five Kingdoms. An Illustrated Guide to the Phyla of Live on Earth*, 3rd edn., Freeman & CO, New York, 1998.
19. Luria, S. and Delbrück, M., Mutations of bacteria from virus sensitivity to virus resistance, *Genetics*, 28, 491, 1943.
20. López-Rodas, V. et al., Resistance of microalgae to modern water contaminants as the result of rare spontaneous mutations, *Eur. J. Phycol.*, 36, 179, 2001.
21. Costas, E. et al., Mutation of algae from sensitivity to resistance against environmental selective agents: The ecological genetics of *Dictyosphaerium chlorelloides* (Chlorophyceae) under lethal doses of 3-(3,4-dichlorophenyl)-1,1 dimethylurea herbicide, *Phycologia*, 40, 391, 2001.
22. Jones, M.E., Thomas, S.M., and Rogers, A., Luria-Delbrück fluctuation experiments: Design and analysis, *Genetics*, 136, 1209, 1994.
23. Rosche, W.A. and Foster P.L., Determining mutation rates in bacterial populations, *Methods*, 20, 4, 2000.
24. Kimura, M. and Maruyama, T., The mutational load with epistatic gene interactions in fitness, *Genetics*, 54, 1337,1966.
25. Reboud, X. et al., *Chlamydomonas reinhardtii* as a model system for pro-active herbicide resistance evolution research, *Biol. J. Linn. Soc.*, 91, 257, 2007.
26. Orellana, G. et al., Biosensores basados en microalgas para la deteccion de contaminantes medioambientales, CT / ES2008 / 000465, 2008, P200701905, 2008.
27. Huertas, I.E. et al., Estimating capability of different phytoplankton groups to adapt to contamination: Herbicides will affect phytoplankton species differently, *New Phytol.*,188, 478, 2010.
28. Novick, A. and Szilard, L., Chemostat on spontaneous mutations of bacteria, *Proc. Natl Acad. Sci. U.S.A.*, 36, 708, 1950.
29. Lopez-Rodas, V., Flores-Moya, A., Maneiro, E., Perdigones, N., Marva, F., Garcia, M.E., and Costas, E., Resistance to glyphosate in the cyanobacterium *Microcystis aeruginosa* as a result of pre-selective mutations, *Evol. Ecol.*, 21, 535, 2007.
30. Garcia-Villada, L. et al., Occurrence of copper resistant mutants in the toxic cyanobacteria *Microcystis aeruginosa*: Characterization and future implications in the use of copper sulphate as algaecide, *Water Res.*, 38, 2207, 2004.
31. Flores-Moya, A. et al., Adaptation of *Spirogyra insignis* (Chlorophyta) to an extreme natural environment (sulphureous waters) through pre-selective mutations, *New Phytol.*, 166, 655, 2005.
32. López-Rodas, V. et al., Microalgal adaptation in the stressful acidic, metal-rich mine waters from Mynydd Parys (N Wales, U.K.) could be due to selection of preselective mutants, *Environ. Exp. Bot.*, 61, 43, 2008a.
33. Lopez-Rodas, V. et al., Adaptation of the chlorophycean *Dictyosphaerium chlorelloides* to the stressful acidic, mine metal-rich waters from Aguas Agrias Stream (SW Spain) as result of pre-selective mutations, *Chemosphere*, 72, 703, 2008b.
34. Costas, E. et al., How eukaryotic algae can adapt to Spain's Rio Tinto: A neo-Darwinian proposal for rapid adaptation to an extremely hostile ecosystem, *New Phytol.*, 175, 334, 2007.
35. Costas, E., Flores-Moya, A., and Lopez-Rodas, V., Rapid adaptation of algae to extreme environments (geothermal waters) by single mutation allows "Noah's Arks" for photosynthesizers during the Neoproterozoic "snowball Earth"? *New Phytol.*, 189, 922, 2008.
36. Cousteau, C., Chevillon, C., and French-Constant, R., Resistance to xenobiotics and parasites: Can we count the cost? *Trends Ecol. Evol.*, 15, 378, 2000.
37. Vila-Auib, M.M., Neve, P., and Powles, S.B., Fitness costs associated with evolved herbicide resistance alleles in plants, *New Phytol.*, 184, 751, 2009.
38. Crow, J.F., and Kimura, M., *An Introduction to Population Genetics Theory*, Harper & Row, New York, 1970, p. 591.
39. Rai, L.C. and Gaur, J.P., *Algal Adaptation to Environmental Stresses*, 1st edn., Springer, Berlin, Germany, 2001.

40. Brock, T.D., Lower pH limit for the existence of blue-green algae: Evolutionary and ecological implications, *Science*, 179, 480, 1973.
41. Brock, T.D., *Thermophilic Microorganisms and Life at High Temperatures*, 1st edn., Springer-Verlag, New York, 1978.
42. Knoll, A.H. and Bauld, J., The evolution of ecological tolerance in prokaryotes, *Trans. R. Soc. Edinburgh Earth Sci.*, 216, 227, 1989.
43. Albertano, P., Microalgae from sulphuric acid environments, in *Algae, Environments and Human Affairs*, Wiesneener, W., Schepf, E., and Starr, R.C., Eds., Biopress, Bristol, U.K., 1995, p. 19.
44. Gimler, H., Acidophilic and acidotolerant algae, in *Algal Adaptation to Environmental Stresses, Physiological, Biochemical and Molecular Mechanisms*, Rai, L.C. and Gaur, J.P., Eds., Springer, Berlin, Germany, 2001, p. 291.
45. Nixdorf, B., Fyson, A., and Krumbeck, H., Review: Plant life in extremely acidic waters, *Environ. Exp. Bot.*, 46, 203, 2001.
46. Steinberg, C.E.W., Schäfer, H., and Beisker, W., Do acid tolerant cyanobacteria exist? *Acta Hydrochim. Hydrobiol.*, 26, 13, 1998.
47. Rouco, M., Mecanismos genéticos y estrategias adaptativas a productores primarios (microalgas y cianobacterias) en un escenario de cambio global, PhD thesis, Complutense University, Madrid, Spain, 2011.
48. Fournadzhieva, S. et al., Influence of the herbicide simazine on *Chlorella*, *Scenedesmus* and *Arthrospira*, *Archiv. fuer Hydrobiol. Suppl. Band*, 106, 97, 1995.
49. Bañares-España, E. et al., Inter-strain variability in the photosynthetic use of inorganic carbon, exemplified by the pH compensation point, in the cyanobacterium *Microcystis aeruginosa*, *Aquat. Bot.*, 85, 159, 2006.
50. Campbell, D. et al., Chlorophyll fluorescence analysis of cyanobacterial photosynthesis and acclimation, *Microbiol. Mol. Biol. Rev.*, 62, 667, 1998.
51. Blanck, H. and Dahl, B., Pollution-induced community tolerance (PICT) in marine periphyton in a gradient of tri-n-butyltin (TBT) contamination, *Aquat. Toxicol.*, 35, 59, 1996.
52. Marvá, F. et al., Adaptation of green microalgae to the herbicides simazine and diquat as result of pre-selective mutations, *Aquat. Toxicol.*, 96, 130, 2010.

17 Benthic *Microcystin* and Climatic Change

*Marina Aboal**

CONTENTS

17.1 INTRODUCTION

Microcystins are some of the most common and most toxic cyanobacterial toxins in freshwaters. These cyclic peptides are synthesized nonribosomally and are powerful inhibitors of eukaryotic protein phosphatases (type 1 and 2A), which have several regulatory roles in eukaryotes [1,2]. From the chemical point of view, microcystins are cyclic heptapeptides with variable aminoacids at seven different positions [3]. All microcystins contain two amino acids that are unique to this type of compound: Adda ((2S, 3S, 8S) S)-3-amino-9-methoxy-2, 6, 8-trimethyl-10-phenyldeca-4, 6 dienoic acid) and ADMAdda (O-acetyl-O-demethylAdda) [4]. The toxicity of microcystins in vertebrates is mediated through their active transport into hepatocytes by the bile acid organic anion transport system, and acute poisoning may provoke death due to massive hepatic hemorrhage in both animals and humans [5]. The main target of acute intoxication is the liver, although a role for these compounds in tumor promotion and carcinogenesis has also been revealed in laboratory animals [6]. The exposure routes are several: skin contact, inhalation, hemodialysis, ingestion, drinking and

* Corresponding author: maboal@um.es

recreational water, food, or dietary supplements [7]. It should be borne in mind that more than one exposure route may operate simultaneously and the implications of the presence of the toxins for industry (food manufacture, agricultural products, aquaculture, etc.), especially in the predicted scenario of increased aridity, are not to be taken lightly.

Microcystins were first isolated from planktic cyanobacteria (*Microcystis aeruginosa*) and the study of such planktic communities has attracted most of the interest of researchers and public administrations because of their implications for human health [3,7]. Much progress has been made in the development of extraction and quantification methods and many recommendations have been published and/or national legislation has come into force regulating the concentration of toxins or the total cyanophyte cell numbers permissible in drinking and recreational waters [3,8–11].

Cyanobacteria constitute an important part of benthic communities in almost all kinds of aquatic and subaerial systems and are usually dominant on calcareous substrates [12,13]. Interest in the potential toxicity of benthic mats arose later than in the case of phytoplankton, and, even now, the study of benthic toxicity represents a very small portion of the research in the field [14]. The same is true for subaerial cyanobacteria. However, both kinds of communities and species have a worldwide distribution. They may be dominant in shallow habitats and arid areas, and some are even edible [15].

Cyanobacteria evolved in the Precambrian period when temperatures were much higher than at present [16], and, although widespread in both oceans and freshwaters, even in Antarctica, they show particular association with warm habitats [17]. Taking into account the present situation of global warming and all the predictions related to climate change, benthic cyanobacteria will most probably end up predominating extensive areas of shallow habitats [17]. As semiarid and arid areas will very likely increase, benthic toxicity could well become a serious concern, especially in developing countries with little or no control over sewage.

The implementation of water-quality-monitoring programs has shown that in shallow Mediterranean streams there is an inverse correlation between macroinvertebrate species richness and the diversity and total toxicity of benthic microcystin-producing mats. Moreover, even in the absence of any kind of perturbation or contamination, the highest values of macroinvertebrate water quality indexes (BMWPc) are never reached [18]. The presence of microcystins in the extracts of these benthic communities was confirmed later in the same rivers and, more recently, in several other regions and habitats [14,19,20]. This has led to many questions being asked about the role that these substances could have in benthic community dynamics. This contribution presents a review of benthic microcystin toxicity, especially in shallow aquatic habitats in a scenario of global warming, emphasizing the importance of environmental studies in the prevention of microcystin intoxication.

17.2 BIOLOGICAL EFFECTS OF MICROCYSTINS

The biological role of the microcystins has been a matter of controversy for many years. These toxic compounds evolved in environmental conditions very different from those observed now in the planet and some authors even hypothesize that the extinction of dinosaurs could be related with the proliferation of toxic cyanobacteria [21]. Whatever the original function of these molecules, they may well have different roles in nature nowadays.

While some authors proposed a potential role as grazer deterrents [22–25], others have found no evidence for the same [26]. Some authors mention allelopathic effects [27–29], or consider microcystins as siderophores [30] or quorum-sensing compounds [31–32]. Yet others suggest a role in protection against photooxidation [34].

In contrast with other secondary metabolites, microcystins are produced from early exponential phase to stationary phase [3]. The gene cluster for microcystin biosynthesis includes an ABC-transporter protein [1] that may play a role in the active export of microcystins from the cells. The existence of this protein may contradict the general assumption that microcystins are released only after cell death and decay. Moreover, the existence of transporters points to an extracellular role for microcystins, at least under conditions of high light intensity [35], when significant extracellular amounts of microcystins are detectable [36], suggesting that an ABC-transporter might be active under these conditions. Similar conclusions were also drawn from transcription data of the corresponding gene [37]. Since bloom-forming cyanobacteria may sense light intensity and/or quality, they could well use this mechanism to control production and possibly export microcystins, which may even act as intercellular signals [37]. A considerable body of data has been accumulated on the effects of toxins in the different components of aquatic systems.

17.2.1 Macroinvertebrates

Aquatic macroinvertebrates, especially groups like Ephemeroptera, are very good indicators of good water quality in aquatic systems. Most of them are absent from headwaters in calcareous streams, even when there is no indication of pollution or perturbation of any kind [18]. A clear correlation was observed between toxicity or microcystin concentration and the diversity or species richness of macroinvertebrate communities in these calcareous Mediterranean streams [18,38]. Some macroinvertebrate groups may feed on toxic cyanobacteria and may accumulate them in their tissues, producing alterations in the same and affecting reproduction, although their elimination and detoxification is relatively rapid [39,40] and the effects are reversible [41].

17.2.2 Fish

There are a great number of references in the literature concerning the pathological effects of microcystins in different species of fish [42]: from the degeneration of hepatocytes [43] to their accumulation in the liver or muscle [44,45]. Some of the information comes from samples collected in the field [45] but others come from experiments with live fish in aquaria [46,47] or from cell cultures [48]. It seems that these toxins enhance oxidative stress in most fish tissues [49] but the influence of toxins in fish survival or reproduction in nature is not known because the concentration of toxins used in experiments tends to be much higher than those found in the environment. Microcystins may penetrate fish bodies via the gills, through the skin, or through the gut. Some fishes (e.g., *Tylapia*) may consume plankton and others (*Cyprinus*) may feed on bottom mud, and both may accidentally ingest toxic material in the process [50]. Fish may contain relatively high concentrations of microcystins in the liver without showing any external symptoms or effects [51], although this may present a risk for populations that consume contaminated samples, especially given the data on microcystin accumulation in different species [52]. Cases of food poisoning and death, presumably after the consumption of fish, particularly fish liver, taken from lagoons with cyanobacterial blooms, have been documented in Poland and Sweden [7].

17.2.3 Birds

Bird deaths linked to cyanobacterial blooms have been reported since the early 1990s [53]. In some cases, the production of several different toxins has been confirmed, but in others, for example, in the United States, Canada, Japan, or South Africa, the death of wintering birds has been linked to microcystins [54–56]. There are no data on the impact of the human consumption of affected birds.

17.2.4 Algae and Bacteria

Allelopathic effects between cyanobacteria and diatoms were reported as long ago as the 1970s [29], and it was later demonstrated that the presence of *Spirogyra* induces the production of microcystins by *Oscillatoria* [57]. However, there is no general agreement on the existence of allelopathy in environmental conditions. Some authors did not consider proved the allelopathy between cyanobacteria and other algae cohabiting in plankton or benthos communities but, recently, much work has been

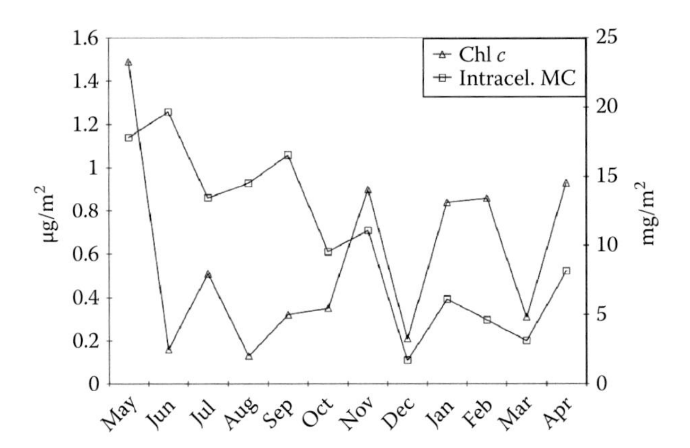

FIGURE 17.1 Relationship between benthic intracellular microcystin and chlorophyll *c* in Alhárabe river throughout the year.

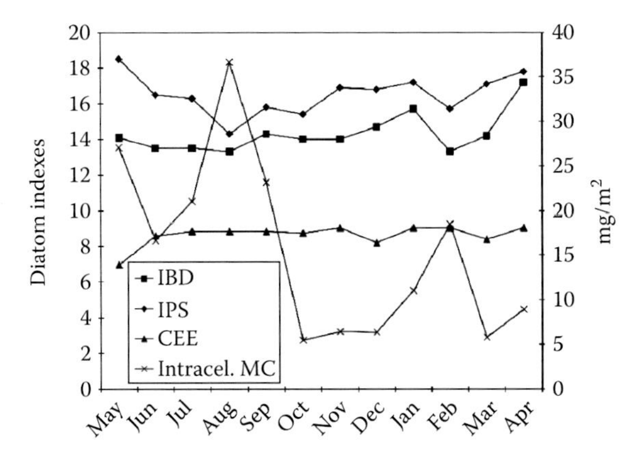

FIGURE 17.2 Relationship between the most common European diatom index (IPS, IBD, CEE) and benthic intracellular microcystin (IPS = pollution sensitivity index, IBD = biological diatom index, CEE = European diatom index).

carried out on biotic relationships [58,59] and clear allelopathic effects have been demonstrated in *Cylindrospermopsis* [28].

An inverse relationship between the growth of cyanobacteria and diatoms is frequently observed in calcareous habitats, ranging from reservoirs to streams (Figure 17.1), and also an almost inverse relationship between diversity or species richness and toxicity or microcystin concentration has been reported [60]. In some streams and reservoirs, an inverse correlation was also found between microcystins and the values obtained for the most commonly used diatom indexes (Figure 17.2).

The results obtained for inhibitory experiments between algae, bacteria, and microcystins are contradictory. Some authors did not find any relevant effects [61] but others reported clear inhibitory effects of cyanobacteria extracts and microcystins on the growth of some strains of algae and bacteria [62]. However, such inhibitory effects were fairly rapidly reversible (Figure 17.3). The importance or significance of such phenomena in nature is unknown.

In some developing countries (China, Peru), the direct consumption of cyanobacteria is common (different *Nostoc* species). However, as the presence of microcystins has been demonstrated in *Nostoc* [63], there should clearly be a stricter sanitary control of this algae food. The ingestion of dietary supplements containing cyanobacterial cells (*Spirulina/Arthrospira*) is also popular, and their daily consumption would represent long-term exposure to bacterial cells and their products. In these cases, there is a need for quality control to monitor the potential presence of toxins [7].

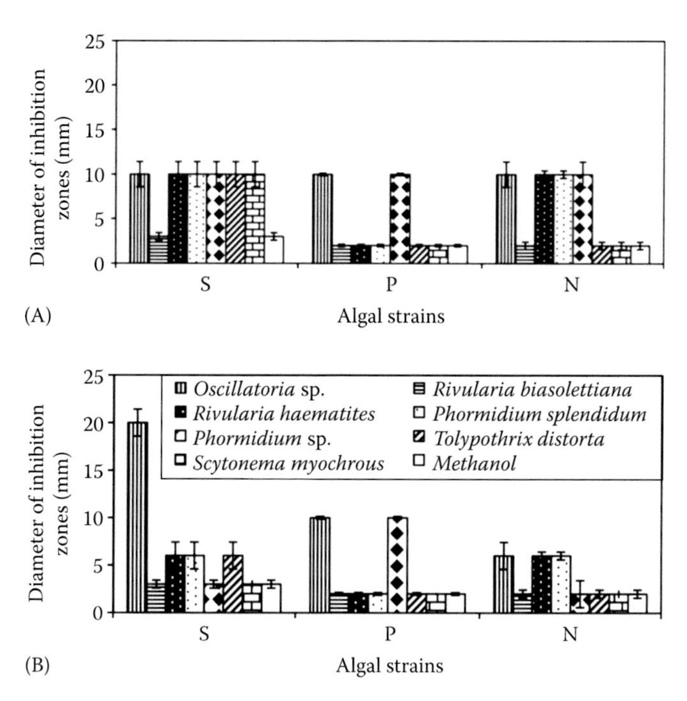

FIGURE 17.3 Antialgal agar diffusion bioassay with extracts of cyanophytes (A) 4 days and (B) 8 days after their inoculation. Target algal strains: *Scytonema sp.* (S), *Phormidium sp.* (P), and *Nostoc sp.* (N). Extracts used from *Oscillatoria sp., Rivularia biasolettiana, Rivularia haematites, Geitlerinema splendidum, Phormidium sp., Tolypothrix distorta*, and *Scytonema myochrous*.

17.2.5 Angiosperms

Several papers have shown the inhibitory effects of crude extracts of cyanobacteria and of microcystins on germination, including the reduction of seedling height and weight, root formation, leaf and root length, and productivity in several plants [64–70]. At the same time, some histological modifications of primary root tissue were observed [69], accompanied by a delay in primary xylem formation and the absence of fibers. Finally, microcystins seem to impair the photosynthetic activity of all plants exposed to them [70]. The promotion of oxidative stress due to exposure to cyanobacterial toxins has also been shown in aquatic macrophytes [71] and terrestrial plants [65,71–74].

The accumulation of microcystins in terrestrial plants [65] and the uptake of some microcystins (MC-LR and MC-LF) by seedling roots of several agricultural plants and their translocation to shoots have been reported [75]. The consumption of edible plants exposed to microcystins via irrigation may represent a health hazard because colonies and cells of cyanobacteria may be lodged in the leaves and are not removed by washing with water [75–76]. The possibility of human exposure via the consumption of crops and plant products has been confirmed in several greenhouse experiments and environmental analyses [7]. The inhibition of photosynthesis in *Phaseolus vulgaris* topically exposed to water containing microcystin-LR has been demonstrated [76]. For this reason, a certain degree of sanitary control should be exercised if crop plants are spray-irrigated with water from reservoirs hosting cyanobacterial blooms. The possibility of cyanobacterial toxins being secreted in milk should also be taken into account in cases where dairy cattle drink from water containing large cyanobacterial populations and/or when dairy cattle pasture is contaminated with water containing the toxins [7]. Even though no direct information about this kind of transfer is available, it is well established that antibiotics used in cattle health care may be present in milk [77], and these antibiotics include cyclic molecules of similar size to cyanobacterial toxins.

17.3 TOXIN PRODUCTION IN BENTHIC COMMUNITIES

17.3.1 Dynamics of Benthic Communities

Hydrological disturbances, light availability, and nutrients are the most relevant factors determining the structure and functioning of the biological communities in Mediterranean rivers [78]. It is widely accepted that the release of intracellular microcystins into water occurs after the decay or senescence of cyanophyte cells [2]. In plankton, the growth and senescence of communities have a clear seasonal basis and can be very clearly observed over short periods of time, but benthic communities may last several months or even be present throughout the year [12,79]. A series of layers can be observed in colonies, with the deepest layers being composed of senescent cells and the upper ones containing actively growing cells. This fact could explain why in calcareous streams and reservoirs, at least in warm areas, the presence of dissolved microcystins in the water may be detected throughout the year [19–80].

The occurrence of cyanobacteria lacking microcystins but containing the *mcy* gene cluster has repeatedly been reported in natural plankton populations and it seems that there is a linear relation between the frequency of inactive *mcy* genotypes and population density [81]. However, there are no data about the proportion of inactive genotypes in natural benthic communities or the factors that increase or decrease these proportions.

17.3.2 Limnological Characteristics of Water and Relation with Toxin Production

In almost all strains cultured in the laboratory, water temperature and the production of planktic microcystins have been seen to be related [4], although such correlation is more difficult to identify in environmental conditions. Most frequently, the environmental data available refer only to the blooming periods, and annual series of data are usually lacking. Moreover, water temperature is a very variable parameter and varies greatly with the time of sampling. For that reason, the mean temperature values (climatic data) should be used when trying to interpret the dynamics of water systems or aquatic communities [82]. The microcystin content was also found to be enhanced with high [35,83–85] or low [4] irradiance. In experimental studies, the same positive correlation between light and toxicity was found [86].

17.3.3 Nutrients

Even if a clear relation with nutrients cannot always be found or if luxurious growth is not always linked to increased toxicity, a correlation between the presence of microcystins and eutrophication is widely accepted in the case of planktic communities [2,30,87,88]. Some evidence strongly suggests that nutrient ratios cannot explain cyanobacteria blooms and that N-fixing by heterocystous cyanobacteria, light limitation, and vertical migration should be taken into account [88,89]. Regression analyses determined that individual nutrient concentrations were better predictors than the ratio between them [90].

As nitrogenated compounds, a certain concentration of nitrogen is needed for microcystin synthesis but, in some cases, it seems that high nitrogen concentrations can also act as an inhibitor for their growth or synthesis (Figure 17.4). Some of the results obtained in cultures could be questioned because the real requirements of the species are unknown; very rich culture media are frequently used, forcing the expression of characteristics that are not usual or perhaps not even present in nature [91]. Moreover, most cyanobacteria (with or without heterocytes) may be independent of the dissolved nitrogen content because they can fix it from the atmosphere [93–94]. The relevance of this ability for microcystin synthesis is not known.

Unlike the data obtained for planktic microcystins, most reports on benthic microcystins are linked to oligotrophic rather than polluted ecosystems [14,18,19,95]. However, the environmental data reported are usually scarce and the relation of mats with nutrients from sediments is unknown.

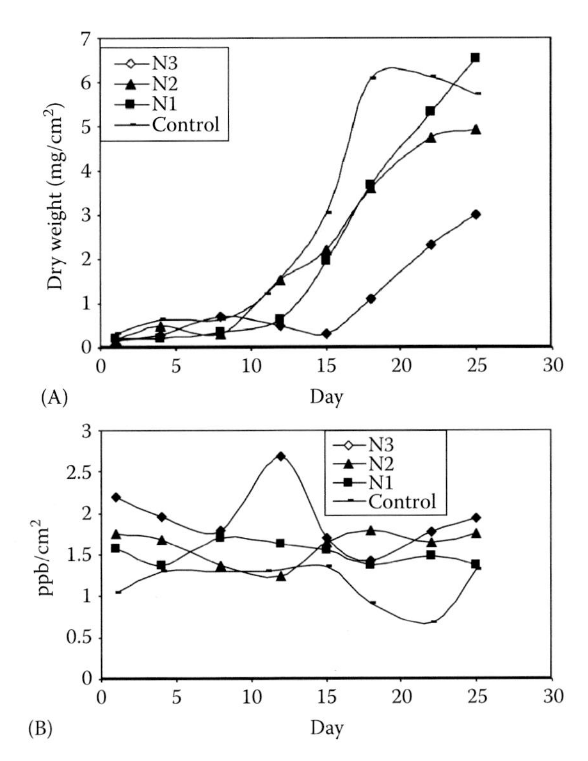

FIGURE 17.4 Effect of increasing nitrogen concentrations in growth (A) and MC production (B) of *Geitlerinema splendidum* in culture (x1, x2, x3 the concentration of BG11 culture medium). Each point represents the mean value of three replicates. The control has ½ de concentration of BG11 culture medium. The values were expressed per surface area. Before each experiment, cultures were maintained 2 weeks in a medium without the considered nutrient in order to force cells to use their reserves.

Some authors report that an increase in phosphate levels enhances microcystin concentrations [36,96] but others point to an inverse correlation between phosphorus and microcystins [96]. Phosphorus is usually the nutrient limiting the growth of algae and cyanobacteria in freshwater, which is why soluble reactive phosphorus (or total phosphorus) is included in all water-monitoring programs. However, most algae, including cyanobacteria, transform the organic phosphorus into inorganic compounds, which are rapidly incorporated into cells. Several authors [97–99] have emphasized the importance of enzymes (phosphatases) specialized in this transformation in the annual dynamics of water chemistry and the growth of algae and cyanobacteria and their respective communities.

17.3.4 Role of Mucilage

The production of extra polysaccharidic substances (EPS) is fairly common among marine and freshwater organisms [100]. Mucilaginous colonies are formed by organisms of all taxonomic groups of algae and cyanobacteria, both planktic and benthic [101]. The production of these compounds has been frequently related with nutrient deficiency [102], although it has often been suggested that mucilage may have other secondary functions.

Benthic algae and cyanobacteria in oligotrophic habitats very frequently produce mucilaginous colonies, and, in calcareous streams, some may contain microcystins [19,61]. Several authors provide evidence that suggests that microcystins could be retained and accumulated in mucilage [103–105] and may play a role in antigrazing protection, as in the case of some marine sponges that use algal toxins (okadaic acid) to eliminate parasites by apoptosis [106].

17.4 GEOGRAPHICAL DISTRIBUTION

The first report of benthic microcystins referred to alpine lakes from Switzerland [95]. More recently, mats were identified in several countries and environmental conditions, from warm areas to high mountains, from temperate zone to the tropics or Antarctica (Table 17.1). Microcystins have been detected in reservoirs, lakes, sources, wetlands, rivers, in aerophytic species, and even in phycosymbionts of lichens [107,108]. This is not surprising because it seems that the capacity

TABLE 17.1
Geographical Distribution of Benthic Microcystin Reports

Water Body	Microcystin Concentration	Taxa	References
Reservoirs			
SE Spain reservoirs	0.00–12.41 ppb	*Phormidium, Lyngbya*	Aboal et al.[14]
Metropolitan Water District of Southern California, USA	1.15–4.15 mg/mg DW.	*Phormidium*	Izaguirre et al.[9]
Agriculture reservoir (Varanasi, India)	0.025 μg/g	*Nostoc*	Bajpai et al.[131]
Contraparada (Murcia, Spain)	0.10–8.50 ppb	*Phormidium, Lyngbya*	Hurtado et al.[28]
Water storage reservoirs	—	—	Mohamed and Shehri[134]
Lakes			
Alpine lakes (Switzerland)	0.01–17.5 nM	*Oscillatoria limosa, O. tenuis*	Metz et al.[16]
Lake Bogoria (Kenia)	221–845 μg/g	*Phormidium terebriforme*	
		O. millei, Spirulina subsalsa	Krienitz et al.[132]
Lake Myall (Australia)	—	—	Dasey et al.[133]
Lakes (Nueva Zelanda)	0.032 mg/g	*Planktothrix* sp.	Wood et al.[134]
Wetlands			
Everglades (Florida, USA)	—	—	Bellinger et al.[26]
Mats (Antarctica)	0.01–0.016	*Nostoc* sp.	Wood et al.[135]
Sources			
Thermal sources (Arabia Saudí)	0.468–0.512	*Oscillatoria limosa Synechococcus lividus*	Mohamed[136]
Sources (Brazil)	0.036	*Fischerella* sp.	Fiore et al.[137]
Subaerial habitats			
Aerophytic, (Finland)	0.004	*Nostoc phycosimbiont*	Kaasalainen et al.[138]
Aerophytic (Murcia, Spain)	0003	*Microcoleus vaginatus*	Dat. Inéd.
Aquacultures tanks			
Aquaculture tanks (Nigeria)	—	—	Chia et al.[139]
Rivers, streams			
Alhárabe river (Murcia, SE Spain)	20.45 mg/m²	*Rivularia biasolettiana Tolypothrix distorta Geitlerinema splendidum*	Aboal et al.[14]
Nile river and irrigation channels (Egypt)	1.6–4.1 μg/mg	*Calothrix parietina Phormidium tenue*	Mohamed et al.[54]
Ouka Meden river (Morocco)	0.139 μg/mg	*N. muscorum*	Oudra et al.[15]

for synthesizing these compounds is fairly old, although it has been lost in some cases during the evolution [109,110]. The concentration of intracellular microcystins detected varied greatly between species and between different strains of the same species. It is sometimes of the same order of magnitude as planktonic species (Table 17.1), although exact comparison of the data is complicated because of the different extraction and quantification methods used by the different researchers.

Some cyanobacteria do not grow well under standard conditions and/or do not express the characteristic morphology of the taxa, complicating their identification. Species living in oligotrophic conditions are especially difficult to culture, because the culture media used for routine study usually contain high concentrations of nitrogen and phosphorus, which are usually found in very low concentrations or are even undetectable in oligotrophic habitats. Moreover, the effects of maintaining cultures for long periods, especially in conditions very different from those that characterize their natural environment, may produce mutations [111,112] and the effects of such mutations on microcystin synthesis are unknown. However, it is now possible to verify the presence of the operon involved in microcystin synthesis (*mcy*) in field material by genomic methods without the need for culturing [113].

17.4.1 Standing and Flowing Aquatic Systems of Mediterranean Spanish Semiarid Regions

Mediterranean aquatic systems are characterized by their great variability and great spatial and temporary heterogeneity [114]. Most rivers and streams are regulated with dams built not only to regulate flow and prevent the risks of flooding but also to satisfy the need for drinking water or for irrigation. Depending on the uses of the water in the reservoirs, the level of water may remain stable over long periods of time, permitting the growth of more or less thick benthic mats. Due to the high alkalinity of the water and the precipitation of phosphorus associated to carbonates, the presence of dissolved inorganic phosphorus is very limited in these habitats.

Organisms living in shallow Mediterranean aquatic systems have evolved a variety of responses to high light conditions and reactive oxygen species (ROS) production. While, the gene expression and physiological responses of cyanobacteria to these environmental conditions have been studied [115], the potential relation of these stressful environments with toxin synthesis has not been addressed.

17.4.2 Reservoirs

In arid and semiarid areas, the increasing need for water for the human population and agriculture has led to the construction of a large number of reservoirs. Most of the reservoirs in the semiarid area of Iberian Peninsula are shallow and must be considered as oligo-mesotrophic (Table 17.2).

TABLE 17.2
Main Physical and Chemical Variables of SE Spain Reservoirs (Only the Mean Annual Values Are Indicated)

Reservoir	Conductivity (μS/cm²)	pH	Temperature (°C)	Silicates (μM/L)	DIN (μM/L)	P-PO4 (μM/L)
Alfonso XIII	7497	8.3	19.3	73.9	14.6	0.4
Argos	1847	8.5	20.9	57.4	53.4	0.9
Cenajo	1574	8.6	16.5	66.4	50.1	0.01
Cierva	1019	8.7	17.4	25.2	35.3	0.1
Ojós	1056	8.2	18.2	84.8	68.4	0.1
Pedrera	1131	8.6	18.7	43.5	71.5	0.05
Puentes	3555	8.2	18.3	218.9	58.6	0.36
Valdeinfierno	4150	8.1	18.7	64.6	19.6	0.6

TABLE 17.3
Cell Density of Phytoplanktic Communities and Benthic Intracellular and Dissolved Microcystins from Contraparada WTP, SE Spain

Cell Density (Cell/mL)	Apr	May	Jun	Jul	Aug	Sep	Oct	Nov	Dec
Surface water									
Oscillatoria limosa	6	61	0	0	0	0	0	0	0
Synechococcus sp.	6	39	0	0	0	0	0	0	0
Pseudanabaena catenata	0	0	779	0	0	0	0	0	0
Microcystis sp.	0	0	0	340	0	0	0	0	0
Planctonema sp.	0	0	0	236	42	0	0	0	0
Aphanocapsa holsatica	0	0	0	0	2682	303	0	0	0
Oscillatoria sp.	0	0	0	0	0	0	0	0	273
Bottom water									
Synechococcus sp.	6	9	0	0	0	0	0	0	0
Pseudanabaena catenata	0	0	5354	0	0	0	0	0	0
Oscillatoria tenuis	0	0	3	0	152	0	91	0	0
Microcystis sp.	0	0	0	212	0	0	0	0	0
Planctonema sp.	0	0	0	303	6	0	0	0	0
Woronichinia delicatula	0	0	0	0	1606	152	0	0	0
Planktothrix agardhii	0	0	0	0	0	0	152	0	0
Oscillatoria sp.	0	0	0	0	0	0	0	0	91
Microcystins									
Benthic intracellular MCs (ppb)	0.152	0.183	0.123	0.462	n.d.	n.d.	n.d.	0.257	n.d.
Dissolved MCs (ppb)	3.629	0.905	0.652	2.209	0.865	2.032	2.263	8.446	2.470

Phytoplankton communities are low in density and cyanobacteria are never the predominant group as regards taxon number or cell density (Table 17.3). However, some benthic communities are well developed in littoral areas and are dominated by cyanobacteria. Dissolved microcystins have been detected in water in almost all the reservoirs analyzed in one or more seasons and, in some cases, their presence was detected throughout the year. Intracellular microcystins (11 different varieties) have also been observed in benthic communities [18–19]. Similar observations have been made for shallow reservoirs or wetlands of California and Florida [14,116].

The shallow character of these reservoirs favors the development of benthic mats that may cover the whole bottom, representing a large area and a fairly high concentration of microcystins [14,49,53,116]. In most cases, the benthic community is mainly composed of Oscillatoriales, with *Phormidium, Leptolyngbya*, and *Pseudanabaena* genera predominating. The presence of planktic cyanobacteria is usually low, both in taxon number and in cell density, although dissolved microcystins are detected throughout the year in these reservoirs (Table 17.4), and a positive and significant correlation between intracellular microcystins in benthic mats and dissolved microcystins has been reported [49]. Although it is very unusual for microcystin concentrations to reach the legal limit (1 ppb), the quantity of microcystins present is very high and could potentially have great impact on animal and vegetal communities [117].

TABLE 17.4
Significant Correlations between Environmental Variables and Benthic Intracellular and Dissolved Microcystins in Shallow Reservoirs from SE Spain

	Depth	P-PO4	Rainfall	Conductivity	Dissolved MCs	Benthic Intracellular MCs
Dissolved MCs	**−0.38**	0.19	0.13	−0.02	1.00	**0.29**
Benthic intracellular MCs	−0.44	0.33	**0.36**	**0.44**	**0.29**	1.00

Bold values represent significant correlations at $p < 0.05$.

Some water treatment plants (WTPs) may use shallow ponds, which suffer the aforementioned problems of cyanophyte growth and microcystin production. Usually, this involves no risk for the human population as long as the water treatment includes ozonization and active carbon steps [49,118]. However, small groups of people in areas where WTPs do not have this technology or who use water without any treatment may well be exposed to the carcinogenic effects of these toxins.

17.4.3 Streams

Mediterranean calcareous streams suffer great variation of flow throughout the year, with periods of drought followed by floods. The highest flows are usually observed in autumn–winter and the lowest in summer. In these calcareous streams, benthic communities are dominated by cyanobacteria throughout the year (Figure 17.5), forming permanent mats and colonies, multilayered most of the time, encrusted

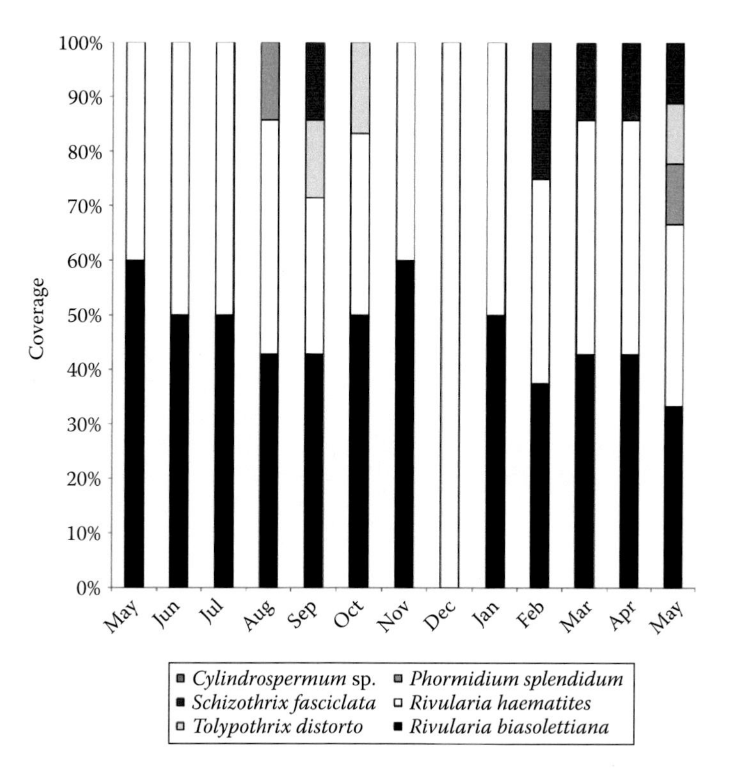

FIGURE 17.5 Annual dynamics of predominant species in Alhárabe river benthic communities.

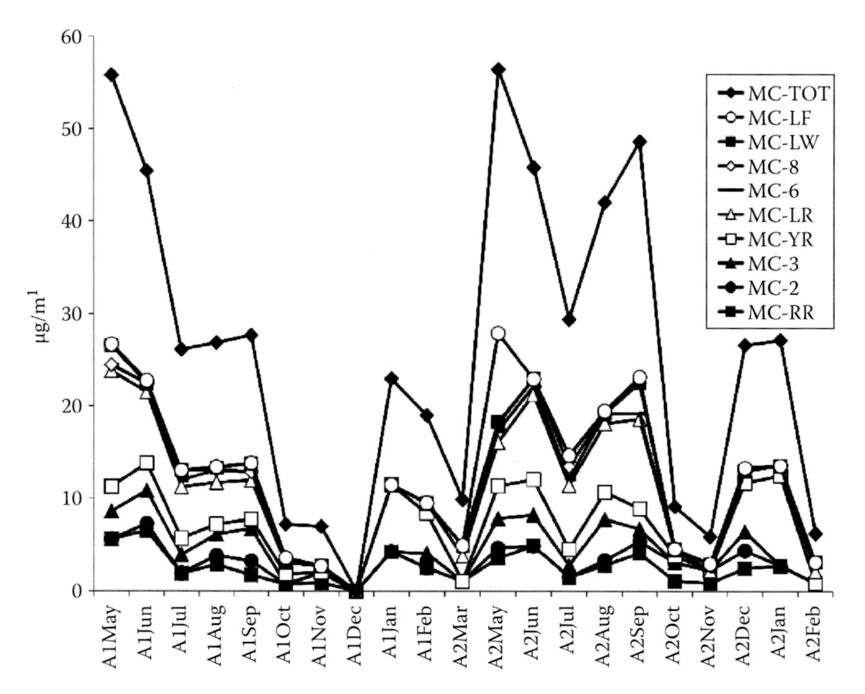

FIGURE 17.6 Annual variation of intracellular microcystin in benthic mats of Alhárabe river (A1, A2 sampling points).

sometimes, with basal layers of more or less senescent material and upper layers of active growing cells [12]. In all cases, the highest concentration of intracellular toxins is detected in summer and the lowest in winter (Figure 17.6). The dry summer period is followed by a rainy autumn–winter and sometimes by floods that may eventually destroy benthic mats and colonies and liberate toxins into the water.

17.5 CLIMATE CHANGE AND THE IMPACT ON PREVALENCE OF TOXIN PRODUCERS

In all projections of climate change and the associated increase in carbon dioxide levels, several scenarios are possible, involving increases in temperature, eutrophication, and the frequency and intensity of floods and drought periods [119]. The spread of what has until now been considered a tropical taxon, *Cylindrospermopsis*, across America and Europe and the recent colonization of Australia by African strains of the genus are considered clear effects of climate change [120,121]. The very scarce data that are available for calcareous streams point to a clear correlation between the mean maximum and minimum temperature and intracellular microcystins in benthic mats (Figure 17.7).

Climate also influences the seasonal hydrological changes of streams, affecting the growth of species, the resilience of communities, the recolonization after floods, and the release of cyanotoxins. In some cases, a clear and significant correlation was found with two important hydrological variables that reflect the annual dynamics of communities: depth and flow. An inverse correlation was observed between dissolved microcystins and flow in all the streams studied (Figure 17.8). A similar correlation with depth has been found in shallow reservoirs [50].

The concentration of inorganic phosphorus in water may be low in nonpolluted habitats but there are usually seasonal inputs of organic phosphorus related with increases in flow or floods or the degradation of decaying materials. In these cases, a negative and significant correlation was found between total phosphorus and chlorophyll a and dissolved microcystins (Table 17.5). A possible explanation for this is that nutrient stress promotes the release of the toxins or that the stress first produces the senescence of a part of the community and then the release of toxins. While the

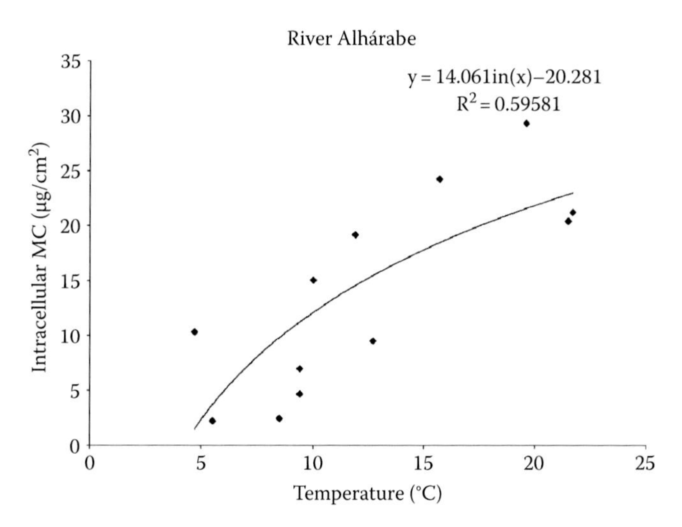

FIGURE 17.7 Relationship between mean month temperature and intracellular microcystin of benthic mats from Alhárabe river.

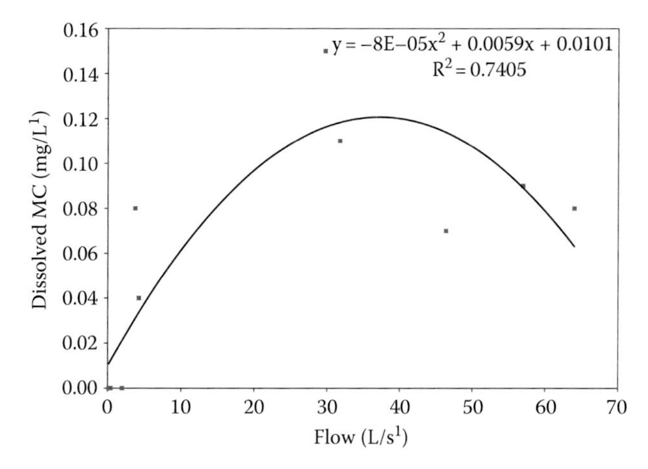

FIGURE 17.8 Relationship between flow and dissolved microcystin of benthic mats in several southeastern Spain shallow rivers.

intracellular microcystins are negatively and significantly correlated with flow and positively with temperature, the dissolved microcystins are positively correlated with rainfall, indicating that intracellular microcystins accumulate during summer when the flow is low and the temperature is high, while the concentration of dissolved microcystins is higher in autumn–winter with the highest rainfall. Interestingly, the different variants of microcystins correlate differently with environmental variables. MC-RR correlates positively with temperature and carotenoid concentration and negatively with nitrate and nitrite; MC-YR correlates negatively with PRS and alkalinity; and MC-LR correlates positively with silicate and temperature and negatively with nitrite, nitrate, flow, and rainfall. This points to a dynamics in toxin production, which decreases from drier and sunny periods (MC-RR) to rainy periods (MC-LR), and explains the difference in the toxin composition of the communities. The highest production of MC-RR was observed with intense light by Kaebernick et al. [35] in laboratory experiments and a negative correlation with nitrate and total nitrogen was observed in some lakes [122], although most data refer to MC-LR or MC-LR equivalents and do not discriminate between the different variants. Recently, it was reported that nitrogen concentration does not influence *mcyD* expression [123].

TABLE 17.5
Correlations between Physical, Chemical and Biological Variables and Benthic Intracellular and Dissolved Microcystins in River Alhárabe, SE Spain

	MC-RR	MC-YR	MC-LR	Total Intracellular MC	Dissolved MC
Silicate	0.35	0.40	***0.42***	***0.59***	−0.14
PRS	0.01	***−0.54***	0.09	−0.19	−0.35
TP	−0.01	0.12	0.20	0.11	***−0.59***
Nitrite	***−0.50***	−0.02	***−0.71***	***−0.67***	0.17
Nitrate	***−0.58***	−0,04	***−0,86***	***−0.83***	0.22
Alkalinity	0.38	***−0,53***	0.34	0.23	0,35
Flow	−0.35	0.37	***−0.76***	***−0.56***	0.08
Rainfall	−0,10	0,33	***−0,49***	−0,17	***0,41***
Temperature max.	***0.56***	0.18	***0.70***	***0.77***	−0.19
N/P Sup.	0.15	***0.54***	−0.06	0.28	0.33
Chlorophyll a	0.30	−0.03	0.17	0.18	***−0.57***
Carotenoids	***0.44***	0.08	0.04	0.16	−0.17
Total intracellular MC	***0.84***	0.23	***0.86***	1.00	−0.20
Dissolved MC	−0.15	0.23	***−0.47***	−0.20	1.00

Significant correlations in bold and italics (p = 0.05).

Some authors suggest that the different variants of microcystin respond differently to environmental factors. Temperature enhances microcystin-RR synthesis [36], while high phosphorus concentrations increase microcystin-LR concentration [124]. These differences may represent a different geographic distribution of variants according to mean temperature. MC-YR is occasionally much more abundant than MC-LR in irrigation channels in Egypt [125] or in streams of southeastern Spain (Aboal, unpublished data). In this last case, a positive correlation with water conductance was found, indicating a potential affinity for the salt-rich waters of arid–semiarid areas.

In all cases, the synthesis of microcystins by cyanobacteria seems to be enhanced in growth-limiting conditions in reservoirs and streams from arid areas. Brand et al. [126] suggest that the benefits of producing microcystins are particularly important under growth-limiting conditions, although this benefit is outweighed by microcystin production costs when environmental conditions favor cell growth. Even in eutrophicated areas, a high nitrogen concentration may be associated with phosphorus deficiency.

17.6 PROPOSALS FOR THE FUTURE

The WHO recommends a limit of 1 ppb for drinking water [2]. Although the European Directive for Drinking Water (98/93/CE) does not regulate the concentration of microcystins in drinking water, some countries have their own regulations and consider 1–1.5 ppb as the limit for drinking water (R.D. 140/2003 [10], Health Canada [11], NHMRC [8], NHNRC [9]). Even though most water agencies regularly control the presence of microcystins, the law only considers toxicological control to be compulsory when eutrophication is evident (R. D.140/2003 [10]). Moreover, controls are only made in reservoirs containing water destined for drinking but not in rivers or reservoirs used for irrigation or fishing.

Present-day molecular tools are not sufficient to prevent toxic events. Although they offer the possibility of knowing the presence of potential toxic strains, it is also necessary to know the biology and requirements of the species involved in the field and in culture in order to predict when they will express their potential toxicity. Microcystin production is not only a problem in the phytoplankton of eutrophicated water systems, and there is increasing evidence of the presence of these toxins in benthic communities, especially in shallow habitats from warm areas with high rates of light absorption and where water is usually a limited resource [127]. In these cases, there are increased health risks for the human population, although modern water treatment plants with ozonization and organic carbon can eliminate them [50] or, at least, reduce them to safer levels [128]. However, the facilities needed for this kind of treatment are not common in most developing countries. If the water of these systems is used without adequate treatment (e.g., in small towns and villages or when water is used for irrigation or farming), the human population may be exposed to health hazards. However, epidemiological studies are still needed to find relationships between the rates of gastrointestinal cancer in the developed world or liver cancer in developing countries and the cyanobacterial contamination of water [6,129]. Taking into consideration the cancer risk, some authors have proposed lowering the maximum tolerable microcystin concentration to 0.3 μg/L, a value three times lower than that permitted nowadays [130].

As arid and semiarid areas will undoubtedly increase in number and extent on a global scale, it is highly recommendable that global collaboration be increased to test the toxicity of water systems and to gain better geographical information concerning toxin production, thereby providing greater insight into the effects of climate change.

ACKNOWLEDGMENTS

Financial support for the research involved in this chapter was obtained from the Ministry of Science and Technology and Development (REN2000-1021-C03-02/HID, CGL2004-06339-C02-02) and Fundación Séneca of Murcia Region (05762/PI/07).

REFERENCES

1. Tillett, D. et al., Structural organization of microcystin biosynthesis in *Microcystis aeruginosa* PCC 7806: An integrated peptide-polyketide synthase system, *Chem. Biol.*, 7, 753–764, 2000.
2. Bell, S.G. and Codd, G.A., Cyanobacterial toxins and human health, *Rev. Med. Microbiol.*, 5, 256–264, 1994.
3. WHO, *Toxic Cyanobacteria in Water: A Guide to Their Public Health Consequences, Monitoring and Management*, Routledge, London, U.K., 1999.
4. Sivonen, K., Effects of light, temperature, nitrate, orthophosphate and bacteria on growth and hepatotoxin production by *Oscillatoria agardhii* strains, *Appl. Environ. Microbiol.*, 56, 2658–2666, 1990.
5. Jochimsen, E.M. et al., Liver failure and death after exposure to microcystins at a hemodialysis center in Brazil, *N. Engl. J. Med.*, 338 (13), 873–878, 1998.
6. Falconer, I., Toxic cyanobacterial bloom problems in Australian waters: Risk and impacts on human health, *Phycology*, 40 (3), 228–233, 2001.
7. Codd, G. et al., Cyanobacterial toxins, exposure routes and human health, *Eur. J. Phycol.*, 34, 405–415, 2010.
8. National Health and Medical Research Council of Australia (NHMRC), National Water Quality Management Strategy: Australian Drinking Water Guidelines, 2004.
9. NHMRC, Guideline for Managing Risks in Recreational Water. Australian Government, 2006.
10. Ministerio de Sanidad y Consumo del Reino de España, Criterios sanitarios de la calidad del agua de consumo humano (R.D. 140/2003, BOE n° 45 de 21 de febrero de 2003), 2003.
11. Health Canada, Summary of guidelines for Canadian drinking water quality, Federal-Provincial-Territorial Committee on Environmental and Occupational Health, Health Canada, Ottawa, Ontario, Canada, 2003.
12. Aboal, M., Epilithic algal communities from River Segura basin, southeastern Spain, *Arch. Hydrobiol.*, 116, 113–124, 1989.
13. Whitton, B.A. and Potts, M., Eds., *The Ecology of Cyanobacteria: Their Diversity in Time and Space*, Kluwer Academic Publishers, Boston, MA, 2002.

14. Izaguirre, G., Jungblut, A.D., and Neilan, B.A., Benthic cyanobacteria (*Oscillatoriaceae*) that produce Microcystin-LR, isolated from four reservoirs in Southern California, *Water Res.*, 41, 492–498, 2007.
15. Gao, K., Chinese studies on the edible blue-green alga, *Nostoc flagelliforme*: A review, *J. Appl. Phycol.*, 10, 37–49, 1998.
16. Falkowski, P., On the evolution of the carbon cycle, in *Phytoplankton Productivity. Carbon Assimilation in Marine and Freshwater Ecosystems*, Williams, P.J.L., Thomas, D.N., and Reynolds, C.S., Eds., pp. 318–349, Blackwell Science, Oxford, U.K., 2002.
17. Moss, B. et al., How important is climate? Effects of warming, nutrient addition and fish on phytoplankton in shallow lake microcosms, *J. Appl. Ecol.*, 40, 782–792, 2003.
18. Aboal, M. et al., Relationship between macroinvertebrate diversity and toxicity of *Cyanophyceae* (Cyanobacteria) in some streams from Eastern Spain, *Verh. Intl. Verein. Limnol.*, 27, 550–555, 2000.
19. Aboal, M., Puig, M.A., and Asencio, A.D., Production of microcystins in calcareous Mediterranean streams: The Alharabe river, Segura river basin in south-east Spain, *J. Appl. Phycol.*, 17, 231–243, 2005.
20. Oudra, B., Dadi-El Andaloussi, M., and Vasconcelos, V., Identification and quantification of microcystins from a *Nostoc muscorum* bloom occurring in Oukaïmeden River (High-Atlas mountains of Marrakech, Morocco), *Environ. Monit. Assess.*, 149, 437–444, 2009.
21. LeBlanc, S., Pick, F.R., and Aranda-Rodriguez, R., Allelopathic effects of the toxic cyanobacterium *Microcystis aeruginosa* on duckweed, *Lemna gibba* L., *Environ. Toxicol.*, 20 (1), 67–73, 2005.
22. Jia, X.H. et al., Alellopathic inhibition by *Scenedesmus obliquus* of photosynthesis and growth of *Microcystis aeruginosa*, *Photosynthesis. Energy from the Sun*, 22, 1339–1342, 2008.
23. Jang, M.H., Jung, J.-M., and Takamura, N., Changes in microcystin production in cyanobacteria exposed to zooplankton at different population densities and infochemical concentrations, *Limnol. Oceanogr.*, 52 (4), 1454–1466, 2007.
24. Smith, J.L., Boyer, G.L., and Zimba, P.V., A review of cyanobacterial odorous and bioactive metabolites: Impacts and management alternatives in aquaculture, *Aquaculture*, 280, 5–20, 2008.
25. Berry, J.P. et al., Cyanobacterial toxins as allelochemicals with potential applications as algaecides, herbicides and insecticides, *Mar. Drugs*, 6, 117–146, 2008.
26. Wilken, S. et al., Microcystins do not provide anti-herbivore defence against mixotrophic flagellates, *Aquat. Microb. Ecol.*, 59, 207–216, 2010.
27. Pflugmacher, S., Possible allelopathic effects of cyanotoxins, with reference to microcystin-LR, in aquatic ecosystems, *Environ. Toxicol.*, 17 (4), 407–413, 2002.
28. Leao, P.N., Vasconcelos, M.T.S., and Vasconcelos, V., Allelopathic activity of cyanobacteria on green microalgae at low densities, *Eur. J. Phycol.*, 44(3), 347–355, 2009.
29. Keating, K.I., Blue-green algal inhibition of diatom growth: The transition from mesotrophic to eutrophic community structure, *Science*, 199, 971–973, 1978.
30. Utkilen, H. and Gjolme, N., Iron-stimulated toxin production in *Microcystisaeruginosa*, *Appl. Environ. Microbiol.*, 61(2), 797–800, 1995.
31. Sedmark, B. and Elersek, T., Microcystins induce morphological and physiological changes in selected representative phytoplanktons, *Microb. Ecol.*, 50 (2), 298–305, 2005.
32. Shatz, D. et al., Towards clarification of the biological role of microcystins, a family of cyanobacterial toxins, *Environ. Microbiol.*, 9 (4), 965–970, 2007.
33. Vassilakaki, M. and Pflugmacher, S., Oxidative stress response of *Synechocystis* sp. (PCC 6803) due to exposure to microcystin-LR and cell-free cyanobacterial crude extract containing microcystin-LR, *J. Appl. Phycol.*, 20, 219–225, 2008.
34. Phelan, R.R. and Downing, T.G., A growth advantage for microcystin production by *Microcystis* PCC7806 under high light, *J. Phycol.*, 47 (6), 1241–1246, 2011.
35. Kaebernick, M. et al., Light and the transcriptional response of the microcystin biosynthesis gene cluster, *Appl. Environ. Microbiol.*, 66, 3387–3392, 2000.
36. Rapala, J. et al., Variation of microcystins, cyanobacterial hepatotoxin, in *Anabaena* spp. as a function of growth stimuli, *Appl. Environ. Microbiol.*, 63 (6), 2206–2212, 1997.
37. Dittman, E. et al., Altered expression of two light-dependent genes in a microcystin-lacking mutant of *Microcystis aeruginosa* PCC 7806, *Microbiology*, 147, 3113–3119, 2001.
38. Romero González-Quijano, C., Methodological and ecological studies on running water and benthic extract toxicities: Cyanobacterial toxicity versus general toxicity in Muga and Matarranya river basins. PhD thesis, Universidad Autónoma de Madrid, Madrid, Spain, 2010.

39. Lance, E. et al., Interactions between cyanobacteria and gastropods. I. Ingestion of toxic *Planktothrix agardhii* by *Lemnaea stagnalis* and the kinetics of microcystin bioaccumulation and detoxification, *Aquat. Toxicol.*, 79, 140–148, 2006.
40. Xie, L. et al., Accumulation of microcystins in various organs of the freshwater snail *Sinotaia histrica* and three fishes in a temperate lake, the eutrophic Lake Suwa, Japan, *Toxicon*, 49, 646–652, 2007.
41. Lance, E. et al., Interactions between cyanobacteria and gastropods. II. Impact of toxic *Planktothrix agardhii* on the life-history traits of *Lymnaea stagnalis*, *Aquat. Toxicol.*, 81, 389–396, 2007.
42. Malbrouck, C. and Kestemont, P., Effects of microcystins on fish, *Environ. Toxicol. Chem.*, 25 (1), 72–86, 2006.
43. Li, X.Y. et al., Toxicity of microcystins in the isolated hepatocytes of common carp (*Cyprinus carpio*), *Ecotoxicol. Environ. Saf.*, 67, 447–451, 2007.
44. Magalhaes, V.F., Soares, R.M., and Azevedo, S.M.F.O., Microcystin contamination in fish from the Jacarepaguá Lagoon (Río de Janeiro, Brazil): Ecological implication and human health, *Toxicon*, 39 (7), 1077–1085, 2001.
45. Atencio, L. et al., Acute Effects of microcystins MC-LR and MC-RR on acid and alkaline phosphatase activities and pathological changes in intraperitoneally exposed tilapia fish (*Oreochromis* sp.), *Toxicol. Pathol.*, 36 (3), 449–458, 2008.
46. Paulíková, M. et al., Accumulation of microcystins in Nile tilapia, *Oreochromis niloticus* L., and effects of a complex cyanobacterial bloom on the dietetic quality of muscles, *Bull. Environ. Contam. Toxicol.*, 87 (1), 26–30, 2011.
47. Puerto, M. et al., Oxidative stress induced by microcystin-LR on PLHC-1 fish cell line, *Toxicology in Vitro*, 23 (8), 1445–1449, 2009.
48. Puerto, M. et al., Differential oxidative stress to pure Microcystin-LR and Microcystin-containing and non-containing cyanobacterial crude extracts on caco-2 cells, *Toxicon*, 55 (2–3), 514–522, 2010.
49. Hurtado, I. et al., Significance of microcystin production by benthic communities in water treatment systems of arid zones, *Water Res.*, 42, 1245–1253, 2008.
50. Molina, R. et al., Acid and alkaline phosphatase activities and pathological changes induced in tilapia fish (*Oreochromis* sp.) exposed subchronically to microcystins from toxic cyanobacterial blooms under laboratory conditions, *Toxicon*, 46, 725–735, 2005.
51. Romo, S. et al., Assessment of microcystins in lake water and fish (Mugilidae, *Liza sp.*) in the largest Spanish coastal lake, *Environ. Monit. Assess.*, 184(2), 939–949, 2011. DOI 10.1007/s10661-011-2011-0.
52. Landsberg, J.H., The effects of harmful algal blooms on aquatic organisms, *Reviews in Fisheries Science*, 10 (2), 113–390, 2002.
53. Carmichael, W.W. and Li, R., Cyanobacteria toxins in the Salton Sea, *Sal. Syst.*, 2, 5, 2006.
54. Driscoll, C.P. et al., Case report: Great blue heron (*Ardeaherodias*) morbidity and mortalityinvestigation in Maryland's Chesapeake Bay, in *Proceedings of the Southeast Fish and Wildlife Conference*, Baltimore, MD, 2002.
55. Matsunaga, H. et al., Possible cause of unnatural mass death of wild birds in a pond in Nishinomiya, Japan: Sudden appearance of toxic cyanobacteria, *Nat. Toxins*, 7 (2), 81–84, 1999.
56. Mohamed, Z.A., Allelopathic activity of *Spirogyra*sp.: Stimulating bloom formation and toxin production by *Oscillatoria* in some irrigation canals, Egypt, *J. Plankton Res.*, 24, 137–141, 2002.
57. Stewart, I., Seawright, A.A., and Shaw, G.R., Cyanobacterial poisoning in livestock, wild mammals and birds—An overview, in *Cyanobacterial Harmful Algal Blooms State of the Science and Research Needs*, Hudnell, H.K., Ed., Springer, New York, 2008.
58. Schagerl, M., Unterrieder, I., and Angeler, D.G., Allelopathic interactions among *Anabaena torulosa* (Cyanoprokaryota) and other algae isolated from lake Neusiedlersee (Austria), *Algological Studies*, 103 (Suppl. 140), 117–130, 2001.
59. Sukenik, A. et al., Inhibition of growth and photosynthesis of the dinoflagellate *Peridinium gatunense* by *Microcystis* sp. (cyanobacteria): A novel allelopathic mechanism, *Limnol. Oceanogr.*, 47 (6), 1656–1663, 2002.
60. Aboal, M. et al., Implications of cyanophytes toxicity on biological monitoring of calcareous streams in north-east Spain, *J. Appl. Phycol.*, 14, 49–56, 2002.
61. Casamatta, D.A. and Wickstrom, C.E., Sensitivity of two disjunct bacterioplankton communities to exudates from the cyanobacterium *Microcystis aeruginosa* Kützing, *Microb. Ecol.*, 41, 64–73, 2000.
62. Valdor, R. and Aboal, M., Effects of living cyanobacteria, cyanobacterial extracts and pure microcystins on growth and ultrastructure of microalgae and bacteria, *Toxicon*, 49 (6), 769–779, 2007.
63. Metcalf, J.S. et al., Analysis of the cyanotoxins anatoxin-a and microcystins in Lesser Flamingo feathers, *Toxicol. Environ. Chem.*, 88 (1), 159–167, 2006.

64. Kurmayer, R., The toxic cyanobacterium *Nostoc* sp. strain 152 produces highest amounts of microcystin and nostophycin under stress conditions. *J. Phycol.*, 47, 200–207, 2011.
65. Bajpai, R. et al., Microcystin producing Cyanobacterium *Nostoc* sp. BHU001 from a pond in India, *Toxicon*, 53, 587–590, 2009.
66. McElhiney, J., Lawton, L.A., and Leifert, C., Investigation into the inhibitory effects of microcystins on plant growth and the toxicity of plant tissues following exposure, *Toxicon*, 39(9), 1411–1420, 2001.
67. Chen, J. et al., Effects of microcystins on the growth and the activity of superoxide dismutase and peroxidase of rape (*Brassica napus* L.) and rice (*Oryza sativa* L.), *Toxicon*, 43 (4), 393–400, 2004.
68. Saqrane, S. et al., Effects of cyanobacteria producing microcystins on seed germination and seedling growth of several agricultural plants, *J. Environ. Sci. Health B*, 43(5), 443–451, 2008.
69. Saqrane, S. et al., Physiological changes in *Triticum durum*, *Zea mays*, *Pisum sativum* and *Lens sculenta* cultivars, caused by irrigation with water contaminated with microcystins: A laboratory experimental approach. *Toxicon*, 53, 786–796, 2009.
70. Pflugmacher, S., Promotion of oxidative stress in the aquatic macrophyte *Ceratophyllum demersum* during biotransformation of the cyanobacterial toxin microcystin-LR, *Aquat. Toxicol.*, 70, 169–178, 2004.
71. Chen, T. et al., Induction of apoptosis in mouse liver by microcystin-LR in combined transcryptomic, proteomic and simulation strategy, *Mol. Cell Proteomics*, 4 (7), 958–974, 2005.
72. Pflugmacher, S. et al., Effects of cyanobacterial toxins and cyanobacterial cell-free crude extracts on germination of alfalfa (*Medicago sativa*) and induction of oxidative stress, *Environ. Toxicol. Chem.*, 25 (9), 2381–2387, 2006.
73. Pflugnacher, S., Auhorn, M., and Grimm, B., Influence of cyanobacterial crude extract containing Microcystin-LR on the physiology and antioxidative defense system of different spinach variants, *New Phytol.*, 175 (3), 482–489, 2007.
74. Peuthert, A., Chakrabarti, S., and Pflugmacher, S., Uptake of microcystin-LR and –LF (cyanobacterial toxins) in seedlings of several important agricultural plant species and the correlation with cellular damage (lipid peroxidation), *Environ. Toxicol.*, 22 (4), 436–442, 2007.
75. Codd, G.A., Metcalf, J.S., and Beattie, A., Retention of *Microcystin aeruginosa* and Microcystins by salad lettuce (*Lactuca sativa*) after spray irrigation with water containing cyanobacteria, *Toxicon*, 37, 1181–1185, 1999.
76. Abe, T. et al., Microcystin-LR inhibits photosynthesis of *Phaseolus vulgaris* primary leaves: Implications for current spray irrigation practice, *New Phytol.*, 133, 651–658, 1996.
77. Schenck, F.J. and Callery, P.S., Chromatographic methods of analysis of antibiotics in milk, *J. Chromatogr. A*, 812, 99–109, 1998.
78. Sabater, S. et al., Hydrology, light and the use of organic and inorganic materials as structuring factors of biological communities in the Mediterranean streams, *Limnetica*, 25 (1–2), 335–348, 2006.
79. Sabater, S., Encrusting algal assemblages in a Mediterranean river basin, *Arch. Hydrobiol.*, 114, 555–573, 1989.
80. Aboal, M. and Puig, M.A., Seasonal variation in intracellular and dissolved microcystin levels in reservoirs of the river Segura basin from Murcia, SE Spain, *Toxicon*, 45, 509–518, 2005.
81. Ostermaier, V. and Kurmayer, R., Distribution and abundance of nontoxic mutants of cyanobacteria in lakes of the Alps, *Microb. Ecol.*, 58, 323–333, 2009.
82. Potapova, M.G. and Charles, D.F., Benthic diatoms in USA rivers: Distributions along spatial and environmental gradients, *J. Biogeogr.*, 29, 167–187, 2002.
83. Wiedner, C. et al., Effects of light on the microcystin content of *Microcystis* strain PCC7806, *Appl. Environ. Microbiol.*, 69 (3), 1475–1481, 2003.
84. Lyck, S. and Christoffersen, K., Microcystin quota, cell division and MC net production of precultured *Microcystis aeruginosa* CYA 228 (Chroococcales, Cyanophyceae) under field conditions, *Phycologia*, 42 (6), 667–674, 2003.
85. Willén, T. and Matteson, R., Water-blooming and toxin producing Cyanobacteria in Swedish fresh and brackish waters, 1981–1995, *Hydrobiologia*, 353, 181–192, 1997.
86. Utkilen, H. and Gjolme, N., Toxins production by *Microcystis aeruginosa* as a function of light in continuous culture and its ecological significance, *Appl. Environ. Microbiol.*, 58, 1321–1325, 1992.
87. Orr, P.T. and Jones, G.J., Relationships between microcystin production and cell division rates in nitrogen limited *Microcystis aeruginosa* cultures, *Limnol. Oceanogr.*, 43, 1604–1614, 1998.
88. Ferber, L.R. et al., Do cyanobacteria dominate in eutrophic lakes because they fix atmospheric nitrogen? *Freshwater Biol.*, 49, 690–708, 2004.
89. Havens, K.E. et al., N:P ratios, light limitation, and cyanobacterial dominance in a subtropical lake impacted by non-point source nutrient pollution, *Environ. Poll.*, 122, 379–390, 2003.

90. Trimbee, A.M. and Prepas, E.E., Evaluation of total phosphorus as a predictor of the relative biomass of blue-green algae with emphasis on Alberta Lakes, *Can. J. Fish. Aquat. Sci.*, 44, 1337–1342, 1987.
91. Whitton, B.A., Phosphorus status and the naming of freshwater algae, *Algas* (*Bulletin of the Spanish Phycological Society*), 4, 20–22, 2009.
92. Zevenboom, W. and Mur, L., N_2 fixing cyanobacteria, why they did not become dominant in Dutch hypertrophic lakes, *Dev. Hydrobiol.*, 2, 123–131, 1980.
93. Lundgren, B.V. et al., Unveiling of novel radiation within the *Trichodesmium* cluster by *hetR* gene sequence analysis, *Appl. Environ. Microb.*, 71, 190–196, 2005.
94. Tsygankov, A., Nitrogen fixing cyanobacteria: A review, *Appl. Biochem. Microbiol.*, 43, 250–259, 2007.
95. Metz, K. et al., Identification of a microcystin in benthic cyanobacteria linked to cattle deaths on alpine pastures in Switzerland, *Eur. J. Phycol.*, 32, 111–117, 1997.
96. Kotak, B.G. et al., Variability of the hepatotoxin microcystin-LR in hyper-eutrophic drinking lakes, *J. Phycol.*, 31, 248–263, 1995.
97. Wicks, R.J. and Thiel, P.G., Environmental factors affecting the production of peptide toxins in floating serums of the cyanobacterium *Microcystis aeruginosa* in hypertrophic African reservoir, *Environ. Sci. Technol.*, 24, 1413–1418, 1990.
98. Whitton, B.A. et al., Ecological aspects of phosphatase activity in cyanobacteria, eukaryotic algae and bryophytes, in *Organic Phosphorus in the Environment*, Turner, B.L., Frossard, E., and Baldwin, D.S., Eds., Commonwealth Agricultural Bureau, Wallingford, U.K., pp. 205–241, 2005.
99. Yelloly, J.M. and Whitton, B.A., Seasonal changes in ambient phosphate and phosphatase activities of the cyanobacterium *Rivularia atra* in intertidal pools at Tyne Sands, Scotland, *Hydrobiologia*, 325, 201–212, 1996.
100. Myklestad, S., Production of carbohydrates by marine planktonic diatoms. I. Comparison of nine different species in culture, *J. Exp. Mar. Biol. Ecol.*, 15, 261–274, 1974.
101. Wotton, R.S., The ubiquity and many roles of expolymers (EPS) in aquatic systems, *Scientia Marina*, 68, 13–21, 2004.
102. Staats, N. et al., Oxygenic photosynthesis as driving process in exopolysaccharide production of benthic diatoms, *Mar. Ecol. Prog. Ser.*, 193, 261–269, 2000.
103. Babica, P., Bláha, L., and Marsalek, B., Exploring the natural role of microcystins—A review of effects on photoautotrophic organisms, *J. Phycol.*, 42, 9–20, 2006.
104. Marco, S. et al., Immunolocalization of microcystins in colonies of the cyanobacterium *Rivularia* in calcareous streams, *Mar. Freshwater Res.*, 63 (2), 160–165, 2011.
105. Young, F.M. et al., Quantification and localization of microcystins in colonies of a laboratory strain of *Microcystis* (cyanobacteria) using immunological methods, *Eur. J. Phycol.*, 43, 217–225, 2008.
106. Schröder, H.C. et al., Okadaic acid, an apoptogenic toxin for symbiotic/parasitic annaids in the demosponge *Suberites domuncula*, *Appl. Environ. Microbiol.*, 72 (7), 4907–4916, 2006.
107. Codd, G.A., Cyanobacterial blooms and their toxins: A history of awareness, regressions and advances, *Phycologist*, 53, 2–4, 1999.
108. Genuario, D.B. et al., Characterization of a microcystin and detection of microcystin synthetase genes from a Brazilian isolate of *Nostoc*, *Toxicon*, 55, 846–54, 2010.
109. Rantala, A. et al., Phylogenetic evidence for the early evolution of microcystin synthesis, *Proc. Natl Acad. Sci. U.S.A.*, 101 (2), 568–573, 2004.
110. Fewer, D.P. et al., Recurrent adenylation domain replacement in the microcystin synthetase gene cluster, *BMC Evol. Biol.*, 7, 183, 2007.
111. Day, J.G. et al., Cryopreservation and conservation of microalgae: The development of a pan-European scientific and biotechnological resource (the COBRA project), *CryoLetters*, 26, 231–238, 2005.
112. Müller, J. et al., Assessing genetic stability of a range of terrestrial microalgae after cryopreservation using amplified fragment length polymorphism (AFLP), *Am. J. Bot.*, 94 (5), 799–808, 2007.
113. Baker, P.D. and Humpage, A.R., Toxicity associated with commonly occurring cyanobacteria in surface waters of Murray-Darling Basin, *Aust. J. Mar. Freshwater Res.*, 45, 773–786, 1994.
114. Sabater, S. et al., The Ter: A Mediterranean river cas-study in Spain, in *River Ecosystems of the World*, Cushing, C.E., Cummins, K.W., and Minshall, G.W. Eds., Elsevier, Amsterdam, the Netherlands, pp. 419–438, 1995.
115. Muramatsu, M. and Hihara, Y., Acclimation to high-light conditions in cyanobacteria: From gene expression to physiological responses, *J. Plant Res.*, 125, 11–39, 2012.
116. Bellinger, B.J. and Hagerthey, S.E., Presence and diversity of algal toxins in subtropical peatland periphyton: The Florida Everglades, USA, *J. Phycol.*, 46 (4), 674–678, 2010.
117. Gantar, M. et al., Allelopathic activity among cyanobacteria and microalgae isolated from Florida freshwater habitats, *FEMS Microbiol. Lett.*, 64, 55–64, 2008.

118. Hitzfeld, B.C., Höger, S.J., and Dietrich, D.R., Cyanobacterial toxins: Removal during drinking water treatment, and human risk assessment, *Environ. Health Perspect.*, 108 (1), 113–122, 2000.
119. Moss, R. et al., *Towards New Scenarions for Analysis of Emissions, Climate Change, Impacts and Response Strategies*, Intergovernmental Panel on Climate Change (UNEP), Geneva, Switzerland, 2008.
120. Sinha, R. et al., Increased incidence of *Cylindrospermopsis raciborskii* in temperate zones—Is climate change responsible? *Water Res.*, 46 (5), 1408–1419.
121. Gugger, M. et al., Genetic diversity of *Cylindrospermopsis* strains (Cyanobacteria) isolated from four continents, *Appl. Environ. Microbiol.*, 71 (2), 1097–1100, 2005.
122. Rinta-Kanto, J.M. et al., Lake Erie Microcystis: Relationship between microcystin production, dynamics of genotypes and environmental parameters in a large lake, *Harmful Algae*, 8, 665–673, 2009.
123. Sevilla, E. et al., Microcystin-LR synthesis as response tonitrogen: Transcriptional analysis of the *mcy* D gene in *Microcystisaeruginosa* PCC7806, *Ecotoxicology*, 19, 1167–1173, 2010.
124. Kaebernick, M. and Neilan, B.A., Ecological and molecular investigation of cyanotoxin production, *FEMS Microbiol. Ecol.*, 35, 1–9, 2001.
125. Mohamed, Z.A. et al., Microcystin production in benthic mats of cyanobacteria in the Nile River and irrigation canals, Egypt, *Toxicon*, 47, 584–590, 2006.
126. Brand, L.E. et al., Cyanobacterial blooms and the occurrence of the neurotoxin beta-N-methylamino-L-alanine (BMAA), in South Florida aquatic food webs, *Harmful Algae*, 9(6), 620–635, 2010.
127. Kosten, S. et al., Warmer climates boots cyanobacterial dominance in shallow lakes, *Global Change Biol.*, 18, 118–126, 2012.
128. Hoeger, S.J. et al., Occurrence and elimination of cyanobacterial toxins in two Australian drinking water treatment plants, *Toxicon*, 43, 639–649, 2004.
129. Yu, S.-Z., Drinking water and primary liver cancer, in *Primary Liver Cancer*, Tang, Z. Y. et al., Eds., China Academia Publishers, New York, pp. 30–37, 1989.
130. Codd, G.A. et al., Cyanobacterial toxins: Risk management for health protection, *Toxicol. App. Pharmacol.,* 203, 264–272, 2005.
131. Mohamed, Z.A. and Shehri, A.M., Cyanobacteria and their toxins in treated-water storage reservoirs in Abha city, Saudi Arabia, *Toxicon*, 50, 75–84, 2007.
132. Krienitz, L. et al., Contribution of hot spring cyanobacteria to the mysterious deaths of Lesser Flamingos at Lake Bogoria, Kenya, *FEMS Microbiol. Ecol.,* 2, 141–148, 2003.
133. Dasey, M. et al., Investigations into the taxonomy, toxicity and ecology of benthic cyanobacteria accumulations in Myall Lake, Australia, *Mar. Freshwater Res.*, 56(1), 45–55, 2005.
134. Wood, S.A. et al., Identification of a benthic microcystin-producing filamentous cyanobacterium (Oscillatoriales) associated with a dog poisoning in New Zealand, *Toxicon*, 55(4), 897–903, 2010.
135. Wood, S.A. et al., Wide spread distribution and identification of eight novel microcystins in Antarctic cyanobacterial mats, *Appl. Environ. Microbiol.*, 74, 7243–7251, 2008.
136. Mohamed, Z.A., Toxic cyanobacteria and cyanotoxins in public hot springs in Saudi Arabia, *Toxicon*, 51, 17–27, 2008.
137. Fiore, M.F. et al., Microcystin production by a freshwater spring cyanobacterium of the genus, *Fischerella*, *Toxicon*, 53(7–8), 754–761, 2009.
138. Kaasalainen, U. et al., Microcystin production in the tripartite cyanolichen, *Peltigera leucophlebia*, *Mol. Plant Microbe Interact.*, 22, 695–702, 2009.
139. Chia, A.M. et al., A survey for the presence of microcystin in aquaculture ponds in Zaria, Northern Nigeria: Possible public health implication, *Afr. J. Biotechnol.*, 8(22), 6282–6289, 2009.

18 Hepatotoxic Microcystins of Cyanobacteria
Biosynthesis and Degradation in Response to Abiotic Stress

*Ashutosh Kumar Rai, Leanne Andrea Pearson, and Ashok Kumar**

CONTENTS

18.1 INTRODUCTION

Cyanobacteria (blue-green algae) are an ancient group of microorganisms that thrive in a broad spectrum of terrestrial, freshwater, and marine habitats ranging from hot springs to frozen ponds [1]. The successful colonization of this wide range of environments by cyanobacteria is linked to their long evolutionary history and adaptation to a number of environmental stresses such as high solar UV radiation, extreme temperatures, desiccation, oxidative stress, and nutrient fluctuations. Present-day cyanobacteria perform oxygenic photosynthesis and possess chlorophyll *a* and water-soluble phycobilin proteins [2,3]. Additionally, certain members of this order are capable of fixing atmospheric dinitrogen gas (N_2), allowing them to colonize environments lacking a stable supply of fixed nitrogen [2]. A range of environmental conditions including higher temperatures and pH values, low turbulence, and high nutrient levels (eutrophication) can result in the proliferation of planktonic cyanobacteria in lakes and reservoirs, often leading to the formation of huge surface blooms. Certain bloom-forming cyanobacteria are also capable of producing secondary metabolites, which have a range of bioactivities, some of which are highly toxic to eukaryotic organisms including humans. Cyanotoxins are very diverse in terms of their chemical structure as well as their toxicity [4–7]. Based on the toxic effects

* Corresponding author: kasokbt@rediffmail.com

TABLE 18.1
Major Toxin Producing Genera of Cyanobacteria

Cyanotoxin	Producing Genera
Neurotoxins	
Anatoxin-a homo-anatoxin-a	*Anabaena*, *Aphanizomenon*, *Oscillatoria*
Anatoxin-a(s)	*Anabaena*, *Oscillatoria* (*Planktothrix*)
Paralytic shellfish poisons (saxitoxins)	*Anabaena*, *Aphanizomenon*, *Cylindrospermopsis*- and *Lyngbya*
Hepatotoxins	
Cylindrospermopsin	*Aphanizomenon*, *Cylindrospermopsis*, *Raphidiopsis*, *Umezakia*
Microcystins	*Microcystis* (*M. aeruginosa*, *M. wesenbergii*, *M. viridis*), *Anabaena flos-aquae*, *Planktothrix* (*Osc. agardhii*, *Osc. rubescens*, *Osc. tenuis*), *Hapalosiphon hibernicus*, *Aphanocapsa cumulus*, *Cyanobium bacillare*, *Arthrospira fusiformis*, *Limnothrix redekei*, *Phormidium formosum*, *Nostoc* sp., *Anabaenopsis* sp., and *Synechocystis* sp.
Nodularins	*Nodularia*
Contact irritant-dermal toxins	
Debromoaplysiatoxin, Lyngbyatoxin	*Lyngbya*
Aplysiatoxin	*Schizothrix*

Source: From Carmichael, W.W. and Li, R.H., *Sal. Syst.*, 2, 5, 2006; Tyagi, M.B. et al., *J. Microbiol. Biotechnol.*, 9, 9, 1999; Zegura, B. et al., *Mutat. Res.*, 727, 16, 2011.

they elicit, they may be classified as dermatotoxins (lipopolysaccharides, lyngbyatoxin-a, and aplysiatoxins), neurotoxins (anatoxin-a, homoanatoxin-a, anatoxin-a(s), and saxitoxins), and hepatotoxins (microcystins, nodularin, and cylindrospermopsin) [8–11]. Generally, these toxins are present within cells but can be released in high concentrations during cell lysis [10] or via active transport mechanisms [12].

The various classes of cyanotoxins and the genera by which they are produced are listed in Table 18.1. Extensive research efforts over the past three decades have provided a wealth of biochemical, genetic, and toxicological information for the cyanotoxins [11,13,14]. It is beyond the scope of this mini review to cover all aspects of cyanotoxin research; however, herein we attempt to provide readers with an overview of the occurrence, biosynthesis, and degradation of the most globally significant cyanotoxin, microcystin. The impact of various stresses on microcystin production is discussed, as are the future implications of global climate change on microcystin-producing cyanobacteria.

18.2 MICROCYSTIN

Microcystins are low molecular weight cyclic heptapeptides produced by numerous cyanobacterial genera (Table 18.1). The various microcystin isoforms range in molecular weight from 909 to 1115 Da [14] and have differing levels of toxicity. They are water-soluble, thermostable compounds, which undergo hydrolysis only when exposed to highly acidic conditions [7]. As potent inhibitors of eukaryotic protein phosphatases (PP) 1 and 2A, microcystins have been linked to the deaths of wild animals and livestock worldwide [10].

Microcystins have the general structure: cyclo (D-Ala^1-X^2-D-MeAsp^3-Z^4-Adda^5-D-Glu^6-Mdha^7), where X and Z are variable L-amino acids; MeAsp stands for erythro-β-methyl aspartate; Adda for a unique β-amino acid (2S,3S,8S,9S)-3-amino-9-methoxy-2,6,8-trimethyl-10-phenyldeca-4,6-dienoic acid; and Mdha is *N*-methyl-dehydroalanine (Figure 18.1) [11]. Adda causes hepatotoxicity by forming a covalent linkage to the catalytic site of PP1 and 2A [12]. Besides hepatotoxicity, Adda also

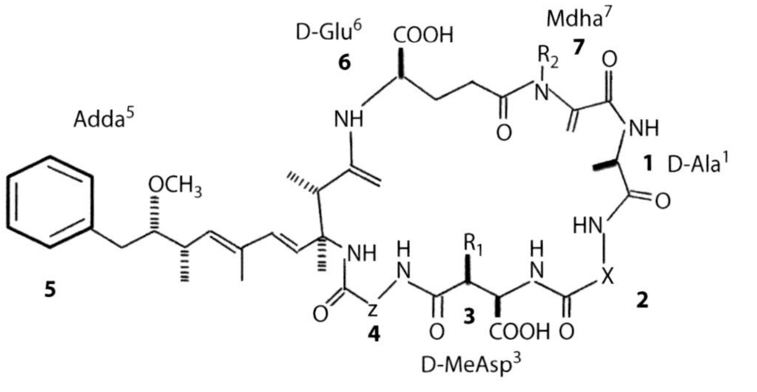

FIGURE 18.1 Generalized structure of microcystin. Variable L-amino acid residues are found at positions X and Z.

provides the characteristic absorption spectrum of microcystin at 238 nm due to the presence of a conjugated diene group in the long carbon chain [10]. Generally, microcystins are produced by planktonic strains of the distantly related cyanobacterial genera, *Microcystis*, *Planktothrix*, and *Anabaena* [10]. However, planktonic, benthic, and terrestrial strains of the genera *Nostoc*, *Anabaenopsis*, *Hapalosiphon*, and *Phormidium* are also capable of producing the toxin (although at relatively low levels) [15–17].

18.2.1 Biosynthesis

Microcystins are synthesized non-ribosomally by a multifunctional enzyme complex via a thiotemplate mechanism (a non-ribosomal, non-RNA-dependent mechanism for the synthesis of various oligopeptides). Due to non-ribosomal mechanism of synthesis, the process is not affected by protein synthesis inhibitors. The enzyme complex (microcystin synthetase) includes non-ribosomal peptide synthetases (NRPS), polyketide synthases (PKS), and tailoring enzymes [18–21]. NRPSs possess a highly conserved modular structure with each module comprising catalytic domains responsible for the adenylation, thioester formation, and, in most cases, condensation of specific amino acids [20–22]. The arrangement of these domains within the multifunctional enzymes determines the number and order of the amino acid constituents of the peptide product. Additional domains for the modification of amino acid residues such as epimerization, heterocyclization, oxidation, formylation, reduction, or N-methylation may also be included in the module [20,22].

Microcystin biosynthesis is encoded by the microcystin synthetase (*mcy*) gene cluster. The *mcy* gene cluster has been characterized in detail for *Microcystis*, *Planktothrix*, and *Anabaena* [18–25]. Interestingly, the size of the *mcy* gene cluster varies between genera. For example, in *Microcystis*, it spans 55 kb (Figure 18.2); in *Planktothrix*, 55.6 kb; and in *Anabaena* sp. 90, 55.4 kb. In *Microcystis aeruginosa*, the *mcy* gene cluster comprises 10 bidirectionally transcribed open reading frames (*mcyABCDEFGHIJ*) arranged in two putative operons (*mcyA-C* and *mcyD-J*) [20]. Altogether, 48 sequential catalytic reactions are involved in microcystin synthesis, in which 45 have been assigned to catalytic domains within six large multienzyme synthases/synthetases (mcyA-E, G), which incorporate the phenylpropanoid starter unit (probably phenyllactate) [26,27], as well as malonyl-CoA, *S*-adenosyl-L-methionine, glutamate, serine, alanine, leucine, D-methyl-isoaspartate,

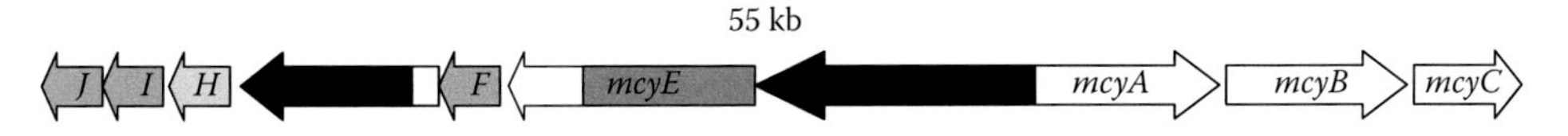

FIGURE 18.2 Typical organization of microcystin synthetase gene cluster from *M. aeruginosa* showing genes encoding non-ribosomal peptide synthetases, polyketide synthases, tailoring enzymes, and ABC transporters (*mcyH*). (Modified from Pearson, L.A. and Neilan B.A., *Curr. Opin. Biotechnol.*, 19, 281, 2008.)

and arginine [20]. The additional four monofunctional microcystin tailoring enzymes are putatively involved in the *O*-methylation of Adda (McyJ) [23], the racemization of L-aspartate (McyF) [28], and the production of D-methyl aspartate (McyI) [29]. The ABC transporter, McyH, is putatively involved in the export of the toxin [20,30]. In *Microcystis* and *Anabaena*, the genes are transcribed from a central bidirectional promoter region [25], whereas in *Planktothrix* all *mcy* genes, except *mcyT*, seem to be transcribed unidirectionally from a promoter located upstream of *mcyD* [23].

18.2.2 Regulation

Regulation of microcystin production has been investigated by numerous researchers and is thought to be influenced by a number of different physical and environmental parameters, including nitrogen, phosphorous, trace metals, growth temperature, light, and pH [31–33]. Unfortunately, as most experiments have been performed using different strains and the data have not been standardized using the same set of controls, it is difficult to draw accurate conclusions regarding microcystin regulation.

Early batch culture experiments have suggested that microcystin production is positively influenced by high nitrogen and phosphorus concentrations and low iron concentrations [31,33]. However, subsequent experiments using continuous cultures indicated that toxin production is actually directly related to the specific growth rate of cells [34].

Transcriptional regulation studies by Kaebernick et al. [35] demonstrated that *mcy* transcripts were upregulated when *M. aeruginosa* cultures were exposed to light intensity of 68 μmol of photons m^{-2} s^{-1} (even at a high intensity of 400 μmol of photons m^{-2} s^{-1}). Importantly, if cultures were exposed to medium (31 μmol of photons m^{-2} s^{-1}) and/or low (16 μmol of photons m^{-2} s^{-1}) light intensities or placed in the dark, transcription rates decreased considerably. Furthermore, Zilliges et al. [36] have recently shown that under light (70 μmol of photons m^{-2} s^{-1}) and oxidative stress conditions, microcystin binds strongly to enzymes of the Calvin cycle, phycobiliproteins, and certain NADPH-dependent reductases. Zilliges et al. [36] also demonstrated that under conditions of oxidative stress and high light, microcystin-producing *M. aeruginosa* cells are more robust than their nontoxic mutant counterparts. These results demonstrate a strong link between toxin production and stress response.

18.2.3 Structural Variants

One of the most fascinating features of microcystin is the high number of isoforms, which can arise due to changes in the specificity of microcystin synthetases or as a result of mutations in the corresponding *mcy* genes [26]. Around 90 variants of microcystin have been identified, which arise mainly due to the variable amino acids present at positions X and Z [10,37,38]. Several independent investigations on *Microcystis* and *Planktothrix* strains have concluded that the natural variation in the microcystin synthetase gene cluster among genera and species is often caused by recombination [26,39]. Mikalsen et al. [26] reported the presence of different variants of the *mcyABC* gene cluster in 11 strains of *Microcystis* species and its absence in 7 strains. They suggested that the recombination between imperfect repeats, gene loss, and horizontal gene transfer may be responsible for the patchy distribution of *mcyABC* gene cluster. In general, *mcyA* and *mcyB* genes show maximum genetic variability among various strains of *Microcystis*.

18.2.3.1 Toxicity

Microcystins are toxic to a range of eukaryotic organisms ranging from planktonic crustaceans to vertebrates such as fish and mammals [40–42]. In mice, the intraperitoneal (i.p.) LD_{50} value for microcystin-LR is usually around 50 μg/kg body weight but it can range from 25 to 125 μg/kg body weight [12,40,41], while the oral LD_{50} value is typically around 5 mg/kg body weight [41]. The toxic effects of microcystin are mediated via the inhibition of protein serine–threonine phosphatases (PP) [43–45], which occurs via noncovalent interactions between the residues Trp206 and Ile130 in the hydrophobic groove of PP1 and the unusual amino acid side-chain, Adda [46]. High concentrations of microcystin promote liver cell death via reactive oxygen species (ROS) signaling without the involvement of typical apoptotic proteins, while intermediate doses activate classical apoptotic pathways. Chronic low concentrations of microcystin promote liver cell survival and proliferation and can cause primary liver cancer [47,48] by inducing oxidative DNA damage [49,50]. Mice exposed to a sublethal dose of microcystin-LR by i.p. injections developed multiple neoplastic nodules in the liver [48].

The WHO has established a drinking water safety guideline of 1 $\mu g\ L^{-1}$ for microcystin-LR, which many countries have adhered to (with slight variation). This guideline is also recommended for food that may be contaminated with the toxin [41]. It should be noted that microcystin-LR is the most-studied variant of microcystin where variable amino acids present at positions X and Z are leucine (L) and arginine (R) (Figure 18.1). In the case of microcystin-RR, which is another variant of microcystin, variable amino acids present at positions X and Z of the ring are arginine (R).

18.2.4 Removal and Physicochemical Degradation

Since microcystins are highly resistant to most physicochemical factors, their removal and transformation from aquatic systems is challenging. However, a few methods have been developed to this end [51–54]. Photodegradation of microcystins as well as cylindrospermopsin under ambient conditions has been tested using different radiation bands within the natural solar spectrum [55]. While photodegradation of cylindrospermopsin was dependent on ultraviolet-A radiation (UV-A) and was very low under natural conditions, microcystin degradation occurred at a high rate with all three radiation bands (photosynthetically active radiation [PAR], UV-A, and UV-B), although PAR and UV-A were more efficient most probably due to their high natural irradiance. The role of visible light in microcystin destruction was also investigated by Graham et al. [56], who demonstrated that rhodium-doped material was the most effective material for degrading microcystin, followed by carbon-modified titania.

Methods such as chlorination and ozonation have been widely used for the destruction of dissolved microcystins [52]; however, continuous treatment with these agents is rarely cost-effective as contamination of water bodies with microcystins is unpredictable and their concentration shows seasonal fluctuation. Advanced oxidation technologies (AOTs) such as Fenton, UV/O_3, UV/TiO_2, and UV/H_2O_2 have been successfully tested for the removal of microcystins from drinking water [53]. Of all the systems, UV/H_2O_2 is the most efficient and preferred, since in this system there is no phase transfer and sludge formation, no secondary pollution, and the system is simple and cost-effective. Qiao et al. [53] reported that after a 60 min reaction, 94.83% of microcystin-RR was removed under optimal experimental conditions. Antoniou et al. [57] explored the potential use of sulfate radical-based advanced oxidation technologies (SR-AOTs) for the degradation of the naturally occurring microcystin-LR. Degradation of microcystins using chlorine dioxide (ClO_2), a strong oxidizing and sanitizing agent, has also been tested [54]. Additionally, powdered activated carbon (PAC), iron oxide nanoparticles, and ultrafiltration (UF) membranes have been explored as potential treatment technologies [51,58]. They are effective but have a limited lifetime and are expensive. Specific strains of probiotic bacteria may also be capable of removing cyanobacterial toxins from solution [59]. Despite recent technological advances, flocculation or slow sand filtration steps are still the preferred (and probably the most effective) methods for the removal of cell-bound microcystins [60]. However, these methods cannot remove toxins that have been exported from cells or released following cell lysis.

18.2.5 Microbial Degradation

Hepatotoxins, including microcystins and nodularin, resist degradation mainly due to their cyclic structure and inclusion of several unusual amino acids. The cyclic structure makes them resistant to many common bacterial proteases. However, during the last two decades, microcystin-degrading bacteria have been isolated from sewage effluents, lakes, rivers, bed sediments, and infiltration soil [61–65]. Jones et al. [61] for the first time reported the isolation of a microcystin-degrading bacterium, *Sphingomonas* sp. strain ACM-3962 (MJ-PV), from an Australian river. Lahti et al. [66] similarly isolated 17 microcystin-degrading bacterial strains including one species of *Sphingomonas* from the bed sediments and water of eutrophic lakes. This was followed by the isolation of a few more *Sphingomonas* sp. or sphingomonad-like strains potentially capable of microcystin degradation [62,63,67]. To date, the majority of microcystin-degrading bacteria belong to the genus *Sphingomonas* sp.; however, a few other bacteria, namely, *Sphingopyxis* sp. USTB-05, *Paucibacter toxinivorans*, *Delftia acidovorans*, *Burkholderia* sp., *Lactobacillus rhamnosus*, *Bifidobacterium*, and *Morganella morganii*, have been reported to degrade microcystins [59,64,65,68,69].

Holst et al. [70] reported that aerobic as well as anaerobic microorganisms in the sediments of a water recharge facility could efficiently remove microcystin. Their conclusion was based on the mineralization of ^{14}C-labeled microcystin-LR (via accumulation of $^{14}CO_2$). In a similar study by Hyenstrand et al. [71], it was reported that carbon dioxide is the major end product of microcystin degradation in eutrophic lake water spiked with ^{14}C-labeled microcystin-LR.

Bourne et al. [67] carried out cloning and gene library screening of *Sphingomonas* strain ACM-3962 and detected the microcystin-degrading gene cluster *mlrABCD* on a 5.8 kb genomic fragment (Figure 18.3). Mlr-mediated microcystin degradation pathways in *Sphingomonas* have been studied in detail [67]. The first enzyme in the degradation pathway, microcystinase (MlrA), is encoded by the *mlrA* gene and is a 336-residue endopeptidase that cleaves the Adda-arginine peptide bond in microcystin-LR and opens the cyclic ring. The linear microcystin-LR is subsequently degraded by peptidases encoded by the genes *mlrB* and *mlrC*. MlrB (402 residues) cleaves linear microcystin LR to a tetrapeptide product and MlrC (507 residues) mediates further peptidolytic degradation. The *mlrD* gene encodes the transporter protein, MlrD (442 residues), that permits uptake of microcystin into the cell [67].

The elucidation of gene clusters involved in microcystin degradation has enabled rapid screening of microcystin-degrading bacteria by polymerase chain reaction (PCR). For example, Saito et al. [63] used *mlrA* as a target sequence to detect microcystin-degrading bacteria. Similarly, Ho et al. [72] used PCR-based methods for the characterization of microcystin-degrading strains within a biologically active sand filter, while Okano et al. [68] identified an *mlrA* homolog in *Sphingopyxis* sp. C-1.

PCR screening for *mlrA* may identify potential microcystin degraders; however, a positive result does not necessarily indicate the presence of the entire *mlr* cluster. In order to reduce the incidence of false positives, it would be desirable to screen candidate strains for all the four *mlr* genes. This could be achieved either by the use of a single set of primers for amplification of each gene separately in PCR assay or by designing four sets of primer (having almost identical annealing temperature) for respective genes and multiplexing them for amplification in one PCR tube (multiplex approach). The multiplex approach saves time and results obtained are identical to those where amplification is performed individually using a single set of primers. Employing the aforementioned approaches of the PCR assay, if the presence of the *mlr* cluster is confirmed, microcystin degradation could be easily assessed by routine analytical techniques.

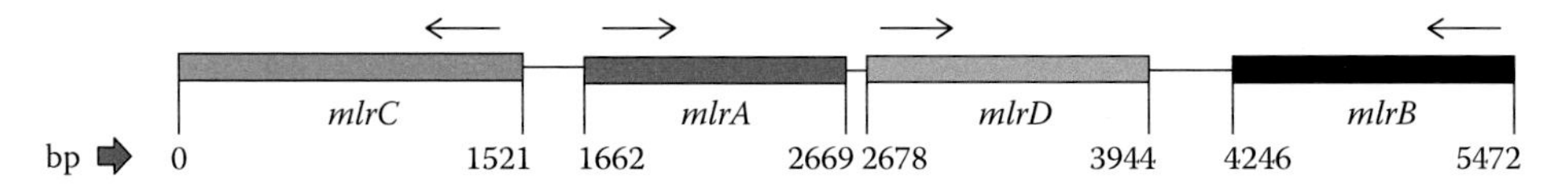

FIGURE 18.3 Organization of 5.8 kb *mlr* gene cluster present only in those isolates of *Sphingomonas* that are strictly involved in microcystin biodegradation. (Bourne, D.G. et al.: Characterisation of a gene cluster involved in bacterial degradation of the cyanobacterial toxin microcystin LR. *Environ. Toxicol.*, 2001, 16, 523. Copyright Wiley-VCH Verlag GmbH & Co. KGaA. With permission.)

18.3 CONCLUSIONS

The global distribution of toxic cyanobacteria is becoming a matter of great human concern owing to the fact that most communities rely on surface water and reservoirs for potable use. It is also presumed that the abundance of toxic blooms shall increase due to the ecological consequences of global warming such as rising temperatures, changes in rainfall patterns, increasing UV-B radiation, and increasing eutrophication. Although knowledge concerning the chemical nature, detection methods, and genetics of microcystin biosynthesis has increased enormously during the last decade, the impact of abiotic stresses on cyanotoxin synthesis and regulation is poorly understood. Generally, abiotic stresses are known to enhance secondary metabolite production in diverse groups of microorganisms. Blooms of cyanobacteria, which are known to produce a variety of secondary metabolites including hepatotoxins (microcystin and nodularin) and neurotoxins (anatoxin-a and saxitoxins), may have a negative impact on water quality and phytoplankton communities. Studies conducted on *M. aeruginosa* have revealed that microcystin is synthesized non-ribosomally by a multifunctional enzyme complex encoded by 55 kb microcystin synthetase (*mcy*) gene cluster in which six structural genes (*mcyA*, *B*, *C*, *D*, *E*, and *G*) are responsible for its biosynthesis. Molecular methods are proving increasingly important for the detection of toxigenic cyanobacteria and for monitoring toxin gene expression in response to a variety of environmental factors. However, a multidisciplinary approach is needed to manage risks associated with these organisms. In line with this, the development of toxin removal and/or degradation methods is an important issue for water quality today, particularly in the face of a changing global climate.

ACKNOWLEDGMENTS

Ashutosh Kumar Rai is thankful to ICAR for his senior research fellowship. This work was partly supported by a research grant sanctioned to Ashok Kumar by ICAR, New Delhi, India (NBAIM/AMAAS/MD/(19)/AK/BG).

REFERENCES

1. Whitton, B., Diversity, ecology and taxonomy of the cyanobacteria, in *Photosynthetic Prokaryotes*, Mann, H. and Carr, N., Eds., Plenum Press, New York, 1992, p. 1.
2. Stewart, W.D.P., Some aspects of structure and function in N_2-fixing cyanobacteria, *Annu. Rev. Microbiol.*, 34, 497, 1980.
3. Adam, D.G., Cyanobacterial phylogeny and development: Question and challenges, in *Prokaryotic Development*, Brun, Y.Y. and Shimkets, L.J., Eds., American Society of Microbiology, Washington, DC, 2000, p. 51.
4. Börner, T. and Dittman, E., Molecular biology of cyanobacterial toxins, in *Harmful Cyanobacteria*, Vol. 3, Huisman, J., Matthijs, H.C.P., and Visser, P.M., Eds., Springer, Norwell, the Netherlands, 2005, p. 25.
5. Carmichael, W.W., Health effects of toxin-producing cyanobacteria: "The cyanoHABs," *Hum. Ecol. Risk Assess.*, **7**, 1393, 2001.
6. Devlin, J.P. et al., Anatoxin-*a*, a toxic alkaloid from *Anabaena flos-aquae* NRC-44h, *Can. J. Chem.*, 55, 1367, 1977.
7. Botes, D.P. et al., The structure of cyanoginosin-LA, a cyclic heptapeptide toxin from the cyanobacterium *Microcystis aeruginosa*, *J. Chem. Soc. Perkin Trans.*, 1, 2311, 1984.
8. Rinehart, K.L., Namikoshi, M., and Choi, B.W., Structure and biosynthesis of toxins from blue-green algae (cyanobacteria), *J. Appl. Phycol.*, 6, 159, 1994.
9. Harada, K.-I. et al., Comprehensive analysis system using liquid chromatography-mass spectrometry for the biosynthetic study of peptides produced by cyanobacteria, *J. Chrom.*, 1033, 107, 2004.
10. Sivonen, K. and Jones, G., Cyanobacterial toxins, in *Toxic Cyanobacteria in Water*, Chorus, I. and Bartram, J., Eds., E & FN Spon, London, U.K., 1999, p. 41.

11. Carmichael, W.W. et al., Naming cyclic heptapeptide toxins of cyanobacteria (blue-green algae), *Toxicon*, 26, 971, 1988.
12. Dawson, R.M., The toxicology of microcystins, *Toxicon*, 36, 953, 1998.
13. Carmichael, W.W. and Li, R.H., Cyanobacteria toxins in the Salton Sea, *Sal. Syst.*, 2, 5, 2006.
14. Kurmayer, R. and Christiansen, G., The genetic basis of toxin production in cyanobacteria, *Freshwater Rev.*, 2, 31, 2009.
15. Prinsep, M.R. et al., Microcystin-LA from a blue-green alga belonging to the Stigonematales, *Phytochemistry*, 31, 1247, 1992.
16. Honkanen, R.E. et al., Protein phosphatase inhibitory activity in extracts of cultured blue-green-algae (cyanophyta). *J. Phycol.*, 31, 478, 1995.
17. Sivonen, K. and Börner, T., Bioactive compounds produced by cyanobacteria, in *The Cyanobacteria: Molecular Biology, Genomics and Evolution*, Herrero, A. and Flores, E., Eds., Caister Academic Press, Norfolk, U.K., 2008, p. 159.
18. Dittmann, E. et al., Insertional mutagenesis of a peptide synthetase gene that is responsible for hepatotoxin production in the cyanobacterium *Microcystis aeruginosa* PCC 7806, *Mol. Microbiol.*, 26, 779, 1997.
19. Nishizawa, T. et al., Polyketide synthase gene coupled to the peptide synthetase module involved in the biosynthesis of the cyclic heptapeptide microcystin, *J. Biochem.*, 127, 779, 2000.
20. Tillett, D. et al., Structural organization of microcystin biosynthesis in *Microcystis aeruginosa* PCC7806: An integrated peptide polyketide synthetase system, *Chem. Biol.*, 7, 753, 2000.
21. Nishizawa, T. et al., Genetic analysis of the peptide synthetase genes for a cyclic heptapeptide microcystin in *Microcystis* spp., *J. Biochem.*, 126, 520, 1999.
22. Pearson, L.A. and Neilan, B.A., The molecular genetics of cyanobacterial toxicity as a basis for monitoring water quality and public health risk, *Curr. Opin. Biotechnol.*, 19, 281, 2008.
23. Christiansen, G. et al., Microcystin biosynthesis in *Planktothrix*: Genes, evolution and manipulation, *J. Bacteriol.*, 185, 564, 2003.
24. Rouhiainen, L. et al., Genes coding for hepatotoxic heptapeptides (microcystins) in the cyanobacterium *Anabaena* strain 90, *Appl. Environ. Microbiol.*, 70, 686, 2004.
25. Kaebernick, M. et al., Multiple alternate transcripts direct the biosynthesis of microcystin, a cyanobacterial nonribosomal peptide, *Appl. Environ. Microbiol.*, 68, 449, 2002.
26. Mikalsen, B. et al., Natural variation in the microcystin synthetase operon *mcyABC* and impact on microcystin production in *Microcystis* strains, *J. Bacteriol.*, 185, 2774, 2003.
27. Hicks, L.M. et al., Structural characterization of in vitro and in vivo intermediates on the loading module of microcystin synthetase, *ACS Chem. Biol.*, 1, 93, 2006.
28. Sielaff, H. et al., The mcyF gene of the microcystin biosynthetic gene cluster from *Microcystis aeruginosa* encodes an aspartate racemase, *Biochem. J.*, 373, 909, 2003.
29. Pearson, L.A., Barrow, K.D., and Neilan, B.A., Characterization of the 2-hydroxy-acid dehydrogenase McyI, encoded within the microcystin biosynthesis gene cluster of *Microcystis aeruginosa* PCC7806, *J. Biol. Chem.*, 282, 4681, 2007.
30. Pearson, L.A. et al., Inactivation of an ABC transporter gene, mcyH, results in loss of microcystin production in the cyanobacterium *Microcystis aeruginosa* PCC7806, *Appl. Environ. Microbiol.*, 70, 6370, 2004.
31. Lukac, M. and Aegerter, R., Influence of trace metals on growth and toxin production of *Microcystis aeruginosa*, *Toxicon*, 31, 293, 1993.
32. Van der Westhuizen, A.J. and Eloff, J.N., Effect of temperature and light on the toxicity and growth of the blue-green alga *Microcystis aeruginosa* (UV-006), *Planta*, 163, 55, 1985.
33. Sivonen, K., Effects of light, temperature, nitrate, orthophosphate, and bacteria on growth of and hepatotoxin production by *Oscillatoria agardhii* strains, *Appl. Environ. Microbiol.*, 56, 2658, 1990.
34. Long, B.M., Jones, G.J., and Orr, P.T., Cellular microcystin content in N-limited *Microcystis aeruginosa* can be predicted from growth rate, *Appl. Environ. Microbiol.*, 67, 278, 2001.
35. Kaebernick, M. et al., Light and the transcriptional response of the microcystin biosynthetic gene cluster, *Appl. Environ. Microbiol.*, 66, 3387, 2000.
36. Zilliges, Y. et al., The cyanobacterial hepatotoxin microcystin binds to proteins and increases the fitness of microcystis under oxidative stress conditions, *PLoS ONE*, 6, e17615, 2011.
37. Welker, M. and von Döhren, H., Cyanobacterial peptides—Nature's own combinatorial biosynthesis, *FEMS Microbiol. Rev.*, 30, 530, 2006.
38. Feurstein, D. et al., Microcystin congener and concentration dependent induction of murine neuron apoptosis and neurite degeneration, *Toxicol. Sci.*, 124(2), 424, 2011. doi:10.1093/ toxsci/kfr243, 2011.

39. Tanabe, Y., Kaya, K., and Watanabe, M.M., Evidence for recombination in the microcystin synthetase (*mcy*) genes of toxic cyanobacteria *Microcystis* spp., *J. Mol. Evol.*, 58, 633, 2004.
40. Tyagi, M.B. et al., Cyanobacterial toxins: The current status, *J. Microbiol. Biotechnol.*, 9, 9, 1999.
41. WHO, *Toxic Cyanobacteria in Water: A Guide to Their Public Health Consequences, Monitoring and Management*, Chorus, I. and Bartram, J., Eds., E & FN Spon, London, U.K., 1999, p. 416.
42. Deng, D-F. et al., Toxic threshold of dietary microcystin (-LR) for quart medaka, *Toxicon*, 55, 787, 2010.
43. MacKintosh, C. et al., Cyanobacterial microcystin-LR is a potent and specific inhibitor of phophatases 1 and 2A from both mammals and higher plants, *FEBS Lett.*, 264, 187, 1990.
44. Yoshizawa, S. et al., Inhibition of protein phosphatases by microcystins and nodularin associated with hepatotoxicity, *J. Cancer Res. Clin. Oncol.*, 116, 609, 1990.
45. Carmichael, W.W., The toxins of cyanobacteria, *Sci. Am.*, 270, 78, 1994.
46. Holmes, C.F. et al., Molecular enzymology underlying regulation of protein phosphatase-1 by natural toxins, *Curr. Med. Chem.*, 9, 1981, 2002.
47. Ueno, Y. et al., Detection of microcystins, a blue green algal hepatotoxin, in drinking water sampled in Haimen and Fusui, endemic areas of primary liver cancer in China, by highly sensitive immunoassay, *Carcinogenesis*, 17, 1317, 1996.
48. Ito, E. et al., Neoplastic nodular formation in mouse liver induced by repeated intraperitoneal injections of microcystin-LR, *Toxicon*, 35, 1453, 1997.
49. Zegura, B., Lah, T.T., and Filipic, M., Alteration of intracellular GSH levels and its role in microcystin-LR-induced DNA damage in human hepatoma HepG2 cells, *Mutat. Res.*, 611, 25, 2006.
50. Zegura, B., Btraser, A., and Filipic, M., Genotoxicity and potential carcinogenicity of cyanobacterial toxins—A review, *Mutat. Res.*, 727, 16, 2011.
51. Lambert, T.W., Holmes, C.F.B., and Hrudey, S.E., Adsorption of microcystin-LR by activated carbon and removal in full scale water treatment, *Water Res.*, 30, 1411, 1996.
52. Tsuji, K. et al., Stability of microcystins from cyanobacteria IV. Effect of chlorination on decomposition, *Toxicon*, 35, 1033, 1997.
53. Qiao, R-P. et al., Degradation of microcystin-RR by UV radiation in the presence of hydrogen peroxide, *Toxicon*, 45, 745, 2005.
54. Ying, J. et al., Degradation of microcystin-RR in water by chlorine dioxide, *J. China Univ. Mining Technol.*, 18, 623, 2008.
55. Wormer, L. et al., Natural photodegradation of the cyanobacterial toxins microcystin and cylindrospermopsin, *Environ. Sci. Technol.*, 44, 3002, 2010.
56. Graham, D. et al., The degradation of microcystin-LR using doped visible light absorbing photocatalysts, *Chemosphere*, 78, 1182, 2010.
57. Antoniou, M.G., de la Cruz, A.A., and Dionysiou, D.D., Degradation of microcystin LR using sulfate radicals generated through photolysis, thermolysis and e-transfer mechanisms, *Appl. Catal. B: Environ.*, 96, 290, 2010.
58. Keijola, A.M. et al., Removal of cyanobacterial toxins in water treatment processes: Laboratory and pilot-scale experiments, *Toxic. Assess.*, 3, 643, 1998.
59. Nybom, S.M., Salminen, S.J., and Meriluoto, J.A., Specific strains of probiotic bacteria are efficient in removal of several different cyanobacterial toxins from solution, *Toxicon*, 52, 214, 2008.
60. Bourne, D.G. et al., Biodegradation of the cyanobacterial toxin microcystin LR in natural water and biologically active slow sand filters, *Water Res.*, 40, 1294, 2006.
61. Jones, G.J. et al., Degradation of the cyanobacterial hepatotoxin microcystin by aquatic bacteria, *Nat. Toxins*, 2, 228, 1994.
62. Park, H.D. et al., Degradation of the cyanobacterial hepatotoxin microcystin by a new bacterium isolated from a hypertrophic lake, *Environ. Toxicol.*, 16, 337, 2001.
63. Saito, T. et al., Detection and sequencing of the microcystin LR-degrading gene, *mlr* A, from new bacteria isolated from Japanese lakes, *FEMS Microbiol. Lett.*, 229, 271, 2003.
64. Ishii, H., Nishijima, M., and Abe, T., Characterization of degradation process of cyanobacterial hepatotoxins by a gram-negative aerobic bacterium, *Water Res.*, 38, 2667, 2004.
65. Eleuterio, L. and Batista, J.R., Biodegradation studies and sequencing of microcystin-LR degrading bacteria isolated from a drinking water biofilter and a fresh water lake, *Toxicon*, 55, 1434, 2010.
66. Lahti, K. et al., Biodegradation of cyanobacterial hepatotoxins-characterization of toxin degrading bacteria, in *Harmful Algae*, Reguera, B. et al., Eds., Xunta de Galicia and Intergovernmental Oceanographic Commission of UNESCO, Santiago de Compostela, Spain, 1998, p. 363.
67. Bourne, D.G. et al., Characterisation of a gene cluster involved in bacterial degradation of the cyanobacterial toxin microcystin LR, *Environ. Toxicol.*, 16, 523, 2001.

68. Okano, K. et al., Characteristics of a microcystin-degrading bacterium under alkaline environmental conditions, *J. Toxicol.*, 2009, 954291, 2009.
69. Zhang, M., Pan, G., and Yan, H., Microbial biodegradation of microcystin-RR by bacterium *Sphingopyxis* sp. USTB-05, *J. Environ. Sci.*, 22, 168, 2010.
70. Holst, T. et al., Degradation of microcystin in sediments at oxic and anoxic, denitrifying conditions, *Water Res.*, 37, 4748, 2003.
71. Hyenstrand, P. et al., Laboratory studies of dissolved radiolabelled microcystin-LR in lake water, *Water Res.*, 37, 3299, 2003.
72. Ho, L. et al., Bacterial degradation of microcystin toxins within a biologically active sand filter, *Water Res.*, 40, 768, 2006.

19 Structural, Physiological, and Ecological Adaptations in Cyanobacterial Mats under Stressful Environment

*Pushpendra Kumar Mishra, John L. Sailo, and Surya Kant Mehta**

CONTENTS

* Corresponding author: skmehta12@rediffmail.com

19.1 INTRODUCTION

In natural environments, microorganisms grow rarely in isolation as a single species or population. Highly specific and less understood interactions among unrelated species play a crucial role in the establishment of a successful microbial assemblage in the form of microbial biofilms or mats. Multilayered microbial assemblage, generally with a coherent structure, growing on or within some substratum is called a microbial mat. There are many varieties of microbial mats including mats of diatoms and biofilms. Microbial mats can be defined as stratified communities of microorganisms that develop along physiochemical gradients at the interface of water and solid substances held together by a dense matrix. In a mat, different groups of microorganisms are protected from deleterious effects of harsh environments like high and low temperatures, hypersalinity, temperature extremes, acidic pH, and many other factors.

Microorganisms have the ability to colonize in different types of environment and interact with each other forming a complex community. Microbial biofilms are formed by bacteria and other microorganisms when they attach to a surface, multiply, and produce exopolymers for mutual attachment [1]. Biofilms are responsible for most of the microbial activities in nature [2]. Microbial mats are multilayered microbial communities where different components (microorganisms) are distributed according to their metabolic specialty and adaptations. Microbial mats are also considered as a microscale ecosystem dominated by phototrophic bacteria. They have the ability to colonize in a variety of environments, forming complex communities consisting of cyanobacteria, purple sulfur bacteria, sulfate-reducing bacteria, etc., along with physicochemical gradients.

Cyanobacterial mats are generally formed by filamentous species that produce visible mat-like structures. Cyanobacterial mats grow on solid surfaces of abiogenic and biogenic substrates in submerged or semi-aquatic habitats under a wide variety of environmental conditions. Similar to other types of microbial mats, cyanobacterial mats are very common in extreme and highly fluctuating environments because in normal environment the development of the mat is often repressed by predators but in extreme conditions mostly predators are excluded. Because cyanobacteria are a very ancient group of microbes and have faced different changing environmental conditions during the course of their evolution, they are well adopted for stressful conditions and are abundant in most harsh environments.

19.2 CYANOBACTERIA: THE PRINCIPAL COMPONENT OF MICROBIAL MATS

19.2.1 Most Successful Photoautotrophs on Earth

Cyanobacteria are the first-known microbes on earth that invented an efficient mechanism for splitting (light-driven photolysis) of water (for sequestration of electron), resulting in oxygenic photosynthesis that laid the foundation for the evolution of aerobic microorganisms and eukaryotes of the past and the present [3]. Cyanobacteria have invented the PSII reaction center that extracts electrons from water. Due to PSII, cyanobacteria are not dependent on other scarcer reduced electron donors as are other non-oxygenic photosynthetic prokaryotes. Cyanobacteria contain a highly organized system of internal membranes for photosynthesis. Chlorophyll *a* and other accessory pigments (phycocyanin and phycoerythrin) found in photosynthetic lamellae are analog to eukaryotic lamellae.

19.2.2 Plant-Like Photosynthesis Together with N_2 Fixation

Many multicellular cyanobacteria irreversibly transform their normal cells into specialized and nitrogen-fixing heterocysts. Heterocyst structure and metabolic activity enable aerobic cyanobacteria to accommodate the oxygen-sensitive process of nitrogen fixation. Heterocysts have evolved to separate plant-like photosynthesis (which produces oxygen) and fixing N_2 to ammonia (a process that involves an enzyme that is inactivated by O_2) in the same filament. The main function of a heterocyst is to fix N_2, while the rest of the cells of the filament divide and perform photosynthesis, thus keeping the two processes separate. Heterocysts are impermeable to gases and have large amounts of nitrogenase. However, heterocysts maintain some permeability to the cells on either side. Neighboring cells take N_2 from the atmosphere and pass it to the heterocyst, which reduces it to NH_3 and returns fixed nitrogen to its neighbors. The neighboring cells also take up and utilize O_2, but they prevent it from reaching the heterocyst. Heterocysts are essentially specialized organs for the "multi-cellular organism" represented by a chain of cyanobacterial cells and they are only formed when nitrogen is limiting. The regulation of this developmental cycle is fascinating and serves as a simple example of multicellular development in a unicellular organism. Details of heterocyst development, morphogenesis, transport of molecules between cells in the filament, differential gene expression, and pattern formation are out of the scope of this chapter and can be found elsewhere.

19.2.3 Morphological Diversity

Cyanobacteria display various morphological forms such as unicellular, colonial, and filamentous (unbranched/branched). The nonfilamentous forms are either in a single cell (*Aphanothece*) or in groups (*Merismopedia*). During cell division, daughter cells separate but in some cases they remain together for shorter or longer periods and form a colony of several cells. The arrangement of cells in colonial forms may be tubular, cubical, large reticulate colonies as in *Microcystis*, or a branched system of mucilage stalk as seen in *Dictyosphaerium*. The trichome may be made up of the same type of cells called the homocystous form (e.g., *Oscillatoria, Lyngbya*) or different types of cells where some cells are enlarged and modified in heterocysts called the heterocystous form. Trichomes are straight or spirally coiled (e.g., *Spirulina*). Some filamentous forms show the ability to differentiate into several different cell types: vegetative cells, the normal, photosynthetic cells that are formed under favorable growing conditions; akinetes, the climate-resistant spores that may form when environmental conditions become harsh; and thick-walled heterocysts. The heterocyst has a strong external thickening of the cell wall. These thickenings are composed of three distinct layers: the outermost is fibrous, the middle homogeneous, and the innermost laminated. The position of the heterocyst in a filament may be terminal or intercalary. Trichomes may be of uniform diameter throughout or taper from base to apex, usually from basal heterocyst (e.g., *Calothrix* sp.) (Figure 19.1).

19.2.4 Environments Inhabited by Cyanobacteria

Due to their wide range of tolerance over extreme and unfavorable conditions, cyanobacteria are found in nearly all ecological habitats. They frequently grow in habitats like freshwater, hot springs, terrestrial environments and soils, deep ocean, hypersaline conditions, hypothermal vent, Antarctic pond sea ice, deserts, etc. [4–9] and are cosmopolitan in distribution. Freshwater and wet conditions like ponds, lakes, and rivers are the most favorable habitats for cyanobacterial growth. Many cyanobacteria (e.g., *Microcystis* sp., *Anabaena* sp., *Lyngbya* sp., *Oscillatoria* sp., *Gloeotrichia* sp.) form freshwater blooms due to their overgrowth.

19.3 CYANOBACTERIAL MAT

Cyanobacteria are major elements of most types of mat ecosystems found in hypersaline, hot springs, and marine environments because of their ability to tolerate and grow in a wide range of extreme conditions [10]. They are common in all microbial mats where light is available, and perform photosynthesis that plays an important role in mat biochemistry by production of O_2 and exopolymeric substances (EPSs), carbon fixation, etc. Cyanobacteria introduce organic compounds in microbial mats, and the subsequent decomposition of organic compounds supports the growth and development of other microbial communities in the mat. The organic compounds synthesized

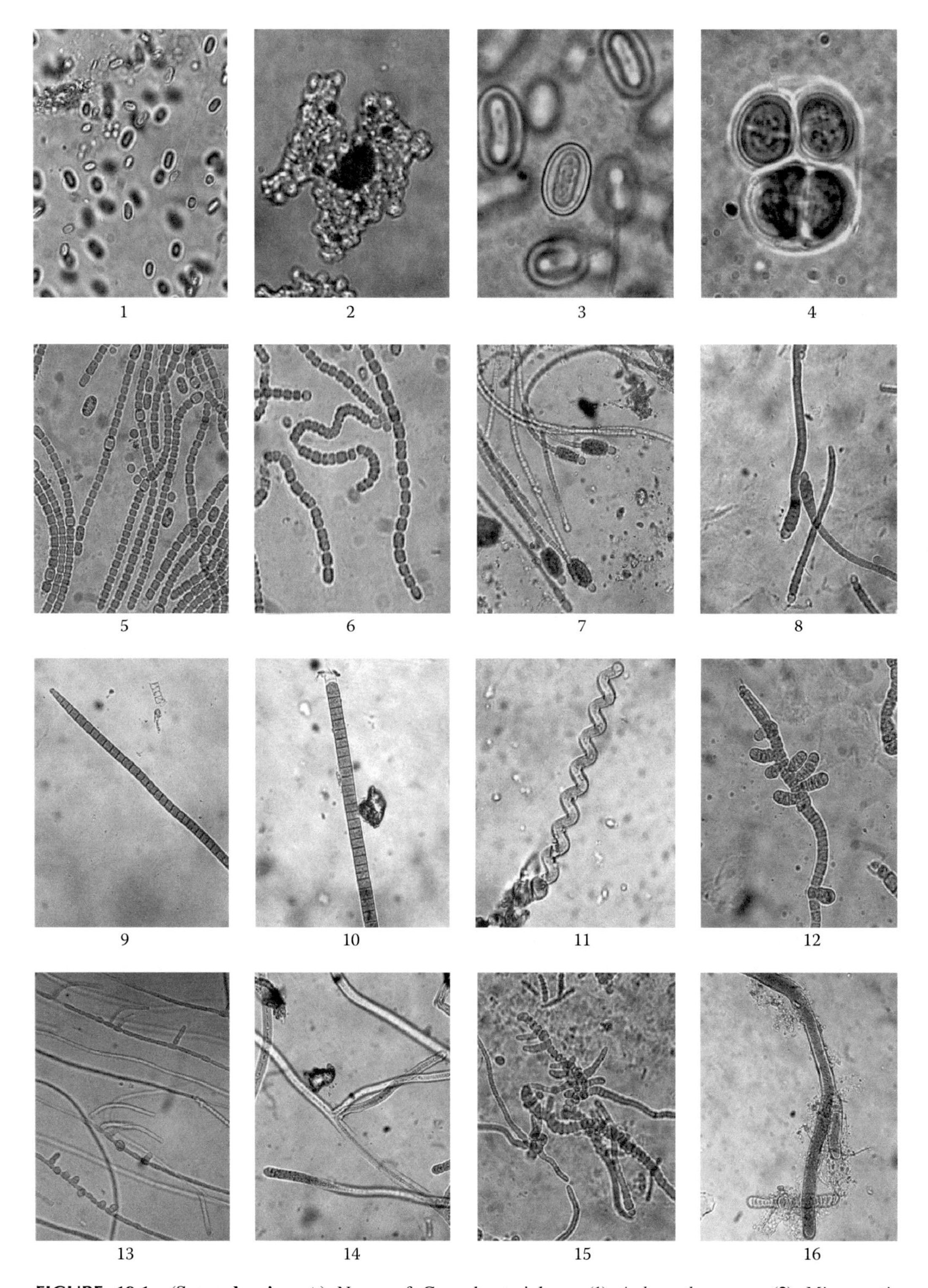

FIGURE 19.1 **(See color insert.)** Name of Cyanobacterial sp. (1) *Aphanothece* sp., (2) *Microcystis* sp., (3) *Gloeothece* sp., (4) *Chrococcus* sp., (5) *Anabaena* sp., (6) *Nostoc* sp., (7) *Cylindrospermum* sp., (8) *Calothrix* sp., (9) *Oscillatoria* sp., (10) *Lyngbya* sp., (11) *Spirulina* sp., (12)*Westiella* sp., (13) *Nostochopsis* sp., (14) *Scytonema* sp., (15) *Fischerella* sp., and (16) *Tolypothrix* sp.

during photosynthesis are released as a result of photorespiration or by the fermentation and secretion of an EPS. EPSs, mainly produced by cyanobacteria, play a crucial role in the formation and stability of mats. Mats inhabiting microorganisms remain embedded in gelatinous matrix and protect themselves against a multitude of stresses (high temperature, desiccation, UV damage, etc.). Cyanobacterial EPSs enhance the survival of community, facilitate the accumulation of nutrients, cell motility, and act as a barrier to the entry of toxic metals in mats. Cyanobacteria excrete a massive amount of slime that helps in the formation of cohesive biofilm structure where all microbial communities of mats colonize together. Purple sulfur and green sulfur bacteria are characteristic components of cyanobacterial mats formed in marine environments and hot springs.

19.3.1 Characteristics

Due to a wide range of metabolic capabilities, cyanobacteria are the most common organisms that form mats that are well adapted to highly fluctuating environments while most eukaryotes cannot survive in such environments. Cyanobacteria are the most successful mat-forming organisms due to a number of factors. Cyanobacteria are the only oxygenic phototrophic prokaryotes and, as such, have the capacity for net carbon fixation using only light and water. Further, cyanobacteria can fix N_2 and thereby add to the nutrient status of microbial communities of mats.

Microbial mats vary in shape, size, width, color, and diversity, depending upon prevailing environmental conditions. The most common cyanobacteria found in mats of perennial streams belong to family Oscillatoriaceae. Filamentous and non-heterocystous cyanobacteria are more common than heterocystous cyanobacteria while unicellular cyanobacteria are also found in some thermal mats. Filamentous cyanobacterium *Microcoleus chthonoplastes* is the most common in hypersaline and hot springs mats worldwide. Other common cyanobacteria found in mats are *Lyngbya aestuarii*, *Synechococcus lividus*, *Spirulina labyrinthiformis*, *Aphanothece* sp, *Phormidium laminosum*, *Calothrix* sp. *Oscillatoria amphigraulata*, *Oscillatoria terebriformis*, *Oscillatoria limosa*, etc. The dominant cyanobacteria found in various microbial mat types are listed in Table 19.1.

19.3.2 Community Structure and Function

Microbial mats are laminated, cohesive microbial communities composed of a consortium of bacteria dominated by photoautotrophic cyanobacteria [11]. There are two more or less clear zones in microbial mats: (a) oxic photosynthetic zone and (b) anoxic zone. The anoxic zone harbors anoxygenic photoautotrophs (purple bacteria) and green sulfur bacteria followed by anaerobic chemoorganotrophic bacteria, sulfate reducers, and methanogens [12]. Table 19.2 shows the various mat types and their microbial community composition diversity.

A unique property of mats is the occurrence of different microbial communities in well-defined and stratified layers. The arrangement of different microbial communities in mats determines the stability, growth, and function of mats. In most photosynthetic mats, cyanobacteria are present in the top layer because they require light to perform photosynthesis. Sometimes diatoms are also present with cyanobacteria in top layers of the mat. The top layer of the mat, covered by dense polymeric mucilage sheath, contains pigments secreted by cyanobacteria, which protect underlying communities from UV damages. The organic matter produced through photosynthesis can be degraded by chemotropic bacterial population. The complete digestion of the photosynthetic product (organic compounds) occurs under anoxic conditions prevailing in the deeper regions of cyanobacterial mats. So the complete decomposition of organic matter in cyanobacterial mats is completed by strictly anaerobic sulfate-reducing bacteria.

Below the cyanobacterial layer, a pink-colored layer, made up of purple sulfur bacteria, may be observed in many cyanobacterial mats. Common to all cyanobacterial mats, the purple sulfur bacteria are anaerobic and perform anaerobic photosynthesis in low intensity of light. A distinct layer of

TABLE 19.1
Dominant Cyanobacterial Species Found in Different Types of Photosynthetic Mats

Type of Mat	Dominant Cyanobacteria	Reference
Hot Spring	*Microcoleus chthonoplastes, Synechococcus* sp., *Phormidium laminosum*	[65]
Marine	*Aphanocapsa littoralis. Oscillatoria sp., Microcoleus chthonoplastes*	[66]
Marine, North Sea, Germany	*Microcoleus chthonoplastes, Aphanocapsa* sp. *Oscillatoria* sp. *Lyngbya sp.*	[86]
Sheltered Beaches	*Phormidium laminosum Synechococcus* sp. *Oscillatoria sp.*	[67]
Marine, Shackleford Banks (USA)	*Microcoleus chthonoplastes, Spirulina* sp., *Oscillatoria sp.*	[68]
Marine (Protected by Mangroves)	*Scytonema* sp., *Oscillatoria* sp., *Lyngbya sp.*	[69]
Marine, Mexico (USA)	*Microcoleus chthonoplastes, Spirulina larynthiformis*	[26]
Marine, Sipperwisset (USA)	*Microcoleus chthonoplastes, Oscillatoria amphigranulata, Lyngbya sp.*	[70]
Hypersaline Sulfur Spring	*Microcoleus chthonoplastes, Spirulina* sp., *Lyngbya sp, Aphanocapsa hylophytica*	[71]
Deep Sea	*Phormidium* sp., *Spirulina larynthiformis, Oscillatoria sp.*	[72]
Inland hypersaline desert stream, Saltanate of Oman	*Microcoleus chthonoplastes, Spirulina subsalsa, Johannesbaptistia pellucida, Chroococcidiopsis* sp., *Aphanocapsa* sp., *Chroococcus* sp., *Gloeocapsa* sp., *Schizothrix* sp. and *Leptolyngbya* sp.	[73]
Hypersaline	*Microcoleus, Oscillatoria, Leptolyngbya, Phormidium, Pleurocapsa, Gloeothece*	[25]

oxidized iron is often present between sulfate reducing bacteria. A sulfate-reducing bacterial layer present in deep layers of the mat is gray in color due to the presence of iron sulfide (FeS). Although these bacteria are distributed throughout the mat, they are quite active in the upper cyanobacterial layer in the dark when the mat turns anoxic. Sulfate-reducing bacteria are strictly aerobic and get energy by the oxidation of sulfur. These are autotrophic and have a high affinity for sulfur products; their occurrence results in highly dynamic sulfur and oxygen gradients.

The oxic photosynthetic zone is occupied by photosynthetic cyanobacteria mixed with some green algae and diatoms. Filamentous non-heterocystous cyanobacterium *Microcoleus chthonoplastes* is most common in mats developed in marine, hypersaline, and hot conditions all over the world [13]. *Microcoleus* trichomes are found in bundles and are twisted like rope and covered by a common dense polysaccharide sheath, a unique feature of any cyanobacteria found in mats. The dense sheath provides an anoxygenic environment for oxygen-sensitive nitrogen-fixing enzyme nitrogenase.

Another widely distributed mat forming cyanobacterium is *Oscillatoria limosa* that is capable of aerobic nitrogen fixation and covered by dense sheath, its motility help the cyanobacteria to bring at appropriate position according to environmental conditions like in high light intensity its trichomes moves toward down. Nitrogen-fixing cyanobacteria *Lyngbya aestuarii, Phormidium laminosum, Plectonema* sp. are also very common in mats. *Scytonema* sp. is common in the mangrove region [14]. In hot springs, the mat is usually dominated by thermophilic cyanobacteria *Synechococcus lividus, Phormidium laminosum, Oscillatoria amphigranulata, Spirulina labrynthiformis*, and *Calothrix* sp.

TABLE 19.2
Microbial Community Composition of Various Mat Types

Types	Species Composition	Reference
Hypersaline (Sultanate of Oman)	*Microcoleus chthonplastes, Chroococcidiopsis* sp., *Spirulina* sp., *Aphanocapsa* sp., *Chrococcus* sp., *Gleocapsa* sp., *Shizothrix* sp., *Leptolyngbya* sp.,	[73]
Hypersaline (Ebro Delta)	*Synechococcus* sp., *Oscillatoria* sp., *Peudoanabaena* sp., *Phormidium* sp., *Beggiatoa* sp., *Spirillum* sp., *Chloroflexus sp.*	[74]
Hypersaline	*Desulfovibrio* sp., *Aphanothece* sp., *Cyanothece* sp., *Chlorobium* sp., *Nitzschia* sp., *Halothece sp.*	[75]
Hot	*Artemia salina, Chloromatium* sp., *Chromatium* sp., *Lyngbya aestuerii, Microcoleus chthonplastes, Oscillatoria sp.*	[76]
Marine (North Carolina)	*Chromatium buderi, Azotobacter vinelondii, Klebsiella pneumoniae, Rhodospirillum rubrum, Rhodobacter capsulatus, Rhizobium meliloti, Clostridium pasteuranum, Azospirillum brasilense.*	[27]
Hypersaline (Bahamas Island)	Chromatium sp., *Desulfovibrio* sp., *Microcoleus chthonplastes, Lyngbya sp.*	[77]
Hypersaline	*Thiocapsa roseopersicina, Thicystis violacea, Allochromatium vinosum, Lamprobacter* sp., *Marichromatium gracile, Ectothiorhodospira* sp., *Halochromatium sp.*	[78]
Marine	*Lyngbya aestuerii, Microcoleus chthonplastes, Leptolyngbya* sp., *Spirulina* sp., *Thalassobaculum* sp., *Sulfitobacter* sp., *Roseobacter* sp., *Methylomonas* sp., *Gleothece* sp. *and Phormidium sp.*	[30]

19.3.3 Cyanobacterial Mat Types

19.3.3.1 Terrestrial Cyanobacterial Mats

These are found in different environmental conditions like deserts, rocks, sand dunes, and carbonate caves. Due to scarcity of water, these habitats support low microbial diversity. The thick sheath on the mat's surface plays an important role in protecting the mat microbial communities from desiccation and low water potential conditions. Some cyanobacteria like *Microcoleus chthonoplastes* and *Chrococcus* sp. have evolved adaptations to these environments and are found in desert rocks. Cyanobacterium *Crinalium epipsammum* is highly adapted to desiccation [15]. The green alga *Klebsormidium flaccidum* is common in some terrestrial microbial mats. In caves, where light intensity is very poor, mats are dominated by unicellular nitrogen-fixing cyanobacterium *Gloeothece* [16] and *Leptolyngbya*. The *Nostoc* is found in a variety of terrestrial mats (Figure 19.2).

19.3.3.2 Hot Springs Cyanobacterial Mats

These environments contain high temperature and acidic conditions that decrease biodiversity and grazers. The concentration of H_2S is very high in these areas, so the growth of most cyanobacteria is limited by a combination of high temperature and high sulfide concentration. Mat-forming *Oscillatoria* sp., frequently growing in hot springs, performs anoxygenic photosynthesis and lowers the sulfide concentration. Anoxygenic photosynthesis degrades the sulfide and protects the lower layers of the mat from oxygen.

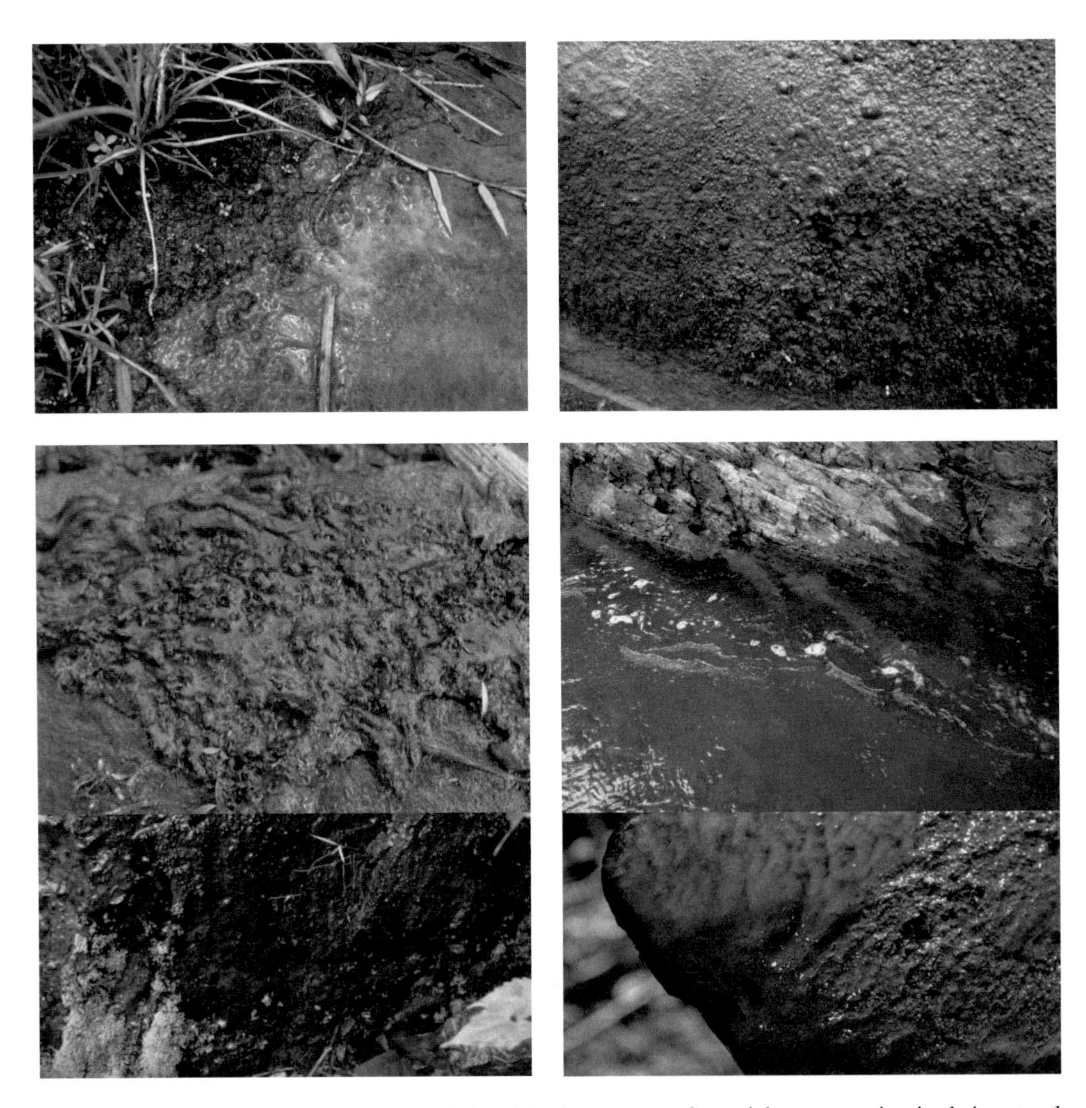

FIGURE 19.2 **(See color insert.)** Terrestrial and freshwater cyanobacterial mats growing in their natural habitats in Mizoram, India.

19.3.3.3 Coastal Cyanobacterial Mats

These are excellent areas for microbial mat development where the slope is low and a large area is filled by water for only short periods during the tide. So these areas experience large fluctuations in water content and temperature, which leads to extreme conditions that inhibit the growth of predators and grazers. Coastal sand flats are nutritionally poor but cyanobacteria are autotrophic, so nutrient demand is low and they can tolerate long periods of drought and high fluctuations in salinity and temperature. Filamentous cyanobacteria that form a dense trichome are more common in coastal habitats. Examples of this habitat are tidal sand flats of the southern North Sea of North America.

19.4 MOLECULAR DIVERSITY OF CYANOBACTERIAL MATS

Microbial mats are micro-scale ecosystems known for their large diversity and metabolic potential. Microbial mats are dominated mainly by filamentous cyanobacteria, although in some cases unicellular cyanobacteria have been shown to be the dominant component. In addition, microbial mats may accommodate a variety of other groups of microorganisms. There exists a good deal of information on microbial biodiversity and composition, structure, and function of communities of cyanobacterial mats of various habitats. The detection of mat diversity based on pure culture and morphology is many times not conclusive. Modern molecular techniques have greatly added to our understanding of microbial community structure, pattern, and dynamics of microbial mats. Molecular approaches have provided a new insight to understand the microbial distribution and ecology of microbial mats by characterization of community structure [17]. The use of 16S rRNA sequences (or the genes encoding them) may avoid the need to cultivate microorganisms to recognize its presence and measure its distribution in a community. Terrestrial hot spring microbial communities were among the first to be analyzed with this technology [18,19] and thus were among the first in which the impressive diversity of uncultivated microbial populations in nature was revealed. The 16S rRNA gene surveys have been carried out with the aim of studying microbial diversity in numerous habitats [20–24].

Investigations using a variety of molecular techniques yielded a more complete picture of the genetic composition of microbial mat communities. For instance, denaturing gradient gel electrophoresis (DGGE) revealed a diverse community in a hypersaline microbial mat comprising filamentous non-heterocystous cyanobacteria *Microcoleus*, *Oscillatoria*, *Leptolyngbya*, and *Phormidium* and the unicellular *Pleurocapsa* and *Gloeothece* [25]. In addition, sulfate-reducing bacteria (SRB) and sulfur-oxidizing and anoxygenic bacteria were present. Analyses of small-subunit rRNA genes of a hypersaline microbial mat revealed 4700 genotypes belonging to over 40 bacterial phyla, emphasizing the high diversity within these microecosystems [26].

In a number of studies, *nifH*, the gene coding for dinitrogenase reductase, which is one of the two proteins that constitute the nitrogenase complex, has been used to identify diazotrophic organisms in complex microbial consortia. These studies revealed a high diversity of diazotrophic organisms in microbial mats, which was not limited to cyanobacteria [27–29]. Sequences that were common in various microbial mats belonged to the cyanobacteria as well as to purple sulfur bacteria and SRB. An investigation based on 41,400 16S rRNA genes and *nifH* sequences from two microbial mats situated in the intertidal zone of the Dutch barrier island Schiemonnikoog showed dominance of cyanobacteria in both mats; however, the composition of the total bacterial community differed [30]. In the aforementioned study, proteobacteria-related sequences were retrieved as the second most abundant group higher up in the littoral (Station I), whereas bacteroidetes were the second most abundant group at the low water mark (Station II). The diazotrophic (nitrogen-fixing) communities at both stations were also different.

19.5 ECOLOGICAL ADAPTATIONS AND STRESS TOLERANCE IN MATS

A typical property of microbial mats is their laminated or stratified structure in which different groups of organisms occur in particular vertical layers and cyanobacteria are associated with the upper layers receiving sufficient light for photosynthesis. By oxygenic or sometime anoxygenic photosynthesis, they fix the CO_2 in the form of organic matter and enrich the mats. The carbon sequestered by cyanobacteria is used by other communities of the mat in a variety of ways. In addition, cyanobacteria may also fix atmospheric N_2 to amino acid and add to the nutrient status of the mat. Of all photosynthetic organisms, cyanobacteria inhibit the widest range of ecological habitats. The 3.5-billion-year survival of mats testifies to their capacity in adapting to and altering hostile environments through cellular and community-mediated activities [31]. Cyanobacteria can survive in cold and hot, alkaline and acidic, marine, freshwater, saline, terrestrial, and symbiotic environments. Cyanobacteria are the primary producers of most of the mats. Cyanobacteria are able to establish competitive growth in almost any environment that has, at least, water and sunlight. The diversity and adaptability of cyanobacteria is due to the fact that they have withstood the challenges of evolutionary environmental changes since their appearance at least 2.5 billion years ago [32]. During the course of their evolution, cyanobacteria have experienced periods of high and low temperatures, high and low CO_2, and low and high O_2 levels. These temporal and spatial variations proved to be a driving force in the evolution and acquisition of many stress-responsive (tolerance) genes and physiological adaptations, which helped in the successful establishment of cyanobacteria in a diverse range of environments today.

Although several metabolic processes are affected by diverse environmental factors, photosynthesis is affected the most. Many habitats, like hot spring, saline, arctic, and deserts, affect photosynthesis by restricting the acquisition of Ci. Major challenges to photosynthesis include (a) the restricted diffusion of Ci in water; (b) variability in the level of Ci and HCO_3^-, or CO_2, which is available; (c) wide fluctuations in light and temperature; and (d) fluctuations in O_2 levels, which can vary from anaerobic to supersaturated. The adverse environmental factors frequently encountered in many habitats further exacerbate the inherent inefficiencies of CO_2-fixing enzyme rubisco. Cyanobacteria have evolved flexible but sophisticated CO_2-concentrating mechanism (CCM) to match habitats' requirements. With the evolution of CCM, cyanobacteria have also evolved a rubisco that is adapted to optimal performance under elevated CO_2 conditions [33]. Importantly, cyanobacterial rubiscos have much lower affinities for both CO_2 and O_2 when compared with other algal or higher plant rubiscos. Due to a CCM, the rubisco operates close to V_{max} and with lowest N input.

The CCM consists of many active Ci transporters that may transport CO_2 or HCO_3^- from an external environment and deliver it as HCO_3^- to the interior of the cell, and the internal HCO_3^- pool is much higher than the external environment. A protein microbody known as carboxysome, containing rubisco, is the subcellular compartment where HCO_3^- is converted into CO_2 by carboxysomal anhydrases (CAs). Although CA is required to generate CO_2 from HCO_3^-, it is identified only in β-Cyanobacteria [34], and none of the α-cyanobacteria sequenced so far has a carboxysomal CA homologue. The carboxysomes are found in all cyanobacteria characterized to date. Polypeptides making up the protein coat of α-carboxysomes from a chemoautotrophic proteobacterium *Halothiobacillus neapolitanus* are coded by cso-type genes (*cso*S1A, *cso*S1B, *cso*S1C, *cso*S2, *cso*S3, *orf*A, *orf*B) [35]. Microbial mats of all kinds are characterized by a number of factors that have great impact on photosynthesis. In a shallow surface zone, light is rapidly attenuated. The diffusion of CO_2 and O_2 is limited, allowing O_2 to build up during the day to over 1000 μM; Ci level is greatly depleted and high pH prevails putting extreme pressure on CO_2 acquisition and photosynthesis.

In adverse conditions like darkness when photosynthesis declines in microbial mats, the mat quickly turns anoxygenic. Many cyanobacteria die in dark anoxygenic conditions. But mat-forming cyanobacteria can survive in such conditions for several days because they can ferment the stored glycogen. So fermentation proves to be an important aspect of cyanobacterial mats, as fermentation products supply substrates to other microorganisms of the mat, like sulfur-reducing bacteria. There is a great diversity in fermentation pathways in cyanobacteria. In majority of pathways, EMP pathway is involved; however, heterolactate fermentation uses the part of the oxidative pentose phosphate pathway found in *Oscillatoria limosa*. The fermentation pathway may be homoacetate, mixed acid, homolactate, or heterolactate, depending upon which enzyme is involved in fermentation [36]. In heterolactate, glycogen serves as a substrate and equimolar amounts of ethanol and lactate are formed that support many other microbial communities of the mat. Homoacetate fermentation, which is mainly found in anaerobic acetogenic bacteria, is also reported in many cyanobacterial mats. The homoacetate fermentation pathway that produces acetate and traces of lactate, CO_2, and H_2 from glycogen has been demonstrated in *Nostoc* sp. The mixed fermentation pathway is found in several cyanobacteria such as *Spirulina platensis*, *Microcoleus chthonoplastes*, etc. Fermentation has a number of advantages such as enhancing the reactivity of the cyanobacteria. Fermentation occurs in cyanobacterial mats when there is much fluctuation in light and oxygen levels. So in these conditions when oxygen declines rapidly, the cyanobacteria ferment the glycogen and provide energy to other organisms for survival and maintenance.

Various types of high molecular weight EPSs are secreted by cyanobacteria and are used for the attachment of cyanobacteria to the mat and also to form a dense matrix in which organisms are aggregated in the form of a mat. Besides, an EPS is a source of polysaccharides for the mat and is required in cell envelope formation. EPSs mainly are heteropolysaccharides that contain polysaccharides and proteins but their composition varies among different microorganisms. An EPS often contains uronic acid–D-glucuronic acid and D-mannuronic acids with carboxyl as the main functional group. EPS molecules produced by bacteria and diatoms are composed of a highly hydrated matrix associated with the cell. EPSs are secreted during the exponential growth phase of cells and provide protection by the binding of toxic ions, thereby preventing toxicants in cells. EPSs have a variety of important functions; these include addition of organisms in mats, protection of organisms against desiccation because it contains large amounts of water such that organisms can withstand long periods of drought, protection from toxic substances, scavenging of trace metals, calcification of mats, etc. Mat-forming cyanobacteria that secrete EPSs produce a matrix that stabilizes the organisms in the mat and contains uronic acid that interacts with sediment particles. EPSs, with a large amount of uronic acid, help in the stabilization of mats in sediments. Further, uronic acid can interact with a number of metals leading to the immobilization of toxic metals or to the scavenging of trace metals that are important (as micronutrients) for microbial mats. EPSs contain sulfated sugars; the sulfate group is also important to the tertiary structure of polysaccharides and is responsible for the stability of the matrix in mats. The EPSs synthesized by cyanobacteria in mats form the matrix for exonucleases, DNA, and plasmid [37], which protect the extracellular DNA against DNAse in mats [38] and provide natural transformation in mats.

Nitrogen is one of the most important elements found in living organisms; most of the nitrogen exists in proteins and amino acids. However, it is also present in significant amounts in nucleic acids, chlorophyll, vitamins, etc. Cyanobacteria produce a unique nitrogenous compound cyanophycin that serves as a nitrogen storage site [39]. Cyanobacteria can use a number of nitrogen sources like amino acid asparagines, ammonia, glutamine, nitrate, nitrite, etc. Some cyanobacteria have the ability to fix the atmospheric nitrogen. Nitrogen fixation is carried out mainly by some filamentous cyanobacteria where some vegetative cells are modified in heterocysts. But some nonfilamentous and non-heterocystous cyanobacteria can also fix the nitrogen. Cyanobacterial photosynthesis is oxygenic, cyanobacterial filaments are highly aerobic, and enzyme nitrogenase does not function under aerobic conditions. Cyanobacteria have developed special strategies to overcome this incompatibility. The most specific adaptation is the differentiation of normal cell to special cell type, heterocyst, which is the site of nitrogen fixation. Most non-heterocystous cyanobacteria also can fix the nitrogen in anaerobic conditions.

Some filamentous and unicellular cyanobacteria (e.g., *Synechococcus* sp, *Plectonema boryanum*, *Oscillatoria limnetica*, etc.) where heterocysts are absent are able to carry out nitrogen fixation. These cyanobacteria have the ability to produce nitrogen-fixing enzyme nitrogenase but cannot protect the enzyme inactivation by oxygen, and hence can fix nitrogen only in anoxygenic environments. The microbial mats found in such environments where steep fluctuations in sulfide and O_2 occur, a high level of sulfide inhibits the oxygenic photosynthesis providing opportunity for anaerobic nitrogen-fixing cyanobacteria to grow diazotrophically. In some microbial mats, aerobic nitrogen-fixing non-heterocystous cyanobacteria (*Microcoleus* sp. *Lyngbya* sp, *Oscillatoria* sp. *Gloeothece* sp.) are known to exist [40,41].

Nitrogen fixation is an expensive process and requires 16 ATP molecules to fix one N_2; cyanobacteria fulfill this high energy demand by oxygenic photosynthesis where they use light as a source of ATP. Thus, in cyanobacterial mats, nitrogen fixation is directly linked to light. Consequently, the nitrogen fixation in microbial mats is not the same every day; the daily variation in nitrogen fixation depends upon the type of cyanobacteria and the fluctuation of light and oxygen in the mats. The daily pattern of N_2 fixation in non-heterocystous cyanobacteria is a result of the combined effect of light, oxygen, and sometimes sulfide. In such mats, photosynthesis occurs during the day and nitrogen fixation at night, as in the case of mats of *Oscillatoria*. Anoxic condition during the night is ideal for nitrogen fixation in non-heterocystous cyanobacteria but there is a problem for energy and reduced substances. Most mat-forming cyanobacteria overcome this problem as most of them are capable of fermentation of endogenous storage. Dark anoxic condition enhances the nitrogenase activity of mat-forming cyanobacteria *Oscillatoria limosa* [42].

19.6 NUTRIENT CYCLES IN MICROBIAL MATS

19.6.1 Nitrogen Cycle

Nitrogen occurs in different oxidation states in mats like ammonia, nitrous oxide, nitric oxide, nitrite, nitrate, hydroxylamine, dinitrogen, etc. Microorganisms carry out the oxidation and reduction reactions by changing from one form to another. In microbial mats, all nitrogen cycle steps are present and cyanobacteria play a significant role in the nitrogen cycle of many microbial mats. Ammonia is converted into amino acids that are used for the synthesis of proteins. In the presence of oxygen, ammonium can be oxidized to nitrate by nitrifying bacteria of the mat via nitrite, a process called nitrification. Nitrate can be taken up by cyanobacteria and assimilated. Under anoxic conditions, nitrate is reduced to ammonium by denitrifying bacteria. Nitrogen can be taken up by cyanobacteria and may be stored as cyanophycin. In cyanobacterial mats, some organic compounds are not fully oxidized so that nitrogen cannot enter into the nitrogen cycle in the microbial mat. The importance of this strategy is not understood, as in mats around 99% of organic matter is recycled within the mat.

19.6.2 Sulfur Cycle

The sulfur cycle also has the significant effect on microbial mats found in sulfur-rich environments like coastal areas. The sulfate-reducing cyanobacteria are found only in euphotic depth in the microbial mats, where conditions are strictly anaerobic and ideal for sulfate-reducing bacteria. The sulfate-reducing bacteria in mats are essentially anaerobic and oxidize simple and small organic matter by using sulfate as electron acceptor leading to the synthesis of sulfide. This sulfide can be oxidized back to sulfate where sulfur produced as an intermediate can be metabolized by anaerobic phototrophic bacteria like purple sulfur and green sulfur bacteria of mats. Some cyanobacteria also can carry out sulfide-dependent anoxygenic photosynthesis but can form thiosulfate or sulfur. The other reactions of the sulfur cycle in mats include oxidation and reduction of sulfite, thiosulfate, and sulfur where one part of the molecule is oxidized and the other reduced [43].

Cyanobacteria in mats release a large number of organic matters by different mechanisms. Compounds produced by cyanobacterial fermentation and photorespiration (glycolate) serve as substrates for sulfate-reducing bacteria. Glycolate can be used by sulfate-reducing bacteria of mats [44]. Thus, the metabolic activity of cyanobacteria has a direct effect on sulfate reduction. Most of sulfate-reducing bacteria are strictly anaerobic and are positioned in the deep layer of microbial mats, but some oxygenic sulfate-reducing bacteria may also be present on the top layer of mats. Sulfate-reducing bacteria present in the top layer of mats face the problem of oxygen stress in cyanobacterial mats. However, they have some strategies, like migration to deeper anoxic zone, formation of clump, and oxygen removal, to escape from oxygen stress [45].

Sulfide is an extremely toxic compound that reacts with iron containing compounds of organisms like hemoproteins, as these compounds are important in electron transport thus interfere the photosynthesis. Sulfide has an irreversible inhibition effect on the PSI of cyanobacteria, but some mat-forming cyanobacteria like *Oscillatoria amphigranulata* and *Oscillatoria limnetica* can overcome sulfide stress and carry out photosynthesis in high sulfur concentrations as found in hot springs and coastal areas.

19.6.3 Phosphorus Cycle

Phosphorus is involved in a variety of physiochemical reactions and is important for the growth and metabolism of organisms. The phosphorus cycle in microbial mats is not well known. Some cyanobacteria can store phosphate as poly-phosphate deposits usually found in coastal waters or shallow seas. H_3PO_4 (ortho-phosphate) is the most common form of inorganic phosphate. In addition, phosphate can also form an insoluble precipitate with iron ($FePO_4$). The free ortho-phosphate in microbial mats is very low. Although phosphorus occurs in other oxidation states (+5 to −3), it is not as important in redox reactions as nitrogen or sulfur. Bacteria can readily oxidize the phosphorus in mats that is taken up by cyanobacteria. The microbial phosphorus cycle includes uptake of inorganic phosphate and the release of organic phosphate. Phosphate deposits are formed when phosphorus reacts with calcium to form calcium phosphates [46].

19.7 APPLICATIONS OF CYANOBACTERIAL MATS

The ecological success of microbial mats and a broad array of microbial activities have proved that these microbial systems are useful in the bioremediation of environmental pollutants and in the production of useful products. Mats (generated) showed very high biomass production rate, which has been demonstrated to be 14.96 g m^{-2} day^{-1} [47]. Mature microbial mats are durable microbial community that may find application in a variety of uses in bioremediation, energy production, agriculture, and aquaculture. Following are some potential applications of microbial mats.

19.7.1 Bioremediation

The increasing concern with environmental pollution significantly motivates the investigation and development of safe technologies. The use of cyanobacterial mats as biosorbent is a fast-growing field in the treatment of waste water due to their size and their ability to grow under controlled conditions and in a wide range of environmental conditions [48]. Although seaweeds and microalgae have shown immense ability to sorb metal ions, confinement of seaweeds to coastal areas and problems in harvesting microalgae during the metal sorption process limit their commercial application [49]. In contrast, cyanobacterial mats are ubiquitous, self-immobilized, rich in exopolysaccharides with tremendous metal-binding capability, and are easy to use during metal sorptive processes [12]. Microbial mats have been shown to remove heavy metals [50,51] and radionuclides [52]. Due to the large diversity of functional groups, microbial mats have great potential in the sequestration of metals, metalloids, radionuclides, and oxyanions. The metal sorption capacities of some cyanobacteria are shown in Table 19.3.

In a study carried out with a tub-and-sink type bioreactor containing 100 L of water recycled at 5 L min^{-1}, 10 mg L^{-1} of chromium (Cr^{3+}), cadmium (Cd^{2+}), and lead (Pb^{2+}), a 100% removal within 24 h was demonstrated [53]. In another study, silica-mat particles (microbial mats immobilized in silica particles) were used to treat water containing 2.4 mg L^{-1} uranium [52]. The study showed 88% reduction of U^{6+} to U^{4+}. Further, it was shown that spent silica-mat particles can be used for subsequent U removal by adding minimal culture media and allowing a 6–8 days growth.

In biofilm environments, heavy metals and radioactive pollutants are removed by a variety of mechanisms including adsorption on surface, precipitation as sulfides or phosphates, and microbial reductive precipitation [54]. The EPSs excreted by microorganisms, found in many cyanobacterial mats, play a role in the sorption of heavy metals in biofilms; as EPS molecules are anionic, they can assist in biomineralization of metal ions that are cationic. EPSs are mainly composed of polysaccharides and proteins that have many charged functional groups. The binding affinity of metal by an EPS depends upon size/charge ratio, EPS composition, pH, etc. A lower pH results in the release of bound ions from an EPS while higher pH favors [55]. The mats can tolerate high metal concentrations because of their ability to precipitate insoluble metal salts outside the cell in the form of sulfides, phosphates, carbonates, or hydroxides [47,56]. The anaerobic sulfate-reducing bacteria of mats carry out the bio-precipitation of metals by microbiologically produced sulfide; this property may be used as an effective method to remove and concentrate metals from waste water [57–59].

19.7.2 Organic Compound Degradation

Microbial mats are capable of degrading various organic compounds like petroleum distillates, trichloroethylene (TCE), tetrachloroethylene, PCB, 2,4,6-trinitrotolune (TNT), pesticides carbofuron,

TABLE 19.3
Metal Adsorption Capacity of Mat-Forming Cyanobacteria

Cyanobacteria	Metals	Adsorption Capacity (m mol g^{-1})	Reference
Phormidium sp.	Pb (II)	1.38	[51]
	Cu (II)	0.88	
	Cd (II)	0.78	
Oscillatoria sp.	Pb (II)	0.48	[51]
	Cu (II)	0.42	
	Cd (II)	0.39	
Aphanothece sp.	Hg (II)	2.27	[79]
Spirulina platensis	Hg (II)	2.12	[79]
Synechococcus sp.	Cu (II)	0.177	[80]
	Pb (II)	0.15	
	Cd (II)	0.064	
	Cr (III)	0.104	
Anabaena spiroides	Mn (II)	0.15	[81]
Cyanothece sp.	Cu (II)	2.25	[82]
Gloeothece magna	Cd (II)	0.0037	[83]
	Mn (II)	0.0043	
Calothrix machica	Pb (II)	0.357	[84]
Cyanospira capsulata	Cu (II)	1.15	[85]

paraquat and prophos, etc. [60–63]. Mat treatment of 42,4,6-TNT, applied to a water column at a concentration of 100 mg L^{-1}, showed 99% removal of TNT in 6 days. Eighty-two percent decrease in carbofuron concentration was found after mat treatment of a banana farm.

19.7.3 Bioenergy

The use of microbial mats in energy production has great potential in view of increasing prices of fossil fuels. Cyanobacteria and purple bacteria of microbial mats are capable of producing hydrogen gas via anoxygenic photosynthesis [64]. Hydrogen production in mats is driven by nitrogen-fixing apparatus (nitrogenase activity); the process is highly ATP consuming and is less efficient than hydrogenase-mediated production. Although hydrogenase occurs in all cyanobacteria, only nitrogenase systems are involved in hydrogen production. The highest H_2 production rate in natural mats is estimated at 6 nmol cm^{-3} day^{-1} [12].

ACKNOWLEDGMENT

This work was supported by the Department of Biotechnology, New Delhi, through a project granted to SKM.

REFERENCES

1. Gilbert, G.S. et al., Effect of an introduced bacterium on bacterial communities on roots, *Ecology*, 74, 40, 1993.
2. Marshall, K., Microbial adhesion in biotechnological process, *Curr Opin Biotechnol.*, 5, 296, 1994.
3. Whitton, B.A., Diversity, ecology and taxonomy of the cyanobacteria, in Mann, N.H. and Carr, N.G. (Eds), *Photosynthetic Prokaryotes*, Plenum Press, New York, pp. 1–51, 1992.
4. Otte, S. et al., Nitrogen, carbon and sulfur metabolism in natural thioploca samples, *Appl Environ Microbiol.*, 65, 3148, 1999.

5. Nuebel, U. et al., Diversity and distribution in hypersaline microbial mats of bacteria related to *Chloroflexus* sp., *Appl Environ Microbiol.*, 67, 4365, 2001.
6. Brunberg, A.K. et al., Characteristics of oligotrophic hardwater lakes in a postglacial land rise area in mid-Sweden, *Freshwater Biol.*, 47, 1451, 2002.
7. Nakagawa, T. and Fukui, M., Phylogenetic characterization of microbial mats and streamers from a Japanese alkaline hot spring with a thermal gradient, *J Gen Appl Microbiol.*, 48, 211, 2002.
8. Lutz, R.A. et al., Life after death in the deep sea, *Am Sci.*, 89, 422, 2001.
9. Madigan, M.T. et al., *Rhodoferax antarcticus* sp. nov.: A moderately Psychrophilic purple non sulfur bacterium isolated from an Antarctic microbial mat, *Arch Microbiol.*, 173, 269, 2000.
10. Stal, L.J., Microbial Mats in coastal environment, in Stal, L.J. and Caumette, P. (Eds), *Microbial Mats. Structure, Development and Environmental Significance*, Springer Verlag, Heidelberg, Germany, p. 21, 1994.
11. Nisbet, E.G. and Fowler, C.M.R., Archaean metabolic evolution of microbial mats, *Proc R Soc.*, 266, 2375, 1999.
12. Bender, J. and Phillips, P., Microbial mats for multiple applications in aqua culture and bio remediation, *Bioresour Technol.*, 94, 229, 2004.
13. Overmann, J. and Garcia-Pichel, F., The phototrophic way of life, in Darwin, M., Falkow, S., Rosenberg, E., and Scheiferk, K.H. (Eds), *The Prokaryote: A Handbook of Biology of Bacteria. Ecophysiology and Biochemistry*, Springer, Heidelberg, Germany, pp. 32–85, 2006.
14. Potts, M., Nitrogen fixation associated with communities of heterocystous and non heterocystous blue green algae in mangrove forest of Sinai, *Oecologia*, 39, 359, 1979.
15. De Winder, B., Stal, L.J., and Mur, L.R., *Crinalium epipsammum* sp. nov.: A filamentous cyanobacterium with trichomes composed of elliptical cells and containing poly-β-(1,4) glucar (cellulose), *Microbiology*, 136, 1645, 1990.
16. Griffiths, M.H., Gallon, J.R., and Chaplin, A.E., The diurnal pattern of dinitrogen fixation by cyanobacteria *in situ*, *New Phycol.*, 107, 649, 1987.
17. Ranchou-Peyruse, A. et al., Vertical migration of phototrophic bacterial populations in a hypersaline microbial mat from salins-de-giraud (amargue France), *FEMS Microbiol Ecol.*, 57, 367, 2006.
18. Stahl, D.A. et al., Characterization of a Yellowstone hot spring microbial community by 5S rRNA sequences, *Appl Environ Microbiol.*, 49, 1379, 1985.
19. Ward, D.M.,. Weller, R., and Bateson, M.M., 16S rRNA sequences reveal numerous uncultured microorganisms in a natural community, *Nature*, 344, 63, 1990.
20. Fuhrman, J.A., McCallum, K., and Davis, A.A., Phylogenetic diversity of subsurface marine microbial communities from the Atlantic and Pacific Oceans, *Appl Environ Microbiol.*, 59, 1294, 1993.
21. Mullins, T.D. et al., Genetic comparisons reveal the same unknown bacterial lineages in Atlantic and Pacific bacterioplankton communities, *Limnol Oceanogr.*, 40, 148, 1995.
22. Ueda, T., Suga, Y., and Matsuguchi, T., Molecular phylogenetic analysis of a soil microbial community in a soybean field, *Eur J Soil Sci.*, 46, 415, 1995.
23. Borneman, J. and Triplett, E.W., Molecular microbial diversity in soils from eastern Amazonia: Evidence for unusual microorganisms and microbial population shifts associated with deforestation, *Appl Environ Microbiol.*, 63, 2647, 1997.
24. Cecilia Callejas., Phyloptype diversity in a benthic cyanobacterial mat community on King George Island, Antartica, *World J Microbiol Biotechnol.*
25. Fourcans A. et al., Characterization of functional bacterial groups in a hypersaline microbial mat community (Salins-de-Giraud, Camargue, France), *FEMS Microbiol Ecol.*, 51, 55, 2004.
26. Ley R.E. et al., Unexpected diversity and complexity of Guerrero Negro hypersaline microbial mats, *Appl Environ Microbiol.*, 72, 3685, 2006.
27. Zehr, J.P. et al., Diversity of heterotrophic nitrogen fixation genes in a marine cyanobacterial mat, *Appl Environ Microbiol.*, 61, 2527–2532, 1995.
28. Steppe T.F. et al., Diazotrophy in modern marine Bahamian stromatolites, *Microb Ecol.*, 41, 36, 2001.
29. Falcon, L.I. et al., Nitrogen fixation in microbial mat and stromatolite communities from Cuatro Cienegas, Mexico, *Microb Ecol.*, 54, 363, 2007.
30. Stal, L.J., Severin, I., and Acinas, S.G., Diversity of nitrogen fixing bacteria in cyanobacterial mats, *FEMS Microbiol Ecol.*, 73, 514, 2010.
31. Hoehler, T.M., Bebout, B.M., and Des Marais, D.J., The role of microbial mats in the production of reduced gases on the early earth, *Nature*, 412, 324, 2001.
32. Giordano, M., Beardall, J., and Raven, J.A., CO concentrating mechanisms in algae: Mechanisms, environmental modulation, and evolution, *Annu Rev Plant Biol.*, 56, 99, 2005.

33. Price, G.D. et al., The functioning of the CO concentrating mechanism in several cyanobacterial strains: A review of general physiological characteristics, genes, proteins and recent advances, *Can J Bot.*, 76, 973, 1998.
34. Price, G.D., Coleman, J.R., and Badger, M.R., Association of carbonic anhydrase activity with carboxysomes isolated from the cyanobacterium *Synechococcus* PCC7942, *Plant Physiol.*, 100, 784, 1992.
35. Cannon, G.C. et al., Carboxysome genomics: A status report, *Funct Plant Biol.*, 29, 175, 2002.
36. Stal, L.J. and Moezelaar, R., Fermentation in cyanobacteria, *FEMS Microbiol Rev.*, 21, 179, 1997.
37. Decho, A.W., Microbial exopolymer secretions in ocean environments: Their roles in food webs and marine processes, *Oceanogr Mar Biol Annu Rev.*, 28, 73, 1990.
38. Romanowski, G., Lorengz, M.G., and Wakernagel, W., Adsorption of plasmid DNA to mineral surfaces and protection against DNAse 1, *Appl Environ Microbiol.*, 57, 1057, 1991.
39. Mackerras, A.H., De Chazal, N.M., and Smith, G.D., Transient accumulation of Cyanophycin in *Anabaena cylindrica* and *Synechocystis* 6308, *J Gen Microbiol.*, 136, 2057, 1990.
40. Pearson, H.W. et al., Aerobic nitrogenase activity associated with a non heterocystous filamentous cyanobacterium, *FEMS Microbiol Lett.*, 5, 163, 1979.
41. Paerl, H.W., Prufert, L.E., and Ambrose, W.W., Contemporeneous N_2 fixing and oxygenic photosynthesis in the nonheterocystous mat forming cyanobacterium *Lyngbya aestuarii*, *Appl Environ Microbiol.*, 57, 3086, 1991.
42. Stal, L.J. and Heyer, H., Dark anaerobic nitrogen fixation in cyanobacteria *Oscillatoria* sp., *FEMS Microbiol Ecol.*, 45, 227, 1987.
43. Bak, F. and Pfennig, N., Chemolithotrophic growth of *Desulfovibrio sulfodismutans* sp. nov. by disproportionation of inorganic sulfur compounds, *Arch Microbiol.*, 147, 184, 1987.
44. Friedrich, M. and Schink, B., Isolation and characterization of desalforubidin containing sulfate reducing bacterium growing with glycolate, *Arch Microbiol.*, 164, 271, 1995.
45. Teske, A., Krekeler, D., and Cypionca, H., Strategies of sulfate reducing bacteria to escape oxygen stress in cyanobacterial mat, *FEMS Microbiol Ecol.*, 25, 89, 1998.
46. Elrich, H.L., *Geomicrobiology*, Marshal Dekker Inc, New York, p. 719, 1996.
47. Bender, J., Lee, R.F., and Philips, P., Uptake and transformation of metals and metalloids by microbial mats and their use in bioreactors, *J Ind Microbiol.*, 14, 113, 1995.
48. Urrutia, M.M., General bacterial sorption processes, in Wase, J. and Forster, C. (Eds). *Biosorbent for Metal Ions*, Taylor & Francis Group, London, U.K., p. 39, 1997.
49. Singh, A., Kumar, D., and Gaur, J.P., Copper(II) and lead(II) sorption from aqueous solution by nonliving *Spirogyra neglecta*, *Bioresour Technol.*, 98, 3622, 2007.
50. Phillips, P., Bender, J., and Thornton, C.F., Field-scale evaluation of microbial mats for metal removal in a funnel-and-gate configuration, in Leeson, A. and Alleman, B.C. (Eds), *Bioremediation of Metals and Inorganic Compounds*, Battelle Press, Columbus, OH, p. 109, 1999.
51. Gaur, J.P. and Kumar, D., Metal biosorption by two cyanobacterial mats in relation to PH, biomass concentration, pretreatment and reuse, *Bioresour Technol.*, 102, 2529, 2011.
52. Bender, J. et al., Bioremediation and bioreduction of U(VI) in groundwaters by microbial mats, *Environ Sci Technol.*, 15, 3235, 2000.
53. Bender, J. et al., Rapid heavy metal removal in a continuous-flow batch reactor by microbial mat, *In situ and On-Site Biorem.*, Vol. 3, p. 373, 1997.
54. Mehta, S.K. and Gaur, J.P., Use of algae for removing heavy metal ions from wastewater: Progress and prospects, *Crit Rev Biotechnol.*, 25, 113, 2005.
55. Wilson, A.R. et al., The effect of pH and surface composition on lead adsorption to neutral freshwater biofilms, *Environ Sci Technol.*, 35, 3182, 2001.
56. Douglas, S. and Beveridge, T.J., Mineral formation by bacteria in natural microbial communities, *FEMS Microbiol Ecol.*, 26, 79, 1998.
57. White, C. and Gadd, G.M., Accumulation and effects of cadmium on sulfate reducing bacterial biofilms, *Microbiology*, 144, 1407, 1998.
58. Utgikar, V.P. et al., Inhibition of sulfate reducing bacteria by metal sulfide formation in bioremediation of acid mine drainage, *Environ Toxicol.*, 17, 40, 2002.
59. White, C., Dennis, J.S., and Gadd, G.M., A mathematical process model for cadmium precipitation by sulfate-reducing bacterial biofilms, *Biodegradation*, 14, 139, 2003.
60. Bender, J. and Phillips, P., Implementation of microbial mats for bioremediation, in Means, J.L. and Hinchee, R.E. (Eds), *Emerging Technology for Bioremediation of Metals*, Lewis Publishers, Boca Raton, FL, p. 85, 1994.
61. Phillips, P. and Bender, J., Biological remediation of mixed waste by microbial mats, *Fed Facilities Environ J Autumn.*, 6, 77, 1995.

62. Phillips, P.C. et al., Biodegradation of naphthalene, phenanthrene, chrysene and hexadecane with a constructed silage–microbial mat, in Hinchee, R.E., Anderson, D.B., Metting, F.B., and Sayles, G.D. (Eds), *Applied Biotechnology for Site Remediation*, Lewis Publishers, Boca Raton, FL, p. 305, 1994.
63. Murray, R., Phillips, P., and Bender, J., Degradation of pesticides applied to St. Vincent banana farm soils comparing native bacteria and microbial mat, *Environ Toxicol Chem.*, 16, 84, 1997.
64. Sasikala, K. et al., Anoxygenic phototrophic bacteria, physiology and advances in hydrogen production technology, *Adv Appl Microbiol.*, 38, 211, 1993.
65. Castenholz, R.W., Composition of hot spring microbial mat: A summary, in Cohen, Y. et al. (Eds), *Microbial Mats Stromatolites*, Alan R Liss, New York, p. 101, 1984.
66. Krumbein, W.E., Cohen, Y., and Shilo, M., Solar lake (Sinai). 4. Strometolitic cyanobacterial mats, *Limnol Oceanogr.*, 22, 635, 1977.
67. Herbert, R.A., Development of mass bloom of photosynthetic bacteria on sheltered beaches in Scopa Flow, Okney Islands, *Proc R Soc Edinburgh*, 87, 15, 1985.
68. Pearl, H.W., Bebout, B.M., and Prufert, L.E., Naturally occurring patterns of oxygenic photosynthesis and N_2 fixation in marine microbial mats: Physiological and ecological ramifications, in Cohen, Y. and Rosenberg (Eds), *Microbial Mats, Physiological Ecology of Benthic Microbial Communities*, American Society for Microbiology, Washington, DC, p. 326, 1989.
69. Margulis, L. et al., Community living long before man: Fossil and living microbial mats and early life, *Sci Total Environ.*, 56, 379, 1986.
70. Gibson, J., Leadbetter, E.R., and Jannasch, H.B., Great Sippewest marsh, in Cohen, Y., Castenholz, R.W., and Halvorson, H.O. (Eds), *Microbial Mats Stromatalised*, Alan R Liss, New York, p. 95, 1984.
71. Orena, A., Photosynthetic and heterophyllic and benthic microbial communities of a hypersaline sulphur spring on the shore of dead sea (Hamei Mazor), in Cohen, Y., Castenholz, R.W., and Halvorson, H.O. (Eds), *Microbial Mats Stromatalized*, Alan R Liss, New York, p. 64, 1989.
72. Belkin, S. and Jannasch, H.W., Microbial mat at deep sea hydrothermal vents: New observations, in Cohen, Y. and Rosenberg, E. (Eds), *Microbial Mats Physiological Ecology of Benthic Microbial Communities*, American Society for Microbiology, Washington, DC, p. 16, 1989.
73. Abed Raeid, M.M. et al., Cyanobacterial diversity and bioactivity of inland hypersaline microbial mats from the desert stream in the Sultanate of Oman, *Fottea*, 11, 215, 2011.
74. Guerrero, M.C. and Wit, R., De Microbial mats in the island saline lake of Spain, *Limnetica*, 8, 197, 1992.
75. Buhring, S.I. et al., A hypersaline microbial mat from the Pacific Atoll Kiritimati: Insights into composition and carbon fixation using biomarker analyses and a ^{13}C-labeling approach, *Geobiology*, 7, 308, 2009.
76. Guerrero, R. et al., Distribution, typology and structure of microbial mat communities in Spain: A preliminary study, *Limnetica*, 8, 185, 1992.
77. Pinckney, J.L. and Paerl, H.W., Anoxygenic photosynthesis and nitrogen fixation by a microbial mat community in a Bahamian hypersaline lagoon, *Appl Environ Microbiol.*, 63, 420–426, 1997.
78. Hubas, C. et al., Tools providing new insight into coastal anoxygenic purple bacterial mats: Review and perspectives, *Res Microbiol.*, 162, 858, 2011.
79. Cain, A., Vannela, R., and Woo, L.K., Cyanobacteria as a biosorbent for mercuric ion, *Bioresour Technol.*, 99, 6578, 2008.
80. Webb, R. et al., Ability of immobilized cyanobacteria to remove metal ions from solution and demonstration of the presence of metallothionein genes in various strains, *J Hazard Subst Res.*, 1, 1998.
81. Freire-Nordi, C.S. et al., The metal binding capacity of *Anabaena spiroides* extracellular polysaccharide: An EPR study, *Process Biochem.*, 40, 2215, 2005.
82. Colica, G. et al., Selectivity in the heavy metal removal by exopolysaccharide producing cyanobacteria, *J. Appl Microbiol.*, 105, 88, 2008.
83. Mohammed, Z.A., Removal of cadmium and manganese by a non toxic strain of the fresh water cyanobacterium *Gloeothece magna*, *Water Res.*, 35, 4405, 2001.
84. Chidthaisong, A. et al., Lead removal from waste water by the cyanobacterium *Calothrix marchica*, *Kasetsart J.*, 40, 784, 2006.
85. Philippis, R. et al., Assessment of metal removal capability of two capsulated cyanobacteria *Cyanospira capsulata* and Nostoc PCC 7936, *J Appl Phycol.*, 15, 155, 2003.
86. Stal, L.J. and Brumbein, W.E., Oxygen protection of nitrogenase in the aerobically nitrogen fixing non heterocystous cyanobacterium, *Oscillatoria* sp., *Arch Microbiol.*, 143, 72, 1985.

Index

F

G

H

J

K

L

P

R

S

T

U

W

Z